Millions of Years Ago	Period	Life Forms	Climates and Major Physical Events
Paleozoic Era			
248–286	Permian	Reptiles diversify. Origin of conifers, cycads, and ginkgos; possible origin of flowering plants; earlier forest types wane.	Extensive glaciation in Southern Hemisphere. Seas drain from land; worldwide aridity. Appalachians formed by end of Paleozoic.
286–360	Carboniferous Pennsylvanian Mississippian	Age of amphibians. First reptiles. Variety of insects. Sharks abundant. Great swamps, forests of ferns, gymnosperms, and horsetails.	Warm; conditions like those in temperate or subtropical zones—little seasonal variation, water plentiful. Lands low, covered by shallow seas or great coal swamps. Mountain building in eastern U.S., Texas, Colorado.
360–408	Devonian	Age of fishes. Amphibians appear. Mollusks abundant. Lunged fishes. Extinction of primitive vascular plants. Origin of modern groups of vascular plants.	Europe mountainous with arid basins. Mountains and volcanoes in eastern U.S. and Canada. Rest of North America low and flat. Sea covers most of land.
408–438	Silurian	Rise of fishes and reef-building corals. Shell-forming sea animals abundant. Invasion of land by arthropods. Earliest vascular plants. Modern groups of algae and fungi.	Mild. Continents generally flat. Again flooded. Mountain building in Europe.
438–505	Ordovician	First primitive fishes. Invertebrates dominant. First fungi. Possible invasion of land by plants.	Mild. Shallow seas, continents low; sea covers U.S. Limestone deposits.
505–590	Cambrian	Shelled marine invertebrates. Explosive diversification of eukaryotic organisms.	Mild. Extensive seas, spilling over continents.
Precambrian Era			
590–4,500		Origin of life. Prokaryotes. Eukaryotic cells and multicellularity by close of era. Earliest known fossils, including soft-bodied marine invertebrates.	Dry and cold to warm and moist. Planet cools. Formation of Earth's crust. Extensive mountain building. Shallow seas. Accumulation of free oxygen.

Invitation to Biology

Summer is brief within the Arctic Circle, where these snowy owl chicks have made their appearance. The sun never sets, moving in a circle, never directly overhead but never dipping below the horizon. Its glancing rays melt the snow and the upper inches of frozen soil, but just beneath the surface the soil remains permanently frozen. Water cannot penetrate the solid ice and so the tundra—as this land of permafrost is called—stays boggy wet all summer long, even though there is as little rainfall as in a desert. The roots of plants cannot penetrate the permafrost either, so all the plants, even the trees, are only a few inches tall. When you walk across the tundra, you are walking on a forest of birches and willows, some hundreds of years old. Flowering plants, racing to set seed before the darkness returns, sometimes bloom when they are only an inch in height. Snowy owls and other land birds nest on the treeless expanse, where the bright, short-stemmed flowers and stunted vegetation barely conceal these curious chicks.

Among the largest of the owls, snowies are almost two feet tall, with a five-foot wingspan. As you can see, the chicks are soot-colored, but most adults are a dazzling white. When courting, it is reported, the male dances stiffly on the ground while holding a dead lemming. Who could resist?

Every year, birds of many species migrate by the millions to the Arctic Circle to take advantage of the abundant fish, the swarms of insects that emerge each season, and most of all, the 24-hour daylight for feeding their young. When fall comes, most head south. The snowy owl is one that remains, seeking lemmings and Arctic hares rustling beneath the snow cover. During the long winter night, the air is cold and clear, and the starlight and moonlight reflect off the surface of the snow. By then, these young snowies will be full grown and airborne—huge white owls gliding as silent as the snow through the cold air.

Invitation to Biology

Fifth Edition

HELENA CURTIS

N. SUE BARNES

Illustrated by Shirley Baty

WORTH PUBLISHERS

Invitation to Biology, Fifth Edition—Complete book (casebound)
Invitation to Biology, Fifth Edition—Part 1: Sections 1, 2, 3, 4, 8 (paperback)
Invitation to Biology, Fifth Edition—Part 2: Sections 5, 6, 7 (paperback)

Copyright 1994, 1985, 1981, 1977, 1972 by Worth Publishers, Inc.

Printed in the United States of America

Library of Congress Catalog Card Number 93-61158—Complete book (casebound)
Library of Congress Catalog Card Number 94-60032—Parts 1 and 2 (paperbacks)

ISBN: 0-87901-679-5—Complete book (casebound)
ISBN: 0-87901-734-1—Part 1: Sections 1, 2, 3, 4, 8 (paperback)
ISBN: 0-87901-735-X—Part 2: Sections 5, 6, 7 (paperback)

Printing: 1   2   3   4   5—98   97   96   95   94

Editor: Sally Anderson

Design: Malcolm Grear Designers

Art Director: George Touloumes

Production Supervisor: Sarah Segal

Layout: Matthew Dvorozniak

Picture Editor: Lana Berkovich

Illustration Consultant: John Norton

Typographer: CRWaldman Graphic Communications

Color Separations and Stripping: Creative Graphic Services Corporation

Printing and Binding: Von Hoffmann Press, Inc.

Covers and Frontispiece: Snowy owlets, © Art Wolfe

Worth Publishers
33 Irving Place
New York, New York 10003

Preface

Preface to the Instructor

As even a glance at the Table of Contents or a few minutes spent thumbing through the text itself will reveal, this Fifth Edition of *Invitation to Biology* is a major revision, the result of a thorough reexamination and rethinking of all aspects of the book.

The most obvious changes are in the design and the illustration program. Our goal was twofold: to make the book as inviting as possible to today's students, who live in a world dominated by visual cues, and, simultaneously, to use art and photographs not simply to enhance the text but also to clarify difficult concepts and actively teach. With the able assistance of our art consultant, John Norton, who provided us with a multitude of suggestions, and the dedicated labors of our gifted illustrator, Shirley Baty, who translated ideas into reality, more than 500 new pieces of art now grace the pages of *Invitation*. A similar effort, involving both our photo researcher, Lana Berkovich, and our editor, Sally Anderson, has resulted in the many marvelous new photographs that appear throughout the book. Of particular importance are the full-page photographs with which each chapter begins. Finding exactly the right photograph for each chapter—a photograph that would, in and of itself, appeal to students and, at the same time, lend itself to a brief essay that would intrigue students and set the stage for the chapter—was a daunting but rewarding task.

Adding more illustrations and increasing their size in order to maximize their appeal and pedagogical effectiveness does, of course, require more space. In order to accommodate the new illustration program without greatly increasing the length of the book—or having to compress explanatory material at the expense of clarity—we have, with this edition, moved from our traditional one-column format to a two-column format. The new design, created by Pat Appleton of Malcolm Grear Designers, has provided great flexibility in both the use of color and the layout of the individual pages. This flexibility and the enormous effort devoted by art director George Touloumes and his colleague, Matthew Dvorozniak, to the detailed layout of the book, page by page, has ensured that the two-column format has not become so tight or "busy" that it is intimidating to students. Working with the new design has been exciting for us, and we believe that it will help draw students into the wonders of biology.

As in all previous editions, the text itself has been subjected to an intense reconsideration and reworking—section by section, chapter by chapter, and sentence by sentence—and then once again thoroughly reviewed. As always, we have had to face the questions of how to organize the book, what to put in, and, most difficult of all, what to leave out. In terms of the depth and breadth of the coverage, it has seemed to us wisest to maintain—and where possible, to strengthen—the firm emphasis on the basic principles of biology that has characterized previous editions of *Invitation*. Specific topics for detailed treatment have been selected on the basis of their centrality to modern biology, their utility in illuminating basic principles, their importance as part of the requisite store of knowledge of an educated adult, and their inherent appeal to students. Throughout, we have tried to lay the foundations and arouse student curiosity so that the ground is prepared for those areas—very diverse—in which individual instructors may wish to give more extensive coverage than is possible in any "short" text.

The essential foundation of biology is, of course, evolution, the major organizing theme of this and all other modern biology texts. The stage is set in the Introduction, which begins with the fascinating organisms of one particular "living laboratory of evolution"—the island of Madagascar—and then moves on to the development and key features of the Darwinian theory. The Introduction concludes with a discussion of the nature of science, where the story of the peppered moth provides both a fully worked-out example of scientific methodology and a concrete example of natural selection in action.

With this edition, we have moved the detailed treatment of evolution much earlier in the book, so that it is now Section 4, following immediately after genetics. This repositioning makes it possible for students to consider the complexities of population genetics while genetics itself is still fresh in their minds. It also makes possible a stronger and more explicit evolutionary theme in Section 5 (The Diversity of Life), Section 6 (Biology of Animals), and Section 7 (Biology of Plants), all of which, in previous editions, preceded the evolution section.

Moving evolution earlier in the book led to other changes, some of which we anticipated from the outset and some of which we discovered only as we worked our way through the revision. In previous editions, for example, human evolution was covered in the evolution section. With the new organization, however, the necessary foundation for a discussion of human evolution—namely the evolution of the vertebrates in general and the mammals in particular—is not covered until after the evolution section. Thus, we moved human evolution into the diversity section. What was previously one chapter on animal diversity is now two chapters: the first covers the invertebrates, and the second covers the vertebrates, culminating with the story of human evolution.

With that change made, it then seemed wise to move directly into animal physiology, for which our representative organism is *Homo sapiens*. Having Section 6, Biology of Animals, follow immediately after the two chapters on animal diversity should make the various examples of comparative animal physiology in that section more meaningful to students, once again because the relevant material will be fresh in their minds. We have also added two new chapters to Section 6: Chapter 33, The Immune Response, and Chapter 38, Development. Both immunology and developmental biology are currently undergoing an explosive growth of knowledge and are also, we believe, of great interest to students.

Section 7, Biology of Plants, which now follows the section on the biology of animals, begins with a chapter on reproduction in the flowering plants, which is followed, in turn, by a chapter on the plant body and its development. This provides a clear and immediate counterpoint to the last two chapters of Section 6, and it should enable students to see more easily the similarities and differences in the ways in which the organisms of the two largest kingdoms accomplish the fundamental tasks of reproduction and development. Beginning Section 7 with the flower and reproduction—rather than with a chapter on plant structure—also enables us to draw students into the fascinating world of plants with that which is familiar and beautiful.

From the plants, we then move into Section 8, Ecology. Here too, the new organization should enable students to have fresh in their minds information that will enhance their exploration of a new topic. This section has been extensively revised, as we attempt to track the continual shifts, rethinkings, and controversies that characterize this most vibrant science. As in the Fourth Edition, the section moves from population dynamics, through the interactions of populations in communities and ecosystems, to the overall organization and distribution of life on Earth. The book ends with a consideration of the tropical forests—the most complex and most seriously threatened of all ecological systems.

All of the feedback we have received—in letters and in surveys—has indicated that, with a few exceptions, the previous organization and sequence of the first three sections of the book—Section 1 (The Unity of Life), Section 2 (Energetics), and Section 3 (Genetics)—was sound and should not be altered. The exceptions were the placement of the chapter on the reproduction of cells (mitosis and cytokinesis), the sequence of chapters at the beginning of the genetics section, and the placement of human genetics in a separate chapter.

In response to your suggestions, we have made several changes. First, the chapter on cell reproduction has been moved out of the first section and is now Chapter 10, the first chapter of genetics. Second, we have resequenced the beginning of Section 3 so that it moves, in three consecutive chapters, from mitosis to meiosis to Mendel. Third, we have eliminated the separate chapter on human genetics and integrated that material throughout Section 3. Because of the inherent interest of human genetics, this change should make the other chapters of the section more appealing. Also, it avoids the problems that are involved in positioning a human genetics chapter that, if it is to be thorough, must range from Mendelian genetics to the cutting edge of molecular genetics. If such a chapter is positioned immediately after the chapters on classical genetics, either it must omit new developments in the diagnosis and treatment of human genetic disorders, or, alternatively, it must attempt to cover those developments before a proper foundation has been laid. If, on the other hand, such a chapter is positioned at the end of the genetics section, students are compelled to make a sudden mental switch from molecular genetics back to classical genetics—and then back again—all within the same chapter. The new organization should lead to both increased student understanding and increased student comfort.

One of the most striking aspects of the enormous burst of new discoveries in molecular and cell biology in recent years is the power of these discoveries to explain processes that previously could only be described.

The immune response, conduction of the nerve impulse, events at the synapse, the summing of information by individual neurons, and differentiation and morphogenesis in animal development are just a few of the many phenomena whose secrets are being revealed by studies at the molecular and cellular level. These revelations depend, in large part, on what is now a flood tide of reports identifying specific membrane proteins, their amino acid sequences, their three-dimensional structures, and, in many cases, the sequences and locations within the genome of the genes coding for these proteins. Because these discoveries, although fascinating in and of themselves, are of such value in explaining organismal phenomena, we have generally chosen to defer their discussion to later sections of the text, where their significance will be most readily grasped. Molecular and cell biology have, in many ways, come of age, and it seems to us that the essential task in the early sections of the book is the clear communication of the underlying principles on which so much is now being built—rather than a catalog of the latest new discoveries, which will soon be superseded by other new discoveries. We have endeavored to explain the fundamental principles as clearly as possible, and the many new illustrations should make these early sections easier to teach and more understandable for students.

Increased student understanding of the phenomena of the living world—coupled with the sheer pleasure of discovering the wonder of it all—is, of course, the whole point of creating a textbook and of teaching biology. As an aid toward the accomplishment of that goal, a number of supplements have been revised or newly created to accompany this edition of *Invitation to Biology*. The excellent *Study Guide* by David J. Fox of the University of Tennessee, Knoxville, makes students active participants in the learning process. A new edition of the well-received *Biology in the Laboratory*, by Doris R. Helms of Clemson University, engages students in critical thinking and problem solving; it is accompanied by a revised *Preparator's Guide* and *Answer Guide* that will make the laboratory course easier to plan and to teach. A set of 200 full-color acetate transparencies and a videodisc provide excellent visual enhancement for lectures. A *Test Bank*, of more than 2,500 multiple-choice, true-false, and brief essay questions, prepared by David Fox, is available on both paper and computer disks.

Of particular interest is *Instructor's Resources* by Robert C. Evans of Rutgers University, Camden. This new supplement includes suggestions for using all of the materials in the package, lecture guides, learning objectives, answers to the questions in the textbook, and a list of appropriate resources for each chapter. Instruc-

tions for outlining and concept mapping, as well as listings of available films and videos, computer programs, and videodiscs, with sources, are also included. The *Lecture Guides* from this supplement plus *Focus Points*, prepared by Bruce Chorba of Mercer County Community College, are also available together on computer disk for preparing your own handouts. Another supplement, *Student Activities*, provides three-hole-punched master copies of critical thinking, extension, and application activities for each chapter.

As with previous editions, we have been greatly assisted by the advice of our consultants and reviewers. In addition to his work on *Instructor's Resources*, Bob Evans reviewed the entire manuscript. His sensitivity to students, his good judgment, and his steady, calm counsel have been a great blessing. The entire manuscript was also expertly reviewed by both Janet Wagner, whose suggestions were invaluable, particularly in the reworking of the genetics section, and Linda Hansford, who has, in addition, prepared the Index. Manuel C. Molles, Jr., of the University of New Mexico, and Andrew Blaustein, of Oregon State University, both made major contributions to our revision of the evolution and ecology sections. Rick Weiss, formerly of *Science News*, and Judith Wilson, of Worth Publishers, contributed to the new essays in Section 6.

In addition, we are grateful for the constructive criticism and many helpful suggestions we received from the following reviewers of this edition:

William L. Bischoff, University of Toledo

Richard Blazier, Parkland College

Maren Brown, Onondaga Community College

Rita Calvo, Cornell University

Mary Colavito, Santa Monica College

Jean DeSaix, University of North Carolina, Chapel Hill

Mark Dubin, University of Colorado

Ray F. Evert, University of Wisconsin

William Greenhood, Santa Fe Community College

Mart Gross, University of Toronto

Thomas Hanson, Temple University

Jean B. Harrison, Mount St. Mary's College

George Hennings, Kean College of New Jersey

Merrill Hille, University of Washington

Ronald Hoham, Colgate University

Norma Johnson, University of North Carolina, Chapel Hill

Stephen W. Kress, National Audubon Society

Melanie Loo, California State University, Sacramento

Joyce Maxwell, California State University, Northridge

Dorothy McMeekin, Michigan State University

Richard Milner, American Museum of Natural History

Douglas W. Morrison, Rutgers University, Newark

David Netzly, Hope College

Bette Nicotri, University of Washington

Frank E. Price, Hamilton College

Julia Riggs, Victoria College

Edward S. Ross, California Academy of Sciences

Anthony P. Russell, University of Calgary

Robert W. Schuhmacher, Kean College of New Jersey

Ian Tattersall, American Museum of Natural History

Robin Tyser, University of Wisconsin, La Crosse

Helen Young, Barnard College

The preparation of this new edition has been a complex and all-consuming task, involving the collaborative efforts of a large number of talented people, many of whom have been noted already. Particular thanks are also due to Bob Worth, whose vision and support through the years have made it all possible; to Sarah Segal, who has managed the production process and somehow kept us all on course—often a thankless and frustrating task; to Demetrios Zangos, who labeled all of the new art and prepared and incorporated structural formulas into the illustrations in which they appear; to Timothy Prairie, who coordinated and edited the supplements; to Brian Donnelley and Treë, both of whom assisted with the photo research; and to Lindsey Bowman, Mary Mazza, and Alexandra Swieconek, who provided expert editorial assistance at different stages of the project. Our greatest debt, however, is to our extraordinary editor, Sally Anderson. Her dedication, her thorough knowledge of biology, her editorial expertise, and her long experience in working with us both have once again seen us through to the successful completion of an enormous project.

Finally, we want to thank all of the instructors and students who have written to us, some with criticisms, some with suggestions, some with questions, and some simply because they enjoyed the book. Those letters remind us of how privileged we are to be writing for students. We continue to appreciate their curiosity, their energy, their imaginativeness, their sense of humor, and their dislike of the pompous and pedantic. We hope we serve them well.

Helena Curtis
Sue Barnes

Long Island, New York
January, 1994

Leaf-tailed gecko, Madagascar

Contents in Brief

Contents

SECTION 1

SECTION 2

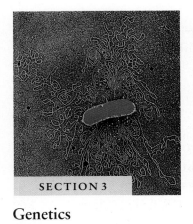

SECTION 3

SECTION 4

Evolution   318

SECTION 5

SECTION 6

Biology of Animals · 506

SECTION 8

Distant islands are the homes of many strange inhabitants—plants and animals found nowhere else on our planet. Lemurs, such as the sifakas shown here, are found only on the island of Madagascar, which lies off the southeast coast of Africa. Some 45 different species of lemurs have been discovered, of which 17 are now extinct. No monkeys, or apes, or other primates (except for humans, who were late arrivals) have been found on Madagascar—only the lemurs. Why are lemurs here and nowhere else?

This same sort of question was repeated over and over in different forms as the naturalists of the eighteenth and nineteenth centuries explored widely separated regions of the world—South America, the Galapagos Islands, Australia, the Malay Archipelago, and Madagascar. These travelers—Charles Darwin among them—were astonished by the vast numbers and extraordinary diversity of the organisms they encountered. The doctrine of special creation held that each kind of plant and animal was created in its present form and placed in the location in which it was subsequently found. But, the naturalist explorers wondered, was it possible, perhaps, that groups of plants and animals, separated from one another—as on islands—change and diversify over time into new kinds of organisms?

The lemur shown here, known to scientists as *Propithecus verreauxi,* is about the size of a small house cat. It is among the largest of the surviving lemurs. Sifakas are the acrobats of this primate world, using their long tails for balance as they fling themselves through the air. They can cover a distance as great as 25 feet in a single leap. These two sifakas are resting in a tamarind tree, found in the dry forest region of Madagascar. When faced with an enemy, they hiss the alarm call that gives them their name: Shifakh!

Introduction

Madagascar, the world's fourth largest island, lies in the Indian Ocean 250 miles east of the African mainland. Slightly smaller than the state of Texas, the island is located almost entirely within the tropics. Two principal types of vegetation are found on Madagascar: dry forest to the west and moist forest to the east, separated by a treeless mountain chain that runs the length of the island from north to south.

Some 200 million years ago, when today's continents were joined together in a single supercontinent, the fragment that would become Madagascar was located between the regions that would ultimately form Africa and India. When Africa and India began to separate, about 160 million years ago, Madagascar was split off. As it gradually moved away from the larger land masses, its plants and animals became increasingly isolated from those on the mainland of both Africa and India. Although occasional plant and animal colonists may have arrived on floating clumps of vegetation carried by the ocean currents, by about 50 million years ago the isolation was complete. The plants and animals of the island remained undisturbed until the arrival of the first human colonists some 2,000 years ago.

In 1771, the French naturalist Philibert de Commerson wrote of his first visit to Madagascar: "Nature seems to have retreated there into a private sanctuary, where she could work on models different from any she has used elsewhere. There, you meet bizarre and marvellous forms at every step." On Madagascar today there are literally thousands of **species,** or kinds, of plants and animals found nowhere else on Earth. For example, of the 8,000 species of flowering plants found on Madagascar, 80 percent grow only on this island. Of the world's 85 species of chameleons, two-thirds are here alone. Similarly, all but two of the 150 species of frogs on Madagascar are unique to the island. Here are found 3,000 different species of butterflies and moths, 97 percent of which live only on Madagascar.

I–1   *The island of Madagascar broke off from Africa some 160 million years ago and then, over a period of approximately 70 million years, moved slowly south to its present position. Madagascar is about 1,000 miles long and 350 miles across at its widest point. Conservationist Gerald Durrell recently described its shape as resembling "a badly made omelet."*

Among the most unusual inhabitants of Madagascar are the small, furry animals known as lemurs. Lemurs are classified by biologists with the primates, a group of mammals that includes monkeys, apes, and ourselves. At one time there were at least 45 species of lemurs, all of which inhabited Madagascar and Madagascar alone. In the last 2,000 years, however, almost half of the lemur species have vanished. Some of the larger species, with adults about the size of a Saint Bernard, were hunted to extinction before the first European explorers arrived on Madagascar. Other species have become extinct over the course of the past two centuries, as the rapid growth of the human population has led to the destruction of the island's forests. Today less than 15 percent of the lemurs' original habitat remains, and only 28 species survive (Figure I–2).

Madagascar has long been of interest to biologists not only because of its array of unique plants and animals but also because of the plants and animals that are *not* found on the island. Totally absent are the monkeys, apes, antelope, gazelles, zebras, giraffes, wildebeest, elephants, and lions of the neighboring African mainland. As European naturalists explored widely sep-

arated regions of the world in the eighteenth and nineteenth centuries, they found that other islands and island chains resembled Madagascar in two important respects: (1) distant islands were typically home to an enormous variety of organisms unique to that island or island group, and (2) such islands often lacked many of the plants and animals characteristic of the nearest mainland.

These observations raised an important question for the explorers: How does one make sense of the bewildering diversity of living things and the patterns in which they are scattered in such profusion over the surface of the Earth? One answer was that each species—each different kind of living organism—was specially created and placed by the creator in its present location. This was the traditional explanation. Another idea was in the air, however, and that idea is the subject of this Introduction.

The Road to Evolutionary Theory

In 1831, the young Charles Darwin set sail from England on what was to prove the most consequential voyage in the history of biology. Not yet 23, Darwin had already abandoned a proposed career in medicine—he described himself as once fleeing a surgical theater in which an operation was being performed on an unanesthetized child—and was a reluctant candidate for the clergy, a profession deemed suitable for the younger son of an English gentleman. An indifferent student, Darwin was an ardent hunter and horseman, a collector of beetles, mollusks, and shells, and an amateur botanist and geologist. When the captain of the surveying ship H.M.S. *Beagle,* himself only a little older than Darwin, offered passage for a young gentleman naturalist who would volunteer to go without pay, Darwin eagerly seized the opportunity to pursue his interest in natural history. The voyage, which lasted five years, shaped the course of Darwin's future work. He returned to an inherited fortune, an estate in the English countryside, and a lifetime of independent work and study that radically changed our view of life and of our place in the living world.

That Darwin was the founder of the modern theory of evolution is well known. Although he was not the first to propose that organisms **evolve**—or change—through time, he was the first to gather a large body of supporting evidence and the first to propose a valid mechanism by which evolution might occur. In order to understand the meaning and significance of Darwin's theory, it is useful to look at the intellectual climate in which it was formulated.

I–2 *Six of the 28 surviving lemur species on Madagascar.* (**a**) *Indri, the largest of the living lemurs. In Malagasy, the principal language of Madagascar, the name of this lemur is "babakoto," the word for "grandfather." However, when one of the local inhabitants used the expression "indri," which means "look at that!" to point out the animal to an early European explorer, the explorer misinterpreted it as the animal's name. With the bodies of monkeys, the black-and-white fur of pandas, the heads of dogs, and no tails, the indri are indeed a sight to behold. They are also noted for their eerie and deafening wail, in which a single song may last as long* as 4 minutes. (**b**) *An aye-aye, the rarest of the lemurs. This small, solitary, and nocturnal animal may have been the original source for the name lemur, which means "ghost." Aye-ayes have bat-like ears, large eyes, and a long third finger that they drum against trees to determine whether they are hollow. If the sound is right, the aye-aye (like the familiar woodpecker) proceeds to extract insects from the decaying wood. Aye-ayes also have a taste for sweet foods, such as sugar cane and coconuts. Malagasy legends say that if an aye-aye should point its long third finger at you, you are destined to die, horribly and* swiftly. (**c**) *The tiny mouse lemur, which is thought to resemble the ancestral primate from which the diverse species of lemurs arose. (**d**) The golden bamboo lemur, first discovered in 1986, is one of the world's most seriously endangered species. The number of surviving individuals is unknown but is thought to be only a few dozen. (**e**) A female black lemur. Unlike the males, for which the species is named, the females are a rusty brown. (**f**) A ring-tailed lemur. The second toe of each foot has a special grooming claw with which the lemur combs its long tail.*

I–3   *Charles Darwin in 1840, four years after he returned from his five-year voyage on H.M.S.* Beagle. *In his autobiography, written in 1876 for his children, Darwin made the following comments about his selection for the voyage: "Afterwards, on becoming very intimate with Fitz Roy [the captain of the* Beagle], *I heard that I had run a very narrow risk of being rejected on account of the shape of my nose! He . . . was convinced that he could judge of a man's character by the outline of his features; and he doubted whether anyone with my nose could possess sufficient energy and determination for the voyage. But I think he was afterwards well satisfied that my nose had spoken falsely."*

Aristotle (384–322 B.C.), the first great biologist, believed that all living things could be arranged in a hierarchy. This hierarchy became known as the *Scala Naturae,* or ladder of nature, in which the simplest creatures had a humble position on the bottommost rung, man occupied the top rung, and all other organisms had their proper places in between. Until the late nineteenth century, many biologists believed in such a natural hierarchy. But whereas to Aristotle living organisms had always existed, the later biologists (at least those of the Western world) believed, in harmony with the teachings of the Old Testament, that all living things were the products of a divine creation. They believed, moreover, that most were created for the service or pleasure of mankind.

That each type of living thing came into existence in its present form—specially and specifically created—was a compelling idea. How else could one explain the astonishing extent to which every living thing was adapted to its environment and to its role in nature? It was not only the authority of the Bible but also, so it seemed, the evidence before one's own eyes that gave such strength to the concept of special creation.

Among those who believed in divine creation was Carolus Linnaeus (1707–1778), the great Swedish naturalist who devised our present system for naming species of organisms. In 1753, Linnaeus published two encyclopedic volumes describing every species of plant known at the time. Even as Linnaeus was at work on this massive project, explorers were returning to Europe from Africa and the New World with thousands of previously undescribed plants and animals. Linnaeus revised edition after edition to accommodate these findings, but he did not change his opinion that all species now in existence were created by the sixth day of God's labor and have remained the same ever since. During Linnaeus's time, however, it became clear that the pattern of creation was far more complex than had been previously imagined.

Evolution before Darwin

The idea that organisms might evolve through time, with one type of organism giving rise to another type of organism, is an ancient one, predating Aristotle. A school of Greek philosophy, founded by Anaximander (611–547 B.C.) and culminating in the writings of the Roman Lucretius (99–55 B.C.), developed not only an atomic theory but also an evolutionary theory, both of which are strikingly similar to modern conceptions. However, as the science of biology began to take form in the eighteenth and nineteenth centuries, the work of this school was largely unknown, as were the ideas of other, non-European cultures.

In the eighteenth century, the French scientist Georges-Louis Leclerc de Buffon (1707–1788) was among the first to propose that species might undergo changes in the course of time. He suggested that, in addition to the numerous creatures that were produced by divine creation at the beginning of the world, "there are lesser families conceived by Nature and produced by Time." Buffon believed that these organisms were produced by a process of degeneration. For example, although an "ideal" type of feline had been created, its descendants had degenerated into a variety of forms, including lions, jaguars, cheetahs, and house cats. Buffon's hypothesis, although vague as to the way in which such changes might occur, did attempt to explain the bewildering variety of creatures in the modern world.

Another doubter of fixed and unchanging species was Erasmus Darwin (1731–1802), Charles Darwin's grandfather. Erasmus Darwin was a physician, a gentleman naturalist, and a prolific writer, often in verse, on both botany and zoology. He suggested, largely in asides and footnotes, that species have historical connections with one another, that animals may change in response to their environment, and that their offspring may inherit these changes. He maintained, for instance, that a polar bear is an "ordinary" bear that, by living in the Arctic, became modified and passed the modifications along to its cubs. These ideas were never convincingly formulated but are interesting because of their possible effects on Charles Darwin.

The Age of the Earth

It was geologists, more than biologists, who paved the way for modern evolutionary theory. One of the most influential of these was James Hutton (1726–1797). Hutton proposed that the Earth had been molded not by sudden, violent events but by slow and gradual processes—wind, weather, and the flow of water—the same processes that can be seen at work in the world today. This theory of Hutton's, which later became known as **uniformitarianism,** was important for three reasons. First, it implied that the Earth has a long history, which was a new idea to eighteenth-century Europeans. Christian theologians, by counting the successive generations since Adam (as recorded in the Bible), had calculated the maximum age of the Earth at about 6,000 years. Yet 6,000 years is far too short for major evolutionary changes to take place, by any theory. Second, the theory of uniformitarianism stated that change is itself the *normal* course of events, as opposed to a static system interrupted by an occasional unusual event, such as an earthquake. Third, although this was never explicit, uniformitarianism suggested that there might be alternatives to the literal interpretation of the Bible.

The Fossil Record

During the latter part of the eighteenth century, there was a revival of interest in **fossils,** which are the preserved remains of organisms long since dead (Figure I–4). In previous centuries, fossils had been collected as curiosities, but they had generally been regarded either as accidents of nature—stones that somehow looked like shells—or as evidence of great catastrophes, such as the Flood described in the Old Testament.

(a)

(b)

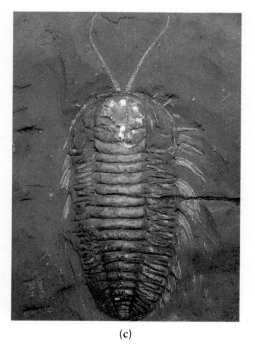

(c)

I–4   *A fossil is a remnant or trace of an organism that died many, many years earlier. (a) Minerals may seep into and fill the hollows left by the decay of soft tissues, as in this cycad leaf dating from the Jurassic period (144 to 213 million years ago). A great diversity of cycads were contemporary with the dinosaurs; about 100 species survive today. (b) A primitive mosquito caught in amber formed from the resin of a tree that lived some 25 to 40 million years ago. The blood from this mosquito's last meal is still visible in her digestive tract. (c) Among the most common early fossils are the outer skeletons of trilobites, marine animals that first appear in the fossil record during the Cambrian period (505 to 590 million years ago). This fossil, with its well-preserved antennae and legs, is filled with fool's gold (pyrite). Trilobites flourished for more than 300 million years before they became extinct.*

I–5   *An outcrop at Olduvai Gorge, in Tanzania, that clearly shows different layers, or strata, in the Earth's surface. The work of geologists who discovered that different strata contain different types of fossils was important in the development of evolutionary theory. The strata at Olduvai have yielded a number of fossils that have contributed to our growing understanding of human evolution.*

The English surveyor William Smith (1769–1839) was among the first to study the distribution of fossils scientifically. Whenever his work took him down into a mine or along canals or cross-country, he carefully noted the order of the different layers of rock, known as **strata,** and collected fossils from each layer. He eventually established that each stratum, no matter where he came across it in England, contained characteristic kinds of fossils. Moreover, these "index fossils" were actually the best way to identify a particular stratum in a number of different geographic locations. Smith, who was nicknamed "Strata" by his contemporaries, did not interpret his findings. However, the implication that the present surface of the Earth had been formed layer by layer over the course of time was an unavoidable one.

Like Hutton's world, the world seen and described by William Smith was a very ancient one. A revolution in geology was beginning; earth science was becoming a study of time and change rather than a mere cataloging of types of rocks. As a consequence, the history of Earth itself became inseparable from the history of living organisms, as revealed in the fossil record (Figure I–5).

Catastrophism

Although the way was being prepared by the revolution in geology, the time was not yet ripe for a parallel revolution in biology. The dominating force in European natural science in the early nineteenth century was Georges Cuvier (1769–1832). Cuvier was the founder of **vertebrate paleontology,** the scientific study of the fossil record of **vertebrates** (animals with backbones). An expert in anatomy and zoology, he applied his knowledge of the way in which animals are constructed to the study of fossil animals. From a few fragments of bone in a rock slab, he could identify an entire animal. To his students, Cuvier's brilliant deductions seemed almost magical.

Today we think of paleontology and evolution as so closely connected that it is surprising to learn that Cuvier was a staunch and powerful opponent of evolutionary theories. He did, however, recognize that many species that had once existed no longer did—a startling conclusion at the time. (In fact, according to modern estimates, considerably less than one percent of all species that have ever lived are represented on Earth today.) Cuvier explained the extinction of species by postulating a series of catastrophes. After each catastrophe, the most recent of which was the Flood, new species filled the vacancies.

Cuvier hedged somewhat on the source of the new animals and plants that appeared after the extinction of older forms. He was inclined to believe they moved in from parts unknown. Another major opponent of evolution, Louis Agassiz (1807–1873), America's leading nineteenth-century biologist, was more straightforward. According to Agassiz, the fossil record revealed 50 to 80 total extinctions of life, followed by an equal number of new, separate creations.

The Concepts of Lamarck

The first European scientist to work out a systematic concept of evolution was Jean Baptiste Lamarck (1744–1829). "This justly celebrated naturalist," as Darwin referred to him, boldly proposed in 1801 that all species, including *Homo sapiens,* are descended from other species. Lamarck, unlike most of the other zoologists of his time, studied one-celled organisms and **invertebrates** (animals without backbones). Undoubtedly it was his long study of these forms of life that led him to think of living things in terms of constantly increasing complexity, each species derived from an earlier, less complex one.

Like Cuvier and others, Lamarck noted that older rocks generally contained fossils of simpler forms of life. Unlike Cuvier, however, Lamarck interpreted this as meaning that the more complex forms had arisen from the simpler forms by a kind of progression. According to his hypothesis, this progression, or evolution, to use the modern term, is dependent on two main forces. The first is the inheritance of acquired characteristics, an idea that did not originate with Lamarck but was, in fact, widely accepted at the time by both scientists and the general public. It was believed that organs in animals became stronger or weaker, larger or smaller, through use or disuse, and that these changes were transmitted from the parents to the offspring. Lamarck's most famous example was the evolution of the giraffe. According to Lamarck, the modern giraffe evolved from ancestors that stretched their necks to reach leaves on high branches. These ancestors transmitted the longer necks—acquired by stretching—to their offspring, which stretched their necks even longer, and so on.

The second important force in Lamarck's concept of evolution was a universal creative principle, an unconscious striving upward on the *Scala Naturae* that moved every living creature toward greater complexity. Amoebas were on their way to *Homo sapiens.* Some might get waylaid—the orangutan, for instance, had been diverted from its course by being caught in an unfavorable environment—but the will was always present. Life in its simplest forms emerged spontaneously to fill the void left at the bottom of the ladder. In Lamarck's formulation, Aristotle's ladder of nature had been transformed into a steadily ascending escalator powered by a universal will.

Lamarck's contemporaries generally did not object to the idea of the inheritance of acquired characteristics, which we, with our present knowledge of genetics, know to be false. Nor did they criticize his belief in an upward-striving metaphysical force, which was a common element in many of the concepts of the time.

But these vague, untestable postulates provided a very shaky foundation for the radical proposal that more complex forms of life evolved from simpler forms. Moreover, Lamarck personally was no match for the brilliant and prestigious Cuvier, who relentlessly attacked his ideas. As a result, Lamarck's career was ruined, and scientists as well as the public became even less prepared to accept any evolutionary doctrine.

Development of Darwin's Theory

Earth Has a History

The person who most influenced Darwin, it is generally agreed, was Charles Lyell (1797–1875), a geologist who was Darwin's senior by 12 years. One of the books that Darwin took with him on his voyage was the first volume of Lyell's newly published *Principles of Geology,* and the second volume was sent to him while he was on the *Beagle.* On the basis of his own observations and those of his predecessors, Lyell opposed the theory of catastrophism. Instead, he produced new evidence in support of Hutton's earlier theory of uniformitarianism.

According to Lyell, the slow, steady, and cumulative effect of natural forces had produced continuous change in the course of the Earth's history. Since this process is demonstrably slow, its results being barely visible in a single lifetime, it must have been going on for a very long time. What Darwin's theory needed was time, and it was time that Lyell gave him. In the words of biologist Ernst Mayr of Harvard University, the discovery that the Earth was ancient "was the snowball that started the whole avalanche."

The Voyage of the *Beagle*

This, then, was the intellectual climate in which Charles Darwin set sail from England. As the *Beagle* moved down the Atlantic coast of South America, through the Strait of Magellan, and up the Pacific coast (Figure I–6), Darwin often left the ship and traveled the interior. He explored the rich fossil beds of South America (with the theories of Lyell fresh in his mind) and collected specimens of the many new kinds of plant and animal life he encountered. He was impressed most strongly during his long, slow trip down one coast and up the other by the constantly changing varieties of organisms he saw. The birds and other animals on the west coast, for example, were very different from those on the east coast, and even as he moved slowly up the western coast, one species would give way to another.

I–6   *The* Beagle's *voyage around South America. The ship left England in December of 1831 and arrived at Bahia, Brazil, in late February of 1832. About 3½ years were spent along the coast of South America, surveying and making inland explorations. The stop at the Galapagos Islands was for slightly more than a month, and, during that brief time, Darwin made the wealth of observations that were to change the course of the science of biology. The remainder of the voyage, across the Pacific to New Zealand and Australia, across the Indian Ocean to the Cape of Good Hope, back to Bahia once more, and at last home to England, occupied another year.*

I–7   *A distinguishing feature of the Galapagos tortoise is the shape of its shell, or carapace, which varies according to its island of origin.* (**a**) *Tortoises found on islands with comparatively lush vegetation are characterized by a domed shell. The domed shell protects the tortoise's soft parts as it makes its way through the thick undergrowth. This tortoise is a native of Isabela Island.* (**b**) *Tortoises found on arid islands, on which the typical vegetation is thornbush and cactus, are characterized by a saddleback shell. The high arch at the front of the shell enables the tortoise to reach upward in search of food. This tortoise was photographed on the island of Española.*

Most interesting to Darwin were the animals and plants that inhabited a small, barren group of islands, the Galapagos, which lie some 580 miles off the coast of Ecuador. The Galapagos were named after the islands' most striking inhabitants, the tortoises (*galápagos* in Spanish), some of which weigh 200 pounds or more. Each island has its own type of tortoise; sailors who took the tortoises on board and kept them as convenient sources of fresh meat on their sea voyages could generally tell which island any particular tortoise had come from (Figure I–7).

(a)

(b)

I–8 *The woodpecker finch is a rare phenomenon in the bird world because it is a tool user. Like the true woodpecker, it feeds on grubs, which it digs out of trees using its beak as a chisel. Lacking the woodpecker's long, barbed tongue, however, it resorts to an artificial probe, a twig or cactus spine, to remove the grubs from their holes. The woodpecker finch shown here has selected a cactus spine, which it has (a) inserted into a grub hole on a dead tree and (b) used to remove the grub from the hole. If the particular tool is an efficient one, as this cactus spine appears to be, the bird will carry it from tree to tree in its search for grubs.*

(a)   (b)

In addition to the tortoises, there was a group of finchlike birds, 13 species in all, that differed from one another in the sizes and shapes of their bodies and beaks, and particularly in the type of food they ate. In fact, although clearly finches, they had many characteristics seen only in completely different types of birds on the mainland. One finch, for example, feeds—as does the familiar woodpecker and also the aye-aye of Madagascar (Figure I–2b)—by routing insects out of the bark of trees. The finch is not fully equipped for this, however, having neither the long tongue of the woodpecker nor the long third finger of the aye-aye. Instead, the woodpecker finch (Figure I–8) uses a small stick or cactus spine to pry the insects loose.

From his knowledge of geology, Darwin knew that the Galapagos, clearly of volcanic origin, were much younger than the mainland. Yet the plants and animals of the islands were different from those of the mainland. Moreover, the inhabitants of the different islands in the archipelago differed from one another. Were the living things on each island the product of a separate special creation? "One might really fancy," Darwin mused at a later date, "that from an original paucity of birds in this archipelago one species had been taken and modified for different ends." This problem continued, in his own word, to "haunt" him.

The Darwinian Theory

Not long after his return to England, Darwin read an essay by the Reverend Thomas Malthus that had first appeared in 1798. In this essay, Malthus warned, as economists have warned ever since, that the human population was increasing so rapidly that it would soon be impossible to feed all of Earth's inhabitants.

Darwin saw that Malthus's conclusion—that food supply and other factors hold populations in check—is true for all species, not just the human one. For example, Darwin calculated that a single breeding pair of elephants, which are among the slowest reproducers of all animals, would, if all their offspring lived and, in turn, reproduced the normal number of offspring over a normal life span, produce a standing population of 19 million elephants in 750 years. Yet the average number of elephants generally remains the same over the years. So, although a single breeding pair could have, in theory, produced 19 million descendants, it did, in fact, produce an average of only two. But why these particular two? The process by which the survivors are "chosen" Darwin called **natural selection.**

Natural selection, according to Darwin, was a process analogous to the type of selection exercised by breeders of cattle, horses, dogs, and pigeons (Figure I–9). In **artificial selection,** we humans choose individual specimens of plants or animals for breeding on the basis of characteristics that seem to us desirable. In natural selection, the action of the environment takes the place of human choice. As individuals with certain hereditary characteristics survive and reproduce, and as individuals with other hereditary characteristics are eliminated, the population will slowly change. If some horses were swifter than others, for example, these individuals would be more likely to escape predators and survive, and their offspring, in turn, might be swifter, and so on.

I–9   *As a pigeon fancier, Darwin knew of the wide variety of exotic breeds that had been produced through artificial selection. All of the birds shown in this illustration from a book published in Germany in the early 1900s were descended from the common rock dove, the familiar city pigeon. As Darwin wrote, "If these varieties can be produced by man's hand, what might Nature not achieve?"*

According to Darwin, inherited variations among individuals, which occur in every natural population, are a matter of chance. They are not produced by the environment, by a "creative force," or by the unconscious striving of the organism. In themselves, they have no goal or direction. They may, however, turn out to be more or less useful to an organism, enhancing or diminishing its survival and reproduction. It is the operation of natural selection—the interaction of individual organisms with their environment—over a series of generations that gives direction to evolution. A variation arises by chance, but if it gives an organism even a slight advantage, it makes it more likely that the organism will leave surviving offspring.

To return to the giraffe, an animal with a slightly longer neck may have an advantage in feeding and thus be healthier, survive longer, and leave more offspring than one with a shorter neck. If neck length is an inherited trait, some of these offspring will also have long necks, and if the long-necked animals in this generation have an advantage, the next generation will include more long-necked individuals. In time, the population of short-necked giraffes will have become a population of longer-necked giraffes (although there will still be variations in neck length).

The essential difference between Darwin's formulation and that of any of his predecessors is the central role he gave to variation. Others had thought of variations as mere disturbances in the overall design, whereas Darwin saw that variations among individuals are the raw material of the evolutionary process. Species arise, he proposed, when differences among individuals within a group are gradually converted, over many generations, into differences between groups. This can occur as subsets of the original group become separated in space and time and are subjected to different environmental forces.

Darwin's *The Origin of Species*, which resulted from more than 20 years of thought after his return to England, is, in his own words, "one long argument." Fact after fact, observation after observation, culled from the most remote Pacific island to a neighbor's pasture, is recorded, analyzed, and commented upon. Every objection is weighed, anticipated, and countered. *The Origin of Species* was published on November 24, 1859, and the Western world has not been the same since.

Acceptance of Darwin's argument revolutionized the science of biology. As Ernst Mayr wrote in 1963:

> The theory of evolution is quite rightly called the greatest unifying theory in biology. The diversity of organisms, similarities and differences between kinds of organisms, patterns of distribution and behavior, adaptation and interaction, all this was merely a bewildering chaos of facts until given meaning by the evolutionary theory. There is no area of biology in which that theory has not served as an ordering principle.*

* From Ernst Mayr, *Animal Species and Evolution*, Harvard University Press, Cambridge, Mass., 1963.

Darwin's Long Delay

Darwin returned to England with the *Beagle* in 1836. Two years later, he read the essay by Malthus, and in 1842 he wrote a preliminary sketch of his theory, which he revised in 1844. On completing the revision, he wrote a formal letter to his wife requesting her, in the event of his death, to publish the manuscript. Then, with the manuscript and letter in safekeeping, he turned to other work. For more than 20 years following his return from the Galapagos, Darwin mentioned his ideas on evolution only in his private notebooks and in letters to his scientific colleagues.

In 1856, urged on by his friends Charles Lyell and botanist Joseph Hooker, Darwin set slowly to work preparing a manuscript for publication. In 1858, some 10 chapters later, Darwin received a letter from the Malay archipelago from another English naturalist, Alfred Russel Wallace, who had corresponded with Darwin on several previous occasions. Wallace presented a theory of evolution that exactly paralleled Darwin's own. Like Darwin, Wallace had traveled extensively and also had read Malthus's essay. Wallace, tossing in bed one night with a fever, had a sudden flash of insight. "Then I saw at once," Wallace recollected, "that the ever-present variability of all living things would furnish the material from which, by the mere weeding out of those less adapted to the actual conditions, the fittest alone would continue the race." Within two days, Wallace's 20-page manuscript was completed and in the mail to Darwin.

When Darwin received Wallace's letter, he turned to his friends for advice, and Lyell and Hooker, taking matters into their own hands, presented the theory of Darwin and Wallace at a scientific meeting just one month later. At that meeting, Lyell and Hooker read to the au-

Alfred Russel Wallace (1823–1913). As a young man, Wallace explored the Malay archipelago for eight years, covering about 14,000 miles by foot and native canoe. During his stay there, he collected 125,000 specimens of plants and animals, many of them previously unknown. His book about his Malay travels bears this inscription: "To Charles Darwin, Author of 'The Origin of Species,' I dedicate this book, not only as a token of personal esteem and friendship but also to express my deep admiration for his genius and his works."

dience four papers from Darwin's notes of 1844, excerpts from two letters written by Darwin, and Wallace's manuscript. Their presentation received little attention, but for Darwin the floodgates were opened. He finished his long treatise in little more than a year, and the book was finally published. The first printing was a mere 1,250 copies, but they were sold out the same day.

Why Darwin's long delay? His own writings, voluminous though they are, shed little light on this question. But perhaps his background does. He came from a conventionally devout family, and he himself had been a divinity student. Perhaps most important, his wife, to whom he was deeply devoted, was extremely religious. It is difficult to avoid the speculation that Darwin, like so many others, found the implications of his theory difficult to confront.

As we shall see throughout the pages that follow, evolution is the thread that links together all the diverse phenomena of the living world.

The theory of evolution has also deeply influenced our way of thinking about ourselves. With the possible exception of the new astronomy of Copernicus and Galileo in the sixteenth and seventeenth centuries, no revolution in scientific thought has had as much effect on human culture as this one. One reason is, of course, that evolution is in contradiction to the literal interpretation of the Bible. Another difficulty is that it seems to diminish human significance. The new astronomy had made it clear that Earth is not the center of the universe or even of our own solar system. Then the new biology asked us to accept the proposition that, as far as science can show, we are not fundamentally different from other organisms in either our origins or our place in the natural world.

Evolutionary Theory Today

Later on, in Chapters 19–27, we shall consider in more detail the process of evolution and the diversity of organisms it has produced. As you will see then, evolutionary theory is itself still evolving. Lively debates about various aspects of evolution occur almost constantly among its students (biologists, like all scientists, are students). Should we, for instance, measure evolutionary relationships—as between humans and chimps—by physical appearance, by the evidence of the fossil record, or by minute variations in the genetic information encoded in the chromosomes? Are animals capable of unselfish behavior, or is seemingly unselfish behavior always—in the evolutionary sense—really self-serving? (One biologist, calculating rapidly, announced that he would lay down his life "for two brothers or eight first cousins.") There will always be controversies of this sort, because that is the way science works.

That evolution has occurred is not, however, a matter of controversy among biologists. There is virtually complete agreement that Earth has had a long history—some 4.5 billion years is the current estimate—and that, during the course of that history, the complex living things present today, including ourselves, arose from simpler forms. There is also agreement that the particular features that suit an organism to its environment—the swiftness of the cheetah, the fragrance of the flower, the camouflage of the katydid, our own opposable thumb—have arisen by the process of natural selection as proposed almost 140 years ago by Charles Darwin.

There are, however, controversies between scientists and religious fundamentalists of various faiths. The fundamentalists seek to explain the world in supernatural or spiritual terms, in accordance with their particular religious doctrines. Scientists do not deny the possibility of supernatural phenomena or the importance of religious belief. They maintain, however, that phenomena such as special creation, whether or not they exist, lie outside the realm of science. Such questions cannot be answered, or even argued, by scientific means because the answers depend ultimately on one's inner faith, rather than on observations or experiments involving the external, material world.

The Nature of Science

Science, including biology, is a way of seeking principles of order. Art is another way, as are religion and philosophy. Science differs from these others in that it limits its search to the natural world, the physical universe. It also differs in the way it conducts the search for understanding.

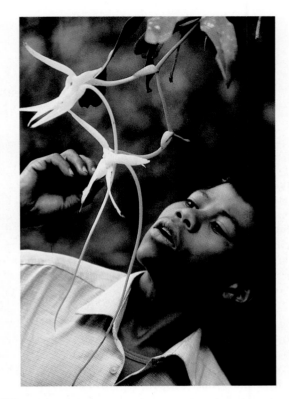

I–10   *The Christmas star orchid of Madagascar is characterized by a tubular nectary, generally about 11½ inches long. A little more than an inch of nectar is found at the very bottom of the tube. Darwin's study of this orchid led him to predict the existence of a moth with a tongue long enough to reach the nectar. Although he was ridiculed at the time by biologists who specialized in the study of insects, such a moth, which feeds on the nectar and pollinates the flowers, was later observed and identified. The young man in the photograph is Bedo Jaosolo, a talented and largely self-taught naturalist and guide who worked closely with scientists studying the rain forest of Madagascar. In 1989, at the age of 20, he was murdered, possibly because of his conservation activities.*

Observation, Hypothesis, Prediction, and Testing

The scientific process begins with an observation that leads to a question. (One of the most important in the history of biology was the question with which we began this Introduction: How does one make sense of the bewildering diversity of living things and the patterns in which they are scattered over the surface of the Earth?) A scientist seeks to answer the question by accumulating further information. Such information can be the product of concentrated thought. Or the additional information may be in the form of further observations, either original observations or reports from other observers (usually a combination of the two). Darwin, for example, spent decades gathering information—and thinking about that information—before proposing his concept of evolution by means of natural selection.

Early on, a scientist usually formulates a **hypothesis**—a tentative explanation. On the basis of the hypothesis, the scientist makes a **prediction** as to what will or will not be observed if the hypothesis is correct. The hypothesis is then subjected to a series of tests. For instance, birds migrate south in the fall, often flying at night, sometimes over open water, to a place they've never been before. How do they find their way? A tentative explanation: They navigate by the stars. A prediction: If birds navigate by the stars, they will not fly on cloudy nights. Testing by observation: Do migrating birds fly on cloudy nights? You might want to look for yourself.

Another example of prediction is provided by Darwin's study of the Christmas star orchid of Madagascar (Figure I–10). The long tubular structure trailing from each flower contains, at the bottom of the tube, a small amount of sugary nectar. On the basis of this structure, Darwin predicted that a moth would be found on Madagascar with a tongue between 10 and 11 inches long—because that length would be required to reach the nectar. Such a moth was discovered, some 40 years later, confirming Darwin's prediction.

Often, the tests of a hypothesis take the form of experiments. These are carefully structured observations that involve some form of manipulation or intervention by the investigator. As an example of the sequence from puzzling observation to hypothesis to prediction to testing, let us look at a study in England that also provides an example of the process of natural selection.

The Case of the Peppered Moth

The peppered moth, known scientifically as *Biston betularia*, was well known to British naturalists of the nineteenth century, who remarked that the moths were usually found on lichen-covered trees and rocks. Against this background, the light coloring of the moths made them practically invisible (Figure I–11). Until 1845, all reported specimens of *Biston betularia* were light-colored, but in that year one black moth of this species was captured in the growing industrial center of Manchester.

With the increasing industrialization of England, smoke particles began to pollute the foliage in the vicinity of industrial towns, killing the lichens and leaving the tree trunks bare. In heavily polluted districts, the trunks and even the rocks and ground became black. During this period, more and more black *Biston betularia* were found. Replacement of light-colored moths by dark ones proceeded briskly. By the 1950s, only a few of the light-colored population could be found, and these were far from industrial centers. Because of the prevailing westerly wind in England, pollutants were carried to the east of industrial towns, and the moths tended to be of the black variety right up to the east coast of England. The few light-colored populations were concentrated in the west, where lichens still grew.

Where did the black *Biston betularia* come from? Eventually, it was demonstrated that the black color was a natural variation in the population. The black moths had always been there, in very small numbers. But why had their numbers increased so dramatically? In the late 1950s, H. B. D. Kettlewell, a physician who was also an amateur collector of moths and butterflies, hypothesized that the color of the moths protected them from insect-eating birds (Figure I–12).

In the face of strong opposition from biologists who specialized in the study of insects—all of whom claimed they had never seen a bird eat a *Biston betularia* of any color—Kettlewell set out to test his hypothesis. He predicted that if he released both light and dark moths in a polluted area and then recaptured them later, he would recapture more of the dark moths than of the light moths. Why? Conversely, he predicted that if he released both types of moths in an unpolluted area, he would later recapture more of the light moths than of the dark ones.

As an experimental test, Kettlewell marked a sample of moths of each color by carefully putting a spot on the underside of the wings, where it could not be seen by birds. Then he released known numbers of marked individuals into a bird reserve near Birmingham, an industrial area where 90 percent of the local *Biston betularia* population consisted of black moths. Another

I–11   Biston betularia, *the peppered moth, resting on a lichen-covered tree trunk in an unpolluted English countryside. Once the moth has positioned itself on a patch of lichen, it remains absolutely motionless, further enhancing its camouflage.*

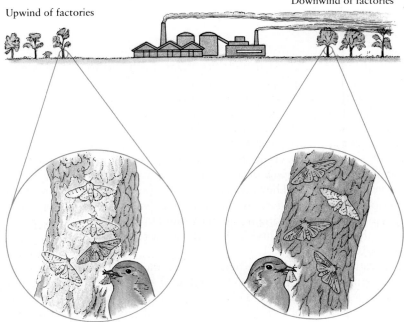

Upwind of factories   Downwind of factories

I–12 *Kettlewell hypothesized that the distribution and abundance of the light and dark varieties of the peppered moth were determined by the visibility of the moths to insect-eating birds. In unpolluted areas upwind of industrial zones, where the trees were covered with lichens, the light-colored moths would be less visible than the black moths and thus would be less likely to be eaten. In polluted areas downwind, however, where the trees were covered with soot, the black moths would be less visible and thus less likely to be eaten.*

sample was released into an unpolluted Dorset countryside, where no black moths were ordinarily found. He returned at night with light traps to recapture the marked moths. From the area around Birmingham, he recovered 40 percent of the black moths, but only 19 percent of the light ones. From the area in Dorset, 6 percent of the black moths and 12.5 percent of the light moths were recaptured.

To clinch the argument, Kettlewell placed moths on tree trunks in both locations, focused hidden movie cameras on them, and was able to record birds actually selecting and eating the moths. As his films revealed, they do this so rapidly that it is not surprising it was not previously observed. The photographic record showed that when equal numbers of black and light moths were available near Birmingham, the birds seized 43 light-colored moths and only 15 black ones; in Dorset, they took 164 black moths but only 26 light-colored forms.

Note that Kettlewell did not just think about the puzzling observation (the increase in the number of black moths) and a possible explanation (in heavily polluted districts, black moths were less likely to be seen and eaten by birds). He also devised tests of his hypothesis. And, when his first test, which was an indirect one, supported the hypothesis, he devised a second, more direct test. Repeated testing, typically using several different methods, is an essential characteristic of the scientific process.

When a scientist has collected enough information to support a particular hypothesis, he or she then reports the results to other scientists. Such a report may take place at a scientific meeting, or in a scientific publication, such as a journal or book (for example, Darwin's

The Origin of Species). If the information is sufficiently interesting or the hypothesis is an important one, other scientists will attempt to repeat the observations or experiments in order to confirm, refute, or extend them.

In the case of the moths, investigations by other scientists in England and the United States have demonstrated that the replacement of light-colored moths by dark-colored ones is a common occurrence in heavily polluted areas. It has been observed among some 70 other moth species in England and some 100 species of moths in the area of Pittsburgh, Pennsylvania. It has also been observed in many species of butterflies.

From Hypothesis to Theory

Once a hypothesis of broad, fundamental importance has survived a number of tests, it is generally referred to as a **theory**. Thus a theory in science has a different meaning from the word "theory" in everyday usage, in which "just a theory" carries with it the implication of a flight of fancy, a hunch, or some abstract, impractical notion. In science, a theory is a carefully formulated, well-tested proposition that describes the relationship among a number of different observations. The theory of evolution, for example, is built on an enormous number and variety of observations.

As new observations are made, however, theories are challenged. For example, various theories describing the structure of the atom have superseded one another as powerful new instruments have made new observations possible. However, the existence of atoms has not been questioned. Similarly, over the past 135 years, the theory of evolution has been tested, retested, amended, and fine-tuned. Some of the new discoveries in genetics—how traits are passed from parents to offspring—seemed at first to challenge the theory of evolution, but, as we shall see in later chapters, they have now become an important part of it. Such challenges will always continue, because, as long as there are curious human beings, there will always be new observations leading to new hypotheses, new predictions, and new tests. That is the way science works.

The Limits of Science

The raw materials of science are our observations of the phenomena of the natural universe. Science—unlike art, religion, or philosophy—is limited to what is observable and measurable and, in this sense, is rightly categorized as materialistic. Hunches are abandoned, hypotheses superseded, theories revised—and occasionally, shattered—but observations endure, and, moreover, they are used over and over again, sometimes in wholly new ways. It is for this reason that scientists stress and seek objectivity. In the arts, by contrast, the emphasis is on subjectivity—experience as filtered through the individual consciousness.

Because of this emphasis on objectivity, value judgments cannot be made in science in the way that such judgments are made in philosophy, religion, and the arts, and indeed in our daily lives. Whether or not something is good or beautiful or right in a moral sense, for example, cannot be determined by scientific methods.

Such judgments, even though they may be supported by a broad consensus, are not subject to scientific testing.

At one time, the sciences, like the arts, were pursued solely for their own sake, for pleasure and excitement and satisfaction of the insatiable curiosity with which we are both cursed and blessed. In the twentieth century, however, the sciences have spawned a host of giant technological achievements—polio vaccine, pesticides, electronic communications, indestructible plastics, nuclear power, perhaps even ways to manipulate our genetic heritage. Yet science cannot give us insight into how to use these achievements wisely or guidance in how to solve the problems they have generated.

The reason that science cannot and does not solve many of the problems for which we seek solutions is inherent in its nature. Most of the problems we now confront can be solved only by value judgments. For example, science has given us nuclear power and can give us predictions as to the extent of the biological damage that would result from accidents that allowed radioactivity to escape into the environment. Yet science cannot help us, as citizens, in weighing our fear of nuclear accidents against our energy needs and our desire for cleaner air. Scientists can develop a synthetic hormone that will produce early abortions, but they are not the ones to decide whether such a hormone will be made available to pregnant women seeking abortions. Science can give us information on which to base our judgments, but it cannot make these judgments for us.

 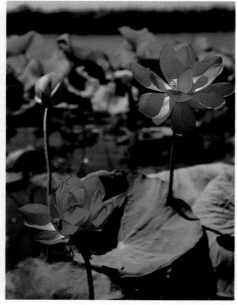

I–13 *In science, great stress is placed on objectivity. As a consequence, scientists, speaking as scientists, refrain from making value judgments. Thus, for example, they would not identify any organism as the ugliest or the most beautiful. As a private individual, however, a scientist may have his or her own opinion about (a) the fishing bat and (b) the sacred lotus.*

(a) (b)

It is one of the ironies of this so-called "age of science and materialism" that probably never before have ordinary individual men and women, including scientists, been confronted with so many moral and ethical dilemmas. In this text, we shall discuss some of the dilemmas that have grown out of the achievements of modern science and technology. Our greater concern, however, is to provide you with the biological knowledge necessary to understand the relevant information as you make your own value judgments regarding the problems that confront us now and that will do so in the future.

You may have decided to study biology because of an interest in environmental problems or a desire to know more about the workings of your own body or a concern with the implications of genetic engineering—in short, because it is "relevant." The study of biology is, indeed, pertinent to many aspects of our day-to-day existence, but do not make this your main focus. Above all other considerations, study biology because it is "irrelevant"—that is, study it for its own sake, because, like art and music and literature, it is an adventure for the mind and nourishment for the spirit.

Summary

Until the eighteenth century, it was generally accepted that species (kinds of organisms) were the products of a special (divine) creation and had remained unchanged ever since. This concept came into question as a result of several developments: (1) the discovery of an enormous number of new species by the European naturalists who fanned out around the globe; (2) studies in geology that indicated a process of gradual, constant change in the surface of the Earth; (3) the accompanying recognition that the Earth has had a long history (without which evolution would have been impossible); and (4) the discovery of the real nature of fossils.

Darwin was not the first to propose a theory of evolution. His most notable predecessor was Lamarck, whose theory of evolution—now known to be in error—was based on the inheritance of characteristics acquired by an organism during its lifetime. Darwin's theory differed from others in that it envisioned evolution as a two-part process, depending upon (1) the existence in nature of inheritable variations among organisms and (2) the process of natural selection by which some organisms, by virtue of their inheritable variations, leave more offspring than others. Darwin's theory is regarded as the greatest unifying principle in biology.

Science is a way of seeking knowledge of the natural world. The essential components of the scientific process are an observation that leads to a question, the formulation of a hypothesis (a tentative answer to the question), the prediction of what would be observed under a particular set of circumstances if the hypothesis were correct, and the testing of the prediction through further observations. Hypotheses that describe the relationship among a number of different observations and that have been repeatedly tested in a variety of different ways are known as theories. Theories provide a framework for further observations, questions, hypotheses, predictions, and tests. They are, as a result, always subject to further revision.

Questions

1. What is the essential difference between Darwin's theory of evolution and that of Lamarck?

2. The chief predator of an English species of snail is the song thrush. Snails that inhabit woodland floors usually have dark shells, whereas those that live on grass have yellow shells, which are less clearly visible against the lighter background. Explain, in terms of Darwinian principles.

3. The phrase "chance and necessity" has been used to describe the Darwinian theory of evolution. Relate this to the fact that snails living on grass do not have green shells, but there are, for example, green frogs and green insects.

4. What tool-using animals other than the woodpecker finch can you think of?

5. When scientists report new findings, they are expected to reveal their methods and raw data as well as their conclusions. Why is such reporting considered essential?

Suggestions for Further Reading

Berra, Tim M.: *Evolution and the Myth of Creationism: A Basic Guide to the Facts in the Evolution Debate,* Stanford University Press, Stanford, Calif., 1990.*

An excellent presentation of the modern theory of evolution, the evidence in its support, and examples of its explanatory power. Written for the general reader, this well-illustrated book also provides an overview of the major events in evolutionary history. It concludes with a discussion of recent political controversies over evolution.

Darwin, Charles: *The Origin of Species by Means of Natural Selection, or The Preservation of Favored Races in the Struggle for Life,* New American Library, Inc., New York, 1986.*

Darwin's "long argument." Every student of biology should, at the very least, browse through this book to catch its special flavor and to begin to understand its extraordinary force.

Darwin, Charles: *The Voyage of the Beagle,* New American Library, Inc., New York, 1988.*

Darwin's own chronicle of the expedition on which he made the discoveries and observations that eventually led him to his theory of evolution. The sensitive, eager young Darwin that emerges from these pages is very unlike the image many of us have formed of him from his later portraits.

Kettlewell, H. B. D.: "Selection Experiments on Industrial Melanism in the Lepidoptera," *Heredity,* vol. 10, pages 287–301, 1955.

The journal article in which Kettlewell reported his now classic studies of natural selection in the peppered moth.

Lanting, Frans: *A World out of Time: Madagascar,* Aperture Foundation, Inc., New York, 1990.

This beautiful book is filled with more than 100 of Lanting's photographs of the peoples, plants, and animals of Madagascar. Essays by Alison Jolly, one of the world's leading authorities on lemurs, John Mack, a cultural anthropologist, and conservationist Gerald Durrell make clear not only the uniqueness of this island paradise but also its imminent peril. Highly recommended.

Lewin, Roger: *Thread of Life: The Smithsonian Looks at Evolution,* Smithsonian Books, Washington, D.C., 1991.*

An absorbing account of the historical development of evolutionary thought, with current applications from biochemistry, paleontology, and geology. Well written and beautifully illustrated with a rich assortment of color photographs.

Mayr, Ernst: *The Growth of Biological Thought: Diversity, Evolution, and Inheritance,* Harvard University Press, Cambridge, Mass., 1982.*

The first of two volumes on the history of biology and its major ideas, written by one of the leading figures in the study of evolution. The introductory chapters provide an outstanding analysis of the philosophy and methodology of the biological sciences. This book, like Darwin's The Origin of Species, should at least be sampled by every serious student of biology.

Milner, Richard: *The Encyclopedia of Evolution: Humanity's Search for Its Origins,* W. W. Norton & Company, New York, 1993.*

With more than 600 entries, this engagingly written reference covers a wide range of topics on the theory of evolution and its influence on Western culture—from Darwin and Wallace to sperm banks, Chief Red Cloud, and Fantasia. *Well illustrated, authoritative, and rich with fascinating detail.*

Moorehead, Alan: *Darwin and the Beagle,* Penguin Books, New York, 1978.*

A delightful narrative of Darwin's journey, beautifully illustrated with contemporary or near contemporary drawings, paintings, and lithographs.

Simpson, George Gaylord: *Fossils and the History of Life,* W. H. Freeman and Company, New York, 1983.

A beautifully written and illustrated summary of the fossil evidence on which much of our knowledge of evolution rests, by one of the major figures in twentieth-century evolutionary biology.

Tattersall, Ian: "Madagascar's Lemurs," *Scientific American,* January 1993, pages 110–117.

Wallace, Alfred Russel: *The Malay Archipelago,* Oxford University Press, New York, 1990.*

A reprint of Wallace's classic work, accompanied by an excellent introduction that provides much additional information about Wallace's life and scientific work.

* Available in paperback.

The Unity of Life

The universe in which we live is composed of stars that, like our Sun, are born and die. In the birth and death of stars, all of the atoms that make up our solar system, our planet, and ourselves had their beginnings.

The birth of a star begins in clouds of gases, mainly hydrogen and helium, that surge and eddy through the spaces between previously formed stars. Moving at random and accompanied by very fine dust, the gases form accidental pockets, their atoms held together by gravity. As a cloud contracts, its interior becomes more dense and its temperature rises. From this mass, a star will form.

As the embryonic star continues to contract, the temperature rises millions of degrees, and nuclear fusion reactions take place. In these reactions, hydrogen is transformed to helium and energy is released. At this moment, the star is born. "Fireflies tangled in a silver braid," wrote the poet Tennyson about the Pleiades, the group of young stars shown here. The haloes and blue patches of haze surrounding the stars are remnants of the dust and gases from which they formed. The Pleiades can be seen in the "shoulder" of the constellation Taurus (the bull).

During a star's lifetime, hydrogen is transformed to helium, and helium to carbon and oxygen, which themselves fuse to form the heavier elements. These reactions release enormous quantities of energy, visible to us as the light of distant stars and as the light and heat of the nearest star—our Sun— which formed some 5 billion years ago.

Eventually a star uses up its fuel, the hydrogen atoms at its core. At its death, there is a sudden great explosion in the sky. During the few minutes of the explosion, the old, dying star releases energy comparable to the total amount emitted by our Sun in a billion years. This enormous blast blows the outer layers of the star into space, creating new clouds of gases and dust. Over time, these will cool and merge with the interstellar clouds of dust and gas from which new stars will be born.

Atoms and Molecules

Our universe began, according to current theory, with the "big bang," a tremendous explosion that filled all space. Prior to this, all of the energy and matter of the present universe is thought to have existed in the form of pure energy, compressed into an infinitesimally small point. This energy was released by the "big bang," and every particle of matter formed from the energy was hurled away from every other particle. The temperature at the time of the explosion—some 10 to 20 billion years ago—was about 100,000,000,000 degrees Celsius (10^{11} °C). At this temperature, not even atoms could hold together; all matter was in the form of subatomic particles. Moving at enormous velocities, these particles had fleeting lives. Colliding with great force, they annihilated one another, creating new particles and releasing more energy.

As the universe expanded and gradually cooled, more matter began to form from energy. When the universe reached a temperature of 2500°C (about the temperature of a white-hot wire in an incandescent light bulb), two types of stable particles, previously present only in relatively small amounts, began to assemble. These particles—protons and neutrons—are very heavy as subatomic particles go. Held together by forces that are still not completely understood, they formed the central cores, or nuclei, of atoms. These nuclei, with their positively charged protons, attracted small, light, negatively charged particles—electrons—which moved rapidly around them. Thus, atoms came into being.

It is from these atoms—blown apart, formed, and re-formed over the course of several billion years—that all the stars and planets of our universe are formed, including our particular star and planet. And it is from the atoms present on this planet that living systems assembled themselves and evolved. Each atom in our own bodies had its origin in that enormous explosion so long ago. You and I are flesh and blood, but we are also stardust.

This text begins where life began, with the atom. At first, it might appear that lifeless atoms have little to do with biology. Bear with us, however. A closer look reveals that the activities we associate with being alive depend on combinations and exchanges between atoms. Moreover, the force that binds the electron to the atomic nucleus stores the energy that powers living systems.

Atoms

All matter, including the most complex living organisms, is made up of combinations of **elements.** Some 92 elements occur naturally on Earth. Many of them are familiar to you; for example, carbon, found in its purest forms in diamond and graphite; oxygen, present in the air we breathe; calcium, incorporated by a great variety of organisms into sea shells, egg shells, bones, and teeth; and iron, the element responsible for the red color of our blood. Elements are, by definition, substances that cannot be broken down into other substances by ordinary chemical reactions. The smallest particle of an element that possesses the properties of that element is an **atom.**

Every atom has a core, or **nucleus,** that contains one or more positively charged particles, called **protons.** The number of protons distinguishes the atoms of different elements from one another. For example, an atom of hydrogen, the simplest of the elements, has 1 proton in its nucleus; an atom of carbon has 6 protons in its nucleus. The number of protons in the nucleus of a particular atom is called its **atomic number.** The atomic number of hydrogen is therefore 1, and the atomic number of carbon is 6.

Outside the nucleus of an atom are negatively charged particles, the **electrons,** which are attracted by the positive charge of the protons. The number of electrons in an atom is determined by and is equal to the number of protons in its nucleus. The electrons, in turn, determine whether an atom will react with other atoms and how it will do so. As we shall see, chemical reactions involve changes in the numbers and energy of electrons.

Atoms also contain **neutrons,** which are uncharged particles of about the same weight as protons. These, too, are found in the nucleus of the atom, where they seem to play a role in stabilizing the nucleus. The **atomic weight** of an element is essentially equal to the number of protons plus neutrons in the nuclei of its atoms. As you can calculate from the data in Table 1–1, the atomic weight of carbon is 12, whereas that of hydrogen, which contains no neutrons, is 1. Electrons are so

light by comparison that they are usually disregarded in calculations of atomic weight. When you weigh yourself, only about 30 grams—approximately 1 ounce—of your total weight is made up of electrons.

Isotopes

All atoms of a particular element have the same number of protons in their nuclei. Sometimes, however, different atoms of the same element contain different numbers of neutrons. These atoms, which differ from one another in their atomic weights but not in their atomic numbers, are known as **isotopes** of the element.

Hydrogen, for example, exists in three different isotopic forms (Table 1–2). The common form of hydro-

Table 1–1	Atomic Structure of Some Familiar Elements			
		Nucleus		
Element	**Symbol**	**Number of Protons**	**Number of Neutrons***	**Number of Electrons**
Hydrogen	H	1	0	1
Helium	He	2	2	2
Carbon	C	6	6	6
Nitrogen	N	7	7	7
Oxygen	O	8	8	8
Sodium	Na	11	12	11
Magnesium	Mg	12	12	12
Phosphorus	P	15	16	15
Sulfur	S	16	16	16
Chlorine	Cl	17	18	17
Potassium	K	19	20	19
Calcium	Ca	20	20	20

* In most common isotope.

Table 1–2	Isotopes of Hydrogen					
Isotope		**Atomic Number**	**Atomic Weight**	**Number of Protons**	**Number of Neutrons**	**Number of Electrons**
Name	**Symbol**					
Hydrogen	^{1}H	1	1	1	0	1
Deuterium	^{2}H	1	2	1	1	1
Tritium	^{3}H	1	3	1	2	1

gen, with its one proton, has an atomic weight of 1 and is symbolized as ^{1}H, or simply H. A second isotope of hydrogen contains one proton and one neutron and so has an atomic weight of 2; this isotope is symbolized as ^{2}H. A third, extremely rare isotope, ^{3}H, has one proton and two neutrons and so has an atomic weight of 3. The chemical behavior of the two heavier isotopes is essentially the same as that of ordinary hydrogen. All three isotopes have only one electron each, and it is the electrons that determine chemical properties.

Most elements have several isotopic forms. The differences in weight, although very small, are sufficiently great that they can be detected with modern laboratory apparatus. Moreover, many, but not all, of the less common isotopes are radioactive. This means that the nucleus of the atom is unstable and emits energy as it changes to a more stable form. The energy given off by the nucleus of a radioactive isotope can be detected with a Geiger counter or on photographic film.

Isotopes have a number of important uses in biological research and in medicine. They can be used, for example, to determine the age of fossils. Each type of radioactive isotope emits energy and changes into another kind of isotope at a characteristic and fixed rate. As a result, the relative proportions of different isotopes in a fossil or in a nearby rock sample give a good indication of how long ago it was formed.

Another use of radioactive isotopes is as "tracers." For example, an isotope of the element thallium, which is unreactive in the human body, can be used to identify blocked blood vessels in persons with symptoms of heart disease. Isotopes have other diagnostic uses in medicine, as well as an important role in the treatment of many forms of cancer.

Since isotopes of the same element all have the same chemical properties, a radioactive isotope will behave in an organism just as its more common nonradioactive isotope does. As a result, biologists have been able to use isotopes of a number of elements—especially carbon, nitrogen, and oxygen—to trace the course of many essential chemical reactions in living organisms.

Electrons and Energy

Atoms—the indivisible units of the elements—are the fundamental building blocks of all matter, both living and nonliving. Yet they are mostly empty space. As the electrons move around the nucleus at almost the speed of light, the distance from electron to nucleus is, on average, about 1,000 times the diameter of the nucleus. The electrons are so exceedingly small that the space is almost entirely empty.

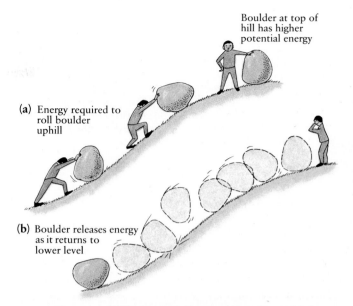

Boulder at top of hill has higher potential energy

(a) Energy required to roll boulder uphill

(b) Boulder releases energy as it returns to lower level

1–1   (a) *The energy used to push a boulder to the top of a hill (less the heat energy resulting from friction between the boulder and the hill) becomes potential energy, stored in the boulder as it rests at the top of the hill. (b) When the boulder rolls back down the hill, this potential energy is converted to energy of motion and released.*

The distance between an electron and the nucleus is determined by the amount of potential energy the electron possesses. **Potential energy,** often called "energy of position," is the stored energy that any object—not just an electron—possesses because of its position. The greater the amount of energy possessed by the electron, the farther it will be from the nucleus. Thus, an electron with a relatively small amount of energy is found close to the nucleus and is said to be at a low **energy level.** An electron with more energy is farther from the nucleus, at a higher energy level.

An analogy may be useful. A boulder resting on flat ground neither gains nor loses potential energy. If, however, you change its position by pushing it up a hill, you increase its potential energy. As long as it sits on the peak of the hill, the boulder once more neither gains nor loses potential energy. If it rolls down the hill, however, potential energy is converted to energy of motion and released (Figure 1–1). The electron is like the boulder in that an input of energy can move it to a higher energy level—farther away from the nucleus. As long as it remains at the higher energy level, it possesses the added energy. And, just as the boulder is likely to roll downhill, the electron also tends to go to its lowest possible energy level.

1–2   (a) *When an atom, such as the hydrogen atom in this diagram, receives an input of energy, an electron (shown here in red) may be boosted to a higher energy level. The electron thus gains potential energy. (b) When the electron returns to its previous level, this energy is released in the form of heat or light.*

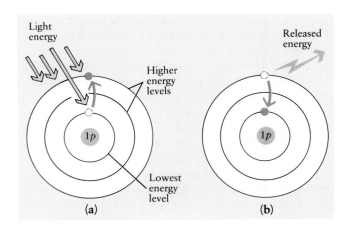

It takes energy to move a negatively charged electron farther away from a positively charged nucleus, just as it takes energy to push a boulder up a hill. However, unlike the boulder on the hill, the electron cannot be pushed part of the way up. With a sufficient input of energy, an electron can move from a lower energy level to any one of several higher energy levels, but it cannot move to an energy state somewhere in between. For an electron to move from one energy level to a higher one, it must absorb a discrete amount of energy, equal to the difference between the two particular energy levels. When the electron returns to its original energy level, that same amount of energy is released (Figure 1–2).

In the green cells of plants and algae, the radiant energy of sunlight raises electrons to higher energy levels. In a series of reactions, which will be described in Chapter 9, these electrons are passed "downhill" from one energy level to another until they return to their original energy level. During these transitions, the radiant energy of sunlight is transformed into the chemical energy on which life on Earth depends.

The Arrangement of Electrons

The way an atom reacts chemically is determined by the number and arrangement of its electrons. An atom is most stable when all of its electrons are at their lowest possible energy levels. Therefore the electrons of an atom fill the energy levels in order—the first is filled before the second, the second before the third, and so on.

The first energy level, closest to the nucleus, can hold a maximum of two electrons. Thus the single electron of hydrogen (atomic number 1) moves around the nucleus within the first energy level. Similarly, the two electrons of helium (atomic number 2) move within the first energy level.

1–3   *The leaves of these corn plants contain chlorophyll, which gives them their green color. When light strikes a molecule of chlorophyll, electrons in the molecule are raised to higher energy levels. As each electron returns to its previous energy level, a portion of the energy released is captured in the bonds of carbon-containing molecules.*

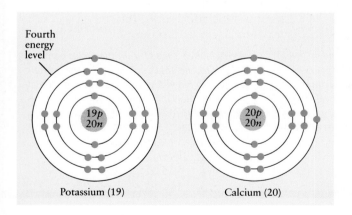

1–4 *Schematic diagrams of the electron arrangements in some familiar elements. In each of these diagrams, the electrons are at their lowest possible energy levels. Note how the energy levels are filled in sequence as the atomic number increases.*

The first energy level can hold a maximum of 2 electrons. The second energy level can hold a maximum of 8 electrons, as can the third energy level of the elements through calcium (atomic number 20). As you can see, potassium (atomic number 19) has one electron in the fourth energy level, and calcium (atomic number 20) has two electrons in the fourth energy level. In atoms of higher atomic number, additional electrons (up to a maximum of 10) fill the third energy level before any more electrons are added to the fourth energy level.

Atoms of higher atomic number than helium have more than two electrons. Since the first energy level is filled by two electrons, the additional electrons must occupy higher energy levels, farther from the nucleus. The second energy level can hold a maximum of eight electrons—and so can the third energy level of elements through atomic number 20 (calcium).

Figure 1–4 illustrates the electron arrangements in the atoms of 14 representative elements, 11 of which play major roles in living systems. Each diagram shows the energy levels occupied by the electrons when the atom is in its ground state—that is, when all of its electrons are at their lowest possible energy levels. You will find it helpful to refer to these diagrams as we consider the chemical behavior of specific elements.

Models of Atomic Structure

The concept of the atom as the indivisible unit of the elements is almost 200 years old. However, our ideas about atomic structure have undergone many changes over the years. These ideas are usually presented in the form of models.

The earliest model of the atom, emphasizing its indivisibility, portrayed the atom as a sphere like a billiard ball. When electrons were discovered and it was realized that they could be removed from the atom, the billiard-ball model gave way to the plum-pudding model. This model represented the atom as a solid, positively charged mass with negatively charged particles, the electrons, embedded in it.

Subsequently, however, physicists found that the electrons are located outside the nucleus, moving around it at very high speed. This discovery gave rise to the planetary model of the atom **(a)**, which depicted the electrons as moving in orbits around the nucleus. Although this model was of limited scientific use,

it gained wide currency as a symbol for atomic energy.

A more useful model is derived from the work of the Danish physicist Niels Bohr (1885–1962), who made the key discovery that different electrons have different amounts of energy and are at different distances from the nucleus. In the Bohr model **(b)**, the energy lev-

(a) *A planetary model of the carbon atom.*

els are depicted as concentric circles surrounding the nucleus. The Bohr model is not, in any sense, a true "picture" of the atom, and it has been superseded by another model. However, schematic diagrams derived from the Bohr model can help us visualize the energy levels of an atom, keep track of the number of electrons in each energy level, and see how atoms are likely to interact with one another.

The most recent model of atomic structure provides a more accurate picture of the atom. An electron is so small and moves so rapidly that it is theoretically impossible to determine, at any given moment, both its precise location and the exact amount of energy it possesses. As a result of this difficulty, the current model describes the pattern of an electron's motion rather than its position. The volume of space in which an electron will be found 90 percent of the time is defined as its **orbital**.

In any atom, the electrons at the lowest energy level—the first energy level—occupy a single spherical orbital, which can contain a

(b) *A Bohr model of the carbon atom.*

(c) *The four orbitals of the second energy level. Each orbital can hold a maximum of two electrons. One orbital is spherical and encloses the spherical orbital of the first energy level, which, in turn, encloses the nucleus, located at the intersection of the axes. The other three orbitals*

are dumbbell-shaped, with their axes perpendicular to one another. For clarity, the orbitals are shown individually in this diagram. In reality, the orbitals influence one another and determine the overall shape of the atom.

maximum of two electrons. At the second energy level, which can contain a maximum of eight electrons, there are four orbitals **(c)**. At the third energy level of elements through atomic number 20 (calcium), there are also four orbitals. In elements of higher atomic number, the pattern becomes more complex.

The electrons in each orbital traverse a particular region of three-dimensional space. Taken together, these patterns of electron movement give the atom as a whole a particular three-dimensional shape **(d)**. The atom, however, has no rigid boundaries but is instead defined by regions of charge. Even these are not always the same. About 10 percent of the time, the electrons of any given orbital are somewhere else in space, farther from the nucleus.

Such a "soft" structure would appear to provide an unlikely foundation for the material world that surrounds us and for the organisms with which it is filled. And, yet, according to modern physics, such is the case.

(d) An orbital diagram of the carbon atom. In this atom, each orbital of the second energy level contains only one electron. Each electron moves in a pattern that provides maximum separation from the other electrons. The resulting shape of the atom resembles four teardrops, extending out from the nucleus.

The Basis of Chemical Reactivity

As we have noted previously, an atom is most stable when all of its electrons are at their lowest possible energy levels. Moreover, an atom in which the outermost energy level is completely filled with electrons is more stable than one in which the outer energy level is only partially filled. For example, helium (atomic number 2) has two electrons at the first energy level, which means that its outer energy level (in this case, also its lowest energy level) is completely filled. Helium is therefore extremely stable and does not ordinarily participate in chemical reactions. Similarly, neon (atomic number 10) has two electrons at the first energy level and eight at the second energy level. Both energy levels are completely filled, and neon is unreactive.

In the atoms of most elements, however, the outer energy level is only partially filled. Atoms in which the outer energy level is partially filled tend to interact with other atoms in such a way that after the reaction both atoms have completely filled outer energy levels. Some atoms lose electrons. Others gain electrons. And, in many of the most important chemical reactions that occur in living systems, atoms share their electrons with each other.

Bonds and Molecules

When atoms interact with one another, resulting in filled outer energy levels, new, larger particles are formed. Discrete particles, consisting of two or more atoms, are known as **molecules.** The forces that hold atoms together within molecules are known as **chemical bonds.** Two principal types are ionic and covalent.

Ionic Bonds

For many atoms, the simplest way to attain a completely filled outer energy level is either to gain or to lose one or two electrons. For example, a chlorine atom (atomic number 17) needs one electron to complete its outer energy level. By contrast, a sodium atom (atomic number 11) has a single electron in its outer energy level. This electron is strongly attracted by the chlorine atom and jumps from the sodium to the chlorine (Figure 1–5). As a result of this transfer, both atoms have outer energy levels that are completely filled, and all the electrons are at their lowest possible energy levels.

(a) Sodium atom (Na)
($11p^+ + 11e^- = 0$ charge)

(b) Chlorine atom (Cl)
($17p^+ + 17e^- = 0$ charge)

(c)

Sodium ion (Na$^+$)
($11p^+ + 10e^- = +1$ charge)

Chloride ion (Cl$^-$)
($17p^+ + 18e^- = -1$ charge)

1–5   *The interaction of sodium and chlorine. (a) The sodium atom has only one electron in its outer energy level. (b) The chlorine atom, by contrast, needs to gain one electron in order to complete its outer energy level. (c) When these two atoms come close to one another, the single electron in the outer energy level of the sodium atom jumps to the outer energy level of the chlorine atom, thereby completing that energy level. When sodium loses the electron, its second energy level—with a full complement of 8 electrons—becomes its outer energy level. Thus both atoms now have filled outer energy levels and are more stable than they were before the electron transfer. However, the atoms have become electrically charged. Sodium now has a charge of +1, and chlorine a charge of −1.*

In this process, however, the original atoms have become electrically charged. Such charged atoms are known as **ions.** The chlorine atom, having accepted an electron from sodium, now has one more electron than proton and is a negatively charged chloride ion, symbolized as Cl$^-$. Conversely, the sodium atom now has one less electron than proton and is a positively charged sodium ion, symbolized as Na$^+$.

Because of their charges, positive and negative ions attract one another. Thus the sodium ion (Na$^+$) with its single positive charge is attracted to the chloride ion (Cl$^-$) with its single negative charge. The resulting substance, sodium chloride (NaCl), is ordinary table salt (Figure 1–6). Similarly, when a calcium atom (atomic number 20) loses two electrons, the resulting calcium ion (Ca^{2+}) can attract and hold two Cl$^-$ ions. Calcium chloride is identified in chemical shorthand as CaCl$_2$, with the subscript 2 indicating that two chloride ions are present for each ion of calcium.

Bonds that involve the mutual attraction of ions of opposite charge are known as **ionic bonds.** Such bonds can be quite strong, but, as we shall see in the next chapter, many ionic substances break apart easily in water, producing free ions. Small ions such as Na$^+$ and Cl$^-$ make up less than 1 percent of the weight of most living matter, but they play crucial roles. Potassium ion (K$^+$) is the principal positively charged ion in most organisms, and many essential biological processes occur only in its presence. Calcium ion (Ca^{2+}), K$^+$, and Na$^+$ are all involved in the production and propagation of the nerve impulse. In addition, Ca^{2+} is required for the contraction of muscles and for the maintenance of a normal heartbeat. Magnesium ion (Mg^{2+}) forms a part of the chlorophyll molecule, the molecule in green plants and algae that traps radiant energy from the sun.

1–6   (a) *Oppositely charged ions, such as sodium and chloride ions, depicted here as spheres, attract one another. Table salt is crystalline NaCl, a latticework of alternating Na$^+$ and Cl$^-$ ions held together by their opposite charges. Such bonds between oppositely charged ions are known as ionic bonds.*

(b) *The regularity of the latticework is reflected in the structure of salt crystals, magnified here about 30 times.*

Na$^+$ Cl$^-$

(a)

(b)

1–7   *The acrobatic leap of the European tree frog, like the movement of all complex animals, is the result of coordinated muscle contractions triggered by nerve impulses. Sodium, potassium, and calcium ions are involved in producing and propagating nerve impulses, and calcium ions are required for the contraction of muscle fibers.*

Covalent Bonds

Another way for atoms to complete their outer energy levels is by sharing electrons with each other. Chemical bonds that result from the sharing of one or more pairs of electrons are known as **covalent bonds.** In a covalent bond, each electron spends part of its time around one nucleus and part of its time around the other. Thus the sharing of electrons completes the outer energy level of each atom and neutralizes the positive charge of its nucleus.

Atoms that need to gain electrons to achieve a filled, and therefore stable, outer energy level have a strong tendency to form covalent bonds. Thus, to take the simplest example, a hydrogen atom forms a covalent bond with another hydrogen atom (Figure 1–8), producing a molecule of hydrogen gas (H_2). A hydrogen atom can also form a covalent bond with any other atom that needs to gain an electron to complete its outer energy level.

(a)

Hydrogen atoms (2H)

H — H
Hydrogen molecule (H_2)

(b)

1–8   (a) *When a molecule of hydrogen is formed, each atom shares its single electron with the other atom. As a result, both atoms effectively have a filled first energy level, containing two electrons—a highly stable arrangement. This type of bond, in which electrons are shared, is known as a covalent bond. In structural formulas, it is represented as a single line:* —.

(b) *A representation of the volume of space traversed by the two electrons of the covalent bond as they move around the hydrogen nuclei. Because the electrons move so rapidly—at nearly the speed of light—the charges of both nuclei are effectively neutralized at all times.*

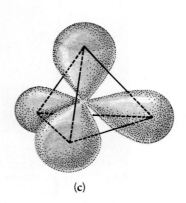

(c)

1–9   *A carbon atom, with four electrons in its outer energy level, can form covalent bonds with as many as four other atoms.*
(a) When a carbon atom reacts with four hydrogen atoms, each of the electrons in its outer energy level forms a covalent bond
with the single electron of one hydrogen atom, producing **(b)** *a molecule of methane (CH₄).*
(c) A three-dimensional representation of the methane molecule. The electrons forming the covalent bonds move rapidly in complex
patterns that encompass the hydrogen nuclei and also bring the electrons close to the carbon nucleus. The positive charge of each nucleus is neutralized, and the outer energy level of each atom is filled.

1–10   *A schematic diagram of the water molecule (H₂O). Each of the two single covalent bonds in the molecule consists of one electron contributed by oxygen and one electron contributed by hydrogen.*

Of extraordinary importance in living systems is the capacity of carbon atoms to form covalent bonds. A carbon atom has four electrons in its outer energy level (see Figure 1–4). It can share each of those electrons with another atom, forming covalent bonds with as many as four other atoms (Figure 1–9). The covalent bonds formed by a carbon atom may be with different atoms (most frequently hydrogen, oxygen, and nitrogen) or with other carbon atoms. As we shall see in Chapter 3, this tendency of carbon atoms to form covalent bonds with other carbon atoms gives rise to the large molecules that form the structures of living organisms and that participate in essential life processes.

Single and Double Covalent Bonds

There are various ways in which atoms can participate in covalent bonds and fill their outer energy levels. Oxygen, for example, has six electrons in its outer energy level (see Figure 1–4). Four of these electrons are grouped into two pairs and are generally unavailable for covalent bonding. The other two electrons are unpaired, and each can be shared with another atom in a covalent bond. In the water molecule (H_2O), one of these electrons participates in a covalent bond with one hydrogen atom, and the other in a covalent bond with a different hydrogen atom (Figure 1–10). Two **single bonds** are formed, and all three atoms have filled outer energy levels.

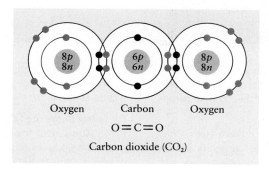

Oxygen   Carbon   Oxygen

$$O = C = O$$

Carbon dioxide (CO_2)

1–11   *A schematic diagram of the carbon dioxide molecule (CO_2). The carbon atom in the center of the molecule participates in two double covalent bonds, one with each oxygen atom. Each double bond consists of two pairs of electrons shared by the two atoms that participate in the bond. In structural formulas, a double bond is represented by two parallel lines:* =.

The bonding situation is different in another familiar substance, carbon dioxide (CO_2). In this molecule, the two available electrons of each oxygen atom participate with two electrons of a *single* carbon atom in the formation of *two* covalent bonds (Figure 1–11). Each oxygen atom is joined to the central carbon atom by two pairs of electrons (four electrons). Such bonds are called **double bonds.** Carbon atoms can form double bonds with each other as well as with other atoms, and so the variety of kinds of molecules that carbon can form is very large.

Single bonds are flexible, leaving atoms free to rotate in relation to one another. Double bonds, however, hold the atoms relatively rigid in relation to one another. The presence of double bonds in a molecule can make a significant difference in its properties. For example, both fats and oils are composed of carbon and hydrogen atoms covalently bonded together, but in fats the bonds are all single and in oils some of the bonds are double. The rigidity caused by these double bonds prevents the oil molecules from packing together, and, as a consequence, oils are liquids at room temperature. In fats, by contrast, the molecules can bend and twist, fitting closely together in a structure that is solid at room temperature.

Polar Covalent Bonds

The atomic nuclei of different elements have different degrees of attraction for electrons. As a consequence, in covalent bonds formed between atoms of different elements, the electrons are not shared equally between the atoms involved. Instead, the shared electrons tend to spend more time around the nucleus with the greater attraction. The atom around which the electrons spend

more time has a slightly negative charge. The other atom has a slightly positive charge, since the electrons spend less time around it and its nuclear charge is not entirely neutralized.

Covalent bonds in which electrons are shared unequally are known as **polar covalent bonds.** Such bonds often involve oxygen atoms, to which electrons are strongly attracted. In molecules that are perfectly symmetrical, such as carbon dioxide, the unequal charges cancel out and the molecule as a whole is **nonpolar.** However, in asymmetrical molecules, such as water, the molecule as a whole is **polar,** with regions of partial positive charge and regions of partial negative charge. Many of the special properties of water, upon which life depends, derive largely from its polar nature, as we shall see in the next chapter.

Chemical Reactions

Chemical reactions—exchanges of electrons among atoms—can be compactly described by **chemical equations.** For example, the equation for the formation of sodium chloride from sodium and chloride ions is

$$\underset{\text{Reactants}}{Na^+ + Cl^-} \longrightarrow \underset{\text{Product}}{NaCl}$$

The arrow in the equation means "forms" or "yields," and it shows the direction of chemical change. Like algebraic equations, chemical equations "balance." The number and kinds of atoms in the products of the reaction must equal the number and kinds of atoms in the original reactants.

To take a slightly more complex example, hydrogen gas can combine with oxygen gas to produce water. As you know, hydrogen gas is H_2; similarly, oxygen gas is O_2. Each molecule of water contains two atoms of hydrogen and one of oxygen. Therefore the reactants must be in the proportion of two hydrogens to one oxygen:

$$\underset{\text{Reactants}}{2H_2 + O_2} \longrightarrow \underset{\text{Product}}{2H_2O}$$

Two molecules of H_2 plus one molecule of O_2 yield two molecules of water. The equation for a chemical reaction thus tells us the kinds of atoms that are present, their proportions, and the direction of the reaction.

A substance that contains atoms of two or more different elements, held together in a definite and constant proportion by chemical bonds, is known as a chemical **compound.** Examples of chemical compounds include water (H_2O), sodium chloride ($NaCl$), carbon dioxide (CO_2), methane (CH_4), and glucose ($C_6H_{12}O_6$).

The Signs of Life

What do we mean when we speak of "life" or "the evolution of life"? Actually, there is no simple definition. Life does not exist in the abstract. There is no "life," only living things. Moreover, there is no single, simple way to draw a sharp line between the living and the nonliving. Certain characteristics, however, taken together, distinguish living things from inanimate (that is, nonliving) objects.

(a) Living things are highly organized, as in this cross section of a stem of a sycamore sapling. This stem reflects the complicated organization of many different kinds of atoms into molecules and of molecules into complex structures. Such complexity of form, which is never found in inanimate objects of natural origin, makes possible the specialization of different parts of a living organism for different functions.

(b) Living organisms are homeostatic, which means simply "staying the same." Although they constantly exchange materials with the outside world, they maintain a relatively stable internal environment quite unlike that of their surroundings. Even this tiny, apparently fragile animal, a rotifer, has an internal chemical composition that differs from its changing environment.

(c) Living things reproduce themselves. They make more of themselves, generation after generation, with astonishing fidelity (and yet, as we shall see, with just enough variation to provide the raw material for evolution). Flowers, the familiar symbols of spring and romance, are the reproductive structures of the largest and most diverse group of plants.

(d) Living organisms grow and develop. Growth and development are the processes by which a single living cell, the fertilized egg, becomes, for example, a tree, or an elephant, or a young koala, shown here with its mother.

(e) Living things take energy from the environment and change it from one form to another. The processes of energy conversion are highly specialized and remarkably efficient. This European kingfisher has converted chemical energy stored in its body to the energy of motion used in catching a fish. After the kingfisher has eaten and digested the fish, the chemical energy stored in the body of the fish will be available for the kingfisher's use.

(f) Living things are adapted to their environment. The Arctic fox is protected from the harsh conditions of winter by a luxuriant coat of soft fur and a thick layer of body fat just below the skin. This insulation is so effective that the fox can comfortably tolerate actual or wind chill temperatures as low as $-60°$ Fahrenheit. When it is inactive or sleeping, the fox curls into a tight ball, providing added protection for tender areas, such as the paws, the belly, and the muzzle.

(g) Living organisms respond to stimuli. When this sphinx moth caterpillar sensed a threat—an approaching photographer—it inflated the front portions of its body, released its front legs from the branch on which it was resting, hung upside down, and waved its body at the camera. Although the caterpillar is only 6 centimeters long, its "fake snake" disguise is sufficiently convincing to protect it from attacks by would-be predators, such as hungry birds.

(a)

(b)

(c)

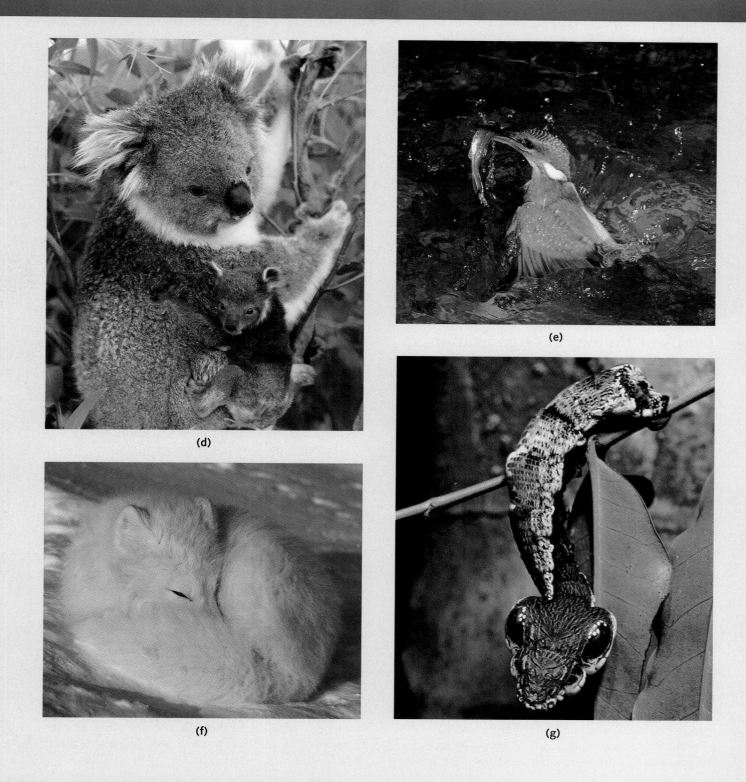

(d)

(e)

(f)

(g)

The Biologically Important Elements

Of the 92 naturally occurring elements, only six make up some 99 percent of all living tissue (Table 1–3). These six elements are carbon, hydrogen, nitrogen, oxygen, phosphorus, and sulfur, conveniently remembered as CHNOPS. These are not the most abundant of the elements of the Earth's surface. Why, as life assembled and evolved from stardust, were these of such importance?

One clue is that the atoms of all of these elements need to gain electrons to complete their outer energy levels (see Figure 1–4). Thus they generally form covalent bonds. Because these atoms are small, the shared electrons in the bonds are held closely to the nuclei, producing very stable molecules. Moreover, with the exception of hydrogen, atoms of these elements can all form bonds with two or more atoms. This makes possible the formation of the complex molecules essential for the structures and functions of living systems.

Levels of Biological Organization

In beginning our study of biology with an examination of how atoms and molecules are put together and how they interact, we have made two fundamental assumptions: (1) Living things are made up of the same chemical components—atoms and molecules—as nonliving things, and (2) living things obey the same chemical and physical laws as nonliving things. This does not mean, however, that organisms are "nothing but" the atoms and molecules of which they are composed. Indeed, there are recognizable differences between living and nonliving systems (see essay, pages 30–31). The key to these differences is organization.

A proton, as we have seen, is a positively charged particle, and an electron a negatively charged particle. Put them together and you have a hydrogen atom, an entity with quite different properties from those of an electron or a proton. Combine hydrogen atoms with each other in hydrogen molecules, and they form a colorless, highly flammable gas. Combine hydrogen atoms with the atoms of oxygen (whose molecules form a different colorless gas), and liquid water is produced—not a gas at all and with very different and remarkable properties, not a bit like those of isolated electrons and protons or of elemental hydrogen and oxygen.

Take just a few more kinds of atoms—carbon, nitrogen, a little sulfur—and put them together in a cer-

Table 1–3 Atomic Composition of Three Representative Organisms			
Element	Human	Alfalfa	Bacterium
Carbon	19.37%	11.34%	12.14%
Hydrogen	9.31	8.72	9.94
Nitrogen	5.14	0.83	3.04
Oxygen	62.81	77.90	73.68
Phosphorus	0.63	0.71	0.60
Sulfur	0.64	0.10	0.32
CHNOPS total:	97.90%	99.60%	99.72%

tain way, and you have the contractile machinery of a muscle cell, the basis for all the voluntary movements of our bodies. Add a little phosphorus, combine the atoms in a different way, and you can spell out the genetic code. Put together enough of these atoms in the right way, enclose them in a membrane (made up of the same kinds of atoms), and they form a living cell.

Each of these states represents a new level of organization. The characteristics of each level are not simply a combination of the characteristics of its components but are, in fact, totally different. At each new level of organization, new and different properties emerge.

The **cell** is the level of organization at which life can unarguably be said to appear as a new, emergent property. Other properties emerge when individual, specialized cells are organized at still higher levels, in the tissues and organs of a multicellular organism. Organized in one way, the cells form a liver; in another way, an intestinal tract; in yet another, a human brain, which represents an extraordinary degree of organizational complexity. Yet it is, in turn, only part of a larger entity whose characteristics are different from those of the brain, although they depend on them.

Nor is the individual organism the ultimate level of biological order. Living organisms interact with each other, individually and in groups. Finally, groups of living organisms are themselves part of an even vaster system of organization. This ultimate level of organization, the **biosphere**, involves not only the great diversity of plants and animals and microorganisms and their interactions with each other but also the physical characteristics of the environment and the planet Earth itself.

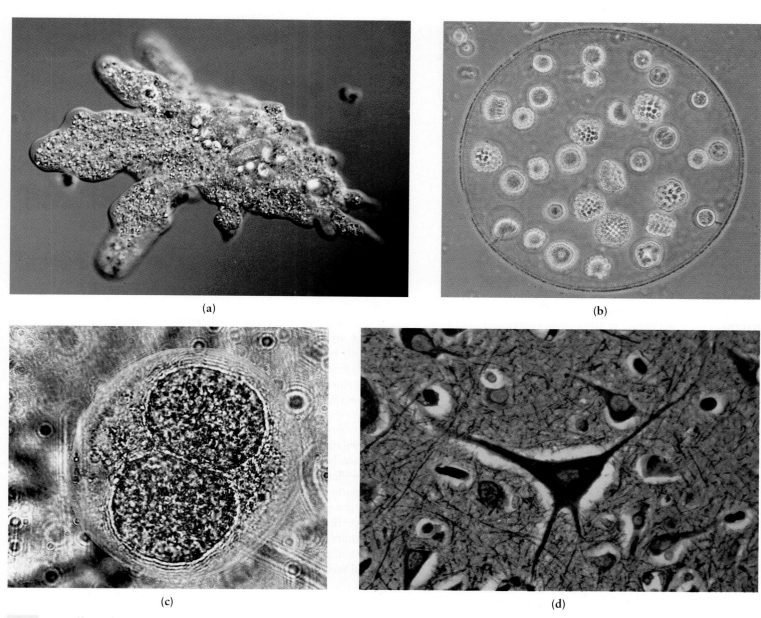

(a)

(b)

(c)

(d)

1–12   *A gallery of cells.* (a) Amoeba pro-teus, *a single-celled organism named for Proteus, a Greek god capable of changing his shape at will. Extensions of the cell, known as pseudopods, enable amoebas to move and to capture prey.*

(b) *This simple organism, called* Eudorina, *is made up of 32 cells, held together by a jelly-like substance, whose outlines you can see. When the organism reproduces, some—but* not all—of the cells divide repeatedly. If you look closely you can see these cells, which appear to be miniature versions of the parent colony. When the cell divisions are completed in the reproductive cells, the parent colony breaks apart and the daughter colonies begin their independent existence.

(c) *The embryo of a mouse at the two-celled stage. Within each cell is a nucleus* that carries all the genetic information needed for every cell in the mature mouse.

(d) *These cells are from the cerebral cortex of a human brain—the most highly organized structure on Earth. The actions of the cells of the cerebral cortex and the interconnections among them are responsible for consciousness, intelligence, dreams, and memory.*

Summary

Matter is composed of atoms, the smallest units of chemical elements. The nucleus of an atom contains positively charged protons and (except for hydrogen, ^1H) neutrons, which have no charge. The atomic number of an atom is equal to the number of protons in its nucleus. The atomic weight of an atom is the sum of the number of protons and neutrons in its nucleus. How an atom will react with other atoms is determined by its electrons—small, negatively charged particles found outside the nucleus. The number of electrons in an atom equals the number of protons.

The nuclei of different isotopes of the same element contain the same number of protons but different numbers of neutrons. Thus the isotopes of an element have the same atomic number but different atomic weights.

The electrons of an atom have differing amounts of energy. Electrons closer to the nucleus have less energy than those farther from the nucleus and thus are at a lower energy level. An electron tends to occupy the lowest available energy level, but with an input of energy, it can be boosted to a higher energy level. When the electron returns to a lower level, energy is released.

The chemical behavior of an atom is determined by the number and arrangement of its electrons. An atom is most stable when all of its electrons are at the lowest available energy levels and those energy levels are completely filled with electrons. The first energy level can hold two electrons, the second energy level can hold eight electrons, and so can the third energy level of the atoms of greatest interest in biology. Chemical reactions between atoms result from the tendency of atoms to reach the most stable electron arrangement possible.

Molecules consist of two or more atoms held together by chemical bonds. Two principal types of bonds are ionic and covalent. The charged atoms produced when electrons jump from one atom to another are known as ions. Ionic bonds are formed by the mutual attraction of ions of opposite charge. In covalent bonds, pairs of electrons are shared between atoms. In single covalent bonds, one pair of electrons is shared; in double covalent bonds, two pairs of electrons are shared. In some covalent bonds, known as polar covalent bonds, pairs of electrons are shared unequally, giving the molecule regions of positive and negative charge.

Chemical reactions—exchanges of electrons among atoms—can be represented by chemical equations. Substances that consist of the atoms of two or more different elements, in definite and constant proportions, are known as chemical compounds.

Six elements (CHNOPS) make up 99 percent of all living matter. The atoms of all of these elements are small and form tight, stable covalent bonds. With the exception of hydrogen, they can all form covalent bonds with two or more atoms, giving rise to the complex molecules that characterize living systems.

Living things are made up of the same chemical components—atoms and molecules—as nonliving things, and they obey the same chemical and physical laws. However, new properties emerge with each increasing level of organization. The properties of a complex molecule depend upon the organization of the atoms within the molecule. Similarly, the properties of a living cell depend upon the organization of molecules within the cell, and the properties of a multicellular organism depend upon the organization of the cells within its body. The ultimate level of biological organization, the biosphere, results from the interactions of the plants, animals, and microorganisms of the Earth with each other and with physical factors in the environment.

Questions

1. Describe the three types of particles of which atoms are composed. What is the atomic number of an atom? The atomic weight?

2. For each of the following isotopes, determine the number of protons and neutrons in the nucleus: (a) ^{11}C, ^{12}C, ^{14}C; (b) ^{31}P, ^{32}P, ^{33}P; (c) ^{32}S, ^{35}S, ^{38}S.

3. Consider the isotopes of phosphorus listed in Question 2. Would you expect all three of these isotopes to exhibit the same chemical properties in a living organism? Why or why not?

4. The street lights in many cities contain bulbs filled with sodium vapor. When electrical energy is passed through the

bulb, a brilliant yellow light is given off. What is happening to the sodium atoms to cause this?

5. Determine the number of protons, the number of neutrons, the number of energy levels, and the number of electrons in the outermost energy level in each of the following atoms: oxygen, nitrogen, carbon, sulfur, phosphorus, chlorine, potassium, and calcium.

6. How many electrons does each of the atoms in Question 5 need to share, gain, or lose to acquire a completed outer energy level?

7. Magnesium has an atomic number of 12. How many electrons are in its first energy level? Its second energy level? Its third energy level? How would you expect magnesium and chlorine to interact? Write the formula for magnesium chloride.

8. Explain the differences between ionic, covalent, and polar covalent bonds. What tendency of atoms causes them to interact with each other, forming bonds?

9. Molecules that contain polar covalent bonds typically have regions of positive and negative charge and thus are polar. However, some molecules containing polar covalent bonds are nonpolar. Explain how this is possible.

10. Knowing that chemical reactions have to be balanced, fill the appropriate numbers into the underlined spaces (*hint:* from 1 to 3 in all cases):

(a) ____ $H_2CO_3 \longrightarrow$ ____ H_2O + ____ CO_2
      Carbonic
      acid

(b) ____ H_2 + ____ $N_2 \longrightarrow$ ____ NH_3
                               Ammonia

(c) ____ NaOH + ____ $H_2CO_3 \longrightarrow$
      Sodium
      hydroxide

                        ____ Na_2CO_3 + ____ H_2O
                                  Sodium
                                  carbonate

11. What six elements make up the bulk of living tissue? What characteristics do the atoms of these six elements share?

Suggestions for Further Reading

Atkins, P. W.: *Molecules,* W. H. Freeman and Company, New York, 1987.*

> *In this beautifully illustrated book, Professor Atkins explores the structure and properties of 160 of the most familiar molecules of our everyday world.*

Powell, Corey S.: "The Golden Age of Cosmology," *Scientific American,* July 1992, pages 17–22.

Silk, Joseph: *The Big Bang: The Creation and Evolution of the Universe,* 2d ed., W. H. Freeman and Company, New York, 1988.

> *A discussion of modern evidence concerning the formation of the solar system and the planet Earth. An excellent, well-written introduction to cosmology.*

* Available in paperback.

Weinberg, Steven: *The Discovery of Subatomic Particles,* W. H. Freeman and Company, New York, 1990.*

> *In this handsome book, an introduction to the structure of the atom is combined with a lively history of twentieth-century physics. This revolution in physics profoundly influenced modern biology.*

Weinberg, Steven: *The First Three Minutes: A Modern View of the Origin of the Universe,* revised ed., Basic Books, Inc., New York, 1988.*

> *A wonderful story, written for the intelligent nonscientist (characterized by the author as a smart old attorney who expects to hear some convincing arguments before he makes up his mind).*

Weisskopf, Victor F.: "The Origin of the Universe," *American Scientist,* vol. 71, pages 473–480, 1983.

Life on Earth depends on the presence of water. The first living systems came into being, according to present hypotheses, in the warm primitive seas, and for many organisms, ourselves included, each new individual begins life bathed and cradled in water.

These are developing embryos of the rain frog, which is native to the rain forests of Costa Rica. The eggs are laid on any convenient surface, including the mosses and other plants that grow on the trunks of the large trees of the forest. Within its gelatinous egg case, which contains an ample supply of water and nutrients, each embryo develops directly into a miniature adult without passing through the free-living tadpole stage commonly seen in frogs. Rain frogs can therefore lay their eggs on terrestrial surfaces, well away from puddles, ponds, and other bodies of water. These two froglets are almost ready to emerge from their egg cases.

Water is composed of atoms of hydrogen and oxygen, and yet its properties are very different from those of either hydrogen or oxygen. It is these special properties of water that are vital to life here on Earth.

Water

In this chapter and the next, we are going to examine the molecules of which living organisms are composed. By far the most abundant of these molecules is water, which makes up 50 to 95 percent of the weight of any functioning living system.

Life on this planet began in water, and today, wherever liquid water is found, life is also present. There are one-celled organisms that eke out their entire existence in no more water than can cling to a grain of sand. Some kinds of algae are found only on the melting undersurfaces of polar ice floes. Certain bacteria thrive in the near-boiling water of hot springs. In the desert, plants race through an entire life cycle—seed to flower to seed—following a single rainfall. In the tropical rain forest, the water cupped in the leaves of a plant forms a miniature world in which a multitude of small organisms are born, mature, reproduce, and die.

Water is the most common liquid on Earth. Three-fourths of the surface of the Earth is covered by water. In fact, if the Earth's land surface were absolutely smooth, all of it would be under 2.5 kilometers* of water. But do not mistake "common" for "ordinary." Water is not an ordinary liquid at all. Compared with other liquids it is, in fact, quite extraordinary. If it were not, it is unlikely that life on Earth could ever have evolved.

The Structure of Water

In order to understand why water is so extraordinary and how, as a consequence, it can play its unique and crucial role in relation to living systems, we must look at its molecular structure. As you know, each water molecule is made up of two atoms of hydrogen and one

---

* A metric table with English equivalents can be found inside the back cover.

atom of oxygen (Figure 2–1). Each of the hydrogen atoms is held to the oxygen atom by a covalent bond; that is, the single electron of each hydrogen atom is shared with the oxygen atom, which also contributes an electron to each bond.

The water molecule as a whole is neutral in charge, having an equal number of electrons and protons. However, the molecule is polar (page 29). Because of the very strong attraction of the oxygen nucleus for electrons, the shared electrons of the covalent bonds spend more time around the oxygen nucleus than they do around the hydrogen nuclei. As a consequence, the region near each hydrogen nucleus has a weak positive charge. Moreover, the oxygen atom has four other electrons in its outer energy level. These electrons, which are not involved in covalent bonding to hydrogen, are paired in two orbitals (see essay, page 24). Each of these orbitals has a weak negative charge. Thus, the water molecule, in terms of its polarity, is four-cornered, with two positively charged "corners" and two negatively charged ones (Figure 2–2a).

When one of these charged regions comes close to an oppositely charged region of another water molecule, the force of attraction forms a bond between them, which is known as a **hydrogen bond.** Hydrogen bonds are found not only in water but also in many large molecules, where they help maintain structural stability. A

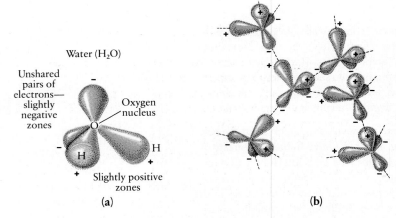

2–2   *The polarity of the water molecule and its consequences.* (a) *As shown in this orbital model of water, four orbitals branch off from the oxygen nucleus. Two of the orbitals are formed by the shared pairs of electrons bonding the hydrogen atoms to the oxygen atom. These two orbitals have a slightly positive charge because their electrons spend more time around the oxygen nucleus than around the hydrogen nuclei. The other two orbitals, containing electrons that are not involved in covalent bonds, have a slightly negative charge.*

(b) *As a result of these positive and negative zones, each water molecule can form hydrogen bonds (dashed lines) with as many as four other water molecules. Under ordinary conditions of pressure and temperature, the hydrogen bonds are continually breaking and re-forming in a shifting pattern. Thus water is a liquid.*

2–1   *The structure of the water molecule (H₂O) can be depicted in several different ways, illustrating different aspects of the structure. The schematic diagram in Figure 1–10 (page 28), for example, shows the electron structure of the atoms and the electron sharing within the molecule. By contrast, the space-filling model (a), in which the atoms are represented by color-coded spheres, provides a compact three-dimensional representation of the molecule. Because of its simplicity, this model is often used as a convenient symbol of the water molecule.*

(b) *The ball-and-stick model emphasizes the covalent bonds that join the atoms. It also gives some indication of the geometry of the molecule. A more accurate description of the molecule's shape is provided by the orbital model in Figure 2–2a.*

hydrogen bond can form between any hydrogen atom that is covalently bonded to an atom that has a strong attraction for electrons—usually oxygen or nitrogen—and the oxygen or nitrogen atom of another molecule. In water, a hydrogen bond forms between a negative "corner" of one water molecule and a positive "corner" of another. Every water molecule can establish hydrogen bonds with as many as four other water molecules (Figure 2–2b).

Any single hydrogen bond is significantly weaker than either a covalent or an ionic bond. Moreover, it has an exceedingly short lifetime. On average, each hydrogen bond in liquid water lasts approximately 1/100,000,000,000th of a second. But, as one is broken, another is made. All together, the hydrogen bonds have considerable strength, causing the water molecules to cling together as a liquid under ordinary conditions of temperature and pressure.

Now let us look at some of the consequences of these attractions among water molecules, especially as they affect living organisms.

(a)   (b)

2–3   *Some consequences of the surface tension of water.*
(a) A water strider, at rest on the calm surface of a pond. The weight of the insect's body is easily supported by the continuous sheet formed by the surface of the water. Note the elasticity of the surface, revealed by the slight depression below each of the insect's legs. The green objects are duckweed, a tiny flowering plant commonly found growing on the surface of ponds and lakes.

(b) When a drop of water strikes a water surface, or a rain drop falls on a pond, the water splashed into the air at impact forms a cylinder of water topped by tiny spheres, each held together by the cohesion of the water molecules. The point of impact, captured here by high-speed photography, resembles a jewel-studded crown.

Consequences of the Hydrogen Bond

Surface Tension

Look at water dripping from a faucet. Each drop clings to the rim and dangles for a moment by a thread of water. Then, just as the tug of gravity breaks it loose, its outer surface is drawn taut, to form a sphere as the drop falls free. Gently place a needle or a razor blade flat on the surface of the water in a glass. Although the metal is denser than water, it floats. Look at a pond in spring or summer. You will see water striders and other insects walking on its surface almost as if it were solid (Figure 2–3a). Rain drops create small crowns as they splash on the pond's surface (Figure 2–3b). These phenomena are all the result of **surface tension**—a taut yet elastic "skin" at the water's surface. Surface tension is produced by the clinging together of the water molecules, caused by their hydrogen bonding. Such a holding together of molecules of the same substance is known as **cohesion.**

Water, because of its negative and positive charges, also clings tightly to other charged molecules and to

charged surfaces. Such a holding together of molecules of different substances is known as **adhesion.** The "wetting" capacity of water—that is, its ability to adhere to a surface—is another result of its polar structure.

Capillary Action and Imbibition

If you hold two dry glass slides together and dip one corner in water, the water will spread upward between the two slides. This movement of liquid through a narrow passage is known as **capillary action.** It results from the combined effects of the cohesion of water molecules to each other and their adhesion to the surface of the glass (Figure 2–4). Capillary action causes water to rise in very fine glass tubes, to creep up a piece of blotting paper, or to move slowly through the tiny spaces between soil particles and so become available to the roots of plants.

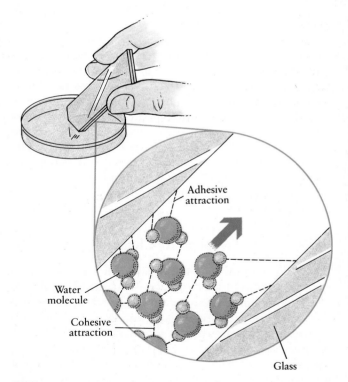

2–4   *The capillary movement of water between two glass slides. Because of the polarity of the water molecules, the molecules at the leading edge of the water surface are attracted to the charged surface of the glass. As the water molecules are pulled forward by this adhesive attraction, their cohesive attraction for other water molecules—a direct result of hydrogen bonding—pulls those molecules along behind. The combined effect of these two forces will continue to move the water forward until the forces of attraction are balanced by the pull of gravity on the increasing weight of the column of water.*

2–5 *The germination of seeds begins with changes in the seed coat that permit a massive uptake of water. The embryo and surrounding structures then swell, bursting the seed coat. In this acorn, photographed on a forest floor, the embryonic root emerged after the tough outer layers of the fruit split open.*

Imbibition ("drinking up") is the capillary movement of water molecules into substances such as wood or gelatin, which swell as a result. The pressures developed by imbibition can be astonishingly great. It is said that stone for the ancient Egyptian pyramids was quarried by driving wooden pegs into holes drilled in the rock face and then soaking the pegs with water. The swelling of the wood created a force great enough to break the stone slab free. Seeds imbibe water as they begin to germinate, swelling and bursting their seed coats (Figure 2–5).

Resistance to Temperature Changes

If you go swimming in the ocean or a lake on one of the first hot days of summer, you will quickly be aware of a striking difference between the air temperature and the water temperature. This difference occurs because a greater input of energy is required to raise the temperature of water than to raise the temperature of air. The amount of heat a given substance requires for a given increase in temperature is known as its **specific heat** (also called heat capacity). One calorie* is defined as the amount of heat that will raise the temperature of 1 gram (1 milliliter or 1 cubic centimeter) of water 1°C. The specific heat of water is about twice the specific heat of oil or alcohol; that is, only 0.5 calorie is needed to

---

* Dietary Calories are actually kilocalories (kcal); 1 kilocalorie equals 1,000 calories.

raise the temperature of 1 gram of oil or alcohol 1°C. The specific heat of water is four times the specific heat of air and 10 times that of iron. Only liquid ammonia has a higher specific heat.

Heat is a form of energy—the **kinetic energy,** or energy of motion, of molecules. Molecules are always moving. They vibrate, rotate, and shift position in relation to other molecules. Heat, which is measured in calories, reflects the *total* kinetic energy in a collection of molecules. It includes not only the magnitude of the molecular movements but also the mass and number of moving molecules present. By contrast, temperature, which is measured in degrees, reflects the *average* kinetic energy of the molecules in question. Thus, heat and temperature are not identical. For example, a lake may have a lower temperature than does a bird flying over it, but the lake contains more heat because it has many more molecules in motion.

The high specific heat of water is a consequence of hydrogen bonding. The hydrogen bonds in water tend to restrict the movement of the molecules. In order for the kinetic energy of water molecules to increase sufficiently for the temperature to rise 1°C, it is necessary first to rupture a number of the hydrogen bonds holding the molecules together. When you heat a pot of water, much of the heat energy added to the water is used in breaking the hydrogen bonds between the water molecules. Only a relatively small amount of heat energy is therefore available to increase molecular movement.

What does the high specific heat of water mean in biological terms? It means that for a given rate of heat input, the temperature of water will rise more slowly than the temperature of almost any other material. Conversely, the temperature will drop more slowly as heat is removed. Because so much heat input or heat loss is required to raise or lower the temperature of water, organisms that live in the oceans or large bodies of fresh water live in an environment where the temperature is relatively constant. Also, the high water content of terrestrial plants and animals helps them to maintain a relatively constant internal temperature. This constancy of temperature is critical because biologically important chemical reactions take place only within a narrow temperature range.

Vaporization

Vaporization—or evaporation, as it is more commonly called—is the change from a liquid to a gas. Water has a high **heat of vaporization.** At water's boiling point (100°C at sea level), it takes 540 calories to change 1 gram of liquid water into vapor, almost 60 times as much as for ether and twice as much as for ammonia.

(a)

(b)

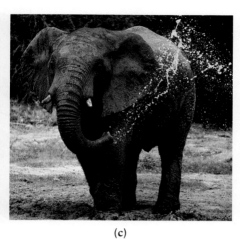
(c)

2-6   *Animals have various devices for utilizing the heat-absorbing properties of evaporating water. (a) Dogs unload heat by panting, which involves short, shallow breaths and the production of a copious saliva that evaporates from the tongue. (b) Among the animals that can sweat over their entire body surface are horses and humans. (c) Elephants, which cannot perspire at all, hose themselves down to keep cool.*

Hydrogen bonding is also responsible for water's high heat of vaporization. Vaporization comes about because some of the most rapidly moving molecules of a liquid break loose from the surface and enter the air. The hotter the liquid, the more rapid the movement of its molecules and, hence, the more rapid the rate of evaporation. But, whatever the temperature, so long as a liquid is exposed to air that is less than 100 percent saturated with the vapor of that liquid, evaporation will take place, down to the last drop.

In order for a water molecule to break loose from its fellow molecules—that is, to vaporize—the hydrogen bonds have to be broken. This requires heat energy. As a consequence, when water evaporates, as from the surface of your skin or a leaf, the escaping molecules carry a great deal of heat away with them. Thus evaporation has a cooling effect. Evaporation from the surface of a land-dwelling plant or animal is one of the principal ways in which these organisms "unload" excess heat and so stabilize their temperatures.

Freezing

Water exhibits another peculiarity when it undergoes the transition from a liquid to a solid (ice). In most liquids, the **density**—that is, the weight of the material in a given volume—increases as the temperature drops. This greater density occurs because the individual molecules are moving more slowly and so the spaces be-tween them decrease, leading to more molecules in the same volume. The density of water also increases as the temperature drops, until it nears 4°C. Then the water molecules come so close together and are moving so slowly that *every one* of them can form hydrogen bonds simultaneously with four other molecules—something they could not do at higher temperatures. However, the geometry of the water molecule is such that, as the temperature drops below 4°C, the molecules must move slightly apart from each other to maintain the maximum number of hydrogen bonds in a stable structure. At 0°C, the freezing point of water, this creates an open latticework (Figure 2-7). Thus water as a solid takes up more volume than water as a liquid. Ice is less dense than liquid water and therefore floats in it.

This increase in volume has occasional disastrous effects on water pipes but, on the whole, turns out to be enormously beneficial for life forms. If water continued to contract as it froze, ice would be heavier than liquid water. As a result, lakes and ponds and other bodies of water would freeze from the bottom up. Once ice began to accumulate on the bottom, it would tend not to melt, season after season. Spring and summer might stop the freezing process, but laboratory experiments have shown that if ice is held to the bottom of even a relatively shallow tank, water can be boiled on the top without melting the ice. Thus if water did not expand when it froze, it would continue to freeze from the bottom up, year after year, and never melt again. Eventually, the body of water would freeze solid and any life in it would be destroyed. By contrast, the layer of floating ice that actually forms tends to protect the organisms in the water. The ice layer effectively insulates the liquid water beneath it, keeping its temperature at or above the freezing point of water.

Oxygen
Hydrogen
Hydrogen bond

(a)

(b)

2–7   (a) *In the crystalline structure of ice, each water molecule is hydrogen-bonded to four other water molecules in a
three-dimensional open latticework. As the water molecules
link up in a hexagonal arrangement, the bond angles in some
of the molecules are distorted. The hexagonal arrangement,
shown here in a small section of the latticework, is repeated
throughout the crystal and is responsible for the beautiful patterns seen in snowflakes and frost. The water molecules are actually farther apart in ice than they are in liquid water.*

   (b) *An ice "lens," photographed on the Alaskan tundra.
When snow melts during the Arctic spring and summer, the
meltwater seeps below the soil surface. It is unable to penetrate deeply, however, because of a thick, permanently frozen
layer of subsoil known as permafrost. In the autumn, when the
surface temperatures once more drop below 0°C, the trapped
water expands as it freezes. The force generated by this expansion heaves the soil upward, revealing the ice lens.*

When ice melts, it draws heat from its surroundings.
For example, ice cubes in a glass of water gradually
melt, cooling the water in the process. The heat energy
absorbed by the ice breaks the hydrogen bonds of the
latticework. Conversely, as water freezes, it releases
heat into its surroundings. In this way, ice and snow
also serve as temperature stabilizers, particularly during
the transition periods of fall and spring. Moderation of
sudden changes in temperature gives organisms time to
make seasonal adjustments essential to survival.

Water as a Solvent

Many substances within living systems are found in
aqueous solution. A **solution** is a uniform mixture of
the molecules of two or more substances. The substance
present in the greatest amount—usually a liquid—is
called the **solvent,** and the substances present in lesser
amounts are called **solutes.**

   The polarity of water molecules is responsible for
water's capacity as a solvent. The polar water molecules
tend to separate ionic substances, such as sodium chloride (NaCl), into their constituent ions. As shown in
Figure 2–8, the water molecules cluster around and segregate the charged ions.

Na⁺   Salt / Water
         crystal   Cl⁻ ion

Cl⁻   Cl⁻

                                              Water
                                              molecules

                                              Na⁺ ion

2–8   *Because of the polarity of water molecules, water can
serve as a solvent for ionic substances and polar molecules.
This diagram shows table salt, sodium chloride (NaCl), dissolving in water. The polar water molecules cluster around the
individual sodium ions (positively charged) and chloride ions
(negatively charged), separating them from one another. Note
the difference between the way the water molecules are arranged around the sodium ions and the way they are arranged
around the chloride ions.*

The Four Seasons of a Lake

As we have seen, water increases in density as its temperature drops, until it reaches 4°C, the temperature of maximum density. Water either colder or warmer than 4°C is less dense and floats above water at 4°C. As a result, the water of temperate-zone lakes is stratified in the summer and winter but undergoes considerable mixing in the fall and spring. The stratifications of summer and winter enable lake-dwelling organisms to avoid life-threatening temperature extremes, while the mixing that occurs in fall and spring provides nutrients and oxygen to organisms at all levels of the lake.

In the summer (a), the top layer of water, called the epilimnion, is heated by the sun and the surrounding air, becoming warmer than the lower layers. Since it becomes less dense as it becomes warmer, this water remains at the surface. Only the water in the epilimnion circulates. In the middle layer, there is an abrupt drop in temperature, known as the thermocline. Since the water in this layer is progressively colder and therefore progressively more dense, it does not mix with the lighter water above. The water of the middle layer effectively cuts off the circulation of oxygen-laden water from the surface into the third layer, the hypolimnion. As the organisms of the hypolimnion gradually use up the available oxygen, the summer stagnation results.

In the fall (b), the temperature of the epilimnion drops until it is the same as that of the hypolimnion. As the surface water becomes more dense, it sinks, and the warmer water in the middle layer rises to the surface, producing the fall overturn. Aided by the fall winds (c), water begins to circulate throughout the lake. Oxygen is returned to the depths, and nutrients released by the activities of bottom-dwelling bacteria are carried to the upper layers of the lake.

As winter deepens (d), the surface water cools below 4°C, becoming lighter as it expands. This water remains on the surface and, in many areas, freezes. The result is the winter stratification.

In the spring (e), as the ice melts and the water on the surface warms to 4°C, it sinks to the bottom, producing the spring overturn. Aided by the spring winds (f), another thorough mixing of the water in the lake occurs.

Many of the covalently bonded molecules important in living systems—such as sugars—are polar; that is, they have regions of weak positive and negative charge. (Such charged regions arise, as you will recall, because the nuclei of atoms of different elements exert differing degrees of attraction for electrons.) Because of their polarity, these molecules attract water molecules and dissolve in water. Polar molecules that readily dissolve in water are often called **hydrophilic** ("water-loving"). Such molecules slip into aqueous solution easily because their charged regions attract water molecules as much as or more than they attract each other. The polar water molecules thus compete with the attraction between the solute molecules themselves.

Molecules, such as fats, that lack polar regions tend to be very insoluble in water. The hydrogen bonding between the water molecules acts as a force to exclude the nonpolar molecules. As a result of this exclusion, nonpolar molecules tend to cluster together in water, just as droplets of fats tend to coalesce, for example, on the surface of chicken soup. Such molecules are said to be **hydrophobic** ("water-fearing"), and the clusterings are known as **hydrophobic interactions.**

We will encounter the properties of hydrophilic and hydrophobic molecules again in later chapters. These weak forces—hydrogen bonds and hydrophobic interactions—play crucial roles in determining the shape of large, biologically important molecules and, as a consequence, in dictating their properties.

The Ionization of Water

In liquid water, there is a slight tendency for the nucleus of one of the hydrogen atoms in a water molecule to leave the oxygen atom to which it is covalently bonded and jump to the oxygen atom to which it is hydrogen-bonded (Figure 2–9). In this reaction, two ions are produced: the hydronium ion (H_3O^+) and the hydroxide ion (OH^-). In any given volume of pure water, a small but constant number of water molecules will be ionized in this way. The number is constant because the tendency of water to ionize is offset by the tendency of the ions to reunite. Thus, even as some molecules are ionizing, an equal number of other molecules are being formed as ions reunite. This state, in which two opposite and equal processes are occurring simultaneously, is known as dynamic **equilibrium.**

The ionization of water is expressed by the following chemical equation:

$$2H_2O \rightleftharpoons H_3O^+ + OH^-$$

The two arrows indicate that the reaction goes in both directions. The fact that the arrow pointing toward $2H_2O$ is longer indicates that, at equilibrium, most of the water is not ionized. As a consequence, in any sample of pure water, only a small fraction exists in ionized form.

2–9   *When water ionizes, a hydrogen nucleus (that is, a proton) shifts from the oxygen atom to which it is covalently bonded to the oxygen atom to which it is hydrogen-bonded. The resulting ions are the negatively charged hydroxide ion and the positively charged hydronium ion.*

Water (H_2O)   +   Water (H_2O)   ⇌   Hydroxide ion (OH^-)   +   Hydronium ion (H_3O^+)

Although the positively charged ion formed when water ionizes is the hydronium ion (H_3O^+), it is, by convention, usually represented as a hydrogen ion (H^+). This convention, which we shall use in the remainder of this chapter and in subsequent chapters, simplifies the bookkeeping involved in balancing chemical equations.

Acids and Bases

In pure water, the number of H^+ ions exactly equals the number of OH^- ions. This is necessarily the case since neither ion can be formed without the other when only H_2O molecules are present. However, when an ionic substance or a substance with polar molecules is dissolved in water, it may change the relative numbers of H^+ and OH^- ions. For example, when hydrogen chloride (HCl) dissolves in water, it is almost completely ionized into H^+ and Cl^- ions. As a result, an HCl solution (hydrochloric acid) contains more H^+ ions than OH^- ions. Conversely, when sodium hydroxide (NaOH) dissolves in water, it forms Na^+ and OH^- ions. Thus, in a solution of sodium hydroxide in water, there are more OH^- ions than H^+ ions.

A solution is acidic when the number of H^+ ions exceeds the number of OH^- ions. Conversely, a solution is basic (alkaline) when the number of OH^- ions exceeds the number of H^+ ions. Thus, an **acid** is a substance that causes an increase in the relative number of H^+ ions in a solution, and a **base** is a substance that causes an increase in the relative number of OH^- ions.

Strong acids and bases are substances, like HCl and NaOH, that ionize almost completely in water, resulting in relatively large increases in the concentrations of H^+ and OH^- ions, respectively. Weak acids and bases, by contrast, are substances that ionize only slightly, resulting in relatively small increases in the concentration of H^+ or OH^- ions.

Chemists express degrees of acidity by means of the **pH scale** (Figure 2–10). (The symbol "pH" is derived from the German *potenz Hydrogen*, "power of hydrogen.") At pH 7, the concentrations of H^+ and OH^- are exactly the same, as they are in pure water. This is a neutral state. Any pH below 7 is acidic, and any pH above 7 is basic. A difference of one pH unit represents a tenfold difference in the concentration of H^+ ion. Most of the chemical reactions of living systems take place within a narrow range of pH that hovers around neutrality.

The Water Cycle

Most of the water on Earth—almost 98 percent—is in liquid form, in the oceans, lakes, and streams. Of the remaining 2 percent, some is frozen in polar ice and glaciers, some is in the soil, some is in the atmosphere in the form of vapor, and some is in the bodies of living organisms.

Water is made available to land organisms by processes powered by the sun. Solar energy evaporates water from the oceans, leaving the salt behind. Water is also evaporated, but in much smaller amounts, from lakes and ponds, from rivers and streams, from moist soil surfaces, from the leaves of plants, and from the bodies of other organisms. These molecules—now water vapor—are carried up into the atmosphere by air currents. At the cooler temperatures of the atmosphere, the molecules condense into clouds of liquid water or ice.

Eventually this water returns to the Earth's surface in the form of rain or snow. Most of the water falls on the oceans, since they cover most of the Earth's surface.

2–10   *The pH scale provides a measure of the degree of acidity of solutions. On this scale, 7 represents the neutral state, in which the concentrations of H^+ and OH^- ions are equal. Lower numbers represent increasingly acidic solutions, and higher numbers represent increasingly basic solutions. Notice that as the H^+ concentration increases, the OH^- concentration decreases, and vice versa.*

A difference of one pH unit reflects a tenfold difference in the concentration of H^+ ions. Cola, for instance, is 10 times as acidic as tomato juice, and gastric juices are about 100 times more acidic than cola drinks.

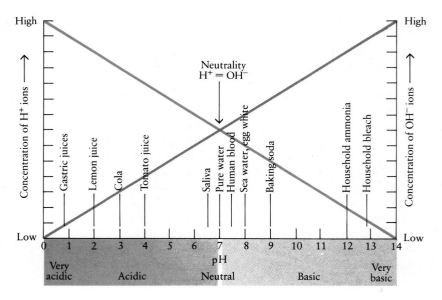

The water that falls on land is pulled toward the oceans by the force of gravity. Some of this runoff, reaching low ground, forms ponds or lakes and streams or rivers, which pour water back into the oceans.

Some of the water that falls on the land soaks in and then percolates down through the soil until it reaches a zone of saturation. In the zone of saturation, all pores and cracks in the rock are filled with water. The upper surface of the zone of saturation is known as the water table; the water within this zone is known as groundwater. Below the zone of saturation is solid rock, through which the water cannot penetrate. The deep groundwater, moving extremely slowly, eventually also reaches the ocean, thereby completing the water cycle.

As we have seen in this chapter, water, essential for life, is a most extraordinary substance. The Earth's supply of water is the permanent possession of our planet, held to its surface by the force of gravity. Through the movements of the water cycle, it is perpetually available to living organisms.

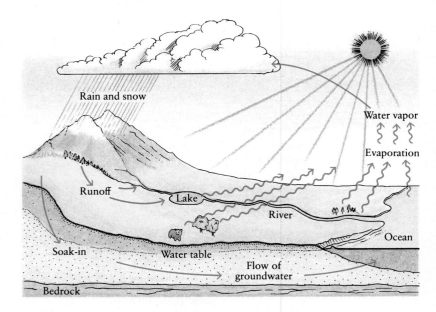

2–11   *The water cycle, powered by solar energy, provides a continuous circulation of the Earth's precious supply of water.*

Summary

Water, the most common liquid on the Earth's surface and the major component, by weight, of all living things, has a number of remarkable properties. These properties are a consequence of its molecular structure and are responsible for water's "fitness" for its roles in living systems.

Water is made up of two hydrogen atoms and one oxygen atom held together by covalent bonds. The water molecule is polar, with two weakly negative zones and two weakly positive zones. As a consequence, weak bonds form between water molecules. Such bonds, which link a somewhat positively charged hydrogen atom that is part of one molecule to a somewhat negatively charged atom that is part of another molecule, are known as hydrogen bonds. Each water molecule can form hydrogen bonds with as many as four other water molecules. Although individual bonds are weak and constantly shifting, the total strength of the bonds holding the molecules together is very great.

Because of the hydrogen bonds holding the water molecules together (cohesion), water has a high surface tension and a high specific heat (the amount of heat that a given amount of the substance requires for a given increase in temperature). It also has a high heat of vaporization (the heat required to change a liquid to a gas). Just before water freezes, it expands; thus, ice has a lower density than liquid water. As a result, ice floats in water.

The polarity of the water molecule is responsible for water's adhesion to other polar substances and hence its tendency for capillary movement. Similarly, water's polarity makes it a good solvent for ions and polar molecules. Molecules that dissolve readily in water are known as hydrophilic. Water molecules, as a consequence of their polarity, actively exclude nonpolar molecules from solution. Molecules that are excluded from aqueous solution are known as hydrophobic.

Water has a slight tendency to ionize, that is, to separate into H^+ ions and OH^- ions. In pure water, the number of H^+ ions and OH^- ions is equal. A solution that contains more H^+ ions than OH^- ions is acidic; one that contains more OH^- ions than H^+ ions is basic. The pH scale reflects the proportion of H^+ ions to OH^- ions. An acidic solution has a pH lower than 7; a basic solution has a pH higher than 7. Almost all of the chemical reactions of living systems take place within a narrow range of pH around neutrality.

Through the water cycle, the water above, on, and below the Earth's surface is recirculated. As a result, it is continuously available to living organisms.

Questions

1. (a) Sketch the water molecule and label the regions of weak positive and negative charge. (b) What are the major consequences of the polarity of the water molecule? (c) How are these effects important to living systems?

2. Distinguish among the following terms: cohesion/adhesion; capillary action/imbibition; heat/temperature; solution/solvent/solute; hydrophilic/hydrophobic; hydrogen bond/hydrophobic interaction; acidic/neutral/basic.

3. The trick with the razor blade (page 39) works better if the blade is a little greasy. Why?

4. Surfaces such as glass or raincoat cloth can be made "nonwettable" by application of silicone oils or other substances that cause water to bead up instead of spreading flat. What do you suppose is happening, in molecular terms, when a surface becomes nonwettable?

5. Generally, coastal areas have more moderate temperatures (not as cold in winter, nor as hot in summer) than inland areas at the same latitude. What explanation can you give for this phenomenon?

6. What is vaporization? Describe the changes that take place in water as it vaporizes. What is heat of vaporization? Why does water have an unusually high heat of vaporization?

7. Many of the molecules important in living systems contain the carboxyl group (—COOH). When these molecules are dissolved in water, the carboxyl group ionizes, as shown by the following equation:

$$-COOH \rightleftharpoons H^+ + -COO^-$$

At equilibrium, is the carboxyl group mostly ionized or only partially ionized? (*Hint:* look at the relative lengths of the two arrows.) Is the carboxyl group an acid or a base? Is it weak or strong?

8. Another group found in many molecules in living systems is the amino group (—NH$_2$). The amino group reacts with water as follows:

$$-NH_2 + H_2O \rightleftharpoons -NH_3^+ + OH^-$$

Is the amino group an acid or a base? Is it weak or strong?

9. Hydrochloric acid (HCl) is a major component of gastric juice, and the digestive processes in the human stomach take place at a pH of about 2. When the food being digested reaches the small intestine, a weak base, sodium bicarbonate (NaHCO$_3$), is released from the pancreas into the small intestine. What effect would you expect this to have on the pH of the partially digested food mass?

Suggestions for Further Reading

Gould, James L., and Carol Grant Gould (eds.): *Life at the Edge,* W. H. Freeman and Company, New York, 1989.*

A collection of articles from Scientific American *on organisms that thrive in extraordinarily difficult environments—such as the frigid waters of the Antarctic or boiling volcanic vents in the ocean depths.*

La Rivière, J. W. Maurits: "Threats to the World's Water," *Scientific American,* September 1989, pages 80–94.

Milne, Lorus J., and Margery Milne: "Insects of the Water Surface," *Scientific American,* April 1978, pages 134–142.

Peterson, Ivars: "A Biological Antifreeze," *Science News,* November 22, 1986, pages 330–332.

Stillinger, Frank H.: "Water Revisited," *Science,* vol. 209, pages 451–457, 1980.

Storey, Kenneth B., and Janet M. Storey: "Frozen and Alive," *Scientific American,* December 1990, pages 92–97.

*Available in paperback.

Atoms, forged in the stars, are incorporated into living organisms through the process of photosynthesis. In this process, green plants extract carbon atoms from the carbon dioxide gas in the atmosphere and harness the energy of the sun to build the complex molecules of which living organisms are formed. These molecules, known as organic molecules, are all constructed on backbones of carbon atoms.

Among the principal types of organic molecules are sugars, such as those contained in the syrupy nectar that has lured this honey bee to this flower. Sugar molecules are a source of energy. Plants use this energy to power the synthesis of other kinds of molecules, such as the proteins and oils present in the pollen grains clinging to the foraging bee. Pollen grains also contain molecules of nucleic acid, which carry the hereditary blueprint of the plant. Rewarded by the energy-rich nectar offered by the flower, the bee transports the pollen grains—and the genetic instructions they contain—from flower to flower.

Soon after the emergence of flowering plants on Earth, an event that occurred more than 120 million years ago, flowers and insects began to adapt to one another. Competition among plants for more efficient means of pollination led to the evolution of nectar and of the brightly colored flowers that advertise its presence. Competition for nectar led to evolutionary changes among insects. The bee's long mouthparts, for example, enable her to probe deeply within the flower for its sugary nectar. Clusters of specialized hairs on the rear legs form baskets for carrying the nutritious pollen grains back to the hive to share with her queen and her sister bees.

3

Organic Molecules

In this chapter, we present some of the types of **organic molecules**—molecules containing carbon—that are found in living things. The molecular drama is a grand spectacular with, literally, a cast of thousands. A single bacterial cell contains some 5,000 different kinds of organic molecules, and an animal or plant cell has about twice that many. These thousands of molecules, however, are composed of relatively few elements (CHNOPS). Similarly, relatively few types of molecules play the major roles in living systems. Consider this chapter, if you will, an introduction to the principal characters. The plot begins to unfold in Chapter 4.

The Central Role of Carbon

As we noted in the previous chapter, water makes up from 50 to 95 percent of a living organism, and small ions, such as sodium (Na^+), potassium (K^+), and calcium (Ca^{2+}), account for no more than 1 percent. Almost all the rest of an organism, chemically speaking, is composed of organic molecules.

Four different types of organic molecules are found in large quantities in organisms. These four are **carbohydrates** (composed of sugars), **lipids** (nonpolar molecules, most of which contain fatty acids), **proteins** (composed of amino acids), and **nucleic acids** (composed of complex molecules known as nucleotides). All of these molecules—carbohydrates, lipids, proteins, and nucleic acids—contain carbon, hydrogen, and oxygen. In addition, proteins contain nitrogen and sulfur. Nucleic acids, as well as some lipids, contain nitrogen and phosphorus.

The Carbon Backbone

As we saw in Chapter 1, a carbon atom can form four covalent bonds with as many as four other atoms. Methane (CH_4), which is natural gas, is an example (see Figure 1–9, page 28). Even more important, in terms

$$\begin{array}{c} H \\ | \\ H\!-\!C\!-\!H \\ | \\ H \end{array}$$

Methane (CH$_4$)

$$\begin{array}{c} H\ \ H \\ |\ \ \ | \\ H\!-\!C\!-\!C\!-\!H \\ |\ \ \ | \\ H\ \ H \end{array}$$

Ethane (C$_2$H$_6$)

$$\begin{array}{c} H\ \ H\ \ H\ \ H \\ |\ \ \ |\ \ \ |\ \ \ | \\ H\!-\!C\!-\!C\!-\!C\!-\!C\!-\!H \\ |\ \ \ |\ \ \ |\ \ \ | \\ H\ \ H\ \ H\ \ H \end{array}$$

Butane (C$_4$H$_{10}$)

3–1   *Ball-and-stick models and structural formulas of methane, ethane, and butane. The gray spheres represent carbon atoms,* *and the smaller blue spheres represent hydrogen atoms. The sticks in the models—and the lines in the structural formulas—represent covalent bonds, each of which consists of a shared pair of electrons. Note that every carbon atom forms four covalent bonds.*

of carbon's biological role, carbon atoms can form bonds with each other. Ethane, for example, contains two carbons; propane, three; butane, four; and so on, forming long chains (Figure 3–1). In general, an organic molecule derives its overall shape from the arrangement of the carbon atoms that form the backbone, or skeleton, of the molecule. The shape of the molecule, in turn, determines many of its properties and its function within the living organism.

In the molecules shown in Figure 3–1, every carbon bond that is not occupied by another carbon atom is taken up by a hydrogen atom. Such compounds, consisting of only carbon and hydrogen, are known as **hydrocarbons.** Structurally, they are the simplest kind of organic compounds. Although hydrocarbons are relatively unimportant in living organisms, they are of great economic importance. The fuels on which we depend— natural gas, heating oil, gasoline, and diesel fuel—are all hydrocarbons. They are derived from the remains of organisms that died millions of years ago. Conditions were such that the bodies did not decay but were instead converted into coal and petroleum.

Carbohydrates: Sugars and Polymers of Sugars

Carbohydrates are the primary energy-storage molecules in most living things. In addition, they form a variety of structural components of living cells. The walls of young plant cells, for example, are about 40 percent cellulose—a carbohydrate that is the most common organic compound in the biosphere.

Carbohydrates are formed from small molecules known as **sugars.** There are three principal kinds of carbohydrates, classified according to the number of sugar subunits they contain. **Monosaccharides** ("single sugars"), such as ribose, glucose, and fructose, consist of only one sugar molecule. **Disaccharides** ("two sugars") contain two sugar subunits linked covalently. Familiar examples are sucrose (table sugar), maltose (malt sugar), and lactose (milk sugar). **Polysaccharides,** such as cellulose and starch, contain many sugar subunits linked together. Large molecules, such as polysaccharides, that are made up of similar or identical subunits are known as **polymers** ("many parts"). The subunits are called **monomers** ("single parts").

Monosaccharides: Ready Energy for Living Systems

Monosaccharides are organic compounds composed of carbon, hydrogen, and oxygen. The proportion is one carbon atom to two hydrogen atoms to one oxygen atom, as indicated by the shorthand formula for monosaccharides: $(CH_2O)_n$. These proportions gave rise to the term "carbohydrate" (meaning "carbon with water added") for sugars and the larger molecules formed from sugar subunits. In the formula, n may be as small as 3, as in $C_3H_6O_3$, or as large as 8, as in $C_8H_{16}O_8$.

Monosaccharides are the building blocks from which living cells construct polysaccharides and other essential molecules. Moreover, they are the principal energy source for most organisms. To understand how monosaccharides provide energy for living systems, we must look again at the covalent bond and consider what happens in a chemical reaction.

Representations of Molecules

As we saw in Chapters 1 and 2, chemists have developed various models to represent the structures of atoms and molecules. Each of these models is a way of organizing a set of scientific data and of focusing attention on particular characteristics of atoms and molecules.

Because the properties of a molecule depend on its three-dimensional shape, physical models are often the most useful. For example, ball-and-stick models of the kind shown in Figure 3–1 emphasize the geometry of a molecule and, in particular, the bonds between atoms. These models, however, fail to suggest the overall shape of the molecule created by the movement of electrons around the atomic nuclei.

A closer approximation of molecular shape is provided by space-filling models. Each atom is represented by the outermost edge of the region occupied by its electrons. Space-filling models are misleading, however, in that molecules do not fill space in the same way that we think of a table or a rock as filling space. The atoms that make up molecules consist mostly of empty space. If the outer perimeter of the region traversed by the electrons in an oxygen atom were the size of the perimeter of the Astrodome in Houston, the nucleus would be a ping-pong ball in the center of the stadium.

The space in molecules is "filled" with regions of charge, associated with the movements of the electrons around the nuclei. One molecule "sees" another molecule in terms of these regions of charge. As a consequence, for instance, a protein that transports glucose molecules into the living cell will not transport fructose molecules because of the differences in the shape of the regions of charge. All the intricate biochemistry that goes on in the cell is based on this ability of molecules to "recognize" one another.

Ball-and-stick and space-filling models are often used in the laboratory, but they are less useful on paper because it is necessary to see them from all angles to see all of the atoms and their bonds. The most accurate two-dimensional representations of molecular structure are orbital models, such as those shown in Figure 2–2 (page 38). For molecules containing more than a few atoms, however, orbital models become extremely complicated. Thus, when representing complex molecules, such as those found in living systems, chemists usually use molecular formulas or structural formulas. A **molecular formula** indicates the number of atoms of each kind within the molecule, while a **structural formula** shows how the atoms are bonded to one another.

Glucose, for example, has 6 carbon atoms, 12 hydrogen atoms, and 6 oxygen atoms. Its molecular formula is $C_6H_{12}O_6$. However, fructose also contains 6 carbons, 12 hydrogens, and 6 oxygens and has a similar structure—a chain of carbon atoms to which hydrogen and oxygen atoms are attached. The differences between glucose and fructose are determined by which carbon atoms the other atoms are attached to. The molecules can therefore be distinguished by their structural formulas:

Glucose
(chain form)

Fructose
(chain form)

In these formulas, the symbol — represents a single covalent bond, and the symbol = represents a double covalent bond.

When glucose and fructose are in solution, however, they tend to form rings and so are more accurately represented by these structural formulas:

Glucose
(ring form)

Fructose
(ring form)

The lower edges of the rings are drawn thicker to hint at a three-dimensional structure. The ring is perpendicular to the page, with the thick edges projecting toward you; the thin edges project behind the page. By convention, the carbon atoms at the intersections of the links in an organic ring structure are "understood" to be present and are not labeled. Although it is not necessary to number the carbon atoms, doing so often makes it easier to interpret the structural formulas.

Structural formulas give us less information than physical models, but you will find them a convenient tool as we examine the molecules involved in the structures and processes of living systems.

Space-filling models of the sugars glucose and fructose. The gray spheres, almost completely hidden at the center of each molecule, represent the carbon atoms. The red spheres at the surface of each molecule represent oxygen atoms, while the blue spheres represent hydrogen atoms.

Glucose   Fructose

Covalent bonds—the bonds commonly found in organic molecules—are strong, stable bonds consisting of shared electrons moving around two or more atomic nuclei. As you will recall from the last chapter, molecules are always in motion—vibrating, rotating, and shifting position in relation to other molecules. The atoms within molecules are also in motion—vibrating and, often, rotating about the axes of their bonds. If this motion becomes great enough (that is, if the atoms acquire enough kinetic energy), the bond will "break" and the atoms will become separated from each other.

Different bonds have different characteristic strengths. Bond strengths are conventionally measured in terms of the amount of energy that must be supplied to break the bond under standard conditions of temperature and pressure. The units in which this energy is expressed are kilocalories per mole.* Table 3–1 compares the strengths of the covalent bonds that occur most frequently in organic molecules. The stronger the bond, the greater the amount of energy required to break it. As you can see, double bonds are considerably stronger than single bonds. Moreover, the single bond between carbon and hydrogen is stronger than the single bond between carbon and oxygen or between two carbon atoms.

When a covalent bond breaks, atoms (or, in some cases, groups of atoms) are released, and each atom usually takes its own complement of electrons with it. This results in atoms whose outer energy levels are only partially filled with electrons. For example, when the atoms of a methane molecule are vibrating and rotating so rapidly that the four carbon-hydrogen bonds break, one carbon atom and four hydrogen atoms are produced—and each of these atoms needs to gain one or more electrons to complete its outer energy level. Thus, the atoms tend to form new covalent bonds quite rapidly, restoring the stable condition of filled outer energy levels. Whether the new bonds that form are identical to those that were broken or are different depends on a number of factors—the temperature, the pressure, and, most important, what other atoms are available in the immediate vicinity.

Chemical reactions that produce new combinations of atoms—that is, new molecules—always involve the

*The mole is the principal unit of measure for quantities of substances involved in chemical reactions. The number of particles (whether ions, atoms, or molecules) in 1 mole of any substance is always exactly the same: 6.023×10^{23}. For example, 1 mole of hydrogen ions contains 6.023×10^{23} ions, 1 mole of carbon atoms contains 6.023×10^{23} atoms, and 1 mole of the monosaccharide glucose contains 6.023×10^{23} molecules.

Table 3–1	A Comparison of Bond Strengths				
Bond	C=O	C=C	—C—H	—C—O—	—C—C—
Energy needed to break bond (kcal/mole)	171	147	99	84	83

rupture of existing chemical bonds and the formation of new bonds. If the new bonds are weaker than the bonds that were broken, energy will be released. However, if the new bonds are stronger than the bonds that were broken, energy will be taken up from the surroundings. For example, in the process of **photosynthesis,** which we shall discuss in Chapter 9, radiant energy from sunlight is taken up by the green cells of plants and algae. These cells use the energy to break and rearrange covalent bonds in carbon dioxide and water, making possible the formation of new chemical combinations—sugar molecules and oxygen molecules. This process, which stores the energy captured from sunlight in the covalent bonds of the sugar molecules, involves many chemical reactions. It can, however, be summarized as follows:

Carbon dioxide + Water + Energy $\longrightarrow$

Sugar + Oxygen

In another series of reactions, living cells can break the covalent bonds of the sugar molecules. In the presence of oxygen, new combinations of atoms—carbon dioxide and water—are formed, and the energy stored in the bonds of the sugar molecules is released:

Sugar + Oxygen $\longrightarrow$

Carbon dioxide + Water + Energy

Other organic substances, such as wood, gasoline, and alcohol, also release their stored energy when they are oxidized (broken down in the presence of oxygen), as occurs when they are burned for fuel. The amount of stored energy an organic compound contains is measured in terms of the amount of heat energy, expressed in kilocalories per mole, released when the compound is oxidized.

The monosaccharide glucose is the principal energy source for most living cells. When 1 mole of glucose is broken down to carbon dioxide and water, 686 kilocalories of energy are released. The yield is the same

3–2 *Over the millennia, many organisms have evolved sensitive mechanisms for detecting sugars. Houseflies, for instance, have sugar detectors in their feet. When a housefly lights on droplets of a sugar solution, its tubular mouthparts are automatically extended and the fly begins to feed. We also have sensory receptors tuned for the detection of sugar, although our receptors, in keeping with our eating habits, are in our tongues. A housefly reacts positively to about the same types of sugars that we do, although its detection mechanism is about 10 million times more sensitive.*

In the course of consuming sugary plant products, houseflies, humans, and other animals obtain not only a rich energy supply but also other essential nutrients, such as plant proteins, lipids, vitamins, and minerals.

whether the reaction takes place in a living cell or in the laboratory. Inside the cell, however, almost 40 percent of this energy is repackaged in the bonds of other organic molecules, as we shall see in Chapter 8. In the laboratory, all of the released energy is given off as heat.

Disaccharides: Transport Forms

Although glucose is the common transport sugar for humans and other vertebrate animals, sugars are often transported in other organisms as disaccharides. Sucrose, often called cane sugar, is the form in which sugar is transported in plants from the photosynthetic cells (mostly in the leaves), where it is produced, to other parts of the plant body. Sucrose is composed of the monosaccharides glucose and fructose. Sugar is transported through the bodies of many insects in the form

of another disaccharide, known as trehalose, which consists of two glucose units linked together. Another common disaccharide is lactose, a sugar that occurs only in milk. Lactose is made up of glucose combined with another monosaccharide, galactose.

In the synthesis of a disaccharide molecule from two monosaccharide molecules, a molecule of water is removed in the process of forming the new bond between the two monosaccharides (Figure 3–3). This type of chemical reaction, which occurs in the formation of most organic polymers from their subunits, is known as **dehydration synthesis.**

When a disaccharide is split into its monosaccharide subunits, which happens when it is used as an energy source, the molecule of water is added again. This splitting is known as **hydrolysis,** from *hydro,* meaning "water," and *lysis,* "breaking apart." Hydrolysis is an energy-releasing reaction. The hydrolysis of sucrose, for example, yields 5.5 kilocalories per mole. Conversely, the formation of sucrose from glucose and fructose requires an energy input of 5.5 kilocalories per mole of sucrose.

3–3 *Sucrose is a disaccharide made up of two monosaccharide subunits, one glucose and one fructose. The formation of sucrose involves the removal of a molecule of water (dehydration synthesis). The new chemical bond forged in the course of this reaction is shown in blue. The reverse reaction—splitting sucrose back into its constituent monosaccharides—requires the addition of a water molecule (hydrolysis).*

(a) Amylose—linear chain of repeated glucose subunits

(b) Amylopectin—branched chain of repeated glucose subunits

Branch point

Branch point

(c)

3–4 *In plants, sugars are stored in the form of starch. Starch is composed of two different types of polysaccharides, amylose and amylopectin. (a) A single molecule of amylose may contain 1,000 or more glucose subunits in a long unbranched chain, which winds to form a uniform coil. (b) A molecule of amylopectin may contain from 1,000 to 6,000 or more glucose subunits. Short chains of glucose subunits branch off from the main chain at frequent intervals. These side branches prevent the formation of a tight coil and produce a looser, more circular structure.*

(c) Starch molecules, perhaps because of their coiled nature, tend to cluster into granules. In this electron micrograph of a single storage cell of a potato, the spherical and egg-shaped objects are starch granules. They are magnified about 1,000 times.

Liver

3–5 *In animal cells, sugars are commonly stored as glycogen, primarily in the liver and in skeletal muscle. Glycogen resembles the amylopectin found in plant cells, but glycogen is more highly branched and the individual branches are shorter. When glucose is needed, hydrolysis can occur at the ends of many branches simultaneously, providing abundant supplies of glucose. The dark granules in this liver cell, magnified about 55,000 times, are glycogen.*

Polysaccharides

Polysaccharides are made up of monosaccharides linked together in long chains. They constitute storage forms for sugars. The principal storage polysaccharide in plants is **starch** (Figure 3–4), and in animals and fungi it is **glycogen** (Figure 3–5). Both starch and glycogen are built up of many glucose units. The differences between them are in the length of the polysaccharide chains and in the ways in which the glucose molecules are linked. Polysaccharides must be hydrolyzed to monosaccharides or disaccharides before the sugar subunits can be used as energy sources or transported through living systems.

Polysaccharides also play structural roles. In plants, the principal structural molecule is the polysaccharide **cellulose.** Although cellulose, like starch, is a polymer of glucose, the bonds linking the glucose units in cellulose are slightly different from those in starch or glycogen. This small difference has a profound effect on the three-dimensional structure of the molecules and thus on their properties. Rather than forming granules,

(a) Cellulose—linear chain of repeated glucose subunits

(b)

3–6 (a) *In the cellulose molecule, the bonds linking the glucose subunits are such that —OH groups (highlighted in blue) project from both sides of the chain. These —OH groups form hydrogen bonds with the —OH groups on adjacent cellulose molecules, producing bundles of cross-linked parallel chains. In the starch molecule (Figure 3–4), by contrast, most of the —OH groups capable of forming hydrogen bonds face toward the exterior of the coil, making it more readily soluble in the surrounding water.*

(b) *The wall of a young plant cell is about 40 percent cellulose. Hundreds of cellulose molecules are bundled together to form strands known as microfibrils, magnified here about 30,000 times. The microfibrils, as strong as an equivalent amount of steel, are embedded in other polysaccharides.*

3–7 *A cicada molting. The relatively hard outer coverings, or exoskeletons, of insects contain chitin, a modified polysaccharide that contains nitrogen. Because exoskeletons do not grow as the insect grows, they must be molted periodically. The discarded exoskeleton is at the bottom, below the insect, which is drying out and waiting for its new exoskeleton to harden. This cicada, a native of Australia, is commonly known as the "greengrocer."*

as do starch and glycogen molecules, cellulose molecules form long, rigid bundles (Figure 3–6).

Because of its structure, cellulose plays a biological role quite different from that of starch or glycogen. Cellulose is a major constituent of plant cell walls, but its glucose units are not readily available as an energy source, either for the plant or for other organisms. In fact, cellulose can be hydrolyzed by only a few microorganisms. Cows and other ruminants, termites, and cockroaches can use cellulose for energy only because of the microorganisms that inhabit their digestive tracts (see page 108).

Chitin, which is a major component of the exoskeletons (the hard, exterior coverings) of insects and other arthropods, is a tough, resistant, modified polysaccharide (Figure 3–7). Chitin is also found in the cell walls of fungi.

Lipids

Lipids are a group of organic substances that are insoluble in polar solvents, such as water, but that dissolve readily in nonpolar organic solvents, such as benzene. Typically, lipids serve as energy-storage molecules—usually in the form of fats or oils—and for structural purposes. Some lipids, however, play major roles as chemical "messengers," both within and between cells.

Fats and Oils: Energy in Storage

Unlike many plants, such as the potato, animals have only a limited capacity to store carbohydrates. In vertebrates, sugars consumed in excess of what can be stored as glycogen are converted into fats. Some plants also store food energy as oils, especially in seeds and fruits. Fats and oils contain a higher proportion of energy-rich carbon-hydrogen bonds than carbohydrates do and, as a consequence, contain more chemical energy. On average, the complete oxidation of fats yields about 9.3 kilocalories per gram* as compared to 3.8 kcal per gram of carbohydrate, or 3.1 kcal per gram of protein. Also, because fats are nonpolar, they do not attract water molecules and hence are not "weighted down" by them, as glycogen is. Taking into account the water factor, fats store six times as much energy, gram for gram, as glycogen, which is undoubtedly why in the course of evolution they came to play a major role in energy storage.

---

* 1,000 grams = 1 kilogram = 2.2 pounds, so oxidation of a pound of fat would yield about 4,200 kilocalories, more than the 24-hour requirement for a moderately active adult.

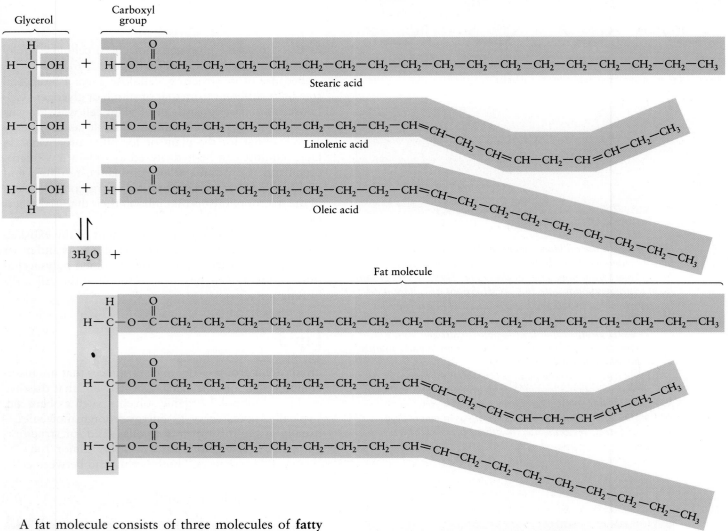

3–8 *A fat molecule consists of three fatty acids joined to a glycerol molecule (hence the term "triglyceride"). The long hydrocarbon chains of the fatty acids terminate in carboxyl (—COOH) groups, which become covalently bonded to the glycerol molecule. Each bond (dark blue) is formed when a molecule of water is removed (dehydration synthesis).*

Three different fatty acids are shown here. Stearic acid is saturated, whereas oleic acid and linolenic acid are unsaturated, as you can see by the double bonds in their structures. Oleic acid, with one double bond, is monounsaturated; linolenic acid, with three double bonds, is polyunsaturated.

The straight chains of saturated fatty acids allow saturated fat molecules to pack together, producing a solid, such as butter or lard. The kinks at the double bonds in unsaturated fatty acids tend to separate the fat molecules, producing a liquid, such as corn oil or peanut oil.

A fat molecule consists of three molecules of **fatty acid** covalently bonded to one **glycerol** molecule. As with the disaccharides and polysaccharides, each bond is formed by the removal of a molecule of water (dehydration synthesis), as shown in Figure 3–8. Fat molecules, which are also known as triglycerides, are said to be neutral because they contain no polar groups. As you would expect, they are extremely hydrophobic.

You have undoubtedly heard a lot about "saturated" and "unsaturated" fats. A fatty acid in which there are no double bonds between carbon atoms is said to be **saturated.** Each carbon atom in the chain has formed covalent bonds to four other atoms, and its bonding possibilities are therefore complete. By contrast, a fatty acid that contains carbon atoms joined by double bonds is said to be **unsaturated.** The double-bonded carbon atoms have the potential to form additional bonds with other atoms.

The physical nature of a fat is determined by the length of the carbon chains in the fatty acids and by whether the acids are saturated or unsaturated. Unsaturated fats tend to be liquid at room temperature. They are more common in plants than in animals; examples

Polar head

Nonpolar tails

$$O^-$$
$$R-O-P-O-^3CH_2$$
$$O$$

Phosphate group

$$H-^2C-O-C-CH_2CH_2CH_2CH_2CH_2CH_2CH_2CH=CHCH_2CH_2CH_2CH_2CH_2CH_3$$
$$O$$

$$H-^1C-O-C-CH_2CH_2CH_2CH_2CH_2CH_2CH_2CH_2CH_2CH_2CH_2CH_2CH_2CH_2CH_3$$
$$O$$
$$H$$

Glycerol

Fatty acid

3–9 *A phospholipid molecule consists of two fatty acids linked to a glycerol molecule, as in a fat, with a phosphate group (shown in yellow) linked to the glycerol's third carbon. It also usually contains an additional chemical group (shown in red) attached to the phosphate group. This group is symbolized by the letter R; its chemical composition varies in different molecules.*

The fatty acid "tails" of a phospholipid molecule are nonpolar and therefore insoluble in water (hydrophobic). The polar "head" containing the phosphate and R groups is water-soluble (hydrophilic).

are olive oil, peanut oil, and corn oil. Animal fats and their derivatives, such as butter and lard, contain saturated fatty acids and are usually solid at room temperature.

Phospholipids and Glycolipids

Lipids, especially phospholipids and glycolipids, also play extremely important structural roles. Like fats, both phospholipids and glycolipids are composed of fatty acid chains attached to a glycerol backbone. In the **phospholipids,** however, the third carbon of the glycerol molecule is occupied not by a fatty acid but by a negatively charged phosphate group, PO_4^{3-} (Figure 3–9), to which another polar group is usually attached. The phosphate end of the molecule is hydrophilic, and the fatty acid portions are hydrophobic.

When phospholipids are added to water, they tend to form a film along its surface, with their hydrophilic "heads" under the water and their hydrophobic fatty acid "tails" protruding above the surface (Figure 3–10a). In the watery interior of the cell, phospholipids tend to align themselves in double rows, with their fatty acid tails oriented toward one another and their phosphate heads directed outward (Figure 3–10b). As we shall see in Chapter 5, this arrangement of phospholipid molecules, with their hydrophilic heads exposed and their hydrophobic tails clustered together, forms the structural basis of cell membranes.

In the **glycolipids** ("sugar lipids"), the third carbon of the glycerol molecule is occupied not by a phosphate group but by a short carbohydrate chain. Depending on the particular glycolipid, this chain may contain anywhere from 1 to 15 sugar subunits. Like the phosphate head of a phospholipid, the carbohydrate head of a glycolipid is hydrophilic, and the fatty acid tails are, of course, hydrophobic. In aqueous solution, glycolipids behave in the same fashion as phospholipids, and they are also important components of cell membranes.

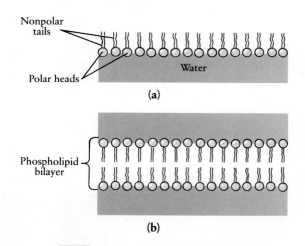

Nonpolar tails

Polar heads

Water

(a)

Phospholipid bilayer

(b)

3–10 (a) *Because phospholipids have polar heads and nonpolar tails, they tend to form a thin film on a water surface. The hydrophilic (water-loving) heads are in the water, and the hydrophobic (water-fearing) tails extend above the water.* (b) *Surrounded by water, phospholipid molecules spontaneously arrange themselves in two layers with their hydrophilic heads extending outward and their hydrophobic tails inward. This arrangement—a phospholipid bilayer—forms the structural basis of cell membranes.*

Regulation of Blood Cholesterol

Although cholesterol plays a number of essential roles in the animal body, it is also a principal villain in heart disease. Deposits containing cholesterol can narrow the arteries carrying blood to the heart muscle, and people with unusually large amounts of cholesterol in their blood have a high risk of heart attacks. How does the body regulate cholesterol levels? What goes wrong to cause elevated cholesterol levels? How does cholesterol cause heart attacks? Given the fact that heart disease is the major cause of death in this country, these questions are not only of biological interest but are also of personal importance to just about every one of us.

The key organ in cholesterol regulation is the liver, which not only synthesizes needed cholesterol from saturated fatty acids but also degrades excess cholesterol circulating in the blood—as a result, for example, of a diet rich in meat, cheese, and egg yolks. Cholesterol is transported to and from body cells, including those of the liver, by way of the bloodstream. Like other lipids, it is insoluble in plasma, the fluid portion of the blood. It is carried in particles consisting of a cholesterol interior and a lipid "wrapper" that has water-soluble proteins embedded in its outer surface. These large complexes exist in two principal forms: low-density lipoproteins (LDLs) and high-density lipoproteins (HDLs).

LDLs function as the delivery trucks of the system, carrying dietary cholesterol and newly synthesized cholesterol to various destinations in the body, including both the liver and the hormone-synthesizing organs. HDLs, however, function more like garbage trucks, carrying excess cholesterol on a one-way trip to the liver for degradation and excretion.

Normally the system is in balance, and the liver synthesizes or degrades cholesterol depending on the body's current needs and the amount of cholesterol in the blood. It can, however, be thrown out of balance by a number of factors. If, for example, the dietary intake of cholesterol is high, the liver becomes swamped and cannot degrade all of the excess. If the dietary intake of saturated fats is high, even in the absence of a high intake of cholesterol itself, the liver increases its synthesis of cholesterol.

Current evidence indicates that the liver monitors the level of cholesterol in the blood through its uptake of LDLs. If the quantities of LDLs taken up by the liver cells are low—a sign that blood cholesterol levels are low—the cells increase their synthesis and export of cholesterol. If the quantities of LDLs taken up by the cells are high—a sign that blood cholesterol levels are high—the cells stop synthesizing cholesterol.

When the quantities of circulating LDLs are greater than can be taken up by the liver and hormone-synthesizing organs, they are taken up by the cells lining the arteries supplying the heart. This can ultimately lead to total blockage of an artery and thus to a heart attack.

Heart disease often runs in families, suggesting that hereditary factors are involved in some cases. Normal liver cells have specialized receptors on their surface that participate in the transport of LDLs into the cells. In one type of hereditary heart disease, these receptors are absent. No LDLs can be taken up by the liver cells, and they continuously synthesize and export cholesterol. Persons with this disease have six to eight times the normal amount of cholesterol in their blood, usually have their first heart attack in childhood, and die of heart disease in their early twenties. Other families seem to be protected against heart disease, apparently because the individuals' bodies synthesize large quantities of HDLs, ensuring that all excess cholesterol makes a speedy one-way trip to the liver.

For most of us, however, the degree of risk depends on our behavior: whether we exercise regularly, which seems to increase HDL levels and thus protect against cholesterol buildup; whether we smoke cigarettes, which seems to decrease HDL levels; and the quantities of cholesterol and saturated fats that we consume.

(a)

(b)   (c)

In one type of heart disease, atherosclerosis, cholesterol and other fatty substances accumulate in the walls of the arteries supplying the heart muscle. This accumulation triggers abnormal growth and the production of fibrous tissues by the cells of the walls. (a) A normal cor- *onary artery. Note the wide channel through which the blood can flow. (b) A coronary artery in which moderate atherosclerosis has developed. Fatty deposits have formed, and the space left for blood flow is significantly decreased. (c) A coronary artery in which the de-* *posits have become so great that only a very narrow channel remains open. Such a narrow channel can be completely blocked by a blood clot. The result is a heart attack and the death of the heart muscle supplied by the artery.*

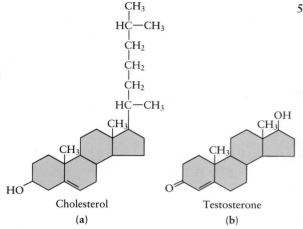

3–11 *All groups of land plants synthesize waxes, which protect exposed plant surfaces from water loss. This scanning electron micrograph reveals the wax-producing glands on the outer surface of a flower bud of a primrose. The stalked structures are the glands, and the fluffy material at the top of the stalks is the wax they have produced. The wax prevents water loss from the flower bud and insulates its delicate internal tissues from the cold. The wax deposits, magnified here 300 times, are sometimes so voluminous that the primrose plant appears to be dusted with flour.*

3–12 *Two examples of steroids. (a) The cholesterol molecule consists of four carbon rings and a hydrocarbon chain. (b) Testosterone, a male sex hormone synthesized from cholesterol by cells in the testes, also has the characteristic four-ring structure but lacks the hydrocarbon tail.*

Waxes

Waxes are another type of structural lipid. They form protective, waterproof coatings on skin, fur, and feathers, on the leaves and fruits of land plants (Figure 3–11), and on the exoskeletons of many insects.

Cholesterol and Other Steroids

Cholesterol belongs to an important group of compounds known as the **steroids** (Figure 3–12). Although steroids do not resemble the other lipids structurally, they are grouped with them because they are insoluble in water. All the steroids have four linked carbon rings, and several of them have a hydrocarbon tail.

Cholesterol is found in animal cell membranes. About 25 percent (by dry weight) of the cell membrane of a red blood cell is cholesterol. It is synthesized in the liver from saturated fatty acids. Cholesterol is also obtained in the diet, principally in meat, cheese, and egg yolks. High concentrations of cholesterol in the blood are associated with atherosclerosis, in which cholesterol is found in fatty deposits on the interior lining of diseased blood vessels (see essay).

A number of hormones are also steroids. These include cortisol and the sex hormones estrogen and testosterone. In the body steroid hormones are synthesized from cholesterol.

Proteins

Proteins are among the most abundant organic molecules. In most living organisms they make up 50 percent or more of the dry weight. Only plants, with their high cellulose content, are less than half protein. Proteins perform an incredible diversity of functions in living systems. In their structure, however, they all follow the same simple blueprint: they are all polymers of nitrogen-containing molecules known as **amino acids**, arranged in a linear sequence. Some 20 different kinds of amino acids are used by living systems to build proteins.

Protein molecules are large, often containing several hundred amino acid monomers. Thus the number of different amino acid sequences, and therefore the possible variety of protein molecules, is enormous—about as enormous as the number of different sentences that can be written with our own 26-letter alphabet. Organisms, however, synthesize only a very small fraction of the proteins that are theoretically possible. The single-celled bacterium *Escherichia coli,** for example, contains 600 to 800 different kinds of proteins at any one time, and the cell of a plant or animal has several times that number. A typical cell in your body contains about 100 million protein molecules of about 10,000 different kinds. Each has a special function, and each, by its unique chemical nature, is specifically suited for that function.

---

* Biologists use a binomial ("two-name") system for designating organisms. Every different kind of organism has a unique two-part name. The first part of the name refers to the genus to which the organism belongs. The second part, in combination with the first part, refers to the particular species, a subdivision of the genus category. In this name, for example, *Escherichia* denotes the genus, while *coli* designates a particular kind, or species, of *Escherichia*, distinguished from all others by certain characteristics. By convention, with the second mention of a scientific binomial, it is permissible to abbreviate the first (genus) name. This is fortunate, particularly when dealing with such names as *Escherichia*.

| Glycine (gly) | Valine (val) | Cysteine (cys) | Serine (ser) | Phenylalanine (phe) | Glutamic acid (glu) |

3–13 *Eight of the 20 different kinds of amino acids used in making proteins. As you can see, the fundamental structure is the same in all of the molecules, but the R groups differ. The other 12 amino acids—* *not shown here—are alanine (ala), glutamine (gln), histidine (his), leucine (leu), asparagine (asn), methionine (met), isoleucine (ile), aspartic acid (asp), lysine (lys), threonine (thr), tyrosine (tyr), proline (pro).*

Arginine (arg)   Tryptophan (trp)

Amino Acids: The Building Blocks of Proteins

Amino acids are made up of carbon, hydrogen, and oxygen, as are sugars, and, in addition, they contain nitrogen. Every amino acid contains an amino group ($-NH_2$), a hydrogen atom, and a carboxyl group ($-COOH$) bonded to a central carbon atom:

$$\text{H---N---C---C---OH}$$

This is the fundamental structure of the molecule, and it is the same in all amino acids.

Different amino acids differ in the side group that occupies the fourth covalent bond of the central carbon atom. The side group is symbolized by the letter "R" (for "rest of molecule"). It has a different chemical structure in each different kind of amino acid (Figure 3–13). These differences are very important because they determine the different biological properties of the individual amino acids and, as a consequence, of the various kinds of proteins.

In yet another example of a dehydration synthesis, the amino "head" of one amino acid can be linked to the carboxyl "tail" of another by the removal of a molecule of water (Figure 3–14). The resulting covalent linkage is known as a **peptide bond,** and the molecule that is formed by the linking of many amino acids is called a **polypeptide** (Figure 3–15).

In order to assemble amino acids into proteins, a cell must have not only a large enough quantity of amino acids but also enough molecules of each kind. This fact is of great importance in human nutrition (see essay).

The Levels of Protein Organization

In a living cell, a protein is assembled in a long polypeptide chain, one amino acid at a time. In this process, the head of one amino acid is linked to the tail of an-

3–14 *A peptide bond is a covalent bond between two amino acids, formed by dehydration synthesis. The amino group of one amino acid is joined to the carboxyl group of the adjacent amino acid, and a molecule of water is split out.*

other, like a line of boxcars. The linear sequence of amino acids, which is dictated by the information stored in the cell for that particular protein, is known as the **primary structure** of the protein (Figure 3–16). Each different protein has a different primary structure. It determines the structural features of the molecule as a whole and thus its biological function. Even one small variation in the sequence may alter or destroy the way in which the protein functions.

Amino Acids and Nitrogen

Like fats, amino acids are formed within living cells using sugars as starting materials. But while fats contain only carbon, hydrogen, and oxygen atoms, all available in the sugar and water of the cell, amino acids also contain nitrogen. Most of the Earth's supply of nitrogen exists in the form of gas in the atmosphere. Only a few organisms, all microscopic, are able to incorporate nitrogen from the air into compounds—ammonia, nitrites, and nitrates—that can be used by living systems. Hence the proportion of the Earth's nitrogen supply available to the living world is very small.

Plants incorporate the nitrogen from ammonia, nitrites, and nitrates into carbon-hydrogen compounds to form amino acids. Animals are able to synthesize some of their amino acids, using ammonia as a nitrogen source. The amino acids they cannot synthe-size, the so-called **essential amino acids,** must be obtained in the diet, either from plants or from the meat of animals that have eaten plants. For adult human beings, the essential amino acids are lysine, tryptophan, threonine, methionine, phenylalanine, leucine, valine, and isoleucine.

People who eat meat usually get enough protein and the correct balance of amino acids. People who are vegetarians, whether for philosophical, esthetic, or economic reasons, have to be careful that they get enough protein and, in particular, all of the essential amino acids.

For many years, agricultural scientists concerned with the world's hungry people concentrated on developing plants with a high caloric yield. More recently, increasing recognition of the role of plants as a major source of amino acids for human populations has led to emphasis on the development of high-protein strains of food plants. Of particular importance has been the development of plants, such as "high-lysine" corn, with increased levels of one or more of the essential amino acids.

Another approach to obtaining the right balance of amino acids from plant sources is to combine certain foods. Beans, for instance, are likely to be deficient in tryptophan and in the sulfur-containing amino acids cysteine and methionine, but they are a good-to-excellent source of isoleucine and lysine. Rice is deficient in isoleucine and lysine but provides an adequate amount of the other essential amino acids. Thus rice and beans in combination make just about as perfect a protein menu as eggs or steak, as some nonscientists seem to have known for quite a long time.

3–15   *Polypeptides are polymers of amino acids linked together, one after another, by peptide bonds. The polypeptide chain shown here contains six different amino acids, but some chains may contain as many as 1,000 linked amino acid monomers.*

3–16   *The primary structure of a relatively small protein, human adrenocorticotropic hormone (ACTH). This was one of the first proteins for which the primary structure was determined. As you can see, it consists of a single polypeptide chain containing 39 amino acids. This hormone, secreted by the pituitary gland, stimulates the production of cortisol and related steroid hormones by the adrenal glands.*

3-17   *In many different types of proteins, the polypeptide chain forms a secondary structure known as a helix. (a) The helix is held in shape by hydrogen bonds, indicated by the dashed lines. The hydrogen bonds form between the double-bonded oxygen atom in one amino acid and the hydrogen atom of the amino group in another amino acid that occurs four amino acids farther along the chain. The R groups, which appear flattened in this diagram, actually extend out from the helix, as shown in (b). In some proteins, virtually all of the molecule is in the form of a helix. In other proteins, only certain regions of the molecule have this secondary structure.*

As a polypeptide chain is assembled, interactions among the various amino acids along the chain cause it to fold in a simple pattern, known as its **secondary structure.** One common secondary structure is a coil or, as it is more commonly called by biochemists, a **helix.** The shape of the helix is maintained by hydrogen bonds (Figure 3-17). Because the hydrogen bonds break and re-form easily, helical proteins—or helical portions of proteins—are stretchable (elastic). Examples of such elastic proteins are myosin, which is one of the major components of muscle, and keratin, the protein that forms hair. The changes that occur when hair is washed and dried involve breaking hydrogen bonds and making new ones.

Other proteins, such as the silk fibers of cocoons and spider webs, are made of extended polypeptide chains lined up in parallel and linked to one another by hydrogen bonds. Proteins with this type of secondary structure, known as a **pleated sheet** (Figure 3-18), are smooth and supple but nonelastic.

Collagen, which is a principal component of cartilage, bones, and tendons, has yet another type of secondary structure. Collagen molecules are made by specialized cells called fibroblasts. As the long polypeptide chains are synthesized, three of them wrap around each other to form a cablelike molecule. These molecules are then packed into long, thin fibrils.

Proteins that exist for most of their length in a helical, pleated-sheet, or cablelike secondary structure are known as **fibrous proteins.** They play a variety of important structural roles in organisms.

3-18   *Another common secondary structure of proteins is the pleated sheet. (a) The pleats result from the alignment of the zigzag pattern of the atoms that form the backbone of polypeptide chains. The sheet is held together by hydrogen bonding between adjacent chains. The R groups extend above and below the pleats, as shown in (b). In some proteins, two or more polypeptide chains are aligned with one another to form a pleated sheet. In other proteins, a single polypeptide chain loops back and forth in such a way that adjacent portions of the chain form a pleated sheet.*

(a)   (b)

3–19   *The fibrous protein keratin is found in all vertebrates. It is the chief component of scales, wool, nails, and feathers. (a) The horn of a rhinoceros consists of tightly packed strands of keratin. Solid rhino horn is used for dagger handles, and powdered horn is sold as an aphrodisiac. A single horn can net a poacher far more than the average annual wage in many parts of Africa. (b) Feathers, such as these spectacularly colored peacock feathers, are made up of a shaft to which thousands of barbs—each with many tiny barbules—are attached.*

In other proteins, known as **globular proteins,** the secondary structure folds back on itself to make a complex **tertiary structure.** The tertiary structure forms as a result of complex interactions among the R groups in the individual amino acids. These interactions include attractions and repulsions among amino acids with polar R groups and repulsions between nonpolar R groups and the surrounding water molecules. In addition, the sulfur-containing R groups of molecules of the amino acid cysteine can form covalent bonds with each other. These bonds, known as **disulfide bridges,** lock portions of the molecule into a particular position.

A great diversity of globular proteins are synthesized by living organisms. They include enzymes, which regulate the chemical reactions occurring within organisms; antibodies, which are important components of the immune system; and receptors on the surface of the cell membrane that interact with a variety of other molecules. As we shall see in subsequent chapters, the three-dimensional structures of all of these proteins are of critical importance in determining their biological functions.

Many proteins are composed of more than one polypeptide chain. The polypeptide chains are held together by hydrogen bonds, disulfide bridges, attractions between positive and negative charges, and hydrophobic forces. This level of organization of proteins, which involves the interaction of two or more polypeptides, is called a **quaternary structure.** The hormone insulin, for example, has an intricate quaternary structure formed by the two polypeptide chains of which the molecule is composed (Figure 3–20).

The secondary, tertiary, and quaternary structures of a protein all depend on the primary structure—the linear sequence of amino acids.

3–20   *A computer-generated model of the insulin molecule, which consists of two short polypeptide chains, folded together in an intricate three-dimensional structure. The individual atoms are identified by color. Carbon atoms are white, sulfur atoms are yellow, slightly hydrophilic nitrogen atoms are light blue, strongly hydrophilic nitrogen atoms are dark blue, slightly hydrophilic oxygen atoms are pink, and strongly hydrophilic oxygen atoms are red. Hydrogen atoms are the same color as the atoms to which they are bonded.*

ala **gly** **leu** **val** **lys** **lys** **gly** **his** **ala** **lys**

Primary
structure

(a)

Secondary
structure

(b)

Tertiary
structure

(c)

Polypeptide chain   Polypeptide chain

Polypeptide chain   Heme   Polypeptide chain

Quaternary
structure

(d)

3–21 *Levels of organization in the hemo-globin molecule.* **(a)** *The sequence of the amino acids in each chain is its primary structure.* **(b)** *The helical form assumed by any part of the chain as a consequence of hydrogen bonding between nearby*

$C{=}O$ *and* $—NH$ *groups is its secondary structure.* **(c)** *The folding of the chains in three-dimensional shapes is the tertiary struc-ture, and* **(d)** *the combination of the four chains into a single functional molecule is the quaternary structure. The outside of the molecule and the hole through the middle are lined by charged amino acids, and the uncharged amino acids are packed inside. Each of the four chains surrounds a heme group (red), which can hold a single oxygen molecule. A hemoglobin molecule is therefore capable of transporting four oxy-gen molecules.*

Hemoglobin: An Example of Specificity

Fibrous proteins usually consist of a relatively small va-riety of amino acids in a repetitive sequence. Many globular proteins, by contrast, have extremely complex, irregular amino acid sequences, as complex and irreg-ular as the sequence of letters in a sentence on this page. Just as these sentences make sense (if they do) because the letters are the right ones and in the right order, the proteins make sense, biologically speaking, because their amino acids are the right ones in the right order.

Hemoglobin, for example, is a protein that is man-ufactured and carried in the red blood cells. Its mole-cules have the special property of being able to combine loosely with oxygen, collecting it in the lungs and re-leasing it elsewhere in the body. The hemoglobin mol-ecule has a quaternary structure that consists of four polypeptide chains, each of which is combined with an iron-containing group known as heme. Hemoglobin has two identical chains of one type, designated alpha, and two identical chains of another type, designated beta. Each type of chain has a unique primary structure containing about 150 amino acids, for a total of about 600 amino acids in all (Figure 3–21).

Sickle-cell anemia is a disease in which the hemoglo-bin molecules are defective. When oxygen is released from them, these molecules change shape and combine with one another to form stiffened rodlike structures.

(a)

(b)

3–22 *Scanning electron micrographs of* **(a)** *a human red blood cell containing nor-mal hemoglobin, and* **(b)** *a red blood cell containing the abnormal hemoglobin associ-ated with sickle-cell anemia. When the oxy-gen concentration in the blood is low, the abnormal hemoglobin molecules stick to-gether, distorting the shape of the cells. As a result, the cells cannot pass readily through small blood vessels. These cells are magnified about 35,000 times.*

Normal hemoglobin

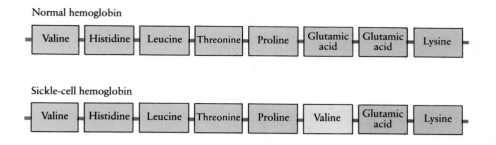

Sickle-cell hemoglobin

3–23   *An example of the remarkable precision of the "language" of proteins. Portions of the beta chains of the normal hemoglobin molecule and the sickle-cell hemoglobin molecule are shown. The entire structural difference between the normal molecule and the sickle-cell molecule (literally, a life-and-death difference) consists of one change in the sequence of each beta chain: one glutamic acid is replaced by one valine.*

Red cells containing large proportions of such hemoglobin molecules become stiff and deformed, taking on the characteristic sickle-cell shape (Figure 3–22). The deformed cells are more fragile than normal red blood cells and rupture easily. They also tend to clog small blood vessels, causing blood clots and depriving vital organs of their full supply of blood. The result is pain, intermittent illness, and, in many cases, a shortened life span.

Analysis of the hemoglobin molecules has revealed that the only difference between normal and sickle-cell hemoglobin is that at a precise location in each beta chain, one glutamic acid is replaced by one valine (Figure 3–23). When the beta chains assume their tertiary structure and become associated with the alpha chains in the quaternary structure of the completed protein, the location at which the substitution occurs is on the outside of the molecule. As you can see in Figure 3–13 (page 60), glutamic acid—present in the normal beta chain—is polar and thus hydrophilic. Valine, by contrast, is nonpolar and thus hydrophobic. The substituted valines in the two beta chains create two "sticky" regions on the surface of the hemoglobin molecule. It is the mutual attraction of these regions in neighboring molecules that causes sickle-cell hemoglobin to form rodlike structures.

In hemoglobin, a difference of only two amino acids in a total of almost 600 can be the difference between life and death. This gives you an idea of the precision and the importance of the arrangement of amino acids in a particular sequence in a protein. In living systems, which must perform many different activities simultaneously, the specificity of function that results from the structural precision of different protein molecules is of crucial importance.

Nucleic Acids

The information dictating the structures of the enormous variety of protein molecules found in living organisms is encoded in molecules known as nucleic acids. Just as proteins consist of long chains of amino acids, nucleic acids consist of long chains of molecules known as **nucleotides.** A nucleotide, however, is a more complex molecule than an amino acid.

As shown in Figure 3–24, a nucleotide consists of three subunits: a phosphate group, a five-carbon sugar, and a **nitrogenous base**—a molecule that has the properties of a base and contains nitrogen. The phosphate group (PO_4^{3-}) is an ion of phosphoric acid (H_3PO_4); it

3–24   *A nucleotide is made up of three different subunits: a phosphate group, a five-carbon sugar, and a nitrogenous base. The nitrogenous base in this nucleotide is adenine, and the sugar is ribose. The nucleotide is adenosine monophosphate, abbreviated AMP.*

3–25   *The sugar subunit of a nucleotide may be either*
(a) ribose or (b) deoxyribose. The structural difference between
the two sugars is highlighted in blue. As you can see, it is very
slight. RNA is formed from nucleotides that contain ribose,
and DNA from nucleotides that contain deoxyribose.

is the source of "acid" in the term "nucleic acid." The
sugar subunit of a nucleotide may be either ribose or
deoxyribose, which contains one less oxygen atom than
ribose (Figure 3–25). Five different nitrogenous bases
occur in the nucleotides that are the building blocks of
nucleic acids. In the nucleotide in Figure 3–24, the nitrogenous base is adenine. In adenine and other nitrogenous bases, each of the nitrogen atoms in the molecule has an unshared pair of electrons in the outer
energy level. These electrons exert a weak attraction for
hydrogen ions (H^+). Thus the molecule is a base, capable of combining with H^+ ions and thereby increasing the relative number of OH^- ions in a solution (see
page 45).

Two types of nucleic acid are found in living organisms. In **ribonucleic acid (RNA),** the sugar subunit in
the nucleotides is ribose. In **deoxyribonucleic acid**
(DNA), it is deoxyribose. Like polysaccharides, fats,
and proteins, RNA and DNA are formed from their
subunits in dehydration synthesis reactions. The result
is a linear molecule, consisting of one nucleotide after
another (Figure 3–26).

3–26   *Nucleic acid molecules are long chains of nucleotides*
in which the sugar subunit of one nucleotide is linked to the
phosphate group of the next nucleotide. The covalent bond
linking one nucleotide to the next—shown here in blue—is
forged by a dehydration synthesis reaction. This reaction is re-
versible; with the addition of a molecule of water in hydrolysis,
the bond can be broken.

RNA molecules consist of a single chain of nucleotides, as
shown here. DNA molecules, by contrast, consist of two
chains of nucleotides, coiled around each other in a double
helix.

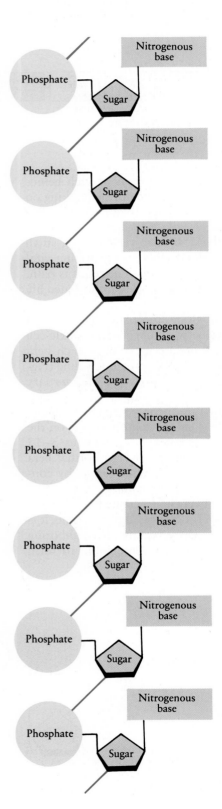

Although their chemical components are very similar, DNA and RNA generally play different biological roles. DNA is the carrier of the genetic message. It contains the information, organized in units known as **genes,** that we and other organisms inherit from our parents. RNA is a copy, or transcript, of the genetic message that serves as a blueprint for protein synthesis.

The discovery of the structure and function of DNA and RNA is undoubtedly the greatest triumph thus far of the molecular approach to the study of biology. In Section 3, we shall trace the events leading to the key discoveries and shall consider in some detail the marvelous processes—the details of which are still being worked out—by which the nucleic acids perform their functions.

ATP: The Cell's Energy Currency

In addition to their role as the building blocks of nucleic acids, nucleotides have an independent and crucial function in living systems. When modified by the addition of two more phosphate groups, they are the carriers of the energy needed to power the numerous chemical reactions occurring within the living cell. The energy in storage carbohydrates, such as starch and glycogen, and in lipids is like money in savings bonds or certificates of deposit—not readily accessible. The energy in glucose is like money in a checking account—accessible but not very handy for everyday transactions. The energy in modified nucleotides, however, is like money in your pocket—available in convenient amounts and universally accepted.

The principal energy carrier for most processes in living organisms is a molecule known as **adenosine triphosphate,** or **ATP.** This molecule is shown schematically in Figure 3–27. Note the three phosphate groups.

Adenosine triphosphate (ATP)

3–27   *A schematic diagram of the ATP (adenosine triphosphate) molecule. The only difference between this molecule and the AMP (adenosine monophosphate) molecule shown in Figure 3–24 is the addition of two phosphate groups. Although this structural difference may seem minor, it is the key to the function of ATP in living systems.*

The bonds linking these groups, designated by a squiggle, are relatively weak, and they can be broken quite readily by hydrolysis. The products of the most common reaction, illustrated in Figure 3–28, are **ADP** (adenosine diphosphate), a phosphate group, and energy. As this energy is released, it can be used to power other chemical reactions.

ADP is "recharged" to ATP when glucose is oxidized to carbon dioxide and water, just as the money in your pocket is "recharged" when you cash a check or visit an automatic teller machine. In Chapter 8, we shall consider this process in more detail. For now, however, the important thing to remember is that ATP is the molecule that is directly involved in providing energy for the living cell, to which we shall turn our attention in the next three chapters.

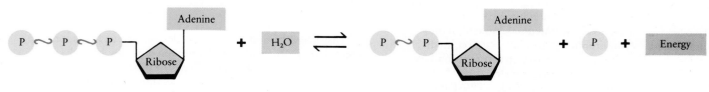

Adenosine triphosphate (ATP)   Adenosine diphosphate (ADP)

3–28   *The hydrolysis of ATP. With the addition of a molecule of water to ATP, one phosphate group is removed from the molecule. The products of this reaction are ADP, a free* *phosphate group, and energy. About 7 kilocalories of energy are released for every mole of ATP hydrolyzed. With an energy input of 7 kilocalories per mole, the reaction can be reversed.*

Summary

The chemistry of living organisms is, in essence, the chemistry of organic compounds—that is, compounds containing carbon. Carbon is uniquely suited to this central role by its ability to form multiple covalent bonds. Because of this ability, carbon can combine with carbon and other atoms to form a great variety of strong and stable chain and ring compounds. Organic molecules derive their three-dimensional shapes primarily from their carbon skeletons. Among the major types of organic molecules important in living systems are carbohydrates, lipids, proteins, and nucleic acids.

Carbohydrates serve as a primary source of chemical energy for living systems. The simplest carbohydrates are the monosaccharides ("single sugars"), such as glucose and fructose. Monosaccharides can be combined to form disaccharides ("two sugars"), such as sucrose, and polysaccharides (chains of many monosaccharides). The polysaccharides starch and glycogen are storage forms for sugar, whereas cellulose, another polysaccharide, is an important structural material in plants. Disaccharides and polysaccharides are formed by dehydration synthesis reactions in which monosaccharide units are covalently bonded with the removal of a molecule of water. They can be broken apart again by hydrolysis, with the addition of a water molecule.

Lipids are hydrophobic organic molecules that, like carbohydrates, play important roles in energy storage and as structural components. Compounds in this group include fats and oils, phospholipids, glycolipids, waxes, and cholesterol and other steroids. Phospholipids and glycolipids are major components of cell membranes.

Proteins are very large molecules composed of long chains of amino acids; these chains are known as polypeptides. The 20 different amino acids used in making proteins vary according to the properties of their side (R) groups. From these relatively few amino acids, an extremely large variety of different kinds of protein molecules can be synthesized, each of which has a highly specific function in living systems. Some proteins are fibrous and have important structural roles. Other proteins are globular and have regulatory and protective functions. The principal levels of protein organization are (1) primary structure, the linear amino acid sequence; (2) secondary structure, often a coiling of the polypeptide chain; (3) tertiary structure, the complex folding of the chain into various shapes; and (4) quaternary structure, the folding of two or more polypeptide chains around each other. The precise structure of a protein molecule is of critical importance in its capacity to carry out a specific biological function.

The building blocks of the nucleic acids are complex molecules known as nucleotides. A nucleotide consists of a phosphate group, a five-carbon sugar, and a nitrogenous base. The nucleic acid known as deoxyribonucleic acid (DNA) is the carrier of the genetic information. Another molecule, ribonucleic acid (RNA), is a copy of that information that functions as a blueprint for protein synthesis. Nucleotides also play key roles in the energy exchanges accompanying chemical reactions within living systems. The principal energy carrier for most processes is adenosine triphosphate (ATP).

Questions

1. Distinguish among the following: hydrocarbon/carbohydrate; glucose/fructose/sucrose; monomer/polymer; glycogen/starch/cellulose; saturated/unsaturated; monounsaturated/polyunsaturated; phospholipid/glycolipid; polysaccharide/polypeptide; primary structure/secondary structure/tertiary structure/quaternary structure; nitrogenous base/nucleotide/nucleic acid; ATP/ADP.

2. Many of the synthetic reactions in living systems take place by dehydration synthesis. What is a dehydration synthesis reaction? What types of molecules undergo dehydration synthesis reactions to form disaccharides and polysaccharides? To form fats? To form proteins? To form nucleic acids?

3. Disaccharides and polysaccharides, as well as lipids, proteins, and nucleic acids, can be broken down by hydrolysis. What is hydrolysis? What two types of products are released when a polysaccharide such as starch is hydrolyzed? How are these products important for the living cell?

4. What do we mean when we say that some polysaccharides are "energy-storage" molecules and that others are "structural" molecules? Give an example of each. In what

sense should any polysaccharide be regarded as an "energy-storage" molecule?

5. Plants usually store energy reserves as polysaccharides, whereas, in most animals, lipids are the principal form of energy storage. Why is it advantageous for animals to have their energy reserves stored as lipids rather than as polysaccharides? (Think about the differences in "life style" between plants and animals.) What kinds of storage materials would you expect to find in seeds?

6. Sketch the arrangement of phospholipids when they are surrounded by water.

7. In pioneer days, soap was made by boiling animal fat with lye (potassium hydroxide). The bonds linking the fatty acids to the glycerol molecule were hydrolyzed, and the po-

tassium hydroxide reacted with the fatty acid to produce potassium stearate. A typical soap available today is sodium stearate. In water, it ionizes to produce sodium ions (Na^+) and stearate ions:

Explain how soap functions to trap and remove particles of grease and dirt.

8. Consider the R groups of the eight amino acids illustrated in Figure 3–13. Which of these R groups would you expect to be nonpolar? Polar? Are there any of the R groups that you would expect to function as weak acids in water solution? As weak bases?

Suggestions for Further Reading

Brown, Michael S., and Joseph L. Goldstein: "How LDL Receptors Influence Cholesterol and Atherosclerosis," *Scientific American*, November 1984, pages 58–66.

Karplus, Martin, and J. Andrew McCammon: "The Dynamics of Proteins," *Scientific American*, April 1986, pages 42–51.

Lehninger, Albert L., David L. Nelson, and Michael M. Cox: *Principles of Biochemistry*, 2d ed., Worth Publishers, Inc., New York, 1993.

This is an updated version of a classic introductory text, outstanding for both its clarity and its consistent focus on the living cell.

Mathews, Christopher K., and K. E. van Holde: *Biochemistry*, The Benjamin/Cummings Publishing Company, Redwood City, Calif., 1990.

Chapters 4–9 of this beautifully illustrated text describe the structure and properties of nucleic acids, proteins, carbohydrates, and lipids.

Olson, Arthur J., and David S. Goodsell: "Visualizing Biological Molecules," *Scientific American*, November 1992, pages 76–81.

Richards, Frederic M.: "The Protein Folding Problem," *Scientific American*, January 1991, pages 54–63.

Scientific American: *The Molecules of Life*, W. H. Freeman and Company, New York, 1986.*

This reprint of the October 1985 issue of Scientific American includes 11 articles on molecules that play key roles in the living cell. The articles on proteins are of particular interest at this point in your study of biology. You will find the other articles in this collection useful at later points in the course.

Sharon, Nathan: "Carbohydrates," *Scientific American*, November 1980, pages 90–116.

Stryer, Lubert: *Biochemistry*, 3d ed., W. H. Freeman and Company, New York, 1988.

An introductory text, with many examples of medical applications of biochemistry. Handsomely illustrated.

* Available in paperback.

Cells: An Introduction

Life first appeared on Earth more than 3.5 billion years ago, when the planet was still very young. The first step in the origin of life was the formation of complex organic molecules from the atoms and simple molecules present in the Earth's crust, waters, and atmosphere. This process required energy. One likely source was the internal heat of the Earth itself. Because of the high temperatures (2200°C to 2800°C) and pressures, rock in the interior of the Earth is molten. It is this molten material that erupts from volcanoes, such as Sakurajima in Japan, shown here in a photograph taken on May 18, 1991.

Volcanic eruptions, releasing enormous amounts of energy, are thought to have been far more frequent on the early Earth than they are today. The energy released takes several forms: dry heat, from the molten rock; damp heat, in the form of steam; and shock waves that travel through the air, water, and rock of the Earth's surface. Volcanic eruptions are also frequently accompanied by lightning. The lightning is caused by the imbalance of electric charge produced by the clouds of volcanic ash, steam, and other gases spewing up into the atmosphere. All of these forms of energy associated with volcanoes are thought to have played a role in the events leading to the origin of life.

The Earth's crust and the upper portion of its mantle are made up of separate plates on which the continents and oceans rest. Volcanoes occur most frequently in areas where different plates converge and one plate slips below another. Sakurajima and the other volcanoes of the Pacific "Ring of Fire" are located along the line where the Asian plate and the Pacific plate converge.

In the last three chapters, we have progressed from subatomic particles through atoms and molecules to large, complex molecules, such as proteins and nucleic acids. At each level of organization, new properties appear. For instance, water is not the sum of the properties of elemental hydrogen and oxygen. It is something more and also something different. In proteins, amino acids become organized into polypeptides, and polypeptide chains are arranged in new levels of organization, the secondary, tertiary, and, in some cases, quaternary structure of the complete protein molecule. Only at the final level of organization do the complex properties of the protein emerge, and only then can the molecule assume its function.

The characteristics of living systems, like those of atoms or of molecules, do not emerge gradually as the degree of organization increases. They appear quite suddenly and specifically, in the form of the living cell—something that is more than and different from the atoms and molecules of which it is composed. No one knows exactly when or how this new level of organization—the living cell—first came into being. However, increasing knowledge of the history of our planet and the results of numerous laboratory experiments provide strong evidence to support the hypothesis that living cells spontaneously self-assembled from molecules present in the primitive seas.

The Formation of the Earth

About 5 billion years ago, cosmologists calculate, the star that is our sun came into being. According to current theory, it formed, like other stars, from an accumulation of dust and hydrogen and helium gases whirling in space among the older stars.

The immense cloud that was to become the sun condensed gradually as the hydrogen and helium atoms were pulled toward one another by the force of gravity, falling into the center of the cloud and gathering speed

4–1   *Of the nine planets in our solar system, only one, as far as we know, has life on it. This planet, Earth, is visibly different from the others. From a distance, it appears blue and green and it shines a little. The blue is water, the green is chlorophyll, and the shine is sunlight reflected off the layer of gases surrounding the planet's surface. Life, at least as we know it, depends on these visible features of Earth.*

as they fell. As the cluster grew denser, the atoms moved more rapidly. More atoms collided with each other, and the gas in the cloud became hotter and hotter. As the temperature rose, the collisions became increasingly violent until the hydrogen atoms collided with such force that their nuclei fused, forming additional helium atoms and releasing nuclear energy. This thermonuclear reaction is still going on at the heart of the sun and is the source of the energy radiated from its turbulent surface.

The planets, according to current theory, formed from the remaining gas and dust moving around the newly formed star. At first, particles would have collected at random, but as each mass grew larger, other particles began to be attracted by the gravity of the largest masses, which would become the planets. The whirling dust and forming planetary spheres continued to revolve around the sun until finally each sphere had swept its own path clean, picking up loose matter like a giant snowball. The orbit nearest the sun was swept by Mercury, the next by Venus, the third by Earth, the fourth by Mars, and so on out to Neptune and Pluto,

the most distant of the planets. The planets, including Earth, are calculated to have come into being about 4.6 billion years ago.

During the time Earth and the other planets were being formed, the release of energy from radioactive materials kept their interiors very hot. When Earth was still so hot that it was mostly liquid, the heavier materials sank to the interior, forming a dense inner core whose diameter is about half that of the planet. As Earth's surface cooled, an outer crust, a skin as thin by comparison as the skin of an apple, solidified. The oldest known rocks in this layer have been dated by isotopic methods as about 4.1 billion years old. A mere 600 million years after these rocks formed—and perhaps even earlier—life on Earth began.

The Beginning of Life

Until very recently, biologists regarded the earliest events in the history of life as chapters that would probably remain forever closed to scientific investigation. Two developments, however, have greatly improved our long-distance vision. The first was the formulation of a testable hypothesis about the events preceding life's origins. This hypothesis generated questions for which answers could be sought experimentally. The results of the initial experimental tests led to the formulation of further hypotheses and to additional experiments, a process that continues today as scientists in many laboratories explore the question of life's origins. The second development was the discovery of fossilized cells more than 3 billion years old.

The testable hypothesis was offered by the Russian biochemist A. I. Oparin. According to Oparin, the appearance of life was preceded by a long period in which complex molecules formed and began to accumulate, a process that is sometimes called **chemical evolution.** The identity of the substances, particularly gases, present in the atmosphere and seas during this period is a matter of controversy. There is general agreement, however, on two critical issues. First, little or no free oxygen was present. The Earth's supply of oxygen was locked in simple compounds such as water, carbon monoxide, carbon dioxide, and oxides of silicon and various metals. Second, the four elements—hydrogen, oxygen, carbon, and nitrogen—that make up more than 95 percent of living tissues were available in some form in the atmosphere and waters of the primitive Earth.

In addition to these raw materials, energy abounded on the young Earth. Radioactive elements within the

Earth released their energy into the atmosphere, as did the frequent volcanic eruptions. There was heat energy, both boiling (moist) heat and baking (dry) heat. Water vapor spewed out of the primitive seas, cooled in the upper atmosphere, collected into clouds, fell back on the crust of the Earth, and steamed up again. Violent rainstorms were accompanied by lightning, which provided electrical energy. And, most important, the sun continuously bombarded the Earth's surface with high-energy particles and ultraviolet radiation.

Oparin hypothesized that under such conditions organic molecules were formed from the atmospheric gases and collected in a thin soup in the Earth's seas and lakes. In the absence of free oxygen, these organic molecules could persist. (Oxygen, as we noted in Chapter 1, has a powerful attraction for electrons; thus it tends to react with and break down organic molecules into simple substances such as carbon dioxide and water.) Some of the newly formed organic molecules might have become locally more concentrated by the drying up of a lake or by the adhesion of the molecules to solid surfaces, such as clay particles.

Oparin published his hypothesis in 1922, but the scientific community ignored his ideas. In the 1950s, the first test of the hypothesis (Figure 4–2) was performed by Stanley Miller, then a graduate student at the University of Chicago. Experiments of this sort, now repeated many times, have shown that almost any source of energy—lightning, ultraviolet radiation, or hot volcanic ash—would have converted molecules believed to have been present on the Earth's surface into a variety of complex organic compounds. With various modifications in the experimental conditions and in the mixture of gases placed in the reaction vessel, almost all of the common amino acids have been produced, as well as the nucleotides that are the essential components of DNA and RNA.

These experiments have not proved that such organic compounds were formed spontaneously on the primitive Earth, only that they could have formed. The accumulated evidence is nevertheless very great, and most biochemists now believe that, given the conditions existing on the young Earth, chemical reactions producing amino acids, nucleotides, and other organic molecules were inevitable.

As the concentrations of such molecules increased, bringing them into closer proximity to each other, they would have been subject to the same chemical forces that act on organic molecules today. As we saw in the last chapter, small organic molecules react with each other, typically in dehydration synthesis reactions, to form larger molecules. Moreover, such forces as hydro-

4–2 *In Miller's experiment, conditions believed to have existed on the primitive Earth were simulated in the laboratory. Hydrogen gas (H_2), water vapor, methane (CH_4), and ammonia (NH_3) were continuously circulated between a lower "ocean" and an upper "atmosphere." The "ocean" was heated, vaporizing the water and forcing the gases into the "atmosphere," through which electric discharges were transmitted. As the gases moved down the portion of the tubing surrounded by cooling water, the water vapor condensed to liquid water and carried with it any organic molecules that had formed. These molecules accumulated in the portion of the tubing leading back to the "ocean." At the end of 24 hours, about half of the carbon originally present in the methane gas was converted to amino acids and other organic molecules. This was the first test of Oparin's hypothesis.*

gen bonds and hydrophobic interactions cause these molecules to assemble themselves into more complex aggregates. In modern chemical systems—either in the laboratory or in the living organism—the more stable molecules and aggregates tend to survive, and the least stable are transitory. Similarly, the compounds and aggregates that had the greatest chemical stability under the prevailing conditions on the primitive Earth would have tended to survive. Hence a form of natural selection played a role in chemical evolution as well as in the biological evolution that was to follow.

(a)   ⊢─────⊣ 1 μm

(b)   ⊢─────⊣ 25 μm

4–3   (a) *When dry mixtures of amino acids are heated at moderate temperatures, polymers known as thermal proteinoids are formed. Each of these polymers may contain as many as 200 amino acid subunits. When the polymers are placed in water solution and maintained under suitable conditions, they spontaneously form proteinoid microspheres, shown here. The microspheres are separated from the surrounding solution by a membrane that appears to be two-layered.*

(b) *These membrane-bound structures, separated from the surrounding solution by a lipid bilayer, formed spontaneously from organic molecules extracted from the Murchison meteorite, which struck Australia in 1969. Samples were taken from the meteorite on the day it landed and then stored under conditions that would prevent their contamination by terrestrial materials. Since that time, samples of the meteorite have been made available to scientists around the world for a variety of investigations.**

The First Cells

Three characteristics distinguish the living cell from a mere aggregate of complex molecules: (1) the capacity to replicate itself, generation after generation; (2) the presence of enzymes, the complex proteins that are essential for the chemical reactions on which life depends; and (3) a membrane that separates the cell from the surrounding environment and enables it to maintain a distinct chemical identity. Which of these characteristics appeared first—and made possible the development of the others—remains an open question. However, as we shall see in Chapter 16, recently discovered functions of RNA suggest that a critical step in the evolution of the first cells may have been the self-assembly of RNA molecules from nucleotides produced by chemical evolution.

In studies simulating conditions during the Earth's first billion years, Sidney W. Fox and his co-workers at the University of Miami have produced membrane-bound structures, known as proteinoid microspheres, that can carry out a few chemical reactions resembling those of living cells. The microspheres (Figure 4–3a) grow slowly by the addition of proteinoid material from the solution and eventually bud off smaller microspheres. These structures are not living cells. Their formation, however, suggests the kinds of processes that could have given rise to self-sustaining protein entities, separated from their environment and capable of carrying out the chemical reactions necessary to maintain their physical and chemical integrity.

Although cell membranes contain large amounts of protein, their basic structure is formed by lipids. It has been hypothesized that comets and meteorites, which struck the primitive Earth with great frequency and sometimes contain carbonaceous material, might have been a source of organic compounds, including lipid-like substances. David W. Deamer of the University of California, Davis, and his co-workers extracted and analyzed organic molecules from carbon-containing meteorites. Some of the molecules spontaneously self-assemble into membrane-bound spheres (Figure 4–3b).

It is not known when the first living cells appeared on Earth, but we can establish some sort of time scale. The earliest fossils found so far (Figure 4–4), which resemble present-day bacteria, have been dated at 3.5 billion years—about 1.1 billion years after the formation of the Earth itself. These fossilized cells are sufficiently complex that it is clear that some little aggregation of chemicals had moved through the twilight zone separating the living from the nonliving millions of years before.

---

* The short straight lines at the bottom of these micrographs and those that follow are scale markers. A scale marker provides a reference for size. The same system is used to indicate distances on road maps. A micrometer, abbreviated μm, is 1/1,000,000 meter.

4-4   *This microfossil of a filament of bacteria-like cells was found in Western Australia in a deposit of a flintlike rock known as black chert. Dated at 3.5 billion years of age, it is one of the oldest fossils known.*

Why on Earth?

Of all the planets in our solar system, only Earth has been found to support life, a fact that holds clues to the requirements for life. A major factor is that Earth is neither too close to nor too distant from the sun. The chemical reactions on which life depends require liquid water, and they virtually cease at very low temperatures. Jupiter, where temperatures are about −150°C, and Saturn, where they are even lower—about −180°C—are far too cold to support life. Conversely, at high temperatures, water, which boils at 100°C, exists only in the form of vapor, and the complex chemical compounds essential for life become unstable. The upper limit for the stability of amino acids, for example, is between 200°C and 300°C. Life is clearly impossible on Mercury and Venus, where temperatures reach 430°C and 480°C, respectively.

Earth's size and mass are also important factors. Planets much smaller than Earth do not have enough gravitational pull to hold a protective atmosphere, and any planet much larger than Earth is likely to have so dense an atmosphere that light from the sun cannot reach its surface. Earth's atmosphere blocks out the most energetic radiations from the sun, which are capable of breaking the covalent bonds between carbon atoms. It does, however, permit the passage of visible light, which made possible the process of photosynthesis, one of the most significant steps in the evolution of complex living systems.

Heterotrophs and Autotrophs

The energy that produced the first organic molecules came from a variety of sources on the young Earth and in its atmosphere—heat, ultraviolet radiation, and electrical disturbances. When the first primitive cells or cell-like structures evolved, they required a continuing supply of energy to maintain themselves, to grow, and to reproduce. The manner in which these cells obtained energy is the subject of continuing discussion.

Modern organisms—and the cells of which they are composed—can meet their energy needs in one of two ways. **Heterotrophs** are organisms that are dependent upon outside sources of organic molecules for both their energy and their small building-block molecules. (*Hetero* comes from the Greek word meaning "other," and *troph* comes from *trophos,* "one that feeds.") All animals and fungi, as well as many single-celled organisms, are heterotrophs.

Autotrophs, by contrast, are "self-feeders." They do not require organic molecules from outside sources for energy or to use as small building-block molecules. They are, instead, able to synthesize their own energy-rich organic molecules from simple inorganic substances. Most autotrophs, including plants and several different types of single-celled organisms, are **photosynthetic,** meaning that the energy source for their synthetic reactions is the sun. Certain groups of bacteria, however, are **chemosynthetic.** These organisms capture the energy released by specific inorganic reactions to power their life processes, including the synthesis of needed organic molecules.

It has long been postulated that the first living cells were heterotrophs, utilizing organic molecules available in the primordial soup—the same molecules from which the cells were themselves assembled. As the primitive cells increased in number, according to this hypothesis, they began to use up the complex molecules on which their existence depended and which had taken millions of years to accumulate. As the supply of these molecules decreased, competition began. Under the pressures of competition, cells that could make efficient use of the limited energy sources now available were more likely to survive and reproduce than cells that could not. In the course of time, other cells evolved that were able to synthesize organic molecules out of simple inorganic materials.

Recent discoveries, however, have raised the possibility that the first cells may have been either chemosynthetic or photosynthetic autotrophs rather than heterotrophs. First, several different groups of chemosynthetic bacteria have been found that would have been well-suited to the conditions prevailing on the young Earth. Some of these bacteria (Figure 4–5) live in swamps, while others have been found in deep ocean trenches in areas where gases escape from fissures in the Earth's crust. There is evidence that these modern bacteria are the surviving representatives of very ancient groups of unicellular organisms. Second, some of the experiments simulating conditions on the young Earth have produced molecules that are the chemical precursors of chlorophyll. When these molecules are mixed with simple organic molecules in an oxygen-free environment and illuminated, primitive photosynthetic reactions occur. These reactions resemble the ones that occur in some types of photosynthetic bacteria.

Both heterotrophs and autotrophs seem to be represented among the earliest microfossils, and biologists are presently unable to resolve the question of which came first. However, it is certain that without the evolution of autotrophs, life on Earth would soon have come to an end. In the more than 3.5 billion years since life first appeared on Earth, the most successful autotrophs (that is, those that have left the most offspring and diverged into the greatest variety of forms) have been those that evolved a system for making direct use of the sun's energy in the process of photosynthesis. With the advent of photosynthesis, the flow of energy in the biosphere came to assume the predominant pattern seen today: radiant energy from the sun is channeled through photosynthetic autotrophs to all other forms of life.

1 μm

4–5   *Methanogens, such as the cells shown here, are chemosynthetic bacteria that produce methane (CH_4) and water from carbon dioxide and hydrogen gas. The chemosynthetic reaction releases about 33 kilocalories of energy for each mole of methane that is formed. This energy is used by the cells to power their life processes. Methanogens can live only in the absence of oxygen—a condition prevailing on the young Earth but occurring today only in isolated environments, such as the muck and mud at the bottom of swamps.*

The Cell Theory

One of the fundamental principles of biology is that all living organisms are composed of one or more similar cells. This concept is of tremendous and central importance to biology because it emphasizes the underlying sameness of all living systems. It therefore brings unity to widely varied studies involving many different kinds of organisms.

The word "cell" was first used in a biological sense some 300 years ago. In the seventeenth century, the English scientist Robert Hooke, using a microscope of his own construction, noticed that cork and other plant tissues are made up of small cavities separated by walls (Figure 4–6). He called these cavities "cells," meaning "little rooms." However, "cell" did not take on its present meaning—the basic unit of living matter—for more than 150 years.

In 1838, Matthias Schleiden, a German botanist, reported his observation that all plant tissues consist of organized masses of cells. In the following year, zoologist Theodor Schwann extended Schleiden's observa-

tion to animal tissues and proposed a cellular basis for all life. In 1858, the idea that all living organisms are composed of one or more cells took on an even broader significance when the pathologist Rudolf Virchow generalized that cells can arise only from preexisting cells: "Where a cell exists, there must have been a preexisting cell, just as the animal arises only from an animal and the plant only from a plant."

From the perspective provided by Darwin's theory of evolution, published in the following year, Virchow's concept takes on an even larger significance. There is an unbroken continuity between modern cells—and the organisms they compose—and the first primitive cells that appeared on Earth.

In its modern form, the **cell theory** states simply that (1) all living organisms are composed of one or more cells; (2) the chemical reactions of a living organism, including its energy-releasing processes and its biosynthetic reactions, take place within cells; (3) cells arise from other cells; and (4) cells contain the hereditary information of the organisms of which they are a part, and this information is passed from parent cell to daughter cell.

(a)

(b)

(c)

25 μm

4–6 (a) *Robert Hooke's drawing of his microscope, reproduced from a book he published in 1665. Light from an oil lamp was directed to the specimen through a water-filled glass globe that acted as a condenser. The specimen was mounted on a pin, just below the tip of the microscope. The microscope was focused by moving it up and* *down, using a screw held to the stand by a clamp. Among the many other illustrations in the book was a drawing of two slices of a piece of cork (*b*). The structure of cork is revealed in more detail in a modern electron micrograph (*c*).*

Hooke was the first to use the word "cells" to describe the tiny compartments *that together make up an organism. The cells in these pieces of cork have died—all that remain are the outer walls. The living cell, however, is filled with a variety of substances, organized into distinct structures and carrying out a multitude of essential processes.*

Prokaryotes and Eukaryotes

All cells—ancient and modern—share two essential features. One is an outer membrane, the **cell membrane** (known more precisely as the **plasma membrane**), that separates the cell from its external environment. The other is the genetic material—the hereditary information—that directs a cell's activities and enables it to reproduce, passing on the cell's characteristics to its offspring.

The organization of the genetic material is one of the characteristics that distinguish two fundamentally distinct kinds of cells, **prokaryotes** and **eukaryotes.** In prokaryotic cells, the genetic material is in the form of a large, circular molecule of DNA, with which a variety of proteins are loosely associated. This molecule is known as the **chromosome.** In eukaryotic cells, by contrast, the DNA is linear, forming a number of distinct chromosomes. Moreover, it is tightly bound to special proteins known as **histones,** which are an integral part of the chromosome structure.

Within the eukaryotic cell, the chromosomes are surrounded by a double membrane, the **nuclear envelope,** that separates them from the other cell contents in a distinct **nucleus** (hence the name, *eu,* meaning "true," and *karyon,* meaning "nucleus" or "kernel"). In prokaryotes ("before a nucleus"), the chromosome is not contained within a membrane-bound nucleus, although it is localized in a distinct region that is known as the **nucleoid.**

The remaining components of a cell (that is, everything within the plasma membrane except the nucleus or nucleoid and its contents) constitute the **cytoplasm.** The cytoplasm contains a large variety of molecules and molecular complexes. For example, both prokaryotes and eukaryotes contain complexes of protein and RNA, known as **ribosomes,** that play a crucial role in the assembly of protein molecules from their amino acid subunits. Eukaryotic cells also contain a variety of membrane-bound structures and compartments called **organelles.** These specialized structures carry out particular functions within the cell.

The plasma membrane of prokaryotes is surrounded by an outer **cell wall** that is manufactured by the cell itself. Some eukaryotic cells, including plant cells and fungi, have cell walls, although their composition and structure is different from that of prokaryotic cell walls. Other eukaryotic cells, including those of our own bodies and of other animals, do not have cell walls. Another feature distinguishing eukaryotes and prokaryotes is size: eukaryotic cells are usually larger than prokaryotic cells.

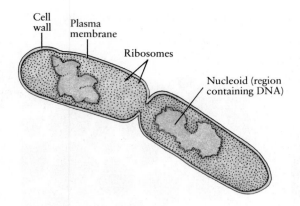

Cell wall
Plasma membrane
Ribosomes
Nucleoid (region containing DNA)

Modern prokaryotes include the bacteria (Figure 4–7) and the cyanobacteria (Figure 4–8), a group of photosynthetic prokaryotes that were formerly known as the blue-green algae.* According to the fossil record, the earliest living organisms were comparatively simple cells, resembling present-day prokaryotes. Prokaryotes were the only forms of life on this planet for almost 2 billion years until eukaryotes evolved.

Many biologists believe that the transition from the prokaryotic cell to the eukaryotic cell, a topic we shall explore in Chapter 24, was the most significant event in the history of life, second only in biological importance to the first appearance of living systems. Figure 4–9 gives an example of a modern single-celled photosynthetic eukaryote, the alga *Chlamydomonas*. It is a common inhabitant of freshwater ponds. These organisms are small and bright green (because of their chlorophyll), and they move very quickly with a characteristic darting motion. Being photosynthetic, they are usually found near the water's surface, where the light intensity is greatest.

The Origins of Multicellularity

The first multicellular organisms, as far as can be told by the fossil record, made their appearance a mere 750 million years ago (Figure 4–10). The cells of modern multicellular organisms closely resemble those of single-celled eukaryotes. They are bound by a plasma membrane identical in appearance to the plasma membrane

0.5 μm

4–7 *Electron micrograph of cells of* Escherichia coli, *the heterotrophic prokaryote that is the most thoroughly studied of all living organisms. The genetic material (DNA) is in the lighter-appearing area in the center of each cell. This region, which is not enclosed by a membrane, is known as the nucleoid. The small, dense bodies in the cytoplasm are ribosomes. The two cells in the center have just finished dividing and have not yet separated completely.*

* The Latin term *alga*, plural *algae*, means "seaweed." Algae is a general term applied to eukaryotic single-celled photosynthetic organisms and to many simple multicellular forms. Until recently, it was also applied to some photosynthetic prokaryotes.

4–8 *Electron micrograph of a single cell from a filament of cells of a photosynthetic prokaryote, the cyanobacterium* Anabaena azollae. *In addition to the genetic material, this cell contains a series of membranes in which chlorophyll and other photosynthetic pigments are embedded.* Anabaena *synthesizes its own energy-rich organic compounds in chemical reactions powered by the radiant energy of the sun.*

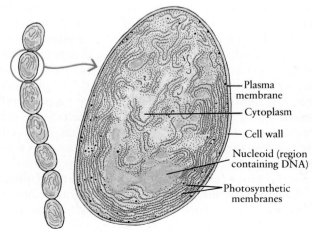

Plasma membrane
Cytoplasm
Cell wall
Nucleoid (region containing DNA)
Photosynthetic membranes

2 μm

1 μm

4–9 *Electron micrograph of* Chlamydomonas, *a photosynthetic eukaryotic cell, which contains a membrane-bound ("true") nucleus and numerous organelles. The most prominent organelle is the single, irregularly shaped chloroplast that fills most of the cell. It is surrounded by a double membrane and is the site of photosynthesis. Other membrane-bound organelles, the mitochondria, provide energy for cellular functions, including the flicking movements of the two flagella. These movements propel the cell through the water. The organism's food reserves are in the form of starch granules, stored around and within a structure known as the pyrenoid. The cytoplasm is enclosed by the plasma membrane, outside of which is a cell wall composed of polysaccharides.*

In the diagram labels:

- Cell wall
- Plasma membrane
- Chloroplast
- Starch
- Pyrenoid
- Base of a flagellum
- Mitochondrion
- Nucleus
- Nuclear envelope

4–10 *The clockface of biological time, which shows when important events in the Earth's past would have occurred if the Earth's 4.6-billion-year history were condensed into one day. Life first appears relatively early, before 6:00 A.M. on a 24-hour time scale. The first multicellular organisms do not appear until the early evening of that 24-hour day, and* Homo, *the genus to which humans belong, is a late arrival—at about 30 seconds to midnight.*

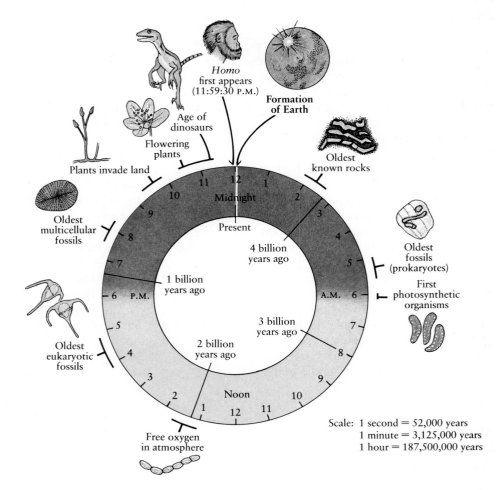

Homo first appears (11:59:30 P.M.)

Formation of Earth

Age of dinosaurs

Flowering plants

Plants invade land

Oldest multicellular fossils

Oldest eukaryotic fossils

Free oxygen in atmosphere

Oldest known rocks

Oldest fossils (prokaryotes)

First photosynthetic organisms

Present

Midnight

4 billion years ago

1 billion years ago

3 billion years ago

2 billion years ago

Noon

P.M.

A.M.

Scale: 1 second = 52,000 years
1 minute = 3,125,000 years
1 hour = 187,500,000 years

Cell wall

Plasma membrane

Nuclear envelope

Nucleus

Nucleolus

Mitochondrion

Starch grain

Chloroplast

Vacuole

0.5 μm

4–11 *Electron micrograph of cells from the leaf of a corn plant. The nucleus can be seen on the right side of the central cell. The granular material within the nucleus is chromatin. It contains DNA associated with his-* *tone proteins. The nucleolus is the region within the nucleus where the RNA components of ribosomes are synthesized. Note the many mitochondria and chloroplasts, all enclosed by membranes. The vacuole, a fluid-* *filled region enclosed by a membrane, and the cell wall are characteristic of plant cells but are generally not found in animal cells. As you can see, this cell closely resembles* Chlamydomonas, *shown in Figure 4–9.*

of a single-celled eukaryote. Their organelles are constructed according to the same design. The cells of multicellular organisms differ from single-celled eukaryotes in that each type of cell is specialized to carry out a relatively limited function in the life of the organism. However, each remains a remarkably self-sustaining unit.

Notice how similar a cell from the leaf of a corn plant (Figure 4–11) is to *Chlamydomonas.* This plant cell is also photosynthetic, supplying its own energy needs from sunlight. However, unlike the alga, it is part of a multicellular organism and depends on other cells for water, minerals, protection from drying out, and other necessities.

The human body, made up of trillions of individual cells, is composed of at least 200 different types of cells, each specialized for its particular function but all working as a cooperative whole. Figure 4–12 shows cells from the lining of an animal trachea (windpipe). They are part of an elaborate organ system involved in delivering oxygen to other cells in the body.

The Forms of Life

Although all organisms are made up of cells, there is incredible diversity in the living world. It is known that

we share this planet with at least 5 million different species of living organisms, and the actual number may be much higher. These different organisms exhibit great variety in the organization of their bodies, in their patterns of reproduction, growth, and development, and in their behavior.

Despite the seemingly overwhelming diversity of organisms, it is possible to group them in ways that reveal not only patterns of similarity and difference but also the evolutionary relationships among different groups. In Section 5, we shall consider these patterns and relationships in some detail. Before reaching that point, however, we shall encounter a marvelous variety of organisms. Thus you will find it helpful to be aware of the five major categories, or **kingdoms,** into which we group organisms in this text.

The first kingdom, **Monera,** includes all of the prokaryotes—the bacteria, the cyanobacteria, and their relatives. The second kingdom, **Protista,** includes the single-celled eukaryotes, as well as a number of simple multicellular eukaryotic organisms. The protists are an extremely varied collection of organisms, including both heterotrophs and autotrophs. Examples of protists are amoebas, paramecia, and the many forms of algae.

The three kingdoms of multicellular organisms— **Fungi, Plantae,** and **Animalia**—are thought to have

2.5 µm

4–12 *Cells from the surface of the trachea (windpipe) of a bat. The free surface of the larger cell is covered with cilia, which are essentially the same in structure as the flagella of Chlamydomonas. Next to the ciliated cells are cells that secrete mucus onto the cell surface. Mucus currents, swept by cilia, remove foreign particles from the trachea.*

evolved from different types of single-celled eukaryotes (Figure 4–13). The fungi include such organisms as molds, yeasts, and mushrooms. They are heterotrophs, absorbing nutrient molecules from the surrounding environment.

Plants are most concisely defined as multicellular photosynthetic autotrophs. They transform the energy of sunlight into the complex molecules of which their bodies are composed. These molecules, which include carbohydrates, proteins, and lipids, are the energy sources for animal life.

Animals are multicellular heterotrophs that must ingest other organisms—mostly plants or other animals—for their sustenance. From our anthropocentric point of view, animal usually means mammal, but actually most animals are invertebrates, animals without backbones. More than 1.5 million different kinds of animals have been recorded, of which 95 percent are invertebrates and more than a million of these are insects.

In the past century, our knowledge of the forms of life, past and present, of the processes occurring within their bodies, and of their interrelationships with one another has rapidly outstripped that gained in all previous centuries of human inquiry. Among the factors contributing to this explosion of knowledge have been significant improvements in microscopy.

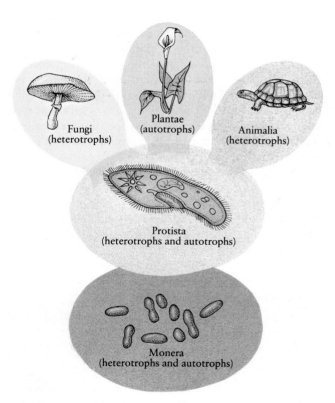

4–13 *The five kingdoms of living organisms. The organisms of kingdom Monera are all prokaryotes, whereas the organisms of the other four kingdoms are all eukaryotes. The first eukaryotic cells were derived from prokaryotes, and the three kingdoms of multicellular eukaryotes are thought to have evolved from different single-celled eukaryotes.*

81

Viewing the Cellular World

In the three centuries since Robert Hooke first observed the structure of cork through his simple microscope (Figure 4–6, page 77), our capacity to view the cell and its contents has increased dramatically. Unaided, the human eye is able to distinguish objects that are separated by a distance of at least 1/10 millimeter, or 100 micrometers (Table 4–1). This capacity is known as **resolving power**. It is expressed in terms of the minimum distance that must be between two objects for them to be perceived as separate objects. For example, if you look at two lines that are less than 100 micrometers apart, you will see a single, somewhat thickened line. Similarly, two dots less than 100 micrometers apart look like a single blurry dot. Conversely, if you look at two lines (or two dots) that are 120 micrometers apart, you can distinguish them from one another.

Most eukaryotic cells are between 10 and 30 micrometers in diameter—some 3 to 10 times below the resolving power of the human eye—and prokaryotic cells are smaller still. In order to distinguish individual cells, to say nothing of examining the structures of which they are composed, we must use instruments that provide greater resolution than does the naked eye. Most of our current knowledge of cell structure has been gained with the assistance of three different types of instruments: the **light microscope**, the **transmission electron microscope**, and the **scanning electron microscope** (Figure 4–14).

The best light microscopes have a resolving power of about 0.2 micrometer, or 200 nanometers, and so improve on the naked eye about 500 times. It is impossible to build a light microscope that will do better than this. The limiting factor is the wavelength of light; the shorter the wavelength, the greater the resolution. The shortest wavelength of visible light is about 0.4 micrometer, and this sets the limit of resolution with the light microscope.

Notice that resolving power and magnification are two different things. If you take a picture through the best light microscope of two lines that are less than 0.2 micrometer, or 200 nanometers, apart, you can enlarge that photograph indefinitely, but the two lines will continue to blur together. Similarly, by using more powerful microscope lenses, you can increase magnification, but this will not improve resolution; you will just see a larger blur.

With the transmission electron microscope, resolving power has been increased about 1,000 times over that provided by the light microscope. This is achieved by using "illumination" of a much shorter wavelength, consisting of electron beams instead of light rays. Areas in the specimen that permit more electrons to pass through—"electron-transparent" regions—show up bright, and areas that do not allow electrons to pass through—"electron-opaque" regions—are seen as dark. Transmission electron microscopy at present affords a resolving power of about 0.2 nanometer, roughly 500,000 times greater than that of the human eye. This is about twice the diameter of a hydrogen atom.

Although the resolving power of the scanning electron microscope is only about 10 nanometers, this instrument has become a valuable tool for biologists. In scanning electron microscopy, the electrons do not pass through the specimen but are instead deflected from its surface, providing an image of the surface contours. The electron beam is focused into a fine probe, which is rapidly passed back and forth over the specimen. Complete scanning from top to bottom usually takes a few seconds. Variations in the surface of the specimen affect the pattern in which the electrons are deflected from it. Holes and fissures appear dark, and knobs and ridges are light. The pattern produced by the electrons is amplified and transmitted to a television monitor, providing a visual image of the specimen. Scanning electron microscopy provides vivid three-dimensional representations of cells and cellular structures that more than compensate for its limited resolution.

Studies with these three types of microscopes, complemented by a variety of biochemical techniques, have revealed a wealth of knowledge about the structure of cells and the dynamic processes that characterize them. We shall begin to examine this knowledge in the next chapter.

Table 4–1   Measurements Used in Microscopy

1 centimeter (cm) = 1/100 meter = 0.4 inch*

1 millimeter (mm) = 1/1,000 meter = 1/10 cm

1 micrometer (μm)† = 1/1,000,000 meter = 1/10,000 cm

1 nanometer (nm) = 1/1,000,000,000 meter = 1/10,000,000 cm

or

1 meter = 10^2 cm = 10^3 mm = 10^6 μm = 10^9 nm

* A metric-to-English conversion table can be found inside the back cover.
† Micrometers were formerly known as microns (μ), and nanometers as millimicrons (mμ).

(d)

4–14   *A comparison of (**a**) the light micro-scope, (**b**) the transmission electron micro-scope, and (**c**) the scanning electron micro-scope. The focusing lenses in the light microscope are glass or quartz; those in the*

electron microscopes are magnetic coils. In both the light microscope and the transmis-sion electron microscope, the illuminating beam passes through the specimen. In the scanning electron microscope, it is deflected

*from the surface of the specimen. (**d**) A mod-ern electron microscope. Depending on how the controls are adjusted, it can function as either a transmission electron microscope or a scanning electron microscope.*

(a)   10 μm

(b)   0.5 μm

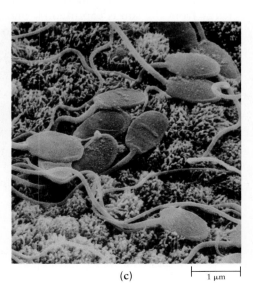

(c)   1 μm

4–15   *Rabbit sperm cells, as seen in (**a**) a light micrograph, (**b**) a transmission electron micrograph, and (**c**) a scanning electron mi-crograph. Note the dramatic increase in the resolution of structural detail in the electron micrographs.*

Electron microscopes produce images only in black and white. However, color is often added to electron micrographs to enhance their visual appeal, as in these micrographs, or to highlight specific structures.

Summary

The properties associated with living systems emerge at the cellular level of organization. Microfossils of bacteria-like cells have been discovered that are 3.5 billion years old. The complexity of these cells suggests that the first primitive cells arose early in the Earth's existence—sometime during the first billion years.

The primitive atmosphere held the raw materials of living matter—hydrogen, oxygen, carbon, and nitrogen—contained in molecules of water vapor and other simple compounds. The energy required to break apart the molecules and re-form them into more complex molecules was present in heat, lightning, radioactive elements, and high-energy radiation from the sun. Laboratory experiments have shown that under these conditions, the types of organic molecules characteristic of living systems can be formed. Other experiments have suggested the kinds of processes by which aggregations of organic molecules could have formed cell-like structures, separated from their environment by a membrane and capable of maintaining their structural and chemical integrity.

The earliest cells may have been heterotrophs (organisms that depend on outside sources for their energy-rich organic molecules) or autotrophs (organisms that can make their own organic molecules from inorganic substances). The first autotrophs may have been chemosynthetic (using the energy released by specific inorganic reactions to synthesize their own organic molecules) or photosynthetic (using the sun's energy to power their synthetic reactions). With the advent of photosynthesis, the flow of energy through the biosphere assumed its dominant modern form—radiant energy from the sun is captured by photosynthetic autotrophs and channeled through them to heterotrophic organisms.

One of the fundamental principles of biology is the cell theory, which states that (1) all living organisms consist of one or more cells; (2) the chemical reactions of a living organism, including energy-releasing processes and biosynthetic reactions, take place within cells; (3) cells arise from other cells; and (4) cells contain the hereditary information of the organisms of which they are a part, and this information is passed from parent cell to daughter cell.

There are two fundamentally distinct types of cells—prokaryotes and eukaryotes. Prokaryotic cells lack the membrane-bound nuclei and organelles found in eukaryotic cells. Prokaryotes were the only form of life on Earth for almost 2 billion years, and then, about 1.5 billion years ago, eukaryotic cells evolved. Multicellular organisms, which are composed of eukaryotic cells specialized to perform particular functions, evolved comparatively recently—only about 750 million years ago.

Five kingdoms of organisms are recognized in this text. The organisms of the kingdom Monera, which includes the bacteria and the cyanobacteria, are prokaryotes. The organisms of the other four kingdoms—Protista, Fungi, Plantae, and Animalia—are eukaryotes. The protists include single-celled organisms, both heterotrophic and autotrophic, as well as a number of simple multicellular forms. Fungi are multicellular heterotrophs that obtain organic molecules by absorption. Plants are multicellular photosynthetic autotrophs, and animals are multicellular heterotrophs that ingest other organisms to obtain needed organic molecules.

Because of the small size of cells and the limited resolving power of the human eye, microscopes are required to visualize cells and subcellular structures. The three principal types are the light microscope, the transmission electron microscope, and the scanning electron microscope.

Questions

1. Distinguish among the following: heterotroph/autotroph; chemosynthetic autotroph/photosynthetic autotroph; prokaryote/eukaryote; light microscope/transmission electron microscope/scanning electron microscope.

2. Why would energy sources have been necessary for the synthesis of simple organic molecules on the primitive Earth?

3. Although there is some uncertainty as to the exact mixture of gases that constituted the early atmosphere, there is general agreement that free oxygen was not present. What properties of oxygen would have made chemical evolution unlikely in an atmosphere containing O_2?

4. A key event in the origin of life was the formation of a membrane that separated the contents of primitive cells from their surroundings. Why was this so critical?

5. Some scientists think that other planets in our galaxy may well contain some form of life. If you were seeking such a planet, what characteristics would you look for?

Suggestions for Further Reading

Cloud, Preston: *Oasis in Space: Earth History from the Beginning*, W. W. Norton & Company, New York, 1989.*

> *An accurate and up-to-date history of Earth's 4.6 billion years, written by a master storyteller. A major theme is the ongoing interplay between physical, chemical, and biological processes. Particular emphasis is placed on the early periods of Earth's history, during which life originated and many of the great diversifications of organisms occurred.*

Groves, David I., John S. R. Dunlop, and Roger Buick: "An Early Habitat of Life," *Scientific American*, October 1981, pages 64–73.

Horgan, John: "In the Beginning . . .," *Scientific American*, February 1991, pages 116–125.

Kasting, James F., Owen B. Toon, and James B. Pollack: "How Climate Evolved on the Terrestrial Planets," *Scientific American*, February 1988, pages 90–97.

Oparin, A. I.: *The Origin of Life*, Dover Publications, Inc., New York, 1953.*

> *In 1938, Oparin, a Russian biochemist, was the first to argue that life arose spontaneously in the oceans of the primitive Earth. Although his concepts have since been somewhat modified in detail, they form the basis for the present scientific theories on the origin of living things.*

Scientific American: *Molecules to Living Cells*, W. H. Freeman and Company, New York, 1980.*

> *A collection of outstanding articles from* Scientific American. *The first two chapters cover the origin of life and the evolution of the earliest cells. Highly recommended.*

Taylor, D. Lansing, Michel Nederlof, Frederick Lanni, and Alan S. Waggoner: "The New Vision of Light Microscopy," *American Scientist*, vol. 80, pages 322–335, 1992.

Vidal, Gonzalo: "The Oldest Eukaryotic Cells," *Scientific American*, February 1984, pages 48–57.

York, Derek: "The Earliest History of the Earth," *Scientific American*, January 1993, pages 90–96.

* Available in paperback.

5

How Cells Are Organized

In multicellular organisms, cells are specialized to carry out the different functions necessary for the survival of the organism. This cell is from the human liver, a three-pound chemical "factory" that synthesizes, processes, and degrades many different types of molecules.

Color has been added to this scanning electron micrograph to highlight the various structures of the cell. The plasma membrane—that is, the outer membrane of the cell—has been colored green. It has been partially peeled away, revealing the contents of the cell.

The large body near the center of the cell, tinted yellow, is the nucleus. It contains the cell's chromosomes and is the control center for the cell's many activities.

Like other cells that produce proteins for export, this cell has a complex system of internal membranes known as the endoplasmic reticulum. These membranes, to which a greenish-tan color has been added, are involved in the synthesis of lipids and proteins.

Embedded in the endoplasmic reticulum and supplying the ATP molecules that power its activities are organelles known as mitochondria, highlighted in a reddish-brown. Mitochondria are also divided into specialized compartments by a complex membrane system.

The history of the evolution of cells can be traced in the increasing role of membranes in the life of the cell. This evolution began with a single, crucial outer membrane separating the earliest prokaryote from the warm seas in which it originated. From that simple beginning have come the complex membrane systems of modern eukaryotic cells.

There are many, many different kinds of cells. In a drop of pond water, you are likely to find a variety of protists, and in even a small pond, there are probably several hundred different kinds of these one-celled eukaryotes, plus a variety of prokaryotes. Plants are composed of cells that appear quite different from those of our bodies, and insects have many kinds of cells not found in either plants or vertebrates. Thus, the first remarkable fact about cells is their diversity.

The second, even more remarkable fact is their similarity. All cells have DNA as the genetic material, they perform the same types of chemical reactions, and they are all surrounded by an external membrane that conforms to the same general design in both prokaryotic and eukaryotic cells. Every cell must carry out essentially the same processes—acquire and assimilate nutrients, eliminate wastes, synthesize new cellular materials, and in many cases, be able to move and to reproduce. Just as the various organs of your body have a structure that enables them to carry out their functions, so your cells have a complex internal architecture that enables them to carry out their functions.

Although we can look at only one structure or process at a time, a cell is not a random assortment of parts. It is a dynamic, integrated entity. Most activities of a cell go on simultaneously and influence one another. For example, *Chlamydomonas* (Figure 4–9, page 79) is swimming, photosynthesizing, absorbing nutrients from the water, building its cell wall, making proteins, converting starch to sugar (or vice versa), and oxidizing food molecules for energy, all at the same time. It is also likely to be orienting itself in the sunlight, it is probably preparing to divide, it is possibly "looking" for a mate, and it is undoubtedly carrying out at least a dozen or more other important activities, many of which may still be unknown.

	One 4-centimeter cube	Eight 2-centimeter cubes	Sixty-four 1-centimeter cubes
Total surface area (square centimeters)	96	192	384
Total volume (cubic centimeters)	64	64	64
Surface-to-volume ratio	1.5 to 1	3 to 1	6 to 1

5–1 *The single 4-centimeter cube, the eight 2-centimeter cubes, and the sixty-four 1-centimeter cubes all have the same total volume. As the volume is divided up into smaller units, however, the total amount of surface area increases, as does the ratio of surface area to volume. For example, the sixty-four 1-centimeter cubes shown here have four times the total surface area of the single 4-centimeter cube. Moreover—and this is the important point—the ratio of surface area to volume in each 1-centimeter cube is four times that in the 4-centimeter cube. Similarly, smaller cells have a higher ratio of surface area to volume than larger cells. This means not only more membrane surface through which materials can move into or out of the cell but also less living matter to be serviced and shorter distances through which materials must move within the cell.*

Cell Size and Shape

Prokaryotic cells are, on average, about 2 micrometers long. By contrast, eukaryotic cells, such as the cells that make up a plant or animal body, are typically between 10 and 30 micrometers in diameter. A principal restriction on cell size is imposed by the relationship between volume and surface area. Figure 5–1 shows that as volume decreases, the ratio of surface area to volume increases rapidly.

It is through the membrane-bound surface of a cell that materials—such as oxygen, carbon dioxide, ions, food molecules, and waste products—enter and leave the cell. These substances are the raw materials and products of the cell's **metabolism,** which is the total of all of the chemical activities in which it is engaged. The more active the cell's metabolism, the more rapidly materials must be exchanged with the environment if the cell is to continue to function. In smaller cells, the ratio of surface area to volume is higher than in larger cells. Thus proportionately greater quantities of materials can move into, out of, and through smaller cells in a given period of time. A larger cell, by contrast, requires the exchange of greater quantities of materials in order to meet the needs of the larger volume of living matter—and yet, the larger the cell, the smaller the ratio of surface area to volume. It is not surprising, therefore, that the most metabolically active cells are usually small.

Another limitation on cell size appears to involve the capacity of the nucleus, the cell's control center, to provide enough copies of the information needed to regulate the processes occurring in a large, metabolically active cell. The exceptions seem to "prove" the rule. In certain large, complex one-celled protists, each cell has two or more nuclei, the additional ones apparently copies of the original.

Like drops of water and soap bubbles, cells have a tendency to be spherical. Often, however, cells have other shapes. For example, cell walls, found in plants, fungi, and many one-celled organisms, are generally in-flexible. The shape of the enclosed cell is determined by the shape of the cell wall, which, depending on the organism and the cell type, may take any number of different forms. The shape of some cells is determined by attachments to and pressure from neighboring cells. This is the case, for example, with the columnar cells that line the surface of the trachea (Figure 4–12, page 81). And, as we shall see later in this chapter, the shape of still other cells is determined by arrays of structural filaments within the cell.

Cell Boundaries

The Plasma Membrane

When you look at a solitary cell—an amoeba, for instance—through a microscope, you can barely distinguish it from its watery surroundings. Although almost invisible and seemingly fragile, it is very different from the vast sea around it: it is alive. One of the crucial factors making possible this extraordinary difference is an external cell membrane—the **plasma membrane**—that separates the living material of the cell from the surrounding environment.

|← 20 nm →|

5–2   *Electron micrograph of a portion of a human red blood cell, showing a cross section of the plasma membrane (indicated by the arrows). In essence, the membrane is a "molecular sandwich." It consists of two layers of phospholipid molecules arranged with their electron-transparent (light) hydrophobic tails pointing inward, forming an inner "filling." The dark line on either side of the "filling" is formed by the electron-opaque hydrophilic heads of the phospholipid molecules. The darker material at the left of the micrograph is hemoglobin, which fills the red blood cell.*

The plasma membrane is only about 7 to 9 nanometers thick and therefore cannot be resolved by the light microscope. With the electron microscope, however, it can be visualized as a continuous, thin double line (Figure 5–2).

The fundamental structure of the plasma membrane is formed by a **phospholipid bilayer,** that is, a double layer of phospholipid molecules arranged with their hydrophobic fatty acid tails pointing inward (Figure 5–3). In animal cells, large numbers of cholesterol molecules are embedded in the hydrophobic interior of the bilayer.

5–3   *Model of the plasma membrane of an animal cell. The fundamental structure is formed from a network of phospholipid molecules, in which cholesterol molecules and large protein molecules are embedded. The phospholipid molecules are arranged in a bilayer with their hydrophobic tails pointing inward and their hydrophilic phosphate heads pointing outward. Cholesterol molecules nestle among the hydrophobic tails.*

The proteins embedded in the bilayer are known as integral membrane proteins. On the cytoplasmic face of the membrane, peripheral membrane proteins are bound to some of the integral proteins. The portion of a protein molecule's surface that is within the lipid bilayer is hydrophobic. The portion of the surface that is exposed outside the bilayer is hydrophilic. It is believed that pores with hydrophilic surfaces pass through some of the protein molecules.

Interspersed among the phospholipid molecules of the outer layer of the bilayer are glycolipid molecules. Their carbohydrate chains and the carbohydrate chains attached to the proteins protruding on the outside of the membrane are thought to be involved in the adhesion of cells to each other and with the "recognition" of molecules at the membrane surface.

Outside of cell

Inside of cell (cytoplasm)

Carbohydrate

Glycolipid head (polar)

Phospholipid head (polar)

Lipid bilayer

Fatty acid tails (nonpolar)

Cholesterol

Hydrophilic region

Hydrophobic region

Integral proteins

Peripheral proteins

In prokaryotes and some protists and in the cells of plants and fungi, cholesterol is generally absent.

The plasma membrane also contains numerous protein molecules, suspended in the lipid bilayer. These proteins, known as **integral membrane proteins,** generally span the bilayer and protrude on either side. The portions embedded in the bilayer have hydrophobic surfaces, whereas the surfaces of those portions that extend beyond the bilayer are hydrophilic. On the cytoplasmic side of the membrane, additional protein molecules, known as **peripheral membrane proteins,** are bound to some of the integral proteins protruding from the bilayer.

Although many of the integral proteins appear to be anchored in place, either by peripheral proteins or by cytoplasmic protein filaments that are concentrated near the plasma membrane, the structure of the bilayer is generally quite fluid. The lipid molecules and at least some of the protein molecules can move laterally within the bilayer, forming different patterns (mosaics) that vary from time to time and place to place. Consequently, this widely accepted model of membrane structure is known as the **fluid-mosaic model.**

Membrane Specificity

The two surfaces of the plasma membrane—one facing the interior of the cell and the other facing the exterior of the cell—differ considerably in chemical composition. The two layers of the bilayer generally have different concentrations of specific types of lipid molecules. In many types of cells, the outer layer is particularly rich in glycolipid molecules. The carbohydrate chains of these glycolipid molecules are, like the phosphate heads of the phospholipid molecules, exposed on the surface of the membrane. The hydrophobic fatty acid tails of both types of molecules are within the membrane.

The protein composition of the outside and inside layers of the membrane also differ. The integral membrane proteins have a definite orientation within the bilayer, and the portions extending on either side are completely different in both amino acid composition and tertiary structure. On the outside of the membrane, short carbohydrate chains are attached to the protruding integral proteins, forming molecules known as **glycoproteins.** The carbohydrate chains of the glycoproteins, along with the carbohydrate chains of the glycolipids, form a carbohydrate coat on the outer surface of the membranes of many types of cells. The carbohydrates are thought to play a role in the adhesion of cells to one another and in the "recogni-

tion" of molecules (such as hormones, antibodies, and viruses) that interact with the cell.

In eukaryotes, all the membranes of a cell, including those surrounding the various organelles, have the same general structure as the plasma membrane. There are, however, differences in the types of lipids and, particularly, in the number and types of proteins and carbohydrates, which vary from membrane to membrane and also from place to place on the same membrane. These differences give the membranes of various cells and organelles unique properties that can be correlated with differences in function.

Membrane proteins, which are extremely diverse structurally, perform a variety of essential functions. Some are enzymes, regulating particular chemical reactions. Others are receptors, influencing the cell's uptake of hormones and other regulatory molecules. Still others are transport proteins, playing critical roles in the movement of ions and molecules across the membrane. As we shall see repeatedly in the course of this text, discoveries concerning the structure and function of specific membrane proteins are shedding new light on a wide variety of processes, ranging from photosynthesis to the response of the human body to infection to the transmission of the nerve impulse.

The Cell Wall

A principal distinction between plant and animal cells is that plant cells are surrounded by a cell wall. The wall is outside the plasma membrane and is constructed by the cell.

As a plant cell divides, a thin layer of gluey material forms between the plasma membranes of the two new cells. This layer becomes the **middle lamella.** Composed of pectins (the compounds that make jellies gel) and other polysaccharides, it holds adjacent cells together. Next, on either side of the middle lamella, each plant cell constructs its **primary cell wall** (Figure 5–4). This wall contains cellulose molecules bundled together in microfibrils and laid down in a matrix of gluey polymers. As you can see in Figure 3–6b on page 55, successive layers of microfibrils are oriented at right angles to one another in the completed cell wall. (Those of you familiar with building materials will note that the cell wall thus combines the structural features of both fiberglass and plywood.)

In plants, growth takes place largely by cell elongation. Studies have shown that the cell adds new materials to its walls throughout this elongation process. The cell, however, does not simply expand in all directions.

2 μm

5–4 *Two adjacent cell walls of tracheids, the cells through which water is conducted in plants. In the electron micrograph, you can see the middle lamella, the primary cell walls, and the layered secondary cell walls, deposited inside the primary walls. The cells, which are from the wood of a ground hemlock, have died. The transparent areas in the top left and lower right portions of the micrograph represent empty space, once filled by the living matter of the cells.*

Its final shape is determined by the structure of its cell wall (Figure 5–5).

As the cell matures, a **secondary wall** may be constructed. This wall is not capable of expansion, as is the primary wall. It often contains molecules that have stiffening properties. In such cells, the living material often dies, leaving only the outer wall, a monument to the cell's architectural abilities.

Cellulose-containing cell walls are also found in many algae. Fungi and prokaryotes also have cell walls, but they usually do not contain cellulose.

The Nucleus

In eukaryotic cells, the **nucleus** is a large, often spherical body, usually the most prominent structure within the cell. It is surrounded by the **nuclear envelope,** which is made up of two concentric membranes, each of which is a phospholipid bilayer. These two membranes are fused together at frequent intervals, creating small **nuclear pores** through which materials pass between the nucleus and the cytoplasm (Figure 5–6).

The chromosomes are found within the nucleus. When the cell is not dividing, the chromosomes are visible only as a tangle of fine threads, called **chromatin.** The most conspicuous body within the nucleus is the **nucleolus.** As we shall see in Chapter 16, the nucleolus is the site at which ribosomal subunits are constructed. Viewed with the electron microscope, the nucleolus appears to be a collection of fine granules and tiny fibers (Figure 5–7). These are thought to be parts of ribosomal subunits and threads of chromatin.

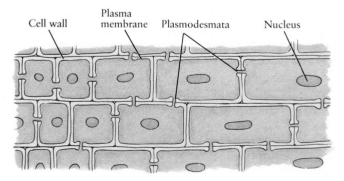

5–5 *Growth of plant cells is limited by the rate at which the cell walls expand. The walls control both the rate of growth and its direction. They do not expand in all directions but elongate in a single dimension. Cells at the left are newly formed. Cells farther to the right are older and have started to elongate. Plasmodesmata are channels connecting the cytoplasm of adjacent cells.*

(a)

$0.1 \mu m$

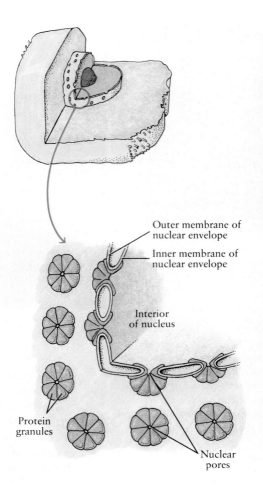

5–6   *The nuclear envelope is formed from two concentric lipid bilayers. A surface view of the nuclear envelope of a guinea pig sperm cell is shown in (a). Nuclear pores, which are very dense in these cells, are clearly visible on its surface. Biochemical studies and electron micrographs of sections through the plane of the envelope (b) have revealed that the structure of each pore consists of eight protein-containing granules. The opening is a narrow channel in the center of the octagonal array. Passage of molecules into and out of the nucleus is thought to be regulated by the nuclear pores.*

Outer membrane of nuclear envelope

Inner membrane of nuclear envelope

Interior of nucleus

Protein granules

Nuclear pores

(b)

$0.1 \mu m$

$0.5 \mu m$

5–7   *The nucleus of a root tip cell from* Arabidopsis thaliana, *as seen in an electron micrograph to which color has been added. The dark area in the lower left portion of the nucleus is the nucleolus. The major components of ribosomes are produced in the nucleolus. The dark granules around its periphery are partially formed ribosomes. Notice also the nuclear envelope with its nuclear pores, two of which are indicated by arrows. The structures highlighted in green are mitochondria.* Arabidopsis *is a small weed that has become an important experimental organism in the study of plant molecular genetics.*

The Functions of the Nucleus

Our current understanding of the role of the nucleus in the life of the cell began with some early microscopic observations. One of the most important of these observations was made more than a hundred years ago. A German embryologist, Oscar Hertwig, was observing the eggs and sperm of sea urchins (Figure 5–8). Sea urchins produce eggs and sperm in great numbers. The eggs are relatively large and so are easy to observe. They are fertilized in the open water, rather than internally, as is the case with land-dwelling vertebrates such as ourselves.

Watching the eggs being fertilized under his microscope, Hertwig observed that only a single sperm cell was required. Further, when the sperm cell penetrated the egg, its nucleus was released and fused with the nucleus of the egg. This observation, confirmed by other scientists and in other kinds of organisms, was important in establishing the fact that the nucleus is the carrier of the hereditary information: the only link between father and offspring is the nucleus of the sperm.

10 µm

5–8   *An egg from a sea urchin, surrounded by sperm cells. Despite the great differences in size of egg and sperm, both contribute equally to the hereditary characteristics of the individual. Since the nucleus is approximately the same size in both cells, the early microscopists postulated that this part of the cell must be the carrier of the hereditary information.*

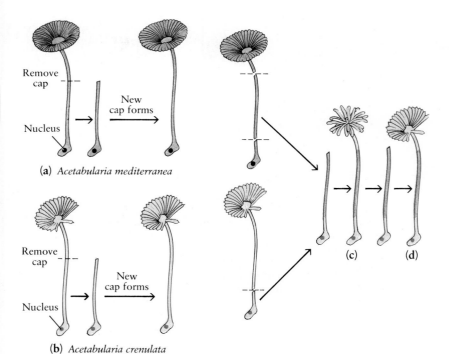

Remove cap

New cap forms

Nucleus

(a) *Acetabularia mediterranea*

Remove cap

New cap forms

Nucleus

(b) *Acetabularia crenulata*

(c)   **(d)**

5–9   (a) *One species of* Acetabularia *has an umbrella-shaped cap, and* (b) *another has a ragged, petal-like cap. If the cap is removed, a new cap forms, similar in appearance to the amputated one. However, if the "foot" (containing the nucleus) is removed at the same time as the cap and a new nucleus from the other species is transplanted, the cap* (c) *that forms will have a structure with characteristics of both species. If this cap is removed, the next cap* (d) *that grows will be characteristic of the cell that donated the nucleus, not of the cell that donated the cytoplasm.*

Since Hertwig's time, a number of experiments have explored the role of the nucleus in the cell. In one simple experiment, the nucleus was removed from an amoeba by microsurgery. The amoeba stopped dividing and, in a few days, it died. If, however, a nucleus from another amoeba was implanted within 24 hours after the original one was removed, the cell survived and divided normally.

In the early 1930s, Joachim Hämmerling studied the comparative roles of the nucleus and the cytoplasm by taking advantage of some unusual properties of the marine alga *Acetabularia*. The body of *Acetabularia* consists of a single huge cell 2 to 5 centimeters in height. Individuals have a cap, a stalk, and a "foot," all of which are differentiated portions of the single cell. If the cap is removed, the cell will rapidly regenerate a new one. Different species of *Acetabularia* have different kinds of caps. *Acetabularia mediterranea*, for example, has a compact umbrella-shaped cap, and *Acetabularia crenulata* has a cap of petal-like structures.

Hämmerling took the "foot," which contains the nucleus, from a cell of *A. crenulata* and grafted it onto a cell of *A. mediterranea*, from which he had first removed the "foot" and the cap. The cap that then formed had a shape intermediate between those of the two species. When this cap was removed, the next cap that formed was completely characteristic of *A. crenulata* (Figure 5–9).

Hämmerling interpreted these results as meaning that certain cap-determining substances are produced under the direction of the nucleus. These substances accumulate in the cytoplasm, which is why the first cap that formed after nuclear transplantation was of an intermediate type. By the time the second cap formed, however, the cap-determining substances present in the cytoplasm before the transplant had been exhausted, and the form of the cap was completely under the control of the new nucleus.

We can see from these experiments that the nucleus performs two crucial functions for the cell. First, it carries the hereditary information that determines whether a particular cell will develop into (or be a part of) a sea urchin, an oak, or a human—and not just any sea urchin, oak, or human, but one that resembles the parent or parents of that particular unique organism. Each time a cell divides, this information is passed on to the two new cells. Second, as Hämmerling's work with *Acetabularia* indicated, the nucleus exerts a continuing influence over the ongoing activities of the cell, ensuring that the complex molecules that the cell requires are synthesized in the number and of the kind needed. The way in which the nucleus performs these functions will be described in Section 3.

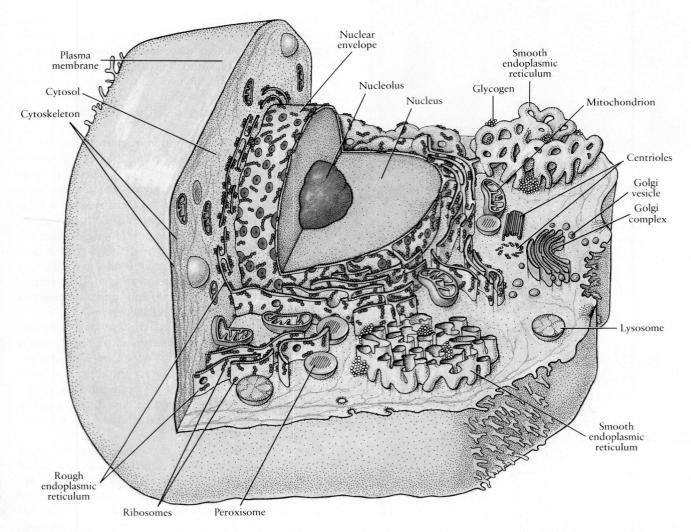

Plasma membrane
Cytosol
Cytoskeleton
Nuclear envelope
Nucleolus
Nucleus
Smooth endoplasmic reticulum
Glycogen
Mitochondrion
Centrioles
Golgi vesicle
Golgi complex
Lysosome
Smooth endoplasmic reticulum
Rough endoplasmic reticulum
Ribosomes
Peroxisome

5–10   *A representative animal cell, as interpreted from electron micrographs. Like all cells, this one is bounded by a* **plasma membrane.** *All materials that enter or leave the cell, including food, wastes, and chemical messages, must pass through this membrane.*

Surrounded by the plasma membrane is the cytoplasm, which consists of a dense solution, the **cytosol,** *and numerous membrane-bound organelles. The cytosol is traversed and subdivided by the membranes of an elaborate organelle system, the* **endoplasmic reticulum,** *a portion of which is shown here. In some areas, the endoplasmic reticulum is covered with ribosomes,* *the structures on which amino acids are assembled into proteins. Ribosomes are also found elsewhere in the cytosol.*

Golgi complexes *are packaging centers for molecules synthesized within the cell.* **Lysosomes** *and* **peroxisomes** *are vesicles in which a number of different types of molecules are broken down to simpler constituents that can either be used by the cell*

or, in the case of waste products, be safely removed from it. The **mitochondria** *are the sites of the chemical reactions that produce large quantities of ATP to power cellular activities.*

The largest body in the cell is the **nucleus.** *It is surrounded by a double membrane, the* **nuclear envelope,** *the outer membrane of which is continuous with the endoplasmic reticulum. The nuclear envelope is studded with nuclear pores, which regulate the movement of substances into and out of the nucleus. Contained within the nucleus is the* **nucleolus,** *the site where the ribosomal subunits are formed, and the chromatin, which consists of long strands of DNA and protein.*

An elaborate, highly structured network of protein filaments, the **cytoskeleton,** *pervades the cytosol. The cytoskeleton, of which only a hint is shown here, maintains the cell's shape, anchors its organelles, and directs the intracellular molecular traffic.*

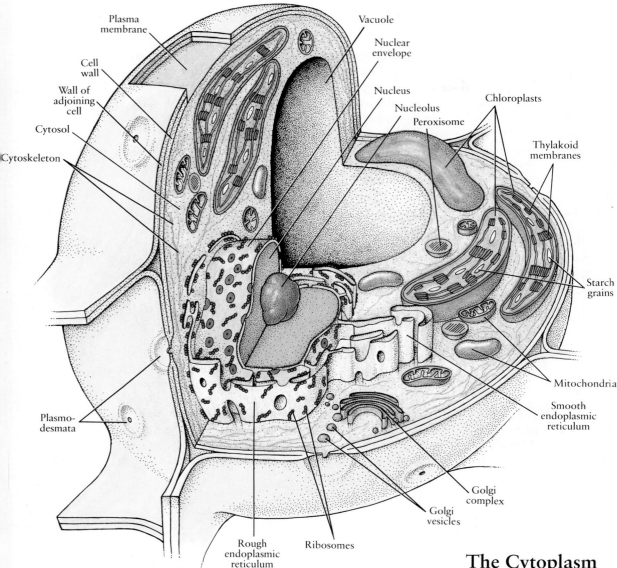

Plasma membrane
Cell wall
Wall of adjoining cell
Cytosol
Cytoskeleton
Plasmodesmata
Vacuole
Nuclear envelope
Nucleus
Nucleolus
Peroxisome
Chloroplasts
Thylakoid membranes
Starch grains
Mitochondria
Smooth endoplasmic reticulum
Golgi complex
Golgi vesicles
Rough endoplasmic reticulum
Ribosomes

5–11   *A relatively young plant cell, as interpreted from electron micrographs. Like the animal cell, it is bounded by a* **plasma membrane.** *Surrounding this membrane is a cellulose-containing* **cell wall. Plasmodesmata,** *which are membrane-lined channels through the cell walls, provide a cytoplasmic connection between adjacent cells.*

The most prominent structure in many plant cells is a large **vacuole,** *filled with a solution of salts and other substances. As we shall see in the next chapter, the vacuole plays a key role in keeping the cell wall stiff and the plant body crisp.*

Chloroplasts, *the large organelles in which photosynthesis takes place, are generally concentrated near the surface of the cell. Molecules of chlorophyll and the other substances involved in the capture of light energy from the sun are located in the* **thylakoid membranes** *within the chloroplasts.*

Like the animal cell, the living plant cell contains a prominent **nucleus,** *extensive* **endoplasmic reticulum,** *and many* **ribosomes** *and* **mitochondria.** *Especially numerous in the growing plant cell are* **Golgi complexes,** *which play an important role in the assembly of materials for the expanding cell wall.*

The Cytoplasm

Not long ago, the cell was visualized as a bag of fluid containing enzymes and other dissolved molecules along with the nucleus and a few other structures. With the development of electron microscopy, an increasing number of complex structures have been identified within the cytoplasm, which is now known to be highly organized. The interior of a typical animal cell is shown in Figure 5–10. Figure 5–11 shows a corresponding view of a typical plant cell.

The fluid portion of the cytoplasm, known as the **cytosol,** is a concentrated solution of ions, small molecules (such as amino acids, sugars, and ATP), and proteins. Most of these solutes are found only in low concentrations—or not at all—outside the cell. The cytosol of both prokaryotic and eukaryotic cells also contains numerous **ribosomes,** the RNA-protein complexes that play a major role in the synthesis of protein molecules from their amino acid subunits.

In most eukaryotic cells, about half of the cytoplasm consists of the membrane-bound compartments known as **organelles.** This name was coined by the early microscopists who viewed these structures, which they could barely distinguish, as comparable to the organs of multicellular animals. On closer inspection, the organelles turned out not to be miniature hearts or kidneys. However, like those organs, each type of organelle carries out its own specialized functions. The amount of membrane surface created within the cell by the organelles is enormous—vastly more than the total surface area of the plasma membrane.

The organelles are anchored within the cytosol by a network of protein filaments that form an internal **cytoskeleton.** The fibers of the cytoskeleton maintain the shape of the cell, enable it to move, and direct its molecular traffic.

In the pages that follow, we shall look at the individual structures of animal and plant cells and discuss their roles in the life of the cell.

Vacuoles and Vesicles: Support and Transport

Many cells, especially plant cells, contain structures known as **vacuoles.** A vacuole is a space in the cytoplasm surrounded by a single membrane and filled with water and solutes. Immature plant cells typically have many vacuoles, but as a plant cell matures, the numerous smaller vacuoles merge, forming one large, central, fluid-filled vacuole. When fully formed, this vacuole is a major supporting element of the cell (Figure 5–12).

Vesicles, which are found in all metabolically active eukaryotic cells, have the same general structure as vacuoles. They are, however, much smaller. One of the principal functions of vesicles is transport. As we shall see, they participate in the transport of materials both within the cell and into and out of the cell.

Ribosomes and Endoplasmic Reticulum: Protein and Lipid Biosynthesis

As we have noted previously, ribosomes are the sites at which amino acids are assembled into proteins. The more protein a cell is making, the more ribosomes it has. In eukaryotic cells, ribosomes are found within the cytosol and also attached to a complex system of internal membranes, the **endoplasmic reticulum.**

The endoplasmic reticulum forms a network of interconnecting flattened sacs, tubes, and channels. It occurs in two general forms, **rough** (with ribosomes attached) and **smooth** (without ribosomes). Rough endoplasmic reticulum (Figure 5–13) is continuous with the outer membrane of the nuclear envelope, which also has ribosomes attached. Similarly, smooth endoplasmic reticulum is continuous with the rough endoplasmic reticulum.

(a)

10 μm

(b)

5–12   (a) *Graphic interpretation and* (b) *electron micrograph of a vacuole. In this cell from the photosynthetic structure of a moss, the vacuole has expanded until it almost entirely fills the cell. The small amount of living cytoplasm, containing the chloroplasts, has been forced to the edges of the cell, up against the plasma membrane.*

(b) |— 0.05 μm —|

(a)

5–13   (a) *Graphic interpretation and* (b) *electron micrograph of rough endoplasmic reticulum. The endoplasmic reticulum is a system of membranes that separates the cell into channels and compartments and provides surfaces on which chemical activities take place. The dense objects on the membrane surfaces are ribosomes. The narrow compartments formed by the membranes of the rough endoplasmic reticulum contain newly synthesized proteins. This endoplasmic reticulum is from a cell of the pancreas, an organ extremely active in the synthesis of digestive enzymes.*

The amount of endoplasmic reticulum in a cell—and the relative proportions of the two forms—are not fixed but vary according to the cell's activities. In most cells, rough endoplasmic reticulum is the predominant form. Only cells specialized for the synthesis or metabolism of lipids—such as the cells that make steroid hormones—have large amounts of smooth endoplasmic reticulum. "Routine" lipid synthesis, such as the synthesis of membrane lipids, generally occurs in the rough endoplasmic reticulum, which also plays a role in the synthesis of many proteins.

The synthesis of proteins always begins on ribosomes that are located within the cytosol. The subsequent location of the ribosomes and the growing polypeptide chains to which they are attached is determined by the way in which the newly synthesized proteins will be used. Some proteins—hemoglobin and certain enzymes, for example—perform their functions within the cytosol. In the case of these molecules, the entire process of protein synthesis, from beginning to end, occurs in the cytosol.

Other proteins, such as digestive enzymes, hormones, or mucus, are released outside the cell, sometimes carrying out their functions at a great distance (on a cellular scale) from their source. Still other proteins are essential components of organelles, organelle membranes, or the plasma membrane. The synthesis of all these proteins begins in the cytosol with the synthesis of a "leader" of hydrophobic amino acids. This portion of the molecule, known as the **signal sequence**, directs the newly forming proteins and their accompanying ribosomes to specific regions of the endoplasmic reticulum. The ribosomes then attach to the endoplasmic reticulum, and the amino acids of the signal sequence assist in the transport of the proteins through the lipid bilayer to the interior cavity of the endoplasmic reticulum.

As synthesis proceeds, the growing proteins continue to move into the endoplasmic reticulum, a process powered by energy from ATP. When the proteins are complete, the ribosomes are detached and return to the cytosol. Subsequently, the newly synthesized proteins move from the rough endoplasmic reticulum through a special transitional endoplasmic reticulum. Here they are packaged, often with newly synthesized membrane lipids, in transport vesicles destined for a Golgi complex.

(a)

0.25 μm

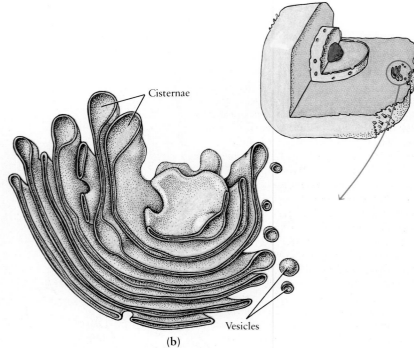

Cisternae

Vesicles

(b)

5–14 (a) *Electron micrograph and* (b) *graphic interpretation of a Golgi complex. A Golgi complex consists of four or more membrane-bound sacs, known as cisternae, arranged in a loose stack. Materials are packaged in membrane-enclosed vesicles at the Golgi complexes and are distributed within the cell or shipped to the cell surface. Note the vesicles pinching off from the edges of the flattened cisternae.*

Golgi Complexes: Processing, Packaging, and Distribution

Each **Golgi complex** consists of flattened, membrane-bound sacs stacked loosely on one another and surrounded by tubules and vesicles (Figure 5–14). The function of the Golgi complex is to accept vesicles from the endoplasmic reticulum, to modify the membranes and contents of the vesicles, and to incorporate these finished products into transport vesicles that deliver them to other parts of the cell and, especially, to the cell surface. Thus Golgi complexes serve as processing, packaging, and distribution centers. They are found in almost all eukaryotic cells. Animal cells usually contain 10 to 20 Golgi complexes, and plant cells may have several hundred.

One of the most critical products processed, packaged, and distributed by the Golgi complexes is new material for the plasma membrane and for the membranes of the cell's organelles. Membrane lipids and proteins, synthesized in the rough endoplasmic reticulum, are delivered to the Golgi complex in vesicles that fuse with it. Within the sacs of the Golgi complex, the final assembly of carbohydrates with proteins (forming glycoproteins) and with lipids (forming glycolipids) occurs. After the chemical processing is completed, the new membrane material is packaged in vesicles that are targeted to the correct location, whether it be the plasma membrane or the membrane of a particular organelle. In plant cells, Golgi complexes also bring together some of the components of the cell walls and export them to the cell surface where they are assembled.

In addition to their function in the assembly of cellular membranes, Golgi complexes have a similar role in the processing and packaging of materials that are released outside the cell. Figure 5–15 summarizes the way in which the ribosomes, the endoplasmic reticulum, and the Golgi complex and its vesicles interact to produce and deliver new material for cellular membranes and molecules for export. One of the intriguing questions about this process is how the various transport vesicles "know" where to go. The speed and efficiency with which each vesicle arrives at the proper location indicates the presence of a sophisticated mechanism for labeling and sorting. The nature of this mechanism is currently under investigation.

Lysosomes and Peroxisomes: Degradation and Recycling

One type of relatively large vesicle commonly formed from the Golgi complex is the **lysosome**. Lysosomes are essentially membranous bags that enclose enzymes involved in hydrolysis reactions. A single lysosome may

5–15   *The interaction of ribosomes, the endoplasmic reticulum, and the Golgi complex and its vesicles.* (**a**) *As proteins are synthesized on the ribosomes, they are fed into the rough endoplasmic reticulum. New membrane lipids are also synthesized in the endoplasmic reticulum.* (**b**) *The proteins then move through a transitional region of the endoplasmic reticulum and are released in vesicles that incorporate the newly synthesized membrane lipids. These vesicles move to the Golgi complex, where they fuse with its sacs.*

(**c**) *In the Golgi complex, carbohydrates, synthesized from sugar subunits available in the cytosol, are added to some of the proteins and lipids, producing glycoproteins and glycolipids. These large molecules are common components of cellular membranes. Molecules destined for export from the cell also undergo chemical processing in the Golgi complex. In some types of cells, this processing includes the addition of lipids to proteins, producing lipoproteins.*

Vesicles containing the finished molecules are released from the Golgi complex and move to other locations within the cell or to its exterior surface. If the vesicles contain new material for the plasma membrane or if their contents are to be secreted from the cell, the vesicle membrane fuses with the plasma membrane.

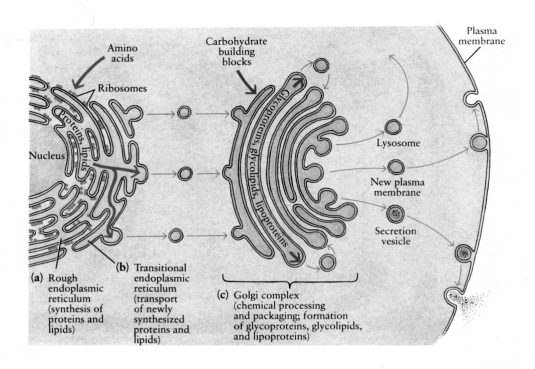

(**a**) Rough endoplasmic reticulum (synthesis of proteins and lipids)

(**b**) Transitional endoplasmic reticulum (transport of newly synthesized proteins and lipids)

(**c**) Golgi complex (chemical processing and packaging; formation of glycoproteins, glycolipids, and lipoproteins)

contain as many as 50 different kinds of hydrolytic enzymes. Within the lysosomes, these enzymes break down proteins, polysaccharides, and lipids from various types of cellular debris. The small molecules that result are then released back into the cytosol for reuse.

Another function of lysosomes is seen among white blood cells, which engulf bacteria in the human body. As the bacteria are taken up by the cell, they are wrapped in membrane-enclosed vacuoles. When this occurs, the lysosomes within the cell fuse with the vacuoles containing the bacteria and release their hydrolytic enzymes. The bacteria are quickly digested.

The membrane of a lysosome separates the hydrolytic enzymes it contains from the rest of the cell. If the lysosomes break open, the cell itself will be destroyed, since the enzymes they carry are capable of hydrolyzing the molecules of the living cell. Why the enzymes do not destroy the membranes of the lysosomes that carry them is an unanswered question. One clue is that the surface of the lysosome membrane facing the interior of the vesicle has a coating of carbohydrates that may perform a protective function.

Another type of enzyme-containing vesicle is the **peroxisome.** Peroxisomes are vesicles in which certain nitrogenous bases and other compounds are broken down by the cell. Some of the reactions that occur in the peroxisomes produce hydrogen peroxide, a compound that is extremely toxic to living cells. The peroxisomes, however, contain another enzyme that immediately breaks hydrogen peroxide into water and oxygen, preventing any damage to the cell.

Chloroplasts and Mitochondria: Cellular Power Plants

The activities of a cell require energy. As we have seen, some cells (autotrophs) manufacture their own energy-rich organic compounds from inorganic molecules. Other cells (heterotrophs) must obtain organic molecules from outside sources.

Photosynthetic autotrophs capture radiant energy from the sun and transform it to chemical energy stored in organic molecules. This process, **photosynthesis,** requires special pigments, of which chlorophyll is the most common. Photosynthesis takes place, however, only when the chlorophyll molecules are embedded in a membrane. In all photosynthetic eukaryotes, the chlorophyll-bearing membranes are organized within a membrane-bound organelle, the **chloroplast** (Figure 5–16).

5–16   (a) *Graphic interpretation and* (b) *electron micrograph of a chloroplast. A chloroplast is surrounded by two membranes, each of which is a lipid bilayer. In addition, it contains an elaborate internal membrane system in which chlorophyll and* *other photosynthetic pigments are embedded. The internal membranes form a series of flattened, interconnected sacs, which are called thylakoids. This chloroplast is from the leaf of a corn plant.*

Stack of thylakoids

Outer membrane

Inner membrane

(a)

(b)   0.25 µm

Inner membrane

Outer membrane

Cristae

(a)

(b)   0.5 µm

5–17   (a) *Graphic interpretation and* (b) *electron micrograph of a mitochondrion. Like the chloroplast, the mitochondrion is surrounded by two distinct membranes. The inner membrane folds inward to make a series of shelves, or cristae. The membrane forming these shelves plays a crucial role in the energy-releasing chemical reactions that occur in the mitochondria.*

Virtually all eukaryotic cells (including photosynthetic ones) have **mitochondria** (singular, mitochondrion), which are also membrane-bound organelles. In the process of **cellular respiration,** which occurs in the mitochondria, energy-rich molecules are broken down. This process uses oxygen and releases energy. The energy is harnessed by the mitochondria to "recharge" ADP molecules to ATP, which is then available to power other cellular activities. As shown in Figure 5–17, mitochondria are always surrounded by two membranes, the inner one of which folds inward. These folds, known as **cristae,** are working surfaces for the reactions that occur in the mitochondria.

Chloroplasts and mitochondria are the essential power plants of eukaryotic cells. Without the energy they make available, most other cellular functions could not be carried out. We shall examine the structure and function of these organelles in more detail when we consider the processes of photosynthesis and respiration in Section 2.

The Cytoskeleton: Structural Support and Motility

The three-dimensional organization of the eukaryotic cell depends, in large part, on a network of filamentous proteins in the cytoplasm. These proteins form an internal cytoskeleton that maintains the shape of the cell, anchors its organelles, directs its traffic, and enables it to move. Three different types of filaments—**actin filaments, intermediate filaments,** and **microtubules**—have been identified as major participants in the cytoskeleton (Figure 5–18).

5–18   *These micrographs of whole cells from the rat kangaroo dramatically reveal the distribution of elements of the cytoskeleton. (a) Actin filaments are bundled together in stress fibers that run the length of the cell and are also concentrated just beneath the plasma membrane. (b) Intermediate filaments surround the nucleus and extend throughout the cytoplasm. (c) Microtubules radiate from an "organizing center" near the nucleus and end near the plasma membrane.*

These cells are from epithelial tissue, which provides a covering for body surfaces and lines the internal organs. Each cell was treated with fluorescent antibodies that would bind to one specific type of cytoskeletal protein. The pattern of fluorescence indicates the location of the protein.

(a)   20 μm

(b)   20 μm

(c)   20 μm

Actin filaments are very fine protein threads, formed from the globular protein actin (Figure 5–19a). Actin is the most abundant protein in many animal cells, and its globular molecules can be rapidly assembled into filaments by the cell—and also rapidly disassembled—depending on the requirements of the cell at any given time. Actin filaments are often concentrated in bundles, known as stress fibers, that are stretched across the cytoplasm like cables or meshed together in the form of belts or webs. In many cells, the plasma membrane is supported by an internal skeleton of actin filaments. The filaments are anchored to particular integral proteins of the plasma membrane, and their assembly begins at these attachment points. Actin filaments also play an important role in cell division.

Intermediate filaments, which are intermediate in size between actin filaments and microtubules, are composed of fibrous proteins. In many cells, the intermediate filaments radiate out from the nuclear envelope.

5–19   *Two components of the cytoskeleton, actin filaments and microtubules, are formed from globular protein subunits. (a) Actin filaments consist of two linear chains of identical actin molecules, coiled around one another to form a helix. (b) Microtubules are hollow tubes composed of two different types of molecules, alpha (α) tubulin and beta (β) tubulin. These molecules come together to form soluble dimers ("two parts"), which then self-assemble into insoluble hollow tubules. Both actin filaments and microtubules can be readily assembled and disassembled by the cell.*

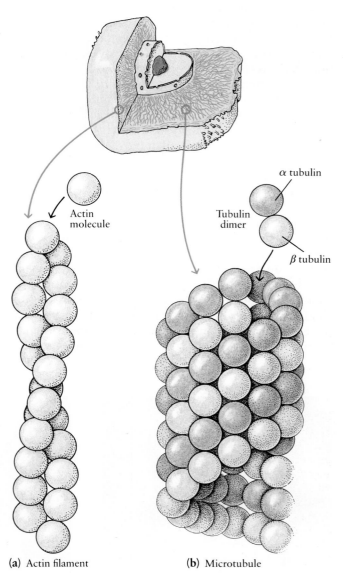

α tubulin

Actin molecule

Tubulin dimer

β tubulin

(a) Actin filament

(b) Microtubule

Their function in the life of the cell is still poorly understood, but they are found in the greatest density in cells, such as those of the skin, that are subject to mechanical stress.

Microtubules, the largest structures of the cytoskeleton, are long, hollow tubes assembled by the cell from globular protein subunits (Figure 5–19b). They usually exist as individual tubules, radiating outward from an "organizing center" near the nucleus and ending near the cell surface. Microtubules determine the position of organelles within the cell, and they provide routes for the movement of both organelles and vesicles. Like actin filaments, microtubules play an important role in cell division. In addition, they seem to provide a temporary scaffolding for the construction of other cellular structures, such as the plant cell wall.

Numerous wisplike fibers interconnect all of the other structures within the cytoplasm. These fibers, formed from accessory proteins of the cytoskeleton, link the various filaments together in specific ways. Although the resulting network gives the cell a highly ordered three-dimensional structure, it is neither rigid nor permanent. The cytoskeleton is a dynamic framework, changing and shifting according to the activities of the cell and making possible its motility.

All cells exhibit some form of movement. Even plant cells, encased in a rigid cell wall, exhibit active movement of the cytoplasm within the cell as well as chromosomal movements and changes in shape during cell division. Embryonic cells migrate in the course of animal development. Amoebas pursue and engulf their prey, and little *Chlamydomonas* cells dart toward a light source.

Two different mechanisms of cellular movement have been identified. The first consists of assemblies of contractile proteins, in which actin filaments are a principal component. In addition to their structural role in the cytoskeleton, actin filaments participate in the internal movement of cellular contents and in the move-

5–20   *A macrophage, a type of white blood cell, on the move across the surface of a culture dish. Its motility is made possible by the interaction of actin filaments with another protein known as myosin. As we shall see in Chapter 33, macrophages perform a number of vital functions in the body's response to invading microorganisms. They are also one of the principal cell types attacked by HIV, the virus that causes AIDS.*

ment of the cell itself (Figure 5–20). The second mechanism involves permanent locomotor structures, cilia and flagella, formed from assemblies of microtubules.

Cilia and Flagella

Cilia (singular, cilium) and **flagella** (singular, flagellum) are long, thin structures extending from the surface of many types of eukaryotic cells. They are essentially the same except for length. (Their names, which are derived from the Latin terms for "eyelash" and "whip," were given before their similarity was realized.) When they are shorter and occur in larger numbers, they are more likely to be called cilia. When they are longer and fewer, they are usually called flagella (Figure 5–21).

(a) |—— 20 μm ——| (b) |—— 5 μm ——|

5–21   (a) *The green alga* Chlamydomonas *propels itself through the water by means of its two flagella.* (b) *The heterotrophic protist* Paramecium, *by contrast, propels itself by means of numerous cilia.*

In one-celled protists and some small animals (such as a few types of flatworms), cilia and flagella are associated with movement of the organism. Similarly, the motile power of the mammalian sperm cell comes from its single powerful flagellum, or "tail."

Many of the cells that line the surfaces within our bodies and the bodies of other animals are also ciliated (see Figure 4–12, page 81). These cilia do not move the cells but, rather, serve to sweep substances across the cell surface. For example, cilia on the surface of cells of the respiratory tract beat upward, propelling a current of mucus that sweeps bits of soot, dust, pollen, tobacco tar—whatever foreign substances we have inhaled either accidentally or on purpose—to our throats, where they can be removed by swallowing.

Only a few large groups of eukaryotic organisms—most notably the flowering plants—have no cilia or flagella on any cells. Some bacteria move by means of flagella, but these prokaryotic flagella are so different in construction from those of eukaryotes that it would be useful if they had a different name.

Almost all eukaryotic cilia and flagella have the same internal structure. Nine fused pairs of microtubules form a cylinder that surrounds two additional, solitary microtubules in the center (Figure 5–22). The movement of cilia and flagella comes from within the structures themselves. If cilia are removed from cells and placed in a medium containing ATP, they beat or swim through the medium. The movement, according to the generally accepted hypothesis, is caused by each outer pair of microtubules moving tractor-fashion with respect to its nearest neighbor. The two "arms" that you can see on one microtubule of each outer pair are enzymes involved in ATP hydrolysis. Other proteins form spokes that extend from the nine pairs of outer microtubules toward the central pair, and still other proteins form more widely spaced links, rather like the hoops of a barrel, connecting the nine outer pairs to each other.

(b)   50 nm

5–22   (a) *Diagram of a cilium with its underlying basal body, and* (b) *an electron micrograph of the flagellum of* Chlamydomonas, *as seen in cross section. Virtually all eukaryotic cilia and flagella, whether they are found on protists or on the surfaces of cells within our own bodies, have this same internal structure, which consists of an outer cylinder of nine pairs of microtubules surrounding two additional microtubules in the center. The "arms," the radial spokes, and the connecting links are formed from different types of protein. The basal bodies from which cilia and flagella arise have nine outer triplets, with no microtubules in the center. The "hub" of the wheel in the basal body is not a microtubule, although it has about the same diameter.*

The spokes are thought to play a role in coordinating the tractorlike movements of the microtubules, whereas the links limit the amount of sliding possible and thus convert the movement into a bending motion.

Cilia and flagella arise from **basal bodies,** which are also made up of microtubules. Their number and arrangement are somewhat different, however, as you can see in Figure 5–22a. Many types of eukaryotic cells also contain structures identical to basal bodies that are known as **centrioles.** The distribution of centrioles within the cell is different from that of basal bodies, and, until recently, it appeared that their function was also different. Thus they were given a different name,

long before electron microscopy revealed their identical structure. Centrioles, like basal bodies, are found only in those groups of organisms that also have cilia and flagella (see Table 5–1). As we shall see in Chapter 10, centrioles also appear to play a role as organizers of microtubules.

The discovery of the complex internal structure of cilia and flagella, repeated over and over again throughout the living world, was one of the spectacular revelations of electron microscopy. For biologists, it is another glimpse down the long corridor of evolution, providing overwhelming evidence, once again, of the basic unity of Earth's living things.

Table 5–1	A Comparison of Cell Characteristics in the Five Kingdoms				
	Monera	**Protista**	**Fungi**	**Plantae**	**Animalia**
Cell type	Prokaryotic	Eukaryotic	Eukaryotic	Eukaryotic	Eukaryotic
Plasma membrane	Present	Present	Present	Present	Present
Cell wall	Present (noncellulose polysaccharide plus protein)	Present in some forms, various types	Chitin and other noncellulose polysaccharides	Cellulose and other polysaccharides	Absent
Nuclear envelope	Absent	Present	Present	Present	Present
Chromosomes	Single, circular DNA molecule	Multiple, consisting of DNA and histone proteins	Multiple, consisting of DNA and histone proteins	Multiple, consisting of DNA and histone proteins	Multiple, consisting of DNA and histone proteins
Ribosomes	Present (smaller)	Present	Present	Present	Present
Endoplasmic reticulum	Absent	Present	Present	Present	Present
Golgi complexes	Absent	Present	Present	Present	Present
Lysosomes	Absent	Often present	Often present	Similar structures (lysosomal compartments) present	Often present
Peroxisomes	Absent	Often present	Absent	Often present	Often present
Vacuoles	Absent	Present	Present	Usually large single vacuole in mature cell	Small or absent
Mitochondria	Absent	Present	Present	Present	Present
Chloroplasts	Absent	Present (some forms)	Absent	Present	Absent
9 + 2 cilia or flagella	Absent	Often present	Absent	Absent (in flowering plants)	Often present
Centrioles	Absent	Often present	Absent	Absent (in flowering plants)	Present

Summary

Cells are the basic units of biological structure and function. The size of cells is limited by proportions of surface to volume; the greater a cell's surface area in proportion to its volume, the greater the quantity of materials that can move into and out of the cell in a given period of time. Cell size is also limited by the capacity of the nucleus to regulate cellular activities.

Cells are separated from their environment by the plasma membrane, which protects the cell's structural and functional integrity. According to the fluid-mosaic model, the plasma membrane and other cellular membranes are formed from phospholipid bilayers in which protein molecules are embedded. Different membrane proteins perform different functions; some are enzymes, others are receptors, and still others are transport proteins.

The two faces of the plasma membrane differ in chemical composition. The cytoplasmic face is characterized by additional protein molecules attached to the proteins embedded in the bilayer. The exterior face of the membrane is characterized by short carbohydrate chains. Some of these chains are the hydrophilic heads of glycolipid molecules that are interspersed among the phospholipid molecules; others are attached to the protruding portions of membrane proteins.

The cells of plants, most algae, fungi, and prokaryotes are further separated from the environment by a cell wall constructed by the cell itself.

The nucleus of eukaryotic cells is surrounded by a double membrane, the nuclear envelope, which contains pores through which molecules pass to and from the cytoplasm. The nucleus contains the chromosomes, which, when the cell is not dividing, exist in an extended form called chromatin. The nucleolus, visible within the nucleus, is involved in the formation of ribosomes. Interacting with the cytoplasm, the nucleus helps to regulate the cell's ongoing activities.

The cytoplasm consists of all the cell contents that are outside the nuclear envelope but within the plasma membrane. The cytosol, the fluid portion of the cytoplasm, is a dense solution of ions, small molecules, and proteins. Dispersed throughout the cytosol in both prokaryotic and eukaryotic cells are numerous ribosomes, complexes of protein and DNA that participate in protein synthesis.

In eukaryotic cells, the cytoplasm contains numerous membrane-bound compartments known as organelles, which are not found in prokaryotic cells (see Table 5–1). The simplest are vacuoles and vesicles, which are bounded by a single membrane.

The cytoplasm of eukaryotic cells is subdivided by a network of membranes known as the endoplasmic reticulum, which serves as a work surface for many of the cell's biochemical activities. Cells that are producing new membrane material or proteins for export have extensive systems of endoplasmic reticulum with ribosomes attached, known as rough endoplasmic reticulum.

Golgi complexes are processing and packaging centers for materials being moved through and out of the cell. Lysosomes, which contain hydrolytic enzymes, are involved in intracellular digestive activities. The enzymes for cellular reactions that produce hydrogen peroxide as a by-product are sequestered in the peroxisomes.

Chloroplasts and mitochondria are the organelles involved in energy capture and energy release, respectively. Photosynthesis occurs within the chloroplasts, and cellular respiration within the mitochondria.

The eukaryotic cytoplasm has a supporting cytoskeleton that includes three principal types of structures: actin filaments, intermediate filaments, and microtubules. The cytoskeleton maintains the shape of the cell, anchors its organelles, directs its traffic, and enables it to move. Assemblies of contractile proteins, including actin filaments, are associated with internal cellular movement and, in some types of cells, with the movement of the cell itself.

Cilia and flagella, which are formed from assemblies of microtubules, are associated with the external movement of cells or the movement of materials along cell surfaces. These whiplike appendages are found on the surface of many types of eukaryotic cells. They have a highly characteristic 9 + 2 structure, with nine pairs of microtubules forming a cylinder around two central microtubules. Cilia and flagella arise from basal bodies, which are cylindrical structures containing nine microtubule triplets with no inner pair. Centrioles have the same internal structure as basal bodies and are found only in those groups of organisms that also have cilia or flagella.

Questions

1. Distinguish among the following terms: plasma membrane/cell wall; nucleus/nucleolus; cytoplasm/cytosol/cytoskeleton; chloroplasts/mitochondria; cilia/flagella; basal body/centriole.

2. Describe the structure of the plasma membrane. How do the two faces of the membrane differ? What is the functional significance of these differences?

3. (a) Sketch an animal cell. Include the principal organelles and label them. (b) Prepare a similar, labeled sketch of a plant cell. (c) What are the major differences between the animal cell and the plant cell?

4. Why is the secondary wall of a plant cell *inside* the primary cell wall? Where is the plasma membrane in relation to the two cell walls?

5. Explain the functions of each of the following structures: ribosomes, endoplasmic reticulum, vesicles, Golgi complexes. How do they interact in the synthesis and delivery of new membrane material and in the export of proteins from the cell?

6. What are the functions of the cytoskeleton? Describe the similarities and differences between actin filaments, intermediate filaments, and microtubules.

7. Use a ruler and the scale marker at the bottom of each micrograph on pages 89 and 103 to determine: (a) the thickness (roughly) of the plasma membrane, (b) the diameter of a cilium, and (c) the diameter of a microtubule within a cilium. (This is how the sizes of cellular components are determined by microscopists.) Would a cilium be resolvable in a light microscope (that is, is its diameter more than $0.2 \ \mu$m)?

8. (a) Sketch a cross section of a cilium. (b) Sketch a cross section of the basal body of a cilium. (c) What are the differences between the two structures?

9. On the basis of what you know of the functions of each of the structures in Table 5–1, what components would you expect to find most prominently in each of the following cell types: muscle cells, sperm cells, green leaf cells, red blood cells, white blood cells?

10. Two brothers were under medical treatment for infertility. Microscopic examination of their semen showed that the sperm were immotile and that the little "arms" were missing from the microtubular arrays. The brothers also had chronic bronchitis and other respiratory difficulties. What is a plausible explanation?

Suggestions for Further Reading

Albersheim, Peter: "The Walls of Growing Plant Cells," *Scientific American*, April 1975, pages 81–95.

Alberts, Bruce, Dennis Bray, Julian Lewis, Martin Raff, Keith Roberts, and James D. Watson: *Molecular Biology of the Cell*, 3d ed., Garland Publishing, Inc., New York, 1994.

> *Progressing from the molecules of which cells are composed, through an examination of cellular structure and function, to the interactions of cells within tissues, this outstanding text describes not only our current knowledge and how it was attained but also the many areas still to be explored. It is clearly written and filled with wonderful micrographs and explanatory diagrams. Highly recommended.*

Allen, Robert Day: "The Microtubule as an Intracellular Engine," *Scientific American*, February 1987, pages 42–49.

Becker, Wayne M., and David W. Deamer: *The World of the Cell*, 2d ed., The Benjamin/Cummings Publishing Company, Redwood City, Calif., 1991.

> *An introductory textbook, designed for a first course in cell biology. Chapters 4–9 provide a concise but clear discussion of cell structure and function.*

Bretscher, Mark S.: "How Animal Cells Move," *Scientific American*, December 1987, pages 72–90.

Darnell, James, Harvey F. Lodish, and David Baltimore: *Molecular Cell Biology*, 2d ed., W. H. Freeman and Company, New York, 1990.

> *A comprehensive treatment of modern cell biology, richly illustrated with diagrams and micrographs. This text places particular emphasis on membrane structure and function, cytoplasmic organelles, and the cytoskeleton, as well as on the techniques used in the contemporary study of cell biology.*

de Duve, Christian: *A Guided Tour of the Living Cell*, W. H. Freeman and Company, New York, 1984.*

> *In this beautifully illustrated two-volume set, de Duve, one of the pioneers of modern cell biology, takes the reader—imagined to be a "cytonaut," a bacterium-sized tourist—on a journey through the eukaryotic cell. The first portion of the journey explores the cellular*

---

* Available in paperback.

membranes; the second, the cytoplasm and its organelles; and the third, the nucleus. At the conclusion of the journey, de Duve considers such key questions of modern biology as the origin of life and the mechanisms of evolution.

Dustin, Pierre: "Microtubules," *Scientific American*, August 1980, pages 67–76.

Ezzell, Carol: "Sticky Situations: Picking Apart the Molecules that Glue Cells Together," *Science News*, June 13, 1992, pages 392–395.

Lazarides, Elias, and Jean Paul Revel: "The Molecular Basis of Cell Movement," *Scientific American*, May 1979, pages 100–113.

Ledbetter, M. C., and Keith R. Porter: *Introduction to the Fine Structures of Plant Cells*, Springer-Verlag, New York, 1970.

An excellent atlas of electron micrographs of plant cells, with detailed explanations.

Porter, Keith R., and Mary A. Bonneville: *An Introduction to the Fine Structures of Cells and Tissues*, 4th ed., Lea & Febiger, Philadelphia, 1973.

An atlas of electron micrographs of animal cells, with detailed commentaries. These are magnificent micrographs, and the commentaries describe not only what the pictures show but also the experimental foundations of our knowledge of cell structure.

Prescott, David M.: *Cells: Principles of Molecular Structure and Function*, Jones and Bartlett Publishers, Boston, 1988.

An up-to-date, yet concise, textbook of cell biology, written for a first course at the undergraduate level. It is a wonderful introduction for any reader wishing to gain an overview of the exciting developments in contemporary cell biology.

Rothman, James E.: "The Compartmental Organization of the Golgi Apparatus," *Scientific American*, September 1985, pages 74–89.

Satir, Peter: "How Cilia Move," *Scientific American*, October 1974, pages 44–52.

Scientific American: *The Molecules of Life*, W. H. Freeman and Company, New York, 1986.*

This reprint of the October 1985 issue of Scientific American *includes 11 articles on molecules that play key roles in the living cell. You will find the articles on the molecules of the plasma membrane and the cytoskeleton of particular interest at this point in your study of biology.*

Scientific American: *Molecules to Living Cells*, W. H. Freeman and Company, New York, 1980.*

A collection of outstanding articles from Scientific American. *Two articles are devoted to cell membranes and their assembly.*

Sharon, Nathan, and Halina Liss: "Carbohydrates in Cell Recognition," *Scientific American*, January 1993, pages 82–89.

Sloboda, Roger D.: "The Role of Microtubules in Cell Structure and Cell Division," *American Scientist*, vol. 68, pages 290–298, 1980.

Stossel, Thomas P.: "How Cells Crawl," *American Scientist*, vol. 78, pages 408–423, 1990.

Unwin, Nigel, and Richard Henderson: "The Structure of Proteins in Biological Membranes," *Scientific American*, February 1984, pages 78–94.

* Available in paperback.

The light brown objects inside this remarkable cell are just what they look like: chips of wood. They were taken into the cell by a process known as phagocytosis—literally "cell-eating."

This organism, a protist bearing the melodic name *Barbulanympha ufalula*, lives in the intestine of the wood roach *Cryptocercus punctulatus*. The roach lives on and among dead logs, feeding on wood chips. Its digestive system breaks up the wood chips into protist-sized pieces, which are then engulfed and digested by *Barbulanympha*. Neither organism can live without the other.

Barbulanympha is about 0.3 millimeter in diameter, barely visible to the unaided eye. The tufts at the top of the cell are flagella— some 13,000 in all—that enable it to move about within the roach's intestine. The blue spherical region, visible below the flagella, is the nucleus. At the opposite end of the cell is a "back door"—a sensitive region of the plasma membrane through which wood chips are brought into the cell.

When the sensitive portion of the plasma membrane makes contact with a wood chip in the roach's intestine, the membrane extends around the wood chip, enclosing it in a vacuole. The vacuole then detaches from the plasma membrane and moves into the interior of the cell. Enzymes synthesized by *Barbulanympha* digest the cellulose in the wood into glucose and then into a smaller breakdown product, acetate. Both digestive enzymes and acetate are exported out of the cell, back into the intestine of the wood roach. In the mitochondria of the roach intestinal cells, the acetate is broken down to carbon dioxide and water. The energy released in this final stage of the process is harnessed to produce ATP.

Although not all cells are as photogenic as *Barbulanympha*, they are all equally dependent on their capacity to transport substances—ions, molecules, food particles, and even other cells—across the plasma membrane.

How Things Get into and out of Cells

Living matter is surrounded on all sides by nonliving matter with which it constantly exchanges materials. Yet living systems differ from their nonliving surroundings in the kinds and amounts of chemical substances they contain. Without this difference, living systems would be unable to maintain the organization on which their existence depends.

In all living systems, from prokaryotes to the most complex multicellular eukaryotes, exchanges of substances between the living organism and the nonliving world occur at the level of the individual cell. These exchanges are regulated by the plasma membrane. In multicellular organisms, the plasma membrane has the additional task of regulating exchanges of substances among the various specialized cells that constitute the organism. Control of these exchanges is essential to (1) protect each cell's integrity, (2) maintain the conditions at which its metabolic activities can take place, and (3) coordinate the activities of the different cells.

In addition to the plasma membrane, which controls the passage of materials between the cell and its environment, internal membranes, such as those surrounding mitochondria, chloroplasts, and the nucleus, control the passage of materials among intracellular compartments. This makes it possible for the cell to maintain the specialized chemical environments necessary for the processes occurring in the different organelles.

Maintenance of the internal environment of the cell and its various parts requires that the plasma membrane and the organelle membranes perform a complex double function: they must keep certain substances out while letting others in, and, conversely, they must keep certain substances in while letting others out. The capacity of a membrane to accomplish this function depends not only on the physical and chemical properties that result from its lipid and protein structure, but also on the physical and chemical properties of the ions and molecules that interact with the membrane.

6–1 *Among the most important ions in many living cells are sodium (Na^+), potassium (K^+), calcium (Ca^{2+}), magnesium (Mg^{2+}), and chloride (Cl^-). This graph shows the relative concentrations of these ions in the cytoplasm of the green alga* Nitella *and in the surrounding pond water. Such differences in ion concentration between the interior of living cells and their nonliving environment indicate that cells regulate their exchanges of materials with the surroundings. This regulation is accomplished by the plasma membrane.*

Of the many kinds of molecules surrounding and contained within the cell, by far the most common is water. Further, the many other molecules and ions important in the life of the cell (Figure 6–1) are dissolved in water. Therefore, let us begin our consideration of transport across cellular membranes by looking again at water, focusing our attention this time on how it moves.

The Movement of Water and Solutes

Except when it is locked in ice, water is constantly moving. It moves across the continents in rivers and streams, from the soil through the bodies of plants into the atmosphere, through the human body by way of the bloodstream, and into and out of living cells. In both the living and nonliving worlds, water molecules move from one place to another because of differences in potential energy. As you will recall (page 21), potential energy is the stored energy an object—or a collection of objects, such as a collection of water molecules—possesses because of its position. The potential energy of water is usually referred to as the **water potential.**

Water moves from a region where water potential is higher to a region where water potential is lower, regardless of the reason for the difference in water potential (Figure 6–2). A simple example is water running downhill in response to gravity. Water at the top of a hill has more potential energy (that is, a higher water potential) than water at the bottom of a hill. As the water runs downhill, its potential energy is converted to kinetic energy. This, in turn, can be converted to mechanical energy doing useful work if, for example, a water wheel is placed in the path of the moving water.

Pressure is another source of water potential. If we fill an eyedropper with water and then squeeze the bulb, water will squirt out. Like water at the top of a hill, this water has been given a high water potential and will move to a lower one. Can we make the water that is running downhill move uphill by means of pressure? Obviously we can move water uphill—but only so long as the water potential produced by the pressure exceeds the water potential produced by gravity.

In solutions, water potential is affected by the concentration of dissolved particles (solutes). As the concentration of solute particles (that is, the number of solute particles per unit volume of solution) increases, the water potential decreases. Conversely, as the concentration of solute particles decreases, the water potential increases. In the absence of other factors (such as pressure) affecting the water potential, water molecules in solutions move from regions of lower solute concentration (higher water potential) to regions of higher solute concentration (lower water potential). This fact is of great importance for living systems.

The concept of water potential is a useful one because it enables us to predict the way that water will move under various combinations of circumstances. Measurements of water potential are usually made in terms of the pressure required to stop the movement of water—that is, the hydrostatic (water-stopping) pressure—under the particular circumstances. The unit usually used to measure this pressure is the atmosphere. One atmosphere is the average pressure of the air at sea level, about 1 kilogram per square centimeter (14.7 pounds per square inch).

Two mechanisms are involved in the movement of water and solutes: bulk flow and diffusion. In living systems, bulk flow is the process that moves water and solutes from one part of a multicellular organism to

6–2   *The three factors that most commonly determine water potential are* (**a**) *gravity,* (**b**) *pressure, and* (**c**) *the concentration of dissolved solutes. Water moves from the region of higher water potential to the region of lower water potential, regardless of the reason for the difference in water potential.*

Often, a combination of factors are involved. In (**b**)*, for example, the pressure applied when the bulb is squeezed is not the only factor contributing to the difference in water potential. Given the angle at which the eyedropper is held, gravity is also a factor.*

another part. Diffusion, by contrast, plays a major role in the movement of many molecules and ions into, out of, and through cells. A particular instance of diffusion—that of water moving through a membrane that separates solutions of different concentration—is known as osmosis.

Bulk Flow

Bulk flow is the overall movement of a fluid. The molecules move all together and in the same direction. For example, water runs downhill by bulk flow in response to the differences in water potential at the top and the bottom of a hill. Blood moves through your blood vessels by bulk flow as a result of the water potential (blood pressure) created by the pumping of your heart. Sap—an aqueous solution of sucrose and other solutes—moves by bulk flow from the leaves of a plant to other parts of the plant body.

Diffusion

Diffusion is a familiar phenomenon. If you sprinkle a few drops of perfumc in one corner of a room, the scent will eventually permeate the entire room even if the air is still. If you put a few drops of dye in one end of a glass tank full of water, the dye molecules will slowly become evenly distributed throughout the tank. The process may take a day or more, depending on the size of the tank, the temperature, and the relative size of the dye molecules.

Why do the dye molecules move apart? If you could observe the individual dye molecules in the tank (Figure 6–3), you would see that each one of them moves individually and at random. Looking at any single molecule—at either its rate of motion or its direction of motion—gives you no clue at all about where the molecule is located with respect to the others. So how do the dye molecules get from one end of the tank to the other?

Imagine a thin section through the tank, running from top to bottom. Dye molecules will move in and out of this section, some moving in one direction, some moving in the other. But you will see more dye molecules moving from the side of greater dye concentration. Why? Simply because there are more dye molecules at that end of the tank. If there are more dye molecules on the left, more dye molecules, moving at random, will move to the right, even though there is an equal probability that any one molecule of dye will move from right to left. Consequently, the *net* movement of dye molecules will be from left to right. Similarly, if you could see the movement of the individual water molecules in the tank, you would see that their *net* movement is from right to left.

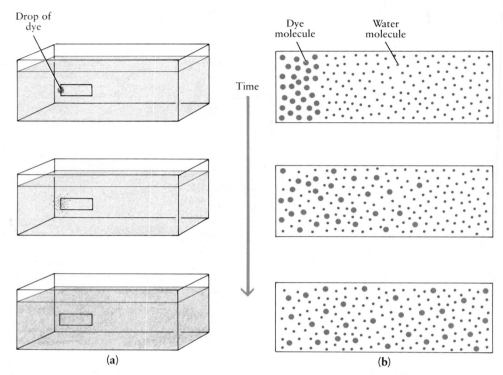

Drop of
dye

Time

Dye
molecule

Water
molecule

(a)

(b)

6–3   *The diffusion process, as revealed by the addition of a drop of dye to a tank of water (top left). The outlined section within each panel in (a) is shown in close-up in (b). Diffusion is the result of the random movement of individual molecules (or ions), which produces a net movement of particles from a region where they are more concentrated to a region where* they are less concentrated. This movement is down *the concentration gradient. Notice that as the dye molecules (red) diffuse to the right, the water molecules (blue) diffuse to the left. The ultimate result is an even distribution of both types of molecules. Can you see why the* net *movement of molecules will slow down as equilibrium is reached?*

Substances that are moving from a region of higher concentration of their own molecules to a region of lower concentration are said to be moving *down a gradient,* which is analogous to flowing downhill. (A substance moving in the opposite direction, toward a region where it is more highly concentrated, moves *against a gradient,* which is analogous to being pushed uphill.) Diffusion occurs only *down* a gradient. The steeper the downhill gradient—that is, the larger the difference in concentration—the more rapid the diffusion.

In our imaginary tank, there are two gradients, one of dye molecules and one of water molecules. The dye molecules are moving in one direction, down their gradient, and the water molecules are moving in the opposite direction, down their gradient. In each case, the molecules are moving from a region of higher potential energy to a region of lower potential energy.

What happens when all the molecules are distributed evenly throughout the tank? The even distribution does not affect the behavior of the molecules as individuals; they still move at random. And, since the movements are random, just as many molecules go to the left as to the right. But because there are now as many molecules of dye and as many molecules of water on one side of the tank as on the other, there is no *net* movement of either. There is, however, just as much random motion as before, provided the temperature has not changed. When the molecules have reached a state of equal distribution, that is, when there are no more gradients, they are said to be in **dynamic equilibrium.**

The essential characteristics of diffusion are (1) that each molecule or ion moves independently of the others and (2) that these movements are random. The net result of diffusion is that the diffusing substances become evenly distributed.

Sensory Responses in Bacteria: A Model Experiment

Concentration gradients are, as we have seen, of vital importance in the diffusion of substances into, out of, and through cells. In the case of many single-celled organisms, they are also important in the movement of the cell itself through the surrounding medium.

Many types of bacteria are able to swim toward a food source or away from a noxious chemical. They accomplish this by moving along a concentration gradient, from a lower concentration of a particular type of molecule to a higher one, or vice versa. Such directed movements in bacteria are extremely sensitive and highly specific. The bacteria can sense only certain molecules and can sense them at very low concentrations. The sensory abilities, it has been shown, depend on receptors in the plasma membrane that detect the molecules in question.

If you watch flagellated bacteria swimming freely, you will see two types of movement. When the flagella are rotating, they drive the cell through the water in much the same way a propeller drives a boat. When the flagella stop, the cell tumbles wildly for perhaps a tenth of a second. Then the propellerlike motion begins again, and the cell moves off in a new direction. When the concentration of chemicals in the water is uniform, the cell tumbles often, changing direction every time. By contrast, when the cell is moving along a gradient, there are fewer tumbles, so the cell continues longer in the same direction.

How do bacterial cells "decide" to move in a particular direction? In other words, how do they detect the concentration gradient? For many years, the most widely held hypothesis was that a bacterial cell could detect the difference in concentration between its front end and its rear end. However, when Daniel E. Koshland, Jr., of the University of California, calculated the concentrations of molecules to which a cell could respond, he began to question this concept. A bacterial cell is so small that, even in a gradient steep enough to produce a strong response, the difference in concentration between the front end of the cell

and the rear end would be only on the order of one molecule in 10,000. Compounding the problem is the probability that the gradient would not be exactly uniform. In short, the analytical task confronting the cell on its journey would seem virtually impossible.

Koshland then formulated an alternative hypothesis and, more important, figured out a way to test between the two. Koshland's hypothesis was that the bacterial cells were making a comparison not in space—between the concentrations at their front end and at their rear end—but in time. In other words, they were comparing the concentration of the solution in which they were currently swimming with the concentration of the solution they had left behind just a few microseconds ago.

In order to choose between the alternatives, Koshland formulated an ingeniously simple experiment. Using a strain of the common bacterium *Salmonella,* he set up an apparatus with which he could transfer a sample of cells almost instantaneously from one liquid medium to another and compare the motility of the cells in the two environments.

First, he put *Salmonella* in a medium that contained no chemical attractants. The cells exhibited their normal tumble-and-run pattern of behavior. He transferred them to a new medium, also containing no chemical attractants. They did not change their pattern of movement. (This part of the experiment—the control—showed that moving the cells, by itself, did not affect their motility.)

Next, in the crucial part of the experiment, he placed *Salmonella* cells in a medium containing a uniform concentration of an attractant, the amino acid serine. The cells behaved just as they had when no attractant was present. Then he transferred them to a medium with a slightly higher concentration of serine. There was an immediate change. For a few seconds, the cells ran more than they tumbled. Then he transferred them to a medium with a lower concentration of serine. For a few seconds, they tumbled more and ran less. In other words, although the bacteria were actually

A *color-enhanced electron micrograph of a* Salmonella *cell.*

moving from one uniform concentration to another, they behaved as if they were moving up or down a gradient. Koshland had tricked the bacteria into revealing their secret and so was able to choose between the alternative hypotheses. The bacteria were analyzing differences in time, not space.

This experiment is a minilesson in how scientists go about their business. They formulate a testable hypothesis and then they challenge it. The test of the hypothesis can take the form of a clever, well-designed experiment, as in this example, of accumulated observations, or of the analysis of reports made by other observers. However, two components are always necessary: the testable hypothesis and the data with which to test it.

Of course, many questions remain. Exactly how do the receptors on the plasma membrane recognize particular substances? How does the cell "remember" the concentration from one moment to the next? How does the sensory response (the detection of the chemical) trigger the motor response (the movement of the flagella)? Here again is a characteristic of the scientific process: the answer to one question nearly always raises still more questions.

Cells and Diffusion

Water, oxygen, carbon dioxide, and a few other simple molecules diffuse freely across the plasma membrane. Carbon dioxide and oxygen, which are both nonpolar, are soluble in lipids and move easily through the lipid bilayer. Despite their polarity, water molecules also move through the membrane without hindrance, apparently through momentary openings created by spontaneous movements of the membrane lipids. Other polar molecules, provided they are small enough, also diffuse through these openings. The permeability of the membrane to these solutes varies inversely with the size of the molecules, indicating that the openings are small and that the membrane acts like a sieve in this respect.

Diffusion is also a principal way in which substances move within cells. One of the major factors limiting cell size is this dependence upon diffusion, which is essentially a slow process, except over very short distances. It becomes increasingly slower and less efficient as the distance traveled by the diffusing molecules increases.

Efficient diffusion requires not only a relatively short distance but also a steep concentration gradient. Cells maintain such gradients by their metabolic activities, thereby hastening diffusion. For example, carbon dioxide is constantly produced as the cell oxidizes fuel molecules for energy. As a result, there is a higher concentration of carbon dioxide inside the cell than out. Thus a gradient is maintained between the inside of the cell and the outside, and carbon dioxide diffuses out of the cell down this gradient. Conversely, oxygen is used up by the cell in the course of its activities, so oxygen present in air or water or blood tends to move into the cells by diffusion, again down a gradient. Similarly, within a cell, molecules or ions are often produced at one place and used at another. Thus a concentration gradient is established between the two regions, and the substance diffuses down the gradient from the site of production to the site of use.

Osmosis: A Special Case of Diffusion

A membrane that permits the passage of some substances, while blocking the passage of others, is said to be **selectively permeable.** The movement of water molecules through such a membrane is a special case of diffusion, known as **osmosis.** Osmosis results in a net transfer of water from a solution that has higher water potential to a solution that has lower water potential. In the absence of other factors that influence water potential (such as pressure), the movement of water in osmosis will be from a region of lower solute concentration (and therefore of higher water potential) to a region of higher solute concentration (lower water potential).

The diffusion of water is not affected by *what* is dissolved in the water, only by *how much* is dissolved— that is, the concentration of particles of solute (molecules or ions) in the water. A small solute particle, such as a sodium ion, counts just as much as a large solute particle, such as a sugar molecule.

Two or more solutions that have equal numbers of dissolved particles per unit volume—and therefore the same water potential—are said to be **isotonic** (from the Greek *isos*, meaning "equal," and *tonos*, "tension"). There is no net movement of water across a membrane separating two solutions that are isotonic to one another, unless, of course, pressure is exerted on one side.

In comparing solutions of different concentration, the solution that has less solute (and therefore a higher water potential) is known as **hypotonic,** and the one that has more solute (a lower water potential) is known as **hypertonic.** (Note that *hyper* means "more"—in this case, more particles of solute; and *hypo* means "less"— in this case, fewer particles of solute.) In osmosis, water molecules diffuse from a hypotonic solution (or from pure water), through a selectively permeable membrane, into a hypertonic solution (Figure 6–4).

Water moves across a selectively permeable membrane

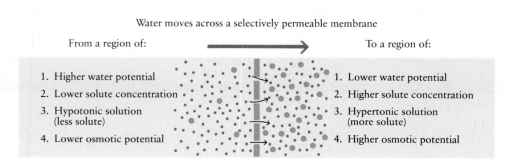

From a region of:

1. Higher water potential
2. Lower solute concentration
3. Hypotonic solution (less solute)
4. Lower osmotic potential

To a region of:

1. Lower water potential
2. Higher solute concentration
3. Hypertonic solution (more solute)
4. Higher osmotic potential

6–4 *The direction of water movement in osmosis.*

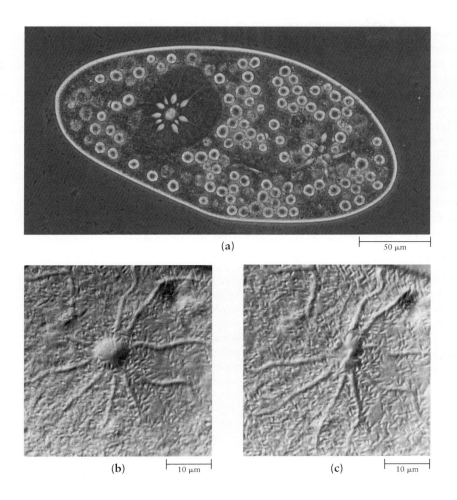

(a)

50 μm

(b)   10 μm

(c)   10 μm

6–5   *A* Paramecium *is hypertonic in relation to its environment, and hence water tends to move into the cell by osmosis. Excess water is expelled through its contractile vacuoles.* (a) *Photomicrograph of a living* Paramecium *in which one of its two rosette-like contractile vacuoles is clearly visible at the left. If you look very closely, you can find the other contractile vacuole at the opposite end of the cell. The numerous small, spherical objects in the cell are food vacuoles, filled with bacteria that are being digested by the* Paramecium.

A contractile vacuole, (b) *filled and* (c) *empty, as revealed by the scanning electron microscope. A group of long, thin collecting tubules converge on a central vacuole. Water moves from the cytosol into these tubules and through them into the central vacuole. When the central vacuole is filled, it contracts, propelling the water out of the cell through a small pore in the plasma membrane. Actin filaments (page 101) are involved in the contraction of the vacuole.*

Osmosis and Living Organisms

The osmotic movement of water across the selectively permeable plasma membrane causes some crucial problems for living systems. These problems vary according to whether the cell or organism is hypotonic, isotonic, or hypertonic in relation to its environment.

The body fluids of many marine fishes, for example, are hypotonic in relation to the surrounding salt water. Thus water tends to move out of their bodies by osmosis. As we shall see in Chapter 32, a combination of mechanisms enable these fishes to drink sea water and then excrete the excess solutes, maintaining their internal solute concentrations at the proper levels.

One-celled organisms that live in salt water are usually isotonic with the medium they inhabit, which is another way of solving the problem. The cells of most marine invertebrates are also isotonic with sea water. Similarly, the cells of vertebrate animals are isotonic with the blood and lymph that constitute the watery medium in which they live.

Many types of cells, however, live in a hypotonic environment. In all single-celled organisms that live in fresh water, such as *Paramecium*, the interior of the cell is hypertonic to the surrounding water. Consequently, water tends to move into the cell by osmosis. If too much water were to move into the cell, it could dilute the cell contents to the point of interfering with function and could even eventually rupture the plasma membrane. This is prevented by a specialized organelle known as a contractile vacuole, which collects water from various parts of the cell and pumps it out with rhythmic contractions (Figure 6–5). As you might expect, this bulk transport process requires an expenditure of energy by the cell.

Osmotic Potential

The water potential of two solutions separated by a selectively permeable membrane will become equal if enough water moves from the hypotonic solution into the hypertonic solution to equalize the solute concentrations—that is, to make the solutions isotonic. If, however, physical barriers prevent the expansion of the hypertonic solution as water moves into it by osmosis, there will be increasing resistance as water molecules continue to move across the membrane. This resistance is caused by a buildup of pressure that gradually increases the water potential of the hypertonic solution, decreasing the gradient of water potential between the two solutions. As the pressure increases, the *net* flow of water molecules will slow and then cease as the gradient of water potential disappears. Individual water molecules continue to move back and forth across the membrane, but these movements are in equilibrium and there is no *net* movement of water.

The pressure that is required to stop the osmotic movement of water into a solution (Figure 6–6) is a measure of the **osmotic potential** of the solution—that is, of the tendency of water to move across a membrane into the solution. The greater the tendency of water to move into the solution, the higher the pressure required to stop its movement—and thus the higher the osmotic potential.

Turgor   Plant cells are usually hypertonic to their surrounding environment, and so water tends to diffuse into them. This movement of water into the cell creates pressure within the cell against the cell wall. In young cells, this pressure causes the cell wall to expand and the cell to enlarge. The elongation that occurs as a plant cell matures (Figure 5–5, page 91) is a direct result of the osmotic movement of water into the cell.

As the plant cell matures, the cell wall stops growing. Moreover, mature plant cells typically have large central vacuoles that contain solutions of salts and other materials. (In citrus fruits, for example, they contain the acids that give the fruits their characteristic sour taste.) Because of these concentrated solutions, plant cells have a high osmotic potential—that is, water has a strong tendency to move into the cells. In the mature cell, however, the cell wall does not expand further. Its resistance to expansion results in an inward directed pressure, analogous to the pressure exerted by the depressed piston in Figure 6–6. This pressure prevents the net movement of additional water into the cell. Consequently, equilibrium of salt concentration is not reached and water continues to "try" to move into the cell, maintaining a constant pressure on the cell wall from the inside (Figure 6–7). This internal pressure on the cell wall is known as **turgor,** and it keeps the cell walls stiff and the plant body crisp. When turgor is reduced, as a consequence of water loss, the plant wilts.

6–6   *Osmosis and the measurement of the osmotic potential of a solution. (a) The tube contains a solution and the beaker contains distilled water. A selectively permeable membrane at the base of the tube permits the passage of water molecules but not of solute particles. (b) The diffusion of water into the solution causes its volume to increase, and thus the column of liquid to rise in the tube. However, the downward pressure created by the force of gravity acting on the column of solution is proportional to the height of the column and the density of the solution. Thus, as the column of solution rises in the tube, the downward pressure gradually increases until it becomes so great that it counterbalances the tendency of water to move into the solution. In other words, the water potential on the two sides of the membrane becomes equal. At this point, there is no further net movement of water. (c) The pressure that must be applied to force the column of solution back to the level of the water in the beaker provides a quantitative measure of the osmotic potential of the solution—that is, of the tendency of water to diffuse across the membrane into the solution.*

6–7   *(a) A turgid plant cell. The central vacuole is hypertonic in relation to the fluid surrounding it and so gains water. The expansion of the cell is held in check by the cell wall. (b) A plant cell begins to wilt if it is placed in an isotonic solution, so that water pressure no longer builds up within the vacuole. (c) A plant cell in a hypertonic solution loses water to the surrounding fluid and so collapses, with its plasma membrane pulling away from the cell wall. Such a cell is said to be plasmolyzed.*

Carrier-Assisted Transport

Cell membranes, as we have seen, are permeable to such substances as water, oxygen, and carbon dioxide, which move through them easily by diffusion. Other molecules used or produced by the cell are unable to diffuse through cell membranes, because of either their size or their polarity. Most ions and polar molecules, for example, cannot pass through the hydrophobic interior of the lipid bilayer. The transport of such substances depends upon integral membrane proteins that act as carriers, ferrying molecules and ions back and forth.

The transport proteins of the plasma membrane and of organelle membranes are highly selective. A particular protein may accept one molecule while it excludes a nearly identical one. It is the configuration of the protein molecule—that is, its tertiary or, in some cases, quaternary structure—that determines what molecules it can transport.

Some transport proteins can move substances across a membrane only if there is a favorable concentration gradient—that is, they can only move substances *down* a concentration gradient. Such carrier-assisted transport is known as **facilitated diffusion.** Facilitated diffusion, like simple diffusion, is driven by the potential energy of a concentration gradient. It is a passive process, requiring no energy outlay by the cell.

Other transport proteins can move molecules *against* a concentration gradient, a process known as **active transport.** Unlike facilitated diffusion, active transport requires the expenditure of energy by the cell (Figure 6–8).

Depending on the situation, the transport of a par-

ticular molecule into a cell may involve either facilitated diffusion or active transport. For example, most cells oxidize glucose to meet their energy needs. A steady supply of glucose is carried into these cells, down the concentration gradient, by facilitated diffusion. Liver cells, however, store glucose, readily converting it to and from glycogen. Although these cells have a high internal concentration of glucose, additional glucose is moved into them, against the concentration gradient, by active transport.

An Example of Active Transport: The Sodium-Potassium Pump

Most animal cells maintain steep concentration gradients of sodium ions (Na^+) and potassium ions (K^+) across the plasma membrane. The concentration of Na^+ outside the cell is as much as 14 times higher than the concentration inside the cell. By contrast, the concentration of K^+ inside the cell is, depending on the cell type, anywhere from 10 to 30 times higher than the concentration on the outside. These concentration gradients, which are important in maintaining osmotic balance and in controlling cell volume, are produced by an active-transport system known as the **sodium-potassium pump.**

The sodium-potassium pump is powered by energy supplied by ATP, which, as you will recall from Chapter 3, is the cell's ready energy currency. A measure of the importance of the sodium-potassium pump to the organism is that more than a third of the ATP used by a resting animal is consumed by this one ion-pumping mechanism.

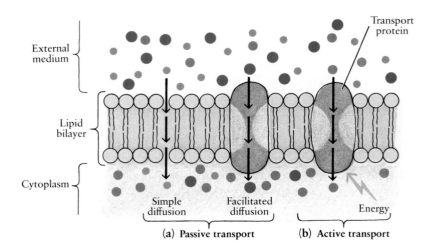

6–8 *Modes of transport through the plasma membrane.* (**a**) *In simple diffusion (tan spheres) and facilitated diffusion (blue-green spheres), molecules or ions move* down *a concentration gradient. The potential energy of the concentration gradient drives these processes, which are, from the standpoint of the cell, passive.* (**b**) *In active transport (red spheres), by contrast, molecules or ions are moved* against *a concentration gradient. Energy, most often supplied by ATP, is required to power active transport. Both facilitated diffusion and active transport require the presence of integral membrane proteins, specific for the substance being transported.*

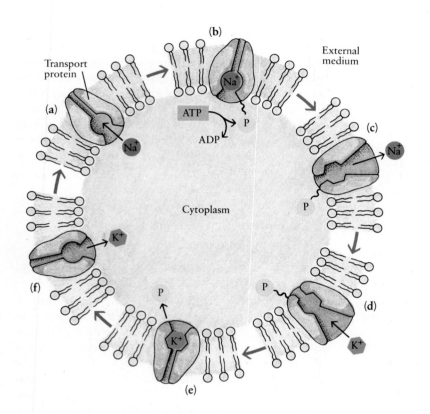

6–9   *A model of the sodium-potassium pump. (a) An Na⁺ ion in the cytosol fits precisely into the transport protein (top left). (b) The hydrolysis of an ATP molecule then attaches a phosphate group (P) to the protein, releasing ADP. This process results in (c) a change in the shape of the protein that causes the Na⁺ to be released outside the cell. (d) A K⁺ ion outside the cell is then bound to the transport protein, which in this form provides a better fit for K⁺ than for Na⁺. (e) The phosphate group is released from the protein, inducing conversion back to the other shape (f) and the release of the K⁺ ion into the cytosol. The protein is now once more ready to transport Na⁺ out of the cell.*

For clarity, only single ions are shown in this diagram. Quantitative studies have shown, however, that each complete pumping sequence, using only a single molecule of ATP, transports three Na⁺ ions out of the cell and two K⁺ ions into the cell. The pump works so rapidly that it can use as many as 100 molecules of ATP per second.

The pumping of Na⁺ and K⁺ ions is accomplished by a transport protein thought to exist in two alternative configurations. One configuration has cavities opening to the inside of the cell, into which Na⁺ ions can fit. The other configuration has cavities opening to the outside, into which K⁺ ions fit. As shown in Figure 6–9, Na⁺ within the cell binds to the transport protein. At the same time, the hydrolysis of ATP to ADP and phosphate (see Figure 3–28, page 67) results in the attachment of a phosphate group to the protein. This triggers its shift to the alternative configuration and the release of the Na⁺ to the outside of the cell. The transport protein is now ready to pick up K⁺, which results in the release of the phosphate group from the protein. This, in turn, causes it to return to the first configuration and to release the K⁺ to the inside of the cell. As this process is repeated, over and over, it maintains the observed gradients of Na⁺ and K⁺ ions across the membrane.

Types of Transport Proteins

Many ingenious models have been proposed to show how transport proteins, such as the sodium-potassium pump, might accept and eject their passengers. A cur-

rent model hypothesizes that transport proteins have hydrophilic cores, or channels, through which the transported molecules are squeezed, propelled by changes in the configuration of the protein. These channels are thought to be regions of the tertiary (or, in some cases, quaternary) structure of the proteins formed by amino acids with hydrophilic R groups.

The changes in configuration that accomplish the transport may be triggered directly, by the binding of the molecule to be transported, or indirectly, by the interaction of a receptor on the membrane surface with some other molecule or ion that is not actually transported through the membrane. The plasma membrane of most nerve cells, for example, contains a complex protein that is a receptor for a molecule known as acetylcholine. When acetylcholine binds to this receptor, a channel is opened in another membrane protein, closely associated with the acetylcholine receptor. This channel allows sodium ions (Na⁺) to flow into the cell, down the concentration gradient created and maintained by the action of the sodium-potassium pump. As we shall see in Chapter 35, this flow of Na⁺ ions into the cell is a key event in the propagation of the nerve impulse.

Current evidence indicates that there are at least three general types of transport proteins (Figure 6–10).

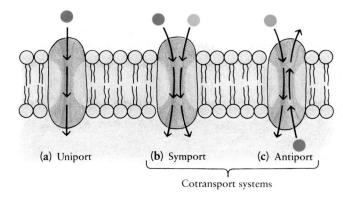

(a) Uniport   **(b)** Symport   **(c)** Antiport

Cotransport systems

6–10   *Three types of transport molecules.* (a) *In the simplest, known as a uniport, one particular solute is moved directly across the membrane in one direction.* (b) *In the type of cotransport system known as a symport, two different solutes are moved across the membrane, simultaneously and in the same direction. Often, a concentration gradient involving one of the transported solutes powers the transport of the other solute.* (c) *In another type of cotransport system, known as an antiport, two different solutes are moved across the membrane, either simultaneously or sequentially, but in opposite directions. The sodium-potassium pump is an example of a cotransport system involving an antiport.*

The simplest transfers one particular kind of molecule or ion directly across the membrane. More complex proteins function as **cotransport systems,** in which the transport of a particular molecule or ion depends upon the simultaneous or sequential transport of a different molecule or ion. In some cotransport systems, both solutes are transported in the same direction. In others, such as the sodium-potassium pump, the two different solutes are transported in opposite directions.

Vesicle-Mediated Transport

The transport proteins that ferry ions and small polar molecules across the plasma membrane cannot accommodate large molecules, such as proteins and polysaccharides, or large particles, such as microorganisms or bits of cellular debris. These large molecules and particles are transported by means of vesicles or vacuoles that bud off from or that fuse with the plasma membrane. For example, many substances are exported from cells in vesicles formed by the Golgi complexes. As we saw in Figure 5–15 (page 99), vesicles move from the Golgi complexes to the surface of the cell. When a vesicle reaches the cell surface, its membrane fuses with the plasma membrane, thus expelling its contents to the outside (Figure 6–11). This process is known as **exocytosis.**

Transport by means of vesicles or vacuoles can also work in the opposite direction. In **endocytosis,** material to be taken into the cell induces the plasma membrane to bulge inward, producing a vesicle enclosing the substance. This vesicle is released into the cytoplasm. Three different forms of endocytosis are known: phagocytosis ("cell-eating"), pinocytosis ("cell-drinking"), and receptor-mediated endocytosis.

0.1 μm

6–11   *Exocytosis. A secretion vesicle, formed by a Golgi complex of the protist* Tetrahymena furgasoni, *discharges mucus at the cell surface. Notice how the membrane enclosing the vesicle has fused with the plasma membrane.*

(a) Phagocytosis

(b) Pinocytosis

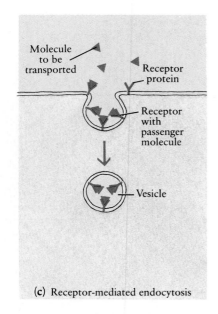

(c) Receptor-mediated endocytosis

6–12   *Three types of endocytosis. (a) In phagocytosis, contact between the plasma membrane and particulate matter, such as a bacterial cell, causes the plasma membrane to extend around the particle, engulfing it in a vacuole. One or more lysosomes then fuse with the vacuole, spilling hydrolytic enzymes into it.*

(b) In pinocytosis, the plasma membrane pouches inward, forming a vesicle around liquid from the external medium that is to be taken into the cell.

(c) In receptor-mediated endocytosis, the molecules to be transported into the cell must first bind to specific receptor proteins. The receptors are either localized in indented areas of the plasma membrane, known as pits, or migrate to such areas after binding the molecules to be transported. When filled with receptors carrying their particular molecules, the pit buds off as a vesicle.

When the substance to be taken into the cell in endocytosis is a solid particle, such as a wood chip (page 108) or a bacterial cell, the process is usually called **phagocytosis** (Figure 6–12a). Many heterotrophic protists, such as amoebas, feed in this way. Similarly, macrophages (Figure 5–20, page 102) and other types of white blood cells in our own bloodstreams engulf bacteria and other invaders in phagocytic vacuoles.

The taking in of liquids, as distinct from particulate matter, is given the special name of **pinocytosis** (Figure 6–12b). It is the same in principle as phagocytosis. Unlike phagocytosis, however, which is carried out only by certain specialized cells and involves specific recognition processes, pinocytosis occurs in all eukaryotic cells, as the cells continually and indiscriminately sip small amounts of fluid from the surrounding medium.

In **receptor-mediated endocytosis,** currently the subject of a great deal of research, particular membrane proteins serve as receptors for specific molecules that are to be transported into the cell (Figure 6–12c). Cholesterol, for example, is carried into animal cells by receptor-mediated endocytosis. As we noted earlier (page 58), cholesterol circulates in the bloodstream in the form of LDL particles, which interact with specific receptors on the cell surface. Binding of LDL particles to the receptor molecules triggers the formation of a vesicle that transports the cholesterol molecules into the cell.

Receptors for some substances, such as the hormone insulin, are apparently free to move laterally in the plasma membrane and, when unoccupied, are scattered at random locations on its surface. As the molecules to be transported into the cell bind to the receptors, the receptors move close together. A vesicle forms, and the hormone-laden receptors are carried into the cell. Receptors for other substances, such as LDL particles, appear to be localized in groups in specific areas of the plasma membrane even before binding of the substance to be transported.

In the areas where specific receptors are localized—or to which they migrate, as in the case of insulin receptors—the inner, or cytoplasmic, face of the plasma membrane is densely coated with a peripheral membrane protein known as clathrin. These areas, which are slightly indented, are known as **coated pits.** The vesicles that form from them, containing receptor molecules and their passengers, thus acquire an external, cagelike coating of clathrin. The formation of such a coated vesicle is illustrated in Figure 6–13.

As you can see by studying Figures 6–12 and 6–13, the surface of the membrane facing the interior of a vesicle or vacuole is equivalent to the surface of the

(a)

(b)

(c)

(d) |— 0.1 μm —|

6–13   *The formation of a coated vesicle in the developing egg cell of a hen. (a) A coated pit in the plasma membrane is covered on the cytoplasmic face with a latticework of clathrin molecules. The large particles clustered in the shallow pit on the external face are lipoprotein molecules, gathered from the surrounding medium and bound to specific membrane receptors that are associated with the underlying clathrin layer. (b) The pit deepens, and then (c) the plasma membrane closes around the pit to form the vesicle. (d) The completed vesicle with its outer coating of clathrin buds off and moves into the cell. The lipoproteins carried by this coated vesicle will be incorporated into the egg yolk.*

plasma membrane facing the exterior of the cell. Similarly, the surface of the vesicle or vacuole membrane facing the cytoplasm is equivalent to the cytoplasmic surface of the plasma membrane. As we noted in the last chapter, new material needed for expansion of the plasma membrane is transported, ready-made, from the Golgi complexes to the membrane by a process similar to exocytosis. Current evidence indicates that the portions of the plasma membrane used in forming endocytic vesicles or vacuoles are also returned to the membrane in exocytosis, thus recycling the membrane lipids and proteins, including the specific receptor molecules.

Cell-to-Cell Communication

Thus far in our consideration of the transport of substances into and out of cells, we have assumed that individual cells exist in isolation, surrounded by a watery environment. In multicellular organisms, however, this is generally not the case. Cells are organized into **tissues,** groups of specialized cells with common functions. Tissues are further organized in concert to form **organs,** each of which has a structure that suits it for a specific function.

As you might imagine, in multicellular organisms it is essential that individual cells communicate with one another so that they can collaborate to create a harmonious tissue or organ. This communication is accomplished by means of chemical signals—that is, by substances that are transported out of one cell and travel to another cell. When they reach the plasma membrane of the target cell, they may be transported into the cell, by any one of the processes we have considered. Alternatively, they may bind to specific membrane receptors at the surface of the target cell, thereby triggering chemical changes within that cell.

Communication in the Cellular Slime Mold

A cellular communication system of particular interest to biologists, because of the comparative ease with which it can be studied, is seen among a group of organisms known as the cellular slime molds. The slime mold *Dictyostelium discoideum* is an example. At one stage in its life cycle, it exists as a swarm of small individual amoeba-like cells, which divide and grow and feed until their food supply (mostly bacteria) gives out.

At this point, the cells alter both their shape and behavior: they become sausage-shaped and begin to migrate toward the center of the group **(a)**. Eventually, they pile up in a heap.

The heap gradually takes on the form of a multicellular mass somewhat resembling a garden slug **(b)**, slowly migrating and depositing a thick slime sheath that collapses behind it.

The sluglike mass soon stops its migration, gathers itself into a mound **(c)**, and sends up a long stalk at the tip of which a small, shimmering fruiting body forms **(d)**. The fruiting body matures **(e)** and eventually bursts open, releasing a new swarm of tiny cells, and the cycle begins again.

The chemical that spreads from cell to cell to initiate this remarkable sequence of events was first called acrasin, after Acrasia, the cruel witch in Spenser's *Faerie Queene* who attracted men and turned them into beasts. Acrasin was later identified as the chemical compound cyclic AMP (adenosine monophosphate). In recent years, it has become clear that many of the communications among cells in the human body also involve cyclic AMP. As we shall see in Chapter 34, the interaction of a number of vertebrate hormones with their specific receptors in the plasma membrane triggers a sequence of events within the cell in which cyclic AMP plays a central role.

(a)

(b)

1 mm

(c)

(d)

(e)

0.5 mm

Primary cell wall

Middle lamella

Primary cell wall

Plasma membrane

Endoplasmic reticulum

Plasmodesmata

(a) |—0.25 μm—| (b)

6–14   (a) *Electron micrograph and* (b) *graphic interpretation of plasmodesmata connecting two leaf cells from a corn plant. The adjacent primary walls of the two cells form the wide gray area running vertically through the micrograph. The walls are traversed by the plasmodesmata, each of which is lined by plasma membrane and filled with cytosol. The dark line extending through the center of each plasmodesma is an extension of the endoplasmic reticulum.*

Cytoplasm   Intercellular space   Cytoplasm

Gap junction

Connexon (6 protein subunits)

Channel

6–15   *A model of a gap junction. Embedded in the plasma membranes of two adjacent cells are structures that have been appropriately named "connexons." Each connexon consists of six identical membrane protein subunits, arranged in a hexagonal pattern with a space through the center. Connexons in adjacent plasma membranes abut each other in perfect alignment, providing a channel connecting the cytosol of the two cells.*

Often, however, the cells within a tissue or organ are tightly packed, allowing for direct and intimate contacts of various types between the cells. Among plant cells, which are separated from one another by cell walls, channels called **plasmodesmata** (singular, plasmodesma) traverse the walls, directly connecting the cytoplasm of adjacent cells (Figure 6–14).

In animal tissues, structures known as **gap junctions** permit the passage of materials between cells. These junctions appear as fixed clusters of very small channels surrounded by an ordered array of proteins (Figure 6–15). Ions and small molecules, such as amino acids and ATP, readily pass through these channels from one cell to another.

The transport of materials into and out of cells through the channels of plasmodesmata or gap junctions, through integral membrane proteins, and by means of endocytosis and exocytosis appear superficially to be three quite different processes. They are fundamentally similar, however, in that they all depend on the precise, three-dimensional structure of a great variety of specific protein molecules. These protein molecules not only form channels through which transport can occur but also endow the plasma membrane with the capacity to "recognize" particular molecules. This capacity is the result of billions of years of an evolutionary process that began, as far as we are able to discern, with the formation of that first fragile film around a few organic molecules.

Summary

The plasma membrane regulates the passage of materials into and out of the cell, a function that makes it possible for the cell to maintain its structural and functional integrity. This regulation depends on interactions between the membrane and the materials that pass through it.

One of the principal substances passing into and out of cells is water. Water potential determines the direction in which water moves; that is, water moves from where the water potential is higher to where it is lower.

Water movement takes place by bulk flow and diffusion. Bulk flow is the overall movement of water molecules and dissolved solutes as a group, as when water flows in response to gravity or pressure. The circulation of blood through the human body is an example of bulk flow.

Diffusion involves the random movement of individual molecules or ions and results in net movement down a concentration gradient. It is most efficient when the surface area through which diffusion is occurring is large in relation to volume, when the distance involved is short, and when the concentration gradient is steep. By their metabolic activities, cells maintain steep concentration gradients of many substances across the plasma membrane and between different regions of the cytoplasm.

Osmosis is the diffusion of water through a membrane that permits the passage of water but inhibits the movement of most solutes. Such a membrane is said to be selectively permeable. In the absence of other forces, the net movement of water in osmosis is from a region of lower solute concentration (a hypotonic medium), and therefore of higher water potential, to one of higher solute concentration (a hypertonic medium), and so of lower water potential. Turgor in plant cells is a consequence of osmosis.

Molecules cross the plasma membrane by simple diffusion or are transported by integral membrane proteins. If carrier-assisted transport is driven by the concentration gradient, the process is known as facilitated diffusion. If the transport requires the expenditure of energy by the cell, it is known as active transport. Active transport can move substances against their concentration gradients. One of the most important active-transport systems is the sodium-potassium pump, which maintains sodium ions at relatively low concentration and potassium ions at relatively high concentration in the cytoplasm.

Controlled movement into and out of a cell may also occur by endocytosis or exocytosis, in which the substances are transported in vacuoles or vesicles composed of portions of the plasma membrane. Three forms of endocytosis are phagocytosis, in which solid particles are taken into the cell; pinocytosis, in which liquids are taken in; and receptor-mediated endocytosis, in which molecules or ions to be transported into the cell are bound to specific receptors in the plasma membrane.

In multicellular organisms, communication among cells is essential for coordination of the different activities of the cells in the various tissues and organs. Much of this communication is accomplished by chemical agents that either pass through the plasma membrane or interact with receptors on its surface. Communication may also occur directly, through the channels of plasmodesmata (in plant tissues) or gap junctions (in animal tissues).

Questions

1. Distinguish among the following: bulk flow/diffusion/ osmosis; hypotonic/hypertonic/isotonic; facilitated diffusion/ active transport; endocytosis/exocytosis; phagocytosis/pinocytosis/receptor-mediated endocytosis; plasmodesmata/gap junctions.

2. What is a concentration gradient? How does a concentration gradient affect diffusion? How does a concentration gradient affect osmosis?

3. When diffusion of dye molecules in a tank of water is complete, random movement of molecules continues at the same rate (as long as the temperature remains the same). However, net movement stops. How do you reconcile these two facts?

4. Why is diffusion more rapid in gases than in liquids? Why is it more rapid at higher temperatures than at lower temperatures?

5. Three funnels have been placed in a beaker containing a solution (see the figure below). What is the concentration of the solution in the beaker? Explain your answer.

| Distilled water | 1% sugar solution | 2% sugar solution | Distilled water | 1% sugar solution | 2% sugar solution |

Initial state   Final state

6. Imagine a pouch with a selectively permeable membrane containing a saltwater solution. It is immersed in a dish of fresh water. Which way will the water move? If you add salt to the water in the dish, how will this affect water movement? What living systems exist under analogous conditions? How do you think they maintain water balance?

7. When you forget to water your house plants, they wilt and the leaves (and sometimes the stems) become very limp. What has happened to the plants to cause this change in appearance and texture? Within a few hours after you water your plants, they resume their normal, healthy appearance. What has occurred within the plants to cause this restoration? Sometimes, if you wait too long to water your plants, they never revive. What do you suppose has happened?

8. What limits the passage of water and other polar molecules and ions through the plasma membrane? How do such molecules get into and out of the cell? Describe four possible routes.

9. In what three ways does active transport differ from simple diffusion? How does it differ from facilitated diffusion?

10. Justify the conclusion that differences in ion concentration between cells and their surroundings (see Figure 6–1) indicate that cells regulate the passage of materials across the plasma membrane.

Suggestions for Further Reading

Brown, Michael S., and Joseph L. Goldstein: "How LDL Receptors Influence Cholesterol and Atherosclerosis," *Scientific American*, November 1984, pages 58–66.

Dautry-Varsat, Alice, and Harvey F. Lodish: "How Receptors Bring Proteins and Particles into Cells," *Scientific American*, May 1984, pages 52–58.

Kessin, Richard H., and Michiel M. Van Lookeren Campagne: "The Development of a Social Amoeba," *American Scientist*, vol. 80, pages 556–565, 1992.

Lienhard, Gustav E., Jan W. Slot, David E. James, and

Mike M. Mueckler: "How Cells Absorb Glucose," *Scientific American*, January 1992, pages 86–91.

Linder, Maurine E., and Alfred G. Gilman: "G Proteins," *Scientific American*, July 1992, pages 56–65.

Satir, Birgit: "The Final Steps in Secretion," *Scientific American*, October 1975, pages 28–37.

Staehelin, L. Andrew, and Barbara E. Hull: "Junctions between Living Cells," *Scientific American*, May 1978, pages 141–152.

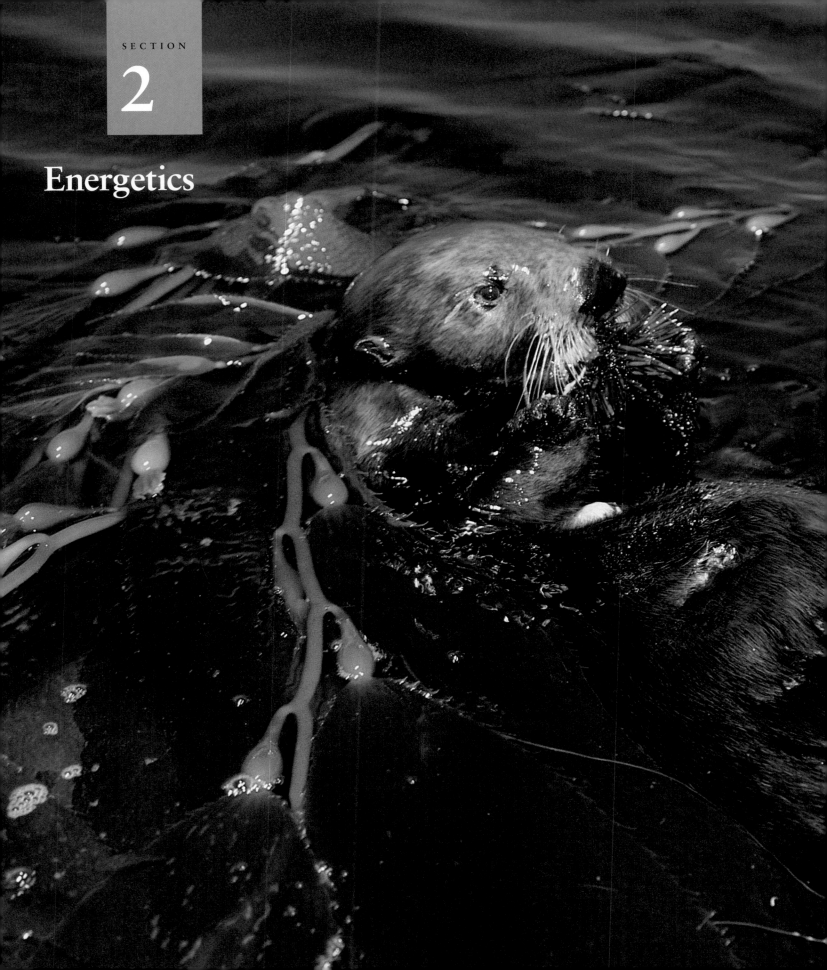

SECTION

2

Energetics

Sea otters are solar-powered. Long strands of kelp, the common marine alga in which this otter is floating, capture the sun's rays and convert them to the energy-rich chemicals of the algal cells. Certain animals, such as the sea urchin seen here cradled in the otter's paws, feed on those algal cells. When the otter eats the urchin—an event about to take place—the sun's energy is converted into the molecules, organelles, cells, organs, and activities that, all together, constitute a sea otter.

It takes a lot of sunshine to make an otter. Only 1 to 3 percent of the sunlight that reaches the alga is captured by its cells. In turn, only about 10 percent of the chemical energy stored in the alga is transferred to and stored in the tissues of the sea urchin. And, only 10 percent of that 10 percent is transferred to and stored in the tissues of the otter. As a consequence, you can always expect to find more algae off the coast of California than sea urchins and more sea urchins than sea otters. Similarly, in the forests of eastern North America, you can expect to find more maple leaves than caterpillars of the maple leaf-cutter moth and more caterpillars than birds that feed upon them. And, in East Africa, more grass than zebras and more zebras than lions.

Each of these—alga, sea urchin, sea otter, leaf, caterpillar, bird, grass, zebra, and lion—owes its existence to continued thermonuclear events taking place at the heart of a middle-aged, mid-sized star—our sun—and to the evolution on Earth of mechanisms and devices to capture and use its energy.

The Flow of Energy

Life on Earth depends on a steady flow of energy from the sun. The amount of energy delivered to the Earth by the sun is about 13×10^{23} (the number 13 followed by 23 zeros) calories per year. It is a difficult quantity to imagine. For example, the amount of solar energy striking the Earth every day is about 1.5 billion times greater than the amount of electricity generated in the United States each year.

About 30 percent of this solar energy is reflected back into space as light, and about 20 percent is absorbed by the atmosphere. Most of the remaining 50 percent, which reaches the surface of the Earth, is absorbed and converted to heat. Some of this absorbed heat energy serves to evaporate the waters of the oceans, producing the clouds that, in turn, produce rain and snow. Solar energy, in combination with other factors, is also responsible for the movements of air and water that help set patterns of climate over the surface of the Earth.

A small fraction—less than 1 percent—of the solar energy reaching the Earth's surface becomes, through a series of operations performed by the cells of plants and other photosynthetic organisms, the energy that drives all the processes of life. Living systems change energy from one form to another, transforming the radiant energy captured from the sun into the chemical and mechanical energy used by everything that is alive.

This flow of energy is the essence of life. In fact, one way of viewing evolution is as a competition among organisms for the most efficient use of energy resources. A cell can be best understood as a complex system for transforming energy. At the other end of the biological scale, the structure of an ecosystem (that is, all the living organisms in a particular locale and the nonliving factors with which they interact) or of the biosphere itself is determined by the energy exchanges occurring among the groups of organisms within it.

In this chapter, we shall look first at the general principles governing all energy transformations. Then we shall turn our attention to the characteristic ways in which cells regulate the energy transformations that take place within living systems. In the chapters that follow, the principal and complementary processes of energy flow through the biosphere will be examined—glycolysis and respiration in Chapter 8 and photosynthesis in Chapter 9.

The Laws of Thermodynamics

"Energy" is such a familiar term today that it is surprising to learn that the word first came into common use less than 200 years ago, at the time of the development of the steam engine. It was only then that scientists and engineers began to understand that heat, motion, light, electricity, and the forces holding atoms together in molecules are all different forms of energy—which can be defined most simply as the capacity to cause change, or, as it is often expressed, to do work. This new understanding led to the study of **thermodynamics**—the science of energy transformations—and to the formulation of its laws.

The First Law

The **first law of thermodynamics** states, quite simply: *Energy can be changed from one form to another, but it cannot be created or destroyed.*

Electricity is a form of energy, as is light. Electrical energy can be changed to light energy (for example, by letting an electric current flow through the tungsten wire in a light bulb). Conversely, light energy can be changed to electrical energy, a transformation that is the essential first step of photosynthesis, as we shall see in Chapter 9.

Energy can be stored in various forms and then changed into other forms. In automobile engines, for example, the energy stored in the chemical bonds of gasoline is converted to heat (kinetic energy), which is then partially converted to mechanical movements of the engine parts. Some of the energy is converted back to heat by the friction of the moving engine parts, and some of it leaves the engine in the exhaust products. Similarly, when organisms break down carbohydrates, they convert the energy stored in chemical bonds to other forms. On a summer evening, for example, a firefly converts chemical energy to mechanical energy, to heat, to flashes of light, and to electrical impulses that travel along the nerves of its body. Birds and mammals convert chemical energy into the heat necessary to maintain their body temperature, as well as into mechanical energy, electrical energy, and other forms of chemical energy. According to the first law of thermo-

dynamics, in these energy conversions, and in all others, energy is neither created nor destroyed.

In all energy conversions, however, some useful energy is converted to heat and dissipates. In an automobile engine, for example, the heat produced by friction and lost in the exhaust, unlike the heat confined in the engine itself, cannot produce work—that is, it cannot drive the pistons and turn the gears—because it is dissipated into the surroundings. But it is nevertheless part of the total equation. In a gasoline engine, about 75 percent of the energy originally present in the fuel is transferred to the surroundings in the form of heat—that is, it is converted to increased motion of atoms and molecules in the air. Similarly, the heat produced by the metabolic processes of animals is dissipated into the surrounding air or water.

Thus, another way of stating the first law of thermodynamics is that *in all energy exchanges and conversions, the total energy of the system and its surroundings after the conversion is equal to the total energy before the conversion.* A "system" can be any clearly defined entity—for example, an exploding stick of dynamite, an idling automobile engine, a mitochondrion, a living cell, a sea otter, a forest, or the Earth itself. The "surroundings" consist of everything outside the system.

7–1 *Electrical energy can be converted to light energy, as in these Hong Kong signs, for example. The energy emitted as an electron falls from one energy level to another is a discrete amount, characteristic for each atom. When electricity is passed through a tube of gas, electrons in the atoms of the gas are boosted to higher energy levels. As they fall back, light energy is emitted, producing, for example, the red glow characteristic of neon and the yellow glow characteristic of sodium vapor.*

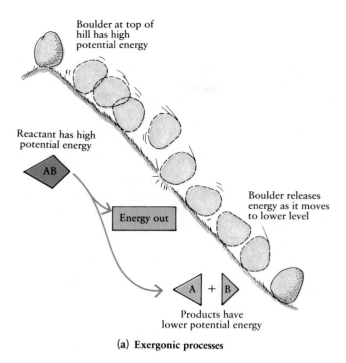

Boulder at top of hill has high potential energy

Reactant has high potential energy

AB

Energy out

Boulder releases energy as it moves to lower level

A + B

Products have lower potential energy

(a) Exergonic processes

Boulder at top of hill has higher potential energy

Product has higher potential energy

XY

Energy in

Energy required to roll boulder uphill

X + Y

Reactants have low potential energy

(b) Endergonic processes

7–2 (a) *In exergonic processes, whether physical or chemical, the potential energy of the final state is lower than the potential energy of the initial state, and energy is released. Exergonic processes occur spontaneously. Although "spontaneously" has an explosive sound, the word says nothing about the speed of the process—just that it can take place* *without an input of energy.* (b) *In endergonic processes, by contrast, the potential energy of the final state is higher than the potential energy of the initial state. Such processes cannot take place spontaneously. For them to occur, an input of energy is required.*

Stated in this form, the first law of thermodynamics can be used as a simple bookkeeping rule: when we add up the energy income and expenditures for any physical process or chemical reaction, the books must always balance.

The Second Law

The energy that is released as heat in an energy conversion has not been destroyed—it is still present in the random motion of atoms and molecules—but it has been "lost" for all practical purposes. It is no longer available to do useful work. This brings us to the **second law of thermodynamics,** which is the more interesting one, biologically speaking. It predicts the direction of all events involving energy exchanges. Thus it has been called "time's arrow."

The second law states that *in all energy exchanges and conversions—if no energy leaves or enters the system under study—the potential energy of the final state will always be less than the potential energy of the initial state.* The second law is entirely in keeping with everyday experience. A boulder will roll downhill but never uphill. Heat will flow from a hot object to a cold one and never the other way. A ball that is dropped will bounce—but not back to the height from which it was dropped.

A process in which the potential energy of the final state is less than that of the initial state is one that releases energy. Such a process is said to be **exergonic** ("energy-out"). Only exergonic processes can take place spontaneously—that is, without an input of energy from outside the system (Figure 7–2a). By contrast, a process in which the potential energy of the final state is greater than that of the initial state is one that requires energy. Such processes are said to be **endergonic** ("energy-in"). In order for them to proceed, there must be an input of energy to the system (Figure 7–2b).

Molecules, as you know, contain potential energy, stored in the chemical bonds holding the atoms together. When these bonds are broken in a chemical reaction, the energy they contain can be used to form new chemical bonds or can be released in the form of heat. If the potential energy of the products of the reaction is lower than the potential energy of the reactants, the reaction is exergonic. Energy will be released. If, however, the potential energy of the products is higher than the potential energy of the reactants, the reaction is endergonic. It will occur only with an input of energy.

One factor determining the difference in potential energy between the reactants and the products of a chemical reaction is their heat content. The lower the heat content, the lower the potential energy. However, another factor—the degree of disorder or randomness of the reactants and the products—is also involved. The greater the disorder or randomness, the lower the potential energy.

Previously, we stated the second law of thermodynamics in terms of the energy change between the initial and final states of a process. The second law can also be stated in another, simpler way: *All natural processes tend to proceed in such a direction that the disorder, or randomness, of the universe increases* (Figure 7–3). This disorder, or randomness, is known as **entropy**.

Living Systems and the Second Law

The universe, according to the present model, is a closed system—that is, neither matter nor energy enters or leaves the system. The matter and energy present in the universe at the time of the "big bang" are all the matter and energy it will ever have. Moreover, after each and every energy exchange and transformation, the universe as a whole has less potential energy and more entropy than it did before. In this view, the universe is running down. The stars will flicker out, one by one. Life—any form of life on any planet—will come to an end. Finally, even the motion of individual molecules will cease. However, even the most pessimistic among us do not believe this will occur for another 20 billion years or so.

In the meantime, life can exist *because* the universe is running down. Although the universe as a whole is a closed system, the Earth is not. It is an open system (Figure 7–4), receiving an energy input of about 13×10^{23} calories per year from the sun. Photosynthetic organisms are specialists at capturing the light energy released by the sun as it slowly burns itself out. They use this energy to organize small, simple molecules (water and carbon dioxide) into larger, more complex molecules (sugars). In the process, the captured light energy is stored as chemical energy in the bonds of sugars and other molecules.

Living cells—including photosynthetic cells—can convert this stored energy into motion, electricity, light, and, by shifting the energy from one type of chemical bond to another, into more convenient forms of chemical energy. At each transformation, energy is lost to the surroundings as heat. But before the energy captured from the sun is completely dissipated, organisms use it to create and maintain the complex organization of structures and activities that we know as life.

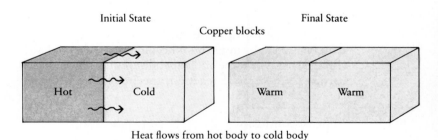

Initial State   Final State
Copper blocks

Hot   Cold   Warm   Warm

Heat flows from hot body to cold body

Solute particles move from region of high concentration to region of low concentration

Order becomes disorder

7–3   *Some illustrations of the second law of thermodynamics. In each case, a concentration of energy—in the hot copper block, in the solute particles on one side of a tank, and in the neatly organized books—is dissipated. In nature, processes tend toward randomness, or disorder. Only an input of energy can reverse this tendency and reconstruct the initial state from the final state. Ultimately, however, disorder will prevail, since the total amount of energy in the universe is finite.*

7–4 (a) *A fish bowl is, like the Earth, an open system—matter and energy enter and leave the system. Sunlight passes through the glass, oxygen diffuses into the water at its surface, and food is added by a human caretaker. Heat leaves the system through the glass and through the opening at the top, carbon dioxide diffuses out of the water at its surface, and the waste products of the animals are removed when the bowl is cleaned. Although energy is lost from the system with each and every energy exchange, a steady supply of energy from outside the system maintains its order.*

(b) *If, however, the fish bowl is placed in an opaque container that is sealed and insulated, it becomes a closed system, consisting of the fish bowl, its contents, and the air within the container. Neither matter nor energy can enter or leave the system. For a period of time after the container is sealed, energy continues to be converted from one form to another by the organisms in the fish bowl. With each and every conversion, however, some of the energy is given off as heat and dissipated into the water, the glass, and the air within the container. In time, the system will run down—the organisms will die and their bodies will disintegrate. The order originally present in the system will have become the disorder of individual atoms and molecules moving at random.*

Sunlight, oxygen, food

Carbon dioxide, waste products

Energy in from the surroundings → Open system: order maintained → Energy out to the surroundings

(a) Open system

No energy in from the surroundings

Closed system: order becomes disorder

(b) Closed system

(a)

(b)

(c)

7–5 *Although energy is dissipated in every conversion from one form to another, considerable work can be accomplished in the process. For example, the transformation of stored chemical energy to mechanical energy can be used by an organism to move into a position to obtain a meal—and thus a new supply of chemical energy. (a) A rainbow nudibranch (right), a type of mollusk, encounters a sea anemone (left) that has fully extended its tentacles from the tube that forms the base of its body. (b) The nudibranch slowly rears up and prepares to strike. When the nudibranch grasps the anemone's tentacles with its large jaws, the anemone rapidly retreats into its tube, pulling the nudibranch down with it (c). In this position, the nudibranch will consume the anemone at its leisure.*

Oxidation-Reduction

You will recall from Chapter 1 that electrons possess differing amounts of potential energy depending on their distance from the atomic nucleus and the attraction of the nucleus for electrons. An input of energy will boost an electron to a higher energy level, but without added energy an electron will remain at the lowest energy level available to it.

Chemical reactions are essentially energy transformations in which the potential energy stored in the chemical bonds of one substance is transferred to other, newly formed chemical bonds of a different substance. In such transfers, electrons shift from one energy level to another. In many reactions, electrons also pass from one atom or molecule to another. These reactions, which are of great importance in living systems, are known as oxidation-reduction (or redox) reactions.

The *loss* of an electron is known as **oxidation,** and the atom or molecule that loses the electron is said to be oxidized. The reason electron loss is called oxidation is that oxygen, which attracts electrons very strongly, is most often the electron acceptor.

Reduction is, conversely, the *gain* of an electron. Oxidation and reduction always take place simultaneously because an electron that is lost by the oxidized atom is accepted by another atom, which is itself reduced in the process.

Redox reactions may involve only a solitary electron, as when sodium loses an electron and becomes oxidized to Na^+, and chlorine gains an electron and is reduced to Cl^- (Figure 7–6a). Often, however, the electron travels with a proton, that is, as a hydrogen atom. In such cases, oxidation involves the removal of hydrogen atoms, and reduction the gain of hydrogen atoms (Figure 7–6b). For example, when glucose is oxidized, hydrogen atoms are lost by the glucose molecule and gained by oxygen:

$$C_6H_{12}O_6 + 6O_2 \longrightarrow 6CO_2 + 6H_2O + \text{Energy}$$

Glucose   Oxygen   Carbon   Water
dioxide

The electrons are moving to a lower energy level, and energy is released.

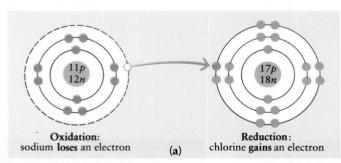

Oxidation:
sodium **loses** an electron
           (a)
Reduction:
chlorine **gains** an electron

Oxidation:
methane **loses**
two hydrogen atoms
$(2e^- + 2p^+ = 2H)$
   (b)
Reduction:
oxygen **gains**
two hydrogen atoms
$(2e^- + 2p^+ = 2H)$

7–6   (a) *In some oxidation-reduction reactions, such as the oxidation of sodium and the reduction of chlorine, a solitary electron is transferred from one atom to another. Such simple reactions typically involve elements or inorganic compounds.* (b) *In other oxidation-reduction reactions, such as the partial oxidation of methane (CH_4), the electrons are accompanied by protons. In these reactions, which often involve organic molecules, oxidation is the loss of hydrogen atoms, and reduction is the gain of hydrogen atoms. When an oxygen atom gains two hydrogen atoms, as shown here, the product is, of course, a water molecule.*

7–7 *The flow of energy through the biosphere. Chloroplasts, present in all photosynthetic eukaryotic cells, capture the radiant energy of sunlight and use it to convert water and carbon dioxide into carbohydrates, such as glucose, starch, and other foodstuff molecules. Oxygen is released as a product of the photosynthetic reactions.*

Mitochondria, present in all eukaryotic cells, carry out the final steps in the breakdown of these carbohydrates and capture their stored energy in ATP molecules. This process, cellular respiration, consumes oxygen and produces carbon dioxide and water, completing the cycling of the molecules.

With each transformation, some energy is dissipated to the environment in the form of heat. Thus the flow of energy through the biosphere is one-way. It can continue only so long as there is an input of energy from the sun.

Conversely, in the process of photosynthesis, hydrogen atoms are transferred from water to carbon dioxide, thereby reducing the carbon dioxide to form glucose:

$$6CO_2 + 6H_2O + Energy \longrightarrow C_6H_{12}O_6 + 6O_2$$

Carbon   Water   Glucose   Oxygen
dioxide

In this case, the electrons are moving to a higher energy level, and an energy input is required to make the reaction occur.

In living systems, the energy-capturing reactions (photosynthesis) and energy-releasing reactions (glycolysis and respiration) are oxidation-reduction reactions. The reduction of carbon dioxide to form a mole of glucose stores 686 kilocalories of energy in the chemical bonds of glucose. Conversely, the complete oxidation of a mole of glucose releases 686 kilocalories of energy.

If this energy were to be released all at once, most of it would be dissipated as heat. Not only would it be of no use to the cell, but the resulting high temperature would be lethal. However, mechanisms have evolved in living systems that regulate these chemical reactions—and a multitude of others—in such a way that energy is stored in particular chemical bonds from which it can be released in small amounts as the cell needs it. These mechanisms, which require only a few kinds of molecules, enable cells to use energy efficiently, without disrupting the delicate balances that characterize a living system. To understand how they work, we must look more closely at the proteins known as enzymes and at the molecule known as ATP.

Enzymes

Most chemical reactions require an initial input of energy to get started. This is true even for exergonic reactions such as the oxidation of glucose or the burning of natural gas. The added energy increases the kinetic energy of the molecules, enabling them to collide with enough force (1) to overcome the repulsion between the electrons surrounding one molecule and those surrounding the other molecule and (2) to break existing chemical bonds within the molecules, making possible the formation of new bonds. The energy that must be possessed by the molecules in order to react is known as the **energy of activation.**

In the laboratory, the energy of activation is usually supplied as heat. But, in a cell, many different reactions are going on at the same time, and heat would affect all of these reactions indiscriminately. Moreover, heat would break the hydrogen bonds that maintain the structure of many of the molecules within the cell and would have other generally destructive effects. Cells get around this problem by the use of **enzymes,** molecules that are specialized to serve as catalysts.

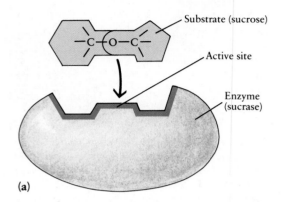

7–8   *In order to react, molecules must possess enough energy—the energy of activation—to collide with sufficient force to overcome their mutual repulsion and to break existing chemical bonds. An uncatalyzed reaction requires more activation energy than a catalyzed one, such as an enzymatic reaction. The lower activation energy in the presence of the catalyst is often within the range of energy possessed by the molecules, and so the reaction can occur at a rapid rate with little or no added energy. Note, however, that the overall energy change from the initial state to the final state is the same with and without the catalyst.*

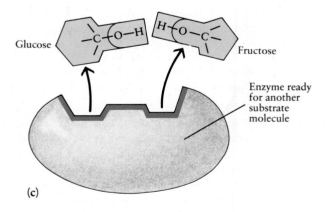

A **catalyst** is a substance that lowers the activation energy required for a reaction by forming a temporary association with the molecules that are reacting (Figure 7–8). This temporary association brings the reacting molecules close to one another and may also weaken existing chemical bonds, making it easier for new ones to form. As a result, little, if any, added energy is needed to start the reaction, and it goes more rapidly than it would in the absence of the catalyst. The catalyst itself is not permanently altered in the process, and so it can be used over and over again. (In Chinese, the word for "catalyst" is the same as the word for "marriage broker," and the functions are indeed similar.)

Because of enzymes, cells are able to carry out chemical reactions at great speed and at comparatively low temperatures. A single enzyme molecule may catalyze the reaction of tens of thousands of identical molecules in a second. Thus enzymes are typically effective in very small amounts.

7–9   *A model of enzyme action.* (**a, b**) *Sucrose, a disaccharide, is hydrolyzed to yield a molecule of glucose and a molecule of fructose* (**c**). *The enzyme involved in this reaction, sucrase, is specific for this process. As you can see, the active site of the enzyme fits the opposing surface of the sucrose molecule. The fit is so exact that a molecule composed, for example, of two subunits of glucose would not be affected by this enzyme.*

7–10 *A computer-generated model of the enzyme lysozyme. The blue lines represent the covalent bonds linking the atoms in the molecule to one another. The purple ribbon represents the backbone of the polypeptide chain, revealing the folding and bending that produce the tertiary structure. The substrate, a complex sugar, is shown in yellow, nestled into the active site.*

Lysozyme acts on the polysaccharides found in the cell walls of many types of bacteria. It was first detected by Sir Alexander Fleming, better known for his discovery of penicillin. Fleming was suffering from a bad cold, and a droplet fell from his nose onto a culture dish containing bacteria. The bacteria in the vicinity of the droplet were destroyed. Lysozyme was subsequently discovered in tears, saliva, milk, egg white, and other body fluids of many different animals.

Almost 2,000 different enzymes are now known, each of them capable of catalyzing a specific chemical reaction. No cell, however, contains all the known enzymes—different types of cells manufacture different types of enzymes. The particular enzymes that a cell manufactures are a major factor in determining the biological activities and functions of that cell. A cell can carry out a given chemical reaction at a reasonable rate only if it has a specific enzyme that can catalyze that reaction.

The molecule (or molecules) on which an enzyme acts is known as its **substrate.** For example, in the reaction diagrammed in Figure 7–9, sucrose is the substrate, and sucrase is the enzyme.

Enzyme Structure and Function

A few enzymes are RNA molecules. All other enzymes, however, are large, complex globular proteins consisting of one or more polypeptide chains. The polypeptide chains of an enzyme are folded in such a way that they form a groove or pocket on the surface (Figure 7–10). The substrate fits into this groove, which is the site of the reactions catalyzed by the enzyme. This portion of the molecule is known as the **active site.**

Only a few amino acids of the enzyme are involved in any particular active site. Some of these may be adjacent to one another in the primary structure, but often the amino acids of the active site are brought close to one another by the intricate folding of the polypeptide chain that produces the tertiary structure. In an enzyme with a quaternary structure, the amino acids of the active site may even be on different polypeptide chains (Figure 7–11).

7–11 *A model of the digestive enzyme chymotrypsin. This enzyme is composed of three polypeptide chains. The three-dimensional shape of the molecule is a result of a combination of disulfide bridges (yellow) and interactions among the chains and between the chains and the surrounding water molecules. These interactions are based on the polarity of the R groups of the various amino acids. As a result of the bending and twisting of the polypeptide chains, particular amino acids come together in a highly specific configuration to form the active site of the enzyme. Three amino acids known to be part of the active site are shown in blue.*

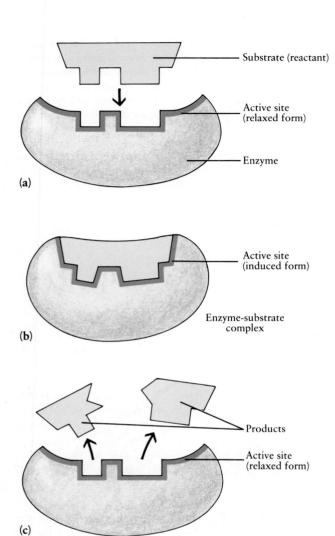

(a)
— Substrate (reactant)
— Active site (relaxed form)
— Enzyme

(b)
— Active site (induced form)
Enzyme-substrate complex

(c)
— Products
— Active site (relaxed form)

7–12 *The induced-fit hypothesis.* (a) *The active site is believed to be flexible and* (b) *to adjust its shape to that of the substrate. This induces a close fit between the active site and the substrate and may also put some strain on the substrate molecule, facilitating the reaction* (c).

Recent studies of enzyme structure have suggested that the active site is flexible. The binding between enzyme and substrate appears to alter the shape of the enzyme, thus inducing a close fit between the active site and the substrate (Figure 7–12). It is believed that this induced fit may put some strain on the reacting molecules and so further facilitate the reaction.

Cofactors in Enzyme Action

The catalytic activity of some enzymes appears to depend only on the interaction between the amino acids of the active site and the substrate. Many enzymes, however, require additional nonprotein substances in order to function. Such substances that are essential for enzyme activity are known as **cofactors.**

Certain ions are cofactors for particular enzymes. For example, the magnesium ion (Mg^{2+}) is required in all enzymatic reactions involving the transfer of a phosphate group from one molecule to another. The phosphate group is usually negatively charged in solution, and its two negative charges are attracted by the two positive charges of the magnesium ion. This attraction holds it in position at the active site of the enzyme.

Nonprotein organic molecules may also function as enzyme cofactors. Such molecules are called **coenzymes.** They are bound, either temporarily or permanently, to the enzyme, usually fairly close to the active site. Many vitamins—the compounds that humans and other animals cannot synthesize and so must obtain in their diets—are coenzymes or parts of coenzymes.

Some coenzymes function as electron acceptors in oxidation-reduction reactions, receiving electrons—often a pair of electrons accompanied by a hydrogen ion (that is, a proton)—and then passing them on to another molecule. There are several different kinds of electron-accepting coenzymes in any given cell, each able to hold electrons at a slightly different energy level. A coenzyme can accept electrons in one particular reaction, and then release them in another reaction. This enables the cell to capture energy efficiently as electrons move from high energy levels to lower ones.

Biochemical Pathways

Enzymes characteristically work in series, like workers on an assembly line. Each enzyme catalyzes one small step in an ordered series of reactions that together form a **biochemical pathway** (Figure 7–13). Different biochemical pathways serve different functions in the life of the cell. For example, one pathway may be involved in the breakdown of the polysaccharides in bacterial cell walls, another in the breakdown of glucose, and another in the synthesis of a particular amino acid.

Step 1   Step 2   Step 3

Enzyme 1   Enzyme 2   Enzyme 3

Initial
reactant: A

Intermediate
product: B

Intermediate
product: C

Final
product: D

7–13   *A schematic representation of a bio-chemical pathway. In order to produce the final product (D) from the initial reactant (A), a series of reactions is required. Each reaction is catalyzed by a different enzyme, and each results in a small but significant modification in the substrate molecule. If any step of the pathway is inhibited—either because of a nonfunctioning enzyme or because a substrate is unavailable—the pathway will shut down, and the subsequent reactions of the series will not occur.*

Cells derive several advantages from this sort of arrangement. First, the groups of enzymes making up a common pathway can be segregated within the cell. Some are found in solution, as in the lysosomes, whereas others are embedded in the membranes of particular organelles. The enzymes located in membranes appear to be lined up in sequence, so the product of one reaction moves directly to the adjacent enzyme for the next reaction of the series. A second advantage is that there is little accumulation of intermediate products, since each product tends to be used up in the next reaction along the pathway. A third advantage is a thermodynamic one. If any of the reactions along the pathway are highly exergonic (that is, energy-releasing), they will rapidly use up the products of the preceding reactions, pulling those reactions forward. Similarly, the accumulation of products from the exergonic reactions will push the subsequent reactions forward by increasing the concentrations of the reactants.

The Cell's Energy Currency: ATP Revisited

Glucose and other carbohydrates are storage forms of energy. They are also the forms in which energy is transferred from cell to cell and organism to organism. As we noted in Chapter 3, they are like money in the bank. **ATP** (adenosine triphosphate), however, is like the change in your pocket—it is the cell's immediately spendable energy currency. Almost every energy-requiring transaction in the cell involves ATP. For this reason, ATP is often referred to as the "universal currency" of the cell.

As you know, ATP is made up of the nitrogenous base adenine, the five-carbon sugar ribose, and three phosphate groups (Figure 7–14). The covalent bonds linking these three phosphates to each other, symbolized by a squiggle, ~, are the key to ATP's function.

Adenine

Phosphates

Ribose

7–14   *The structure of adenosine triphosphate (ATP), the cell's chief energy currency. ATP function depends on the bonds, shown here in blue, between the three phosphate groups in the molecule.*

These particular covalent bonds, often called "high-energy" bonds, are easily broken, releasing an amount of energy adequate to drive many of the essential reactions of the cell.

When one phosphate group is removed by hydrolysis, the ATP molecule becomes ADP, adenosine diphosphate (see Figure 3–28, page 67). About 7 kilocalories of energy per mole are released. If another phosphate group is removed by hydrolysis, the ADP molecule becomes AMP, adenosine monophosphate, and another 7 kilocalories of energy per mole are released. In most cellular reactions, however, only the first phosphate group is removed. The resulting ADP molecule can be "recharged" to ATP with the addition of a phosphate group and an energy input of 7 kilocalories per mole.

ATP in Action

Coupled Reactions

To understand the role of ATP in living systems, we must return briefly to the concept of the chemical bond. Because a chemical bond is a stable configuration of electrons, reacting molecules must possess a certain amount of energy before existing chemical bonds can be broken and new bonds formed. This is the energy of activation. Because of enzymes, which reduce the required energy of activation, the reactions essential to life are able to proceed at an adequate rate. However,

in biosynthetic reactions, in which larger, more complex molecules are synthesized from smaller, simpler molecules, additional energy is required.

In a biosynthetic reaction—for example, the formation of a disaccharide from two monosaccharide molecules—the electrons forming the chemical bonds of the product are at a higher energy level than the electrons in the bonds of the reactants. In other words, the potential energy of the product is greater than the potential energy of the reactants. Thus the reaction is endergonic—it can occur only with an input of energy. Cells get around this difficulty by means of **coupled reactions,** in which endergonic reactions are linked to exergonic reactions that provide a surplus of energy, enabling the overall process to proceed (Figure 7–15). The molecule that most frequently supplies the energy in coupled reactions is ATP.

Hydrolysis and Phosphorylation

In living cells, ATP is sometimes directly hydrolyzed to ADP plus phosphate, releasing energy for a variety of activities. ATP hydrolysis provides, for example, a means for producing heat, as in those animals, such as birds and mammals, that generally maintain a high and constant body temperature.

Enzymes catalyzing the hydrolysis of ATP are known as **ATPases.** A variety of different ATPases have been identified. The protein "arms" on the microtubules in cilia and flagella (page 103), for example, are ATPase

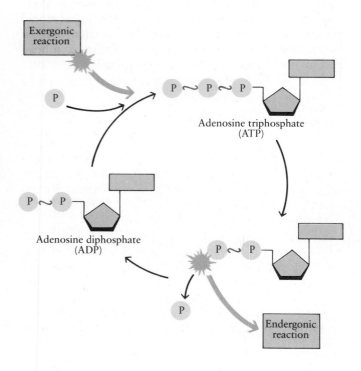

7–15   *In living systems, endergonic reactions, such as biosynthetic reactions, are powered by the energy released in exergonic reactions to which they are coupled. In most coupled reactions, ATP is the intermediate that carries energy from one reaction to the other.*

Exergonic reaction

P

Adenosine triphosphate
(ATP)

P ~ P

Adenosine diphosphate
(ADP)

P ~ P

P

Endergonic reaction

(a)

(b)

7–16 *Living organisms use the energy stored in the phosphate bonds of ATP for a variety of purposes.* (a) *In the skunk cabbage, a common plant in bogs and marshy areas of the northeastern United States, ATP is hydrolyzed to ADP, releasing energy in the form of heat. The plant produces enough heat to melt surrounding snow or ice, while maintaining a nearly constant internal temperature of about 22°C (72°F).* (b) *Flashlight fish contain specialized organs, located just below the eyes, that are filled with bioluminescent bacteria—as many as 10 billion cells in each milliliter of fluid. These bacteria transform ATP energy into light energy, thereby glowing in the dark. The French divers who discovered the first flashlight fish, with its blinking headlights, called it "le petit Peugeot." Flashlight fish live in reefs in the Indian Ocean and the Red Sea. These fish were photographed off the Comoro Islands, which are located at the northern end of the channel separating Madagascar from the African mainland.*

molecules, catalyzing the energy release that causes the microtubules to move past one another. Many of the proteins that move molecules and ions through cell membranes against a concentration gradient (page 117) are not only transport proteins but also ATPases, releasing energy to power the transport process.

Usually, however, the terminal phosphate group of ATP is not simply removed but is transferred to another molecule. This addition of a phosphate group is known as **phosphorylation**. Phosphorylation reactions transfer some of the energy of the phosphate group in the ATP molecule to the phosphorylated compound, which, thus energized, participates in a subsequent reaction.

Consider, for example, the formation of sucrose, a disaccharide synthesized in plants from the monosaccharides glucose and fructose. This process requires an input of energy to raise electrons to the higher energy level of the new chemical bond. The energy is supplied by the phosphorylation of the glucose and fructose molecules:

$$\text{ATP} + \text{Glucose} \longrightarrow \text{Glucose phosphate} + \text{ADP}$$

$$\text{ATP} + \text{Fructose} \longrightarrow \text{Fructose phosphate} + \text{ADP}$$

In each of these reactions, some of the energy made available by the conversion of ATP to ADP is conserved by the transfer of the terminal phosphate group of an ATP molecule to a sugar molecule. The glucose and fructose molecules thus become energized.

Next, glucose phosphate reacts with fructose phosphate to form sucrose:

Glucose phosphate + Fructose phosphate $\longrightarrow$
Sucrose + 2 Phosphates

In this second step, the phosphate groups are released from the glucose and fructose. Much of the energy made available by their release (energy originally derived from ATP) is used to form the bond between glucose and fructose, producing sucrose.

Where does the ATP come from? As we shall see in the next chapter, energy released in the cell's oxidation reactions, such as the breakdown of glucose, is used to "recharge" the ADP molecule to ATP by the addition of a phosphate group. Thus the ATP/ADP system serves as a universal energy-exchange system, shuttling between energy-releasing processes and energy-requiring ones.

Summary

Living systems convert energy from one form to another as they carry out essential functions of maintenance, growth, and reproduction. In these energy conversions, as in all others, some useful energy is lost to the surroundings at each step.

The laws of thermodynamics govern transformations of energy. The first law states that energy can be converted from one form to another but cannot be created or destroyed. The second law of thermodynamics states that in the course of energy conversions, the potential energy of the final state will always be less than the potential energy of the initial state. Stated differently, all natural processes tend to proceed in such a direction that the disorder, or randomness, of the universe increases. To maintain the organization on which life depends, living systems must have a constant supply of energy to overcome the tendency toward increasing disorder. The sun is the original source of this energy.

The energy transformations in living cells involve the movement of electrons from one energy level to another and, often, from one atom or molecule to another. Reactions in which electrons move from one atom to another are known as oxidation-reduction reactions. An atom or molecule that loses electrons is oxidized, and one that gains electrons is reduced.

Enzymes are the catalysts of biological reactions, lowering the energy of activation and thus enormously increasing the rate at which reactions take place. With a few exceptions, enzymes are large, globular protein molecules folded in such a way that particular groups of amino acids form an active site. The reacting molecule or molecules, known as the substrate, fit precisely into this active site. Although the shape of an enzyme may change temporarily in the course of a reaction, it is not permanently altered.

Many enzymes require cofactors, which may be simple ions or nonprotein organic molecules known as coenzymes. Coenzymes often serve as electron carriers, with different coenzymes holding electrons at slightly different energy levels.

Enzymes generally work in series. Each enzyme in the series catalyzes one specific reaction in the sequence of reactions that constitute a biochemical pathway. The stepwise reactions of biochemical pathways enable cells to carry out their chemical activities with remarkable efficiency in terms of both energy and materials.

ATP supplies the energy for most of the activities of the cell. The ATP molecule consists of the nitrogenous base adenine, the five-carbon sugar ribose, and three phosphate groups. The three phosphate groups are linked by two covalent bonds that are easily broken, each yielding about 7 kilocalories of energy per mole. Cells are able to carry out endergonic (energy-requiring) reactions by coupling them with exergonic (energy-yielding) reactions that provide a surplus of energy. Such coupled reactions usually involve ATP.

Questions

1. Distinguish among the following: the first law of thermodynamics/the second law of thermodynamics; exergonic reaction/endergonic reaction; oxidation/reduction; active site/substrate; ATP/ADP/AMP.

2. At present, at least four types of energy conversions are going on in your body. Name them.

3. The laws of thermodynamics apply only to closed systems, that is, to systems in which no energy is entering or leaving. Is a terrarium ordinarily a closed system? Could you convert it to one? A spaceship may or may not be a closed system, depending on certain features of its design. What would these features be? Is the Earth a closed system?

4. Explain why it is that living systems, despite appearances, are not in violation of the second law of thermodynamics.

5. What is there about the orderliness of a living organism that most significantly distinguishes it from the orderliness of a machine, such as a computer or the telephone system?

6. What is the basis for the specificity of enzyme action? What is the advantage to the cell of such specificity? What might be its disadvantages to the cell?

7. When a plant does not have an adequate supply of an essential mineral, such as magnesium, it is likely to become sickly and may die. When an animal is deprived of a particular vitamin in its diet, it too is likely to become ill and may die. What is a reasonable explanation of such phenomena?

8. Some human societies use a barter system for the exchange of goods and services. However, all complex societies have some form of monetary exchange. What are the advantages of a monetary exchange? Relate your answer to the ATP/ADP system.

9. In the photo at the right, why are there more plants than wildebeest and more wildebeest than cheetahs? (Explain in terms of thermodynamics.)

Suggestions for Further Reading

Alberts, Bruce, Dennis Bray, Julian Lewis, Martin Raff, Keith Roberts, and James D. Watson: *Molecular Biology of the Cell*, 3d ed., Garland Publishing, Inc., New York, 1994.

This outstanding cell biology text includes a clear, up-to-date discussion of our current understanding of the processes that occur in mitochondria and chloroplasts. The authors' discussion of the experimental procedures used to study these processes is especially helpful.

Darnell, James, Harvey F. Lodish, and David Baltimore: *Molecular Cell Biology*, 2d ed., W. H. Freeman and Company, New York, 1990.

A comprehensive treatment of modern cell biology, richly illustrated with diagrams and micrographs. Chapters 15 and 16 of this outstanding text are devoted to a thorough explication of our current understanding of the processes that occur in mitochondria and chloroplasts.

Dressler, David, and Huntington Potter: *Discovering Enzymes*, W. H. Freeman and Company, New York, 1990.

This beautifully illustrated volume from the Scientific American Library *provides both a history of the science of biochemistry and a clear presentation of our current knowledge of enzyme structure and function in living organisms.*

Karplus, Martin, and J. Andrew McCammon: "The Dynamics of Proteins," *Scientific American*, April 1986, pages 42–51.

Knowles, Jeremy R.: "Tinkering with Enzymes: What Are We Learning?" *Science*, vol. 236, pages 1252–1258, 1987.

Kornberg, Arthur: *For the Love of Enzymes: The Odyssey of a Biochemist*, Harvard University Press, Cambridge, Mass., 1991.*

Kornberg, who began his scientific career with studies of jaundice in World War II army recruits, went on to make major contributions to our understanding of enzymes. He is a wonderful writer, and this book is enriched by the clarity of his explanations, by his wise reflections on the scientific enterprise, and by his sense of humor.

Lehninger, Albert L., David L. Nelson, and Michael M. Cox: *Principles of Biochemistry*, 2d ed., Worth Publishers, Inc., New York, 1993.

Lehninger was one of the foremost experts on cellular energetics. This updated version of his classic introductory text is enriched by his vast experience and thorough understanding of the processes by which cells provide themselves with energy.

Mathews, Christopher K., and K. E. van Holde: *Biochemistry*, The Benjamin/Cummings Publishing Company, Redwood City, Calif., 1990.

A clearly written and beautifully illustrated textbook. Chapter 3 covers the principles of cellular energetics, and Chapters 10 and 11 are devoted to enzymes and the regulation of their activity.

Neurath, Hans: "Evolution of Proteolytic Enzymes," *Science*, vol. 224, pages 350–357, 1984.

Prescott, David M.: *Cells: Principles of Molecular Structure and Function*, Jones and Bartlett Publishers, Boston, 1988.

An up-to-date, yet concise, textbook of cell biology, written for a first course at the undergraduate level. Chapter 4 is devoted to energy flow and metabolism.

Scientific American: *Molecules to Living Cells*, W. H. Freeman and Company, New York, 1980.*

A collection of articles from Scientific American. *Chapters 5 and 6 are concerned with the structure and function of enzymes.*

Stryer, Lubert: *Biochemistry*, 3d ed., W. H. Freeman and Company, New York, 1988.

A good introduction to cellular energetics, handsomely illustrated.

Weiss, Rick: "Blazing Blossoms: Investigating the Metabolic Machinations of Heat-generating Plants," *Science News*, June 24, 1989, pages 392–394.

* Available in paperback.

How Cells Make ATP: Glycolysis and Respiration

ATP is the universal energy currency in living systems. In eukaryotes, such as this winning example, most of the body's ATP is generated in specialized organelles, the mitochondria. Muscle cells, like other cells with high energy requirements, abound in mitochondria. Powered by the ATP produced in their mitochondria, muscle cells are capable of contraction. It is their contractions that make possible the movements of walking, running, swimming, flying—and diving.

With the remarkable airborne performance shown here, Fu Mingxia of China won the 1992 Olympic gold medal in the women's 10-meter platform diving event. At the time, she was only thirteen years old.

In addition to providing the energy that made Fu Mingxia's dive possible, ATP was also simultaneously powering the many chemical activities taking place within the other cells of her body and providing the heat that kept her internal temperature well above that of the surrounding air and water.

The chemical energy stored in Fu Mingxia's ATP molecules came from the oxidation of glucose and other organic molecules. Ultimately, however, Fu Mingxia, like the rest of us, operates on solar power. And, evolution, somewhat like the Olympics, can be viewed as a contest in which the victors are those individuals who capture and use energy most efficiently.

In living organisms, the principal energy carrier is ATP. It participates in a great variety of cellular events, from the biosynthesis of organic molecules, to the flick of a cilium, the flashing of a firefly on a summer evening, the twitch of a muscle, or the active transport of a molecule across the plasma membrane. In this chapter, we shall show how a cell breaks down carbohydrates and captures a portion of the released energy in the terminal phosphate bonds of ATP. The process begins with the glucose molecule, the form in which carbohydrates usually enter the cell.

An Overview of Glucose Oxidation

The oxidation of glucose (and other carbohydrates) is complicated in detail, but simple in its overall design. **Oxidation,** as we saw in the last chapter, is the loss of electrons. **Reduction** is the gain of electrons. In the oxidation of glucose, the molecule is split apart, and the hydrogen atoms (that is, electrons and their accompanying protons) are removed from the carbon atoms and combined with oxygen, which is thereby reduced. The electrons go from higher energy levels to lower energy levels, and energy is released. The summary equation for this process is

Glucose + Oxygen $\longrightarrow$
$$\text{Carbon dioxide} + \text{Water} + \text{Energy}$$

Or,

$$C_6H_{12}O_6 + 6O_2 \longrightarrow$$
$$6CO_2 + 6H_2O + 686 \text{ kilocalories/mole}$$

Each time we inhale, we bring into our bodies the oxygen required for this process. Each time we exhale, we expel the carbon dioxide that is formed as a by-product.

About 40 percent of the energy released by the oxidation of glucose is used to convert ADP to ATP. As you will recall, about 75 percent of the energy in gasoline is "lost" as heat by an automobile engine, and only 25 percent is converted to useful forms of energy. The living cell is significantly more efficient. Its greater efficiency is largely due to the fact that the energy release occurs over the course of a series of reactions, each of which involves only a small change in energy.

The oxidation of glucose takes place in two major stages (Figure 8–1). The first is known as **glycolysis**—the lysis, or splitting, of glucose. The second is **respiration,** which, in turn, consists of two stages: the **Krebs cycle** and **terminal electron transport.** Glycolysis occurs in the cytosol, and the two stages of respiration take place within the mitochondria.

In glycolysis and the Krebs cycle, hydrogen atoms are removed from the carbon backbone of the glucose molecule and are passed to coenzymes that function as electron carriers. The first of these is nicotinamide adenine dinucleotide, abbreviated **NAD$^+$.** (As you can tell from its name, the molecule consists of two nucleotides, each containing a nitrogenous base, a sugar, and a phosphate group.) NAD$^+$ can accept a proton and two electrons, becoming reduced to **NADH.** The second molecule is flavin adenine dinucleotide, abbreviated **FAD.** FAD can accept two hydrogen atoms (that is, two protons and two electrons), being reduced to **FADH$_2$.** In glycolysis and the Krebs cycle, NAD$^+$ and FAD accept electrons and protons and are thus reduced.

In the final stage of respiration, NADH and FADH$_2$ give up their electrons to an **electron transport chain.**

8–1   *An overview of the oxidation of glucose. In glycolysis, glucose is split into pyruvic acid. A small amount of ATP is synthesized from ADP and phosphate, and a few electrons* (e$^-$) *and their accompanying protons (H$^+$) are transferred to coenzymes that function as electron carriers.*

In the presence of oxygen, the pyruvic acid is fed into the Krebs cycle. In the course of this cycle, additional ATP is synthesized and more electrons and protons are transferred to electron carriers. The electron carriers then transfer the electrons to an electron transport chain in which the electrons drop, step by step, to lower energy levels. As they do so, considerably more ATP is synthesized. At the end of the electron transport chain, the electrons reunite with protons and combine with oxygen to form water.

In the absence of oxygen, the pyruvic acid is converted to either lactic acid or ethanol. This process, known as fermentation, produces no ATP, but it does regenerate the electron carriers that are necessary for glycolysis to continue.

8–2   *Ancient Egyptian wall paintings, such as the one shown here, are the earliest historical record of wine-making. They have been dated to about 5,000 years ago. However, recently discovered pottery fragments, stained with wine, suggest that the Sumerians had mastered the art of wine-making at least 500 years before the Egyptians.*

The electrons are then passed "downhill" along a series of electron carriers. As the electrons drop, step by step, to lower energy levels, the energy released is used to power the formation of ATP from ADP and phosphate. When the electrons reach their lowest energy level, they combine with protons (H^+) and oxygen to form water.

In the presence of oxygen (an **aerobic** environment), the complete oxidation of a molecule of glucose produces about 38 molecules of ATP. These convenient packages of energy are then available to power the many activities of the cell.

In the absence of oxygen (an **anaerobic** environment), respiration cannot occur. Such a situation exists, for example, in muscle cells depleted of oxygen by strenuous exercise and in kegs containing yeast cells and fruit juices. Glycolysis, however, can occur, in conjunction with a process known as **fermentation.** The energy yield is only two molecules of ATP for each molecule of glucose, but this is enough to supply the immediate needs of muscle cells until the oxygen supply can be replenished. And, it is enough to keep yeast cells alive and well while their enzymes carry out the reactions that produce wine, appreciated by *Homo sapiens* (and perhaps other animals as well) since the dawn of history (Figure 8–2).

Now that you have the general outline in mind, let us look at the details of glycolysis, respiration, and fermentation.

Glycolysis

In glycolysis, the six-carbon glucose molecule is split into two molecules of a three-carbon compound known as **pyruvic acid** (Figure 8–3). Four hydrogen atoms (that is, four electrons and four protons) are removed from the glucose molecule in this process. The electrons

and two of the protons are accepted by NAD^+ molecules, while the other two protons remain in solution as hydrogen ions (H^+). Approximately 143 kilocalories are released per mole of glucose during glycolysis. The other 543 kilocalories remain stored in the bonds of the pyruvic acid molecules.

Glycolysis exemplifies the way the biochemical processes of a living cell proceed in small sequential steps. It takes place in a series of nine reactions, each catalyzed by a specific enzyme. Do *not* try to memorize the steps of glycolysis, but, as we describe each step (beginning at the top of the next page), look closely at Figure 8–4 to see how a series of small changes can add up to a major transformation. Notice especially how the carbon backbone of the glucose molecule is dismembered and its atoms rearranged step by step. Note also the formation of ATP from ADP and of NADH and H^+ from NAD^+. *ATP and NADH represent the cell's net energy harvest from glycolysis.*

8–3   *In glycolysis, the six-carbon glucose molecule is split into two molecules of a three-carbon compound known as pyruvic acid. In the course of glycolysis, four hydrogen atoms are removed from the original glucose molecule.*

8–4   *The steps of glycolysis.*

Step 1

Hexokinase

Step 2   Phosphoglucoisomerase

Step 3   Phosphofructokinase

Step 4

Aldolase

(Splits molecule)

Step 1. The first steps in glycolysis require an input of energy. This energy is supplied by coupling these steps to the ATP/ADP system. The terminal phosphate group is transferred from an ATP molecule to the carbon in the sixth position of the glucose molecule, to make glucose 6-phosphate. The reaction of ATP with glucose to produce glucose 6-phosphate and ADP is an exergonic reaction. Some of the energy it releases is conserved in the chemical bond linking the phosphate to the glucose molecule, which thus becomes energized. This reaction is catalyzed by a specific enzyme (hexokinase), and each of the reactions that follows is similarly catalyzed by a specific enzyme.

Step 2. The molecule is reorganized. The six-sided ring characteristic of glucose becomes the five-sided fructose ring. (As you know, glucose and fructose both have the same number of atoms—$C_6H_{12}O_6$—and differ only in the arrangements of these atoms.) This reaction is pushed forward by the accumulation of glucose 6-phosphate and the removal of fructose 6-phosphate as the latter enters Step 3.

Step 3. In this step, which is similar to Step 1, fructose 6-phosphate gains a second phosphate by the investment of another ATP. The added phosphate is bonded to the first carbon, producing fructose 1,6-bisphosphate, that is, fructose with phosphates in the 1 and 6 positions. Note that in the course of the reactions thus far, two molecules of ATP have been converted to ADP and no energy has been recovered.

Step 4. The six-carbon sugar molecule is split in half, producing two three-carbon molecules, dihydroxyacetone phosphate and glyceraldehyde phosphate. These two molecules can be converted from one to the other by the enzyme isomerase. However, because the glyc-

(Continued at top of next page)

eraldehyde phosphate is used up in subsequent reactions, all of the dihydroxyacetone phosphate is eventually converted to glyceraldehyde phosphate. Thus, *the products of all subsequent steps must be counted twice to account for the fate of one glucose molecule.* With the completion of Step 4, the preparatory reactions are complete.

Step 5. Glyceraldehyde phosphate molecules are oxidized—that is, hydrogen atoms with their electrons are removed—and NAD^+ is reduced to NADH and H^+ (a total of two molecules of NADH and two H^+ ions per molecule of glucose). This is the first reaction from which the cell harvests energy. Some of the energy from this oxidation reaction is also conserved in the attachment of a phosphate group to the 1 position of the molecule, forming 1,3-bisphosphoglycerate. The properties of this new bond are similar to those of the phosphate bonds of ATP, as indicated by the squiggle.

Step 6. The phosphate in the 1 position is released from the bisphosphoglycerate molecule and used to recharge a molecule of ADP (a total of two molecules of ATP per molecule of glucose). This is an exergonic reaction, and it pulls all the preceding reactions forward.

Step 7. The remaining phosphate group is enzymatically transferred from the 3 position to the 2 position.

Step 8. In this step, a molecule of water is removed from the three-carbon compound. The resulting internal rearrangement of the molecule concentrates energy in the vicinity of the phosphate group.

Step 9. The phosphate is transferred to a molecule of ADP, forming another molecule of ATP (again, a total of two molecules of ATP per molecule of glucose). This is also an exergonic reaction, and it pulls forward the preceding two reactions (Steps 7 and 8).

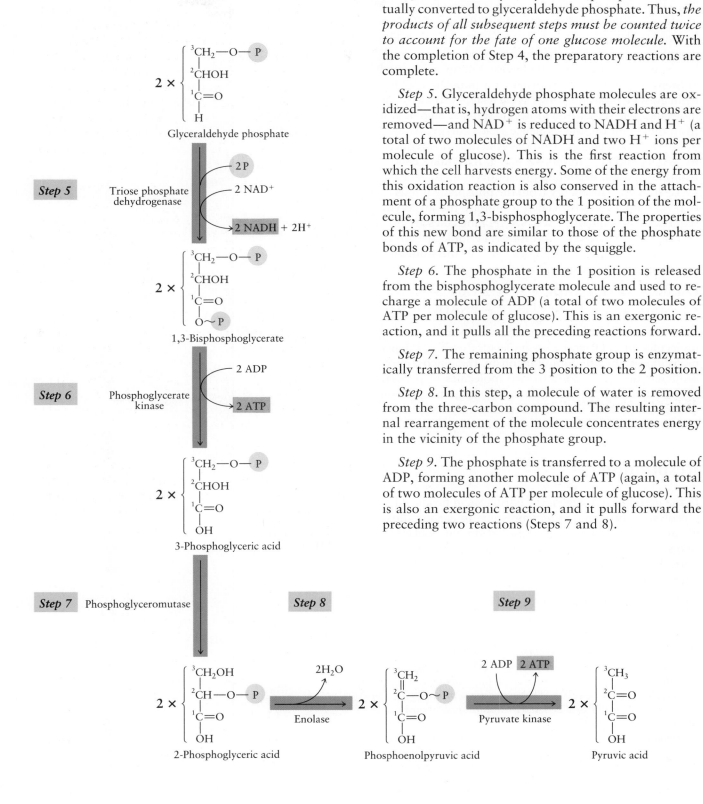

Summary of Glycolysis

The series of reactions we have just described is carried out by virtually all living cells—from the simplest prokaryote to the eukaryotic cells of our own bodies. The complete sequence, which is summarized in Figure 8–5, begins with one molecule of glucose. Energy is invested at Steps 1 and 3 by the transfer of a phosphate group from an ATP molecule—one at each step—to the sugar molecule. The six-carbon molecule splits at Step 4, and from this point on, the sequence yields energy. At Step 5, NAD^+ takes energy from the system and is reduced to NADH and H^+. At Steps 6 and 9, ADP takes energy from the system, becoming phosphorylated to ATP.

To sum up: The energy from the phosphate bonds of two ATP molecules is needed to initiate the glycolytic sequence. Subsequently, two NADH molecules are produced from two NAD^+ and four ATP molecules from four ADP plus four phosphates ("P_i" in the following equation—the "i" stands for "inorganic").

Glucose $+ 2ATP + 4ADP + 2P_i + 2NAD^+ \longrightarrow$
    2 Pyruvic acid $+ 2ADP + 4ATP + 2NADH + 2H^+$

Thus the net harvest—the energy recovered—is two molecules of ATP and two molecules of NADH per molecule of glucose. Simultaneously, the glucose molecule has been converted to two molecules of pyruvic acid. These molecules still contain a large proportion of the energy that was stored in the original glucose molecule.

Respiration

In the presence of oxygen, the pyruvic acid produced in glycolysis is broken down completely, to carbon dioxide and water. This process, respiration, takes place in two stages: the Krebs cycle and terminal electron transport. In eukaryotic cells, these reactions take place within the mitochondria.

8–5   *A summary of the two stages of glycolysis. The first stage requires an energy investment of 2 ATP. This stage ends with the splitting of the six-carbon sugar molecule into two three-carbon sugar molecules. The second stage produces an energy yield of 4 ATP and 2 NADH—a substantial return on the original investment.*

Compounds other than glucose, including glycogen, starch, various disaccharides, and a number of monosaccharides, can undergo glycolysis once they have been converted to glucose 6-phosphate or fructose 6-phosphate.

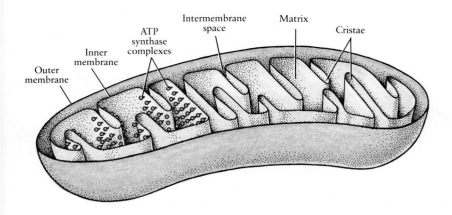

8–6   *A mitochondrion is surrounded by two membranes. The inner membrane folds inward to make a series of shelves, the cristae. Many of the enzymes and electron carriers involved in respiration are built into the membrane forming the cristae. Among these enzymes are the ATP synthase complexes, which, as we shall see, play a critical role in the formation of ATP in the final stage of respiration. The inner membrane encloses a dense solution, the matrix, which contains enzymes involved in the earlier stages of respiration, plus coenzymes, phosphates, and other solutes. The space between the inner membrane and the outer membrane (the intermembrane space) contains a solution of different composition.*

The Structure of the Mitochondrion

A mitochondrion, as we saw in Chapter 5, is surrounded by two membranes, each of which is a phospholipid bilayer. The outer membrane is smooth, but the inner membrane folds inward, forming a convoluted series of shelves known as **cristae** (Figure 8–6). Within the inner compartment of the mitochondrion, bathing the cristae, is a dense solution known as the **matrix**. The matrix contains enzymes, coenzymes, water, phosphates, and other molecules involved in respiration. Some of the enzymes of the Krebs cycle are in solution in the matrix. Other Krebs cycle enzymes and the enzymes and other components of the electron transport chain are built into the cristae.

The outer membrane of the mitochondrion is permeable to most small molecules, and the solution in the space between the inner and outer membranes is similar in composition to the cytosol. The inner membrane, however, permits the passage of only certain molecules, such as pyruvic acid, ADP, and ATP. It restrains the passage of other molecules and ions, including H^+ ions (protons). As we shall see, this selective permeability of the inner membrane is critical to the ability of the mitochondria to harness the power of respiration to the production of ATP.

A Preliminary Step:
The Oxidation of Pyruvic Acid

Pyruvic acid passes from the cytosol, where it is produced by glycolysis, to the matrix of the mitochondrion, crossing the outer and inner membranes. Before entering the Krebs cycle, the three-carbon pyruvic acid molecule is oxidized (Figure 8–7). The first carbon and its attached oxygen atoms are removed in the form of carbon dioxide, and a two-carbon acetyl group (CH_3CO) remains. In the course of this reaction, a molecule of NADH is formed from NAD^+. Because the oxidation

of one glucose molecule produces two pyruvic acid molecules, this step yields two NADH molecules for each glucose molecule. The original glucose molecule has now been oxidized to two CO_2 molecules and two acetyl groups.

Each acetyl group is momentarily accepted by a compound known as **coenzyme A.** Like many other coenzymes, it is a large molecule, a portion of which is a nucleotide. The combination of an acetyl group and coenzyme A is abbreviated as **acetyl CoA.** Its formation is the link between glycolysis and the Krebs cycle. Acetyl groups from sources other than the oxidation of pyruvic acid—for example, from the breakdown of fats and amino acids—can also combine with coenzyme A and enter the respiratory sequence at this point.

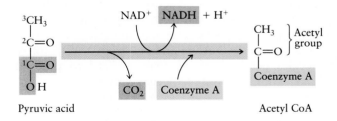

8–7   *The three-carbon pyruvic acid molecule is oxidized to the two-carbon acetyl group, which is combined with coenzyme A to form acetyl CoA. Following this reaction, acetyl CoA enters the Krebs cycle. The oxidation of the pyruvic acid molecule is coupled to the reduction of NAD^+. One molecule of NADH is produced for each molecule of pyruvic acid oxidized. Thus two molecules of NADH are produced for each glucose molecule that enters glycolysis.*

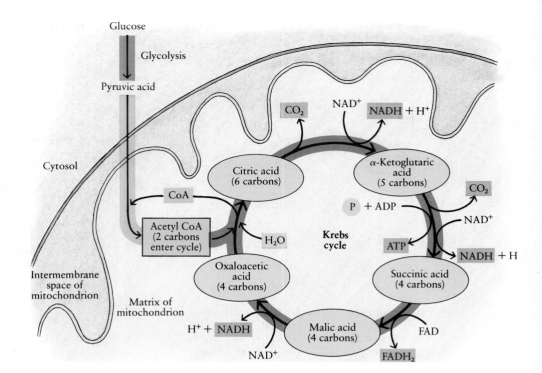

8–8   *The principal steps of the Krebs cycle. In the course of this reaction sequence, the carbons donated by the acetyl group of acetyl CoA are oxidized to carbon dioxide, and the hydrogen atoms are passed to the electron carriers NAD^+ and FAD. As in glycolysis, a specific enzyme catalyzes each step. Coenzyme A shuttles back and forth between the oxidation of pyruvic acid and the Krebs cycle, linking these two stages of respiration.*

One molecule of ATP, three molecules of $NADH$, and one molecule of $FADH_2$ represent the energy yield of one turn of the cycle. Two turns are required to complete the oxidation of one molecule of glucose.

This cycle is named after its discoverer, Sir Hans Krebs. It is also known as the citric acid cycle.

The Krebs Cycle

The Krebs cycle, like other biochemical pathways, consists of a series of reactions in which each step involves only a small change but the process as a whole produces major change. The principal steps of the Krebs cycle are shown in Figure 8–8.

The process begins when the two-carbon acetyl group is combined with a four-carbon compound (oxaloacetic acid) to form a six-carbon compound (citric acid). In the course of the Krebs cycle, two of the six carbons are removed and oxidized to carbon dioxide. And, a molecule of oxaloacetic acid is regenerated—making this series literally a cycle. Each turn around the cycle uses up one acetyl group and regenerates a molecule of oxaloacetic acid, which is then ready to begin the sequence again.

In the course of these reactions, some of the energy released by the oxidation of the carbon atoms is used to convert ADP to ATP (one molecule per cycle), some is used to produce NADH and H^+ from NAD^+ (three molecules per cycle), and some is used to produce $FADH_2$ from FAD (one molecule per cycle).

Oxaloacetic acid + Acetyl CoA + H_2O +

$$ADP + P_i + 3NAD^+ + FAD \longrightarrow$$

Oxaloacetic acid + $2CO_2$ + CoA +

$$ATP + 3NADH + 3H^+ + FADH_2$$

Two turns of the cycle are required to complete the oxidation of one molecule of glucose. Thus the total energy yield of the Krebs cycle for one glucose molecule is two molecules of ATP, six molecules of NADH, and two molecules of $FADH_2$.

No oxygen is required for the Krebs cycle. The electrons and protons removed in the oxidation of carbon are accepted by NAD^+ and FAD. However, oxygen *is* required for the next—and last—stage of respiration, which begins when NADH and $FADH_2$ give up their electrons and protons, regenerating NAD^+ and FAD. If oxygen is not available, the supply of NAD^+ and FAD is rapidly depleted and the Krebs cycle shuts down.

Terminal Electron Transport

The glucose molecule is now completely oxidized. Some of its energy has been used to produce ATP from ADP and phosphate. Most of its energy, however, remains in the electrons removed from the carbon atoms and passed to the electron carriers NAD^+ and FAD. These electrons—from glycolysis, the oxidation of pyruvic acid, and the Krebs cycle—are still at a high energy level.

In terminal electron transport, which is the final stage of respiration, these high-energy-level electrons are passed step-by-step to the low energy level of oxygen. This stepwise passage is made possible by the electron transport chain (Figure 8–9), a series of electron carriers, each of which holds the electrons at a slightly lower energy level. Among the principal components of the electron transport chain are molecules known as **cytochromes.** Although the structures of the individual cytochromes are similar, they differ enough to enable them to hold electrons at different energy levels.

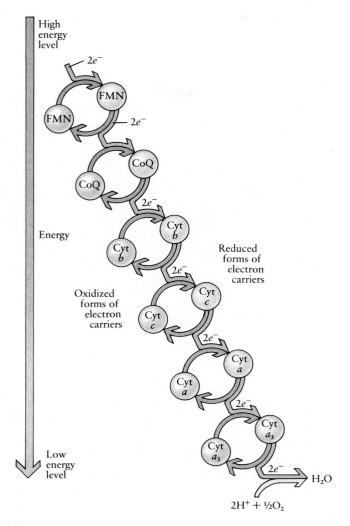

8-9   *A schematic representation of the electron transport chain. The molecules shown here—flavin mononucleotide (FMN), coenzyme Q (CoQ), and cytochromes b, c, a, and a₃—are the principal electron carriers of the chain. At least nine other molecules function as intermediates between these electron carriers.*

Electrons carried by NADH enter the chain when they are transferred to FMN, which is thus reduced (blue). Almost instantaneously, FMN passes the electrons on to CoQ. FMN returns to its oxidized form (gray), ready to receive another pair of electrons, and CoQ is reduced. CoQ then passes the electrons on to the next carrier and returns to its oxidized form, and so on down the line. As the electrons move down the chain, they drop to successively lower energy levels. The electrons are ultimately accepted by oxygen, which combines with protons (hydrogen ions) to form water.

Electrons carried by FADH₂ are at a slightly lower energy level than those carried by NADH. They enter the electron transport chain farther down the line at CoQ.

As the electrons move through the electron transport chain, dropping to lower energy levels, energy is released. This energy is harnessed by the mitochondrion to power the synthesis of ATP from ADP, in a process known as **oxidative phosphorylation.** Quantitative measurements show that for every two electrons that pass from NADH to oxygen, three molecules of ATP are formed from ADP and phosphate. For every two electrons that pass from $FADH_2$, which holds them at a slightly lower energy level than NADH, two molecules of ATP are formed.

With the synthesis of ATP in oxidative phosphorylation, the process that began with the glucose molecule is complete. However, before we sum up the total energy harvest, let us look more closely at how ATP is synthesized in conjunction with terminal electron transport.

The Mechanism of Oxidative Phosphorylation: Chemiosmotic Coupling

The mechanism of oxidative phosphorylation was, until the early 1960s, one of the most baffling puzzles in all of cell biology. As a result of the insight and experimental creativity of the British biochemist Peter Mitchell (1920–1992)—and the subsequent work of many other investigators—much of the puzzle has now been solved. Oxidative phosphorylation depends on a gradient of protons (H^+ ions) across the mitochondrial membrane and the subsequent use of the potential energy stored in that gradient to form ATP from ADP and phosphate.

As shown in Figure 8–10, the components of the electron transport chain are arranged sequentially in the inner membrane of the mitochondrion. Most of the electron carriers are tightly associated with proteins embedded in the membrane, forming three distinct complexes. According to current evidence, these complexes are locked into place in the membrane. Within each complex, the electron carriers are held in the proper positions in relation to one another.

The protein complexes also have another crucial function: they are proton pumps. As the electrons drop to lower energy levels during their transit through the electron transport chain, the released energy is used by the protein complexes to pump protons from the mitochondrial matrix into the intermembrane space. It is thought that for each pair of electrons moving down the electron transport chain from NADH to oxygen, 10 protons are pumped out of the matrix.

The inner membrane of the mitochondrion is, as we noted earlier, impermeable to protons. Thus, the protons that are pumped into the intermembrane space cannot easily move back across the membrane into the

8–10   *The arrangement of the components of the electron transport chain in the inner membrane of the mitochondrion. Three complex protein structures (here labeled I through III) are embedded in the membrane. They contain electron carriers and the enzymes required to catalyze the transfer of electrons from one carrier to the next. Complex I contains the electron carrier FMN and receives electrons from NADH. CoQ, which is located in the lipid interior of the membrane, ferries electrons from complex I to complex II, which contains cytochrome b. From complex II, the electrons move to cytochrome c, which is a peripheral membrane protein that shuttles back and forth between complexes II and III. The electrons then move through cytochromes a and a_3, located in complex III, back into the matrix where they combine with H^+ ions and oxygen, forming water.*

As the electrons make their way down the electron transport chain, protons (H^+ ions) are pumped through the three protein complexes from the matrix to the intermembrane space. This transfer of protons from one side of the membrane to the other establishes the proton gradient that powers the synthesis of ATP.

matrix. The result is a concentration gradient of protons across the inner membrane of the mitochondrion, with a much higher concentration of protons in the intermembrane space than in the matrix.

Like a boulder at the top of a hill or water at the top of a falls, the difference in the concentration of protons between the intermembrane space and the matrix represents potential energy. This potential energy results not only from the actual concentration difference (more hydrogen ions outside the matrix than inside) but also from the difference in electric charge (more + charges outside than inside). The potential energy is thus in the form of an **electrochemical gradient.** It is available to power any process that provides a channel allowing the protons to flow down the gradient back into the matrix.

Such a channel is provided by a large enzyme complex known as **ATP synthase** (Figure 8–11). This enzyme complex, which is embedded in the inner membrane of the mitochondrion, has binding sites for ATP and ADP. It also has an inner channel, or pore, through which protons can pass. When protons flow through this channel, moving down the electrochemical gradient from the intermembrane space back into the matrix, the energy released powers the synthesis of ATP from ADP and phosphate.

This mechanism of ATP synthesis, summarized in Figure 8–12, is known as **chemiosmotic coupling.** The term "chemiosmotic," which was coined by Peter Mitchell, reflects the fact that the production of ATP in

oxidative phosphorylation includes both chemical processes and transport processes across a selectively permeable membrane. As we have seen, two distinct events take place in chemiosmotic coupling: (1) a proton gradient is established across the inner membrane of the mitochondrion, and (2) potential energy stored in the gradient is used to generate ATP from ADP and phosphate.

Chemiosmotic power also has other uses in living systems. For example, it provides the power that drives the rotation of bacterial flagella. In photosynthetic cells, as we shall see in the next chapter, it is involved in the formation of ATP using energy supplied to electrons by the sun. And, it can be used to power other transport processes. In the mitochondrion, for example, the energy stored in the proton gradient is also used to carry other substances through the inner membrane. Both phosphate and pyruvic acid are carried into the matrix by membrane proteins that simultaneously transport protons down the gradient.

We noted earlier (page 145) that "about" 38 molecules of ATP are formed for each molecule of glucose oxidized to carbon dioxide and water. The exact amount of ATP formed depends on how the cell apportions the energy made available by the proton gradient. When more of this energy is used in other transport processes, less of it is available for ATP synthesis. The needs of the cell vary according to the circumstances, and so does the amount of ATP synthesized.

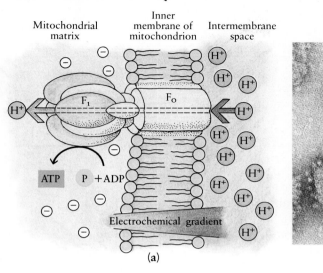

Mitochondrial matrix · Inner membrane of mitochondrion · Intermembrane space

F_1 · F_o · H⁺

ATP   P +ADP

Electrochemical gradient

(a)

(b)   0.1 μm

8–11   (a) *The ATP synthase complex. This enzyme complex consists of two major portions, F_o, which is contained within the inner membrane of the mitochondrion, and F_1, which extends into the matrix. Binding sites for both ATP and ADP are located on the F_1 portion, which consists of nine separate protein subunits. A channel, or pore, connecting the intermembrane space to the mitochondrial matrix, passes through the entire complex. When protons flow through this channel, moving down the electrochemical gradient, ATP is synthesized from ADP and phosphate.*

(b) The knobs protruding from the vesicles in this electron micrograph are the F_1 portions of ATP synthase complexes. The F_o portions to which they are attached are embedded in the membrane and are not visible. The vesicles were prepared by disrupting the inner mitochondrial membrane with ultrasonic waves. When the membrane is disrupted in this way, the fragments immediately reseal, forming closed vesicles. The vesicles are, however, inside-out. The outer surface here is the surface that faces the matrix in the intact mitochondrion.

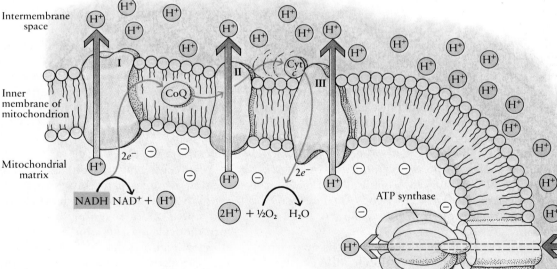

Intermembrane space

Inner membrane of mitochondrion

Mitochondrial matrix

I   CoQ   II   Cyt c   III

$2e^-$ $2e^-$

NADH   NAD⁺ + H⁺

2H⁺ + ½O₂   H₂O

ATP synthase

ATP   P +ADP

8–12   *A summary of the chemiosmotic synthesis of ATP in the mitochondrion. As electrons pass down the electron transport chain, which forms a part of the inner mitochondrial membrane, protons are pumped out of the mitochondrial matrix into the intermembrane space. This creates an electrochemical gradient. The subsequent movement of protons down the gradient as they pass through the ATP synthase complex provides the energy by which ATP is regener-* *ated from ADP and phosphate. Current evidence suggests that three protons flow through the ATP synthase complex for each molecule of ATP formed.*

It is estimated that the inner membrane of each mitochondrion in a liver cell has more than 10,000 copies of the electron transport chain and of the ATP synthase complex. Each mitochondrion in a heart muscle cell, which uses enormous quantities of ATP, is estimated to have more than 30,000 copies.

8–13   *A summary of the maximum energy yield from the complete oxidation of one molecule of glucose.*

		Molecules produced in:			
Process		Cytosol	Matrix of mitochondrion	Electron transport	
Glycolysis		2 ATP			2 ATP
		2 NADH		6 ATP	6 ATP
Respiration	Pyruvic acid to acetyl CoA		2 × (1 NADH)	2 × (3 ATP)	6 ATP
	Krebs cycle		2 × (1 ATP)		2 ATP
			2 × (3 NADH)	2 × (9 ATP)	18 ATP
			2 × (1 FADH$_2$)	2 × (2 ATP)	4 ATP

Total: **38 ATP**

Overall Energy Harvest

We are now in a position to see how much of the energy originally present in the glucose molecule has been recovered in the form of ATP. Because the proton gradient across the inner membrane of the mitochondrion can be used for purposes other than ATP synthesis, the amount of ATP formed varies. The numbers shown in Figure 8–13 represent the maximum energy harvest.

Glycolysis yields two molecules of ATP directly and two molecules of NADH. When oxygen is present, the electrons held by these NADH molecules are "shuttled across" the inner membrane of the mitochondrion and then enter the electron transport chain. In most cells, the energy cost of the shuttle is quite low, and each NADH formed in glycolysis ultimately results in the synthesis of three molecules of ATP. In these cells, the total gain from glycolysis is 8 ATP. In other cells, including those of brain, skeletal muscle, and insect flight muscle, the energy cost of the shuttle is higher. In such cells, the total gain from glycolysis is only 6 ATP.

The conversion of pyruvic acid to acetyl CoA, which occurs in the matrix of the mitochondrion, yields two molecules of NADH for each molecule of glucose. When the electrons held by these two NADH molecules pass down the electron transport chain, 6 ATP are produced.

The Krebs cycle, which also occurs in the matrix of the mitochondrion, yields two molecules of ATP, six of NADH, and two of FADH$_2$. The passage down the electron transport chain of the electrons held by these NADH and FADH$_2$ molecules yields 22 ATP. Thus, for each molecule of glucose, the total yield of the Krebs cycle is 24 ATP.

As Figure 8–13 shows, the complete yield from a single molecule of glucose is a maximum of 38 molecules of ATP. Note that all but 2 of the 38 molecules of ATP have come from reactions taking place in the mitochondrion. And, all but 4 result from the passage down the electron transport chain of electrons carried by NADH or FADH$_2$.

In the course of glycolysis and respiration, 686 kilocalories are released per mole of glucose. About 266 kilocalories (7 kilocalories per mole of ATP × 38 moles of ATP) have been captured in the phosphate bonds of ATP, an efficiency of almost 40 percent.

The ATP molecules, once formed, are exported across the inner membrane of the mitochondrion by a shuttle system that simultaneously brings in one molecule of ADP for each ATP exported.

Fermentation

The pathway of respiration, which we have just examined, is the principal pathway of energy metabolism for most cells in the presence of oxygen. In the absence of oxygen, however, fermentation occurs. The pyruvic acid formed in glycolysis is converted to ethanol (ethyl alcohol) or to one of several organic acids, of which lactic acid is the most common. The reaction product depends on the type of cell.

Yeast cells, which are often present as a "bloom" on the skin of grapes, can grow either with or without oxygen. When the sugar-filled juices of grapes are extracted and stored under anaerobic conditions, the yeast cells turn the fruit juice to wine by converting glucose into ethanol (Figure 8–14). When the sugar is exhausted, the yeast cells cease to function. At this point, the alcohol concentration is between 12 and 17 percent, depending on the variety of the grapes and the season at which they were harvested.

Lactic acid is formed from pyruvic acid by a variety of microorganisms and also by some animal cells when oxygen is scarce or absent (Figure 8–15). As we noted on page 145, it is produced in muscle cells during strenuous exercise, as when an athlete runs a sprint. We breathe hard when we run fast, thereby increasing the supply of oxygen, but even this increase may not be

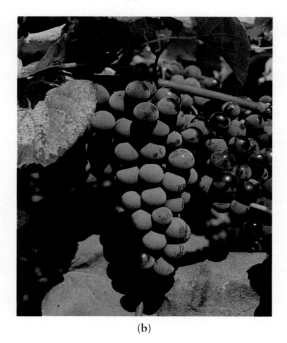

(a) Pyruvic acid (from glycolysis) → Acetaldehyde → Ethanol

(b)

(a) *The steps by which pyruvic acid, formed in glycolysis, is converted to ethanol (ethyl alcohol) in the absence of oxygen. In the first step, carbon dioxide is released. In the second, NADH is oxidized, and acetaldehyde is reduced. Most of the energy in the original glucose molecule remains in the alcohol molecule, which is the end product of the sequence. However, by regenerating NAD$^+$, these steps allow glycolysis to continue, with its small but sometimes vitally necessary yield of ATP.*

(b) *Yeast cells on the skins of these grapes give them their dustlike "bloom." When the grapes are crushed, the yeast cells mix with the juice. Storing the mixture under anaerobic conditions causes the yeast to break down the glucose in the grape juice to alcohol.*

Pyruvic acid (from glycolysis) → Lactic acid

(a)

(b)

enough to meet the immediate needs of the muscle cells. These cells, however, can continue to work by accumulating what is known as an **oxygen debt.** Glycolysis continues, using glucose released from glycogen stored in the muscle, but the resulting pyruvic acid is not oxidized to acetyl CoA. Instead, it is converted to lactic acid, which, as it accumulates, produces the sensations of muscle fatigue.

Later, when oxygen is more abundant (as a result of the deep breathing that follows strenuous exercise) and ATP demand is reduced, the lactic acid is converted back to pyruvic acid. The pyruvic acid is then used in the resynthesis of glucose or glycogen.

Why is pyruvic acid converted to lactic acid, only to be converted back again? The function of the initial conversion is simple: it uses NADH and regenerates the NAD$^+$ without which glycolysis cannot go forward (see Step 5, page 147). Even though the overall process seems to be wasteful in terms of energy consumption, the regeneration of NAD$^+$ may spell the difference between life and death when an animal "out of breath" needs one last burst of ATP to escape from a predator or to capture a prey.

(a) *The enzymatic reaction that produces lactic acid from pyruvic acid anaerobically in muscle cells. In the course of this reaction, NADH is oxidized and pyruvic acid is reduced. The NAD$^+$ molecules produced in this reaction and the one shown in Figure 8–14 are recycled in glycolysis. Without this recycling, glycolysis cannot proceed.*

(b) *Later, when ample oxygen is available and ATP demand is reduced, the reaction shown in (a) is reversed. The accumulated lactic acid is converted back to pyruvic acid, which is then used in the resynthesis of glucose and glycogen. This restores the chemical energy reserves of the muscle cells, which are then ready, once more, to meet whatever demands are placed upon them.*

Ethanol, NADH, and the Liver

The human body can dispose fairly readily of most toxic products of its own manufacture, such as carbon dioxide and the end products of nitrogen metabolism. In contrast, most ingested toxic substances, such as ethanol (beverage alcohol), must first be broken down by the liver, which possesses special enzymes not present in other tissues.

It has been known for many years that heavy drinkers are at great risk for severe, and often fatal, liver disease. Biochemical studies have demonstrated that the origin of the problem lies in the simple chemical steps involved in the breakdown of ethanol. Enzymes in the liver first oxidize ethanol (CH_3CH_2OH) to acetaldehyde (CH_3CHO), removing two hydrogen atoms and reducing a molecule of NAD^+. This is the reverse of the second reaction shown in Figure 8–14a. The acetaldehyde is then oxidized to acetic acid, which is, in turn, oxidized to carbon dioxide and water and eliminated from the body.

The intoxicating effects of alcohol are due mostly to the acetaldehyde, which stimulates the release of adrenalinelike agents. The chief culprits in the development of liver disease, however, are the hydrogen atoms (electrons and protons) removed from ethanol. These "extra" hydrogens—carried by NADH—follow two principal pathways within the cell.

Most of the hydrogen atoms are fed directly into the electron transport chain, producing water and ATP. Because of the high levels of NADH present in the cell from the oxidation of ethanol, the production of NADH by glycolysis and the Krebs cycle is reduced. As a result, sugars, amino acids, and fatty acids are not broken down but are instead converted to fats. The fats accumulate in the liver. The mitochondria also swell, presumably as a result of the distortion of their normal function—the electron transport chain is doing very heavy duty, while the Krebs cycle is effectively shut down.

Other hydrogen atoms are used in the synthesis of fatty acids from the sugars and amino acids that are not being processed in glycolysis and the Krebs cycle. More fats accumulate. It does not take long. In human volunteers fed a good high-protein, low-fat diet, six drinks (about 10 ounces) a day of 86 proof alcohol produced an eightfold increase in fat deposits in the liver in only 18 days. Fortunately, these early effects are completely reversible.

The liver cells work hard to get rid of the excess fats, which are not soluble in blood plasma. Before being released into the bloodstream, the fats are coated with a thin layer of protein in a process carried out on the membranes of the endoplasmic reticulum. The liver cells of heavy drinkers show enormous proliferation of the endoplasmic reticulum.

After a few years—depending on how much alcohol is consumed—liver cells, engorged with fat, begin to die, triggering the inflammatory process known as alcoholic hepatitis. Liver function becomes impaired. Cirrhosis is the next step. It is the formation of scar tissue, which interferes with the function of the individual cells and also with the supply of blood to the liver. This leads to the death of more cells. The liver can no longer carry out its normal activities—such as breaking down nitrogenous wastes—which is why cirrhosis is a cause of death. In fact, cirrhosis of the liver is the ninth leading cause of death in the United States for the population as a whole, and the seventh leading cause of death among those who are 25 to 44 years old.

Not so long ago, it was commonly believed

(a) 20 μm

(b) 20 μm

(a) *Normal cells from a human liver.*
(b) *Abnormal cells from the liver of a person with cirrhosis of the liver. Note the many globular fat droplets that have accumulated in the cells.*

that a good diet was all that was required to protect even a heavy drinker from the deleterious effects of alcohol. In fact, if one were just to add a few vitamins to the alcohol itself, some sophisticates maintained, most of the long-term physical damage of alcohol would disappear. This evidence refutes such comforting notions, and it comes at a time when alcohol is enjoying a resurgence of popularity among persons of high school and college age.

The fact that glycolysis does not require oxygen suggests that the glycolytic sequence evolved early, before free oxygen was present in the atmosphere. Primitive one-celled organisms presumably used glycolysis (or something very much like it) to extract energy from organic compounds they absorbed from their watery surroundings. Although anaerobic glycolysis generates only two molecules of ATP for each glucose molecule processed (a small fraction of the ATP that is generated when oxygen is present), it was and is adequate for the needs of many organisms.

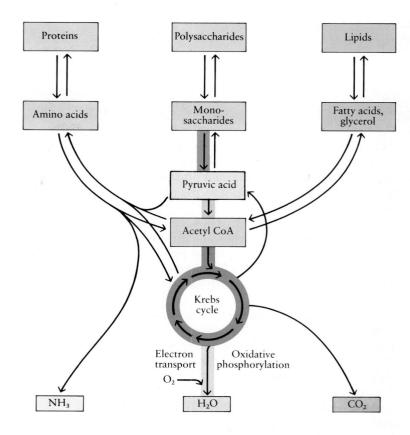

8–16 *An outline of the major pathways of catabolism and anabolism in the living cell. Catabolic pathways (arrows pointing down) are exergonic. A significant portion of the energy released in these pathways is, as we have seen, captured in the synthesis of ATP. Anabolic pathways (arrows pointing up) are endergonic. The energy that powers the reactions in these pathways is supplied primarily by ATP and NADH.*

The Strategy of Energy Metabolism

Most organisms do not feed directly on glucose. How do they extract energy from, for example, fats or proteins? The answer lies in the fact that the Krebs cycle is a Grand Central Station for energy metabolism. Other foodstuffs are broken down and converted into molecules that can feed into this central pathway.

Polysaccharides, such as starch, are broken down to their constituent monosaccharides and phosphorylated to glucose 6-phosphate or fructose 6-phosphate. In these forms, they enter the glycolytic pathway at Steps 2 and 3, respectively (see page 146).

Fats are first split into their glycerol and fatty acid components. Glycerol is subsequently broken down in a series of reactions that ultimately result in the transfer of electrons into the electron transport chain at CoQ. The fatty acids are chopped up into two-carbon fragments and slipped into the Krebs cycle as acetyl CoA.

Proteins are similarly broken down into their constituent amino acids, and the amino groups are removed. The residual carbon skeleton is converted either to an acetyl group, which enters the Krebs cycle as acetyl CoA, or to one of the larger compounds that are intermediates in glycolysis or in the Krebs cycle. If the amino groups are not reused, they are excreted as nitrogenous wastes.

These various pathways by which foods are broken down to yield energy are known collectively as **catabolism.** They are also central to the biosynthetic processes of life. These processes, known collectively as **anabolism,** are the pathways by which cells synthesize the diversity of molecules that constitute a living organism. Since many of these molecules, such as proteins and lipids, can be broken down and fed into the central pathway of glucose metabolism, you might guess that the reverse process can occur—namely, that the various intermediates of glycolysis and the Krebs cycle can serve as precursors for biosynthesis. This is indeed the case, as outlined in Figure 8–16.

For the reactions of the catabolic and anabolic pathways to occur, there must be a steady supply of organic molecules that can be broken down to yield energy and building-block molecules. Without a supply of such molecules, the metabolic pathways cease to function and the life of the organism ends. Heterotrophic cells (including the heterotrophic cells of plants, such as root cells) are dependent on external sources—specifically, autotrophic cells—for the organic molecules that are essential for life. Autotrophic cells, however, are able to synthesize monosaccharides from simple inorganic molecules and an external energy source. These monosaccharides are then used not only to supply energy but also as building-block molecules.

By far the most important autotrophic cells are the photosynthetic cells of algae and plants. In the next chapter, we shall examine how these cells capture the energy of sunlight and use it to synthesize the monosaccharide molecules on which life on this planet depends.

Summary

The oxidation of glucose is a chief source of energy in most cells. As the glucose is broken down in a series of enzymatic reactions, a significant portion of the energy in the molecule is repackaged in the phosphate bonds of ATP molecules.

The first phase in the breakdown of glucose is glycolysis, in which the six-carbon glucose molecule is split into two three-carbon molecules of pyruvic acid. A net yield of two molecules of ATP and two of NADH results from the process. Glycolysis takes place in the cytosol of the cell.

The second phase in the breakdown of glucose and other fuel molecules is respiration. It requires oxygen and, in eukaryotic cells, takes place in the mitochondria. It occurs in two stages: the Krebs cycle and terminal electron transport.

In the course of respiration, the three-carbon pyruvic acid molecules from glycolysis are broken down to two-carbon acetyl groups, which then enter the Krebs cycle. In a series of reactions in the Krebs cycle, the two-carbon acetyl group is oxidized completely to carbon dioxide. In the course of the oxidation of each acetyl group, four electron acceptors (three NAD^+ and one FAD) are reduced, and another molecule of ATP is formed.

The final stage of respiration is terminal electron transport, which involves a chain of electron carriers and enzymes embedded in the inner membrane of the mitochondrion. Along this series of electron carriers, the high-energy electrons carried by NADH from glycolysis and the oxidation of pyruvic acid and by NADH and $FADH_2$ from the Krebs cycle pass downhill to oxygen. During their passage down the electron transport chain, the energy released powers the pumping of protons (H^+ ions) out of the mitochondrial matrix. This creates a gradient of potential energy across the inner membrane of the mitochondrion. When protons pass through the ATP synthase complex as they flow down the gradient back into the matrix, the energy released powers the formation of ATP molecules from ADP and phosphate. This mechanism, by which oxidative phosphorylation is accomplished, is known as chemiosmotic coupling.

In the course of the breakdown of the glucose molecule, a maximum of 38 molecules of ATP can be formed, most of them from processes occurring in the mitochondrion.

In the absence of oxygen, the pyruvic acid produced by glycolysis is converted to either ethanol or lactic acid by the process of fermentation. NAD^+ is regenerated, allowing glycolysis to continue, producing a small but vital supply of ATP for the organism.

Organisms can also obtain energy from food molecules other than glucose. Polysaccharides, fats, and proteins are degraded to molecules that can enter the pathways of glucose metabolism at various steps. The biosynthesis of these substances also originates with molecules derived from intermediates in the respiratory sequence and is driven by the energy derived from those processes.

Questions

1. Distinguish among the following: oxidation of glucose/glycolysis/respiration/fermentation; $NADH^+$/NADH; FAD/$FADH_2$; Krebs cycle/electron transport; aerobic pathways/anaerobic pathways.

2. Sketch the structure of a mitochondrion. Describe where the various stages in the breakdown of glucose take place in relation to mitochondrial structure. What molecules and ions cross the mitochondrial membranes during these processes?

3. (a) As we have seen, a cell can obtain a maximum of 38 molecules of ATP from each molecule of glucose that is completely oxidized. Account for the production of each molecule of ATP. (b) In the course of glycolysis, the Krebs cycle, and electron transport, 40 molecules of ATP are actually formed. Why is the net yield for the cell only 38 molecules? (c) What other factors can reduce the yield of ATP?

4. Cyanide can combine with—and thereby deactivate—cytochrome a and cytochrome a_3. In our bodies, however, cyanide tends to react first with hemoglobin and to make it impossible for oxygen to bind to the hemoglobin. Either way, cyanide poisoning has the same effect: it inhibits the synthesis of ATP. Explain how this is so.

5. Certain chemicals function as "uncoupling" agents when they are added to respiring mitochondria. The passage of electrons down the electron transport chain to oxygen continues, but no ATP is formed. One of these agents, the antibiotic valinomycin, is known to transport K^+ ions through the inner membrane into the matrix. Another, 2,4-dinitro-

phenol, transports H^+ ions through the membrane. How do these substances prevent the formation of ATP?

6. In the cells of a specialized tissue known as brown fat, the inner membrane of the mitochondrion is permeable to H^+ ions. These cells contain large stores of fat molecules, which are gradually broken down to acetyl groups. These acetyl groups are fed into the Krebs cycle. The electrons captured by NADH and $FADH_2$ are, in turn, fed into the electron transport chain and ultimately accepted by oxygen. No ATP is synthesized, however. Why not? Brown fat tissue is found in some hibernating animals and in mammalian infants that are born hairless, including human infants. What do you suppose the function of brown fat tissue is?

7. If aerobic (oxygen-utilizing) organisms are so much more efficient than anaerobes in extracting energy from or-ganic molecules, why are there any anaerobes left on this planet? Why didn't they all become extinct long ago?

8. Describe the process of fermentation. What conditions are essential if it is to occur? With some strains of yeast, fermentation stops before the sugar is exhausted, usually at an alcohol concentration in excess of 12 percent. What is a plausible explanation?

9. Describe how the cell is adapted to the efficient use of a variety of foodstuffs, and to the efficient production of the variety of molecules that it needs to manufacture for its own use.

10. In terms of the cell's economy, what do catabolic processes provide for the cell? What do anabolic processes provide? How are they dependent on each other?

Suggestions for Further Reading

Alberts, Bruce, Dennis Bray, Julian Lewis, Martin Raff, Keith Roberts, and James D. Watson: *Molecular Biology of the Cell*, 3d ed., Garland Publishing, Inc., New York, 1994.

This outstanding cell biology text includes a clear, up-to-date discussion of our current understanding of the processes that occur in the mitochondria.

Darnell, James, Harvey F. Lodish, and David Baltimore: *Molecular Cell Biology*, 2d ed., W. H. Freeman and Company, New York, 1990.

A comprehensive treatment of modern cell biology, richly illustrated with diagrams and micrographs. Chapter 15 of this outstanding text is devoted to a thorough explication of our current understanding of glycolysis and respiration.

Dickerson, Richard E.: "Cytochrome *c* and the Evolution of Energy Metabolism," *Scientific American*, March 1980, pages 136–153.

Kolata, Gina: "How Do Proteins Find Mitochondria?" *Science*, vol. 228, pages 1517–1518, 1985.

Lane, M. Daniel, Peter L. Pedersen, and Albert S. Mildvan: "The Mitochondrion Updated," *Science*, vol. 234, pages 526–527, 1986.

Lehninger, Albert L., David L. Nelson, and Michael M. Cox: *Principles of Biochemistry*, 2d ed., Worth Publishers, Inc., New York, 1993.

Lehninger, who made the key discovery that oxidative phosphorylation occurs in the mitochondria, was one of the foremost experts on cellular energetics. This updated version of his classic introductory text is enriched by his vast experience and thorough understanding of the processes by which cells provide themselves with energy.

Newsholme, Eric, and Tony Leech: *The Runner: Energy and Endurance*, Fitness Books, Roosevelt, N.J., 1984. *

A lively introduction to the energetics of the working human body. Equally useful as a primer for runners who want to know more about their own physiology and as an introduction to the biochemistry of carbohydrate and fat metabolism.

Prescott, David M.: *Cells: Principles of Molecular Structure and Function*, Jones and Bartlett Publishers, Boston, 1988.

An up-to-date, yet concise, textbook of cell biology, written for a first course at the undergraduate level. Glycolysis and respiration are covered in Chapter 4.

Scientific American: *Molecules to Living Cells*, W. H. Freeman and Company, New York, 1980. *

A collection of articles from Scientific American. *Chapter 12 is an excellent presentation of the chemiosmotic synthesis of ATP in both mitochondria and chloroplasts.*

Stryer, Lubert: *Biochemistry*, 3d ed., W. H. Freeman and Company, New York, 1988.

Glycolysis, the Krebs cycle, and oxidative phosphorylation are covered in Chapters 15, 16, and 17 of this outstanding text.

Webb, A. Dinsmoor: "The Science of Making Wine," *American Scientist*, vol. 72, pages 360–367, 1984.

* Available in paperback.

Photosynthesis, Light, and Life

We are here on Earth—you and I—because of events taking place at this very moment in the cells of plants and algae. These events—known as photosynthesis—capture the radiant energy of the sun and use it to form carbohydrates and other organic molecules. These molecules, which are the building blocks of the tissues of plants and algae, are incorporated into the bodies of animals and other organisms that eat plants and algae. The molecules produced in photosynthesis also serve as the power source—the fuel—for the nonphotosynthetic inhabitants of our planet.

The three great staple foods for the human population are wheat, which had its origins in the temperate regions of the Old World; corn, which had its beginnings in the wild plants of the New World; and rice, found in the tropics and subtropics. Much of the world's rice crop—almost 500 million metric tons each year—is harvested by hand, as in these rice paddies on the island of Sumatra, in Indonesia.

Learning to plant, cultivate, and harvest crops such as rice was the most important event in the cultural evolution of *Homo sapiens.* It led to the end of the hunter-gatherer way of life for most of the human population. For better or worse, people began to have permanent dwelling places and to accumulate food, land, and other wealth. This transition led to a human situation that endures to this day—a large and still growing population, divided between rich and poor.

The first photosynthetic organisms probably appeared 3 to 3.5 billion years ago. Before the evolution of photosynthesis, the physical characteristics of Earth and its atmosphere were the most powerful forces shaping the course of natural selection. With the evolution of photosynthesis, however, organisms began to change the face of our planet and, as a consequence, to exert strong influences on each other. Organisms have continued to change the environment, at an ever-increasing rate, up to the present day.

As we saw earlier, the atmosphere in which the first cells evolved lacked free oxygen. These earliest organisms were, of course, adapted to living in that environment. In fact, free oxygen, with its powerful electron-attracting capacities, would have been poisonous to them (as it is to many modern anaerobes). Their energy came from anaerobic processes, most likely glycolysis and fermentation, which would have released carbon dioxide into the atmosphere. The organic molecules they used as fuel may have been formed by nonbiological processes (page 73), or they may have been produced by chemosynthetic autotrophs (page 75) or by primitive photosynthetic cells that, like some modern photosynthetic bacteria, did not release oxygen to the environment.

Then, it is hypothesized, there evolved photosynthetic organisms that used carbon dioxide as their carbon source and released oxygen, as do most modern photosynthetic forms. As these photosynthetic organisms multiplied, they provided a new supply of organic molecules, and free oxygen began to accumulate. In response to these changing conditions, cell species arose for which oxygen was not a poison but rather a requirement for existence.

As we saw in the last chapter, oxygen-utilizing organisms have an advantage over those that do not use oxygen. A higher yield of energy can be extracted per molecule from the aerobic breakdown of carbon-containing compounds than from anaerobic processes, in which fuel molecules are not completely oxidized.

9-1   *The bubbles on the leaves of this sprig of* Elodea, *a common pondweed and aquarium plant, are oxygen. As we shall see, the oxygen is released when water molecules are split apart in the first step of photosynthesis.*

Energy released in cells by reactions using oxygen made possible the development of increasingly active, increasingly complex organisms. Without oxygen, the complex forms of life that now exist on Earth could not have evolved.

Life on Earth continues to be dependent on photosynthesis both for its oxygen and for its carbon-containing fuel molecules. Photosynthetic organisms capture light energy and use it to form carbohydrates and free oxygen from carbon dioxide and water, in a complex series of reactions. The overall equation for photosynthesis can be summarized as:

Carbon dioxide + Water + Light energy $\longrightarrow$
Glucose + Oxygen

Or,

$$6CO_2 + 6H_2O + 686 \text{ kilocalories/mole} \longrightarrow$$
$$C_6H_{12}O_6 + 6O_2$$

To understand how organisms are able to capture light energy and convert it into stored chemical energy, we must look first at the characteristics of light itself.

The Nature of Light

Over 300 years ago, the English physicist Sir Isaac Newton (1642–1727) separated visible light into a spectrum of colors by passing it through a prism. Then by passing the light through a second prism, he recombined the colors, producing white light once again. By this experiment, Newton showed that white light is actually made up of a number of different colors, ranging from violet at one end of the spectrum to red at the other (Figure 9–2). Their separation is possible because light of different colors is bent at different angles in passing through the prism.

In the nineteenth century, through the genius of James Clerk Maxwell (1831–1879), it became known that what we experience as light is in truth a very small part of a vast continuous spectrum of radiation, the **electromagnetic spectrum.** All the radiations in this spectrum act as if they travel in waves. The **wavelengths**—that is, the distances from the crest of one wave to the crest of the next—range from those of gamma rays, which are measured in fractions of a nanometer (1 nanometer $= 10^{-9}$ meter), to those of low-frequency radio waves, which are measured in kilometers (1 kilometer $= 10^3$ meters). Radiation of each particular wavelength has a characteristic amount of energy associated with it. The longer the wavelength, the lower the energy; conversely, the shorter the wavelength, the higher the energy. Within the spectrum of visible light, violet light has the shortest wavelengths, and red light the longest. The shortest rays of violet light have almost twice the energy of the longest rays of red light.

The Fitness of Light

From the physicist's point of view, the difference between radiations we can see and radiations we cannot see—so dramatic to the human eye—is only a few nanometers of wavelength, or, expressed differently, a small amount of energy. Why does this particular group of radiations, rather than some other, make the leaves grow and the flowers burst forth, cause the mating of fireflies and palolo worms, and, when reflecting off the surface of the moon, excite the imagination of poets and lovers? Why is this tiny portion of the electromagnetic spectrum responsible for vision, for the rhythmic day-night regulation of many biological activities, for the bending of plants toward the light, and—most important of all—for photosynthesis, on which life depends? Is it an amazing coincidence that all these biological activities are dependent on these same wavelengths?

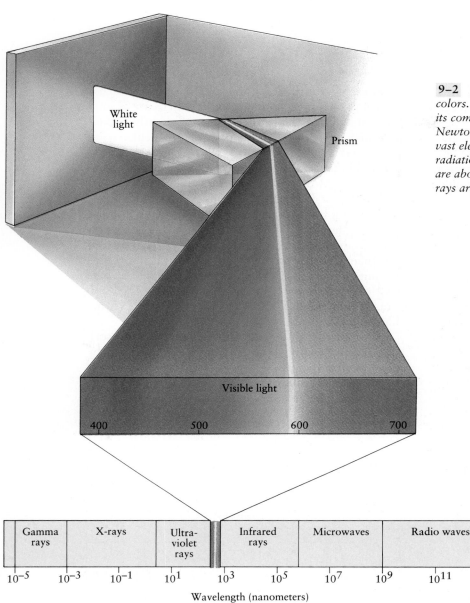

9–2 *White light is actually a mixture of light of different colors. When it is passed through a prism, it is separated into its components—"the celebrated phaenomena of colors," as Newton referred to it. Visible light is only a small portion of a vast electromagnetic spectrum. For the human eye, the visible radiations range from violet light, in which the shortest rays are about 380 nanometers, to red light, in which the longest rays are about 750 nanometers.*

George Wald of Harvard University, an expert on the subject of light and life, says no. He thinks that if life exists elsewhere in the universe, it is probably dependent on the same fragment of the electromagnetic spectrum. Wald bases this conjecture on two points. First, living things are composed of large, complicated molecules held in special configurations and relationships to one another by hydrogen bonds and other weak bonds. Radiation of even slightly higher energies than the energy of violet light breaks these bonds and so disrupts the structure and function of the molecules. DNA molecules, for example, are particularly vulnerable to such disruption. Radiations with wavelengths less than 200 nanometers—that is, with still higher energies— drive electrons out of atoms. On the other hand, light of wavelengths longer than those of the visible band—

that is, with less energy than red light—is absorbed by water, which makes up the great bulk of all living things on Earth. When this light is absorbed by molecules, its energy causes them to increase their motion (increasing heat), but it does not trigger changes in their electron configurations. Only those radiations within the range of visible light have the property of exciting molecules—that is, of moving electrons into higher energy levels—and so of producing chemical and, ultimately, biological change.

The second reason that the visible band of the electromagnetic spectrum has been "chosen" by living things is simply that it is what is available. The bulk of the radiation reaching the surface of the Earth from the sun is within this range. Most of the higher-energy wavelengths are screened out by oxygen and ozone in the atmosphere. Much infrared radiation is screened out by water vapor and carbon dioxide before it reaches the Earth's surface.

This is an example of what has been termed "the fitness of the environment." The suitability of the environment for life and that of life for the physical world are exquisitely interrelated. If they were not, life could not exist.

Chlorophyll and Other Pigments

In order for light energy to be used by living systems, it must first be absorbed. A substance that absorbs light is known as a **pigment.** Some pigments absorb all wavelengths of light and so appear black. Others absorb only certain wavelengths, transmitting or reflecting the wavelengths they do not absorb (Figure 9–3). For example, chlorophyll, the pigment that makes leaves green, absorbs light in the violet and blue wavelengths and also in the red. Because it reflects and transmits green light, it appears green.

Different groups of photosynthetic organisms use various pigments in photosynthesis. There are several different kinds of chlorophyll that vary slightly in their molecular structure. In photosynthetic eukaryotes (plants and algae), **chlorophyll *a*** is the pigment directly involved in the transformation of light energy to chemical energy. Most photosynthetic cells also contain a second type of chlorophyll—in plants and green algae, it is **chlorophyll *b***—and representatives of another group of pigments called the **carotenoids.** The carotenoids are red, orange, or yellow pigments. In the green leaf, their color is masked by the chlorophylls, which are more abundant. In some tissues, however, such as those of a ripe tomato, the carotenoid colors predominate, as they do also when leaf cells stop synthesizing chlorophyll in the fall.

Chlorophyll *b* and the carotenoids are able to absorb light at wavelengths different from those absorbed by chlorophyll *a* (Figure 9–4). They apparently can pass the energy on to chlorophyll *a*, thus extending the range of light available for photosynthesis.

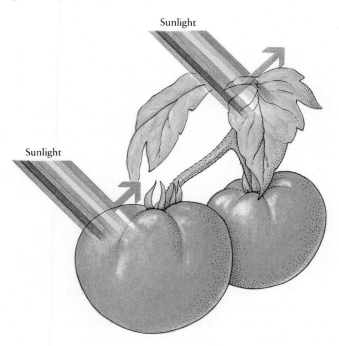

9–3 *When light strikes a pigmented object, some wavelengths are absorbed, whereas others are transmitted or reflected. The colors we perceive are the wavelengths of light that are transmitted and reflected. For example, a ripe tomato appears red because it reflects light in the red portion of the spectrum; light in all the other portions of the visible spectrum is absorbed. Similarly, the leaves of the tomato plant appear green because they reflect light in the green portion of the spectrum.*

9–4 *Different pigments absorb light energy of different wavelengths. The percentage of the light of each wavelength that is absorbed by a pigment can be measured and the results presented in the form of a graph. Combining the graphs for different pigments makes it easy to compare the absorption patterns. The three curves shown here reveal the absorption patterns of the principal pigments found in the chloroplast: chlorophyll a, chlorophyll b, and the carotenoids. The two forms of chlorophyll absorb violet, blue, orange, and red wavelengths of light. The carotenoids, by contrast, absorb violet, blue, and green wavelengths.*

When a pigment absorbs light, electrons within the pigment molecules are boosted to higher energy levels. In most cases, the electrons fall back to their original energy levels almost immediately. The energy released as they fall back may (1) be absorbed by a neighboring molecule, boosting its electrons to higher energy levels, or (2) be dissipated as heat, or (3) be re-emitted as light of a longer wavelength, a phenomenon known as fluorescence.

In other cases, however, the absorbed energy triggers a chemical reaction. The energy absorbed by the pigment drives electrons from the molecule, which is thus oxidized. These high-energy electrons are immediately accepted by another molecule, which is thus reduced. Whether a chemical reaction occurs depends not only on the structure of the particular pigment but also on its relationships with neighboring molecules. Chlorophyll can convert light energy to chemical energy—a process that begins with a simple oxidation-reduction reaction—only when it is associated with certain proteins and is embedded in a specialized membrane.

Chloroplasts

The specialized membranes in which chlorophyll and other pigments are embedded are known as **thylakoids**. They usually take the form of flattened sacs, or vesicles. In eukaryotes, the thylakoids form a part of the internal membrane structure of specialized organelles, the chloroplasts (Figure 9–5). The alga *Chlamydomonas*, for

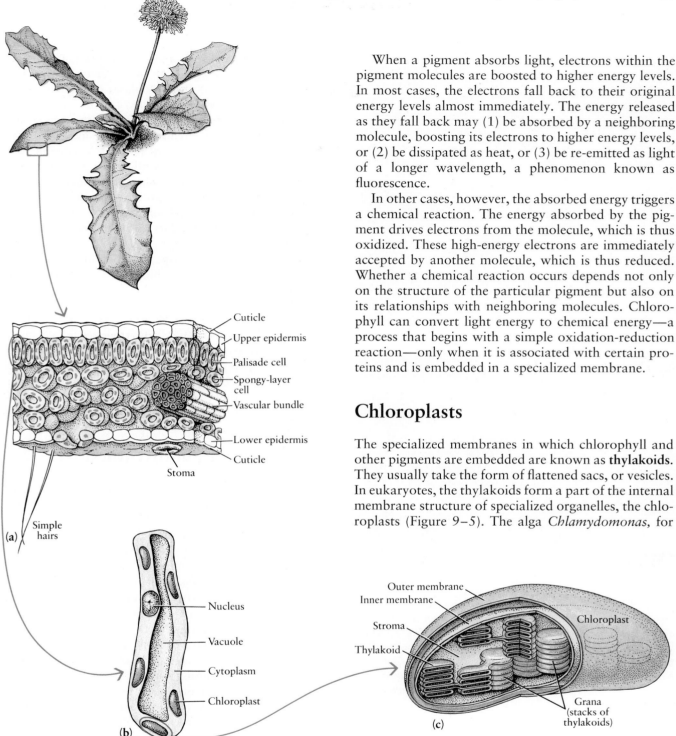

9–5 *Journey into the chloroplast of a dandelion leaf.* (**a**) *The inner tissues of the leaf are completely enclosed by transparent epidermal cells that are coated with a waxy layer, the cuticle. Oxygen, carbon dioxide, and other gases enter the leaf largely through special openings, the stomata (singular, stoma). These gases and water vapor fill the* spaces between cells in the spongy layer, leaving and entering cells by diffusion. Water, taken up by the roots, enters the leaf by way of the vascular bundles, and sugars, the products of photosynthesis, move out of the leaf by this same route, traveling to nonphotosynthetic parts of the plant. Much of the photosynthesis takes place in the palisade *cells* (**b**), *elongated cells directly beneath the upper epidermis. Palisade cells have a large central vacuole and numerous chloroplasts* (**c**) *that move within the cytosol, orienting themselves with respect to the light. Light is captured in the membranes of the disk-shaped thylakoids within the chloroplast.*

(a)   0.5 μm

(b)   1 μm

9–6   *The first stage of photosynthesis takes place in the thylakoid, a flattened sac, whose membranes contain chlorophyll and other pigments. In plants and algae, thylakoids are part of an elaborate membrane system enclosed in a special organelle, the chloroplast. (a) Stacks of thylakoids (grana) in a chloroplast from a leaf cell of a corn plant. The inner compartments of the thylakoids are interconnected, forming the thylakoid space, which contains a solution whose composition differs from that of the stroma and of the cytosol. The black particles are droplets of lipid, available for the synthesis of new membrane material. (b) A chloroplast, showing the elaborate system of internal membranes comprising interconnected stacks of thylakoids. This chloroplast is from the leaf of a soybean plant.*

instance, has a single very large chloroplast. The cell of a leaf characteristically has 40 to 50 chloroplasts, and there are often 500,000 chloroplasts per square millimeter of leaf surface.

Chloroplasts, like mitochondria, are surrounded by two membranes. The thylakoids, in the interior of the chloroplast, constitute a third membrane system. Surrounding the thylakoids, and filling the interior of the chloroplast, is a dense solution, the **stroma,** which is different in composition from the cytosol. The thylakoids enclose an additional compartment, known as the **thylakoid space,** which contains a solution of still different composition.

With the light microscope, it is possible to see little spots of green within the chloroplasts of leaves. The early microscopists called these green specks **grana** ("grains"), and this term is still in use. With the electron microscope, it can be seen that the grana are stacks of thylakoids (Figure 9–6). All the thylakoids in a chloroplast are oriented parallel to each other. Thus, as the chloroplast swings toward the light, all of its millions of pigment molecules can be simultaneously aimed for optimum reception, as if they were miniature electromagnetic antennae (which, indeed, they are).

The Stages of Photosynthesis

The reactions of photosynthesis take place in two stages (Figure 9–7). In the first stage—the **light-dependent reactions**—light strikes chlorophyll *a* molecules that are packed in a special way in the thylakoid membranes. Electrons from the chlorophyll *a* molecules are boosted to higher energy levels, and the chlorophyll *a* molecules are oxidized. In a stepwise series of reactions, the energy carried by these electrons is used to form ATP from ADP and to reduce a molecule known as $NADP^+$. $NADP^+$ closely resembles NAD^+, and it too is reduced by the addition of two electrons and a proton, forming **NADPH.** Water molecules are also broken apart in this stage of photosynthesis, supplying electrons that replace those boosted from the chlorophyll *a* molecules.

In the second stage of photosynthesis—the **light-independent reactions**—the ATP and NADPH formed in the first stage are used to reduce carbon dioxide to a simple sugar. Thus the chemical energy temporarily stored in ATP and NADPH molecules is transferred to molecules suitable for transport and storage in the algal cell or plant body. At the same time, a carbon skeleton is formed, from which other organic molecules can be built. This incorporation of carbon dioxide into organic compounds, known as **carbon fixation,** occurs in the stroma of the chloroplast.

Outer membrane
of chloroplast

Inner membrane
of chloroplast

Light
energy

P + ADP

ATP

CO$_2$

Calvin
cycle

H$^+$ + NADP$^+$

NADPH

2e$^-$

H$_2$O   2H$^+$ + ½O$_2$

Sugars

(a) Light-dependent reactions
    (thylakoid membrane)

(b) Light-independent reactions
    (stroma)

9–7   *An overview of photosynthesis. Photosynthesis takes place in two stages: the light-dependent reactions and the light-independent reactions. (a) In the light-dependent reactions, light energy absorbed by chlorophyll a molecules in the thylakoid membrane is used to power the synthesis of ATP. Simultaneously, in the interior of the thylakoid, water is split into oxygen gas and hydrogen atoms (electrons and protons). The electrons are ultimately accepted by $NADP^+ + H^+$, producing NADPH. (b) In the light-independent reactions, which occur in the stroma of the chloroplast, sugars are synthesized from carbon dioxide and the hydrogen carried by NADPH. This process is powered by the ATP and NADPH produced in the light-dependent reactions. As we shall see, it involves a series of reactions, known as the Calvin cycle, that are repeated over and over.*

The Light-Dependent Reactions

In the thylakoids, chlorophyll and other molecules are, according to the present model, packed into units called **photosystems.** Each unit contains from 250 to 400 molecules of pigment, which serve as light-trapping antennae. Once light energy is absorbed by one of the antenna pigments, it is bounced around (like a hot potato) among the other pigment molecules of the photosystem until it reaches a particular form of chlorophyll *a*, which is the reaction center of the photosystem.

Present evidence indicates that there are two different photosystems. In **Photosystem I,** the reactive chlorophyll *a* molecule is known as P_{700} (P is for pigment) because one of the peaks of its absorption curve is at 700 nanometers, a slightly longer wavelength than the usual chlorophyll *a* peak. P_{700} is not a different kind of chlorophyll but rather two chlorophyll *a* molecules that are bound together. Its unusual properties result from its association with specific proteins in the thylakoid membrane and its position in relation to other molecules. **Photosystem II** also contains a reactive chlorophyll *a* molecule, P_{680}.

Figure 9–8 shows how the two photosystems are thought to work together in photosynthesis. Light energy enters Photosystem II, where it is trapped by the reactive chlorophyll *a* molecule P_{680}, causing an electron from the P_{680} molecule to be boosted to a higher energy level. The electron is then transferred to a primary electron-acceptor molecule that holds electrons at a higher energy level than does chlorophyll *a*. Next, the electron passes downhill along an electron transport chain to Photosystem I. As electrons travel down this chain, the energy they release is used to pump protons from the stroma into the thylakoid space. A proton gradient is thus established across the thylakoid membrane. ATP synthase complexes, embedded in the thylakoid membrane, provide a channel through which protons can flow down the gradient, back into the stroma. As they do so, the potential energy of the gradient drives the synthesis of ATP from ADP. For each molecule of ATP synthesized, two electrons must travel down the electron transport chain from Photosystem II to Photosystem I.

Higher energy level

Photosystem II

Primary electron acceptor

$2e^-$

$2e^-$

$2e^-$

$2e^-$

Electron carriers

Proton gradient

Light energy

$2e^-$

Pigment molecules of antenna complex

Reactive chlorophyll *a* molecule (P_{680})

$2e^-$

H_2O $2H^+ + \frac{1}{2}O_2$

P + ADP ATP

Photosystem I

Primary electron acceptor

$2e^-$

$2e^-$

$2e^-$

Electron carriers

$NADP^+ + H^+$

NADPH

Light energy

$2e^-$

$2e^-$

Pigment molecules of antenna complex

Reactive chlorophyll *a* molecule (P_{700})

To light-independent reactions in stroma

Calvin cycle

9–8 *A schematic diagram of the light-dependent reactions of photosynthesis. In Photosystem II, light energy absorbed by pigment molecules of the antenna complex is transferred to the reactive chlorophyll a molecule (P_{680}). This energy boosts electrons from the P_{680} molecule to a primary electron acceptor at a higher energy level. The electrons are then passed from the primary electron acceptor down an electron transport chain to a lower energy level, the reactive chlorophyll a molecule (P_{700}) of Photosystem I. As the electrons travel down this transport chain, the energy they release is used to power the synthesis of ATP. To generate one molecule of ATP, two electrons must be boosted from Photosystem II and pass down the electron transport chain to Photosystem I. The electrons removed from Photosystem II are replaced by electrons released when water molecules are split into protons (H^+ ions) and oxygen.*

In Photosystem I, light energy boosts electrons from the P_{700} molecule to another primary electron acceptor. From this electron acceptor, they are passed via other carriers to $NADP^+$. Two electrons and one proton (H^+ ion) combine with one molecule of $NADP^+$ to form one molecule of NADPH. The additional H^+ ion released from each water molecule that was split by the reaction in Photosystem II remains in solution in the thylakoid space. The electrons removed from Photosystem I are replaced by those from Photosystem II.

ATP and NADPH represent the net gain from the light-dependent reactions. To generate one molecule of NADPH, two electrons must be boosted from Photosystem II and two from Photosystem I.

This chemiosmotic process, illustrated in Figure 9–9, is known as **photophosphorylation.** It is similar to the oxidative phosphorylation that occurs in the mitochondria (page 151).

During photophosphorylation, three other processes are taking place simultaneously:

1. The P_{680} chlorophyll molecule, having lost two electrons, is avidly seeking replacements. It finds them in a water molecule, which is stripped of two electrons and then broken into protons and oxygen.

2. Additional light energy is trapped in the reactive chlorophyll molecule (P_{700}) of Photosystem I. The molecule is oxidized, and electrons are boosted to a primary electron acceptor, from which they go downhill to $NADP^+$. Two electrons and one proton combine with $NADP^+$ to form NADPH.

3. The electrons removed from the P_{700} molecule of Photosystem I are replaced by the electrons that moved downhill from the primary electron acceptor of Photosystem II.

Thus in the light there is a continuous flow of electrons from water to Photosystem II to Photosystem I to $NADP^+$. In the words of the late Nobel laureate Albert Szent-Györgyi: "What drives life is . . . a little electric current, kept up by the sunshine."

9–9 *Embedded in the thylakoid membrane, or closely associated with it, are the molecules and molecular complexes that participate in the light-dependent reactions of photosynthesis. These include the pigments and proteins of Photosystems I and II, the carriers of their respective electron transport chains, and all necessary enzymes, including ATP synthase.*

The arrangement of these molecules in the thylakoid membrane makes possible the chemiosmotic synthesis of ATP in photophosphorylation. In this process, electrons from the reactive chlorophyll a molecule of Photosystem II are boosted to a high energy level by sunlight. As they flow down a series of electron carriers that lead to the reactive chlorophyll a molecule of Photosystem I, the energy they release is used to pump protons (H$^+$ ions) through one of the carrier complexes. The protons are pumped from the stroma into the thylakoid space. This creates an electrochemical gradient of potential energy. As the protons flow down this gradient through the ATP synthase complexes, moving from the thylakoid space back into the stroma, ADP is phosphorylated to ATP.

The chemical structures of the electron carriers and enzymes of the thylakoid membrane (including ATP synthase) are only slightly different from those of the mitochondrial membrane.

The Light-Independent Reactions

In the first stage of photosynthesis, light energy is converted to electrical energy—the flow of electrons—and the electrical energy is converted to chemical energy stored in the bonds of NADPH and ATP. In the second stage of photosynthesis, this energy is used to reduce carbon and synthesize simple sugars.

Carbon is available to photosynthetic cells in the form of carbon dioxide. Algal cells obtain dissolved carbon dioxide directly from the surrounding water. In plants, however, carbon dioxide reaches the photosynthetic cells through specialized openings, called **stomata** (Figure 9–10), that are found in leaves and green stems.

The reactions in the second stage of photosynthesis require molecules—ATP and NADPH—that are synthesized in the chloroplast only in the presence of light. However, once these molecules are available, the subsequent reactions can take place regardless of whether light is or is not present. Thus, these reactions are said to be "light-independent."

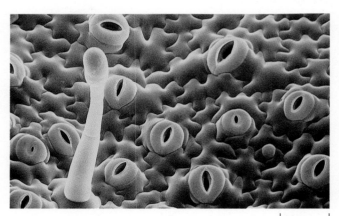

50 μm

9–10 *Scanning electron micrograph of stomata on the lower surface of a tobacco leaf. The plant can open or close its stomata, as needed; in this micrograph, most of them are open. It is through the stomata that the carbon dioxide required for photosynthesis diffuses into the interior of the leaf and the oxygen produced as a by-product of photosynthesis diffuses out. The structure extending from the surface at the lower left is a leaf hair.*

Ribulose
bisphosphate
(RuBP)

2 molecules of phosphoglyceric acid

The Calvin Cycle: The Three-Carbon Pathway

The reduction of carbon takes place in the stroma in a cyclical series of reactions named after its discoverer, Melvin Calvin. The **Calvin cycle** is analogous to the Krebs cycle (page 150) in that, in each turn of the cycle, the starting compound is regenerated. The starting (and ending) compound in the Calvin cycle is a five-carbon sugar with two phosphates attached, **ribulose bisphosphate (RuBP)**.

The cycle begins when carbon dioxide is bound to RuBP, which immediately splits to form two molecules of phosphoglyceric acid, or PGA (Figure 9–11). This reaction is catalyzed by a specific enzyme, **RuBP carboxylase** (known also as "rubisco"), which constitutes more than 15 percent of the chloroplast protein. In fact, RuBP carboxylase is thought to be the most abundant protein on Earth. Each of the PGA molecules produced in this initial reaction contains three carbon atoms; thus the Calvin cycle is also known as the three-carbon pathway.

The complete cycle is diagrammed in Figure 9–12. As in the Krebs cycle, each step is catalyzed by a specific enzyme. At each full turn of the cycle, one molecule of carbon dioxide enters the cycle and is reduced, and one molecule of RuBP is regenerated. Three turns of the cycle introduce three molecules of carbon dioxide, enough to produce one molecule of a three-carbon sugar. Six revolutions of the cycle, with the introduction of six molecules of carbon dioxide, are necessary to produce the equivalent of a six-carbon sugar, such as glucose. The overall equation is

$$6RuBP + 6CO_2 + 12NADPH + 12H^+ + 18ATP \longrightarrow$$
$$6RuBP + Glucose + 12NADP^+ +$$
$$18ADP + 18P_i + 6H_2O$$

The immediate product of the Calvin cycle is glyceraldehyde phosphate. This same three-carbon sugar-phosphate molecule is formed when the fructose bisphosphate molecule is split at the fourth step in glycolysis (page 146).

The Problem of Photorespiration

In the presence of ample carbon dioxide, the enzyme RuBP carboxylase feeds carbon dioxide into the Calvin cycle with great efficiency. However, when the carbon dioxide concentration in the leaf is low in relation to the oxygen concentration, this same enzyme catalyzes a reaction of RuBP with oxygen, rather than with carbon dioxide. This reaction is the first step in a process known as **photorespiration,** in which carbohydrates are oxidized to carbon dioxide and water in the presence of light. Unlike mitochondrial respiration, however, photorespiration is a wasteful process, yielding neither

9–11 *In order to trace the sequence of reactions in which carbon is reduced, Calvin and his collaborators briefly exposed photosynthesizing algae to carbon dioxide in which the carbon atom was radioactive (shown in red). The first product in which they could detect the radioactive carbon atom was phosphoglyceric acid (PGA). This indicated that when carbon dioxide binds to RuBP, the molecule immediately splits to form two molecules of PGA. This is the first step of the Calvin cycle.*

ATP nor NADH. In some plants, as much as 50 percent of the carbon fixed in photosynthesis may be reoxidized to carbon dioxide during photorespiration.

Conditions that can set the stage for photorespiration are quite common. Carbon dioxide is not continuously available to the photosynthesizing cells of a plant. As we have seen, it enters the leaf by way of the stomata, specialized pores that open and close, depending on, among other factors, water stress. When a plant is subjected to hot, dry conditions, it must close its stomata to conserve water. This cuts off the supply of carbon dioxide and also allows the oxygen produced by photosynthesis to accumulate. The resulting low carbon dioxide and high oxygen concentrations lead to photorespiration.

Also, when plants are growing in close proximity to one another, the air surrounding the leaves may be quite still, with little gas exchange between the immediate environment and the atmosphere as a whole. Under such conditions, the concentration of carbon dioxide in the air closest to the leaves may be rapidly reduced to low levels by the photosynthetic activities of the plant. Even if the stomata are open, the concentration gradient between the outside of the leaf and the inside is so slight that little carbon dioxide diffuses into the leaf. Meanwhile, oxygen accumulates, and photorespiration is likely to occur, greatly reducing the photosynthetic efficiency of the plants.

A Solution: The Four-Carbon Pathway

The problem of photorespiration is solved in some plants by an alternative pathway for the capture of carbon dioxide. In these plants, the first step in the fixation of carbon is the binding of carbon dioxide to a compound known as phosphoenolpyruvic acid (PEP), forming the four-carbon compound oxaloacetic acid (Figure

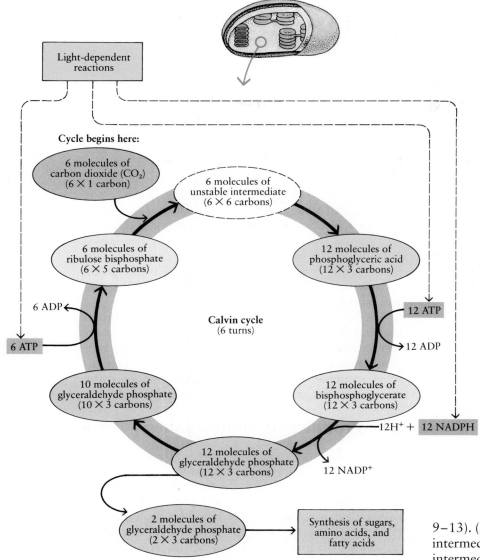

Light-dependent reactions

Cycle begins here:

6 molecules of carbon dioxide (CO_2) (6×1 carbon)

6 molecules of unstable intermediate (6×6 carbons)

6 molecules of ribulose bisphosphate (6×5 carbons)

12 molecules of phosphoglyceric acid (12×3 carbons)

6 ADP

6 ATP

12 ATP

12 ADP

Calvin cycle (6 turns)

10 molecules of glyceraldehyde phosphate (10×3 carbons)

12 molecules of bisphosphoglycerate (12×3 carbons)

$12H^+$ + 12 NADPH

12 molecules of glyceraldehyde phosphate (12×3 carbons)

12 $NADP^+$

2 molecules of glyceraldehyde phosphate (2×3 carbons)

Synthesis of sugars, amino acids, and fatty acids

9–12 *Summary of the Calvin cycle. At each full "turn" of the cycle, one molecule of carbon dioxide enters the cycle. Six turns are summarized here—the number required to make two molecules of glyceraldehyde phosphate, the equivalent of one molecule of a six-carbon sugar. The energy that drives the Calvin cycle is in the form of ATP and NADPH, produced by the light-dependent reactions in the first stage of photosynthesis.*

The cycle begins at the upper left. Six molecules of ribulose bisphosphate (RuBP), a five-carbon compound, are combined with six molecules of carbon dioxide. This produces six molecules of an unstable intermediate that splits apart immediately, yielding twelve molecules of phosphoglyceric acid (PGA), a three-carbon compound. In the next two steps, these molecules are reduced to twelve molecules of glyceraldehyde phosphate. Ten of these three-carbon molecules are combined and rearranged to form six five-carbon molecules of RuBP. The two "extra" molecules of glyceraldehyde phosphate represent the net gain from the Calvin cycle. Using these molecules as the starting point, the cell can synthesize a variety of sugars, amino acids, and fatty acids.

9–13). (Phosphoenolpyruvic acid, you may recall, is an intermediate in glycolysis, and oxaloacetic acid is an intermediate in the Krebs cycle.) Plants that utilize this pathway are commonly called **C_4 plants,** as distinct from the **C_3 plants,** in which carbon is bound first into the three-carbon compound phosphoglyceric acid (PGA).

PEP carboxylase, the enzyme that catalyzes the formation of oxaloacetic acid in C_4 plants, has a much higher affinity for carbon dioxide than does RuBP carboxylase, the enzyme that catalyzes the formation of PGA. Compared with RuBP carboxylase, PEP carboxylase fixes carbon dioxide faster and at lower levels, thereby keeping the carbon dioxide concentration lower within the cells near the surface of the leaf. This maximizes the gradient of carbon dioxide between the outside air and the leaf interior. Thus, when the stomata are open, carbon dioxide readily diffuses down the concentration gradient into the leaf. If the stomata must be closed much of the time—as they must be to conserve water in a hot, dry climate—the plant with C_4 metabolism will take up more carbon dioxide with each gasp (so to speak) than the plant that has only C_3 metabolism.

COOH
|
C—O—P
|
CH_2

Phosphoenol-pyruvic acid (PEP)

CO_2 H_2O

PEP carboxylase

COOH
|
C=O
|
CH_2
|
COOH

Oxaloacetic acid

COOH
|
CHOH
|
CH_2
|
COOH

Malic acid

9–13 *In C_4 plants, carbon dioxide is first combined with phosphoenolpyruvic acid (PEP) to yield the four-carbon oxaloacetic acid. This reaction is catalyzed by the enzyme PEP carboxylase. The oxaloacetic acid is then converted either to malic acid, as shown here, or to aspartic acid. Carbon dioxide is subsequently released from the malic acid (or aspartic acid) for use within the Calvin cycle. Sugarcane, corn, and sorghum are among the best-known C_4 plants.*

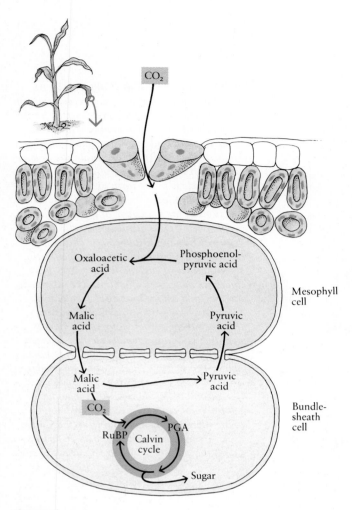

9–14 *A pathway for carbon fixation in C_4 plants. In cells in the leaf interior, known as mesophyll ("middle leaf") cells, CO_2 is initially fixed in oxaloacetic acid, which is then converted to malic acid. By way of the plasmodesmata, the malic acid passes to bundle-sheath cells, which form sheaths around the vascular bundles. Here the malic acid is broken down to pyruvic acid and CO_2, which enters the Calvin cycle. The pyruvic acid moves through the plasmodesmata into the mesophyll cells, where it is converted back to phosphoenolpyruvic acid (PEP), ready to receive another molecule of CO_2.*

The carbon dioxide bound into oxaloacetic acid by C_4 plants is ultimately released and then bound to RuBP, entering the Calvin cycle. This does not occur, however, until it has passed through a series of reactions (Figure 9–14) that transport it to other cells, known as **bundle-sheath cells**, that surround the vascular bundles of the leaf. In the bundle-sheath cells, the carbon dioxide is released at a high concentration, enabling RuBP carboxylase and the other enzymes of the Calvin cycle to function with maximum efficiency.

C_4 plants evolved primarily in the tropics and are especially well adapted to high light intensities, high temperatures, and dryness. The optimal temperature

range for C_4 photosynthesis is much higher than that for C_3 photosynthesis, and C_4 plants flourish at temperatures that would eventually be lethal to many C_3 species. Because of their more efficient use of carbon dioxide, C_4 plants can attain the same photosynthetic rate as C_3 plants but with smaller stomatal openings and, hence, with considerably less water loss.

A familiar example of the competitive capacity of C_4 plants is seen in lawns in the summertime. In most parts of the United States, lawns consist mainly of C_3 grasses such as Kentucky bluegrass. As the summer days become hotter and drier, these dark green, fine-leaved grasses are often overwhelmed by rapidly growing crabgrass, which disfigures the lawn as its yellowish-green, broader-leaved plants slowly take over. Crabgrass, you will not be surprised to hear, is a C_4 plant.

The list of plants known to utilize the four-carbon pathway has grown to over 1,000 species. Current evidence indicates that this pathway has arisen independently many times in the course of evolution. It is another example of the exquisite adaptation of living systems to their environment.

The Products of Photosynthesis

Glyceraldehyde phosphate, the three-carbon sugar produced by the Calvin cycle, may seem an insignificant reward, both for all the enzymatic activity on the part of the cell and for the intellectual effort required to understand photosynthesis. However, this molecule and those derived from it provide (1) the energy source for virtually all living systems, and (2) the basic carbon skeleton from which the great diversity of organic molecules can be synthesized. Carbon has been fixed—that is, it has been brought from the inorganic world into the organic one.

Molecules of glyceraldehyde phosphate may flow into a variety of different metabolic pathways, depending on the activities and requirements of the cell. Often they are built up to glucose or fructose, following a sequence that is in many of its steps the reverse of the glycolytic sequence described in the previous chapter. Plant cells use these six-carbon sugars to make starch and cellulose for their own purposes and sucrose for export to other parts of the plant body. Animal cells store them as glycogen. All cells use sugars, including glyceraldehyde phosphate and glucose, as the starting point for the manufacture of other carbohydrates, fats and other lipids, and, with the addition of nitrogen, amino acids and nitrogenous bases. Finally, as we saw in the preceding chapter, the carbon fixed in photosynthesis is the source of ATP energy for heterotrophic organisms and for the heterotrophic cells of plants.

The Carbon Cycle

In photosynthesis, living systems incorporate carbon dioxide from the atmosphere into organic compounds. In respiration, organic compounds are broken down again into carbon dioxide and water. These processes, viewed on a worldwide scale, result in the carbon cycle.

The principal photosynthesizers in the carbon cycle are plants and microscopic marine algae. These organisms synthesize carbohydrates from carbon dioxide and water and release oxygen into the atmosphere. About 100 billion metric tons of carbon per year are bound into carbon compounds by photosynthetic organisms.

Some of the carbohydrates are used by the photosynthesizers themselves in glycolysis and respiration. Plants release carbon dioxide from their roots and leaves, and marine algae release it into the water. The carbon dioxide dissolved in the waters of the Earth is in equilibrium with the carbon dioxide in the atmosphere. Some 500 billion metric tons of carbon are "stored" as dissolved carbon dioxide in surface waters, and some 740 billion metric tons in the atmosphere.

Some of the carbohydrates produced in photosynthesis are oxidized in the cells of animals that feed on the living plants, on algae, and on one another, releasing carbon dioxide. An enormous amount of carbon, however, is contained in the dead bodies of plants and other organisms plus discarded leaves and shells, feces, and other waste materials that settle into the soil or sink to the ocean floors where they are consumed by small invertebrates, bacteria, and fungi. The metabolic processes of these organisms also release carbon dioxide into the reservoir of the atmosphere and the oceans. Another, even larger store of carbon lies below the surface of the Earth in the form of coal and oil, deposited there some 300 million years ago.

The natural processes of photosynthesis and respiration generally balance one another out. Over the long span of geologic time, the carbon dioxide concentration of the atmosphere has varied, but for the last 10,000 years it has remained relatively constant. By volume, it is a very small proportion of the atmosphere, only about 0.03 percent. It is important, however, because carbon dioxide, unlike most other components of the atmosphere, acts as a blanket, trapping the heat produced when sunlight is absorbed at the Earth's surface. Since 1850, carbon dioxide concentrations in the atmosphere have been increasing, owing in large part to our use of fossil fuels, to our plowing of the soil, and to our destruction of forests, particularly in the tropics.

At the beginning of the 1980s, a major study by the Environmental Protection Agency predicted that this increase in the carbon dioxide "blanket" would significantly increase the average temperatures here on Earth, be-ginning by the turn of the century. A number of subsequent studies, coupled with worldwide temperatures in the 1980s that were higher than in any decade since records have been kept, have now convinced not only members of the scientific community but also political leaders throughout the world that the increase is real and has already begun.

The consequences of this increase cannot be known with certainty. In some parts of the world, there may be lengthened growing seasons, increased precipitation, and, in conjunction with the increased levels of carbon dioxide available to plants, greater agricultural productivity. In other parts of the world, however, it is thought that precipitation will be reduced, lowering crop yields and, in already arid areas, accelerating the spread of the great deserts of the world. Rises in sea level, resulting from the melting of polar ice, pose a potential threat not only to human inhabitants of coastal regions but also to the multitude of marine organisms that live or reproduce in shallow waters at the continents' edges.

Although it now seems apparent that a certain amount of global warming is the inevitable consequence of past and present human activities, national and international efforts are currently underway to develop strategies for agriculture, energy use, and manufacturing that will slow—and perhaps eventually reverse—the process. It remains to be seen if we have awakened in time.

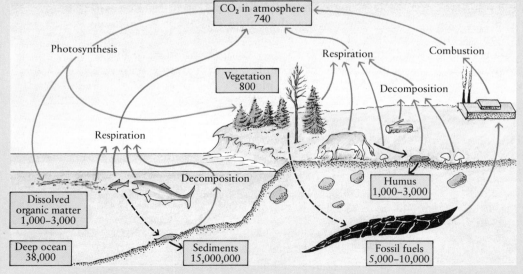

The carbon cycle. The arrows indicate the movement of carbon atoms. The numbers are all estimates of the amount of carbon, expressed in billions of metric tons, stored in various reservoirs as of 1988. The amount of carbon released by respiration and combustion has begun to exceed the amount fixed by photosynthesis. The amount of carbon stored in the atmosphere is currently increasing by 3 billion metric tons each year.

Summary

In photosynthesis, light energy is converted to chemical energy, and carbon is fixed into organic compounds.

Light energy is captured by the living world by means of pigments. The pigments involved in photosynthesis in eukaryotes include the chlorophylls and the carotenoids. Light absorbed by antenna pigments boosts their electrons to higher energy levels. Because of the way the pigments are packed into membranes, they are able to transfer this energy to reactive chlorophyll a molecules.

In eukaryotes, photosynthesis takes place within organelles known as chloroplasts, which are surrounded by two membranes. Contained within the membranes of the chloroplast are a solution of organic compounds and ions known as the stroma and a complex internal membrane system that forms sacs called thylakoids. The pigments and other molecules responsible for capturing light are located in and on the thylakoid membranes.

Photosynthesis takes place in two stages: (1) the light-dependent reactions, in which light energy trapped by chlorophyll is converted to the chemical energy of ATP and NADPH; and (2) the light-independent reactions, in which carbon dioxide is reduced and carbohydrates formed.

In the currently accepted model of the light-dependent reactions, light energy strikes antenna pigments of Photosystem II. Electrons are boosted uphill from the reactive chlorophyll a molecule P_{680} to a primary electron acceptor. As the electrons are removed, they are replaced by electrons from water molecules, with the simultaneous production of free oxygen and protons (H^+ ions). The electrons then pass downhill to Photosystem I along an electron transport chain. This passage generates a proton gradient that drives the synthesis of ATP from ADP (photophosphorylation). Light energy absorbed in antenna pigments of Photosystem I and passed to chlorophyll P_{700} results in the boosting of electrons to another primary electron acceptor. The electrons removed from P_{700} are replaced by the electrons from Photosystem II. The electrons are ultimately accepted by the electron carrier $NADP^+$. The energy yield from this series of reactions is contained in molecules of NADPH and ATP.

In the light-independent reactions, which take place in the stroma, NADPH and ATP produced in the light-dependent reactions are used to reduce carbon dioxide to organic carbon. This is accomplished by means of the Calvin cycle. In the Calvin cycle, a molecule of carbon dioxide is combined with the starting material, a five-carbon sugar called ribulose bisphosphate (RuBP). At each turn of the cycle, one carbon atom enters the cycle. Three turns of the cycle produce a three-carbon molecule, glyceraldehyde phosphate. Two molecules of glyceraldehyde phosphate (six turns of the cycle) can combine to form a glucose molecule. With each turn of the cycle, RuBP is regenerated. The glyceraldehyde phosphate can also be used as a starting material for other organic compounds needed by the cell.

When the carbon dioxide concentration in photosynthetic cells is low in relation to the oxygen concentration, photorespiration occurs. In this process, which requires sunlight, carbohydrates are oxidized to carbon dioxide and water, reducing the photosynthetic efficiency of the plant. In C_4 plants, photorespiration is minimized. Carbon dioxide is initially accepted by a compound known as PEP to yield the four-carbon oxaloacetic acid. Following a series of reactions that transport it to cells that surround the vascular bundles of the leaf, the carbon dioxide is released, binding with RuBP and entering the Calvin cycle. The C_4 pathway enables plants to minimize water loss and maximize photosynthetic efficiency. Under conditions of intense sunlight, high temperatures, or drought, C_4 plants are more efficient than C_3 plants.

Questions

1. Distinguish between the following: grana/thylakoid; stroma/thylakoid space; light-dependent reactions/light-independent reactions; Photosystem I/Photosystem II; C_3 photosynthesis/C_4 photosynthesis.

2. Why is it plausible to argue, as the Nobel laureate George Wald does, that wherever in the universe we find living organisms, we will find them (or at least some of them) to be colored?

3. Predict what colors of light might be most effective at stimulating plant growth.

4. Sketch a chloroplast and label its structures. Compare your sketch with Figure 9–5.

5. Describe in general terms the events of photosynthesis.

6. What is the source of the oxygen released in photosynthesis?

7. The experiment shown in Figure 4–2 on page 73 was attempted in Stockton, California, in 1973, but instead of making amino acids, the electric sparks caused the apparatus to explode. (No one was hurt, fortunately.) What is present in today's atmosphere that was not present in the primitive atmosphere and that would account for the explosion?

8. Some plants have acquired C_4 photosynthesis in the course of evolution. How is this adaptation advantageous to these plants? Many plants, however, have not evolved C_4 photosynthesis. Why is it advantageous to such plants *not* to have C_4 photosynthesis?

9. Trace a carbon atom through a series of biological events, such as those illustrated on page 126.

Suggestions for Further Reading

Alberts, Bruce, Dennis Bray, Julian Lewis, Martin Raff, Keith Roberts, and James D. Watson: *Molecular Biology of the Cell*, 3d ed., Garland Publishing, Inc., New York, 1994.

This outstanding cell biology text includes a clear, up-to-date discussion of our current understanding of the processes that occur in the chloroplast.

Bazzaz, Fakhri A., and Eric D. Fajer: "Plant Life in a CO_2-Rich World," *Scientific American*, January 1992, pages 68–74.

Berner, Robert A., and Antonio C. Lasaga: "Modeling the Geochemical Carbon Cycle," *Scientific American*, March 1989, pages 74–81.

Conant, James Bryant (ed.): *Harvard Case Histories in Experimental Science*, vol. 2, Harvard University Press, Cambridge, Mass., 1964.

Case #5, Plants and the Atmosphere, edited by Leonard K. Nash, describes the early work on photosynthesis, often presented in the words of the investigators themselves. The narrative illuminates the historical context in which the discoveries were made.

Detwiler, R. P., and C. A. S. Hall: "Tropical Forests and the Global Carbon Cycle," *Science*, vol. 239, pages 42–47, 1988.

Govindjee, and William J. Coleman: "How Plants Make Oxygen," *Scientific American*, February 1990, pages 50–58.

Houghton, Richard A., and George M. Woodwell: "Global Climatic Change," *Scientific American*, April 1989, pages 36–44.

Jones, Philip D., and Tom M. L. Wigley: "Global Warming Trends," *Scientific American*, April 1990, pages 84–91.

Kasting, James F., Owen B. Toon, and James B. Pollack: "How Climate Evolved on the Terrestrial Planets," *Scientific American*, February 1988, pages 90–97.

Ledbetter, M. C., and Keith R. Porter: *Introduction to the Fine Structures of Plant Cells*, Springer-Verlag, New York, 1970.

An excellent atlas of electron micrographs of plant cells, with detailed explanations.

Nassau, Kurt: "The Causes of Color," *Scientific American*, October 1980, pages 124–154.

Post, Wilfred M., et al.: "The Global Carbon Cycle," *American Scientist*, vol. 78, pages 310–326, 1990.

Scientific American: *Managing Planet Earth*, W. H. Freeman and Company, New York, 1990.[*]

This reprint of the September 1989 issue of Scientific American *contains 11 articles on the environmental crises currently threatening our planet, the effect of continuing human population growth, and strategies for the future to sustain economic growth without further damage to the environment. Chapters 2 and 3 examine the changes now occurring in the atmosphere and in worldwide climate as a consequence of the imbalance in the carbon cycle caused by human activities.*

Scientific American: *Molecules to Living Cells*, W. H. Freeman and Company, New York, 1980.[*]

A collection of articles from Scientific American. *Chapter 12 is an excellent presentation of the chemiosmotic synthesis of ATP in both mitochondria and chloroplasts, and Chapter 13 examines the photosynthetic membrane.*

Waggoner, Paul E.: "Agriculture and Carbon Dioxide," *American Scientist*, vol. 72, pages 179–184, 1984.

Youvan, Douglas C., and Barry L. Marrs: "Molecular Mechanisms of Photosynthesis," *Scientific American*, June 1987, pages 42–48.

[*]Available in paperback.

Genetics

10

The Reproduction of Cells

"Of everything that creepeth on the Earth, there went in two and two . . ."

Knowledge that like begets like is at least as old as recorded history. Biological inheritance has been perhaps the single most important factor in shaping the social evolution of humankind. Since the beginning of civilization, the ties of maternity and paternity have governed the transfer of power and wealth (and also of powerlessness and poverty). Today, throughout the world, the biological relationships of race and tribe and royalty still play a major role in determining the fates of nations and of individuals.

What is the source of this great power? What exactly do we mean by inheritance? What is passed from generation to generation? Why is it necessary to have two lions—and lions of opposite sexes, at that—in order to ensure a continuity of lions?

The way in which the continuity of life is preserved from generation to generation, and the way that male and female each contribute to the characteristics of their offspring were fundamental questions of biological science from its earliest beginnings. These ancient questions have now been largely answered, as the result of an extraordinary period of research that began in the middle of the nineteenth century and that has continued, at an ever-accelerating pace, to the present.

Our story begins with a seemingly simple event—the reproduction of the individual cell.

Cells reproduce in a process known as **cell division,** in which the cellular contents are divided between two new **daughter cells.** In one-celled organisms, such as bacteria and many protists, cell division increases the number of individuals in the population. In many-celled organisms, such as plants and animals, cell division is the means by which the organism grows, starting from one single cell. It is also the means by which injured or worn-out tissues are repaired and replaced.

An individual cell grows by taking in materials from its environment and synthesizing these materials into new structural and functional molecules. When the cell reaches a certain critical size and metabolic state, it divides. The two daughter cells, each of which has received about half of the mass of the parent cell, then begin growing again. A bacterial cell may divide as often as every six minutes. In a one-celled eukaryote, such as *Paramecium,* cell division may occur every few hours.

The new cells are structurally and functionally similar both to the parent cell and to one another. They are similar, in part, because each new cell usually receives about half of the parent cell's cytosol and organelles. More important, in terms of structure and function, each new cell inherits an exact replica of the hereditary information of the parent cell.

Cell Division in Prokaryotes

The distribution of exact replicas of the hereditary information is relatively simple in prokaryotic cells. In such cells, most of the hereditary material is in the form of a single, long, circular molecule of DNA, with which a variety of proteins are associated. This molecule, the cell's **chromosome,** is replicated before cell division.

According to present evidence, each of the two daughter chromosomes is attached to a different spot on the interior of the plasma membrane. As the membrane elongates, the chromosomes move apart

Cell wall
Plasma membrane
Chromosome

10–1 *Cell division in a bacterium. Attachment of the chromosome to an inward fold of the plasma membrane ensures that one chromosome replica is distributed to each daughter cell as the plasma membrane elongates.*

(Figure 10–1). When the cell has approximately doubled in size and the chromosomes are completely separated, the plasma membrane pinches inward, and a new cell wall forms that separates the two new cells and their chromosome replicas.

Cell Division in Eukaryotes

In eukaryotic cells, the problem of dividing the genetic material is much more complex. A typical eukaryotic cell contains about a thousand times more DNA than a prokaryotic cell, and this DNA is linear, forming a number of distinct chromosomes. For instance, human body cells have 46 chromosomes, each different from the others. When these cells divide, each daughter cell has to receive one copy—and only one copy—of each of the 46 chromosomes.

The solution to this problem is, as you will see, ingenious and elaborate. In a series of steps, known collectively as **mitosis,** a complete set of chromosomes is allocated to each of two daughter nuclei. Mitosis is usually followed by **cytokinesis,** a process that divides the cell into two new cells. Each new cell contains not only a nucleus with a full chromosome complement but also approximately half of the cytosol and organelles of the parent cell.

Although mitosis and cytokinesis are the culminating events of cell division, they represent only two stages of a larger process.

The Cell Cycle

Dividing eukaryotic cells pass through a regular, repeated sequence of cell growth and division, known as the **cell cycle** (Figure 10–2). The cycle consists of five major phases: G_1, S, G_2, mitosis, and cytokinesis. Completion of the cycle requires varying periods of time from a few hours to several days, depending on both the type of cell and external factors, such as temperature or available nutrients.

Before a cell can begin mitosis and actually divide, it must replicate its DNA, synthesize more of the proteins associated with the DNA in the chromosomes, produce a supply of organelles adequate for two daughter cells, and assemble the structures needed to carry out mitosis and cytokinesis. These preparatory processes occur during the G_1, S, and G_2 phases of the cell cycle, which are known collectively as **interphase.**

The key process of DNA replication occurs during the **S phase** (synthesis phase) of the cell cycle, a time in which many of the DNA-associated proteins are also synthesized. G (gap) phases precede and follow the S phase.

The **G_1 phase,** which precedes the S phase, is a period of intense biochemical activity. The cell doubles in size, and its enzymes, ribosomes, mitochondria, and other cytoplasmic molecules and structures also increase in number. Some of the structures can be synthesized entirely *de novo* ("from scratch") by the cell. These include microtubules, actin filaments, and ribosomes, all of which are composed, at least in part, of protein subunits. Membranous structures, such as Golgi complexes, lysosomes, vacuoles, and vesicles, are all apparently derived from the endoplasmic reticulum, which is renewed and enlarged by the synthesis of lipid and protein molecules.

In those cells that contain centrioles (that is, most eukaryotic cells except those of the flowering plants),

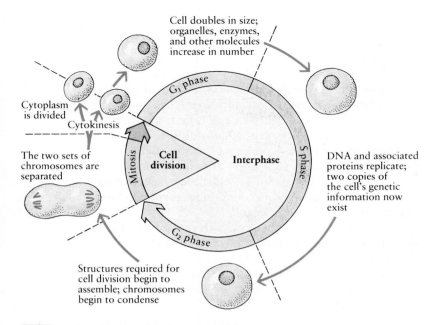

Cell doubles in size; organelles, enzymes, and other molecules increase in number

G₁ phase

S phase

DNA and associated proteins replicate; two copies of the cell's genetic information now exist

G₂ phase

Interphase

Cell division

Mitosis

Cytokinesis

Cytoplasm is divided

The two sets of chromosomes are separated

Structures required for cell division begin to assemble; chromosomes begin to condense

10–2   *The cell cycle. Cell division, which consists of mitosis (the division of the nucleus) and cytokinesis (the division of the cytoplasm), takes place after the completion of the three preparatory phases (G₁, S, and G₂) that constitute interphase. After the G₂ phase comes mitosis, which is usually followed immediately by cytokinesis. In cells of different species or of different tissues within the same organism, the different phases occupy different proportions of the total cycle.*

the two centrioles begin to separate from each other and to replicate. Mitochondria and chloroplasts, which are produced only from previously existing mitochondria and chloroplasts, also replicate. Each mitochondrion and chloroplast has its own chromosome, which is organized much like the single circular chromosome of the bacterial cell.

During the **G₂ phase,** which follows the S phase and precedes mitosis, the final preparations for cell division occur. The newly replicated chromosomes, which are dispersed in the nucleus in the form of threadlike strands of chromatin, slowly begin to coil and condense into a compact form. Replication of the centriole pair is completed, with the two mature centriole pairs lying just outside the nuclear envelope, somewhat separated from each other (Figure 10–3). Also during this period, the cell begins to assemble the structures required for the allocation of a complete set of chromosomes to each daughter cell during mitosis and for the separation of the two daughter cells during cytokinesis.

Plasma membrane

Nucleus

Centriole pair

10–3   *By the end of the G₂ phase of the cell cycle, two centriole pairs lie outside the nucleus. (a) As this cross section of a centriole from a human cell reveals, the structure of a centriole is identical to that of a basal body (page 103). (b) A centriole pair from a human cell, as seen in longitudinal section. Centrioles are formed either from preexisting centrioles, with the newly formed centriole appearing at right angles to the previously existing one, or from basal bodies.*

(a)   0.05 μm

(b)   0.05 μm

Regulation of the Cell Cycle

Some cell types pass through successive cell cycles throughout the life of the organism. This group includes the one-celled organisms and certain types of cells in both plants and animals. An example is provided by the **stem cells** in the human bone marrow that give rise to red blood cells. The average red blood cell lives only about 120 days, and there are about 25 trillion (2.5×10^{13}) of them in an adult. To maintain this number, about 2.5 million new red blood cells must be produced each second by the division of stem cells. The red blood cells themselves, like certain other highly specialized cells, such as nerve cells, lose the capacity to divide once they are mature.

Some cell types retain the capacity to divide after they mature but do so only under special circumstances. Cells in the human liver, for example, do not ordinarily divide. However, if a portion of the liver is removed surgically, the remaining cells (even if as few as a third of the original number) continue to divide until the liver reaches its former size. Then they stop. All told, about 2 trillion (2×10^{12}) cell divisions occur in an adult human every 24 hours, or about 25 million per second.

$\vdash\!\!-\!\!-\!\!\dashv$ 2 μm

10–4   *These dividing malignant cells, of a type known as HeLa cells, are just at the point of separation. HeLa cells are derived from a tissue sample obtained in 1951 from an extremely aggressive cervical cancer that afflicted Henrietta Lacks, a Baltimore mother of four. Since that time, HeLa cells have been studied in laboratories around the world and have continued to divide, over and over and over again. By contrast, when normal human cells are cultured in the laboratory, they typically divide about 50 times and then die.*

In a multicellular organism, it is of critical importance that cells of the various different types divide at a sufficient rate to produce as many cells as are needed for growth and replacement—and only that many. If any particular cell type divides more rapidly than is necessary, the normal organization and functions of the organism may be disrupted as specialized tissues are invaded and overwhelmed by the rapidly dividing cells. Such is the course of events in cancer.

The nature of the control or controls that regulate the cell cycle is currently the subject of intense research, not only because of its biological interest but also because of its potential importance in the control of cancer. A number of different substances that influence the growth and division of cells have been discovered in the last few years. Work is now underway to determine their molecular structure, the structure of the membrane receptors to which they bind, and the events triggered within the cell in response to that binding.

Mitosis

The function of mitosis is to maneuver the replicated chromosomes so that each new cell gets a full complement—one of each. The capacity of the cell to accomplish this distribution depends on (1) the condensed state of the chromosomes during mitosis and (2) an assembly of microtubules known as the **spindle.**

By the beginning of mitosis, the chromosomes have become sufficiently condensed to be visible under the light microscope. Each chromosome consists of two replicas, called **chromatids** (Figure 10–5), joined together by a constricted area common to both chromatids. This region of attachment is known as the **centromere.** Within the constricted region are disk-shaped protein-containing structures, the **kinetochores,** to which microtubules of the spindle are attached.

When completely formed, the spindle (Figure 10–6) is a three-dimensional football-shaped structure, consisting of at least two groups of microtubules: (1) **polar fibers,** which reach from each pole of the spindle (analogous to the ends of the football) to a central region midway between the poles, and (2) **kinetochore fibers,** which are attached to the kinetochores of the replicated chromosomes. These two groups of spindle fibers are responsible for separating the sister chromatids during mitosis.

In those cells that contain centrioles, each pole of the spindle is marked by a pair of newly replicated centrioles. Such cells also contain a third group of shorter spindle fibers, extending outward from the centrioles.

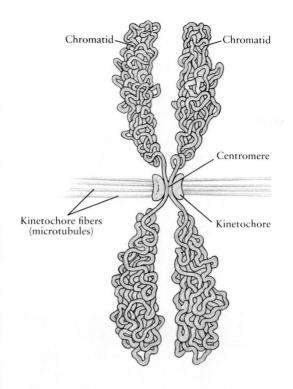

Chromatid   Chromatid

Centromere

Kinetochore fibers
(microtubules)

Kinetochore

10–5   *A replicated, fully condensed chromosome. The DNA and proteins that form the chromosome were replicated during the S phase of the cell cycle. Each chromosome now consists of two identical parts, called chromatids. The centromere, the constricted area at the center, is the site of attachment of the two chromatids. The kinetochores are protein-containing structures, one on each chromatid, associated with the centromere. Attached to the kinetochores are microtubules that form part of the spindle.*

These additional fibers are known collectively as the **aster.** It has been hypothesized that the fibers of the aster may brace the poles of the spindle against the plasma membrane during the movements of mitosis. In cells that lack centrioles and asters, the rigid cell wall may perform a similar function.

Most of the tubulin molecules from which the microtubules of the spindle are formed (see Figure 5–19b, page 101) are apparently borrowed from the cytoskeleton. As a consequence, dividing cells take on a characteristic rounded appearance. Following cell division, the spindle is disassembled, the cytoskeletal network of microtubules is reassembled, and the cell assumes its nondividing shape.

The Phases of Mitosis

The process of mitosis is conventionally divided into four phases: **prophase, metaphase, anaphase,** and **telophase.** Of these, prophase is usually by far the longest. If a mitotic division takes 10 minutes (which is about the minimum time required), during about six of these minutes the cell will be in prophase.

10–6   *The mitotic spindle. (a) This micrograph of a dividing cell from the lung of an amphibian illustrates the spindle's three-dimensional quality. The red fibers are the spindle microtubules. The large blue bodies near the equator of the spindle are the chromosomes.*

The basic framework of the spindle in (b) an animal cell and (c) a plant cell. For clarity, only one replicated chromosome is shown in each cell. In the animal cell, a centriole pair is present at each pole. The polar fibers, which form the bulk of the spindle, are sharply focused on the centrioles, and additional fibers radiate outward from the centrioles, forming the aster. In plant cells, by contrast, centrioles are absent, and the spindle is less sharply focused at the poles. Although there are a few shorter fibers near the poles, no aster is formed.

(a)   10 μm

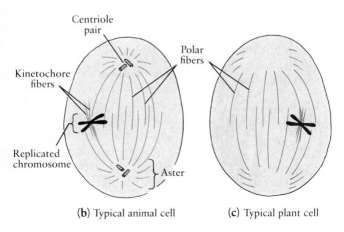

Centriole
pair

Polar
fibers

Kinetochore
fibers

Replicated
chromosome

Aster

(b) Typical animal cell

(c) Typical plant cell

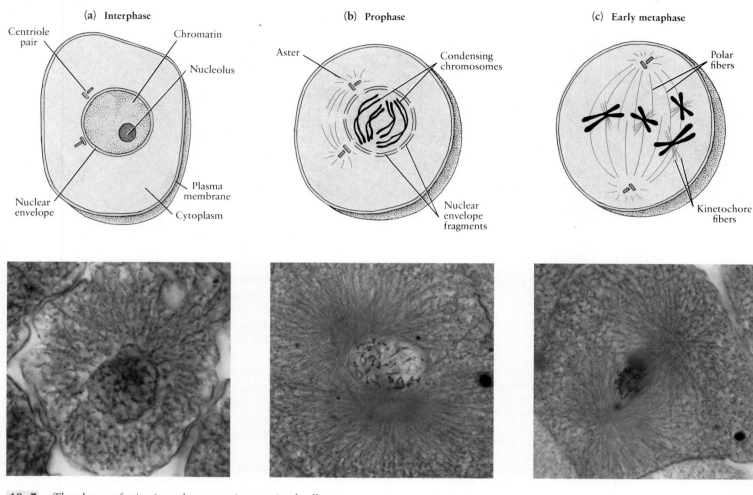

(a) Interphase

Centriole pair
Chromatin
Nucleolus
Nuclear envelope
Plasma membrane
Cytoplasm

(b) Prophase

Aster
Condensing chromosomes
Nuclear envelope fragments

(c) Early metaphase

Polar fibers
Kinetochore fibers

10–7   *The phases of mitosis as they occur in an animal cell. The accompanying micrographs show the phases of mitosis in embryonic cells of a whitefish. (a) Interphase. The chromatin, although replicated, has not yet begun to condense. Two centriole pairs lie just outside the nuclear envelope. (b) Prophase. The centrioles begin to move toward opposite poles of the cell, the condensing chromosomes become visible, the nuclear envelope breaks down, and the spindle apparatus begins to form. In cells with centrioles, the aster becomes visible. (c) Early metaphase. The chromatid pairs are tugged back and forth as they are moved into position, maneuvered by the polar and kinetochore fibers of the spindle. (d) Late metaphase. The chromatid pairs are lined up at the equator of the cell. (e) Anaphase. The two sets of newly separated chromosomes are pulled toward opposite poles of the cell. (f) Telophase. Nuclear envelopes form around the two sets of chromosomes, and the chromosomes uncoil, becoming diffuse once more. The spindle apparatus disappears, and the plasma membrane pinches in, a process that will ultimately separate the two daughter cells.*

During the interphase portions of the cell cycle, little can be seen in the nucleus (Figure 10–7a). The two centriole pairs can be seen at one side of the nucleus, outside the nuclear envelope.

By early prophase (Figure 10–7b), the chromatin has condensed sufficiently that the individual chromosomes, each consisting of two chromatids, become visible under the light microscope. (The threadlike appearance of the chromosomes when they first become visible is the source of the name "mitosis"; *mitos* is the Greek word for "thread.") The cell becomes more spherical and the cytoplasm more viscous at this stage, as the microtubules of the cytoskeleton are disassembled in preparation for the formation of the spindle.

During prophase, the centriole pairs move apart. Between the centriole pairs, forming as the centriole pairs separate (or, more likely, separating the centriole pairs as they form), are the microtubules that become the polar fibers of the spindle. In those cells that have centrioles, the microtubules that form the aster radiate outward from the centrioles. By this time the nucleolus has

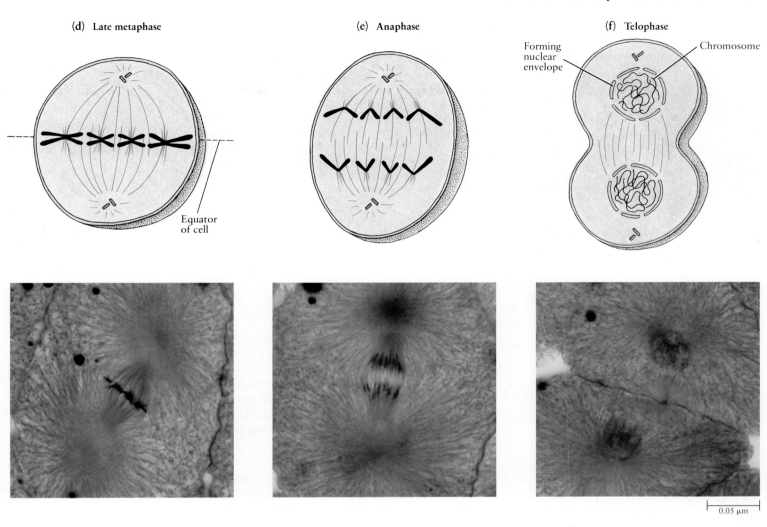

(d) Late metaphase

Equator of cell

(e) Anaphase

(f) Telophase

Forming nuclear envelope

Chromosome

0.05 μm

usually disappeared from view. As the chromosomes continue to condense, the nuclear envelope breaks down, dispersing into membranous fragments similar to fragments of endoplasmic reticulum.

By the end of prophase, the chromosomes are fully condensed. And, with the dispersion of the nuclear envelope, they are no longer separated from the cytoplasm. The centriole pairs have reached the poles of the cell, and the members of each pair are of equal size. The polar fibers of the spindle are fully formed, and the kinetochore fibers, attached to the kinetochores of the chromosomes, have also formed.

During early metaphase (Figure 10–7c), the chromatid pairs move back and forth within the spindle, apparently maneuvered by the spindle fibers. They appear to be tugged first toward one pole and then toward the other. Finally they become arranged precisely at the midplane (equator) of the cell (Figure 10–7d). This marks the end of metaphase.

At the beginning of anaphase, the most rapid stage of mitosis, the two chromatids of each pair, until this point attached at the centromere, separate. This happens simultaneously in all of the chromatid pairs. The chromatids of each pair then move apart, each chromatid now a separate chromosome, each apparently drawn toward the opposite pole by the kinetochore fibers. The centromeres move first, while the arms of the newly separated chromosomes seem to drag behind. As anaphase continues (Figure 10–7e), the two identical sets of chromosomes move rapidly toward the opposite poles of the spindle.

By the beginning of telophase, the chromosomes have reached the opposite poles and the spindle begins to disperse into tubulin molecules. During late telophase (Figure 10–7f), a nuclear envelope re-forms around each set of chromosomes, which once more become diffuse. In each nucleus, the nucleolus reappears. Often, a new centriole begins to form adjacent to each of the previous ones. As we saw earlier, replication of the centrioles continues during the subsequent cell cycle, so that each daughter cell has two centriole pairs by prophase of the next mitotic division.

(a)

(b)

10-8   *Dividing cells in a seed of the African globe lily illustrate the phases of mitosis in a plant cell. (a) Early prophase. The microtubules of the cytoskeleton are in a meshwork surrounding the nucleus and have not yet become reorganized into a spindle. The chromosomes are condensing. (b) Metaphase. The condensed chromosomes are lined up at the equator of the cell. (c) Anaphase. The chromosomes are moving toward the poles. (d) Telophase. Note the unstained region at the equator of the cell. During cytokinesis, the cell plate will form here, completing the separation of the two cells.*

(c)

(d)   ⊢——⊣ 25 μm

Cytokinesis

Cytokinesis, the division of the cytoplasm, usually but not always accompanies mitosis, the division of the nucleus. The visible process of cytokinesis generally begins during telophase of mitosis, and it usually divides the cell into two nearly equal parts.

Cytokinesis differs significantly in plant and animal cells. In animal cells, during early telophase, the plasma membrane begins to constrict along the circumference of the cell in the plane of the equator of the spindle. At first, a furrow appears on the surface, and this gradually deepens into a groove (Figure 10–9). Eventually, the connection between the daughter cells dwindles to a slender thread, which soon parts. The constriction of the plasma membrane is caused by a belt of actin filaments (page 101). These filaments are believed to act as a sort of "purse string," gathering in the plasma membrane of the parent cell at its midline, thus pinching apart the two daughter cells.

In plant cells, the cytoplasm is divided at the midline by a series of polysaccharide-containing vesicles produced from the Golgi complexes (Figure 10–10). These vesicles eventually fuse to form a flat, membrane-bound region, the **cell plate.** As more vesicles fuse, the edges of the growing plate fuse with the plasma membrane, completing the separation of the two daughter cells. During the fusion process, the polysaccharides within the vesicles are deposited between the two cells. This polysaccharide layer becomes impregnated with pectins and ultimately forms the middle lamella (page 91). Each new cell then constructs its own cell wall, laying down cellulose and other polysaccharides against the outer surface of its plasma membrane.

When cell division is complete, two daughter cells are produced, smaller than the parent cell but otherwise indistinguishable from it and from each other.

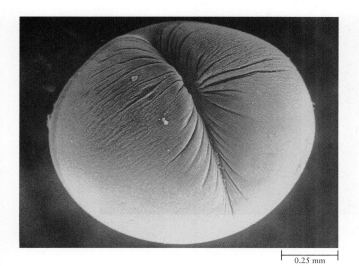

10–9   *Cytokinesis in an animal cell, the egg of a frog. In this scanning electron micrograph, note the puckering of the plasma membrane as the constriction furrow deepens into a groove. Eventually the two daughter cells will be connected by only a slender thread, similar to that connecting the two HeLa cells shown in Figure 10–4 (page 180). When that connection is severed, cytokinesis is complete.*

Cell Division and the Reproduction of the Organism

In one-celled organisms, mitosis is the key event in reproduction. It is the means by which exact replicas of the chromosomes are transmitted from parent to offspring. Mitosis plays the same essential role in the reproduction of some large organisms. Sea anemones and sponges, for example, may break apart to form new sea anemones and sponges. Plants may produce roots or runners from which new individuals arise. Reproduction in which exact replicas of the chromosomes are transmitted from parent to offspring by mitosis is known as **asexual reproduction.** It is also sometimes called vegetative reproduction, since it is particularly common among plants.

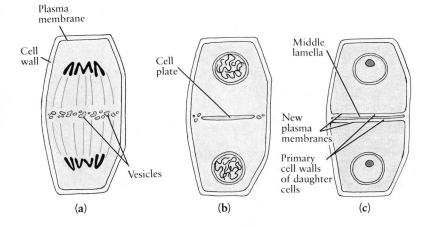

(a)   (b)   (c)

Plasma membrane
Cell wall
Vesicles
Cell plate
Middle lamella
New plasma membranes
Primary cell walls of daughter cells

(d)   1 μm

10–10   *In plants, the separation of the two daughter cells is effected by the formation of a structure known as the cell plate. (a) During telophase, vesicles containing polysaccharides become aligned across the equatorial plane of the cell. (b) The vesicles gradually fuse, forming a flat, membrane-bound region, the cell plate. This region is filled with the polysaccharides originally contained within the vesicles. The cell plate extends outward until it reaches the plasma membrane of the parent cell. As the last vesicles fuse with the plasma membrane, the two daughter cells become completely separated. (c) Each daughter cell then constructs a new primary cell wall on its side of the cell plate, which becomes the middle lamella.*

(d) Cell plate formation in a seedling of the soft maple (Acer saccharinum).

10–11   *This green alga of the genus* Microasterias *can, like many one-celled organisms, reproduce either asexually or sexually. The mature cell consists of two halves separated by a narrow waist, which, in this micrograph, runs from top to bottom. When the cell divides, it splits at the waist. Each half then generates a perfect replica of itself, restoring the original shape. The new half-cell (at the right) is almost full size, and the cell will soon divide again.*

25 μm

All multicellular organisms produce new cells—and thus grow—by mitosis and cytokinesis, and some produce entirely new individuals in this fashion. Most multicellular organisms, however, arise from a unique single cell—the fertilized egg. This cell is formed by the fusion of two cells, the sperm and the egg. How are these cells formed? And what process ensures that the new organism receives a full complement of chromosomes, with exactly one half from the father and one half from the mother? In the next chapter, we shall turn our attention to these questions.

Summary

Cell division in prokaryotes is a relatively simple process, in which two daughter chromosomes are attached to different spots on the interior of the plasma membrane. As the membrane elongates, the chromosomes are separated. The plasma membrane then pinches inward, and a new cell wall forms, completing the division of the daughter cells.

Cell division is a more complex process in eukaryotes, which contain a vast amount of genetic material, organized into a number of different chromosomes. Dividing eukaryotic cells pass through a regular sequence of growth and division known as the cell cycle. The cycle consists of a G_1 phase, during which cytoplasmic molecules and structures increase in number; an S phase, during which the chromosomes are replicated; a G_2 phase, during which condensation of the chromosomes and assembly of the structures required for mitosis and cytokinesis begin; mitosis, during which the replicated chromosomes are apportioned between the two daughter nuclei; and cytokinesis, during which the cytoplasm is divided, separating the parent cell into two daughter cells. The first three phases of the cell cycle are known collectively as interphase. Regulation of the cell cycle is thought to involve a number of interacting chemical factors.

When the cell is in the interphase portions of the cycle, the chromosomes are visible only as thin strands of threadlike material (chromatin) within the nucleus. As prophase of mitosis begins, the condensing chromosomes, previously replicated during the S phase, become visible under the light microscope. They consist of pairs of identical replicas, called chromatids, held together at the centromere. Simultaneously, the spindle is forming. Prophase ends with the breakdown of the nuclear envelope and disappearance of the nucleolus. During metaphase, the chromatid pairs, maneuvered by the spindle fibers, move toward the center of the cell. At the end of metaphase, they are arranged on the equatorial plane. During anaphase, the sister chromatids separate, and each chromatid, now an independent chromosome, moves to an opposite pole. During telophase, a nuclear envelope forms around each group of chromosomes.

The spindle begins to break down, the chromosomes uncoil and once more become extended and diffuse, and nucleoli reappear.

Cytokinesis in animal cells results from constrictions in the plasma membrane between the two nuclei. In plant cells, the cytoplasm is divided by the fusion of vesicles to form the cell plate, within which new cell walls are subsequently laid down. In both cases, the result is the production of two new, separate cells. As a result of mitosis, each cell has received an exact copy of the chromosomes of the parent cell and, as a result of cytokinesis, approximately half of the cytosol and organelles.

In multicellular organisms, mitosis and cytokinesis produce new cells for growth and replacement. In unicellular organisms and some multicellular organisms, mitosis and cytokinesis produce new individuals, a process known as asexual reproduction.

Questions

1. Distinguish among the following: cell cycle/cell division; mitosis/cytokinesis; chromatid/chromosome; centriole/centromere/kinetochore.

2. Describe the activities occurring during each phase of the cell cycle and the role of each phase in the overall process of cell division.

3. What is a chromosome? How is it related to chromatin?

4. Why do we often refer to chromatids as sister chromatids? When are sister chromatids formed? How? When do they first become visible under the microscope?

5. The drawings at the right show stages in mitosis in a plant cell. Identify each stage and describe what is happening.

6. In what ways does cell division in plant cells differ from that in animal cells?

7. What is the function of cell division in the life of an organism? Suppose you, as an organism, were made up of a large single cell rather than trillions of small ones. How would you differ from your present self?

(a)

(b)

(c)

Suggestions for Further Reading

Aaronson, Stuart A.: "Growth Factors and Cancer," *Science,* vol. 254, pages 1146–1153, 1991.

Balter, Michael: "Cell Cycle Research: Down to the Nitty Gritty," *Science,* vol. 252, pages 1253–1254, 1991.

Glover, David M., Cayetano Gonzalez, and Jordan W. Raff: "The Centrosome," *Scientific American,* June 1993, pages 62–68.

Marx, Jean: "The Cell Cycle: Spinning Farther Afield," *Science,* vol. 252, pages 1490–1492, 1991.

McIntosh, J. Richard, and Michael P. Koonce: "Mitosis," *Science,* vol. 246, pages 622–628, 1989.

McIntosh, J. Richard, and Kent L. McDonald: "The Mitotic Spindle," *Scientific American,* October 1989, pages 48–56.

Murray, Andrew W., and Marc W. Kirschner: "What Controls the Cell Cycle," *Scientific American,* March 1991, pages 56–63.

Sloboda, Roger D.: "The Role of Microtubules in Cell Structure and Cell Division," *American Scientist,* vol. 68, pages 290–298, 1980.

This passionate embrace of the female golden toad by the male (from whose brilliant color the species derives its common name) often lasts for several hours. During that time, the male kicks his hind legs furiously to repel rivals. Eventually the female releases a procession of some 200 eggs. As they are laid, the male deposits his sperm on them. The sperm cells, which are too small to be seen in this photograph, then penetrate the outer layers of the eggs and fertilize them.

Each of the egg cells visible here contains just half the number of chromosomes present in the body cells of the female. Similarly, each of the sperm cells contains just half the number of chromosomes present in the body cells of the male. The process by which the chromosome number is halved is known as meiosis. At fertilization, when the sperm and egg unite, the number of chromosomes characteristic of the body cells is restored.

For reasons that remain unclear, these handsome amphibians, whose scientific name is *Bufo periglenes,* have disappeared from the few square miles of the Monteverde Cloud Forest Reserve in Costa Rica where they were previously found in great abundance. In the past, the entire population of golden toads emerged from their hiding places once a year when the first heavy spring rains filled the breeding pools. Within the space of a single week, all of the toads would congregate in the breeding pools, with thousands of eggs laid by the females and fertilized by the males. However, only a few of the fertilized eggs completed the journey to adulthood, largely because the little pools of water in which the eggs were laid dried up so quickly.

Each year from 1972 through 1987, some 1,500 golden toads were observed during the mating period. In 1988 and 1989, however, only a few were seen, and, since 1990, no golden toads have been observed at all. No one knows if the toads are alive and in hiding, perhaps waiting until an unusually long period of severe drought ends and pools of water suitable for breeding form once more—or if they have disappeared forever.

11

Meiosis and Sexual Reproduction

Most multicellular eukaryotic organisms—for example, golden toads, sea urchins, whitefish, pea plants, and human beings—reproduce sexually. Many unicellular eukaryotes—even those that typically reproduce asexually by mitosis and cytokinesis—are also capable of sexual reproduction.

Sexual reproduction generally requires two parents, and it always involves two events: **fertilization** and **meiosis.** Fertilization is the means by which the different genetic contributions of the two parents are brought together to form the genetic identity of the offspring. Meiosis is a special kind of nuclear division that is believed to have evolved from mitosis and that uses much of the same cellular apparatus. As we shall see, however, meiosis differs from mitosis in a number of important respects.

Haploid and Diploid

To understand meiosis, we must look once more at the chromosomes, focusing this time on their numbers. Every organism has a chromosome number characteristic of its particular species. In a mosquito, for example, each **somatic** (body) cell contains 6 chromosomes; in a cabbage, 18; in corn, 20; in a cat, 38; in a human being, 46; in a potato, 46; and in a goldfish, 94. However, in these organisms and most other familiar plants and animals, the sex cells—or **gametes**—have exactly half the number of chromosomes that is characteristic of the somatic cells of the organism.

The number of chromosomes in the gametes (from the Greek word *gamein,* "to marry") is referred to as the **haploid** ("single set") number. And, as you might expect, the number in the somatic cells is known as the **diploid** ("double set") number. Cells that have more than two sets of chromosomes are said to be **polyploid** ("many sets"). Polyploid cells rarely occur in animals, but the somatic cells of a wide variety of flowering plants are polyploid.

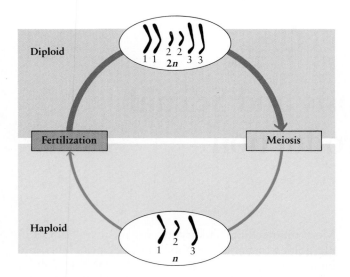

11–1   *Sexual reproduction is characterized by two events: the coming together of the gametes (fertilization) and the halving of the number of chromosomes (meiosis). Following meiosis, there is a single set of chromosomes, that is, the haploid number* (n). *Following fertilization, there is a double set of chromosomes, that is, the diploid number* (2n).

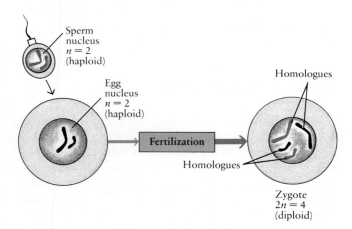

11–2   *During gamete formation, individual homologues (members of a homologous pair of chromosomes) are parceled out by meiosis so that a haploid* (n) *gamete, which is produced from a diploid* (2n) *cell, carries only one member of each homologous pair. At fertilization, the chromosomes in the sperm nucleus and the egg nucleus come together in the zygote, producing, once again, pairs of homologous chromosomes. Each pair consists of one homologue from the father (paternal chromosome) and one from the mother (maternal chromosome). Here, and in subsequent diagrams, red and black are used to indicate the paternal and maternal chromosomes of a homologous pair.*

In biological shorthand, the haploid number is designated as n and the diploid number as $2n$. In humans, for example, $n = 23$ and $2n = 46$. When a sperm fertilizes an egg, the two haploid nuclei fuse, $n + n = 2n$, and the diploid number is restored (Figure 11–1). A diploid cell produced by the fusion of two gametes is known as a **zygote** (from the Greek *zygotos*, "a pair").

In every diploid cell, each chromosome has a partner. These pairs of chromosomes are known as homologous pairs, or **homologues.** The two resemble each other in size and shape and also, as we shall see, in the kinds of hereditary information each contains. One homologue comes from the gamete of one parent, and its partner is from the gamete of the other parent. After fertilization, both homologues are present in the zygote (Figure 11–2).

In meiosis, the diploid set of chromosomes, which contains the two homologues of each pair, is reduced to a haploid set, which contains only one homologue of each pair. Meiosis thus counterbalances the effects of fertilization, ensuring that the number of chromosomes remains constant from generation to generation. Meiosis is also a source of new combinations within the chromosomes themselves.

Meiosis and the Life Cycle

Meiosis occurs at different times during the life cycle of different organisms (Figure 11–3). In many protists, such as the alga *Chlamydomonas*, it occurs shortly after fusion of the mating cells (Figure 11–4). The cells are ordinarily haploid, and meiosis restores the haploid number after fertilization.

In plants, a haploid phase typically alternates with a diploid phase. For example, in ferns (Figure 11–5), the form of the plant commonly seen is the **sporophyte,** the diploid organism. By meiosis, fern sporophytes produce spores, usually on the undersides of their fronds (leaves). These spores have only the haploid number of chromosomes. They germinate to form tiny plants, typically only a few cell layers thick, that are known as **gametophytes.** In these plants, all the cells are haploid. The small, haploid gametophytes produce gametes by mitosis. The gametes fuse and then develop into a new, diploid sporophyte. This process, in which a haploid phase is followed by a diploid phase and again by a haploid phase, is known as **alternation of generations.** As we shall see in Chapter 25, alternation of generations occurs in all sexually reproducing plants, although not always in the same form.

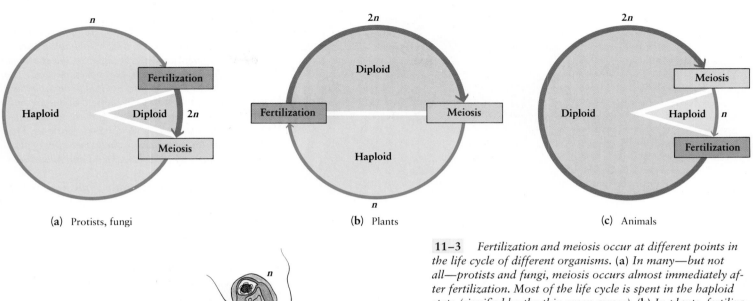

(a) Protists, fungi

(b) Plants

(c) Animals

11–3   *Fertilization and meiosis occur at different points in the life cycle of different organisms.* (**a**) *In many—but not all—protists and fungi, meiosis occurs almost immediately after fertilization. Most of the life cycle is spent in the haploid state (signified by the thin green arrow).* (**b**) *In plants, fertilization and meiosis are separated in time. The life cycle of the organism consists of both a diploid phase and a haploid phase.* (**c**) *In animals, completion of meiosis is followed almost immediately by fertilization. As a consequence, during most of the life cycle the organism is diploid (signified by the thick blue arrow).*

n

n

Fertilization

Haploid

Diploid

Diploid
zygote

Meiosis

Protective coat
of zygote

n

n

n

11–4   *The life cycle of* Chlamydomonas *is of the type shown in Figure 11–3a. The organism is haploid for most of its life cycle. Fertilization, the fusion of cells of different mating strains, produces the diploid zygote. The zygote manufactures a thick coat that allows it to remain dormant during harsh conditions. Following dormancy, the diploid zygote divides meiotically, forming four new haploid cells. Each haploid cell can reproduce asexually (by mitosis and cytokinesis) to form either more haploid cells or, in periods of environmental stress, haploid cells of a particular mating strain. These cells can fuse with cells of a different mating strain, and another sexual cycle is underway.*

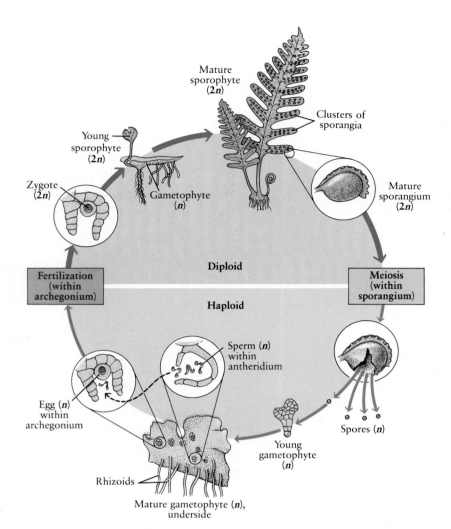

Mature sporophyte (**2n**)

Clusters of sporangia

Young sporophyte (**2n**)

Gametophyte (**n**)

Zygote (**2n**)

Mature sporangium (**2n**)

Diploid

Fertilization (within archegonium)

Meiosis (within sporangium)

Haploid

Sperm (**n**) within antheridium

Egg (**n**) within archegonium

Spores (**n**)

Young gametophyte (**n**)

Rhizoids

Mature gametophyte (**n**), underside

11–5 *The life cycle of a fern is of the type shown in Figure 11–3b. Following meiosis (far right), spores, which are haploid, are produced in the sporangia. As the sporangia dry and split open, the spores are shed. The spores develop into haploid gametophytes. In many species, the gametophytes are only one layer of cells thick and are somewhat heart-shaped, as shown here (bottom). From the lower surface of the gametophyte, filaments, the rhizoids, extend downward into the soil.*

On the lower surface of the gametophyte are borne the flask-shaped archegonia, which enclose the egg cells, and the antheridia, which enclose the sperm. When the sperm are mature and there is an adequate supply of water, the antheridia burst, and the sperm cells, which have numerous flagella, swim to the archegonia and fertilize the eggs.

From the zygote, the diploid (2n) sporophyte develops, growing out of the archegonium within the gametophyte. After the young sporophyte becomes rooted in the soil, the gametophyte disintegrates. The sporophyte matures, develops sporangia, in which meiosis occurs, and the cycle begins again.

Human beings have the typical animal life cycle, in which the diploid individual produces haploid gametes by meiosis shortly before fertilization. Fusion of male and female gametes at fertilization then restores the diploid number. Virtually all of the life cycle is spent in the diploid state (Figure 11–6).

Note that although meiosis in animals produces gametes, meiosis in plants produces **spores**. A spore is a haploid reproductive cell that, unlike a gamete, can develop into a whole organism (a haploid organism) without first fusing with another cell. With the formation of either gametes or spores, however, meiosis has the same result: at some point during the life cycle of a sexually reproducing organism, it reduces the diploid chromosome set to the haploid chromosome set.

The Preparations for Meiosis

As we saw in the last chapter, mitosis consists of a single nuclear division and results in the formation of two daughter nuclei. Each of these nuclei receives an exact copy of the parent cell's chromosomes. Meiosis, by contrast, consists of two successive nuclear divisions, producing a total of four daughter nuclei. Each of these daughter nuclei contains half the number of chromosomes present in the original nucleus. Moreover, each daughter nucleus receives just one member of each pair of homologous chromosomes.

The key events in meiosis—on which all else depends—occur in the interphase preceding meiosis and in prophase of the first meiotic division. During interphase, the chromosomes are replicated, so that by the beginning of meiosis each chromosome consists of two identical sister chromatids held together at the centromere.

Early in prophase, after the chromosomes are replicated, the homologous chromosomes come together in pairs (Figure 11–7a). Once contact is made at any point between two homologues, pairing extends, zipperlike, along the length of the chromatids. Since each chromosome consists of two identical chromatids, the pairing of the homologous chromosomes actually involves four chromatids. Each complex of paired homologous chromosomes is thus known as a **tetrad** (from the Greek *tetra,* meaning "four").

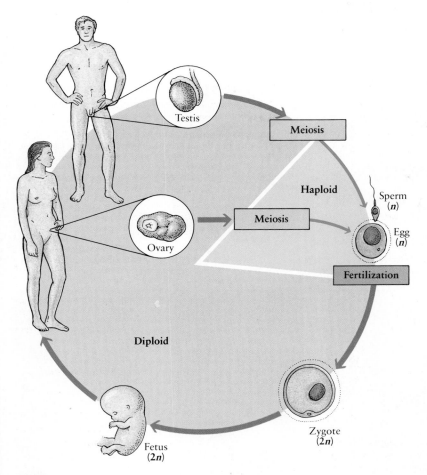

11–6 *The life cycle of Homo sapiens. Gametes—egg cells and sperm cells—are produced by meiosis. At fertilization, the haploid gametes fuse, restoring the diploid number in the fertilized egg. The zygote develops into a mature man or woman, who again produces haploid gametes. As is the case with most other animals, the cells are diploid during almost the entire life cycle, the only exception being the gametes. This is the type of life cycle diagrammed in Figure 11–3c.*

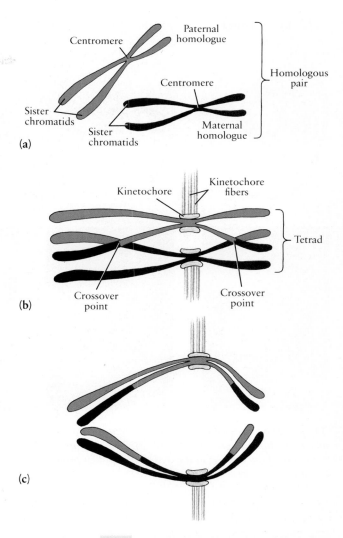

11–7 (a) *A homologous pair of chromosomes, prior to meiosis. One member of the pair is of paternal origin, and the other of maternal origin. Each of these chromosomes has replicated and consists of two sister chromatids.* (b) *In prophase of the first meiotic division, the two homologues of each pair come together and become closely associated with one another. Each homologous pair consists of four chromatids and is therefore also known as a tetrad. Within the tetrad, chromatids of the two homologues intersect at a number of points, making possible the exchange of chromatid segments. This phenomenon is known as crossing over, and the locations at which it occurs are called crossover points.* (c) *The result of crossing over is a recombination of the genetic material of the two homologues. The sister chromatids of each homologue are no longer identical.*

At this point, a crucial process occurs that can alter the genetic makeup of the chromosomes. This process, known as **crossing over,** involves the exchange of segments of one chromosome with corresponding segments from its homologous chromosome (Figure 11–7b). At the sites of crossing over, portions of the chromatids of one homologue are broken and exchanged with corresponding portions of the chromatids of the other homologue. The breaks are resealed, and the result is that the sister chromatids of a single homologue no longer contain identical genetic material (Figure 11–7c). The maternal homologue now contains portions of the paternal homologue, and vice versa. Thus crossing over is an important mechanism for recombining the genetic material from the two parents.

| (a) Late prophase I | (b) Metaphase I | (c) Anaphase I | (d) Telophase I | (e) Interphase II |

11–8   *The phases of meiosis as they occur in a plant cell in which the diploid number is 6 (n = 3). The accompanying micrographs show meiosis in the anther (the pollen-producing structure) of a lily.*

The Phases of Meiosis

The two successive nuclear divisions of meiosis are designated as meiosis I and meiosis II. In meiosis I, homologous chromosomes pair and then separate from one another. In meiosis II, the chromatids of each homologue separate. In this discussion, we shall describe meiosis in a plant cell in which the diploid number is 6 (*n = 3*). Three of the six chromosomes were originally derived from one parent and three from the other parent. For each chromosome from one parent there is a homologous chromosome, or homologue, from the other parent.

Following replication of the chromosomes, the first of the two nuclear divisions of meiosis begins. It proceeds through the stages of prophase, metaphase, anaphase, and telophase. All of these are designated "I" to indicate that they are substages of meiosis I.

During **prophase I** (Figure 11–8a), the condensed,

replicated chromosomes first become visible under the light microscope. The spindle microtubules assemble and can be seen radiating out from the two poles of the cell. During this phase, the nucleolus disappears and the nuclear envelope breaks down. Also, and most important, the pairing of the homologous chromosomes and crossing over take place. If you remember one essential point—that the homologous chromosomes are arranged in pairs by the end of prophase I—you will be able to remember all of the subsequent events with little difficulty.

In **metaphase I** (Figure 11–8b), the homologous pairs line up along the equatorial plane of the cell. (By contrast, in metaphase of mitosis, replicated chromosomes line up at the equator with no pairing of the homologues.) Spindle fibers have become associated with the kinetochores (see page 180).

During **anaphase I** (Figure 11–8c), the homologues, each consisting of two sister chromatids, separate, as if pulled apart by the kinetochore fibers. However, the two sister chromatids of each homologue do not separate as they do in mitosis.

By the end of the first meiotic division, **telophase I**

(f) Prophase II **(g)** Metaphase II **(h)** Anaphase II **(i)** Telophase II **(j)** Cytokinesis

25 μm

(Figure 11–8d), the homologues have moved to the poles. Each chromosome group now contains only half the number of chromosomes as the original nucleus.* Moreover, because of exchanges that took place during crossing over, these chromosomes may be different from any of those present in the original cell. Depending on the species, new nuclear envelopes may or may not form, and cytokinesis may or may not take place.

Meiosis, however, does not end here. Although each nucleus contains the haploid *number* of chromosomes, it nevertheless contains double the haploid *amount* of hereditary material. Why? Because each chromosome consists of two chromatids that have not yet separated.

Meiosis II resembles mitosis except that it is not preceded by replication of the chromosomal material. A short interphase (Figure 11–8e) may occur, during which the chromosomes partially uncoil, but meiosis in many species proceeds from telophase I directly to prophase II.

* In counting, it is often difficult to know whether to count a chromosome that has replicated but has not divided as one or two chromosomes. It is customary to count such a chromosome as one. The trick is to count centromeres.

During **prophase II** (Figure 11–8f), the chromosomes, if dispersed, condense fully again. The nuclear envelopes, if present, disintegrate, and new spindle fibers begin to appear. Remember that in this example there are three chromosomes in each nucleus (the haploid number), and each is still in the form of two chromatids held together at the centromere.

During **metaphase II** (Figure 11–8g), the three chromosomes in each nucleus line up on the equatorial plane. Each consists of two sister chromatids. At **anaphase II** (Figure 11–8h), as in anaphase of mitosis, the sister chromatids of each chromosome separate from one another. Each individual chromatid, which can now be called a chromosome, moves toward one of the poles.

During **telophase II** (Figure 11–8i), the spindle microtubules disappear and a nuclear envelope forms around each set of chromosomes. There are now four nuclei in all, each containing the haploid number of chromosomes. Cytokinesis (Figure 11–8j) proceeds as it does following mitosis. Cell walls form, dividing the cytoplasm, and these haploid plant cells begin to differentiate into spores.

195

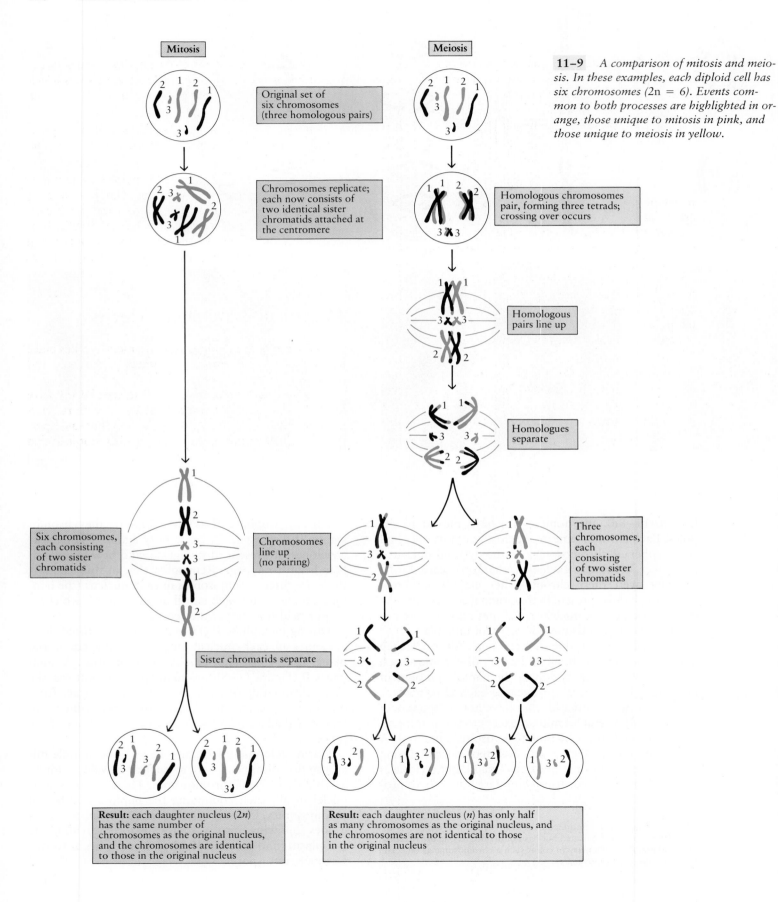

Mitosis

Meiosis

Original set of
six chromosomes
(three homologous pairs)

Chromosomes replicate;
each now consists of
two identical sister
chromatids attached at
the centromere

Homologous chromosomes
pair, forming three tetrads;
crossing over occurs

Homologous
pairs line up

Homologues
separate

Six chromosomes,
each consisting
of two sister
chromatids

Chromosomes
line up
(no pairing)

Three
chromosomes,
each
consisting
of two sister
chromatids

Sister chromatids separate

Result: each daughter nucleus (2*n*)
has the same number of
chromosomes as the original nucleus,
and the chromosomes are identical
to those in the original nucleus

Result: each daughter nucleus (*n*) has only half
as many chromosomes as the original nucleus, and
the chromosomes are not identical to those
in the original nucleus

11–9   *A comparison of mitosis and meiosis. In these examples, each diploid cell has
six chromosomes (2n = 6). Events common to both processes are highlighted in orange, those unique to mitosis in pink, and
those unique to meiosis in yellow.*

Thus, beginning with one cell containing six chromosomes (three homologous pairs), we end with four cells, each with three chromosomes (no homologous pairs). The chromosome number has been reduced from the diploid number to the haploid number.

Meiosis and Mitosis Compared

As we have seen, the events that take place during meiosis resemble those of mitosis. However, a direct comparison of the two processes (Figure 11–9) reveals a number of differences. The most crucial are the pairing of the homologous chromosomes in prophase I of meiosis and their alignment in metaphase I, followed by the separation of the homologues in anaphase I. These events, which do not occur in mitosis, are the key to the reduction in chromosome number.

Other important points to remember:

1. Meiosis can occur only in cells with the diploid (or polyploid) number of chromosomes—that is, in cells in which each chromosome has a partner. By contrast, mitosis, which does not require any pairing of the chromosomes, can occur in either haploid or diploid cells.

2. During meiosis, each diploid nucleus divides twice, producing a total of four nuclei. The chromosomes, however, replicate only once—prior to the first nuclear division.

3. Thus, each of the four nuclei produced contains half the number of chromosomes present in the original nucleus.

4. The haploid nuclei produced by meiosis contain new combinations of chromosomes. The homologous chromosomes, originally derived from the organism's parents, are assorted randomly among the four new haploid nuclei. Moreover, as a result of crossing over, these chromosomes are not identical to those with which meiosis began.

Meiosis in the Human Species

In all vertebrates, including humans, meiosis takes place in the reproductive organs, the **testes** of the male and the **ovaries** of the female.

In males, the diploid cells that undergo meiosis are known as **primary spermatocytes.** Each primary spermatocyte divides twice, producing four haploid **spermatids.** The spermatids then mature into motile sperm cells (Figure 11–10).

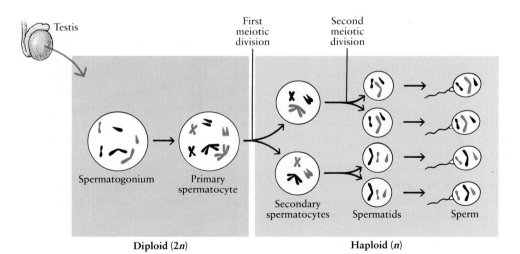

11–10 *The series of changes resulting in the formation of sperm cells begins with the growth of spermatogonia into large cells known as primary spermatocytes. For clarity, only six (n = 3) of the 46 human chromosomes are shown. At the first meiotic division, each primary spermatocyte divides into two haploid secondary spermatocytes.*

The second meiotic division results in the formation of four haploid spermatids. The spermatids then mature into functional sperm cells. This process occurs continuously after puberty; the normal ejaculate of an adult human male contains between 300 and 400 million sperm cells.

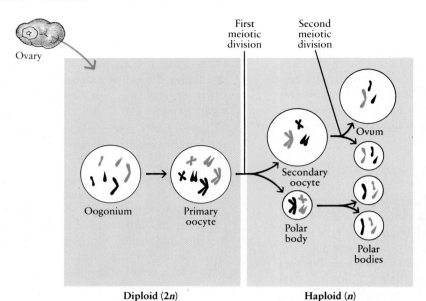

First meiotic division

Second meiotic division

Ovary

Oogonium

Primary oocyte

Secondary oocyte

Polar body

Ovum

Polar bodies

Diploid (2n)

Haploid (n)

11–11   *Formation of the ovum begins with the growth of an oogonium into a primary oocyte. In the first meiotic division, this cell divides into a secondary oocyte and a polar body. In human females, the first meiotic division begins during the third month of fetal development and ends at ovulation, which, in the case of some cells, may not take place until 50 years later. The second meiotic division, which produces the ovum and a second polar body, does not take place until after the fertilizing sperm cell has penetrated the secondary oocyte. The first polar body may also divide.*

In females, diploid cells known as **primary oocytes** undergo meiosis. Haploid nuclei are also produced by the meiotic divisions of these cells, but the cytoplasm is apportioned unequally during cytokinesis in both meiosis I and II. From each primary oocyte, only one **ovum** (egg cell) is produced, along with two or three **polar bodies** (Figure 11–11). The polar bodies contain the other nuclei formed by meiosis and usually disintegrate. As a result of this unequal division of the cytoplasm, the ovum is well supplied with ribosomes, mitochondria, enzymes, and stored nutrients, all important for the development of the embryo.

Mistakes in Meiosis

As you know, the diploid number of chromosomes in the human species is 46, and the haploid number is 23. In all of the homologous pairs except one, the chromosomes appear to be identical in both males and females; these chromosomes are known as **autosomes.** The structure of one pair, however, differs between males and females. The chromosomes of this pair are known as **sex chromosomes.** In females, the two sex chromosomes are identical, but in males they are dissimilar. One of the male sex chromosomes is the same as the female sex chromosomes, but the other is much smaller. The chromosome that is the same in the cells of both males and females is known as the **X chromosome,** and the unlike chromosome characteristic of the cells of males is known as the **Y chromosome** (Figure 11–12)

A number of genetic disorders are caused by abnormalities in the number or structure of either the autosomes or the sex chromosomes. These abnormalities result from "mistakes" during meiosis (or, less often, during mitotic divisions early in embryonic development). For example, from time to time, homologous chromosomes fail to separate during the first division of meiosis. Similarly, the chromatids occasionally fail to separate during the second division of meiosis. This phenomenon is known as **nondisjunction.**

When nondisjunction occurs in meiosis, the results are gametes with one or more chromosomes too many and other gametes with one or more chromosomes too few. A gamete with too few chromosomes (unless the missing chromosome is a sex chromosome) cannot produce a viable embryo. However, sometimes a gamete with too many chromosomes does produce a viable embryo. The result is an individual with one or more extra chromosomes in every cell of his or her body. In the vast majority of such cases, however, the fetus is spontaneously aborted early in pregnancy, an event (commonly known as a miscarriage) that occurs in 15 to 20 percent of recognized pregnancies.

Individuals with additional autosomes always have widespread abnormalities. With the exception of those with Down syndrome, those who are not stillborn typically survive only a few months. Among the few who survive, most are mentally retarded and those who survive to maturity are usually sterile. They frequently have abnormalities of the heart and other organs as well.

(a)

25 μm

(b)

11–12 (a) *The chromosomes of a dividing somatic cell of a human female, photographed after replication and condensation. The chromosomes have been stained, revealing banding patterns that make it relatively easy to identify homologues. The large, round yellow-stained structure is the nucleus from which the chromosomes were released.*

(b) *Chromosomes from a single human somatic cell, arranged in the kind of graphic display known as a karyotype (see essay). In a karyotype, the autosomes are grouped by size (A, B, C, etc.), and then the probable homologues are paired. The normal diploid chromosome number of a human being is 46, 22 pairs of autosomes and two sex chromosomes. A normal woman has two X chromosomes and a normal man has an X and a Y, as shown here.*

Nondisjunction may also produce individuals with unusual numbers of sex chromosomes. A normal XY combination in the twenty-third pair produces maleness, but so do XXY, XXXY, and even XXXXY. These latter males, however, are usually sexually underdeveloped and sterile. XXX combinations sometimes produce normal females, but many of the XXX women and almost all women with only one X chromosome are sterile.

Nondisjunction is not the only type of "mistake" in meiosis that can cause abnormalities. For example, a **deletion** occurs when a segment of a chromosome breaks off and is not replaced by the corresponding segment from its homologue. In some cases, the segment is incorporated into the homologue, in which the segment then occurs twice. This abnormality is known as a **duplication**. Sometimes a deleted segment of one chromosome is transferred to and becomes part of another, nonhomologous chromosome, a phenomenon known as **translocation**. Depending on the size of the segment and the hereditary information it carries, the deletion of a portion of a chromosome can be as lethal as the loss of the entire chromosome. Similarly, duplication and translocation can, in some cases, have the same consequences as the presence of an extra chromosome.

Down Syndrome

One of the most familiar conditions resulting from an abnormality in the number of autosomal chromosomes is Down syndrome, named after the physician who first described it. Because it usually involves more than one defect, it is referred to as a **syndrome,** a group of disorders that occur together. Down syndrome includes, in most cases, a short, stocky body type with a thick neck; mental retardation, ranging from mild to severe in different individuals; a large tongue, resulting in speech defects; an increased susceptibility to infections; and, often, abnormalities of the heart and other organs. Individuals with Down syndrome who survive into their thirties or forties also have a high probability of developing a form of senility similar to Alzheimer's disease (to be discussed in Chapter 36).

Down syndrome arises when an individual has three, rather than two, copies of chromosome 21. In about 95 percent of the cases, the cause of the abnormality is nondisjunction during formation of a parental gamete, resulting in 47 chromosomes, with an extra copy of chromosome 21 in the cells of the affected individual (Figure 11–13).

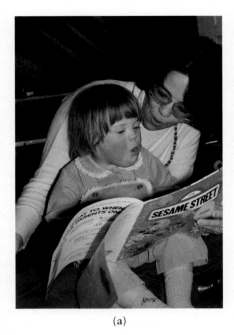

(a)

(b)

11–13   (a) *Although children with Down syndrome share certain physical characteristics, there is a wide range of mental capacity among these individuals.* (b) *The karyotype of a male with Down syndrome caused by nondisjunction. Note that there are three copies of chromosome 21.*

Down syndrome may also result from a translocation in the chromosomes of one of the parents. The person with Down syndrome caused by translocation usually has a third chromosome 21 (or, at least, most of it) attached to a larger chromosome, most often chromosome 14. Such an individual, although he or she has only 46 chromosomes, has the functional equivalent of a third chromosome 21.

When cases of Down syndrome due to translocation are studied, it is usually found that one parent, although normal, has only 45 separate chromosomes. One chromosome is usually composed of most of chromosomes 14 and 21 joined together. Thus, parents who have a child with Down syndrome are advised to have their karyotypes prepared. If either parent has the translocation, they are at a high risk of having another child with Down syndrome and, moreover, half of their children without Down syndrome are likely to be carriers of the translocation.

It has been known for many years that Down syndrome and a number of other disorders involving nondisjunction are more likely to occur among infants born to older women (Figure 11–14). The reasons for this are not known. Recent studies have also indicated that in about 5 percent of the cases of Down syndrome due to nondisjunction, the extra chromosome comes from the father rather than the mother.

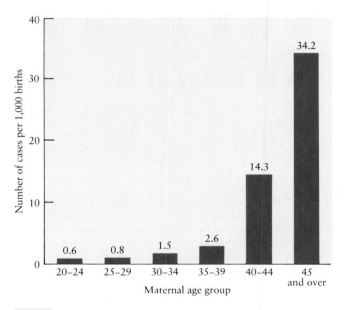

11–14   *The frequencies of births of infants with Down syndrome in relation to the ages of the mothers. The number of cases shown for each age group represents the occurrence of Down syndrome in every 1,000 live births by mothers in that group. As you can see, the risk of having a child with Down syndrome increases rapidly after the mother's age exceeds 40. An increased risk is also thought to occur after the father's age exceeds 55.*

Preparation of a Karyotype

A karyotype is a graphic display of the complete chromosome complement of a particular organism. Although the karyotypes of some organisms have been known since the 1920s, it was not until 1956 that the human karyotype was determined. This established, after a flurry of controversy, that the diploid number of human chromosomes is 46: 44 autosomes and two sex chromosomes. Interest in karyotyping received great impetus three years later with the discovery that the presence of an extra chromosome could be associated with a major medical problem, Down syndrome. Karyotypes are now prepared routinely for genetic counseling of couples at risk for having children with Down syndrome or other conditions associated with gross chromosome abnormalities.

The chromosomes shown in a karyotype are mitotic metaphase chromosomes, each consisting of two sister chromatids held together at the centromere. To prepare a karyotype, cells in the process of dividing are interrupted at metaphase by the addition of colchicine, a drug that interferes with the spindle microtubules and prevents the subsequent steps of mitosis from taking place. After treating and staining, the chromosomes are photographed, enlarged, cut out, and arranged according to size. Chromosomes of the same size are paired according to centromere position, which results in different "arm" lengths. From the karyotype, certain abnormalities, such as an extra chromosome or piece of a chromosome, can be detected.

Add colchicine

Add water

Spread one drop

White cells are transferred

Centrifuge

Fix with alcohol and stain

Red cells settle out

Stops all cells at metaphase

Causes cells to swell

White cells settle out

Cells at metaphase have burst

Photograph and enlarge

Cut out individual chromosomes from the photograph

Paste in order of diminishing size with centromere on line

(a)

(b)

11–15   (a) *Amniocentesis. The position of the fetus is first determined by ultrasound, with the sonogram continuously displayed on a video monitor. Then a needle is inserted into the amniotic cavity, and fluid containing fetal cells is withdrawn into a syringe. The cells are grown in tissue culture and then are analyzed for chromosome abnormalities and other genetic disorders. The procedure is usually not performed until the fourteenth week of pregnancy, to ensure both that there are enough fetal cells in the amniotic cavity to make detection possible and that there is sufficient amniotic fluid that removal of the small amount necessary for the test will not endanger the fetus.*

(b) A sonogram of the uterus of a pregnant woman carrying a four-month-old fetus. The solid black regions enclosed by the muscular uterine walls are the fluid surrounding the fetus. The fetus is lying on its back, with its head at the left, and appears to be sucking its thumb.

Prenatal Detection

A procedure known as **amniocentesis** makes possible the prenatal detection of Down syndrome and a number of other genetic disorders in the fetus. A thin needle is inserted through the mother's abdominal wall and through the membranes that enclose the fetus, and a sample of the amniotic fluid surrounding the fetus is withdrawn (Figure 11–15). The amniotic fluid contains living cells sloughed off from the skin of the fetus. These cells can be induced to divide, thus providing mitotic cells from which a karyotype can be made. Although amniocentesis must be done with great care, it is simple, quick, and usually harmless.

More recently, a technique has been developed for collecting cells from the chorion, one of the fetal membranes. The advantage of this method is that the test can be performed as early as the eighth week of pregnancy.

In most cases, treatment of conditions detected by prenatal testing—particularly those involving gross chromosome abnormalities—is not possible. Parents are faced with the difficult decision of whether or not to abort the affected fetus. However, as we shall see in Chapter 18, current developments in gene therapy are providing the first rays of hope that successful treatment of other types of genetic disorders is not an impossible dream.

The Consequences of Sexual Reproduction

Many organisms can reproduce both asexually (by mitosis and cytokinesis) and sexually. For example, most one-celled eukaryotes have a life cycle in which either sexual or asexual reproduction may take place, often depending on environmental circumstances. Many plants can also reproduce both sexually and asexually. In some animals, asexual reproduction can take place by the breaking off of a fragment of the parent animal, as occurs in sponges and sea anemones. Because of the careful copying process of mitosis, asexually produced individuals are genetically identical to their parents.

By contrast, the potential for genetic variability in sexually produced individuals is enormous. Figure 11–16 shows the possible distributions of chromosomes at meiosis in organisms with relatively few chromosomes. The red chromosomes are of paternal origin, and the black chromosomes of maternal origin. In the course of meiosis, these chromosomes are distributed among the haploid cells. As you can see, chromosomes

(a)   (b)   (c)

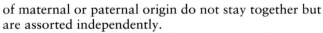

11–16   *The different chromosome combinations possible in the gametes of three organisms with relatively few chromosomes. Red represents chromosomes of paternal origin, and black represents chromosomes of maternal origin. (a) An organism in which the diploid number is 4, and the haploid number is 2. (b) An organism in which the diploid number is 6, and the haploid number is 3. (c) An organism in which the diploid number is 8, and the haploid number is 4.*

of maternal or paternal origin do not stay together but are assorted independently.

The number of possible chromosome combinations in the gametes is 2^n; 2 is the number of homologues in a pair, and n equals the haploid chromosome number. For example, if the haploid number of chromosomes is 2, the number of possible combinations of chromosomes is 2^2, or 4. If the haploid number is 3, the number of possible combinations is 2^3, or 8. If it is 4, then 16 different combinations (2^4) are possible.

A human male with his 46 chromosomes is capable of producing 2^{23} kinds of sperm cells—8,388,608 different combinations of chromosomes, a number larger than the population of New York City. Similarly, a human female is capable of producing 2^{23} kinds of egg

cells—8,388,608 different combinations of chromosomes. And this does not take into account the additional variations that may be introduced by crossing over.

As you know from your own observations, the similarities and differences between parents and children and among brothers and sisters vary greatly from child to child and from trait to trait. Some children closely resemble one parent in certain traits but resemble the other parent in other traits. Other children resemble a grandparent more closely than they resemble either parent. And, in some cases, cousins resemble each other more strongly than they resemble their own brothers and sisters. This variability in inheritance, which results primarily from the enormous diversity of possible gametes and the random meeting of two particular gametes in fertilization, was a complete puzzle until the mid-nineteenth century. Then, as we shall see in the next chapter, an Augustinian monk began a series of experiments that revealed the fundamental principles governing inheritance.

Summary

Sexual reproduction involves a special kind of nuclear division called meiosis. Meiosis is the process by which the chromosomes are reassorted and cells are produced that have the haploid chromosome number *(n)*. The other principal component of sexual reproduction is fertilization, the coming together of haploid cells to form the zygote. Fertilization restores the diploid number *(2n)*. There are characteristic differences among major groups of organisms as to when in the life cycle these events take place.

During interphase preceding meiosis, the chromosomes replicate. During prophase of the first meiotic division, they arrange themselves in homologous pairs. One homologue of each pair is of maternal origin, and one is of paternal origin. Each homologue consists of two identical sister chromatids. While the chromosomes are tightly paired, crossing over occurs between homologues, resulting in exchanges of chromosomal material.

In the first stage of meiosis, meiosis I, the homologues are separated. Two nuclei are produced, each with a haploid number of chromosomes, which, in turn, consist of two chromatids each. The nuclei may enter interphase, but the chromosomal material is not replicated. In the second stage of meiosis, meiosis II, the sister chromatids of each chromosome separate as in mitosis. The two nuclei divide, cytokinesis occurs, and four haploid cells are formed.

Meiosis in the human male produces four haploid spermatids from each diploid primary spermatocyte. The spermatids mature into sperm cells. In meiosis in the human female, however, the cytoplasm is unequally divided. The result is that each diploid primary oocyte yields only one haploid ovum; the other haploid nuclei form polar bodies, which soon disintegrate.

Chromosome abnormalities resulting from "mistakes" in meiosis include extra chromosomes (usually a result of nondisjunction—the failure of two homologues or of two chromatids to separate), translocations, deletions, and duplications. Down syndrome is among the disorders associated with an extra chromosome; it may be caused either by nondisjunction or, less commonly, by translocation. Extra sex chromosomes can also result from nondisjunction and are usually associated with sterility. Many genetic disorders can now be detected in the fetus by the use of amniocentesis, the collection of fetal cells from the amniotic fluid.

Each of the haploid cells produced by meiosis contains a unique assortment of chromosomes due to crossing over and random assortment of homologues. Thus meiosis is a source of variation in the offspring.

Questions

1. Distinguish among the following: haploid/diploid/polyploid; gamete/zygote/spore; sporophyte/gametophyte; homologues/tetrad; meiosis I/meiosis II; sex chromosome/autosome; nondisjunction/translocation; deletion/duplication.

2. (a) Dogs have a diploid chromosome number of 78. How many chromosomes would you expect to find in a gamete? In a kidney cell? (b) Plums have a haploid chromosome number of 24. How many chromosomes would you expect to find in the cell of a leaf? In the nucleus of a pollen grain?

3. Draw a diagram of a cell with six chromosomes $(n = 3)$ at meiotic prophase I. Label each pair of chromosomes differently (for example, label one pair A^1 and A^2, and another B^1 and B^2, etc.).

4. Diagram the eight possible gametes resulting from meiosis in a plant cell with six chromosomes $(n = 3)$. Label each chromosome differently, as in Question 3. Assume that crossing over does not occur.

5. Identify the stages of meiosis in the lily cells shown in the micrographs at the right.

(a)

(b)

(c)

(d)

6. (a) Compare metaphase of mitosis and metaphase II of meiosis. (b) Compare anaphase of mitosis with anaphase I and anaphase II of meiosis. In your answers, consider both the positions and composition of the chromosomes, as well as the consequences.

7. Compare and contrast the processes and the genetic consequences of meiosis and mitosis.

8. In some species, sexual reproduction—that is, fertilization and meiosis—does not require two parents. Explain how this is possible.

9. Nondisjunction can occur at the first meiotic division or the second. How do the effects differ? Include diagrams with your answers.

10. Describe the two types of chromosome abnormalities that can cause Down syndrome. With which type is it possible to unequivocally identify prospective parents who are at a higher-than-average risk of having a child with Down syndrome? How?

11. In our bodies and in those of most other animals, both mitosis and meiosis occur. What are the end products of these two processes? Where in our bodies do these two processes occur?

Suggestions for Further Reading

Alberts, Bruce, Dennis Bray, Julian Lewis, Martin Raff, Keith Roberts, and James D. Watson: *Molecular Biology of the Cell,* 3d ed., Garland Publishing, Inc., New York, 1994.

> *This outstanding cell biology text provides an excellent discussion of our current understanding of meiosis and fertilization.*

Anderson, Alun: "The Evolution of Sexes," *Science,* vol. 257, pages 324–326, 1992.

Mange, Arthur P., and Elaine Johansen Mange: *Genetics: Human Aspects,* 2d ed., Sinauer Associates, Inc., Sunderland, Mass., 1990.

> *A thorough account of human genetics. Chapters 7, 8, and 9 provide extensive coverage of chromosome abnormalities in both autosomes and sex chromosomes.*

Patterson, David: "The Causes of Down's Syndrome," *Scientific American,* April 1987, pages 52–60.

Raghavan, V.: "Germination of Fern Spores," *American Scientist,* vol. 80, pages 176–185, 1992.

Weiss, Rick: "Uneven Inheritance: A Genetic Quirk Leaves Some People with a Chromosomal Odd Couple," *Science News,* July 7, 1990, pages 8–11.

The great contribution of Gregor Mendel, the "father of genetics," was to show that inheritance follows predictable patterns. His experimental subject was very ordinary—the common garden pea—and the traits he selected for study were unremarkable—flower color and seed shape, for example. But the principles he discovered opened the floodgates for the studies that have come to be known as classical genetics.

Mendel's discoveries have important modern applications in medical genetics—the study of inherited disorders. One human disorder that follows a Mendelian pattern of inheritance is Huntington's disease, which causes a progressive destruction of brain cells, followed by death within 10 to 20 years of the onset of symptoms. Most human hereditary disorders are evident at birth or in early infancy, but Huntington's usually first manifests itself among individuals in their 30s and 40s. One tragic consequence of this late onset is that the individual often already has children who, in turn, carry the same lethal, hidden time bomb. According to Mendelian principles, any child of a man or woman who develops Huntington's has a fifty-fifty chance of also developing the disorder.

Research into Huntington's disease has led geneticists to the shores of Lake Maracaibo in Venezuela. Among the inhabitants of a remote village, reachable only by small fishing boats, are more than 100 individuals with Huntington's and another 1,000 who, like the children shown here, are at high risk for the disorder. All are descendants of a single woman, Maria Concepción Soto, who lived in the early nineteenth century and is thought to have inherited the disorder from a German sailor who was her father.

In recent years, geneticists have compiled a pedigree (a detailed chart) tracing the family relationships and incidence of Huntington's among almost 10,000 individuals. This pedigree has made it possible to predict which children in the population are at a high risk of eventually developing the disorder.

From an Abbey Garden: The Beginning of Genetics

For more than 3 billion years, living things have been reproducing, generation after generation, with each parent passing on to its offspring all the biological instructions necessary for the offspring to develop into the same kind of organism as the parent. Although biological inheritance has been the object of wonder—and considerable thought—since early in human history, only rather recently have we begun to understand how it works. In fact, the scientific study of heredity—known as **genetics**—did not really begin until the second half of the nineteenth century.

The questions that are the focus of genetics are among the most fundamental in biology, including not only the transmission of hereditary information from generation to generation but also the translation of that information into the specific characteristics of a particular organism. Our present understanding of these processes is based on the work of some of the most brilliant scientists our civilization has known. In this chapter and the ones that follow, we are going to trace their discoveries.

The Concept of the Gene

In the mid-nineteenth century, well before the developments in microscopy that made possible the discovery of chromosomes and their movements in mitosis and meiosis, biologists were attempting to answer some especially puzzling questions about heredity. For example, why does a child resemble its mother in certain features and its father in other features? Why do some features seem to skip a generation, with the result that a child resembles a grandparent more closely than either parent? Similar questions were of considerable practical importance to breeders of plants and animals, who were attempting to develop varieties with particular desirable characteristics.

At about the same time that Darwin was writing *The Origin of Species* (page 10) and Hertwig was observing the fusion of sea urchin egg and sperm (page 92), Gregor Mendel was beginning a series of experiments that provided the first useful answers to these basic questions about heredity. His work, pursued in a quiet monastery garden in what was then the city of Brünn in the Austro-Hungarian Empire (now Brno in the Czech Republic), and ignored until after his death, marks the beginning of modern genetics.

By Mendel's time, breeding experiments with domestic plants and animals had shown that both parents contribute to the characteristics of their offspring. Further, it was known that these contributions are carried in the gametes—that is, in the sperm and the eggs.

Mendel's great achievement was to demonstrate that inherited characteristics are determined by discrete factors, which are passed from one generation to the next and are parceled out separately (reassorted) in each generation. These discrete factors, which Mendel called *Elemente*, eventually came to be known as **genes.**

Mendel's Experimental Method

For his experiments in heredity, Mendel chose the common garden pea. It was a good choice. The plants were commercially available, easy to cultivate, and grew rapidly. Different varieties had clearly different characteristics that "bred true," appearing unchanged from one crop to the next. Further, the reproductive structures of the pea flower are entirely enclosed by petals, even when they are mature (Figure 12–1). Consequently, the flower normally self-pollinates; that is, sperm cells from the flower's own pollen fertilize its egg cells. Although the plants could be crossbred experimentally, accidental crossbreeding could not occur to confuse the experimental results. As Mendel said in his original paper, "The value and utility of any experiment are determined by the fitness of the material to the purpose for which it is used."

Mendel's choice of the pea plant for his experiments was not original. However, he was successful in formulating the fundamental principles of heredity— where others had failed—because of his approach to the problem. First, he tested a very specific hypothesis in a series of logical experiments. He planned his experiments carefully and imaginatively, choosing for study only clear-cut hereditary differences and avoiding characteristics that could be "more or less" apparent in the offspring. Second, he studied the offspring of not only the first generation but also of the second and subsequent generations. Third, and most important, he counted the different types of offspring resulting from each cross and then analyzed the results mathematically. Even though his mathematics was simple, the idea

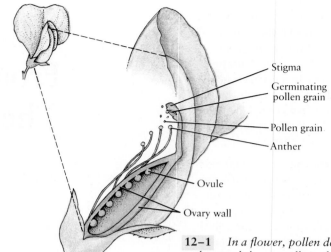

(a) Self-pollination

(b) Artificial cross-pollination

Stigma
Germinating pollen grain
Pollen grain
Anther
Ovule
Ovary wall

Pollen grains from another plant

Anthers removed

12–1   *In a flower, pollen develops in the anthers and the egg cells in the ovules. Pollination occurs when pollen grains land on the stigma, germinate, and grow down to the ovules, where they release the sperm nuclei. The nuclei of egg and sperm unite, and the fertilized eggs develop within the ovules, which are attached to the ovary wall. In the garden pea, the ovules with their enclosed embryos form the peas (the seeds), while the ovary wall becomes the pod.*

Pollination in most species of flowering plants involves the pollen from one plant (often carried by an insect) being caught on the stigma of another plant. This is called cross-pollination.

(a) In the pea flower, however, the stigma and anthers are completely enclosed by petals, and the flower, unlike most, does not open until after fertilization has taken place. Thus self-pollination is the normal course of events in the pea flower—that is, pollen is deposited on the stigma of the same flower in which it is produced. (b) In his crossbreeding experiments, Mendel pried open the flower bud before the pollen matured and removed the anthers with tweezers, preventing self-pollination. Then he artificially cross-pollinated the flower by dusting the stigma with pollen collected from another plant.

that a biological problem could be studied quantitatively was startlingly new. Fourth, and last, he organized his data in such a way that his results could be evaluated simply and objectively. The experiments themselves were described so clearly that they could be repeated and checked by other scientists, as eventually they were.

Table 12–1 Results of Mendel's Experiments with Pea Plants

| Trait | Original Crosses | | Second Filial Generation (F₂) | |
	Dominant × Recessive		Dominant	Recessive
Seed form	Round	× Wrinkled	5,474	1,850
Seed color	Yellow	× Green	6,022	2,001
Flower position	Axial	× Terminal	651	207
Flower color	Purple	× White	705	224
Pod form	Inflated	× Constricted	882	299
Pod color	Green	× Yellow	428	152
Stem length	Tall	× Dwarf	787	277

The Principle of Segregation

Mendel began with 32 different types of pea plants, which he studied for several years before he began his quantitative experiments. As a result of his preliminary observations, Mendel selected for study seven traits, each of which appeared in two conspicuously different forms in different varieties of plants. One variety of plant, for example, always produced yellow peas (seeds), while another always produced green ones. In one variety, the seeds, when dried, had a wrinkled appearance; in another variety, they were smooth. The complete list of traits is given in Table 12–1.

Mendel performed experimental crosses, removing the pollen-containing anthers from flowers and dusting their stigmas with pollen from a flower of another variety. He found that in every case in the first generation (now known in biological shorthand as the F₁, for "first filial generation"), all of the offspring showed only one of the two alternative characteristics. The other characteristic disappeared completely. For example, all of the seeds produced as a result of a cross between true-breeding yellow-seeded plants and true-breeding green-seeded plants were yellow. Similarly, all of the flowers produced by plants resulting from a cross between a true-breeding purple-flowered plant and a true-breeding white-flowered plant were purple. Characteristics that appeared in the F₁ generation, such as yellow seeds and purple flowers, Mendel called **dominant.**

The interesting question was: What had happened to the alternative characteristic—the greenness of the seed or the whiteness of the flower—that had been passed on so faithfully for generations by the parent stock? Mendel let the pea plant itself carry out the next stage of the experiment by permitting the F₁ plants to self-pollinate (Figure 12–2). The characteristics that had disappeared in the first generation reappeared in the second, or F₂, generation. These characteristics, which were present in the parental (**P**) generation and reappeared in the F₂ generation, must also have been present somehow in the F₁ generation, although not apparent there. Mendel called these characteristics **recessive.**

Mendel's actual counts are shown in Table 12–1. If you plug these numbers into your calculator, you will notice, as Mendel did, that the dominant and recessive characteristics appear in the second, or F₂, generation in ratios of approximately 3:1. How do the recessives disappear so completely and then reappear again, and always in such constant proportions? It was in answering these questions that Mendel made his greatest contribution. He saw that the appearance and disappearance of alternative characteristics, as well as their constant proportions in the F₂ generation, could be explained if hereditary characteristics are determined by

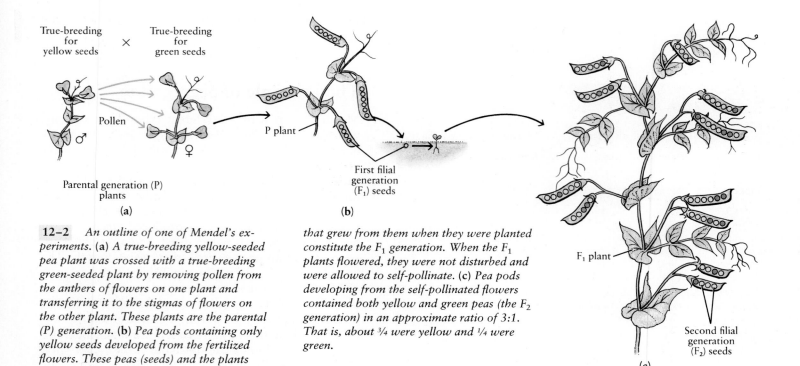

(a)

True-breeding for yellow seeds × True-breeding for green seeds

Pollen

♂

♀

Parental generation (P) plants

P plant

First filial generation (F₁) seeds

(b)

F₁ plant

Second filial generation (F₂) seeds

(c)

12–2 *An outline of one of Mendel's experiments. (a) A true-breeding yellow-seeded pea plant was crossed with a true-breeding green-seeded plant by removing pollen from the anthers of flowers on one plant and transferring it to the stigmas of flowers on the other plant. These plants are the parental (P) generation. (b) Pea pods containing only yellow seeds developed from the fertilized flowers. These peas (seeds) and the plants that grew from them when they were planted constitute the F₁ generation. When the F₁ plants flowered, they were not disturbed and were allowed to self-pollinate. (c) Pea pods developing from the self-pollinated flowers contained both yellow and green peas (the F₂ generation) in an approximate ratio of 3:1. That is, about ¾ were yellow and ¼ were green.*

discrete (separable) factors. These factors, Mendel realized, must have occurred in the F₁ plants in pairs, with one member of each pair inherited from the maternal parent and the other from the paternal parent. The paired factors separated again when the mature F₁ plants produced gametes, resulting in two kinds of gametes, with one member of the pair in each.

The hypothesis that every individual carries pairs of factors for each trait and that the members of a pair segregate (separate from each other) during the formation of gametes has come to be known as Mendel's first law, or the **principle of segregation.**

Consequences of Segregation

We now recognize that any given gene, for instance, the gene for seed color, can exist in different forms. These different forms of a gene are known as **alleles.** For example, yellow seededness and green seededness are determined by different alleles—that is, different forms—of the gene for seed color. The alleles are represented in biological shorthand by letters. By convention, capital letters are used for alleles for dominant characteristics and lowercase letters for alleles for recessive characteristics. Thus, for example, the allele for yellow seededness is represented by Y, and the allele for green seededness by y.

How a given trait is expressed in an organism is determined by the particular combination of two alleles for that trait present in the cells of the organism. If the two alleles are the same (for example, YY or yy), then the organism is said to be **homozygous** for that particular trait. (The term is from the Greek *homos*, meaning "same" or "similar," plus *zygotos*, "a pair.") If the two alleles are different from one another (for example, Yy), then the organism is **heterozygous** for that trait (from the Greek *heteros*, meaning "other" or "different").

When gametes are formed, alleles are passed on to them, but each gamete receives only one allele for any given gene (Figure 12–3). When two gametes combine to form a fertilized egg, the alleles occur in pairs again. If the two alleles in a given pair are the same (a homozygous state), the characteristic they determine will be expressed. If the alleles are different (a heterozygous state), one may be dominant over the other. A dominant allele is one that produces its particular characteristic in the heterozygous as well as in the homozygous state. A recessive allele, by contrast, is one that produces its particular characteristic only in the homozygous state.

The outward appearance and other observable characteristics of an organism constitute its **phenotype.** Even though a recessive allele may not be expressed in the phenotype, each allele of a pair still exists independently and as a discrete unit in the genetic makeup, or **genotype,** of the organism. The two alleles of a pair will separate from each other when gametes are again formed. Only if two recessive alleles come together in the fertilized egg—one from the female gamete and one from the male gamete—will the phenotype show the recessive characteristic.

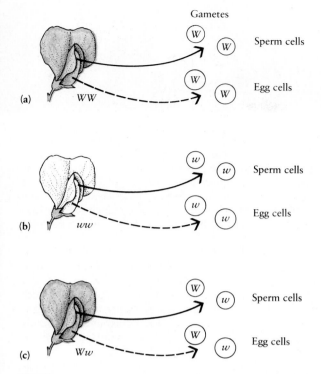

Gametes

12-3 *An example of the segregation of alleles in gamete formation. (a) In a pea plant homozygous for purple flowers (WW), all of the gametes, both sperm and egg, will contain a purple-flower (W) allele. (b) Similarly, in a pea plant homozygous for white flowers (ww), all of the gametes will contain a white-flower (w) allele. (c) However, in a pea plant that is heterozygous for flower color, half of the sperm cells will contain a W allele and half will contain a w allele. Similarly, half of the egg cells will contain a W allele and half will contain a w allele.*

The letters W and w are used here because of a convention by which geneticists derive the allele symbol from the first letter of the name of the form that is less common in nature (in this case, white pea flowers).

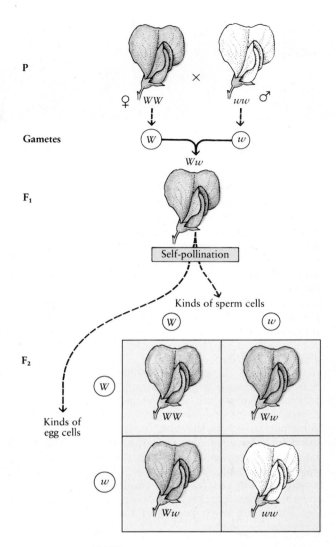

12-4 *Mendel's principle of segregation, as exemplified in the F₁ and F₂ generations following a cross between two homozygous parent (P) pea plants, one with two dominant alleles for purple flowers (WW) and the other with two recessive alleles for white flowers (ww). The female symbol ♀ identifies the plant that contributes the egg cells (female gametes), and the male symbol ♂ indicates the plant that contributes the sperm cells (male gametes).*

The phenotype of the offspring in the F₁ generation is purple, but note that the genotype is Ww. The F₁ heterozygote produces four kinds of gametes, ♀ W, ♀ w, ♂ W, ♂ w, in equal proportions. When this plant self-pollinates, the W and w egg and sperm cells combine randomly to form, on the average, ¼ WW (purple), 2/4 (or ½) Ww (purple), and ¼ ww (white) offspring. It is this underlying 1:2:1 genotypic ratio that accounts for the phenotypic ratio of 3 dominants (purple) to 1 recessive (white).

When pea plants homozygous for purple flowers are crossed with pea plants homozygous for white flowers, only pea plants with purple flowers are produced. However, each plant in this F₁ generation carries both an allele for purple and an allele for white. What happens in the F₂ generation when the F₁ generation self-pollinates? One of the simplest ways to predict the types of offspring that will be produced is to diagram the cross as shown in Figure 12–4. This sort of checkerboard diagram is known as a **Punnett square,** after the English geneticist who first used it for the analysis of genetically determined traits.

As you study Figure 12–4, notice that the result would be exactly the same if an F₁ individual were crossed with another F₁ individual. That is how these breeding experiments are performed with animals and with plants that are not self-pollinating.

From Pea Plants to Humans

The principles of genetics are, of course, the same for humans as they are for pea plants or members of any other diploid eukaryotic species. Historically, however, there have been important differences in the methodology of genetic studies. Breeding experiments, so readily performed with Mendel's pea plants, are not possible with humans. Thus, until very recently, most of our knowledge about human genetics has come from the analysis of inheritance patterns. Family pedigrees, which reveal these patterns, have usually been worked out only when a particular abnormality is associated with a medical problem.

Huntington's Disease

In terms of the numbers of individuals affected, serious medical problems caused by dominant alleles are rare, simply because severely afflicted individuals are typically unable to reproduce. Perhaps the most familiar disorder caused by a dominant allele is Huntington's disease (see page 207). Although the incidence of Huntington's is low (about 5 to 10 persons in every 100,000), the devastating physical and mental decline that it causes, coupled with its late onset, means that it affects the lives of many others.

Phenylketonuria (PKU)

Many inherited disorders are, like the white flowers of Mendel's pea plants, the result of the coming together of two recessive alleles. Individuals heterozygous for the gene are usually symptom-free. One of the best-studied examples of a disorder inherited as a Mendelian recessive is phenylketonuria, or PKU. Individuals with PKU lack an enzyme that plays a critical role in the breakdown of the amino acid phenylalanine. When this enzyme is missing or deficient, phenylalanine and its abnormal breakdown products accumulate in the bloodstream and urine. These substances are harmful to the cells of the developing nervous system. Without treatment, the result is profound mental retardation and a life span of little more than 30 years. About 1 in every 15,000 infants born in the United States is homozygous for this allele, which is most common among individuals of Northern European extraction.

It is not yet known how the high levels of phenylalanine and its abnormal breakdown products bring about the tragic neurological symptoms. However, the knowledge we do have is enough to effectively treat infants with PKU and prevent the symptoms from appearing. Most states now require routine tests of all newborn babies in order to detect PKU homozygotes. Those identified at birth are put on a special diet containing low amounts of phenylalanine—enough to supply dietary needs but not enough to permit toxic accumulations. After more than 30 years of experience, it is clear that low phenylalanine diets, maintained for at least the first six years of life (when brain development is still in progress), allow PKU homozygotes to develop normally.

More recently, a biochemical test has been developed that, in conjunction with amniocentesis, allows the detection of PKU in the developing fetus. In such cases, the mother-to-be is placed on a low phenylalanine diet for the remainder of her pregnancy, thus protecting her child from any early effects of PKU.

Tay-Sachs Disease

Another neurological disorder that occurs only in individuals homozygous for a recessive allele is Tay-Sachs disease. As in PKU, Tay-Sachs homozygotes appear normal at birth and through the early months. However, by about eight months, symptoms of severe listlessness become evident. Blindness usually occurs within the first year, and afflicted children rarely survive past their fifth year. Homozygous individuals lack an enzyme, normally found in the lysosomes of brain cells, that breaks down a specific lipid in the cells. In the child lacking this enzyme, the lysosomes of the brain cells fill with the lipid and swell, and the cells die. There is no therapy yet available for Tay-Sachs disease.

While Tay-Sachs disease is a rare disorder in the general population (1 in 300,000 births), until recently it has had a much higher incidence (1 in 3,600 births) among Jews of Eastern and Central European extraction, who make up more than 90 percent of the American Jewish population. It is estimated that among this population approximately 1 in 28 individuals is a heterozygous carrier of the Tay-Sachs allele. The development of a blood test measuring levels of the enzyme that is deficient in Tay-Sachs disease has made it possible for prospective parents to determine if they are carriers. Since its development, this test has been so extensively utilized by prospective parents in the American Jewish population that the incidence of Tay-Sachs births has now dropped dramatically. In fact, in the United States today, the majority of the Tay-Sachs babies born are to non-Jewish parents.

Sickle-Cell Anemia and Cystic Fibrosis

Other serious conditions caused by a recessive allele in the homozygous state are sickle-cell anemia (page 64) and cystic fibrosis. Sickle-cell anemia has a high incidence among African-Americans (1 in 400 births) and also occurs with some frequency among individuals of Mediterranean extraction. Cystic fibrosis is the most common genetic disorder among individuals of European descent (1 in 2,000 births). Affecting cells specialized for secretion, its most serious consequence is the production of copious amounts of thick mucus in the lungs. Although the resulting congestion can be relieved, the tissues of the lungs are gradually destroyed, leading to death. The average life span of cystic fibrosis victims is presently about 24 years.

As we shall see in subsequent chapters, the underlying basis of both sickle-cell anemia and cystic fibrosis is now understood, and effective treatments are, at long last, being developed.

A number of familiar human characteristics are governed by simple Mendelian inheritance of dominant alleles. Among these are dimples and cleft chin, as illustrated by the actor Kirk Douglas and his son Michael. Although this allele has no medical consequences for the individual who carries it, it has undoubtedly caused the hearts of many others to beat a bit faster.

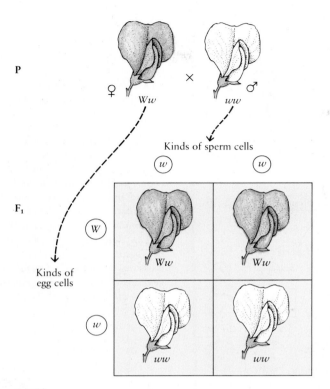

P

♀
Ww   ×   *ww*   ♂

Kinds of sperm cells

Ⓦ *w*   Ⓦ *w*

F₁

Ⓦ *W*   *Ww*   *Ww*

Kinds of
egg cells   Ⓦ *w*   *ww*   *ww*

12–5   *A testcross. In order for a pea flower to be white, the plant must be homozygous for the recessive allele* (ww). *But a purple pea flower can be produced by a plant with either a* Ww *or a* WW *genotype. How can you determine the genotype of a purple-flowered plant? Geneticists solve this problem by crossing such plants with homozygous recessives. This sort of experiment is known as a testcross. As shown here, a phenotypic ratio in the F₁ generation of one purple to one white indicates that the purple-flowering parent used in the testcross must have been heterozygous.*

What would have been the result if the plant being tested had been homozygous for the purple-flower allele?

A Testcross

In order to test the hypothesis that alleles occur in pairs and that the two alleles of a pair segregate during gamete formation, it is necessary to perform an additional experiment: cross purple-flowering F₁ plants (the result of a cross between true-breeding purple- and white-flowering plants) with white-flowering plants. To the casual observer, it would appear as if this were simply a repeat of Mendel's first experiment, crossing plants having purple flowers with plants having white flowers. But if Mendel's hypothesis is correct, the results will be different from those of his first experiment. Can you predict the results of such a cross? Stop a moment and think about it.

The easiest way to analyze the possible results of such a cross is with a Punnett square, as in Figure 12–5. This type of experiment, which reveals the genotype of the parent with the dominant phenotype, is known as a **testcross**. A testcross is an experimental cross between an individual with the dominant phenotype (genotype unknown) for a given trait and another individual with the recessive phenotype (and thus homozygous for the recessive allele). Whether only one phenotype or two different phenotypes are produced in the offspring indicates whether the parent with the dominant phenotype is homozygous or heterozygous for the trait being studied. The testcross shown in Figure 12–5 reveals that the genotype of the plant being tested is *Ww* rather than *WW*.

The Principle of Independent Assortment

In a second series of experiments, Mendel studied crosses between pea plants that differed in two characteristics. For example, one parent plant produced peas that were round and yellow, and the other had peas that were wrinkled and green. Both the round and yellow characteristics are dominant, and the wrinkled and green are recessive (see Table 12–1). As you would expect, all the F₁ seeds produced by a cross between the true-breeding parental types were round and yellow.

When these F₁ seeds were planted and the resulting flowers allowed to self-pollinate, 556 seeds were produced in the F₂ generation. Of these, 315 showed the two dominant characteristics, round and yellow, but only 32 combined the recessive characteristics, green and wrinkled. All the rest of the seeds were unlike either parent; 101 were wrinkled and yellow, and 108 were round and green. Totally new combinations of characteristics had appeared.

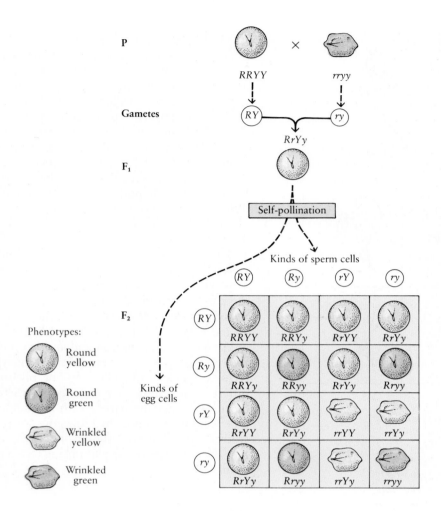

P

RRYY   ×   rryy

Gametes   RY   ry

RrYy

F₁

Self-pollination

Kinds of sperm cells

RY   Ry   rY   ry

F₂

Phenotypes:

Round yellow

Round green

Wrinkled yellow

Wrinkled green

Kinds of egg cells

RY | RRYY | RRYy | RrYY | RrYy
Ry | RRYy | RRyy | RrYy | Rryy
rY | RrYY | RrYy | rrYY | rrYy
ry | RrYy | Rryy | rrYy | rryy

12–6 *One of the experiments from which Mendel derived his principle of independent assortment. A plant homozygous for round (RR) and yellow (YY) peas is crossed with a plant having wrinkled (rr) and green (yy) peas. The F₁ peas are all round and yellow, but notice how the characteristics will, on the average, appear in the F₂ generation. Of the 16 possible combinations in the offspring, nine show the two dominant characteristics (round and yellow), three show one combination of dominant and recessive (round and green), three show the other combination (wrinkled and yellow), and one shows the two recessives (wrinkled and green). This 9:3:3:1 distribution of the phenotypes is always the expected result from a cross involving two independently assorting genes, each with one dominant and one recessive allele in each of the parents.*

This experiment did not, however, contradict Mendel's previous results. If the two traits, seed color and seed shape, are considered independently, round and wrinkled still appeared in a ratio of approximately 3:1 (423 round to 133 wrinkled), and so did yellow and green (416 yellow to 140 green). But the seed shape and the seed color characteristics, which had originally been combined in a certain way (round only with yellow and wrinkled only with green), behaved as if they were entirely independent of one another (yellow could now be found with wrinkled and green with round). From this, Mendel formulated his second law, the **principle of independent assortment.** This principle states that when gametes are formed, the alleles of one gene segregate independently of the alleles of another gene.

Figure 12–6 diagrams Mendel's interpretation of these results. It shows why, in a cross involving two independently assorting genes, each with one dominant and one recessive allele, the phenotypes in the offspring will, on the average, be in the ratio of 9:3:3:1. The fraction 9/16 represents the proportion of the F₂ offspring expected to show the two dominant characteristics, 1/16 the proportion expected to show the two recessive characteristics, and 3/16 and 3/16 the proportions expected to show the two alternative combinations of dominant and recessive characteristics.

The 9:3:3:1 ratio in the F₂ offspring holds true when one of the original parents is homozygous for both recessive characteristics and the other parent is homozygous for both dominant characteristics, as in the experiment just described *(RRYY × rryy)*. It also holds true when each original parent is homozygous for one recessive and one dominant characteristic *(rrYY × RRyy)*.

The Influence of Mendel

Mendel's experiments were first reported in 1865 before a small group of people at a meeting of the Brünn Natural History Society. None of them, apparently, understood the significance of Mendel's results. His paper was, however, published the following year in the *Proceedings* of the Society, a journal that was circulated to libraries all over Europe. In spite of this, his work was ignored for 35 years, during most of which he devoted himself to the duties of an abbot, and he received no scientific recognition until after his death.

It was not until 1900 that biologists were finally able to understand and accept Mendel's findings. Within a single year, his paper was independently rediscovered by three scientists, each working in a different European country. Each of the scientists had done similar breeding experiments and was searching the scientific literature for related work that would provide confirmation of his results. And each found, in Mendel's brilliant analysis, that much of his own work had been anticipated.

During the 35 years that Mendel's work remained in obscurity, great improvements were made in microscopy and, as a consequence, in **cytology**—the study of cell structure. It was during this period that chromosomes were discovered and their movements during mitosis and meiosis were first observed and recorded.

Cytology and Genetics Meet: Sutton's Hypothesis

In 1902, shortly after the rediscovery of Mendel's work, Walter Sutton, a graduate student at Columbia University, was studying the formation of sperm cells in male grasshoppers. Observing the process of meiosis, Sutton noticed that the chromosomes were paired early in the first meiotic division. He also noticed that the two chromosomes of any one pair had physical resemblances to one another. In diploid cells, he noted, chromosomes apparently come in pairs. The pairing was obvious only at meiosis, although the discerning eye might also find the matching, but unpaired, homologues during metaphase of mitosis (see, for example, Figure 11–12a on page 199).

Sutton was struck by the parallels between what he was seeing and the first principle of Mendel—the principle of segregation. Suddenly the facts fell into place. Suppose chromosomes carried genes, the factors described by Mendel, and suppose the alleles of a gene occurred on homologous chromosomes. Then the alleles would always remain independent and so would be separated at meiosis I as homologous chromosomes separated. New combinations of alleles would be formed as gametes fused at fertilization. Mendel's principle of the segregation of alleles could thus be explained by the segregation of the homologous chromosomes at meiosis.

12–7   *Gregor Mendel, holding a fuchsia, is third from the right in this photograph of members of the Augustinian monastery in Brünn in 1862. As a teenager, Mendel received a thorough training in agricultural science. He then spent two years at the University of Vienna, where he studied physics, chemistry, mathematics, and botany. Although Mendel published only two scientific papers during his lifetime, he performed breeding experiments with a variety of plants until his election as abbot of the monastery in 1871. Unfortunately, almost all of Mendel's papers relating to his scientific work were destroyed shortly before or after his death in 1884.*

Mendel and the Laws of Probability

In applying mathematics to the study of heredity, Mendel was asserting that the laws of probability apply to biology as they do to the physical sciences. Toss a coin. The probability that it will turn up heads is fifty-fifty, that is, one chance in two, or ½. The probability that it will turn up tails is also one chance in two, or ½. The probability that it will turn up one or the other is certain, or one chance in one.

Now toss two coins. The probability that one will turn up heads is again ½. The probability that the second will turn up heads is also ½. The probability that *both* coins will turn up heads is ½ × ½, or ¼. The probability that both coins will turn up tails is also ½ × ½. Similarly, the probability of the first turning up tails and the second turning up heads is ½ × ½, and the probability of the second turning up tails and the first heads is ½ × ½. We can diagram these four possible outcomes of a two-coin toss in a Punnett square (see figure), which indicates that the combination in each square has an equal probability of occurring.

The two-coin toss provides an example of what is known as the **product rule of probability.** This rule states that the probability of two independent events occurring together is simply the probability of one occurring alone multiplied by the probability of the other occurring alone. For instance, in Mendel's experiment diagrammed in Figure 12–4, the probability that a gamete produced by an F_1 plant of *Ww* genotype will carry the *W* allele

is ½, and the probability that it will carry the *w* allele is ½. Therefore the probability of any specific combination of the two alleles in the offspring—that is, *WW, Ww, wW,* or *ww*—is ½ × ½, or ¼. It was undoubtedly the observation that one-fourth of the offspring in the F_2 generation showed the recessive phenotype that indicated to Mendel that he was dealing with a simple case of the laws of probability.

Returning to our coin toss, if there were three coins involved, the probability of any given combination would be simply the product of all three individual probabilities: ½ × ½ × ½, or ⅛. Similarly, with four coins, the probability of any specific combination is ½ × ½ × ½ × ½, or 1/16. The Punnett square in Figure 12–6 expresses the probability of each of any one of four possible phenotype combinations.

When there is more than one possible arrangement of the events producing a specified outcome, the individual probabilities are added. For instance, what is the probability of throwing a head and a tail, in either order? There are two ways you could do this: by throwing a head first and then a tail (HT), or a tail first and then a head (TH). The probability of throwing a head first and then a tail (HT) is ½ × ½, or ¼. The probability of throwing a tail first and then a head (TH) is also ½ × ½, or ¼. Thus, the probability that a head and a tail will be thrown, in whichever

order, is the sum of their individual probabilities: (½ × ½) + (½ × ½) = ¼ + ¼ = ½. This is known as the **sum rule of probability.**

In the cross diagrammed in Figure 12–4, a heterozygote is produced by either *Ww* or *wW*. The probability of a heterozygote in the F_2 generation is the sum of the probability of each of the two possible combinations: ¼ + ¼ = ½.

The sum rule of probability, like the product rule, applies in more complex cases as well. For example, if you were asked the probability of throwing two heads and a tail, the answer would be ⅜. Three combinations are possible: HHT, HTH, and THH. For each of these combinations, the probability is ½ × ½ × ½ = ⅛, that is, the product of three independent throws. Thus, the probability of throwing two heads and a tail is the sum of the probability of each of the three possible combinations: ⅛ + ⅛ + ⅛ = ⅜.

Notice that in planning his experiments, Mendel made several assumptions: (1) of the male gametes produced, one-half contain one paternal allele and one-half contain the other paternal allele for each gene; (2) of the female gametes produced, one-half contain one maternal allele and one-half contain the other maternal allele for each gene; (3) the male and female gametes combine at random. Thus, the laws of probability could be employed—an elegant marriage of biology and mathematics.

If you toss two coins 4 times, it is unlikely that you will get the precise results diagrammed here. However, if you toss two coins 100 times, you will come close to the proportions predicted in the Punnett square, and if you toss two coins 1,000 times, you will be very close indeed. As Mendel knew, the ratio of dominants to recessives in the F_2 generation might well not have been so clearly visible if he had been dealing with a small sample. The larger the sample, however, the more closely it will conform to the results predicted by the laws of probability.

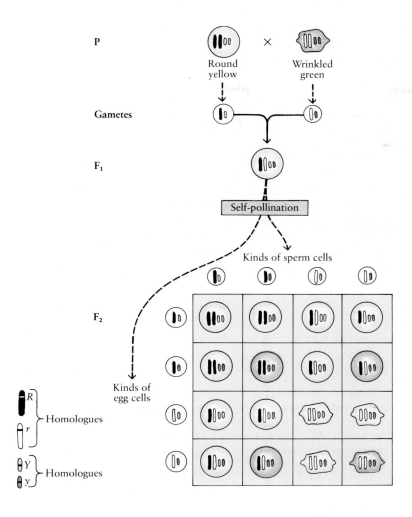

P — Round yellow × Wrinkled green

Gametes

F₁

Self-pollination

Kinds of sperm cells

F₂

Kinds of egg cells

R
r — Homologues

Y
y — Homologues

12–8 *The chromosome distributions in Mendel's cross of round yellow and wrinkled green peas, according to Sutton's hypothesis. Although the pea has 14 chromosomes (n = 7), only four are shown here, the homologous pair carrying the alleles for round or wrinkled and the homologous pair carrying the alleles for yellow or green. (This selection of specific pairs of homologous chromosomes is analogous to Mendel's selection of specific traits to study.) As you can see, one parent plant is homozygous for the dominants, and the other parent plant is homozygous for the recessives. Therefore, the gametes of one parent can contain only R and Y, and the gametes of the other parent can contain only r and y. The F₁ generation, therefore, must be Rr and Yy. When a cell of this generation undergoes meiosis, R is separated from r and Y from y when the members of each homologous pair separate at anaphase I. Because the alleles of the two genes are on different pairs of homologous chromosomes, they are assorted independently in meiosis. Four different types of haploid egg nuclei are possible, as the diagram reminds us, and also four different types of haploid sperm nuclei. These can combine in 4 × 4, or 16 different ways, as illustrated in the Punnett square.*

What about Mendel's second principle in relation to the movement of the chromosomes during meiosis? This principle, as we have seen, states that the alleles of different genes assort independently. We can see that this can be true if—and this is an important point—the genes are on different pairs of homologous chromosomes, as shown in Figure 12–8.

On the basis of these parallels, Sutton proposed that the factors described by Mendel are carried on the chromosomes. Although Sutton's proposal was widely accepted as a working hypothesis, proof of the physical location of the gene depended on further studies, which we shall consider in the next chapter.

Summary

How are hereditary characteristics passed from generation to generation? Gregor Mendel provided the first clear answers to this question. His breeding experiments in pea plants revealed that hereditary characteristics are determined by discrete factors (now called genes) that occur in pairs. According to Mendel's principle of segregation, the members of each pair segregate (that is, separate from one another) during gamete formation. When two gametes come together in fertilization, the offspring receives one member of each pair from each parent.

The members of a given pair may be the same, in which case the individual is homozygous for the trait determined by that gene, or they may be different, in which case the individual is heterozygous for that trait. Different forms of the same gene are known as alleles.

The genetic makeup of an organism is known as its genotype. Its observable characteristics are known as its phenotype. An allele that is expressed in the phenotype of a heterozygous individual to the exclusion of the other allele is a dominant allele. An allele whose effects are concealed in the phenotype of a heterozygous individual is a recessive allele. In crosses involving two individuals heterozygous for the same gene, the expected ratio of dominant to recessive phenotypes in the offspring is 3:1.

Mendel's other great principle—independent assortment—applies to the behavior of two or more different genes. This principle states that, during gamete formation, the alleles of one gene segregate independently of the alleles of another gene. When organisms heterozygous for each of two independently assorting genes are crossed, the expected phenotypic ratio in the offspring is 9:3:3:1.

Sutton was among the first to notice the analogy between the behavior of the chromosomes at meiosis and the segregation and assortment of the factors described by Mendel. On the basis of this observation, Sutton proposed that genes are carried on chromosomes.

Questions

1. Distinguish between the following terms: gene/allele; dominant/recessive; homozygous/heterozygous; genotype/phenotype; the F_1/the F_2.

2. State Mendel's two principles in your own words.

3. In the experiments summarized in Table 12–1, which of the alternative characteristics appeared in the F_1 generation?

4. (a) What is the genotype of a pea plant that breeds true for tall? (Use the symbol T for tall, t for dwarf.) What possible gametes can be produced by such a plant? (b) What is the genotype of a pea plant that breeds true for dwarf? What possible gametes can be produced by such a plant? (c) What will be the genotype of the F_1 generation produced by a cross between a true-breeding tall pea plant and a true-breeding dwarf pea plant? (d) What will be the phenotype of this F_1 generation? (e) What will be the probable distribution of characteristics in the F_2 generation? Illustrate with a Punnett square.

5. PKU, phenylketonuria, is a disorder caused by the presence of two recessive alleles of a particular gene. Individuals who are homozygous for the dominant allele of that gene or who are heterozygous show no signs of PKU. If two parents, neither of whom has the disorder, have a child with PKU, what are their genotypes with respect to PKU? What is the probability that their next child will have the same disorder?

6. The inheritance of sickle-cell anemia is similar to that of PKU—that is, it occurs when two recessive alleles of a particular gene come together in the same individual. If two healthy parents have a child with sickle-cell anemia, what are their genotypes with respect to this allele? (Use H^a to represent the normal hemoglobin allele, and h^s to represent the sickle-cell allele.) Having had one such child, what is the probability that their next child will have the same disorder?

7. What proportion of the children of the parents in Question 6 would be expected to be carriers of the sickle-cell allele (that is, heterozygous)? What proportion of their children would not be expected to carry the allele? (Draw a Punnett square to diagram this problem.)

8. Why is a homozygous recessive always used in a testcross?

9. The ability to taste a bitter chemical, phenylthiocarbamide (PTC), is due to a dominant allele. In terms of tasting ability, what are the possible phenotypes of a man both of whose parents are tasters? What are his possible genotypes? (Use N for the allele for tasting ability, and n for the allele for the absence of tasting ability.)

10. If the man in Question 9 marries a woman who is a nontaster, what proportion of their children could be tasters? Suppose one of the children is a nontaster. What would you know about the father's genotype? Explain your results by drawing Punnett squares.

11. A taster and a nontaster have four children, all of whom can taste PTC. What is the probable genotype of the parent who is a taster? Is there another possibility?

12. A pea plant that breeds true for round, green seeds *(RRyy)* is crossed with a plant that breeds true for wrinkled, yellow seeds *(rrYY)*. Each parent is homozygous for one dominant characteristic and for one recessive characteristic. (a) What is the genotype of the F_1 generation? (b) What is the phenotype? (c) The F_1 seeds are planted and their flowers are allowed to self-pollinate. Draw a Punnett square to determine the ratios of the phenotypes in the F_2 generation. How do the results compare with those of the experiment shown in Figure 12–6?

13. In Jimson weed, the allele for violet petals *(W)* is dominant over the allele for white petals *(w)*, and the allele for prickly capsules *(S)* is dominant over the allele for smooth capsules *(s)*. A plant with white petals and prickly capsules was crossed with one that had violet petals and smooth capsules. The F_1 generation was composed of 47 plants with white petals and prickly capsules, 45 plants with white petals and smooth capsules, 50 plants with violet petals and prickly capsules, and 46 plants with violet petals and smooth capsules. What were the genotypes of the parents?

14. Suppose you have just flipped a coin five times and it has turned up heads every time. What is the probability that the next time you flip it, it will turn up tails?

15. What is the probability of drawing two aces out of a deck of 52 playing cards, one the ace of hearts and the other the ace of spades?

16. (a) Suppose you would like to have a family consisting of two girls and a boy. What are your chances, assuming you have no children now? (**b**) If you already have one boy, what are your chances of completing your family as you would like? (**c**) If you have two girls, what is the probability that the next child will be a boy?

17. Mendel did not know of the existence of chromosomes. Had he known, what change might he have made in his second principle?

18. Segregation of alleles can occur at either of two stages of meiosis. Identify the two stages, and explain what happens in each of them.

19. You do not look exactly like your mother or your father. Why is this so? Explain how you might have inherited some of your maternal grandfather's characteristics. (Start with a gamete produced by your grandfather, and end with one of your somatic cells.)

Suggestions for Further Reading

Jacob, François: *The Logic of Life: A History of Heredity*, Pantheon Books, New York, 1982.*

Jacob's principal theme concerns the changes in the way people have looked at the nature of living beings. These changes, which are part of our total intellectual history, determine both the pace and direction of scientific investigation. The opening chapters are particularly brilliant.

Klug, William S., and Michael R. Cummings: *Concepts of Genetics*, 3d ed., Macmillan Publishing Company, New York, 1991.

In this textbook, designed for a first course in genetics, the emphasis is on fundamental concepts, which are illustrated by carefully and clearly explained examples. It also contains many sample problems and their solutions.

Mange, Arthur P., and Elaine Johansen Mange: *Genetics: Human Aspects,* 2d ed., Sinauer Associates, Inc., Sunderland, Mass., 1990.

A thorough account of human genetics.

Oldroyd, David: "Gregor Mendel: Founding-father of Modern Genetics," *Endeavour*, vol. 8, pages 29–31, 1984.

Peters, James A. (ed.): *Classic Papers in Genetics*, Prentice-Hall, Inc., Englewood Cliffs, N.J., 1959.*

Includes papers by most of the scientists, including Mendel and Sutton, who were responsible for the major developments in genetics from the 1850s through the 1950s. The authors are surprisingly readable, and the papers give a feeling of immediacy that no modern account can achieve.

Pines, Maya: "In the Shadow of Huntington's," *Science 84,* May 1984, pages 30–39.

Strickberger, Monroe W.: *Genetics*, 3d ed., Macmillan Publishing Company, New York, 1986.

A cohesive account of the science, this book provides a broad coverage of classical genetics.

*Available in paperback.

13

Mendel Rediscovered: Classical Genetics

"Anatomy is destiny," according to one widely quoted source. Some of us may disagree with this statement and with its author. However, we are all likely to admit that whether we turned out male or female has played an important role in our life histories.

The determinants of this fundamental biological characteristic are two intensively studied microscopic structures: the human X and Y chromosomes. All human egg cells contain, as a result of meiosis, one X chromosome. Each human sperm cell contains either an X or a Y. Consequently, depending on which sperm wins the race to the egg, the zygote will be either XX (female) or XY (male)—and all else that follows from that.

As you can readily see in these scanning electron micrographs, the X chromosome (on the left) is much larger than the Y. In keeping with its size, the X carries many more genes than the Y. This difference is a significant factor in medical genetics. Suppose you have a defective allele on one X chromosome. If you are female, a normal allele on the other X chromosome can usually compensate for any deficiency. If you are male, however, you have only the one allele and so its effects are exposed. That is why some hereditary conditions, such as color blindness, hemophilia, and some forms of muscular dystrophy, are so much more common in males than in females.

The visible differences in the chromosome complements of males and females in many species provided some of the first evidence that the secrets of inheritance were to be found in the chromosomes. Differences in inheritance patterns in males and females noted by Mendel's successors lent strong support to the developing realization—now a commonplace fact—that particular chromosomes carry particular genes. Today, it is becoming possible to pinpoint the precise location of particular genes on particular chromosomes. An example is the human SRY gene (for "sex-determining region of the Y"), which is responsible for setting in motion the processes in embryonic development that result in maleness. Its location near the tip of the Y chromosome is marked by the arrow.

In the decades following the rediscovery of Mendel's work in 1900, an enormous number of studies were carried out in genetics, as biologists sought to confirm and extend Mendel's observations. The investigations of this period included a wide variety of breeding experiments with many different types of organisms, as well as numerous cytological studies. The methods and the data that resulted from this span of about 50 years form the core of what is now known as **classical genetics.**

Classical genetics not only provided new understandings of the fundamental principles of heredity but also generated a wealth of data on the inheritance of specific traits in a diversity of plant and animal species. This data base has been—and continues to be—of great practical value to breeders of plants and animals and also to physicians concerned with human hereditary disorders. Moreover, both the data base and the methods of classical genetics are assuming a new importance as biologists try to relate the current avalanche of discoveries in molecular genetics to the phenotypic characteristics of living organisms.

Broadening the Concept of the Gene

The new studies that followed the rediscovery of Mendel's 1866 paper confirmed his work in principle. However, many of the studies showed that the patterns of inheritance are not always as simple and direct as Mendel's reported results had indicated. This is not surprising when you recall that Mendel had carefully selected certain traits for study, using only those that showed clear-cut differences.

The phenotypic effects of a particular gene, it was found, are influenced not only by the alleles of that gene present in the organism, but also by other genes and by the environment. And many, indeed most, traits are influenced by more than one gene, just as most genes can influence more than a single trait. Perhaps most surprising was the discovery that genes can change.

Mutations

In 1902, the Dutch botanist Hugo de Vries reported results of his studies on Mendelian inheritance in the evening primrose (Figure 13–1). Heredity in the primrose, he found, was generally orderly and predictable, as in the garden pea. Occasionally, however, a characteristic appeared that was not present in either parent or indeed anywhere in the lineage of that particular plant.

De Vries hypothesized that such characteristics came about as the result of abrupt changes in genes and that the characteristic produced by a changed gene was then passed along like any other hereditary characteristic. De Vries called these abrupt hereditary changes **mutations,** and organisms exhibiting such changes came to be known as **mutants.** Different alleles of a gene, de Vries proposed, arose as a result of mutations. For example, the allele for wrinkled peas is thought to have arisen as a mutation of the gene for round peas.

As it turned out, only about 2 of some 2,000 changes observed by de Vries in the evening primrose were actually mutations. The vast majority of the changes were simply the result of new allele combinations, produced in the course of meiosis, crossing over, and fertilization. However, de Vries's concept of mutation as the source of genetic variation proved of great importance, even though most of his examples were not valid.

Mutations and Evolutionary Theory

An important gap in Darwin's theory of evolution, first published in 1859, was the lack of explanation as to how variations can persist in populations. Mendel's work filled this gap. Segregation of alleles explained how variation is maintained from generation to generation. Independent assortment explained how individuals could have characteristics in combinations not present in either parent and so perhaps be better adapted, in evolutionary terms, than either parent.

Mendelian principles, however, presented new problems to the early evolutionists. If all hereditary variations were to be explained by the reshuffling process proposed by Mendel, there would be little or no opportunity for the kind of change in organisms envisioned by Darwin. As a result of mutations, however, there is a wide range of variability in natural populations. In a complex or shifting environment, a particular variation may give an individual or its offspring a slight edge. Although mutations seldom, if ever, determine the direction of evolutionary change, they are now recognized as the ultimate—and continual—source of the hereditary variations that make evolution possible.

13–1   *A flower of one species* (Oenothera biennis) *of evening primrose. Ironically, most of the genetic changes in the evening primrose observed by Hugo de Vries, one of the rediscoverers of Mendel's work, were not mutations. However, the concept that the basic genetic constitution of an organism is subject to change proved a valid and important one.*

Allele Interactions

Incomplete Dominance and Codominance

As the early studies in genetics proceeded, it soon became apparent that dominant and recessive characteristics are not always as clear-cut as in the seven traits studied by Mendel in the pea plant. Some characteristics appear to blend. For instance, a cross between a red-flowering snapdragon and a white-flowering snapdragon produces heterozygotes that are pink (Figure 13–2).

This phenomenon, in which the phenotype of the heterozygote is intermediate between those of the two homozygotes, is known as **incomplete dominance.** As we shall see in Chapter 15, it is a result of the combined effects of gene products—in this case, pigments synthesized by cells in the flower petals. When the heterozygous pink snapdragons are allowed to self-pollinate, the red and white characteristics are sorted out once again, showing that the alleles themselves, as Mendel had asserted, remain discrete and unaltered.

In other cases, alleles may act in a **codominant** manner, with heterozygotes expressing both homozygous phenotypes simultaneously. A familiar example is the human blood type AB, in which the red blood cells have the distinctive characteristics of both type A and type B.

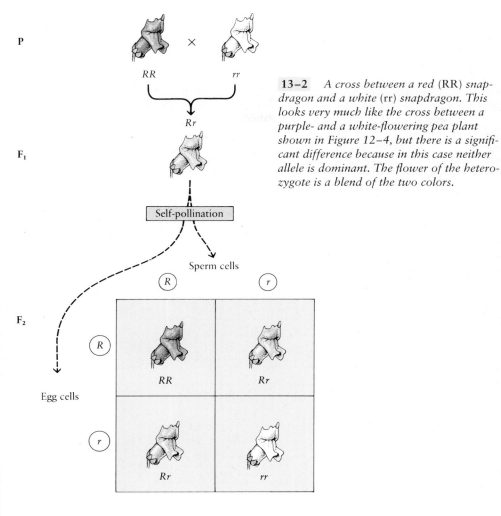

P

RR × rr

Rr

F₁

Self-pollination

Sperm cells

R   r

F₂

Egg cells

R

RR Rr

r

Rr rr

13–2 *A cross between a red (RR) snap-dragon and a white (rr) snapdragon. This looks very much like the cross between a purple- and a white-flowering pea plant shown in Figure 12–4, but there is a signifi-cant difference because in this case neither allele is dominant. The flower of the hetero-zygote is a blend of the two colors.*

(a)

(b)

(c)

Multiple Alleles

Although any individual diploid organism can have only two alleles of any given gene, it is possible that more than two forms of a gene—**multiple alleles**—may be present in a population of organisms (Figure 13–3). Multiple alleles result from different mutations of a single gene.

In humans, the four major blood groups (A, B, AB, and O) are determined by a gene with three alleles—*A*, *B*, and *O*. Phenotypically, the blood groups are characterized by particular polysaccharides on the surface of the red blood cells and by specific antibodies in the blood plasma.

13–3 *Coat color in rabbits is determined principally by a single gene, of which four different alleles are known. An individual rabbit, of course, carries only two alleles for this gene in its somatic (body) cells. Different combinations of alleles produce (a) the wild-type, or agouti, rabbit, (b) the chinchilla rab-bit, (c) the Himalayan rabbit, and (d) the albino rabbit.*

(d)

Table 13–1   The Genetic Basis of the ABO Blood Groups

Phenotype	Genotype (Alleles Present)	Polysaccharides on Surface of Red Blood Cells	Antibodies in Blood Plasma	Reaction with Antibodies	
				Antibody A	Antibody B
O	OO	—	Antibody A / Antibody B	No	No
A	AA, OA	A	Antibody B	Yes	No
B	BB, OB	B	Antibody A	No	Yes
AB	AB	A, B	—	Yes	Yes

The genetic basis of the blood groups is shown in Table 13–1. Alleles *A* and *B* are codominant, whereas allele *O* is recessive. Thus individuals with type A blood have either two *A* alleles or one *A* and one *O*, and their red blood cells bear the A polysaccharide. Their plasma does not contain antibodies against the A polysaccharide found on their own red blood cells, but it does have antibodies against the B polysaccharide. Individuals with type B blood have the B polysaccharide on their red blood cells and antibodies against the A polysaccharide in their plasma. Individuals with type AB blood have both polysaccharides but neither A nor B antibodies; conversely, type O individuals have neither polysaccharide but both A and B antibodies.

Gene Interactions

In addition to the interactions that occur between alleles of the same gene, interactions also occur among the alleles of different genes. Indeed, most of the characteristics (both structural and chemical) that constitute the phenotype of an organism are the result of the interaction of two or more distinct genes. These interactions may take a variety of forms.

Sometimes, when a trait is affected by two or more different genes, a completely novel phenotype may appear. For instance, comb shape in chickens is determined by two different genes, known as rose and pea, each with two alleles *(R, r* and *P, p)*. A genotype of *RR* or *Rr* results in rose comb, whereas *rr* produces single comb. Similarly, *PP* or *Pp* produces pea comb, and *pp* produces single comb. However, when *R* and *P* occur together in the same individual, a novel phenotype, walnut comb, results. Thus, four different types of combs are possible depending on the interaction of the alleles of these two genes (Figure 13–4).

In some cases, however, gene interaction does not produce a novel phenotype. Instead, one gene interferes with or masks the effect of another. This type of interaction is called **epistasis** ("standing upon"). A classic example of epistasis occurs in the sweet pea, a plant in which the flowers may be either purple or white. In order to produce purple flowers, a plant must have at least one dominant allele of each of two different genes. If a plant is homozygous recessive for either gene, the effects of the other gene are hidden and the flowers are white. Similarly, congenital deafness in humans can occur if either of two distinct genes is homozygous recessive.

(a) rrpp

(b) RRpp, Rrpp

(c) rrPP, rrPp

(d) RRPP, RRPp, RrPP, RrPp

13-4 *The four types of combs observed in chickens are single comb, rose comb, pea comb, and walnut comb. These comb types are determined by two different genes, of which the alleles are R, r and P, p. (a) Single comb occurs in chickens homozygous recessive for both genes. (b) Rose comb results from the presence of at least one dominant R allele coupled with two recessive p alleles. (c) Pea comb, by contrast, is produced by at least one dominant P allele coupled with two recessive r alleles. (d) A novel phenotype, walnut comb, is produced when at least one dominant allele of each gene is present in the same individual.*

Normal hearing requires at least one dominant allele of each of the two genes.

Polygenic Inheritance

Some traits, such as size or height, shape, weight, color, metabolic rate, and behavior, are not the result of interactions between one, two, or even several genes. Instead, they are the cumulative result of the combined effects of many genes. This phenomenon is known as **polygenic inheritance.**

A trait affected by a number of genes does not show a clear difference between groups of individuals—such as the differences tabulated by Mendel. Instead, it shows a gradation of small differences, which is known as **continuous variation.** If you make a chart of differences among individuals for any trait affected by a number of genes, for example, the height of adult men, you get a curve such as that shown in Figure 13-5.

One hundred years ago, the average height of men in the United States was less than it is now, but the shape of the curve was the same: the great majority fell within the middle range, and the extremes in height were represented by only a few individuals. Some of these height variations are produced by environmental factors, such as diet, but even if all the men in a population were maintained from birth on the same type of diet, there would still be a continuous variation in height in the population. This is due to genetically determined differences in hormone production, bone formation, and numerous other factors.

Table 13-2 illustrates a simple example of polygenic inheritance, color in wheat kernels, which is controlled by two genes, the four alleles of which exhibit cumulative quantitative effects. Human skin color is believed to be under a similar kind of genetic control (although involving more than two genes), as are many other traits.

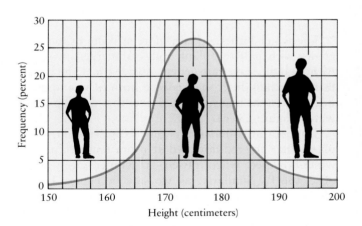

13-5 *Height distribution of males in the United States. Height is an example of polygenic inheritance; that is, it is affected by a number of genes. Such genetic traits are characterized by small gradations of difference. A graph of the distribution of such traits takes the form of a bell-shaped curve, as shown, with the mean, or average, usually falling in the center of the curve. The larger the number of genes involved, the smoother the curve.*

Multiple Effects of a Single Gene

Although most traits are influenced by a number of different genes, a single gene can often have multiple effects on the phenotype of an organism. This phenomenon is known as **pleiotropy** ("many turnings").

A striking example of pleiotropy is provided by a gene in rats that controls the production of a protein involved in the formation of cartilage. A mutation of this one gene causes a whole complex of birth defects, including thickened ribs, a narrowing of the passage through which air moves to and from the lungs, a loss of elasticity in the lungs, blocked nostrils, a blunt snout, and a thickening of the heart muscle. Needless to say, the mutation leads to a greatly increased mortality rate. Since cartilage is one of the most common structural substances of the body, the widespread effects of such a gene are not difficult to understand. Similarly, it is very likely that Mendel's allele for wrinkled peas affects other structural characteristics of the pea plant.

Genes and the Environment

The expression of a gene is always the result of its interaction with the environment. To take a common example, a seedling may have the genetic capacity to be green, to flower, and to fruit, but it will never turn green if it is kept in the dark, and it may not flower and fruit unless certain precise environmental requirements are met.

The water buttercup is a more striking example. It grows with half the plant body submerged in water. Although the leaves are genetically identical, the broad, floating leaves differ markedly in both form and physiology from the finely divided leaves that develop under water (Figure 13–6).

Temperature often affects gene expression. Primrose plants that are red-flowered at room temperature are white-flowered when raised at temperatures above 30°C (86°F). Similarly, as we saw in Figure 13–3c, Himalayan rabbits usually have black ears, forepaws, noses, and tails. However, when rabbits of the same genotype with respect to coat color are raised at high temperatures (above 35°C), their coats are entirely white.

The expression of a gene may be altered not only by factors in the external environment but also by factors in the internal environment of the organism, particularly during embryonic development. These factors include temperature, pH, ion concentrations, hormones, and a multitude of other influences, including the action of other genes.

When the expression of a gene is altered by environmental factors or by other genes, two outcomes are pos-

Table 13–2   The Genetic Control of Color in Wheat Kernels*

Parents:	RRSS (Dark Red)	×	rrss (White)	
F₁:	RrSs (Medium Red)			

F₂:	Genotype		Phenotype	
1		RRSS	Dark red	
2	} 4	RRSs	Medium-dark red	
2		RrSS	Medium-dark red	
4		RrSs	Medium red	15 red
1	} 6	RRss	Medium red	to
1		rrSS	Medium red	1 white
2	} 4	Rrss	Light red	
2		rrSs	Light red	
1		rrss	White	

* Two genes are involved, each of which has two alleles, symbolized here as R and r for gene 1, and S and s for gene 2.

13–6   *The water buttercup,* Ranunculus peltatus, *grows with half the plant body submerged in water. Leaves growing above the water are broad, flat, and lobed. The genetically identical underwater leaves are thin and finely divided, appearing almost rootlike. These differences are thought to be related to differences in the turgor (page 116) of the immature leaf cells in the two environments. The degree of turgor affects the expansion of the cell walls and thus the ultimate size of the cells.*

sible. First, the degree to which a particular genotype is expressed in the phenotype may vary from individual to individual. In humans, this **variable expressivity** is seen in polydactyly, the presence of extra fingers and toes, which is caused by a dominant allele. Often, there is great variability in the expression of this allele among members of a family in which it is present. The result is that some individuals have both extra toes and fingers, while others have only a portion of an extra toe on one foot.

Second, the proportion of individuals that show the phenotype may be less than expected. The genotype shows **incomplete penetrance.** For example, individuals known to carry the allele for polydactyly may have absolutely normal hands and feet. Similarly, some individuals known to carry the dominant allele responsible for Huntington's disease (page 212) never develop the disorder.

These are all examples of a universal verity: The phenotype of any organism is the result of the interaction between its genes and their environment.

The Reality of the Gene

One of the most remarkable features of the earliest work in classical genetics was that the gene had no physical reality. It was a pure abstraction. The work of Sutton (page 215) and other cytologists was known, but, to most researchers, it seemed irrelevant to studies of inheritance. Several lines of study were to provide the evidence that conclusively linked genes and chromosomes.

Sex Determination

As Sutton had observed, the chromosomes of a diploid organism occur in pairs. In all of the pairs except one, the chromosomes in both males and females appear to be the same. As we noted in Chapter 11, these chromosomes are called autosomes. The structure of one pair, however, may differ between males and females. The chromosomes of this pair are known as the sex chromosomes.

In some animal groups, such as birds, the two sex chromosomes in males are structurally the same. These are, by convention, termed Z chromosomes, and the male is designated ZZ. The sex chromosomes of females in these groups consist of one Z chromosome (which is structurally identical to the male Z chromosome) and one W chromosome. Thus the females of these species are designated ZW.

By contrast, in mammals, including humans, and in many other groups of animals, it is the female in which

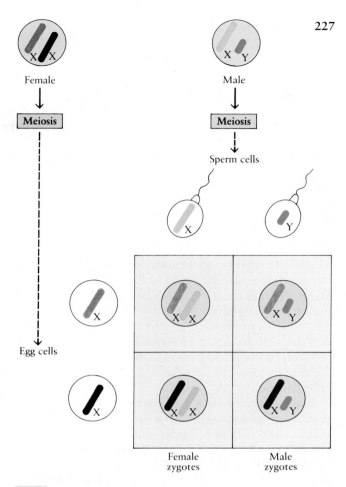

13–7 *Sex determination in organisms (such as humans and other mammals) in which the two sex chromosomes of the male are unlike. At meiosis, each egg cell receives an X chromosome from the mother. A sperm cell, however, may receive either an X chromosome or a Y chromosome.*

If a sperm cell carrying an X chromosome fertilizes an egg cell, the offspring will be female (XX). If, however, a sperm cell carrying a Y chromosome fertilizes an egg cell, the offspring will be male (XY). As the Punnett square predicts, male and female offspring are produced in approximately equal numbers.

the two sex chromosomes are the same. These are, by convention, termed X chromosomes, and the female is designated XX. The sex chromosomes of the male consist of one X chromosome (which is the same as the female X chromosome) and one Y chromosome. Thus the males of these species are designated XY.

In XY males, when the gametes are formed by meiosis, half of the sperm carry an X chromosome and half carry a Y chromosome. All the gametes formed by a female carry the X. The sex of the offspring is determined by whether the female gamete is fertilized by a male gamete carrying an X chromosome or a male gamete carrying a Y chromosome (Figure 13–7). Since equal numbers of X and Y sperm are produced, there is theoretically an even chance of producing male or female offspring.

The correlation of particular chromosomes with a particular trait—sex—gave strength to the hypothesis that genes are located on the chromosomes. Further evidence was to come from a variety of studies.

Sex Linkage

Introducing Drosophila

Much of the early work in classical genetics was carried out at a few research centers in Europe and the United States. Among the most important was the laboratory founded at Columbia University in 1909 by T. H. Morgan, a biologist from the United States who had visited de Vries's laboratory in Holland. By a remarkable combination of insight and good fortune, Morgan selected the fruit fly *Drosophila melanogaster* as his experimental organism.

Biologists have often used for their experiments "insignificant" plants and animals—such as pea plants, sea urchins, and fruit flies. Underlying this approach is the assumption that basic biological principles are universal, applying equally to all living things. As it turned out, the little fruit fly proved to be a "fit material" for a wide variety of genetic investigations. In the decades that followed, *Drosophila* was to become famous as the biologist's principal tool in studying animal genetics.

Drosophila means "lover of dew," although actually this useful animal is not attracted by dew but feeds on the fermenting yeast that it finds in rotting fruit. The fruit fly was an excellent choice for genetic studies since it is easy to breed and maintain. These tiny flies, each only 3 millimeters long, can produce a new generation every two weeks. Each female lays hundreds of eggs during her adult life, and very large numbers of flies can be kept in a half-pint bottle, as they were in Morgan's laboratory. Also, *Drosophila* has only four pairs of chromosomes (Figure 13–8), a feature that turned out to be particularly useful, although Morgan could not have foreseen that.

The White-Eyed Fruit Fly

The investigators in Morgan's laboratory were, at first, looking for genetic differences among individual flies that they could study in breeding experiments similar to those Mendel had carried out with pea plants. Shortly after Morgan established his *Drosophila* colony, such a difference appeared. One of the prominent and readily visible characteristics of fruit flies is their brilliant red eyes. One day, a white-eyed fly, a mu-

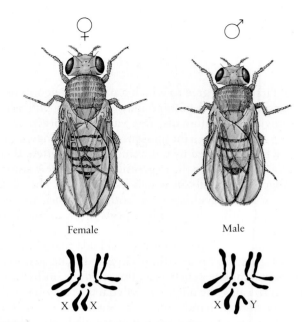

13–8   *The fruit fly* (Drosophila melanogaster) *and its chromosomes. Fruit flies have only four pairs of chromosomes (2n = 8), a fact that simplified Morgan's experiments. Six of the chromosomes (three pairs) are autosomes (including the two small spherical chromosomes in the center), and two are sex chromosomes.*

tant, appeared in the colony (Figure 13–9). This fly, a male, was mated with a red-eyed female, and all the F_1 offspring had red eyes. Apparently the white-eyed phenotype was recessive.

Morgan then crossbred the F_1 offspring, just as Mendel had done in his pea experiments. However, the offspring in the F_2 generation did not conform to the expected 3:1 ratio of dominant to recessive phenotypes (that is, of red-eyed to white-eyed individuals). And, as the detailed counts of the F_2 generation reveal, all of the white-eyed individuals were males:

Red-eyed females	2,459
White-eyed females	0
Red-eyed males	1,011
White-eyed males	782

Why were there no white-eyed females? To explore the situation further, Morgan crossed the original white-eyed male with one of the F_1 females. The following results were obtained from this testcross:

Red-eyed females	129
White-eyed females	88
Red-eyed males	132
White-eyed males	86

13–9 *A mutant white-eyed fruit fly (left) and a normal, or wild-type, red-eyed fruit fly (right). While looking for genetic differences among* Drosophila, *Calvin Bridges, a student in Morgan's laboratory, discovered a single white-eyed fruit fly in a population of thousands. This chance occurrence launched a series of studies that established the chromosomal basis of heredity.*

On the basis of these experiments, diagrammed in Figure 13–10, Morgan and his co-workers formulated the following hypothesis: The gene for eye color is carried only on the X chromosome. (In fact, as it was later shown, the Y chromosome of *Drosophila* carries very little genetic information of any kind.) The allele for white eyes must indeed be recessive, since all of the F_1 flies had red eyes. Thus a heterozygous female would have red eyes—which is why there were no white-eyed females in the F_2 generation. However, a male that received an X chromosome carrying the allele for white eyes would be white-eyed since no other allele would be present.

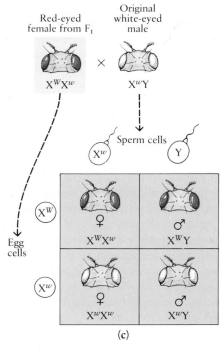

13–10 *Punnett square diagrams of the experiments Morgan performed after discovery of the white-eyed male* Drosophila. *The mutant allele for white eyes is represented here by w, and the wild-type allele for red eyes by W. Alleles located on the sex chromosomes are commonly designated by superscripts to the chromosome symbols.*

(a) Morgan first mated a homozygous red-eyed female to the white-eyed male. All of the F_1 offspring had red eyes. (b) Next, he mated an F_1 red-eyed female to an F_1 red-eyed male. Although both red-eyed and white-eyed males were produced in the F_2 generation, all F_2 females had red eyes, suggesting a relationship between the inheritance of eye color and the inheritance of the sex chromosomes.

(c) A testcross between a red-eyed F_1 female and the original white-eyed male produced both red-eyed and white-eyed flies of both sexes. This led to the conclusion that the gene for eye color must be carried on the X chromosome. The allele for red eyes (W) is dominant, and the allele for white eyes (w) is recessive.

Further experimental crosses, such as the one diagrammed in Figure 13–11, confirmed Morgan's hypothesis. They also revealed that white-eyed fruit flies are more likely to die before they reach adulthood than are red-eyed fruit flies. This suggested that the allele for white eyes has other effects as well, and it explained why the numbers of white-eyed fruit flies were lower than expected in both the F_2 generation and the testcross.

These experiments introduced the concept of **sex-linked traits.** They also convinced Morgan, and many other geneticists as well, that Sutton's hypothesis was correct: Genes *are* on chromosomes. Conclusive demonstration of the physical location of the gene, however, depended on a subsequent series of experiments.

Linkage Groups

An important research tool became available when H. J. Muller, one of Morgan's collaborators, found that exposure to x-rays greatly increases the rate at which mutations occur in *Drosophila*. Other forms of radiation, such as ultraviolet light, and certain chemicals were also shown to act as **mutagens,** or agents that produce mutations. As increasing numbers of mutants were produced in Columbia University's *Drosophila* collection, it became possible to do breeding experiments in which the flies differed in more than one characteristic.

As you will recall, Mendel had demonstrated that certain pairs of alleles, such as those for round and wrinkled peas, assort independently of other pairs, such as those for yellow and green peas. And, as we noted at the end of the last chapter, the alleles of two different genes will always assort independently if the genes are on different pairs of homologous chromosomes (see Figure 12–8, page 217). If, however, the alleles of the two genes are on the same pair of homologous chromosomes, segregation of the alleles of one gene will *not* be independent of the segregation of the alleles of the other gene. In other words, if the alleles of two different genes are on the *same* chromosome, they should both be transmitted to the *same* gamete during meiosis. Genes that tend to stay together because they are on the same pair of homologous chromosomes are said to be in the same **linkage group.**

As the *Drosophila* breeding experiments continued, the mutations began to fall into four linkage groups, in accord with the four pairs of chromosomes visible in the cells. Indeed, in all organisms that have been studied in sufficient genetic detail, the number of linkage groups and the number of pairs of chromosomes have been the same, providing further support for Sutton's hypothesis that genes are on the chromosomes.

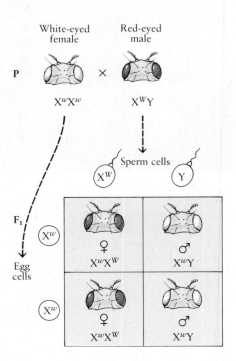

13–11   *Offspring of a cross between a white-eyed female fruit fly and a red-eyed male fruit fly, illustrating what happens when a recessive allele is carried on an X chromosome. The F_1 females, with one X chromosome from the mother and one from the father, are all heterozygous (X^wX^W) and so will be red-eyed. But the F_1 males, with their single X chromosome received from the mother carrying the recessive (w) allele, will all be white-eyed because the Y chromosome carries no gene for eye color. Thus the recessive allele on the X chromosome inherited from the mother will be expressed.*

Recombination

Large-scale studies of linkage groups soon revealed some unexpected difficulties, however. For instance, most fruit flies have light tan bodies and long wings, both of which are dominant characteristics. When individuals homozygous for these characteristics were bred with mutant fruit flies having black bodies and vestigial (short) wings, both of which are recessive characteristics, all the F_1 offspring had tan bodies and long wings, as would be expected. Then the F_1 generation was inbred. Two outcomes seemed possible:

1. The genes for body color and wing length would be assorted independently, giving rise to Mendel's 9:3:3:1 ratio in the phenotypes of the F_2 offspring. This would indicate that the genes for these two traits were on different pairs of homologous chromosomes.

Human Sex-Linked Traits

In humans, as in fruit flies, the Y chromosome carries much less genetic information than the X chromosome. Among the genes carried on the human X chromosome—but not on the Y—are the genes responsible for the capacity to distinguish red and green.

The ability to perceive color depends on three genes, producing three different visual pigments, each responsive to light in a different region of the spectrum. The genes for the pigments responsive to red and green light are both on the X chromosome.

In males, if the gene for green is defective, green cannot be distinguished from red, and, conversely, a defect in the gene for red results in red appearing as green (a). In heterozygous females, the defective alleles are recessive to the normal alleles on the other X chromosome, and so color vision is usually normal. Complete red-green color blindness in females occurs only when both X chromosomes carry the same defective allele.

If a woman carrying a defective allele on one X chromosome (b) transmits that X chromosome to a daughter, the daughter will also have normal color vision if she receives an X chromosome with the normal allele from her father (that is, if he is not color-blind). If, however, the X chromosome with the defective allele is transmitted from mother to son, he will be color-blind since, lacking a second X chromosome, he has only the defective allele.

Another classic example of sex-linked inheritance is the hemophilia that has afflicted some royal families of Europe since the nineteenth century. Hemophilia is a group of disorders in which the blood does not clot normally. Clotting occurs through a complex series of reactions in which each reaction depends on the presence of certain protein factors in the blood. Failure to produce one essential protein, known as Factor VIII, results in the most common form of hemophilia, which is associated with a recessive allele of a gene carried on the X chromosome.

For individuals with this form of hemophilia, even minor injuries carry the risk of bleeding to death. Until very recently, the only effective treatment was Factor VIII extracted from normal human blood. Unfortunately, this carried the risk of transmission of infectious agents, particularly the virus that causes AIDS. However, as we shall see in Chapter 18, a form of Factor VIII produced through genetic engineering is now available, eliminating the risk of contamination with infectious agents.

The original carrier of hemophilia among European royalty was probably Queen Victoria. One of her sons died of hemophilia at the age of 31. Because male-to-male inheritance of the disorder is impossible—as it is with all X-linked characteristics—Prince Albert, Victoria's husband, could not have been the source. Because none of Victoria's relatives other than her descendants were affected, it is thought that the mutation occurred on an X chromosome in one of her parents or in the cell line from which her own eggs were formed.

At least two of Victoria's daughters were carriers, since a number of *their* descendants were hemophiliacs. And so, through various intermarriages, the disease spread from throne to throne across Europe. Tsarevitch Alexis, the only son of Nicholas II and Alexandra, the last Tsar and Tsarina of Russia, inherited the allele for hemophilia from his mother, who was a granddaughter of Queen Victoria. The great concern of Alexis's parents for his health, which apparently distracted them from affairs of state, contributed to the turbulent events surrounding the Russian revolution.

(a) *Individuals with normal color vision can easily read the two-digit number embedded in this dot pattern. Individuals with red-green color blindness—about 8 percent of human males and 0.04 percent of human females—cannot do so.*

(b) *The pedigree of a family in which the mother has inherited one normal and one defective allele for red-green color vision. The normal allele is dominant, and she has normal color vision. However, half of her eggs (on the average) will carry the defective allele and half will carry the normal allele—and it is a matter of chance which kind is fertilized. Since her husband's Y chromosome, the one that determines a son rather than a daughter, carries no gene for color vision, the single allele the* mother contributes will determine whether or not a son is color-blind. Therefore, on average, half of her sons will be color-blind. Assuming that her children marry individuals with X chromosomes with the normal alleles, the expected distribution of the trait among her grandchildren will be as shown on the bottom line. Note that all the daughters of a color-blind man will be carriers of the defective allele, and all his sons will have normal color vision, unless their mother is a carrier or is herself color-blind.*

Normal female

Normal male

Carrier female

Color-blind male

X Normal X chromosome

X Chromosome with allele for color blindness

Mother   Father

X X   X Y

X X   X Y

X X   X Y   X Y   X X   X Y

No alleles for color blindness

No alleles for color blindness

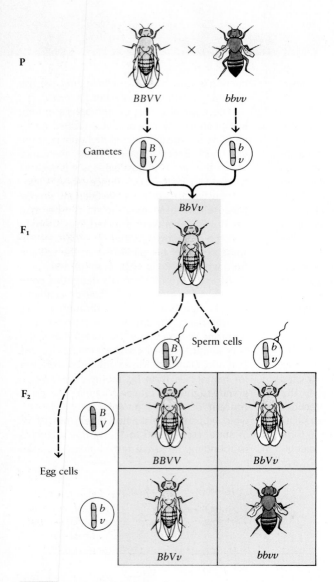

13–12   *The expected results in the F₂ generation of an initial cross between a fruit fly homozygous for tan body color and long wings and a fruit fly homozygous for black body color and short wings, if the two genes are located on the same pair of homologous chromosomes. The allele symbols are B for tan body, b for black body, V for long wings, and v for vestigial, or short, wings.*

When Morgan performed this experiment, the phenotypes in the F₂ generation were close to the 3:1 ratio shown in the Punnett square, but they did not conform exactly. Some of the F₂ flies were tan with vestigial wings, and some were black with long wings.

2. The genes for the two traits would be linked. This possibility is diagrammed in Figure 13–12. As you can see, 75 percent of the F₂ flies would be tan with long wings, and 25 percent, homozygous for both recessives, would be black with vestigial wings (a 3:1 ratio).

In the case of these particular traits, the results closely resembled the second possibility, but they did not conform exactly. In a few of the offspring, the genes for these traits appeared to assort independently. That is, some of the F₂ flies were tan with vestigial wings, and some were black with long wings. How could this be? Somehow alleles that were presumed to be on the same chromosome had become separated.

To find out what was happening, Morgan tried a testcross, breeding a member of the F₁ generation with a fly that was homozygous recessive for both genes. If the alleles for black and tan assorted independently of

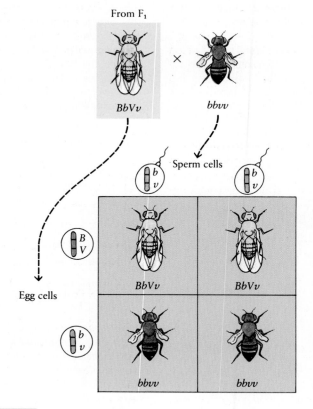

13–13   *The expected results of a testcross of an F₁ fly from the cross in Figure 13–12 with a fly homozygous for black body and vestigial wings, if the two genes are located on the same pair of homologous chromosomes. Once again, the results were close to those shown in the Punnett square, but they did not conform exactly.*

the alleles for long and vestigial—that is, if the two genes were on different pairs of homologous chromosomes—25 percent of the offspring of this cross should be black with long wings, 25 percent tan with long wings, 25 percent black with vestigial wings, and 25 percent tan with vestigial wings (a ratio of 1:1:1:1).

On the other hand, if the genes for color and wing size were on the same pair of homologous chromosomes, the alleles of the two genes should move together. In this case, 50 percent of the testcross offspring should be tan with long wings and 50 percent should be black with vestigial wings, as shown in Figure 13–13. But actually, as it turned out, over and over, in counts of hundreds of fruit flies resulting from such crosses, 42 percent were tan with long wings, 42 percent were black with vestigial wings, 8 percent were tan with vestigial wings, and another 8 percent were black with long wings (Figure 13–14).

Morgan was convinced by this time that genes are located on chromosomes. It now seemed clear that the genes for these two traits were located on a single pair of homologous chromosomes, since the characteristics did not show up in the 1:1:1:1 ratio of independently assorted alleles. The only way in which the observed figures could be explained, Morgan reasoned, was if alleles could sometimes be exchanged between homologous chromosomes—that is, if they could be recombined.

As we noted in Chapter 11, it now has been established that exchange of portions of homologous chromosomes—crossing over—takes place in prophase I of meiosis (Figure 13–15). If crossing over takes place between the positions at which two different genes are located on a pair of homologues, then the alleles of the two genes can become separated as chromatids of the two homologues break and rejoin with each other (Figure 13–16).

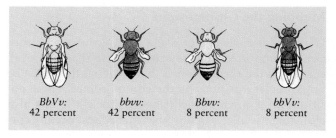

BbVv:
42 percent

bbvv:
42 percent

Bbvv:
8 percent

bbVv:
8 percent

13–14 *The actual results of the testcross of an F_1 fly from the cross illustrated in Figure 13–12 with a fly homozygous for black body and vestigial wings. These results suggested that alleles could sometimes be exchanged between homologous chromosomes. As we now know, this process—crossing over—occurs in prophase I of meiosis.*

2 μm

13–15 *Homologous chromosomes of a grasshopper, as seen in prophase I. All four chromatids are visible. Crossing over—that is, the exchange of genetic material—has probably occurred at the points at which these chromatids intersect.*

13–16 (a, b) *Crossing over takes place when breaks occur in the chromatids of homologous chromosomes early in meiosis I, when the chromosomes are tightly paired. The broken end of each chromatid joins with the chromatid of a homologous chromosome. In this way, alleles are exchanged between chromosomes.* (c) *As a result of crossing over, two of the chromosomes have allele combinations different from those of the original chromosomes.*

(a)

(b)

(c)

Mapping the Chromosome

With the discovery of crossing over, growing evidence clearly supported not only the premise that genes are carried on the chromosomes, but also that they must be at particular positions on the chromosomes. These positions are known as **loci** (singular, locus). It was also clear that the alleles of any given gene must occupy corresponding loci on homologous chromosomes. Otherwise, the exchange of sections of chromosomes would result in genetic chaos rather than in an exact exchange of alleles.

As other traits were studied, the data revealed that the percentage of recombinations between any two genes, such as those for body color and wing length, was different from the percentage of recombinations between two other genes, such as those for body color and leg length. In addition, as Morgan's experiments had shown, these percentages were fixed and predictable. It occurred to A. H. Sturtevant, then a student working in Morgan's laboratory, that the percentage of recombinations probably had something to do with the physical distances between the gene loci, or, in other words, with their spacing along the chromosome. This concept opened the way to the earliest "mapping" of chromosomes.

Sturtevant postulated that (1) genes are arranged in a linear series on chromosomes; (2) genes that are close together will be separated by crossing over less frequently than genes that are farther apart; and (3) it should therefore be possible to use the frequencies of various recombinations to plot the sequence of the genes along the chromosome and the relative distances between them. In Figure 13–16, for example, you can see that in a crossover, the chance that a strand would break and rejoin with its homologous strand somewhere between *A* and *C* should be more likely than this happening somewhere between *C* and *D*. This is the case simply because there is a greater distance between *A* and *C* in which crossing over can occur.

In 1913, Sturtevant began constructing chromosome maps using data from crossover studies in fruit flies. As a standard unit of measure, he arbitrarily defined one map unit as equal to the distance that would give (on the average) one recombinant organism per 100 fertilized eggs. For example, two genes with 10 percent recombination would be 10 units apart; two genes with 8 percent recombination would be 8 units apart. By this method, he and other geneticists constructed chromosome maps locating a variety of genes and their mutants in *Drosophila* (Figure 13–17).

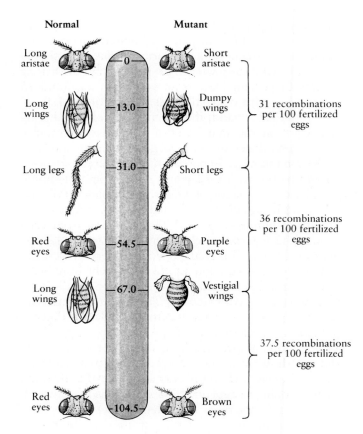

13–17   *A portion of the map of chromosome 2 of* Drosophila melanogaster, *showing the relative positions of some of the genes on that chromosome. The map distance between any two genes represents the number of recombinations of the alleles of those two genes per 100 fertilized eggs. As this map reveals, more than one gene may affect a single trait.*

Although Sturtevant's method of chromosome mapping has now been superseded by much more precise techniques, these studies were of critical importance in the history of genetics. They confirmed not only that genes are located on chromosomes, as Sutton had hypothesized, but also that they have fixed positions in a linear sequence. Moreover, the data from this work constitute a detailed atlas of the *Drosophila* chromosomes—an invaluable reference for current research workers.

Giant Chromosomes

Recombination does not usually affect the order of genes on a chromosome. However, as we noted in Chapter 11, when chromatid segments break during crossing over, they may sometimes be lost or they may join with a different chromosome. They may also rejoin the correct chromosome but in a different orientation. The outcome, in any of these cases, is a change in the sequence of genes on the affected chromosomes.

Evidence for such changes in chromosome structure, giving rise to changes in gene order and altered hereditary patterns, came from the study of giant chromosomes discovered in 1933 in the salivary glands of *Drosophila* larvae (immature forms). In *Drosophila,* as in many other insects, certain cells do not divide during the larval stages of the insect. In such cells, however, the chromosomes continue to replicate, over and over again, but since the daughter chromosomes of the nondividing cells do not separate from one another after replication, they simply become larger and larger. Ultimately, they may be composed of as many as a thousand copies. As you can see in Figure 13–18, giant chromosomes are characterized by very distinctive patterns of dark and light bands.

These banding patterns became another useful tool for geneticists, enabling them to detect structural changes in the chromosomes themselves. By observing changes in the banding patterns in the giant chromosomes, geneticists can see where changes in chromosome structure have occurred. Such changes include the deletions, duplications, and translocations discussed in Chapter 11 (page 199). They also include another type of chromosome abnormality, known as an **inversion.** This results when a double break in a chromosome occurs and a segment is turned 180° and then reincorporated into the chromosome. When one member of a homologous pair contains a large inverted segment, that chromosome must loop inside the other for close pairing to occur. In giant chromosomes, homologues are paired and such loops are greatly magnified, making it possible to readily identify regions with inverted segments.

The correlation of changes in the banding patterns of the giant chromosomes with observed genetic changes in the individual fruit flies provided conclusive confirmation of Sturtevant's hypothesis that genes occur in a linear sequence on the chromosomes.

13–18   *Chromosomes from the salivary gland of a* Drosophila *larva. These chromosomes are hundreds of times larger than the chromosomes in ordinary body cells, and their details are therefore much easier to see. (This micrograph was taken with a light microscope, not an electron microscope.) Because of the distinctive banding patterns, it is possible in some cases to associate genes with specific regions in particular chromosomes. Each visible band represents several genes.*

25 μm

Summary

The rediscovery of Mendel's work in 1900 was the catalyst for many new discoveries in genetics, leading to the modification and extension of some of Mendel's conclusions and to the identification of chromosomes as the carriers of the genetic information.

Mutations are abrupt changes in the genotype. They are the ultimate source of the genetic variations studied by Mendel. Different mutations of a single gene increase the diversity of alleles of that gene in a population. As a consequence, mutation provides the variability among organisms that is the raw material for evolution.

Many traits are inherited according to the patterns revealed by Mendel. However, in others—perhaps the majority—the patterns are more complex. Although many alleles interact in a dominant-recessive manner, some show varying degrees of incomplete dominance and codominance. In a population of organisms, multiple alleles of a single gene may exist, but only two alleles can be present in any given diploid individual. The particular combination present determines both the genotype and the phenotype.

Different genes can also interact with one another. Novel phenotypes may result from these interactions, or genes may affect one another in an epistatic manner, such that one hides the effect of the other. The phenotypic expression of many traits is influenced by a number of genes. This phenomenon is known as polygenic inheritance. Such traits typically show continuous variation, as represented by a bell-shaped curve. Conversely, a single gene can affect two or more superficially unrelated traits. This property of a gene is known as pleiotropy.

Gene expression is also affected by factors in the external and internal environments. Variable expressivity or incomplete penetrance of particular alleles may result from environmental influences, interactions with other genes, or both.

Strong support for the hypothesis that genes are on the chromosomes came from work with the fruit fly *Drosophila*, which has been used in a wide variety of genetic studies. It has four pairs of chromosomes; three pairs (the autosomes) are structurally the same in both sexes, but the fourth pair, the sex chromosomes, is different. In fruit flies, as in many other species (including humans), the two sex chromosomes are XX in females and XY in males.

At the time of meiosis, the sex chromosomes, like the autosomes, segregate. Each egg cell receives an X chromosome, but half of the sperm cells receive an X chromosome and half receive a Y chromosome. Thus it is the sperm cell that determines the sex of the offspring in species with XY males.

In the early 1900s, breeding experiments with *Drosophila* showed that certain traits are sex-linked, that is, their genes are carried on the sex chromosomes. Because the X chromosome carries genes that are not present on the Y chromosome, a single recessive allele on the X chromosome of a male will result in a recessive phenotype—simply because no other allele is present. By contrast, a female heterozygous for a sex-linked trait will show the dominant phenotype. In humans, sex-linked characteristics include color blindness and hemophilia.

Some genes assort independently in breeding experiments, and others tend to remain together. Genes that do not assort independently are said to be in the same linkage group. A linkage group corresponds to a pair of homologous chromosomes.

Alleles are sometimes exchanged between homologous chromosomes as a result of crossing over in meiosis. Such recombinations can take place because: (1) the genes are arranged in a fixed linear array along the length of the chromosomes, and (2) the alleles of a given gene are at corresponding sites (loci) on homologous chromosomes. Chromosome maps, showing the relative positions of gene loci along the chromosomes, have been developed from recombination data provided by breeding experiments.

Genetic studies have shown that chromosome breaks other than those resulting in crossovers may sometimes occur. A portion of a chromosome may be lost, or deleted, it may be duplicated, it may be translocated to a nonhomologous chromosome, or it may be inverted. Studies of the giant chromosomes of *Drosophila* larvae provided visual confirmation of these changes, as well as the final, conclusive evidence that the chromosomes are the carriers of the genetic information.

Questions

1. Distinguish among the following: mutation/mutant/mutagen; incomplete dominance/codominance/epistasis; polygenic inheritance/pleiotropy; variable expressivity/incomplete penetrance.

2. The so-called "blue" (really gray) Andalusian variety of chicken is produced by a cross between the black and white varieties. Only a single pair of alleles is involved. What color chickens (and in what proportions) would you expect if you crossed two blues? If you crossed a blue and a black? Explain.

3. In snapdragons, the allele that produces tall stems is completely dominant to the allele for dwarf stems, while the allele that produces red flowers is only partially dominant to that for white flowers. (a) Describe the phenotype (height and flower color) of the F_1 plants resulting from a cross between a homozygous tall, red-flowered plant and a homozygous dwarf, white-flowered plant. (b) If one of these F_1 plants self-pollinates, what will be the appearance and proportions of phenotypes in the resulting F_2 generation? (c) Which two of these phenotypes will breed true?

4. If you are of blood type O and neither of your parents is, what are their possible genotypes? What is the probability that one of your siblings is also of type O? (The use of a Punnett square here may be helpful.)

5. Although blood types can never prove that someone is the father of a particular child, they can demonstrate that someone could *not* be the father. Fill in the table at the right to see how this is so. In the famous Charlie Chaplin paternity case in the 1940s, the baby's blood was B, the mother's A, and Chaplin's O. If you had been the judge, how would you have decided the case?*

6. In chickens, as we have seen, two pairs of alleles determine comb shape. RR or Rr results in rose comb, whereas rr produces single comb. PP or Pp produces pea comb, and pp produces single comb. When R and P occur together, they produce a new type of comb: walnut. (a) What would be the genotype of the F_1 generation resulting from $RRpp$ × $rrPP$? The phenotype? (b) If F_1 hybrids were crossbred, what would be the probable distribution of genotypes? Of phenotypes? (Illustrate this cross with a Punnett square.)

* As a matter of fact, Chaplin was judged to be the father. In some states at that time, blood-group data were not admitted as evidence in cases of disputed parentage.

7. In Duroc-Jersey pigs, coat color is determined by two genes, R and S. The homozygous recessive condition, $rrss$, produces a white coat. The presence of at least one copy each of R and S produces red. The presence of one or the other allele (either R or S) produces a new phenotype, sandy. Give the phenotypes of the following genotypes:

RRSS	*rrss*
RrSs	*rrSs*
RRSs	*rrSS*
RrSS	*RRss*

8. Mating a red Duroc-Jersey boar to sow A (white) gave pigs in the ratio of 1 red: 2 sandy: 1 white. Mating this same boar to sow B (sandy) gave 3 red: 4 sandy: 1 white. When this boar was mated to sow C (sandy), the litter had equal numbers of red and sandy piglets. Using the information presented in Question 7, give the possible genotypes of the boar and the three sows.

9. In humans, either of two recessive alleles *(a or b)*, when homozygous, can cause congenital deafness. Hence, it is possible that all of the children of parents who are both congenitally deaf could have normal hearing. What would be the genotypes of the parents in such a situation?

Phenotypes of Parents		Phenotypes Possible in Children	Phenotypes Not Possible in Children
A	A		
A	B		
A	AB		
A	O		
B	B		
B	AB		
B	O		
AB	AB		
AB	O		
O	O		

10. A woman homozygous for the dominant alleles *A* and *B*, necessary for normal hearing, marries a man who is congenitally deaf. What are the possible genotypes of the man? What is the probability that this couple will have a deaf child?

11. You and a geneticist are looking at a mahogany-colored Ayrshire cow with a newly born red calf. You wonder if it is male or female, and the geneticist says it is obvious from the color which sex the calf is. She explains that in Ayrshires the genotype *AA* is mahogany and *aa* is red, but the genotype *Aa* is mahogany in males and red in females. What is she trying to tell you—that is, what sex is the calf? What are the possible phenotypes of the calf's father?

12. In one strain of mice, skin color is determined by five different pairs of alleles. The colors range from almost white to dark brown. Would it be possible for some pairs of mice to produce offspring darker or lighter than either parent? Explain.

13. The size of the eggs laid by one variety of hens is determined by three pairs of alleles. Hens with the genotype *AABBCC* lay eggs weighing 90 grams, and hens with the genotype *aabbcc* lay eggs weighing 30 grams. Each of the alleles *A*, *B*, or *C* adds 10 grams to the weight of the egg. When a hen from the 90-gram strain is mated with a rooster from the 30-gram strain, the hens of the F_1 generation lay eggs weighing 60 grams. If a hen and rooster from this F_1 generation are mated, what will be the weight of the eggs laid by hens of the F_2?

14. Height and weight in animals follow a distribution similar to that shown in Figure 13–5. By inbreeding large animals, breeders are usually able to produce some increase in size among their stock. But after a few generations, increase in size characteristically stops. Why?

15. Draw a diagram similar to Figure 13–7 indicating sex determination in a robin.

16. Why is male-to-male inheritance of red-green color blindness impossible? Under what conditions would such color blindness be found in a woman? If she married a man who was not color-blind, what proportion of her sons would be color-blind? Of her daughters?

17. What is the probability that a woman with normal color vision whose father is color-blind but whose husband has normal color vision will have a color-blind son? A color-blind daughter?

18. A woman whose maternal grandfather was hemophiliac has parents who are clinically normal. She too seems normal, as does her husband. What are the chances that her first son will *not* have hemophilia? *(Hint:* Determine the genotype of the woman's mother and then the possible genotypes of the woman herself.)

19. A man with a particular hereditary disorder marries a woman who does not have the disorder. They have six children, three girls and three boys. The girls all have the father's disorder but the boys do not. What type of inheritance pattern is suggested? (Although this situation was not discussed in the text, you should be able to deduce the answer.)

20. A particular diploid organism has 42 chromosomes per somatic cell. How many linkage groups does it have?

21. The first two pairs of genes studied simultaneously in crosses were *Rr* and *Yy* in pea plants. As it turned out, they were on different chromosomes. How do you think the development of the principles of genetics would have been affected had they been linked?

22. Does crossing over necessarily result in a recombination of alleles? Explain.

23. In a series of breeding experiments, a linkage group composed of genes *A*, *B*, *C*, *D*, and *E* was found to show approximately the recombination frequencies in the chart below. Using Sturtevant's standard unit of measure, "map" the chromosome.

		Gene					
		A	*B*	*C*	*D*	*E*	
	A	—	8	12	4	1	
	B	8	—	4	12	9	
Gene	*C*	12	4	—	16	13	Recombinations per 100 Fertilized Eggs
	D	4	12	16	—	3	
	E	1	9	13	3	—	

Suggestions for Further Reading

Klug, William S., and Michael R. Cummings: *Concepts of Genetics,* 3d ed., Macmillan Publishing Company, New York, 1991.

In this textbook, designed for a first course in genetics, the emphasis is on fundamental concepts, which are illustrated by carefully and clearly explained examples. It also contains many sample problems and their solutions.

Mange, Arthur P., and Elaine Johansen Mange: *Genetics: Human Aspects,* 2d ed., Sinauer Associates, Inc., Sunderland, Mass., 1990.

A thorough account of human genetics.

Peters, James A. (ed.): *Classic Papers in Genetics,* Prentice-Hall, Inc., Englewood Cliffs, N.J., 1959.*

A collection of papers by the scientists—including Morgan and his collaborators at Columbia University—who were responsible for the major developments in genetics from the 1850s through the 1950s. The authors are surprisingly readable, and the papers give a feeling of immediacy that no modern account can achieve.

Sapienza, Carmen: "Parental Imprinting of Genes," *Scientific American,* October 1990, pages 52–60.

Strickberger, Monroe W.: *Genetics,* 3d ed., Macmillan Publishing Company, New York, 1986.

A cohesive account of the science, this book provides a broad coverage of classical genetics.

Weiss, Rick: "A Genetic Gender Gap," *Science News,* May 20, 1989, pages 312–315.

* Available in paperback.

This is a photomicrograph of dividing cells in the root tip of an onion. The cells, many of which are undergoing mitosis, have been stained to show the chromosomes.

As we have seen, chromosomes have two important functions in the life of a cell. First, they transmit the genetic information. When a cell undergoes mitosis and cytokinesis, each daughter cell receives a full and precise copy of all of the genes that were present in the parent cell. This everyday occurrence—taking place right now in almost every tissue of your body—is the basis for the survival of cells, of organisms, of generations—that is, of life on Earth.

The second important function of chromosomes is to instruct the cells in their day-to-day, moment-to-moment activities. In pea plants, for example, chromosomes direct cells in the flower petals to make pigment and cells in the roots to make the membrane proteins that allow them to absorb needed minerals from the soil.

The work of Mendel and his successors showed us that there were different genes for different functions and that genes were located on the chromosomes. But their work provided barely a clue as to how these microscopic scraps of matter could transmit information and direct cellular activities.

The answers to these crucial questions began to emerge when scientists turned their attention away from the microscopic examination of chromosomes and the counting of phenotypes toward a wholly new field of research—the chemistry of heredity.

The Chemical Basis of Heredity: The Double Helix

By the early 1940s, the existence of genes and the fact that they are carried in the chromosomes were no longer in doubt. A turning point in the history of genetics came when scientists began to focus on the question of how it is possible for the chromosomes to carry what the scientists realized must be an enormous amount of extremely complex information.

The chromosomes, like all the other parts of a living cell, are composed of atoms arranged into molecules. Some scientists, a number of them eminent in the field of genetics, thought it would be impossible to understand the complexities of heredity in terms of the structure of "lifeless" chemicals. Others thought that if the chemical structure of the chromosomes were understood, we could then come to understand how chromosomes function as the bearers of the genetic information. This insight marked the beginning of the vast range of investigations that we know as **molecular genetics**.

The DNA Trail

Early chemical analyses revealed that the eukaryotic chromosome consists of both **deoxyribonucleic acid (DNA)** and protein, in about equal amounts. Thus, both were candidates for the role of the genetic material. Proteins seemed the more likely choice, because of their greater chemical complexity. As you will recall from Chapter 3, proteins are polymers of amino acids, of which there are some 20 different types in living cells. DNA, by contrast, is a polymer formed from only four different types of nucleotides.

Speculative thinkers in the field of biology were quick to point out that the amino acids, the number of which is so provocatively close to the number of letters in our own alphabet, could be arranged in a variety of different ways. The amino acids were seen as making up a sort of language—"the language of life"—that spelled out the directions for all the many activities of the cell.

241

Many prominent investigators, particularly those who had been studying proteins, believed that the genes themselves were proteins. They thought that the chromosomes contained master models of all the proteins that would be required by the cell and that enzymes and other proteins active in cellular life were copied from these master models. This was a logical hypothesis, but, as it turned out, it was wrong.

Sugar-Coated Microbes: The Transforming Factor

To trace the beginning of the other hypothesis—the one that ultimately proved correct—it is necessary to go back to 1928 and pick up an important thread in modern biological history. In that year, an experiment was performed that seemed at the time to have little relevance to the field of genetics. Frederick Griffith, a public health bacteriologist in England, was studying the possibility of developing vaccines against pneumococci, the bacteria that cause one kind of pneumonia. In those days, before the development of antibiotics, bacterial pneumonia was an extremely serious disease, often referred to as the grim "captain of the men of death."

Pneumococci, as Griffith knew, come in either **virulent** (disease-causing) forms with polysaccharide capsules or **nonvirulent** (harmless) forms without capsules (Figure 14–1). The production of the capsule and its composition are genetically determined—that is, they are inherited properties of the bacteria. Griffith thought that injections of heat-killed virulent bacteria or of liv-

14–2   *Discovery of the transforming factor, a substance that can transmit genetic characteristics from one cell to another, resulted from studies of pneumococci. One strain of these bacteria has polysaccharide capsules; another does not. The capacity to make capsules and cause pneumonia is an inherited characteristic, passed from one bacterial generation to another as the cells divide. (a) Injection into mice of encapsulated pneumococci led to pneumonia, which killed the mice. (b) The nonencapsulated strain produced no infection. (c) If the encapsulated strain was heat-killed before injection, it too produced no infection. (d) If, however, heat-killed encapsulated bacteria were mixed with living nonencapsulated bacteria and the mixture was injected into mice, the mice developed pneumonia and died. (e) Blood samples from the dead mice revealed living encapsulated pneumococci. Something had been transferred from the dead bacteria to the living ones that endowed them with the capacity to make polysaccharide capsules and cause pneumonia. This "something" was later isolated and found to be DNA.*

14–1   (a) *Encapsulated and* (b) *nonencapsulated forms of pneumococci. The capsules—the lighter colored zones surrounding the cells in* (a)—*are made up of polysaccharides deposited outside the cell walls. The encapsulated form, which is resistant to phagocytosis by white blood cells, can cause pneumonia. The nonencapsulated form is harmless.*

ing nonvirulent bacteria (without capsules) might immunize mice against living virulent pneumococci. As a result of his experiments, which are outlined in Figure 14–2, Griffith found that something can be passed from dead to living bacteria that changes their hereditary characteristics. This phenomenon became known as **transformation,** and the "something" that caused it was called the **transforming factor.**

In 1943, after almost a decade of patient chemical isolation and analysis, O. T. Avery and his co-workers at Rockefeller University demonstrated that the transforming factor is DNA. Subsequent experiments showed that a variety of genetic factors can be passed from bacterial cells of one strain to cells of another, similar strain by means of isolated DNA.

The Nature of DNA

DNA had first been isolated by a German physician named Friedrich Miescher in 1869—in the same remarkable decade in which Darwin published *The Origin of Species* and Mendel presented his results to the Brünn Natural History Society. The substance Miescher isolated was white, sugary, slightly acidic, and contained phosphorus. Since it was found only in the nuclei of cells, it was called nucleic acid. This name was later amended to deoxyribonucleic acid, to distinguish it from a similar chemical also found in cells, ribonucleic acid (RNA).

By the time of Avery's discovery, it was known that DNA is made up of nucleotides. Each nucleotide consists of a nitrogenous base, a deoxyribose sugar, and a phosphate group (Figure 14–3a). The nitrogenous bases are of two kinds: **purines** (Figure 14–3b), which have two rings, and **pyrimidines** (Figure 14–3c), which have one ring. There are two kinds of purines in DNA, **adenine (A)** and **guanine (G),** and two kinds of pyrimidines, **cytosine (C)** and **thymine (T).** So DNA is made up of four types of nucleotides, differing only in their nitrogen-containing purine or pyrimidine.

Avery's results offered evidence for DNA as the genetic material, but his discovery was slow to gain full recognition. This was partly because bacteria, which are, of course, prokaryotes, were considered "lower" and "different." And, it was partly because the DNA molecule—made up of only four different nucleotides—seemed too simple for the enormously complex task of carrying the hereditary information.

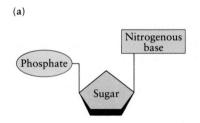

(b) Purine-containing nucleotides

(c) Pyrimidine-containing nucleotides

14–3   (a) *A nucleotide is made up of three different components: a nitrogenous base, a five-carbon sugar, and a phosphate group.* **(b)** *The two purine-containing nucleotides and* **(c)** *the two pyrimidine-containing nucleotides found in DNA. Each nucleotide consists of one of the four possible nitrogenous bases, a deoxyribose sugar, and a phosphate group.*

0.1 μm

14–4 *Bacteriophages of the type used in the studies that provided positive identification of DNA as the genetic material. Notice their highly distinctive shape. Each bacteriophage consists of a head, which appears hexagonal in electron micrographs, and a complex tail assembly. These bacteriophages attach to* E. coli *cells by means of the thin fibers extending from the tail assembly.*

The Bacteriophage Experiments

In 1940, a series of crucial experiments was begun with another "fit material," destined to become as important to genetic research as the garden pea and the fruit fly. The fit material was a group of viruses that attack bacteria and are therefore known as **bacteriophages** ("bacteria eaters"). The bacteriophages originally chosen for study were ones that attack *Escherichia coli,* the familiar bacterium found in the healthy human intestine.

These viruses, commonly called **phages,** were inexpensive, easy to maintain in the laboratory, and required little space or equipment. Moreover, they were phenomenal at reproducing themselves. About 25 minutes after a single virus infected a bacterial cell, the cell would burst open, releasing a hundred or more new viruses, all exact copies of the original virus. Another advantage of these phages was their highly distinctive shape (Figure 14–4), which enabled them to be readily identified with the electron microscope.

Chemical analysis of the bacteriophages revealed

that they consist quite simply of DNA and of protein, the two leading contenders for the role of the genetic material. The question of which type of molecule contains the viral genes—that is, the hereditary material that directs the synthesis of new viruses within the bacterial cells—was answered in 1952 by Alfred D. Hershey and Martha Chase. Their simple but ingenious experiments are summarized in Figure 14–5. Remember that protein contains sulfur (in the amino acids cysteine and methionine) but no phosphorus, and DNA contains phosphorus but no sulfur. You will then see how these experiments demonstrated that only the DNA of the bacteriophages was involved in the replication process and that the protein could not be the hereditary material.

Electron micrographs later confirmed that this type of bacteriophage attaches to the bacterial cell wall by its tail fibers and injects its DNA into the cell. The empty protein coat (the "ghost") is left on the outside (Figure 14–6). In short, the protein is just a container for the phage DNA. It is the DNA that enters the cell and carries the complete hereditary message of the virus, directing the formation of new viral DNA and new viral protein.

Further Evidence for DNA

The role of DNA in bacterial transformation and in viral replication formed very convincing evidence that DNA is the genetic material. Two other lines of work also helped to lend weight to the argument. First, Alfred Mirsky, in a long series of careful studies, showed that, in general, the somatic cells of any given species contain equal amounts of DNA. The gametes, however, contain just half as much DNA as the somatic cells.

A second important series of contributions was made by Erwin Chargaff. Chargaff analyzed the purine and pyrimidine content of the DNA of many different kinds of living things and found that the nitrogenous bases do *not* occur in equal proportions. The proportions of the four nitrogenous bases are the same in all cells of all individuals of a given species, but they vary from one species to another.

These variations suggested that the four bases might indeed provide a "language" in which the instructions controlling cell growth could be written. Some of Chargaff's results are reproduced in Table 14–1. Do you notice anything interesting about the relative proportions of the four bases? These ratios provided important clues to the molecular structure of DNA.

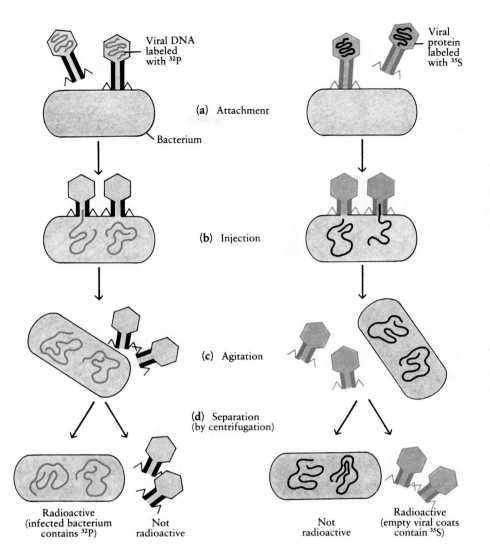

(a) Attachment

Viral DNA labeled with ³²P

Viral protein labeled with ³⁵S

Bacterium

(b) Injection

(c) Agitation

(d) Separation (by centrifugation)

Radioactive (infected bacterium contains ³²P)

Not radioactive

Not radioactive

Radioactive (empty viral coats contain ³⁵S)

14–5 *A summary of the Hershey-Chase experiments. Two separate samples of phages were prepared, one in which the DNA was labeled with a radioactive isotope of phosphorus, ³²P, and the other in which the protein was labeled with a radioactive isotope of sulfur, ³⁵S. (a) One culture of bacteria (left) was infected with the ³²P-labeled phage, and a similar culture (right) was infected with the ³⁵S-labeled phage. After the infection cycles had begun (b), the bacterial cells were agitated in a blender (c) and then spun down in a centrifuge (d) to separate the cells from any viral material remaining on the outside. Samples of extracellular material and of intracellular material from each culture were then tested for radioactivity. Hershey and Chase found that the ³⁵S had remained outside the bacterial cells with the empty viral coats and the ³²P had entered the cells, infected them, and caused the production of new viral particles. It was therefore concluded that the genetic material of the bacteriophage is DNA rather than protein.*

Table 14–1	Composition of DNA in Several Species			
	Purines		**Pyrimidines**	
Source	**Adenine**	**Guanine**	**Cytosine**	**Thymine**
Human being	30.4%	19.6%	19.9%	30.1%
Ox	29.0	21.2	21.2	28.7
Salmon sperm	29.7	20.8	20.4	29.1
Wheat germ	28.1	21.8	22.7	27.4
E. coli	24.7	26.0	25.7	23.6
Sea urchin	32.8	17.7	17.3	32.1

0.1 µm

14–6 *Electron micrograph of bacteriophages attacking a cell of E. coli. The viruses attach to the bacterial cell by their tail fibers. Viral DNA, contained within the head of the virus, is injected through the tail and into the cell. In this micrograph, the heads of some of the viruses are empty, indicating that the injection process has already occurred.*

The Watson-Crick Model

In the early 1950s, a young American scientist, James Watson, went to Cambridge, England, on a research fellowship to study problems of molecular structure. There, at the Cavendish Laboratory, he met physicist Francis Crick. Both were interested in DNA, and they soon began to work together to solve the problem of its molecular structure. They did not do experiments in the usual sense but rather undertook to examine all the data about DNA and to unify them into a meaningful whole.

The Known Data

By the time Watson and Crick began their studies, quite a lot of information on the subject had already accumulated. It was known that the DNA molecule is very large, very long and thin, and contains nucleotides, each consisting of either one purine or one pyrimidine plus a deoxyribose sugar and a phosphate group.

In 1950, Linus Pauling had shown that proteins sometimes take the form of a helix (see page 62) and that the helical structure is maintained by hydrogen bonding between successive turns in the helix. Pauling had suggested that the structure of DNA might be similar. Subsequent x-ray studies by Rosalind Franklin and Maurice Wilkins at Kings College, London, provided strong evidence that the DNA molecule was indeed a giant helix.

Finally, there were Chargaff's data, which indicated that the ratio of DNA nucleotides containing thymine to those containing adenine is approximately 1:1 and that the ratio of nucleotides containing guanine to those containing cytosine is also approximately 1:1.

Building the Model

From these data, Watson and Crick attempted to construct a model of DNA that would fit the known facts and explain the biological role of DNA. In order to carry the vast amount of genetic information, the molecules should be heterogeneous and varied. Also, there must be some way for them to replicate readily and with great precision so that faithful copies could be passed from cell to cell and from parent to offspring, generation after generation.

By piecing together the data, Watson and Crick were able to deduce that DNA is an exceedingly long, entwined double helix. If you were to take a ladder and twist it into a helix, keeping the rungs perpendicular, you would have a crude model of the DNA molecule (Figure 14–7). The two rails, or sides, of the ladder are made up of alternating sugar and phosphate molecules.

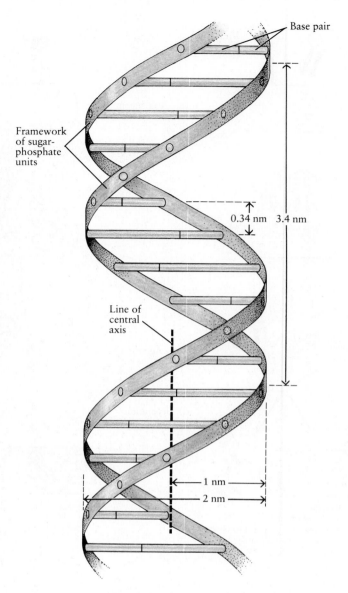

14–7 *The double-stranded helical structure of DNA, as first presented in 1953 by Watson and Crick. The framework of the helix is composed of the sugar-phosphate units of the nucleotides. The rungs are formed by the four nitrogenous bases adenine and guanine (the purines) and thymine and cytosine (the pyrimidines). Each rung consists of a pair of bases. Knowledge of the distances shown here was crucial in establishing the structure of the DNA molecule. These distances were determined from x-ray diffraction photographs of DNA taken by Rosalind Franklin.*

The perpendicular rungs of the ladder are formed by the nitrogenous bases—adenine (A), thymine (T), guanine (G), and cytosine (C)—a pair of bases to each rung. Each base is covalently bonded to a sugar subunit in the "rail" adjacent to it. The paired bases meet in the interior of the helix and are joined by hydrogen bonds, the relatively weak bonds that Pauling had demonstrated in his studies of protein structure.

Then there is the question, what would have happened if Watson and I had not put forward the DNA structure? This is "iffy" history which I am told is not in good repute with historians, though if a historian cannot give plausible answers to such questions I do not see what historical analysis is about. If Watson had been killed by a tennis ball I am reasonably sure I would not have solved the structure alone, but who would? Olby has recently addressed himself to this question. Watson and I always thought that Linus Pauling would be bound to have another shot at the structure once he had seen the King's College x-ray data, but he has recently stated that even though he immediately liked our structure it took him a little time to decide finally that his own was wrong. Without our model he might never have done so. Rosalind Franklin was only two steps away from the solution. She needed to realise that the two chains must run in opposite directions and that the bases, in their correct tautomeric forms, were paired together. She was, however, on the point of leaving King's College and DNA, to work instead on TMV [tobacco mosaic virus] with Bernal. Maurice Wilkins had announced to us, just before he knew of our structure, that he was going to work full time on the problem. Our persistent propaganda for model building had also had its effect (we had previously lent them our jigs to build models but they had not used them) and he proposed to give it a try. I doubt myself whether the discovery of the structure could have been delayed for more than two or three years.

There is a more general argument, however, recently proposed by Gunther Stent and supported by such a sophisticated thinker as Medawar. This is that if Watson and I had not discovered the structure, instead of being revealed with a flourish it would have trickled out and that its impact would have been far less. For this sort of reason Stent had argued that a scientific discovery is more akin to a work of art than is generally admitted. Style, he argues, is as important as content.

I am not completely convinced by this argument, at least in this case. Rather than believe that Watson and Crick made the DNA structure, I would rather stress that the struc-

(a)

(b)

(a) *James Watson (left) and Francis Crick in 1953, with one of their models of DNA. At the time they announced their discovery of the DNA structure, Watson was 23 and Crick was 34.*

(b) *A computer-generated model of a portion of a DNA molecule. The sugar-phosphate backbone is indicated by the blue ribbons and green dots. The purines are shown in yellow, and the pyrimidines in red. The hydrogen bonds linking* the base pairs are represented by blue dashed lines.

In 1993, 40 years after their discovery, Watson remarked, "The molecule is so beautiful. Its glory was reflected on Francis and me. I guess the rest of my life has been spent trying to prove that I was almost equal to being associated with DNA, which has been a hard task." Crick replied, "We were upstaged by a molecule."

ture made Watson and Crick. After all, I was almost totally unknown at the time and Watson was regarded, in most circles, as too bright to be really sound. But what I think is overlooked in such arguments is the intrinsic beauty of the DNA double helix. It is the molecule which has style, quite as much as scientists. The genetic code was not revealed all in one go but it did not lack for impact once it had been pieced together. I doubt if it made all that difference that it was Columbus who discovered America. What mattered much more was that people and money were available to exploit the discovery when it was made. It is this aspect of the history of the DNA structure which I think demands attention, rather than the personal elements in the act of discovery, however interesting they may be as an object lesson (good or bad) to other workers.

Francis Crick: "The Double Helix: A Personal View," *Nature*, vol. 248, pages 766–769, 1974.

14–8   *The double-stranded structure of a small portion of a DNA molecule. Each nucleotide consists of a phosphate group, a deoxyribose sugar, and a purine or pyrimidine base.*

Note the repetitive sugar-phosphate-sugar-phosphate sequence that forms the backbone of each strand of the molecule. Each phosphate group is attached to the 5' carbon of one sugar subunit and to the 3' carbon of the sugar subunit in the adjacent nucleotide. Each strand of the DNA molecule thus has a 5' end and a 3' end, determined by these 5' and 3' carbons. The strands are antiparallel—that is, the direction from the 5' to the 3' end of one strand is opposite to that of the other.

The strands are held together by hydrogen bonds (represented here by dashed lines) between the bases. Notice that adenine and thymine form two hydrogen bonds, whereas guanine and cytosine form three. Because of these bonding requirements, adenine can pair only with thymine, and guanine can pair only with cytosine. Thus the order of bases along one strand determines the order of bases along the other strand.

The sequence of bases varies from one DNA molecule to another. It is customarily written as the sequence in the 5' to 3' direction of one of the strands. Here, using the strand on the left, the sequence is TCAG.

According to x-ray measurements, the distance between the two sides, or rails, is 2 nanometers. Two purines in combination would take up more than 2 nanometers, and two pyrimidines would not reach all the way across. But if a purine paired in each case with a pyrimidine, there would be a perfect fit. The paired bases—the "rungs" of the ladder—would therefore always be purine-pyrimidine combinations.

As Watson and Crick worked their way through the data, they assembled actual tin-and-wire models of the molecules (see essay, page 247), testing where each piece would fit into the three-dimensional puzzle. As they worked with the models, they realized that the nucleotides along any one strand of the double helix could be assembled in any order. Since a DNA molecule may be thousands of nucleotides long, there is a possibility for great variety, one of the primary requirements for the genetic material.

The most exciting discovery, however, came when Watson and Crick set out to construct a matching strand. They encountered an interesting and important restriction. Not only could purines not pair with purines and pyrimidines not pair with pyrimidines, but also, because of the structures of the bases, adenine could pair only with thymine, forming two hydrogen bonds (A=T), and guanine could pair only with cytosine, forming three hydrogen bonds (G≡C). The paired bases were **complementary.** Look at Table 14–1 again and see how well these chemical requirements explain Chargaff's data.

The double-stranded structure of a small portion of a DNA molecule is shown in Figure 14–8. As you can see, the strands have direction. In each strand, each phosphate group is attached to one sugar at the 5′ position (the fifth carbon in the sugar ring) and to the other sugar at the 3′ position (the third carbon in the sugar ring). Thus, each strand has a 5′ end and a 3′ end. And, within the double helix, the two strands run in opposite directions—that is, the direction from the 5′ end to the 3′ end of each strand is opposite. The strands are therefore said to be **antiparallel.**

Although the nucleotides along one chain of the double helix can occur in any order, their sequence then determines the order of nucleotides in the other chain. This is necessarily the case, since the bases are complementary (G pairs only with C, and A pairs only with T).

DNA Replication

An essential property of the genetic material is the ability to provide for exact copies of itself. Implicit in the

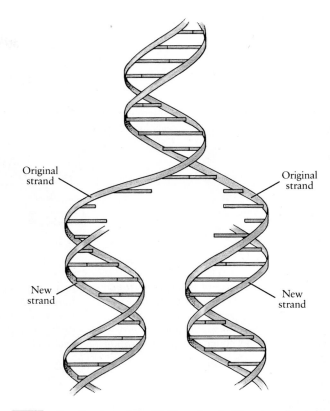

14–9   *Replication of the DNA molecule. The strands separate down the middle as the hydrogen bonds are broken and the paired bases separate. Each of the original strands then serves as a template along which a new, complementary strand forms from nucleotides available in the cell.*

double and complementary structure of the DNA helix is a mechanism by which it can reproduce itself.

At the time of chromosome replication, the molecule "unzips" down the middle, with the paired bases separating as the hydrogen bonds are broken. As the two strands separate, they act as **templates,** or guides. Each strand directs the synthesis of a new complementary strand along its length (Figure 14–9), using the raw materials in the cell. If a T is present on the original strand (the template), only an A can fit in the adjacent location of the new strand; a G will pair only with a C, and so on. In this way, each strand forms a copy of its original partner strand, and two exact replicas of the molecule are produced. The age-old question of how hereditary information is duplicated and passed on, generation after generation, had, in principle, been answered.

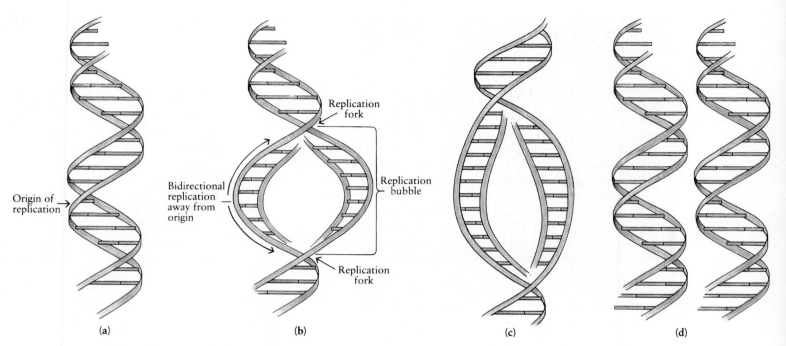

Origin of replication

Bidirectional replication away from origin

Replication fork

Replication bubble

Replication fork

(a) (b) (c) (d)

14–10 *An overview of DNA replication. The two strands of the DNA molecule (a) separate at the origin of replication as a result of the action of special initiator proteins and enzymes. (b, c) The two replication forks move away from the origin of replication in opposite directions, forming a replication bubble that expands in both directions (bidirectionally). (d) When synthesis of the new DNA strands is complete, the two double-stranded chains separate into two new double helixes. Each consists of one old strand and one new strand.*

The Mechanics of DNA Replication

Replication of DNA is a process that occurs only once in each cell generation, during the S phase of the cell cycle (page 178). It is the essential event in the replication of the chromosomes. In most eukaryotic cells, DNA replication leads ultimately to mitosis, but in the cells that give rise to gametes it leads instead to meiosis. It is a remarkably rapid process. For example, in humans and other mammals, the rate of synthesis is about 50 nucleotides per second. In prokaryotes, it is even faster—about 500 nucleotides per second.

The principle of DNA replication, in which each strand of the double helix serves as a template for the formation of a new strand, is relatively simple and easy to understand. However, the actual process by which the cell accomplishes replication is considerably more complex. Like other biochemical reactions of the cell, DNA replication requires a number of different enzymes, each catalyzing a particular step of the process. The identification of the principal enzymes, their precise functions, and the sequence of events in replication has required a number of years and the efforts of many scientists working in different laboratories. Although our understanding is still incomplete, the general outlines of the process are now clear.

Initiation of DNA replication always begins at a specific nucleotide sequence known as the **origin of replication.** It requires special initiator proteins and enzymes that break the hydrogen bonds linking the complementary bases at the replication origin, opening up the helix so replication can occur. As the two strands of the DNA double helix are separated, other proteins attach to the individual strands, holding them apart. This makes possible the next stage, the actual synthesis of the new strands, catalyzed by a group of enzymes known as **DNA polymerases.**

If replicating DNA is viewed under the electron microscope, the region of synthesis appears as an "eye," or **replication bubble.** At either end of the bubble, where the existing strands are being separated and the new, complementary strands are being synthesized, the molecule appears to form a Y-shaped structure. This is known as a **replication fork.** The two replication forks move in opposite directions away from the origin (Figure 14–10), and thus replication is said to be **bidirectional.**

In prokaryotes, there is a single replication origin in the chromosome. In eukaryotes, by contrast, there are many replication origins in each chromosome. Replication proceeds along the linear chromosome as each bubble expands bidirectionally (Figure 14–11) until it meets an adjacent bubble. As the bubbles merge, the entire length of the chromosome is replicated.

14–11 *Replication of eukaryotic chromosomes is initiated at multiple origins. The individual replication bubbles spread until ultimately they meet and join. In this electron micrograph of a replicating chromosome in an embryonic cell of* Drosophila, *the replication bubbles are indicated by the arrows.*

5 kb

Proofreading

One of the essential features of DNA replication is that DNA polymerase has a proofreading function. In the course of synthesis, errors are sometimes made, and the wrong nucleotide is added to the newly forming strand. In other words, the nucleotide added to the strand is not complementary to the nucleotide on the template strand. DNA polymerase, however, is able to add nucleotides to a strand only if the preceding nucleotides are correctly paired with their complementary nucleotides on the template strand. If a mistake has occurred, the enzyme backtracks, removing nucleotides until it encounters a correctly paired nucleotide. At that point, the enzyme begins moving forward once more, adding nucleotides to the growing strand as it goes. This ability to remove incorrectly paired nucleotides provides an important proofreading that ensures the accuracy of DNA replication.

In addition to the proofreading that occurs during DNA replication, other enzymes constantly monitor the DNA double helixes of the cell. Whenever an incorrectly paired nucleotide is encountered, DNA repair enzymes move in to snip it out and replace it with the correct nucleotide. This process is essential in maintaining the integrity of the genetic machinery.

The Energetics of DNA Replication

The nucleotides required for DNA replication are synthesized by the same biochemical pathways as the nucleotides required for other functions of the cell. However, they are supplied not in the form of the monophosphates shown in Figure 14–3, but rather as triphosphates. For example, adenine is supplied as deoxyadenosine triphosphate (dATP). The only differ-

ence between ATP and dATP is in the sugar subunit—in ATP, it is ribose, and in dATP, it is deoxyribose. Similarly, guanine is supplied as dGTP, cytosine as dCTP, and thymine as dTTP.

Just as ATP provides energy to power a variety of biosynthetic reactions, dATP, dGTP, dCTP, and dTTP provide energy to power the synthesis of DNA. As each nucleotide is attached to the growing DNA strand, the "extra" P~P group is removed and immediately split into two phosphate ions. The energy released in this process powers the reactions mediated by the DNA polymerases.

DNA as a Carrier of Information

You will recall that a necessary property of the genetic material is the capacity to carry information. The Watson-Crick model showed how the DNA molecule is able to do this. The information is carried in the sequence of the bases, and *any* sequence of bases is possible. Since the number of paired bases ranges from about 5,000 for the simplest known virus up to an estimated 6 billion in the 46 human chromosomes, the number of possible variations in sequence is astronomical.

The DNA from a single human cell—which if extended in a single thread would be almost 2 meters long—can contain information equivalent to some 600,000 printed pages of 500 words each, or a library of about a thousand books. Obviously, the DNA structure can well account for the endless diversity among living things.

Summary

Classical genetics had been concerned with the mechanics of inheritance—how the units of heredity are passed from one generation to the next and how changes in the hereditary material are expressed in individual organisms. New questions soon arose, however, and geneticists began to explore the nature of the gene—its structure, composition, and properties.

During the 1940s, many investigators believed that genes were proteins, but others were convinced that the hereditary material was deoxyribonucleic acid (DNA). Important, although not widely accepted, evidence for the genetic role of DNA was presented by Avery in his experiments to identify the transforming factor of pneumococci. Confirmation of Avery's hypothesis came from studies with bacteriophages (bacterial viruses) showing that DNA and not protein is the genetic material of the virus.

Further support for the genetic role of DNA came from two more sets of data: (1) Almost all somatic cells of any given species contain equal amounts of DNA, and the gametes contain just half that amount. (2) The proportions of nitrogenous bases are the same in the DNA of all cells of a given species, but they vary in different species.

In 1953, Watson and Crick proposed a structure for DNA. The DNA molecule, according to their model, is a double-stranded helix, shaped like a twisted ladder. The two sides of the ladder are composed of repeating subunits consisting of a phosphate group and the five-carbon sugar deoxyribose. The "rungs" are made up of paired nitrogenous bases, one purine base pairing with one pyrimidine base. There are four bases in DNA—adenine (A) and guanine (G), both of which are purines, and thymine (T) and cytosine (C), which are pyrimidines. A can pair only with T, and G only with C. The four bases are the four "letters" used to spell out the genetic message. The paired bases are joined by hydrogen bonds.

When the DNA molecule replicates, the two strands come apart, breaking at the hydrogen bonds. Each strand acts as a template for the formation of a new, complementary strand from nucleotides available in the cell. The addition of nucleotides to the new strands is catalyzed by DNA polymerases. A variety of other enzymes also play key roles in the replication process.

Replication begins at a particular nucleotide sequence on the chromosome, the origin of replication. It proceeds bidirectionally, by way of two replication forks that move in opposite directions. In the course of DNA replication, DNA polymerase proofreads, backtracking when necessary to remove nucleotides that are not correctly paired with the template strand.

The nucleotides incorporated into the growing DNA strands are supplied in the form of triphosphates. The energy required to power replication is provided by the removal and splitting of the "extra" P~P groups.

On the basis of the structure of the DNA double helix, as revealed by Watson and Crick, the role of DNA as the carrier and transmitter of the genetic information was universally accepted. With the discovery of the mechanism by which the living cell replicates its DNA, the question of how the hereditary information is faithfully transmitted from parent cell to daughter cell, generation after generation, was answered.

Questions

1. What are the steps by which Griffith demonstrated the existence of the transforming factor? Can you think of any implications of Griffith's discovery for modern medicine?

2. What characteristics of bacteriophages make them a useful experimental tool?

3. When the structure of DNA was being worked out, it became apparent that one purine base must be paired with a pyrimidine base, and that the other purine base must be paired with the other pyrimidine base. The evidence for this requirement came from two types of data. What were the data, and how did they indicate this structural requirement?

4. Further consideration of the structures of the four nitrogenous bases indicated that adenine could pair only with thymine, and cytosine only with guanine. What structural features of the bases imposed this requirement on the structure of the DNA molecule?

5. One of the chief arguments for the erroneous hypothesis that proteins constitute the genetic material was that proteins are heterogeneous. Explain why the genetic material must have this property. What feature of the Watson-Crick model of DNA structure is important in this respect?

6. Shown here is the sequence of bases in the 5' to 3' di-

rection in one strand of a hypothetical DNA molecule. Identify the sequence of bases in the complementary strand.

5′ —A—A—G—T—T—T—G—G—T—T—A—C—T—T—G— 3′
3′ — — — — — — — — — — — — — — — — 5′

7. Distinguish among the following terms: origin of replication, replication bubble, and replication fork.

8. Among the functions of DNA polymerase is the proofreading of the newly synthesized strands as they are formed. DNA polymerase can identify any mismatched bases, snip them out, and replace them with the correct bases. What is the biological significance of this activity?

9. Eukaryotic cells are grown for a number of generations in a medium containing thymine labeled with tritium, a radioactive isotope of hydrogen (^3H). The cells are then removed from the radioactive medium, placed in an ordinary, nonradioactive medium, and allowed to divide. Studies of the distribution of the radioactive isotope are made after each generation to determine the presence or absence of radioactive material in the chromatids.

Before the cells are placed in the nonradioactive medium, all the chromatids contain ^3H. After one generation in the nonradioactive medium, all of the chromatids still contain radioactive ^3H. (a) Assuming each chromatid contains a single DNA molecule, explain the results. (b) Is this consistent with the Watson-Crick hypothesis of DNA replication? (c) What would be the distribution of the ^3H after two generations in the nonradioactive medium? Why? *(Hint:* You may find it helpful to look at Figure 14–10 as you think about these questions.)

10. Suppose you are talking to someone who has never heard of DNA. How would you support an argument that DNA is the genetic material? List at least five of the strong points in such an argument.

Suggestions for Further Reading

Crick, Francis: *What Mad Pursuit: A Personal View of Scientific Discovery,* Basic Books, Inc., New York, 1988.*

> *This book is one of a series of autobiographies by major figures in twentieth-century science, commissioned by the Sloan Foundation and intended for the general reader. Crick's account of the discovery of the structure of the DNA molecule provides an interesting counterpoint to Watson's* The Double Helix *(see below). As a physicist who became a biologist, Crick has many valuable insights into the significant differences between biology and the physical sciences.*

Felsenfeld, Gary: "DNA," *Scientific American,* October 1985, pages 58–67.

Judson, Horace F.: *The Eighth Day of Creation: Makers of the Revolution in Biology,* Simon and Schuster, New York, 1979.*

> *A comprehensive study of the human and scientific aspects of molecular biology from the 1930s through the mid-1970s. As events unfold, the story is told from each participant's point of view. We are treated to an enlightening and personal glimpse of the development of scientific thought.*

Kornberg, Arthur: *For the Love of Enzymes: The Odyssey of a Biochemist,* Harvard University Press, Cambridge, Mass., 1991.*

> *Kornberg, who began his scientific career with studies of jaundice in World War II army recruits, went on to discover the DNA polymerases and to make major contributions to our understanding of DNA replication. He is a wonderful writer, and this book is enriched by the clarity of his explanations, by his wise reflections on the scientific enterprise, and by his sense of humor.*

Luria, S. E.: *A Slot Machine, a Broken Test Tube,* Harper & Row, Publishers, Inc., New York, 1985.*

> *Luria was one of the founders of the "phage group" in the 1940s, whose work with the bacteriophages of E. coli led to the beginnings of molecular genetics. A highly reflective book, this autobiography recounts not only his scientific experiences but also many aspects of his personal life, including his flight from Italy through France to the United States at the beginning of World War II.*

Olby, Robert: *The Path to the Double Helix,* University of Washington Press, Seattle, 1975.

> *An account, written by a professional historian of science, of twentieth-century genetics. Olby is interested not only in the scientific concepts and experiments but also in the various personalities involved and their effects on one another and on the course of scientific discovery.*

Radman, Miroslav, and Robert Wagner: "The High Fidelity of DNA Duplication," *Scientific American,* August 1988, pages 40–46.

Watson, James D.: *The Double Helix,* Atheneum Publishers, New York, 1968.*

> *"Making out" in molecular biology. A brash and lively book about how to become a Nobel laureate.*

* Available in paperback.

Escherichia coli, or E. coli, for short, is one of the most common of microscopic organisms. Its preferred residence is your intestinal tract. A prokaryote, it is much smaller than the eukaryotic cells of your intestinal lining, with which it cohabits. In fact, it is only about the size of a mitochondrion in one of those cells. At any given moment, however, a single E. coli cell contains some 5,000 different macromolecules. One of these is a single long molecule of DNA—the bacterial chromosome. In this electron micrograph, an E. coli cell has been exploded by osmotic shock to release its DNA, which can be seen surrounding it.

The revelation by Watson and Crick of the double helical structure of DNA led immediately to an understanding, in principle, of the way in which the molecule duplicates itself. ("It has not escaped our notice," they stated slyly in their original report, "that the special pairing we have postulated immediately suggests a possible copying mechanism for the genetic material.") The next big question, however, remained wide open: How does the genetic material, now identified as DNA, direct the day-to-day activities of living cells?

In the search for the answer to this question, E. coli has turned out to be a major player, comparable in status to the garden pea, Drosophila, and the bacteriophages. Rising from its humble origins, E. coli, it is now generally recognized, is the best understood of all living things on this planet.

The Genetic Code and Its Translation

The Watson-Crick model both established the structure of DNA and indicated the mechanism for its replication. The DNA molecule carries the instructions for the structure and function of the cell and also transmits these instructions to new cells and organisms. Yet two fundamental questions were left unanswered: How are instructions encoded in the DNA molecule? And, how are they carried out?

The studies that led to the answers to these questions had their beginnings in a concept first proposed in 1908. In a series of lectures in that year, an English physician, Sir Archibald Garrod, set forth a new concept of human diseases, which he called "inborn errors of metabolism." Garrod postulated that certain diseases are caused by the body's inability to perform particular chemical processes and are hereditary in nature. With a leap of the imagination that spanned almost half a century, he hypothesized that such diseases are the result of enzyme deficiencies. Implicit in this hypothesis was the idea that genes act by influencing the production of enzymes.

Genes and Proteins

One Gene–One Enzyme

By the 1940s, biologists had come to realize that all of the biochemical activities of the living cell, including the multitude of biosynthetic reactions that produce all of its constituent molecules, depend upon different specific enzymes. Even the synthesis of enzymes depends on enzymes. Further, it was becoming clear that the specificity of different enzymes is a result of their primary structure, the linear sequence of amino acids in the protein molecule.

Meanwhile, George Beadle, a geneticist, was working with the eye-color mutants of Drosophila that had been discovered in Morgan's laboratory (page 228). As a result of his studies, Beadle formulated the hypothesis that each of the various eye colors observed in the mutants is the result of a change in a single enzyme in a biosynthetic pathway. To test this notion—that genes control enzymes—on a broader scale, he teamed up in 1941 with Edward L. Tatum, a biochemist. The organism they chose for their studies was the red bread mold *Neurospora.*

Utilizing x-rays to increase the mutation rate, Beadle and Tatum analyzed the resultant mutants by genetic mapping studies analogous to those performed with *Drosophila.* A number of the mutations proved to affect the activity of single enzymes, and Beadle and Tatum were able to show that a mutation in one gene could be correlated with the loss of function of one enzyme. On the basis of these experiments, they formulated the hypothesis that a single gene specifies a single enzyme.

"One gene–one enzyme" turned out to be an oversimplification, for there are many proteins that are not enzymes. Some, for instance, are hormones, such as insulin, others are structural proteins, such as collagen, and still others are membrane proteins, such as the sodium-potassium pump. These proteins, too, are specified by genes. This expansion of the original concept did not modify it in principle. "One gene–one enzyme" was simply amended to "one gene–one protein." Subsequently, with the realization that many proteins consist of more than one polypeptide chain, it was modified once more, to the less memorable but more precise "one gene–one polypeptide chain."

The Structure of Hemoglobin

What causes the change or loss of function in an enzyme or other protein specified by a gene that has undergone a mutation? Linus Pauling was one of the first to see the implications of the work of Beadle and Tatum with regard to this question. Perhaps, Pauling reasoned, human disorders involving hemoglobin, such as sickle-cell anemia, could be traced to a variation from the normal protein structure of the hemoglobin molecule.

To test this hypothesis, Pauling took samples of hemoglobin from people with sickle-cell anemia (a homozygous recessive condition), from others heterozygous for the allele (Figure 15–1), and from still others homozygous for the normal allele. To try to detect differences in these proteins, Pauling used a process known as **electrophoresis,** in which molecules in solution are exposed to a weak electric field. Very small differences, even in very large molecules, may be reflected in the electric charges of the molecules. These charge differences cause the molecules to move at different rates in the electric field.

Figure 15–2 shows the results of Pauling's experiment. A person who has sickle-cell anemia makes a different sort of hemoglobin than a person who does not have the disorder. A person who is heterozygous (carrying one copy of the allele for sickling and one copy of the allele for normal hemoglobin) makes both kinds of hemoglobin molecules. However, enough normal molecules are produced to prevent anemia. (Notice that the terms "dominant" and "recessive" are becoming more complicated.)

A few years later, it was shown that the actual difference between the normal and the sickle-cell hemoglobin molecules is two amino acids in 600, as we noted previously (page 65).

The Virus Coat

Additional evidence that DNA specifies the structure of proteins came from the bacteriophage studies described in Chapter 14. You will recall that the introduction of viral DNA into a bacterial cell results in the production not only of more viral DNA, but also of the proteins of the virus coat. Clearly, the viral DNA carries the information for the synthesis of the coat proteins.

20 μm

15–1 *Although individuals heterozygous for the sickle-cell allele have no symptoms of anemia, they can be identified by the fact that they have children with sickle-cell anemia. They can also be identified by means of a simple laboratory test in which a sample of blood is treated to remove oxygen from the hemoglobin molecules. As this scanning electron micrograph of deoxygenated blood from a heterozygous individual reveals, most of the red blood cells are normal, but an occasional cell is sickled under the artificial conditions of the test.*

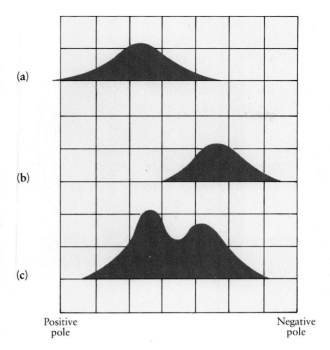

15–2   *The results, as reported by Linus Pauling in 1949, of the electrophoresis of (a) normal hemoglobin, (b) the hemoglobin of a person with sickle-cell anemia, and (c) the hemoglobin of a person who is heterozygous for the sickle-cell allele. The height of the curves indicates the amount of hemoglobin at each point. Because of slight differences in electric charge, normal and sickle-cell hemoglobins move at different rates in an electric field. The normal hemoglobin is more negatively charged. Hence it moves closer to the positive pole than does the sickle-cell hemoglobin. The hemoglobin from the heterozygote separates into the two different positions, indicating that it consists of both normal and sickle-cell hemoglobins.*

Positive pole

Negative pole

From DNA to Protein: The Role of RNA

As a result of all of these studies, there was general agreement that the DNA molecule contains a coded message with instructions for biological structure and function. Moreover, these instructions are carried out by proteins, which also contain a highly specific biological "language." As we saw in Chapter 3, the linear sequence of amino acids in a polypeptide chain determines the three-dimensional structure of the completed protein molecule, and it is the three-dimensional structure that, in turn, determines function. The question thus became one of translation. How did the order of bases in DNA specify the sequence of amino acids in a protein molecule?

The search for the answer to this question led to **ribonucleic acid (RNA)**, a close chemical relative of DNA. As you can see in Figure 15–3, there are three principal differences between RNA and DNA:

1. In the nucleotides of RNA, the sugar component is ribose, rather than deoxyribose.

2. The nitrogenous base thymine, found in DNA, does not occur in RNA. Instead, RNA contains a closely related pyrimidine, **uracil (U)**. Uracil, like thymine, pairs only with adenine.

3. Most RNA is single-stranded.

15–3   *Chemically, RNA is very similar to DNA, but there are two differences in its nucleotides. (a) One difference is in the sugar component. Instead of deoxyribose, RNA contains ribose, which has an additional oxygen atom. (b) The other difference is that instead of thymine, RNA contains the closely related pyrimidine uracil (U). Uracil, like thymine, pairs only with adenine.*

A third, and very important, difference between the two nucleic acids is that RNA is usually single-stranded and does not form a regular helical structure.

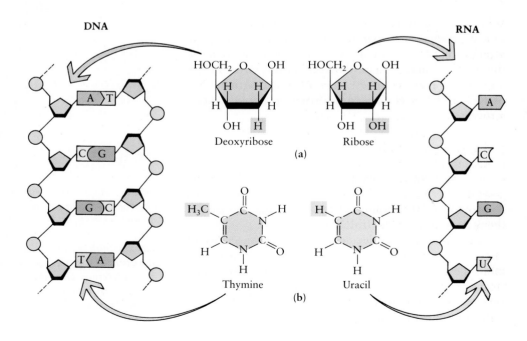

DNA   RNA

Deoxyribose   Ribose

(a)

Thymine   Uracil

(b)

15–4   (a) *Diagram of a very small portion of tobacco mosaic virus (TMV). This virus has a central core not of DNA but rather of RNA. Its outer coat is composed of 2,150 identical protein molecules, each consisting of 158 amino acids. If the RNA is separated from its protein coat and rubbed into scratches on a tobacco leaf, new TMV particles are formed, complete with new protein coats. Studies with TMV were among the first to suggest that RNA could direct the assembly of proteins. (b) When viewed under the electron microscope, TMV particles appear as rod-shaped structures.*

Protein subunits

RNA

(a)   (b)   150 nm

There were several clues that RNA might play a role in the translation of genetic information from DNA into a sequence of amino acids. For instance, cells that are making large amounts of protein are rich in RNA. In eukaryotic cells, the RNA is found mostly in the cytosol, and it is here that most protein synthesis takes place. Also, both prokaryotic and eukaryotic cells making large amounts of protein have numerous ribosomes, and ribosomes are rich in RNA.

Additional evidence came from viruses. When a bacterial cell is infected by a DNA-containing bacteriophage, RNA is synthesized from the viral DNA before viral protein synthesis begins. Also, some viruses consist of only RNA and protein. Tobacco mosaic virus (Figure 15–4) is an example. These clues all indicated that RNA as well as DNA contained information about protein structure.

RNA as Messenger

As it turned out, not one but three kinds of RNA play roles as intermediaries in the steps that lead from DNA to protein. At this point, we shall consider just one of them: **messenger RNA (mRNA).**

Messenger RNA molecules are copies (transcripts) of nucleotide sequences encoded in the DNA. Unlike DNA molecules, however, RNA molecules are usually single-stranded. Each new mRNA molecule is copied, or transcribed, from one of the two strands of DNA by the same base-pairing principle that governs DNA replication (Figure 15–5).

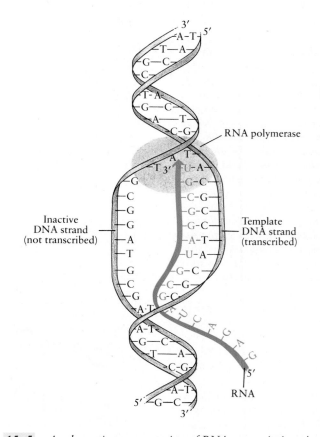

RNA polymerase

Inactive DNA strand (not transcribed)

Template DNA strand (transcribed)

RNA

15–5   *A schematic representation of RNA transcription. At the point of attachment of the enzyme RNA polymerase, the DNA opens up, and, as the RNA polymerase moves along the DNA molecule, the two strands of the molecule separate. Nucleotide building blocks are assembled into RNA in a 5' to 3' direction as the enzyme reads the template DNA strand in a 3' to 5' direction. Note that the base sequence of the RNA strand is complementary—not identical—to that of the template strand from which it is transcribed. It is, however, identical to that of the inactive (untranscribed) DNA strand, except for the replacement of thymine (T) by uracil (U).*

The Elusive Messenger

The cytosol of cells that are synthesizing proteins is full of RNA. This observation was the major clue that RNA plays a role in directing the assembly of proteins. Although the existence of RNA molecules that carry the genetic information from DNA to protein was hypothesized, confirmation of the hypothesis required the detection and isolation of the messenger molecules. The problem was complicated, however, by the fact that most of any cell's RNA is bound into ribosomes and is therefore unlikely to be the carrier of genetic information. This paradox puzzled molecular biologists for almost a decade.

Escherichia coli and its bacteriophages once again provided the tools of discovery. As you know, the genetic material of these phages is DNA, and their coats are made of protein. When they infect a bacterial cell, new coat proteins are synthesized. If the messenger hypothesis were true, new RNA should be formed between the time of cell infection and the synthesis of new virus coats. To test the hypothesis, *E. coli* cells were infected with phage and then exposed briefly to uracil labeled with radioactive carbon. Short-lived RNA molecules with radioactive labels were detected in the cells. They were associated with ribosomes but they were not part of the ribosomes. Were they the long-sought messengers?

To answer this question it was necessary to show that the radioactive RNA was complementary to the phage DNA. The method used was simple but ingenious. If dissolved DNA molecules are heated gently, the hydrogen bonds break apart and the two strands of the double helix separate. Subsequently, when the solution is slowly cooled, complementary strands pair up again and the hydrogen bonds re-form. If the newly formed, radioactive RNA were complementary to the phage DNA, the investigators reasoned, it too should form a duplex (a double-stranded molecule) with that DNA.

As a control, the radioactive RNA molecules were first mixed in a beaker with a solution of *E. coli* DNA and heated. When this mixture was cooled, no duplexes containing radioactivity could be detected. The fact that no hybrid RNA-DNA molecules had formed

(a)

(b)

RNA-DNA hybrids can be used to show the complementarity in nucleotide sequence between an RNA molecule and the DNA molecule from which it was transcribed. **(a)** *The formation of a hybrid molecule between a radioactive*

indicated that the RNA was not complementary to the *E. coli* DNA.

Then the radioactive RNA molecules were mixed with a solution of phage DNA. When this mixture was heated and then cooled, the results were clear-cut. Duplexes containing radioactivity had formed, indicating that the RNA had bound to its complementary strand of phage DNA. The messenger had been disclosed.

RNA molecule (red) and its template DNA strand (black). **(b)** *If the same RNA molecule is mixed with unrelated DNA, no hybrid molecules containing radioactivity are formed.*

Hybridization of DNA with DNA and of RNA with DNA has since become an enormously powerful tool in both molecular genetics and evolutionary biology. As we shall see in later chapters, it is used in a great variety of studies, ranging from the detection of specific genes responsible for particular human disorders to the resolution of evolutionary enigmas.

Like a strand of DNA, each RNA molecule has a 5′ end and a 3′ end. The nucleotides, which are present in the cell as triphosphates, are added, one at a time, to the 3′ end of the growing RNA chain. This process, known as **transcription,** is catalyzed by the enzyme **RNA polymerase.** The enzyme moves along the template DNA strand in a 3′ to 5′ direction, synthesizing a new complementary strand of nucleotides in a 5′ to 3′ direction.

DNA is the master copy of the genetic information, kept permanently "on file" in the chromosomes. Messenger RNA, by contrast, is the working copy of the genetic information. Incorporating the instructions encoded in the DNA molecule, mRNA dictates the sequence of amino acids in proteins. When its job is done, it is degraded into its constituent nucleotides, which are then available for use in the synthesis of other mRNA molecules.

The Genetic Code

The identification of mRNA as the working copy of the genetic instructions still left the big question unresolved. Proteins contain 20 different amino acids, but DNA and RNA each contain only four different nucleotides. Somehow, these nucleotides constituted a **genetic code** for the amino acids.

As it turned out, the idea of a code was useful not only as a dramatic metaphor but also as a working analogy. Scientists, seeking to understand how the sequence of nucleotides stored in the double helix could specify the quite dissimilar structures of protein molecules, approached the problem with methods used by cryptographers in deciphering codes.

As we saw in Chapter 3, the primary structure of each particular kind of protein molecule consists of a specific linear arrangement of the 20 different amino acids. Similarly, there are four different nucleotides, arranged in a specific linear sequence in a DNA molecule. If each nucleotide "coded" for one amino acid, only four amino acids could be specified by the four bases. If two nucleotides specified one amino acid, there could be a maximum number, using all possible arrangements of the nucleotides, of 4×4, or 16—still not quite enough to code for all 20 amino acids. Therefore, following the code analogy, at least three nucleotides in sequence must specify each amino acid. This would provide for $4 \times 4 \times 4$, or 64, possible combinations, or **codons**—clearly, more than enough.

The three-nucleotide, or triplet, codon was widely and immediately adopted as a working hypothesis. Its existence, however, was not actually demonstrated until the code was finally broken, a decade after Watson and Crick first presented their DNA structure. The scientists who performed the initial, crucial experiments toward breaking the code were Marshall Nirenberg and his colleague Heinrich Matthaei, both of the National Institutes of Health.

Breaking the Code

Messenger RNA, then newly discovered, gave Nirenberg the tool he needed. He broke apart *E. coli* cells, extracted their contents, and added radioactively labeled amino acids and crude samples of RNA from a variety of sources to the extracts. All of the RNA samples stimulated protein synthesis. The amounts of radioactively labeled protein produced were small but measurable. In other words, the material extracted from the *E. coli* cells would start producing protein molecules even when the RNA "orders" it received were from a "complete stranger." Even the RNA from tobacco mosaic virus, which naturally multiplies only in cells of the leaves of tobacco plants, could be read as an mRNA by the machinery of the bacterial cell.

Nirenberg and Matthaei then tried an artificial RNA. Perhaps if the *E. coli* extracts could read a foreign message and translate it into protein, they could read a totally synthetic message, one dictated by the scientists themselves. Severo Ochoa of New York University had developed a process for linking nucleotides into a long strand of RNA. With this process, carried out in a test tube, he had produced an RNA molecule that contained only one nitrogenous base, uracil, repeated over and over again. It was called "poly-U."

Nirenberg and Matthaei prepared 20 different test tubes, each of which contained extracts from *E. coli* cells. These extracts included ribosomes, ATP, the necessary enzymes, and all of the amino acids. In each test tube, one of the amino acids, and only one, carried a radioactive label. Synthetic poly-U was added to each test tube. In 19 of the test tubes, no radioactive polypeptides were produced. In the twentieth tube, however, to which radioactive phenylalanine had been added, the investigators were able to detect newly formed, radioactive polypeptide chains.

When these polypeptides were analyzed, they were found to consist only of phenylalanines, one after another. Nirenberg and Matthaei had dictated the message "uracil . . . uracil . . . uracil . . . uracil . . . uracil . . . uracil . . .," and a clear answer had come back, "phenylalanine . . . phenylalanine. . . ." The experiment not only defined the first code word (UUU = phe) but also made available a method for defining the others.

Second letter

		U	C	A	G	
First letter (5′ end)	**U**	UUU ⎤ phe UUC ⎦ UUA ⎤ leu UUG ⎦	UCU ⎤ UCC ⎥ ser UCA ⎥ UCG ⎦	UAU ⎤ tyr UAC ⎦ UAA stop UAG stop	UGU ⎤ cys UGC ⎦ UGA stop UGG trp	U C A G
	C	CUU ⎤ CUC ⎥ leu CUA ⎥ CUG ⎦	CCU ⎤ CCC ⎥ pro CCA ⎥ CCG ⎦	CAU ⎤ his CAC ⎦ CAA ⎤ gln CAG ⎦	CGU ⎤ CGC ⎥ arg CGA ⎥ CGG ⎦	U C A G
	A	AUU ⎤ AUC ⎥ ile AUA ⎦ AUG met	ACU ⎤ ACC ⎥ thr ACA ⎥ ACG ⎦	AAU ⎤ asn AAC ⎦ AAA ⎤ lys AAG ⎦	AGU ⎤ ser AGC ⎦ AGA ⎤ arg AGG ⎦	U C A G
	G	GUU ⎤ GUC ⎥ val GUA ⎥ GUG ⎦	GCU ⎤ GCC ⎥ ala GCA ⎥ GCG ⎦	GAU ⎤ asp GAC ⎦ GAA ⎤ glu GAG ⎦	GGU ⎤ GGC ⎥ gly GGA ⎥ GGG ⎦	U C A G

Third letter (3′ end)

15–6 *The genetic code, consisting of 64 triplet combinations (codons) and the corresponding amino acids, indicated here by their abbreviations (see page 60). The codons shown here are the ones that would appear in the mRNA molecule. Of the 64 codons, 61 specify particular amino acids. The other three codons are stop signals, which cause the synthesis of a polypeptide chain to terminate.*

Since 61 triplets code for only 20 amino acids, there are "synonyms," as many as six different codons for leucine, for example. Most of the synonyms, as you can see, differ only in the third nucleotide. Each codon, however, specifies only one amino acid.

Similar experiments were subsequently performed in a number of laboratories, using artificial RNAs in which either two or three different nucleotides were repeated over and over again in a known sequence. As a result of these experiments, the mRNA codons for all of the amino acids were soon worked out. Of the 64 possible triplet combinations, 61 specify particular amino acids and 3 are stop signals. With 61 combinations coding for only 20 amino acids, you can see that there must be more than one codon for many of the amino acids. As shown in Figure 15–6, codons specifying the same amino acid often differ only in the third nucleotide.

The Universality of the Genetic Code

Three decades have passed since the genetic code was first broken, and the DNA, mRNA, and proteins of many more organisms have been examined. The evidence is now overwhelming: for virtually all organisms, from *Escherichia coli* to *Homo sapiens*, the genetic code is universal. UUA, for example, codes for the amino acid leucine not only in prokaryotes but also in protists, fungi, plants, and animals. The genetic code evolved early, remained constant, and determines the underlying unity of all living things.

A few exceptions to the code shown in Figure 15–6 have been found, however. Most of them involve mitochondria, which contain their own DNA, transcribe their own RNA molecules, and carry out some protein synthesis. In several instances, the mitochondrial code differs from that carried in the chromosomes of both prokaryotes and eukaryotes.

Protein Synthesis

With a knowledge of the genetic code, we can now turn our attention to the question of how the information encoded in the DNA and transcribed into mRNA is subsequently translated into a specific sequence of amino acids in a polypeptide chain. The basic principles of protein synthesis are the same in both prokaryotic and eukaryotic cells, but there are some differences in detail, which will be described in the next chapter. Here we shall focus on the process as it takes place in prokaryotes, particularly *E. coli*.

As we have seen, instructions for protein synthesis are encoded in sequences of nucleotides in the DNA of a cell and are transcribed into mRNA molecules following the same base-pairing rules that govern DNA replication. The single-stranded mRNA molecules produced in this process range from 500 to 10,000 nucleotides in length. Specific nucleotide sequences of the DNA, called **promoters,** are the binding sites for RNA polymerase and thus are the start signals for RNA synthesis. Other nucleotide sequences, called **terminators,** are the stop signals for RNA synthesis, marking the locations at which transcription ends.

The synthesis of proteins requires, in addition to mRNA molecules, two other types of RNA: ribosomal RNA and transfer RNA. These molecules, which are transcribed from their own specific genes in the DNA of the cell, differ both structurally and functionally from mRNA.

Ribosomal RNA and Transfer RNA

Ribosomes are, as you know, the sites of protein synthesis. They are about one-third protein and two-thirds RNA. The form of RNA they contain is called **ribosomal RNA (rRNA).** As shown in Figure 15–7, each ribosome is composed of two subunits, each with its characteristic rRNAs and proteins. The smaller subunit has a binding site for messenger RNA; in *E. coli* and

15–7   (a) *One view of the three-dimensional structure of the* E. coli *ribosome, as revealed by electron micrographs.* (b) *A schematic diagram of another view of the* E. coli *ribosome. In both prokaryotes and eukaryotes, ribosomes consist of two subunits, one large and one small. Each subunit is composed of specific rRNA and protein molecules. In protein synthesis, the ribosome moves along an mRNA molecule that is threaded between the two subunits.*

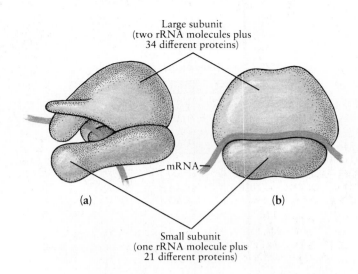

Large subunit
(two rRNA molecules plus
34 different proteins)

mRNA

(a)   (b)

Small subunit
(one rRNA molecule plus
21 different proteins)

Amino acid
attaches here

Invariant
bases

Anticodon

15–8   *The structure of a tRNA molecule. Such molecules consist of about 80 nucleotides linked together in a single chain. The chain always terminates in a CCA sequence at its 3' end. An amino acid links to its specific tRNA molecule at this end. Some nucleotides are the same in all tRNAs; these are shown in light green. The other nucleotides vary according to the particular tRNA. The unlabeled boxes represent unusual modified nucleotides characteristic of tRNA molecules.*

Some of the nucleotides are hydrogen-bonded to one another, as indicated by the dashed lines. In some regions, the unpaired nucleotides form loops. The loop on the right in this diagram is thought to play a role in binding the tRNA molecule to the surface of the ribosome. Three of the unpaired nucleotides in the loop at the bottom of the diagram (bright green) form the anticodon. They serve to "plug in" the tRNA molecule to an mRNA codon.

other prokaryotes, the leading (5') end of the mRNA molecule attaches to this binding site even as the rest of the molecule is still being transcribed. The larger subunit has two binding sites for transfer RNAs.

Transfer RNA (tRNA) molecules are, in effect, a bilingual dictionary by which the language of nucleic acids is translated into the language of proteins. Cells contain more than 20 different kinds of tRNA molecules, at least one for each of the different kinds of amino acids. These molecules, which are comparatively small, all have a characteristic cloverleaf shape (Figure 15–8). At the 3' end of the molecule is the attachment site for its specific amino acid. A second attachment site, at the opposite end of the three-dimensional structure, is exposed by a loop within the molecule. This site consists of three nucleotides that form an **anticodon**, which is complementary to a specific codon in mRNA.

A third region of the tRNA molecule functions as the recognition site for an enzyme known as an **aminoacyl-tRNA synthetase**. There are at least 20 different aminoacyl-tRNA synthetases in the cell, one or more for each amino acid. Each of these enzymes has a binding site for a particular amino acid and for its matching tRNA molecule. In reactions powered by the energy released in ATP hydrolysis, aminoacyl-tRNA synthetases catalyze the attachment of specific amino acids to specific tRNA molecules—in effect, packaging the amino acids for delivery to the correct location in a growing polypeptide chain.

Translation

The synthesis of proteins is known as **translation,** since it is the transfer of information from one language (nucleic acids) to another (proteins). Translation takes place in three stages: initiation, elongation, and termination (Figure 15–9).

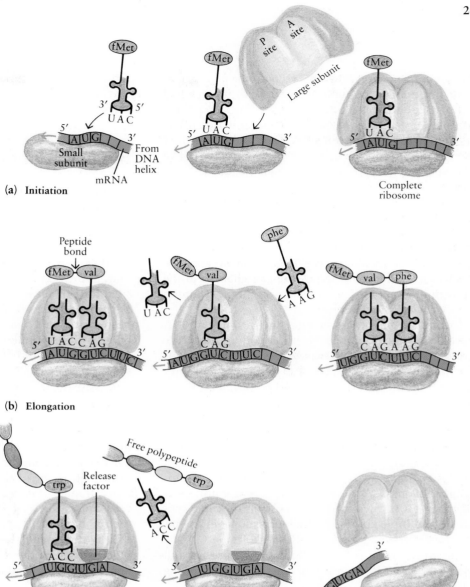

15–9 *The three stages of protein synthesis.* **(a)** *Initiation. The smaller ribosomal subunit attaches to the 5′ end of the mRNA molecule. The first tRNA molecule, bearing the modified amino acid fMet, plugs into the AUG initiator codon on the mRNA molecule. The larger ribosomal subunit then locks into place, with the tRNA occupying the P (peptide) site. The A (aminoacyl) site is vacant. The initiation stage is now complete.*

(b) *Elongation. A second tRNA with its attached amino acid moves into the A site, and its anticodon plugs into the mRNA. A peptide bond is formed between the two amino acids brought together at the ribosome. At the same time, the bond between the first amino acid and its tRNA is broken. The ribosome moves along the mRNA chain in a 5′ to 3′ direction, and the second tRNA, with the dipeptide attached, is moved to the P site from the A site as the first tRNA is released from the ribosome. A third tRNA moves into the A site, and another peptide bond is formed. The growing peptide chain is always attached to the tRNA that is moving from the A site to the P site, and the incoming tRNA bearing the next amino acid always occupies the A site. This step is repeated over and over until the polypeptide is complete.*

(c) *Termination. When the ribosome reaches a termination codon (in this example, UGA), the polypeptide is cleaved from the last tRNA and the tRNA is released from the P site. The A site is occupied by a release factor that triggers the dissociation of the two subunits of the ribosome.*

(a) Initiation

(b) Elongation

(c) Termination

The first stage, **initiation,** begins when the smaller ribosomal subunit attaches to the strand of messenger RNA near its 5′ end, exposing its first, or initiator, codon. Next, the first, or initiator, tRNA comes into place to pair with the initiator codon of mRNA. This initiator codon is usually (5′)–AUG–(3′), which is complementary to the tRNA anticodon (3′)–UAC–(5′). In prokaryotes, the tRNA with this anticodon carries a modified form of the amino acid methionine, known as fMet. Thus, fMet is the first amino acid of the newly synthesized polypeptide chain, but it is usually later removed from the completed polypeptide.

The combination of the small ribosomal subunit, mRNA, and the initiator tRNA is known as the **initiation complex.** Once it has formed, the larger subunit of the ribosome attaches to the smaller subunit, and the initiator tRNA with its attached fMet is locked into the P (peptide) site of the larger subunit—one of the two sites for binding tRNA molecules. The energy for this step, which completes the initiation stage, is provided by the hydrolysis of guanosine triphosphate (GTP).

At the beginning of the **elongation** stage, the second codon of the mRNA is positioned opposite the A (aminoacyl) site of the large subunit. A tRNA with an anticodon complementary to the second mRNA codon plugs into the mRNA molecule and, with its attached amino acid, occupies the A site of the ribosome. When both sites are occupied, a peptide bond (see page 60) is forged between the two amino acids, attaching the first (fMet) to the second. The first tRNA is released. The ribosome then moves one codon down the mRNA molecule. Consequently, the second tRNA, to which is now attached fMet and the second amino acid, is transferred from the A position to the P position. A third tRNA–amino acid moves into the now vacant A position opposite the third codon on the mRNA, and the step is repeated. Over and over, the P position accepts the tRNA bearing the growing polypeptide chain, and the A position accepts the tRNA bearing the next amino acid that will be added to the chain.

Toward the end of the coding sequence of the mRNA molecule are one or more of the three codons that serve as **termination** signals. No tRNAs exist with anticodons that "match" these codons, and so no tRNAs will enter the A site in response to them. When a termination codon is reached, translation stops, the polypeptide chain is freed, and the two ribosomal subunits separate.

As the ribosome moves along the mRNA molecule during the elongation process, the initiator portion of the mRNA is freed, and another ribosome can form an initiation complex with it. A group of ribosomes reading the same mRNA molecule is known as a **polysome** (Figure 15–10). Polysomes make possible the rapid synthesis of multiple copies of a polypeptide from the instructions carried by a single mRNA molecule.

The working out of the details of the precise and elegant process of translation was an awe-inspiring achievement. Even more awe-inspiring is the knowledge that at this very moment a similar process is taking place in virtually every cell of our own bodies.

Biological Implications

Mutations Revisited

With the process of protein synthesis in mind, we can now consider some of the broader implications of the genetic code and its translation. For example, let us take another look at sickle-cell anemia. The mutant allele that, in the homozygous recessive condition, is responsible for this disorder apparently originated in Africa. For reasons that we shall discuss in Chapter 19, it has been maintained by natural selection at a very high frequency in the populations of certain regions of Africa. About 9 percent of African-Americans are heterozygous for the sickle-cell allele, and about 0.2 percent are homozygous for the allele and therefore have the symptoms of sickle-cell anemia.

These symptoms are triggered by low oxygen concentrations in the blood, which cause the sickle-cell hemoglobin molecules to become insoluble. The molecules stick together, forming bundles of stiff fibers that distort the shape of the red blood cells and make them more fragile. Premature degradation of these fragile cells causes the anemia. Also, the loss of flexibility of the red blood cells makes it difficult for them to make their way through small blood vessels. Blocking of the blood vessels in the joints and in vital organs by these abnormal red cells is both painful and life-threatening.

15–10   *This scanning electron micrograph shows a series of ribosomes (highlighted in green) reading a single mRNA molecule in a human brain cell. Such groups of ribosomes are called polyribosomes, or polysomes.*

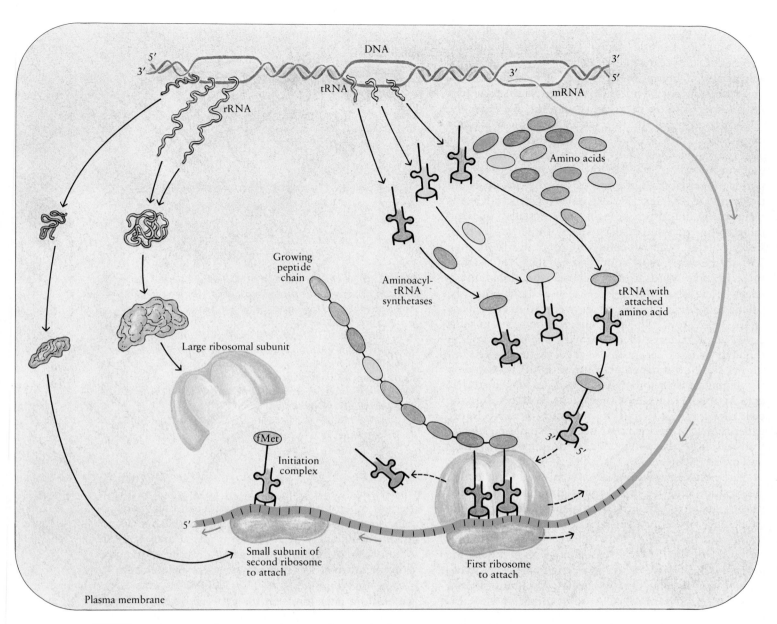

15–11   *A summary of protein synthesis in a bacterial cell. Three different rRNA molecules are transcribed from the DNA of the chromosome. These combine with specific proteins to form the ribosomal subunits. At least 32 different kinds of tRNA molecules are also transcribed from the DNA. These molecules are so structured that each can be attached at one end (by an aminoacyl-tRNA synthetase) to a specific amino acid. Each contains an anticodon that is complementary to an mRNA codon for that particular amino acid.*

The process of protein synthesis begins as an mRNA strand is being transcribed from its DNA template. When the leading (5') end of the mRNA strand attaches to a small ribosomal subunit, an initiator tRNA carrying fMet plugs into the initiator

codon of the mRNA. Addition of a large ribosomal subunit completes the initiation complex. The second amino acid is brought in by a tRNA molecule with an anticodon complementary to the next codon in the mRNA strand. The tRNA plugs in momentarily to that codon, a peptide bond is formed between the first and second amino acids, and the first tRNA molecule is released. As the ribosome moves along the mRNA strand, this process is repeated over and over, as the amino acids are brought into line one by one, following the exact order originally dictated by the DNA from which the mRNA was transcribed.

It is estimated that E. coli can synthesize as many as 3,000 proteins, each different and each assembled in this same way.

The hemoglobin molecule, as you will recall, contains four polypeptide chains—two alpha chains and two beta chains. Normal beta chains contain glutamic acid at a particular position; sickle-cell beta chains contain valine at that position (see Figure 3–23, page 65). As Figure 15–6 reveals, the difference between the mRNA codons for glutamic acid and valine is a single nucleotide. In mRNA, GAA and GAG specify glutamic acid (glu), and GUA and GUG are among the codons that specify valine (val). So the difference between the two is the replacement of one adenine by one uracil. In the template strand of DNA from which the mRNA is transcribed, the difference is the replacement of one thymine by one adenine in a sequence of nucleotides that, since it dictates a polypeptide that contains more than 150 amino acids, must contain more than 450 nucleotides. In other words, the tremendous functional difference—literally a matter of life and death—can be traced to a single "misprint" in over 450 nucleotides.

De Vries, some 90 years ago, defined mutation in terms of characteristics appearing in the phenotype. In the light of more recent knowledge, the current definition is somewhat different: A mutation is a change in the sequence or number of nucleotides in the nucleic acid of a cell. Mutations that occur in gametes or the cells that give rise to gametes are transmitted to future generations. Mutations that occur in somatic cells are transmitted to the daughter cells produced by mitosis and cytokinesis.

Many mutations involve only a single nucleotide substitution and are called **point mutations.** As in the case of sickle-cell anemia, such a substitution can lead to changes in the protein produced by a gene. A mutation that has this effect is known as a **missense** mutation. More than 100 missense mutations involving the genes coding for the alpha and beta chains of the hemoglobin molecule are now known. About 20 of these result in serious disorders. A point mutation can also produce what is known as a **nonsense** mutation. In such cases, the result of the nucleotide substitution is a "stop" codon. This causes protein synthesis to end before the complete polypeptide has been translated.

Changes in the amino acid sequence of a protein can result not only from nucleotide substitutions but also from the deletion or addition of nucleotides within a gene (Figure 15–12). When this occurs, the reading frame of the gene may shift—that is, the way in which the nucleotides are grouped into triplets changes—resulting in the production of an entirely new protein. These **frame-shifts,** as they are known, almost invariably lead to nonfunctional proteins.

Among the agents known to cause mutations are x-rays, ultraviolet rays, radioactive materials, and a va-

15–12   *Deletion or addition of nucleotides within a gene leads to changes in the protein produced. The original DNA molecule, the mRNA transcribed from it, and the resulting peptide are shown in* (a).

In (b) *we see the effect of the deletion of a nucleotide pair (T–A) from the location indicated by the arrow. The reading frame for the gene is altered, and a different sequence of amino acids occurs in the peptide.*

A similar change results from the addition of a nucleotide pair (bright green), as shown in (c).

riety of other chemical substances. Most mutations occur "spontaneously"—meaning simply that we do not know the chemical factors that trigger them. The rate of spontaneous mutation is generally low. For mutations detectable in the phenotype of eukaryotes, the rate varies from 1 in 1,000 to 1 in 1,000,000 gametes per generation, depending on the gene involved. Different genes—and even different alleles of the same gene—have different rates of mutation. These differences are thought to be related to both the chemical composition of the gene (or allele) in question and its position in the chromosome.

Classical Genetics Revisited

The realization that genes code for polypeptides provided a simple and clear explanation for the phenomenon of dominance observed in many traits of pea plants, fruit flies, humans, and other organisms. In heterozygotes, enough of the protein coded by the normal allele is produced to compensate for the recessive allele, which codes either for a defective, poorly functioning

15–13   *A summary of the information flow from DNA to RNA to protein. Replication of the DNA occurs only once in each cell cycle (page 179), during the S phase prior to mitosis or meiosis. Transcription and translation, however, occur repeatedly throughout the interphase portions of the cell cycle.*

polypeptide or for none at all. Only when two recessive alleles come together in the same individual—and none of the normal polypeptide is synthesized—is the recessive phenotype expressed.

The discovery of the genetic code and the new understanding of the nature and consequences of mutations also clarified some of the apparent exceptions to Mendelian principles. Incomplete dominance, for example, results from the combined effect of the proteins translated from both the dominant and the recessive alleles of a gene.

Other exceptions, such as epistasis (page 224), are more complicated. As you know, most of the biochemical reactions in the cell occur in series—biochemical pathways—in which each reaction of the series is catalyzed by a specific enzyme (Figure 7–13, page 137). For the complete series of reactions in a biochemical

pathway to occur, at least one copy of the normal allele coding for each enzyme involved in the pathway must be present. If, for any given enzyme, two copies of a defective allele are present, all steps of the pathway beyond that point will be shut down—and the presence of normal alleles coding for other enzymes of the pathway will be masked. Moreover, the end product of the reaction series will not be formed, and harmful intermediates may accumulate. Such is the case, for example, in PKU and in Tay-Sachs disease (page 212).

Mutations of different genes coding for different enzymes in a pathway can have different effects. In the pathway for the breakdown of phenylalanine, for example, a mutation in the gene for one enzyme leads to PKU, and a mutation in the gene for another enzyme leads to a disorder known as alkaptonuria, in which the urine of affected individuals is jet black. A mutation in the gene for still another enzyme in the same pathway blocks the synthesis of the brown pigment melanin. The result is albinism—the lack of pigmentation in skin, hair, and eyes.

Although the discovery of the mechanism of protein synthesis and the working out of the genetic code shed new light on mutations and answered many questions, they raised many other questions. The genetic material, the DNA, is an enormously long sequence of nucleotides, undifferentiated in form. Yet the information it contains is compartmentalized in the units we call genes. How is this accomplished? And, how does the cell regulate its genes so they produce the right amounts of the right proteins at the right times? In the next chapter, we shall focus on these questions.

Summary

Genetic information is encoded in the sequence of nucleotides in molecules of DNA, and these, in turn, determine the sequence of amino acids in molecules of protein.

The way in which the information carried by DNA is translated into protein has been worked out in considerable detail. The information is transcribed from the DNA into a long, single strand of RNA (ribonucleic acid). This type of RNA molecule is known as messenger RNA, or mRNA. Synthesis of mRNA, catalyzed by the enzyme RNA polymerase, uses one strand of the DNA as a template, following the principles of base pairing first suggested by Watson and Crick. The mRNA therefore is complementary to the template DNA strand. Each group of three nucleotides in the mRNA molecule is the codon for a particular amino acid.

The genetic code has now been "broken," that is, it is known which amino acid is called for by a given mRNA codon. Of the 64 possible triplet combinations of the four-letter nucleotide code, 61 combinations specify amino acids. Each of these combinations codes for one—and only one—of the 20 amino acids that make up protein molecules. The other three triplets serve as stop signals, terminating protein synthesis. The genetic code is universal, the same in all living organisms.

Protein synthesis—translation—takes place at the ribosomes. A ribosome is formed from two subunits, one large and one small, each consisting of characteristic ribosomal RNAs (rRNAs) complexed with specific proteins. Also required for protein synthesis is another group of RNA molecules, known as transfer RNA (tRNA). These small molecules can carry an amino acid on one end, and they have a triplet of bases, the anticodon, on a central loop at the opposite end of the three-dimensional structure. The tRNA molecule is the adapter that pairs the correct amino acid with each mRNA codon during protein synthesis. There is at least one kind of tRNA molecule for each kind of amino acid found in proteins.

In bacteria, even as the mRNA strand is still being transcribed, ribosomes are attaching, one after another, near its free end. At the point where the strand of mRNA is in contact with a ribosome, tRNAs are bound temporarily to the mRNA strand. This binding takes place by complementary base pairing between the mRNA codon and the tRNA anticodon. Each tRNA molecule carries the specific amino acid called for by the mRNA codon to which the tRNA attaches. Thus, following the sequence originally dictated by the DNA, the amino acid units are brought into line one by one and, as peptide bonds form between them, are linked into a polypeptide chain.

Mutations are now defined as changes in the sequence or number of nucleotides in the nucleic acid of a cell or organism. They may take the form of substitutions of one nucleotide for another, or deletions or additions of nucleotides. Nucleotide substitutions can result in missense mutations, in which one amino acid is substituted for another, or in nonsense mutations, in which protein synthesis is terminated prematurely. Deletions or additions of nucleotides can result in frameshift mutations and the production of entirely new—and usually nonfunctional—proteins.

The fact that genes code for proteins—and that mutations result in defective proteins—makes clear the underlying basis of the phenomenon of dominance, first documented by Mendel. It also explains many of the apparent exceptions to Mendelian principles, such as incomplete dominance and epistasis.

Questions

1. Distinguish among the following: mRNA/tRNA/rRNA; code/codon/anticodon; transcription/translation; P site/A site; initiation/elongation/termination; point mutation/missense mutation/nonsense mutation/frame-shift mutation.

2. A person heterozygous for the allele for sickle-cell anemia makes the variant hemoglobin molecules but does not suffer from anemia. In what respect is this situation different from the definitions of "dominant" and "recessive" given in Chapter 12?

3. Most of the bacterial DNA codes for messenger RNA, and most of the RNA produced by the cell is mRNA. Yet analysis of the RNA content of a cell reveals that, typically, rRNA is about 80 percent of the cellular RNA and tRNA makes up most of the rest. Only about 2 percent is normally mRNA. How do you explain these findings? What do you think the functional explanation might be?

4. Given the details of protein synthesis, what further amendment would you make to the principle "one gene–one polypeptide"? (Hint: See page 261, and look again at Figure 15–11.)

5. Explain the term "genetic code." In what respects is it a useful analogy?

6. In a hypothetical segment of one strand of a DNA molecule, the sequence of nucleotides is

$$(3')–AAGTTTGGTTACTTG–(5')$$

(a) What would be the sequence in an mRNA strand transcribed from this DNA segment? (b) What would be the sequence of amino acids coded by the mRNA? (c) Does it matter at what point on the template strand the transcription from DNA to mRNA begins? Explain your answer.

7. Fill in the missing letters.

T G T	_ _ _	_ _ _	
_ _ A	C _ _	_ _ _	DNA
U _ _	_ C A	_ _ _	mRNA codons
_ _ _	_ _ _	G C A	tRNA anticodons

Specify the amino acid for which each mRNA triplet codes.

8. What amino acid is carried by the tRNA molecule shown in Figure 15–8? How do you know?

9. Suppose you have the peptide arg–lys–pro–met, and you know that the tRNA molecules used in its synthesis had the following anticodons:

(3′)–GGU–(5′)
(3′)–GCU–(5′)
(3′)–UUU–(5′)
(3′)–UAC–(5′)

What is the DNA nucleotide sequence for the template strand of the gene that codes for this peptide? *(Hint:* Determine the mRNA codons complementary to each of the tRNA codons and their sequence in the mRNA strand from which the peptide was translated.)

10. Deletion or addition of nucleotides within a gene leads to changes in the protein produced. The original DNA molecule, the mRNA transcribed from it, and the resulting peptide are:

DNA

mRNA

ser ⟶ ser ⟶ ser ⟶ ser   Peptide

Deleting the second T–A pair, as indicated by the arrow, yields the following DNA molecule:

How is the resulting amino acid sequence altered? How does the addition of a C–G pair to the original molecule,

affect the amino acid sequence?

11. In prokaryotes, transcription and translation are "linked" processes. As we shall see in the next chapter, however, this is not the case in eukaryotes. On the basis of your knowledge of cell structure, propose a likely explanation.

Suggestions for Further Reading

Amato, Ivan: "Expanding the Genetic Alphabet," *Science News,* February 10, 1990, pages 88–90, 94.

Darnell, James E., Jr.: "RNA," *Scientific American,* October 1985, pages 68–87.

Dickerson, Richard E.: "The DNA Helix and How It Is Read," *Scientific American,* December 1983, pages 94–111.

Lake, James A.: "The Ribosome," *Scientific American,* August 1981, pages 84–97.

Loupe, Diane E.: "Breaking the Sickle Cycle," *Science News,* December 2, 1989, pages 360–362.

Rich, Alexander, and Sung Hou Kim: "The Three-Dimensional Structure of Transfer RNA," *Scientific American,* January 1978, pages 52–62.

Ross, Jeffrey: "The Turnover of Messenger RNA," *Scientific American,* April 1989, pages 48–55.

Scientific American: *Molecules to Living Cells,* W. H. Freeman and Company, New York, 1980.*

More than half of the articles in this collection from Scientific American *are accounts of major breakthroughs in our understanding of the nucleic acids and their functions. Written for the general reader by the scientists who made the key discoveries, they are highly recommended.*

* Available in paperback.

This cat is a tortoiseshell, so called because of the patches of yellow and black in its fur. You can bet that a tortoiseshell is a female, because yellow and black coat colors are determined by alleles of a gene on the X chromosome. A male cat, with only one X chromosome, can be either black or yellow—but not both. However, if a female, with her two X chromosomes, is heterozygous for the gene, she can be a tortoiseshell.

If you were to examine the somatic cells of a tortoiseshell cat at interphase, you would see a dark spot of chromatin at the edge of the nucleus. This spot, known as a Barr body, is a highly condensed, inactivated X chromosome. Early in development, one X chromosome or the other is inactivated in all of the embryonic cells except those that will later give rise to the gametes. The same X chromosome is inactivated in all of the cells descended from each embryonic cell. Which is why a particular patch of fur is either yellow or black.

Barr bodies are of interest to geneticists as well as to cat fanciers. Found in the somatic cells of all female mammals, they represent one of the ways in which genes can be "turned off." Inactivation of the X chromosome occurs at random. Thus, the somatic cells of a female mammal are not identical but are of two types, depending on which X chromosome is active and which is inactive. Human females who are carriers of the allele for hemophilia, for example, show wide variability in the amount of clotting factor in their blood, ranging from 20 to 100 percent of normal levels. There also appear to be slight variations in the timing. Female carriers of the allele for color blindness are sometimes color blind in one eye but not in the other. In this cat, X-chromosome inactivation occurred relatively late, and the black and yellow patches are small and closely intermingled. When the inactivation occurs earlier in embryonic development, the patches are larger and more distinct—in which case the cats are known as calicos.

You should know that you might lose your bet about the sex of a tortoiseshell. An occasional tortoiseshell is male. Such males, as you might guess, are XXY and are sterile.

Chromosome Structure and the Regulation of Gene Expression

As we saw in the last chapter, the genetic code—three specific nucleotides coding for each amino acid—is universal. It is the same in bacteria, bread molds, pea plants, fruit flies, cats, humans, and all other organisms—awesome evidence that all living things are descended from a common ancestor. However, a variety of studies in molecular genetics have shown that, despite the basic similarities of prokaryotic and eukaryotic chromosomes, there are important differences in their structure. There are also important differences in the mechanisms that turn genes "on" and "off" and thus regulate gene expression. In this chapter, we shall explore some of these differences.

The Prokaryotic Chromosome

The prokaryotic chromosome is a single, continuous (circular) thread of double-stranded DNA, only 2 nanometers wide. In *Escherichia coli*, it contains some 4.7 million base pairs and is approximately 1 millimeter long when fully extended (see page 254). An *E. coli* cell is less than 2 micrometers long, so the chromosome is some 500 times longer than the cell itself. Within the cell, the chromosome is folded into an irregularly shaped body known as the nucleoid (see Figure 4–7, page 78).

Regulation of Gene Expression in Prokaryotes

Transcription in *E. coli* and other prokaryotes takes place, as we saw in the last chapter, by the synthesis of a molecule of mRNA along a template strand of DNA. The process begins when the enzyme RNA polymerase attaches to the DNA at a specific site known as the **promoter**. The RNA polymerase molecule binds tightly to the promoter and causes the DNA double helix to open,

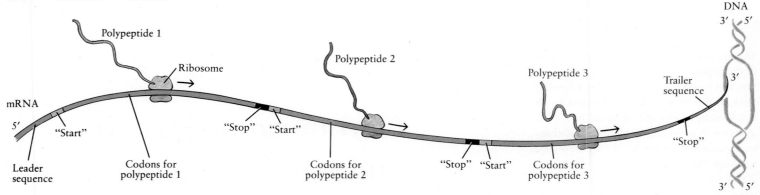

16–1 *In prokaryotes, transcription often results in an mRNA molecule that contains coding sequences for several different polypeptide chains, with the sequences separated by stop and start codons. In this diagram, the stop and start codons are adjacent, but sometimes they are separated by as many as 100 to 200 nucleotides. The 5' end of the mRNA* *molecule has a short leader sequence, and the 3' end has a trailer sequence; neither of these sequences codes for protein. Translation, which starts at the leading (5') end of the mRNA, generally begins while the rest of the molecule is still being transcribed.*

initiating transcription. The growing RNA strand remains hydrogen-bonded to the DNA template only briefly, and then it peels off in a single strand.

A segment of DNA that codes for a polypeptide is known as a **structural gene.** In the bacterial chromosome, structural genes coding for polypeptides with related functions often occur together in sequence. Such functional groups might include, for instance, two polypeptide chains that together constitute a particular enzyme or three enzymes that work in a single biochemical pathway. Groups of genes coding for such molecules are typically transcribed into a single mRNA strand (Figure 16–1). Thus, a group of polypeptides that are needed by the cell at the same time and in the same quantity can be synthesized simultaneously, a simple and efficient inventory-control system.

A cell does not make all of its possible proteins all of the time, but only when they are needed and only in the amounts needed. For example, cells of *E. coli* supplied with lactose as a nutrient require the enzyme beta-galactosidase to split the lactose, a disaccharide, into its two constituent monosaccharides, glucose and galactose. Cells growing on lactose have approximately 3,000 molecules of beta-galactosidase per cell. In the absence of lactose, there is an average of one molecule of the enzyme per cell. In short, the presence of lactose leads to the production of the enzyme molecules needed to break it down.

In other cases, the presence of a particular nutrient may inhibit the synthesis of particular proteins. Like most other bacteria, *E. coli* can synthesize each of its amino acids from ammonium ions (NH_4^+) and a carbon source. The enzymes needed for the biosynthesis of the amino acid tryptophan, for instance, are synthesized continuously in growing cells—unless tryptophan is

present. In the presence of tryptophan, production of the enzymes ceases.

Mutants of *E. coli* sometimes occur that are unable to regulate enzyme production. These cells produce beta-galactosidase even in the absence of lactose, for example, or the enzymes that synthesize tryptophan even when tryptophan is present. These and similar mutants are generally at a disadvantage because they are squandering their energies and resources producing enzymes they don't need. Normal *E. coli* cells rapidly outmultiply them.

Although regulation of protein synthesis could theoretically take place at many points in the process, in prokaryotes it occurs mostly at the level of transcription. Regulation involves interactions between the chemical environment of the cell and special regulatory proteins, coded by regulatory genes. These proteins can work either as negative controls, repressing mRNA transcription, or as positive controls, enhancing transcription. The fact that mRNA is translated into protein so immediately (before transcription is even completed) and then broken down very rapidly further increases the efficiency of this strategy of regulation.

The Operon

Our current understanding of the regulation of transcription in prokaryotes rests upon a model, known as the **operon** model, proposed some years ago by the French scientists François Jacob and Jacques Monod. According to this model, an operon (Figure 16–2) comprises the promoter, one or more structural genes, and another DNA sequence known as the **operator.** The operator is a sequence of nucleotides located between the promoter and the structural gene or genes.

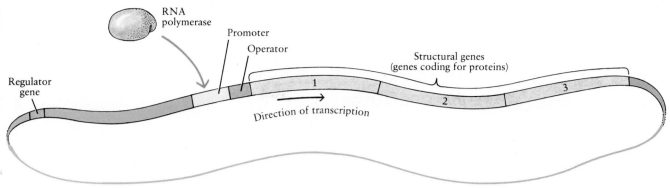

Regulator gene

RNA polymerase

Promoter

Operator

Structural genes (genes coding for proteins)

1

2

3

Direction of transcription

DNA (bacterial chromosome)

16–2 *A schematic representation of an operon. An operon consists of a promoter, an operator, and structural genes (that is, genes that code for proteins, often enzymes that work sequentially in a particular biochemical pathway). The promoter, which precedes the operator, is the binding site for RNA polymerase. The operator is the site at which a repressor can bind; it may overlap the promoter, the first structural gene, or both.*

Another gene involved in operon function is the regulator, which codes for the repressor. Although the regulator may be adjacent to the operon, in most cases it is located elsewhere on the bacterial chromosome.

Transcription of the structural genes often depends on the activity of still another gene, the **regulator,** which may be located anywhere on the bacterial chromosome. This gene codes for a protein called the **repressor,** which binds to the operator. When a repressor is bound to the operator, it obstructs the promoter. As a consequence, RNA polymerase either cannot bind to the DNA molecule or, if bound, cannot begin its movement along the molecule. The result in either case is the same: no mRNA transcription occurs. However, when the repressor is removed, transcription may begin.

The capacity of the repressor to bind to the operator and thus to block protein synthesis depends, in turn, on yet another molecule that either activates or inactivates the repressor for that particular operon (Figure 16–3).

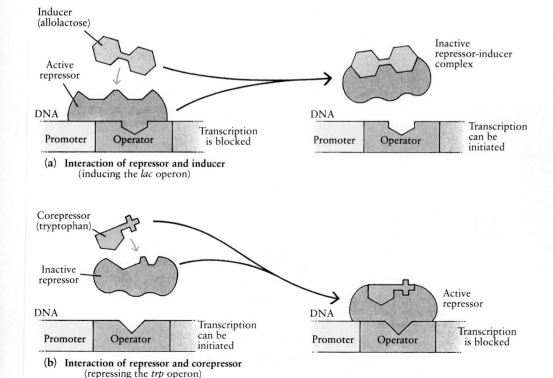

Inducer (allolactose)

Active repressor

DNA

Promoter   Operator

Transcription is blocked

Inactive repressor-inducer complex

DNA

Promoter   Operator

Transcription can be initiated

(a) Interaction of repressor and inducer (inducing the *lac* operon)

Corepressor (tryptophan)

Inactive repressor

DNA

Promoter   Operator

Transcription can be initiated

DNA

Promoter   Operator

Active repressor

Transcription is blocked

(b) Interaction of repressor and corepressor (repressing the *trp* operon)

16–3 *In an operon system, the transcription of mRNA—and thus the synthesis of proteins—is regulated by interactions involving either a repressor and an inducer or a repressor and a corepressor. (a) In some systems, such as the lac operon, the repressor molecule is active until it combines with the inducer. (b) In other systems, such as the trp operon, the repressor is not active until it combines with the corepressor.*

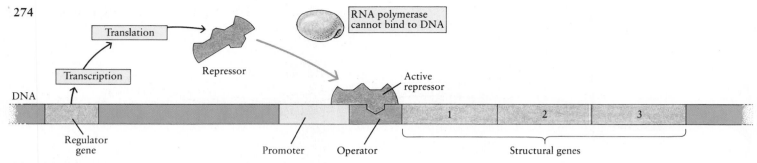

(a) *Lac* operon, operator "off"

16–4 *Regulation of the* lac *operon.* **(a)** *In the absence of lactose, the repressor binds to DNA at the operator and so prevents the RNA polymerase from initiating transcription.* **(b)** *In the presence of lactose, the repressor is inactivated. Thus RNA polymerase can bind to the promoter, and synthesis of mRNA can proceed. The genes of the operon are transcribed onto a single mRNA molecule directing the synthesis of three proteins: the enzyme beta-galactosidase, a transport protein that brings lactose from the external medium into the cell, and the enzyme transacetylase, which transfers an acetyl group from acetyl CoA (page 149) to galactose.*

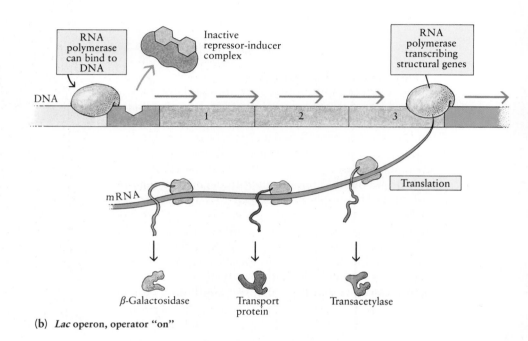

(b) *Lac* operon, operator "on"

16–5 *A computer-generated model of a repressor bound to the DNA double helix. The three-dimensional shape of the repressor mirrors the shape of the nucleotide sequence that forms the operator. When the repressor is bound to the operator, it blocks a sequence of some 10 to 20 base pairs. Repressors function as molecular switches, turning off the synthesis of proteins that are not needed.*

A molecule that activates a repressor is known as a **corepressor,** and one that inactivates a repressor is known as an **inducer.** For example, when lactose is present in the growth medium, the first step in its metabolism produces a closely related sugar, allolactose. Allolactose is an inducer—it binds to and inactivates the repressor, removing it from the operator of the lactose *(lac)* operon. As a consequence, RNA polymerase can begin its movement along the DNA molecule, transcribing the structural genes of the operon into mRNA (Figure 16–4).

An example of a corepressor is provided by the amino acid tryptophan. When present in the growth medium, it activates the repressor for the tryptophan *(trp)* operon. The activated repressor then binds to the operator and blocks the synthesis of the unneeded enzymes. Both tryptophan and allolactose—as well as the molecules that interact with the repressors of other operons—exert their effects by causing a change in the three-dimensional shape of the repressor molecule.

Some 75 different operons have now been identified in *E. coli*, comprising 260 structural genes. As we shall see in Chapter 18, manipulating operons is an essential part of the scientific trick of inducing bacterial cells to produce proteins of medical importance, such as human insulin.

The Eukaryotic Chromosome

In the early days of molecular genetics (the late 1950s and into the 1960s), it was tempting to think that the eukaryotic chromosome might turn out to be simply a large-scale version of the *E. coli* chromosome. However, it soon became clear that there are many important differences between the bacterial chromosome and the chromosomes of eukaryotes—some expected and some very surprising. These differences include (1) a far greater quantity of DNA in eukaryotic chromosomes; (2) a great deal of repetition in this DNA, with much of it lacking any apparent function; (3) a close association of the DNA with proteins that play a major role in chromosome structure; and (4) considerably more complexity in the organization of the protein-coding sequences of the DNA and the regulation of their expression.

DNA is an "exquisitely thin filament," in the words of E. J. DuPraw, who calculated that a length sufficient to reach from the Earth to the sun would weigh only half a gram. The DNA of each eukaryotic chromosome is thought to be in the form of a single, linear molecule. If fully extended, the DNA forming each human chromosome would be from 3 to 4 centimeters long. Each diploid cell, with its 46 chromosomes, thus contains about 2 meters of DNA, and the entire human body contains some 25 billion kilometers of DNA double helix.

The Structure of the Chromosome

In the nucleus of the eukaryotic cell, the DNA is always found combined with proteins. This combination, as we noted in Chapter 5, is known as **chromatin** ("colored threads") because of its staining properties. Chromatin is more than half protein, and the most abundant proteins, by weight, belong to a class of small polypeptides known as **histones**. Histones are positively charged (basic) and so are attracted to the negatively charged (acidic) DNA. They are always present in chromatin and are synthesized in large amounts during the S phase of the cell cycle. The histones are primarily responsible for the folding and packaging of DNA. In a human cell, for instance, the approximately 2 meters of DNA are

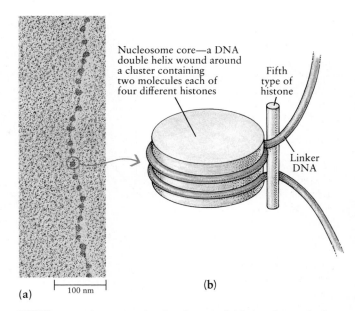

Nucleosome core—a DNA double helix wound around a cluster containing two molecules each of four different histones

Fifth type of histone

Linker DNA

(a) |— 100 nm —| **(b)**

16–6   (a) *Chromatin that has been unfolded and stretched out as a result of chemical treatment, revealing the beadlike nucleosomes. The distance between nucleosomes is about 10 to 11 nanometers, and the diameter of each bead is about 7 nanometers.*

(b) *The structure of a nucleosome. The negatively charged DNA coils twice around the protein core, composed of eight positively charged histone molecules. Another histone molecule (also positively charged) binds to the outer surface of the nucleosome. The DNA of the nucleosome core contains about 140 base pairs. The strand of linker DNA between the nucleosome cores contains another 30 to 60 base pairs.*

packed into 46 cylinders (the 46 chromosomes) that, when condensed at metaphase, have a combined length of only 200 micrometers.

There are five distinct types of histones. They are present in enormous quantities—about 60 million molecules per cell of each of four of the types, and about 30 million molecules per cell of the fifth type. With the exception of the fifth type, the amino acid sequences of the histones are very similar in widely diverse groups of organisms. One histone of the garden pea, for instance, differs from the corresponding histone of the cow by only two amino acids out of a total of 102.

The fundamental packing units of chromatin are the **nucleosomes** (Figure 16–6), which resemble beads on a string. Each nucleosome consists of a core of eight histone molecules, around which the DNA filament is wrapped twice, like thread around a spool. Another histone molecule lies on the DNA, outside the nucleosome core. When a fragment of DNA is tied up in a nucleosome, it is about one-sixth the length it would be if fully extended.

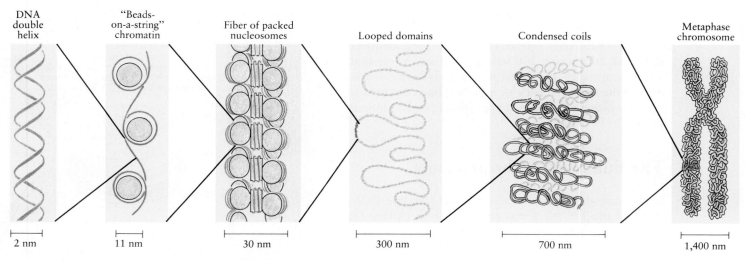

DNA double helix	"Beads-on-a-string" chromatin	Fiber of packed nucleosomes	Looped domains	Condensed coils	Metaphase chromosome
2 nm	11 nm	30 nm	300 nm	700 nm	1,400 nm

16–7 *Stages in the folding of chromatin, culminating in a fully condensed metaphase chromosome. This model is derived from electron micrographs of chromatin at different degrees of condensation. According to present evidence, each chromatid of a replicated chromosome contains a single molecule of double-stranded DNA and is, by weight, about 60 percent protein.*

As shown in Figure 16–7, additional packing of the nucleosomes produces a fiber that is about 30 nanometers in diameter. Further condensation of this fiber produces a series of loops, known as looped domains. The looped domains also coil until, ultimately, clusters of neighboring looped domains condense into the compact chromosomes that become visible during mitosis and meiosis.

Other proteins associated with the chromosome are the enzymes involved in DNA and RNA synthesis, regulatory proteins, and a large number and variety of molecules that have not yet been isolated and identified. Unlike the histones, these molecules vary from one cell type to another.

Regulation of Gene Expression in Eukaryotes

As we saw earlier, regulation of gene expression in prokaryotes typically involves turning genes on and off in response to changes in the nutrients available in the environment. In eukaryotes, especially multicellular eukaryotes, the problems of regulation are very different. A multicellular organism usually starts life as a fertilized egg, the zygote. The zygote divides repeatedly by mitosis and cytokinesis, producing many cells. At some stage these cells begin to differentiate, becoming, in humans, for example, muscle cells, nerve cells, blood cells, intestinal lining cells, and so forth. Each cell type, as it

differentiates, begins to produce characteristic proteins that distinguish it structurally and functionally from other types of cells.

This is nicely illustrated by mammalian red blood cells. In the early stages of fetal life, developing red blood cells synthesize one type of fetal hemoglobin. Red blood cells produced at later stages contain a second type of fetal hemoglobin. Then, sometime after the birth of the organism, the developing red blood cells begin to produce the polypeptide chains characteristic of adult hemoglobin. The genes are expressed in a precisely timed sequence, one after the other. Moreover, the DNA segments that code for these hemoglobin molecules are expressed only in developing red blood cells.

There is evidence, however, that all of the genetic information originally present in the zygote is also present in every diploid cell of the organism (Figure 16–8). Thus, for example, the DNA segments that code for hemoglobin (both the fetal types and the adult type) are present in skin cells and heart cells and liver cells and nerve cells and, indeed, in every one of the nearly 200 different types of cells in the body. Similarly, the DNA sequence that codes for the hormone insulin is present not only in the specialized cells of the pancreas that manufacture insulin but also in all the other cells. Since each type of cell produces only its characteristic proteins—and not the proteins characteristic of other cell types—it becomes apparent that differentiation of the cells of a multicellular organism depends on the inactivation of certain groups of genes and the activation of others.

Before puffing — Tightly condensed chromatin

Open loops of DNA

Chromosome "puff"

16–9 *Observations of chromosome puffs support the concept that the DNA is somehow unwound to make it available for RNA transcription. This puff was observed in a chromosome of the Brazilian gnat, which, like the fruit fly, has giant chromosomes in some of its cells. The red lines between the two views of the chromosome indicate corresponding bands of the chromosome. Puffs occur normally but can also be induced experimentally. This puff occurred in response to a hormone that causes molting.*

Cell from tadpole intestine

Micropipette containing nucleus from tadpole cell

Egg from another frog—nucleus has been destroyed by ultraviolet radiation

Nucleus from tadpole cell is implanted in enucleated egg

Egg develops into frog with the same characteristics as original, nucleus-donating tadpole

16–8 *A number of experiments have shown that early development does not result in permanent inactivation of genes or the loss of functional DNA. In experiments by J. B. Gurdon, illustrated here, nuclei were removed from intestinal cells of a tadpole and implanted into egg cells in which the nucleus had been destroyed. In some cases, the embryo developed normally, indicating that the tadpole intestinal cell nucleus contained all the information required for all the cells of the organism.*

In other experiments, F. C. Steward demonstrated that under certain conditions, a single differentiated cell of a carrot can be persuaded to divide repeatedly, ultimately reconstituting an entire carrot plant.

Condensation of the Chromosome and Gene Expression

Many lines of evidence indicate that the degree of condensation of the DNA of the chromosome, as shown by chromatin staining, plays a major role in the regulation of gene expression in eukaryotic cells. Staining reveals two types of chromatin: **euchromatin,** the more open chromatin, which stains weakly, and **heterochromatin,** the more condensed chromatin, which stains strongly. During interphase, heterochromatin remains tightly condensed, but euchromatin becomes less condensed. Transcription of DNA to RNA takes place only during interphase, when the euchromatin is less condensed—and is therefore accessible to molecules of RNA polymerase.

Some regions of heterochromatin are constant from cell to cell and are never expressed. An example is the highly condensed chromatin located in the centromere region of the chromosome. This region, which does not code for protein, is believed to play a structural role in the movement of the chromosomes during mitosis and meiosis. Similarly, little or no transcription takes place from Barr bodies (page 271), which are tightly condensed and irreversibly inactivated mammalian X chromosomes.

The degree of condensation of other regions of the chromatin, however, varies from one type of cell to another within the same organism. This reflects the fact that different types of cells synthesize different proteins and thus require the transcription of different segments of the DNA. Moreover, as a cell differentiates during embryonic development, the proportion of heterochromatin to euchromatin increases as the cell becomes more specialized. Segments of DNA that will not be needed by the differentiated cell are effectively "silenced."

Further evidence linking the degree of chromosome condensation to gene expression comes from studies of the giant chromosomes of insects (see page 235). At various stages of larval growth in insects, diffuse "puffs" occur in various regions of these chromosomes (Figure 16–9). The puffs are open loops of DNA. Studies with radioactive isotopes have shown that they are the sites of rapid RNA synthesis. When ecdysone, a hormone that produces molting in insects, is injected, the puffs occur in a definite sequence that can be related to the developmental stage of the animal.

Regulation by Specific Binding Proteins

In eukaryotes, as in prokaryotes, transcription is also regulated by proteins that bind to specific sites on the DNA molecule. Many of these proteins and their binding sites have now been identified, and it is increasingly clear that this level of transcriptional control is far more complex in multicellular eukaryotes than in prokaryotes.

A gene in a multicellular organism appears to respond to the sum of many different regulatory proteins, some tending to turn the gene on and others to turn it off. The sites at which these proteins bind may be hundreds or even thousands of base pairs away from the promoter sequence at which RNA polymerase binds and transcription begins. This, as you might expect, adds to the difficulty of identifying the regulatory molecules and also of understanding exactly how they exert their effects. Recent research suggests that changes in the activities of some of these regulators are linked with the development of cancer, a matter we shall explore in the next chapter.

The DNA of the Eukaryotic Chromosome

Early studies of the DNA of eukaryotic cells revealed two surprising facts. First, with a few exceptions, the amount of DNA per cell is the same for every diploid cell of any given species (which is not surprising), but the variations among different species are enormous. *Drosophila* has about 1.4×10^8 base pairs per haploid chromosome set, only about 70 times more than *E. coli*. Humans (with approximately 3.5×10^9 base pairs) have 25 times as much as *Drosophila*, somewhat more than a mouse, but about the same amount as a toad (3.32×10^9 base pairs). The largest amount of DNA so far has been found in a salamander with 8×10^{10} base pairs per haploid set of chromosomes.

Second, in every eukaryotic cell, there is what appears to be a great excess of DNA, or at least of DNA whose functions are unknown. It is estimated that in eukaryotic cells less than 10 percent of all the DNA codes for proteins; in humans, it may even be as little as 1 percent. By contrast, prokaryotes use their DNA very thriftily. Except for regulatory or signal sequences, virtually all of their DNA is expressed.

Continuing research is revealing the organization of eukaryotic DNA, as well as some of the functions of the long skeins of nucleotides that are never translated into protein.

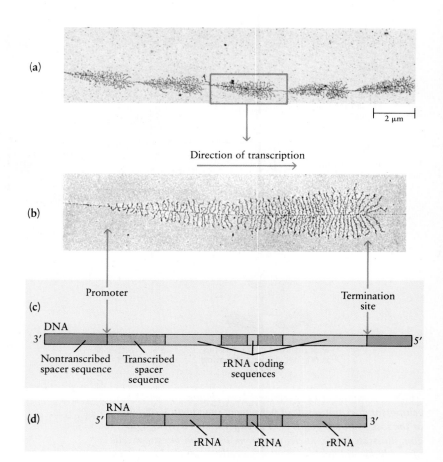

16–10   *A single gene codes for three of the four types of rRNA molecules found in the eukaryotic ribosome. This gene occurs in multiple copies, repeated in tandem (head to tail). (a) Electron micrograph of five copies of this gene separated by nontranscribed spacer sequences. (b) An enlargement of one of the genes and the RNA molecules being transcribed from it. The DNA is the thin horizontal line, and the RNA molecules are the fine fibrils perpendicular to the DNA. (c) A map of the three-rRNA gene. The portions of the DNA coding for the three rRNA molecules are shown in light green; the transcribed spacer sequences are shown as gray. (d) A map of the RNA molecule transcribed from the three-rRNA gene. When transcription is complete, enzymes cleave the molecule and remove the transcribed spacer sequences. This releases the three individual rRNA molecules.*

Classes of DNA: Repeats and Nonrepeats

Simple-Sequence DNA

The DNA of eukaryotic cells can be divided into three general classes. The first, known as **simple-sequence** DNA, consists of short repetitive sequences, arranged in tandem (head to tail). These sequences are typically 5 to 10 base pairs in length, and they are present in enormous quantities. *Drosophila virilis*, for instance, has ACAAACT repeated 12 million times. About 10 percent of the DNA of the mouse, and about 20 to 30 percent of human DNA, is made up of short, highly repetitive sequences.

Simple-sequence DNA is thought to be vital to chromosome structure. Long blocks of short repetitive sequences have been found around the centromere, and, indeed, may *be* the centromere. More recently, the tips of all human chromosomes have been found to consist of some 1,500 to 6,000 nucleotides in which a simple sequence—TTAGGG—is repeated over and over. This same simple, repeated sequence has also been found at the chromosome tips in a wide range of other mammals, birds, reptiles, and even protists. It is thought that the "caps" formed by this repeated sequence help stabilize the chromosome itself.

Intermediate-Repeat DNA

The second class of eukaryotic DNA consists of **intermediate-repeat** sequences, generally 150 to 300 nucleotides in length. Sometimes these sequences are arranged in tandem, one after another, but more often they are dispersed throughout the DNA. About 20 to 40 percent of the DNA of multicellular organisms consists of intermediate-repeat sequences.

Among the most thoroughly studied examples of this class of DNA are the genes coding for histones and for three of the four ribosomal RNAs found in eukaryotes (Figure 16–10). The histone genes are present in multiple copies (from 50 to 500) in the cells of all multicellular eukaryotes. Cells of multicellular eukaryotes, which may contain some 10 million ribosomes per cell, also have from 50 to 5,000 copies of the rRNA genes. The rRNA genes occur in tandem, head to tail; the chromosome regions in which they are located form the structure we recognize as the nucleolus (see essay).

Other intermediate-repeat sequences are similar to one another, rather than identical, and are said to form **gene families.** The best studied of these are the genes of the globin family, which code for the polypeptide chains of fetal and adult hemoglobins. Figure 16–11 summarizes the evolutionary steps that are thought to have led to this gene family. The ancestral gene is thought to have coded for a polypeptide chain resembling that of myoglobin, a relatively small molecule found in muscle cells. Myoglobin is formed by a single polypeptide chain of 153 amino acids and holds one oxygen-binding heme group. It is thought that the ancestral globin gene was accidentally duplicated several times in the course of evolutionary history. Over time, mutations led to divergence of the duplicates, eventually giving rise to the present family of genes. A corresponding divergence of function occurred in the proteins for which the globin genes code, leading to modern myoglobin and to the forerunners of the alpha and beta chains of hemoglobin.

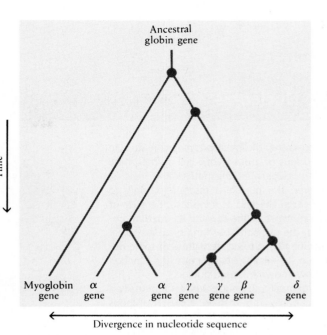

16–11 *A hypothetical evolutionary sequence that could have given rise to the genes of the globin family. The solid dots represent duplications of genes. The divergence of the genes following duplication is the result of mutations. The globin family includes the genes coding for myoglobin and for the alpha (α), beta (β), gamma (γ), and delta (δ) chains of hemoglobin. Fetal hemoglobin molecules contain two alpha chains and two gamma chains, whereas adult hemoglobin contains two alpha and two beta chains. The delta gene, which is found only in primates, codes for another variant of the beta chain.*

In addition to the protein-coding genes, the globin family contains other nucleotide sequences that are never expressed. These sequences, known as **pseudogenes,** are believed to be duplicates that have been disabled by their accumulated mutations.

Single-Copy DNA

The third class of eukaryotic DNA is **single-copy** DNA, in which each sequence is present in only one or a few copies. Depending on the species, it comprises anywhere from 50 to 70 percent of the DNA of the organism. With the exception of the histone genes and the members of gene families, all of the protein-coding genes belong to this class of DNA. However, only a small proportion of single-copy DNA—perhaps as little as 1 percent of the total DNA in the cell—appears to be translated into protein. Some, but by no means all, of the noncoding single-copy DNA is accounted for by long stretches of spacer DNA, which is never transcribed into RNA. The remainder has been found in a most unexpected location.

279

The Nucleolus

As you know, the most prominent feature in the nucleus of a eukaryotic cell in interphase is the nucleolus. During mitosis and meiosis, however, the nucleolus disappears, only to reappear at the end of telophase. Because of its prominence in the cell and its puzzling behavior, the nucleolus has long been an object of scrutiny by cytologists. Although some details are still missing, its structure and function are now known in broad outline.

Structurally, the nucleolus is not actually a distinct entity. It is, instead, a cluster of loops of chromatin, often from different chromosomes. For example, 10 of the 46 human chromosomes contribute chromatin loops to the nucleolus. The loops that form the basic structure of the nucleolus are the DNA segments that contain copies of the gene coding for three of the four types of rRNA molecules found in eukaryotic ribosomes. During the condensation of the chromosomes at the beginning of mitosis or meiosis, these loops are reeled back into their respective chromosomes, and the nucleolus disappears.

1 μm

Functionally, the nucleolus is a ribosome factory in which rRNA molecules are transcribed from the chromatin loops and ribosomal subunits are assembled. Ribosomal proteins, themselves synthesized on ribosomes in the cytosol of the cell, are transported into the nucleus and assembled into subunits. The

Electron micrograph of the nucleus of a type of pancreatic cell that produces and exports many of the enzymes used in digestion. The dark body in the lower center is the nucleolus, where the RNAs of the ribosomes are transcribed and the ribosomal subunits are assembled. Partially formed ribosomal subunits are around its periphery. Notice also the nuclear envelope, with its many nuclear pores (red arrows). Surrounding the nucleus are membranes of the endoplasmic reticulum, as well as a mitochondrion.

nearly completed ribosomal subunits, containing both rRNAs and protein, are then shipped back into the cytosol. There, after a few finishing touches, the ribosomes begin to perform their essential functions in the assembly of amino acids into proteins.

Introns and Exons

One of the great surprises in the study of eukaryotic DNA was the discovery that the protein-coding sequences of genes are usually not continuous but are instead interrupted by noncoding sequences—that is, by nucleotide sequences that are not translated into protein. These noncoding interruptions within a gene are known as intervening sequences, or **introns.** The coding sequences—the sequences that *are* translated into protein—are called **exons.**

Introns were discovered in the course of mRNA-DNA hybridization experiments (page 259). Investigators found that there was not a perfect match between eukaryotic messenger RNA molecules and the genes from which they were transcribed. The nucleotide sequences of the genes were much longer than the complementary mRNA molecules found in the cytosol.

It is now known that most, but not all, structural genes of multicellular eukaryotes contain introns. The introns are transcribed onto RNA molecules, but they are snipped out before translation occurs. The number of introns per gene varies widely. For example, the gene for ovalbumin, a protein found in large quantities in vertebrate egg cells, has seven introns (Figure 16–12). In chickens, the gene for collagen, a very common structural protein, has 50 introns. Introns have also been found in genes coding for transfer RNAs and ribosomal RNAs.

The Functional Significance of Introns

The way in which introns established themselves in the DNA and their functions, if any, are not known. One suggestion is that they promote recombination. Crossing over during meiosis is more likely in genes containing introns than in genes lacking introns, just because of the distances involved.

There are also indications that, in some cases, different exons code for different structural and functional regions of the finished protein (Figure 16–13). For ex-

16–12 *This electron micrograph reveals the results of an experiment in which a single strand of DNA containing the gene coding for ovalbumin was hybridized with the messenger RNA for ovalbumin. The complementary sequences of the DNA and mRNA are held together by hydrogen bonds. There are eight such sequences (the exons labeled 1 through 8 in the accompanying diagram). Some segments of the DNA do not have corresponding mRNA segments and so loop out from the hybrid. These are the seven introns, labeled A through G. Only the exons are translated into protein.*

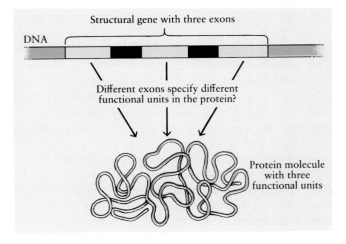

16–13 *An attractive hypothesis: Exons code for discrete functional regions in the protein for which the entire gene codes.*

ample, all of the genes of the globin family consist of three exons and two introns, one large and one small, always in the same position in the gene. The central exon of each gene codes for the portion of the polypeptide that holds the heme group, and the other two exons code for regions of the molecule that fold around this central portion (see Figure 3–21, page 64). It is hypothesized that new combinations of such regions, brought about by the reshuffling of exons, might foster the rapid evolution of new proteins.

Transcription and Processing of mRNA in Eukaryotes

Transcription in eukaryotes is the same, in principle, as in prokaryotes. It begins with the attachment of an RNA polymerase to a particular nucleotide sequence, the promoter, on the DNA molecule. The enzyme then moves along the molecule, using the 3′-to-5′ strand as a template for the synthesis of RNA molecules, as shown in Figure 15–5 (page 258). The transcribed RNA molecules (rRNAs, tRNAs, and mRNA) then play their various roles in the translation of the encoded genetic information into protein.

Despite these basic similarities, there are differences between prokaryotes and eukaryotes in transcription, translation, and the events that occur between these two processes. One important difference is that eukaryotic genes are not grouped in operons in which two or more structural genes are transcribed onto a single RNA molecule. In eukaryotes, each structural gene is transcribed separately, and its transcription is under separate controls.

Another important difference is that in eukaryotes, unlike prokaryotes, transcription and translation are separated in both time and space. After transcription is completed in the nucleus, the mRNA transcripts are extensively modified before they are transported to the cytoplasm, the site of translation.

Even before transcription is completed, while the newly forming mRNA strand is only about 20 nucleotides long, a "cap" of an unusual nucleotide is added to its leading (5′) end. This cap is necessary for the binding of the mRNA to the eukaryotic ribosome. After transcription has been completed and the molecule released from the DNA template, special enzymes add a string of adenine nucleotides to the trailing (3′) end of the molecule. This added segment, known as the poly-A tail, may contain as many as 200 nucleotides.

Before the modified mRNA molecules leave the nu-

RNA and the Origin of Life

The discovery of the role of DNA in heredity launched a debate among biologists as to whether life had its molecular beginnings in protein or in DNA. DNA was a likely choice because it is the repository of the genetic information and provides the template for its own precise replication. Proteins have neither of these qualifications. Proponents of proteins, however, have noted that virtually all of the chemical reactions of the cell depend on proteins that function as catalysts—that is, on enzymes. Even the replication of DNA depends on protein catalysts.

An important clue toward the resolution of this "chicken-and-egg" dilemma has come from an unexpected source. Using the single-celled protist *Tetrahymena,* T. C. Cech and his co-workers at the University of Colorado were studying the removal of introns and the splicing together of exons. This process must be carried out with exquisite precision, since a mistake of one nucleotide could render the entire molecule nonfunctional. In order to isolate the catalysts required for the reaction, Cech and his co-workers set up two experimental systems. One contained a *Tetrahymena* RNA molecule from which an intron was to be re-

The two ringlike structures visible in this electron micrograph are introns removed from a transcribed RNA molecule of the protist Tetrahymena. *These introns have the capacity to catalyze their own removal and to splice together the exons.*

moved, plus proteins that were potential catalysts. The other system, the control, contained the RNA but no proteins. The intron was neatly removed in the first system, as expected, but, to everyone's surprise, the removal and splicing process also took place in the control.

It was subsequently shown that the intron—a segment of RNA containing some 400 nucleotides—has an enzyme-like catalytic activity that carries out its own removal and the splicing of the exons. This RNA segment folds up to form a complex surface that functions

like an enzyme. Although RNA catalysts are not common, a number of them have now been found in other types of reactions and in exon-splicing in other types of cells. Such catalysts, dubbed "ribozymes," have become the subject of intense research.

The discovery that RNA can act as a catalyst makes it easier to imagine how life had its beginnings. According to cell biologist Bruce M. Alberts, "One suspects that a crucial early event was the evolution of an RNA molecule that could catalyze its own replication." These molecules then diversified into a collection of catalysts that could, for example, assemble nucleotides in RNA synthesis or accumulate lipid-like molecules to form the first primitive cell membranes. Gradually, other RNAs evolved and assembled the first proteins. Then, because the proteins were better catalysts, they gradually took over the catalytic functions. In the third step, DNA appeared on the scene, and its more stable double-stranded structure became the ultimate repository of the genetic information. Thus, researchers speculate, the catalytic intron can be regarded as a living fossil, a provocative clue to the events of almost 4 billion years ago.

---

cleus, the introns are removed, and the exons are spliced together to form a single, continuous molecule (Figure 16–14). The splicing mechanism is very precise, preventing the slight errors that could cause a frame-shift in the transcribed message (see page 266).

A number of instances have now been found in which identical mRNA transcripts are processed in more than one way. Such alternative splicing can result in the formation of different functional polypeptides from RNA molecules that were originally identical (Figure 16–15). In such cases, an intron may become an exon, or vice versa. Thus, the more that is learned about eukaryotic DNA and its expression, the more difficult it becomes to define "gene" or "intron" or "exon."

As we noted earlier, molecular biologists once thought that the eukaryotic chromosome might be simply a large-scale version of the prokaryotic chromosome. That turned out not to be the case. As we have

seen, the structure and organization of the chromosome, the regulation of gene expression, and the processing of mRNA molecules are all much more complex in eukaryotes than in prokaryotes. For many years it also seemed reasonable to believe that the chromosomes of eukaryotes were stable. Recombinations were produced by crossing over, of course, but it was thought that—except for occasional mutations and mistakes in meiosis—the structure of each chromosome was essentially fixed and unchanging. Perhaps the greatest surprise of all has been the discovery that this too is not the case. In both prokaryotes and eukaryotes, segments of DNA can move from one place to another within a chromosome, they can move into and out of chromosomes, they can move from one chromosome to another—and sometimes even from one organism to another. In the next chapter, we shall explore some of the ways in which these movements occur and their significance for the organisms involved, including ourselves.

16–14   *A summary of the stages in the processing of an mRNA transcribed from a structural gene of a eukaryote. The genetic information encoded in the DNA (a) is transcribed into an RNA copy. This copy is then edited, with the addition of a cap at the 5' end and a poly-A tail at the 3' end (b). Next, the introns are snipped out, and the exons are spliced together. The mature mRNA (c) then goes to the cytoplasm, where it is translated into protein.*

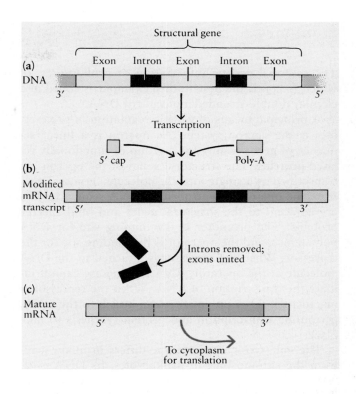

16–15   *Alternative splicing of identical mRNA transcripts results in the synthesis of different polypeptides from the information encoded by a single gene. For example, when the mRNA transcript shown here is processed (a) in the thyroid gland, segments E and F are removed in the initial processing and the poly-A tail is added to the end of exon D. The three introns are then removed, and the mature mRNA molecule is translated into the peptide hormone calcitonin. (b) In the pituitary gland, however, the poly-A tail is added to the end of exon F. Five introns are removed from this transcript, including segment D, which in the thyroid was retained as an exon. The mature mRNA, formed from five exons, is translated into a different hormone, known as calcitonin-gene-related protein (CGRP).*

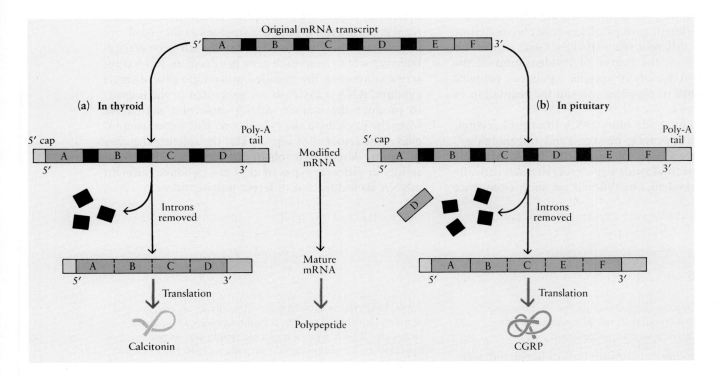

Summary

The essential genetic information of prokaryotes, of which *E. coli* is the best-studied example, is coded in a circular double-stranded molecule of DNA.

A principal means of genetic regulation in prokaryotes is the operon system. An operon is a linear sequence of genes coding for a group of functionally related proteins. The structural genes of the operon are transcribed as a single mRNA molecule. Transcription from the operon is controlled by DNA sequences that are adjacent to the structural genes and bind specific proteins. The promoter is the binding site for RNA polymerase, and the operator is the binding site for the repressor. When the repressor is attached to the DNA molecule at the operator, RNA polymerase cannot initiate the transcription of RNA. When the repressor is inactivated, RNA polymerase can attach to the DNA, permitting transcription and protein synthesis to take place.

The eukaryotic chromosome differs in many ways from the chromosome of prokaryotes. Its DNA is always associated with proteins, which constitute more than half of the chromosome. Most of these proteins are histones, which are relatively small molecules. The DNA molecule wraps around histone cores to form nucleosomes, which are the basic packaging units of eukaryotic DNA.

Regulation of gene expression is more complex in eukaryotes than in prokaryotes. During embryonic development, different groups of genes are activated or inactivated in different types of cells. Gene expression is correlated with the degree of condensation of the chromosome. A variety of specific regulatory proteins are also thought to play key roles in the regulation of gene expression.

Eukaryotes have far more DNA than prokaryotes, much of which appears to be excess and "meaningless." There are three classes of eukaryotic DNA. Multiple repeats of short nucleotide sequences, characteristically arranged in tandem, are known as simple-sequence DNA. Simple-sequence DNA is associated with the tightly condensed chromatin in the region of the centromere and with the tips of the chromosomes.

Longer repeats, usually dispersed throughout the chromosomes, are known as intermediate-repeat DNA. Intermediate-repeat DNA includes multiple copies of the genes coding for the rRNAs and histones, as well as members of gene families. The individual genes of a family differ slightly in their nucleotide sequences. As a consequence, the proteins for which they code differ slightly in structure and biological properties. Some members of gene families are not expressed; these DNA sequences are known as pseudogenes.

The third class of eukaryotic DNA, known as single-copy DNA, makes up from 50 to 70 percent of the chromosomal DNA. Current data indicate that as little as 1 percent of human DNA may be translated into protein.

In eukaryotes, the coding sequence of most structural genes is not continuous but is interrupted by sequences known as introns. Although introns are transcribed into RNA in the nucleus, they are not present in the mRNA in the cytoplasm and thus are not translated into protein. The segments that are present in the cytoplasmic mRNA and are translated into protein are known as exons.

Transcription in eukaryotes differs from that in prokaryotes in a number of respects. In eukaryotes, a multiplicity of regulatory proteins are involved. Also, structural genes are not grouped in operons as they often are in prokaryotes. The transcription of each gene is regulated separately, and each gene produces an RNA transcript containing the encoded information for a single product. RNA transcripts are processed in the nucleus to produce the mature mRNA molecules that move from the nucleus to the cytoplasm. This processing includes the removal of introns and the splicing together of exons. Alternative splicing of identical RNA transcripts in different types of cells can produce different mRNA molecules and different polypeptides.

Questions

1. Distinguish among the following: operon/promoter/operator; repressor/regulator; nucleosome/nucleolus; euchromatin/heterochromatin; intron/exon.

2. Nutrient molecules can be used by cells in two different ways. One type is broken down (usually as an energy source). Another is used as a building block for a larger molecule. Which type of molecule would you expect to function as a corepressor? As an inducer? Do the examples given in the text conform to your expectations?

3. In what ways are the chromosomes of prokaryotes and eukaryotes similar? In what ways are they different?

4. Describe the three classes of eukaryotic DNA. What are some of the functions that have been identified for each class?

5. (**a**) What might be the advantage to an organism of having multiple copies of the genes for the rRNAs and the histones? (**b**) Would you expect an organism also to have multiple copies of the genes for the tRNAs? Explain your answer.

6. Gene expression theoretically can be regulated at the level of transcription, translation, or activation of the protein. In the latter case, the polypeptide produced by protein synthesis is in an inactive form. It undergoes some structural modification (controlled by enzymes) before it can perform its function in the cell. What would be the advantages of each type of regulation, in terms of the cell? Under what circumstances might one type be more useful than another? Which is the more economical?

7. The genes for coat color in cats are carried on the X chromosome. Black (*b*) is the recessive and yellow (*B*) is the dominant. What coat colors would you expect in the offspring of a cross between a black female and a yellow male? What coat colors would you expect in the sons of a tortoiseshell female? (You may find it helpful to diagram these crosses with Punnett squares.)

Suggestions for Further Reading

Alberts, Bruce, Dennis Bray, Julian Lewis, Martin Raff, Keith Roberts, and James D. Watson: *Molecular Biology of the Cell*, 3d ed., Garland Publishing, Inc., New York, 1994.

> *This outstanding cell biology text provides an excellent discussion of chromosome structure and the regulation of gene expression.*

Cech, Thomas R.: "RNA as an Enzyme," *Scientific American*, November 1986, pages 64–75.

Chambon, Pierre: "Split Genes," *Scientific American*, May 1981, pages 60–71.

Darnell, James E., Jr.: "The Processing of RNA," *Scientific American*, October 1983, pages 90–100.

Darnell, James, Harvey Lodish, and David Baltimore: *Molecular Cell Biology*, 2d ed., W. H. Freeman and Company, New York, 1990.

> *Chapters 7–11 provide a thorough review of gene structure and function. Chapters 5 and 6 discuss the techniques used in molecular biology.*

Eigen, Manfred, William Gardiner, Peter Schuster, and Ruthild Winkler-Oswatitsch: "The Origin of Genetic Information," *Scientific American*, April 1981, pages 88–118.

Felsenfeld, Gary: "DNA," *Scientific American*, October 1985, pages 58–67.

Grunstein, Michael: "Histones as Regulators of Genes," *Scientific American*, October 1992, pages 68–74B.

Holliday, Robin: "A Different Kind of Inheritance," *Scientific American*, June 1989, pages 60–73.

Jacob, François: *The Statue Within: An Autobiography*, Basic Books, Inc., New York, 1988.

> *In this gracefully written memoir, François Jacob describes his privileged boyhood in Paris, his turbulent years of military service during World War II, and his postwar return to the laboratory, where, along with André Lwoff and Jacques Monod, he undertook his Nobel Prize–winning work on the operon. The book is rich with the insights of a brilliant and eloquent scientist-poet.*

Kornberg, Roger D., and Aaron Klug: "The Nucleosome," *Scientific American*, February 1981, pages 52–64.

McKnight, Steven Lanier: "Molecular Zippers in Gene Regulation," *Scientific American*, April 1991, pages 54–64.

Moyzis, Robert K.: "The Human Telomere," *Scientific American*, August 1991, pages 48–55.

Ptashne, Mark: "How Gene Activators Work," *Scientific American*, January 1989, pages 40–47.

Rhodes, Daniela, and Aaron Klug: "Zinc Fingers," *Scientific American*, February 1993, pages 56–65.

Steitz, Joan A.: "Snurps," *Scientific American*, June 1988, pages 56–63.

Watson, James D., Nancy H. Hopkins, Jeffrey W. Roberts, Joan A. Steitz, and Alan M. Weiner: *Molecular Biology of the Gene*, 4th ed., The Benjamin/Cummings Publishing Company, Redwood City, Calif., 1987.

> *Molecular Biology of the Gene is a classic. Now in its Fourth Edition, it continues to serve as a celebration of the extraordinary achievements of biology during the second half of the twentieth century. Well written and richly illustrated, the book covers an incredible array of topics, ranging from molecular biology of prokaryotes to that of eukaryotes and multicellular organisms.*

17

Genes on the Move

Barbara McClintock, shown here at Cold Spring Harbor Laboratory on Long Island, New York, spent more than half a century analyzing color differences and other variations in corn kernels (each of which is actually an embryonic corn plant). On the basis of mapping studies, analogous to those carried out with *Drosophila* in Morgan's laboratory, she deduced that the changes she observed were not the result of ordinary mutations. Instead, she postulated, the changes came about as a result of the movement of genes, or "controlling elements," that jumped from one place to another on a chromosome—or even from one chromosome to another.

McClintock's findings, first published in 1951, were largely ignored. Like Mendel before her, she was a lonely traveler with an odd tale that did not fit into the scheme of things as then understood. James Watson, who has also worked at Cold Spring Harbor (first as a nineteen-year-old graduate student, then as its director from 1968 to 1993, and now as its president), once described McClintock as "fiercely independent, beholden to no one." Despite the lack of recognition, she stubbornly pursued her research, sometimes working without pay. "It was fun," she said, "I could hardly wait to get up in the morning."

McClintock was more than vindicated by a rush of discoveries, some 30 years later, of a host of movable genetic elements such as those she had first postulated. "Jumping genes," or transposons, as they came to be known, move from place to place on the chromosome and influence the expression of other genes. When McClintock was 81 years old, she received the Nobel Prize.

Her life did not change much, however. She continued her work at Cold Spring Harbor until shortly before her death in 1992 at the age of 90. "There are really three main figures in the history of genetics," Watson noted at a tribute to her the year before. "The three M's: Mendel, Morgan, and McClintock. Gregor Mendel and Thomas Hunt Morgan showed us how regular the genome is, and Barbara McClintock showed us how irregular it is."

In the mid-1970s, when it seemed as if the mysteries of heredity had been so satisfactorily explained that all that was left was the tidying up of a few details, genetics underwent another revolution. Quite suddenly, biologists became able to manipulate genes in ways never before imagined—copying, dissecting, analyzing, modifying, and recombining portions of DNA molecules from different sources. This new technology has generated an avalanche of studies that are providing new knowledge and new possibilities—particularly in medicine and agriculture—at a stunning pace.

Like the early advances in molecular genetics, the new technology resulted from work with bacteria and viruses—and specifically from the manipulation of the natural mechanisms by which genes move both within and between chromosomes. In the next chapter, we shall look at this technology and some of its most promising and intriguing applications. Before we do so, however, we must look first at the natural phenomena on which it is based. As we shall see in this chapter, the capacity of genes to "jump" has an importance for the organisms in which it occurs—and sometimes for other organisms as well—that is quite independent of the applications made possible by its deliberate manipulation.

Although the ability of genes to move from one location to another in the eukaryotic chromosome was first detected by Barbara McClintock in her studies of corn, other scientists were not convinced until such movement was demonstrated at the molecular level in far simpler organisms. We shall begin our story where they began, with some "extra" DNA molecules found in *E. coli* and other bacteria.

Plasmids and Conjugation

Although the bacterial chromosome (page 271) contains all of the genes necessary for growth and reproduction of the cell, virtually all types of bacteria have

0.05 μm

17–1   *Plasmids from an E. coli cell. The two plasmids that are connected, forming a figure eight, are probably about to complete replication.*

0.5 μm

17–2   *Conjugating E. coli cells. The F^+ (donor) cell at the left is connected to the F^- (recipient) cell at the right by a long pilus. Genes on the F plasmid are responsible for the production of these specialized pili, which are necessary for conjugation. After contact has been made, the long pilus retracts, pulling the two cells close enough to one another that a cytoplasmic bridge can form between them. Numerous shorter pili are also visible on the F^+ cell.*

been found to carry additional DNA molecules known as **plasmids.** Plasmids, which are much smaller than the bacterial chromosome, may carry as few as two genes or as many as thirty.

Like the bacterial chromosome, plasmids are circular and self-replicating (Figure 17–1). Some plasmids replicate at the same time as the chromosome, and each daughter cell has only one copy of the plasmid. Other plasmids replicate more frequently than the chromosome, with the result that the cell may contain multiple copies. In the case of some small plasmids, as many as 50 copies have been detected in a single cell. Alternatively, if the plasmid replicates less frequently than the chromosome, some daughter cells may not receive any copies of the plasmid.

About a dozen different kinds of plasmids have been described in *E. coli* alone. Two of the most important are fertility, or F, plasmids, and drug resistance, or R, plasmids.

The F Plasmid

The first plasmid to be identified was the **F factor** of *E. coli.* This plasmid contains some 25 genes, many of which control the production of long, rod-shaped protein structures, known as **pili** (singular, pilus), that extend from the surface of cells containing the F plasmid. Such cells are known as F^+ cells. Cells that lack the F

plasmid—and thus the pili—are known as F^- cells. F^+ cells can attach themselves to F^- cells by the pili (Figure 17–2) and transfer a copy of the F plasmid to them through cytoplasmic bridges. Such transfer of DNA from one cell to another by cell-to-cell contact is known as **conjugation.**

Transfer of the F factor in conjugation gives the recipient cells the capacity to produce pili and to transfer the F plasmid—that is, the recipient cells become F^+, or donor, cells. The replication of an F plasmid and its transfer from a donor cell to a recipient cell are diagrammed in Figure 17–3.

Like many other plasmids, the F factor can exist independently in the cell or it can become spliced into the bacterial chromosome. When the F factor becomes integrated into the chromosome, it then becomes possible for a portion of the chromosome itself to be transferred from one cell to another. During conjugation, the chromosome opens up within the F factor sequence and begins to replicate. As it replicates, the free end of a single strand of DNA moves into the F^- cell. Recombination may then occur between the chromosome of the recipient cell and those portions of the chromosome of the donor cell that have been transferred, with new material replacing old in the recipient's chromosome. Studies of this process using mutant strains of bacteria have enabled scientists to prepare detailed genetic maps of the *E. coli* chromosome.

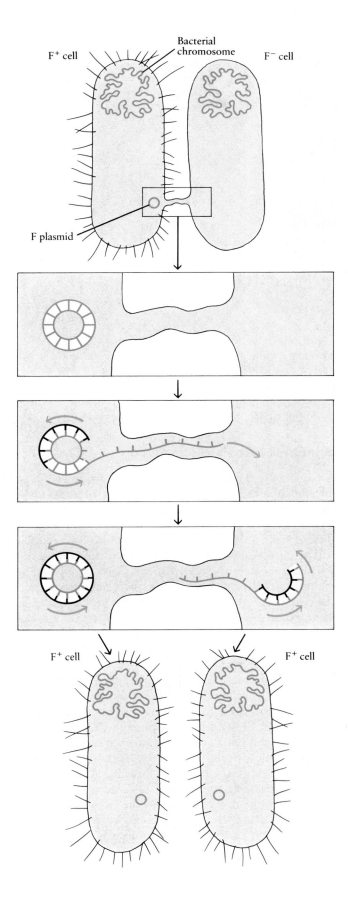

F⁺ cell

Bacterial chromosome

F⁻ cell

F plasmid

F⁺ cell   F⁺ cell

17–3 *Transfer of an F plasmid from an F⁺ cell to an F⁻ cell during conjugation. Both the bacterial chromosome and the plasmid are circular molecules of double-stranded DNA. The plasmid, however, is much smaller than the chromosome and contains far fewer base pairs.*

A single strand of DNA moves from the donor cell, through the cytoplasmic bridge, and into the recipient cell. Its complementary strand (black) is then synthesized from nucleotides available in the recipient cell. As this DNA strand is transferred, the remaining strand in the donor cell "rolls" counterclockwise, exposing its unpaired nucleotides. These serve as a template for the synthesis of a complementary DNA strand (black). As a result of this "rolling-circle" replication, the plasmid in the donor cell continues to be a circle of double-stranded DNA. The transferred plasmid converts the recipient cell to an F⁺ cell.

R Plasmids

In 1959, a group of Japanese scientists discovered that resistance to certain antibiotics and other antibacterial drugs can be readily transferred from one bacterial cell to another. Under experimental conditions, 100 percent of a population of drug-sensitive cells can become resistant within an hour after being mixed with suitable drug-resistant bacteria. It was subsequently found that the genes conveying drug resistance are often carried on plasmids that have come to be known as **R plasmids.**

Drug resistance in bacterial cells is often the result of the synthesis of enzymes that break down the drug or that set up a new enzymatic pathway, circumventing the effects of the drug. Thus, resistance may depend on the synthesis of specific enzymes in high concentrations. The fact that resistance genes are on plasmids allows many copies of those genes to be produced very rapidly within a single cell.

Resistance genes can be transferred not only from one cell to another but also from one R plasmid to another. A single plasmid may collect as many as 10 different resistance genes, making the cell it inhabits (and any cell to which it is transferred) resistant to as many as 10 different antibiotics (Figure 17–4). Typically, only a few copies of these large plasmids exist in a single

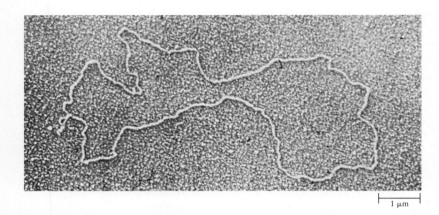

17–4   *This plasmid, known as R6, carries genes that confer resistance to six different drugs, including the antibiotics tetracycline, neomycin, and streptomycin.*

1 μm

cell. They are passed from mother to daughter cells at cell division, are transferred by conjugation, or, in another example of transformation (page 243), they may be simply passed from cell to cell through the plasma membranes.

Resistance genes can also be transferred from plasmids to the bacterial chromosome, to viruses, and to bacteria of other species, including species that cause serious human illness. Thus, for example, the usually innocuous *E. coli* can pick up R plasmids by conjugation and transfer them to *Shigella*, a bacterium capable of causing a sometimes fatal form of dysentery. Infectious drug resistance has been found among an increasing number of types of bacteria, including those responsible for typhoid fever, cholera, pneumonia, meningitis, gonorrhea, and, most disturbing of all, tuberculosis.

The incidence of tuberculosis, once thought to be essentially eradicated in the developed countries, has increased dramatically in the United States since 1985. In 1991, there were 26,000 new cases reported and, in 1992, almost 27,000. In New York City, one-third of the cases were caused by bacteria resistant to one or more of the antibiotics used to treat the disease, and drug-resistant strains of the bacteria were also identified in 35 other states. In otherwise healthy individuals, drug-resistant tuberculosis has a fatality rate of about 50 percent. In individuals with immune systems that are suppressed (often a side effect of cancer chemotherapy) or weakened (for example, by the virus that causes AIDS), the fatality rate is much higher.

For bacterial cells, the ability to transfer drug-resistance genes from one cell to another is clearly beneficial. However, as the resurgence of tuberculosis illustrates, this ability also threatens to negate the great advances that have been made in this century in the control of infectious disease.

Viruses

Some biologists consider the infectious agents known as viruses to be living organisms. Other biologists, however, take a different view. Salvador Luria, one of the pioneers in the study of bacteriophages, once described viruses as "bits of heredity looking for a chromosome." As we shall see, their quest is often successful.

The Structure of Viruses

A virus consists essentially of one or more molecules of nucleic acid enclosed in a protein coat, or **capsid**. As shown in Figure 17–5, the capsid may consist of a single protein molecule repeated over and over, or it may be formed from a number of different kinds of proteins.

The composition of the protein coat determines the specificity of the virus. A virus can infect a cell only if that type of cell has receptor sites on its surface to which the viral proteins can bind. Thus bacteriophages attack bacterial cells, tobacco mosaic virus (page 258) attacks leaf cells of the tobacco plant, and the viruses responsible for the common cold invade cells in the lining of the human respiratory tract.

Viruses contain no cytoplasm or metabolic machinery and can multiply only within a living cell. In some virus infections, the protein coat is left outside the cell while the nucleic acid enters (see page 245). In others, the intact virus enters the cell, but once inside, its protein coat is destroyed by enzymes, freeing the viral nucleic acid. Within the host cell, this nucleic acid directs the production of new viruses. This is accomplished using not only the raw materials of the cell—nucleotides and amino acids—but also the cell's enzymes, tRNA molecules, ribosomes, and ATP molecules and other energy sources.

Depending on the type of virus, its nucleic acid—that

(a)   0.25 μm

(b)

(c)   0.2 μm

(d)

17–5   *Representative virus structures.*
(**a**) *An electron micrograph of the capsid of
an adenovirus, one of the many viruses that
cause colds in humans. Within the capsid is a
core of double-stranded DNA.* (**b**) *A computer-generated model of the capsid of an adenovirus. The capsid has 20 sides, each of
which is an equilateral triangle composed of
identical protein subunits. There are 252
subunits in all. Rodlike protein structures
project from the 12 subunits located at the
points where the vertices of the triangles
meet.*

(**c**) *Electron micrograph of type A influenza viruses. The inner layer of each capsid is
formed from helical proteins, and the outer
layer is a lipoprotein membrane through
which stubby protein spikes protrude. Within
the capsid are eight separate and dis-*

*tinct RNA molecules. For reasons that are
not understood, influenza viruses mutate frequently. The resulting changes in nucleotide
sequence alter the proteins of the outer envelope. As a consequence, previously formed
antibodies circulating in the bloodstream no
longer "recognize" the virus. New strains of
influenza are likely to arise more rapidly
than new vaccines can be produced to com-
bat them.*

(**d**) *A model of the capsid of the type of
bacteriophage used in the Hershey-Chase experiments (see Figures 14–4 and 14–5, on
pages 244 and 245). The double-stranded
DNA of the virus, located within the head
of the capsid, codes for all of the proteins
required to form its different structural
components.*

is, its chromosome—may be either DNA or RNA, single-stranded or double-stranded, circular or linear.
When DNA viruses infect a cell, two processes occur.
The viral DNA is replicated, forming more viral DNA,
and it is transcribed into mRNA that directs the synthesis of viral proteins. With most RNA viruses, the
situation is similar. The RNA replicates, forming more
viral RNA, and also serves directly as mRNA.

The viral chromosome always codes for the coat proteins and for one or more enzymes involved in replication of the viral chromosome. In at least some cases,
it codes for repressors and other regulatory molecules
as well. In most viruses, the chromosome also codes for
enzymes that, once the new virus particles are assembled, enable them to **lyse** (break apart) the host cell and
escape. Some viruses, however, insert themselves into
the plasma membrane of the host cell and are budded
off in a process similar to exocytosis (page 119).

The infection cycle of a virus is complete when the
newly synthesized viral nucleic acid molecules are packaged into the newly synthesized protein coats and the
virus particles leave the host cell.

Viruses as Vectors

The genetic makeup of bacterial cells can be altered, as
we have seen, by the introduction of DNA from other
bacterial cells by means of plasmids. Viruses can also
play a role as **vectors,** or carriers, that move pieces of
DNA from one cell to another.

Early in the study of bacteriophages, it was noted
that a virus infection could suddenly erupt in a colony
of apparently uninfected bacterial cells. The cause of
this phenomenon, it was discovered, was the capacity
of certain viruses to set up a long-term relationship with
their host cell, remaining latent through many cell di-

visions before initiating a **lytic cycle** in which the cell is destroyed. Such viruses became known as **temperate bacteriophages.**

The DNA of temperate phages, like that of the F plasmid, may become integrated at specific sites in the host chromosome, replicating along with the chromosome. Such integrated phages are known as **prophages,** and bacteria that harbor them are known as **lysogenic bacteria.** Prophages break loose from the host chromosome spontaneously about once in every 10,000 cell divisions, initiating a lytic cycle (Figure 17–6). In the laboratory, this process may be triggered by ultraviolet light, x-rays, or other agents that damage nucleic acids.

Temperate phages resemble plasmids in that (1) they are autonomously replicating molecules of DNA, and (2) they may become integrated into the bacterial cell chromosome. Unlike plasmids, however, temperate phages can manufacture a protein coat and thus exist (although they cannot multiply) outside a host cell.

Transduction

The transfer of DNA from one host cell to another by means of viruses is known as **transduction.** During the lytic cycle of many viruses, the host DNA becomes fragmented. When the new virus particles are assembled, some of these DNA fragments, picked up at random, may be enclosed by the coat proteins. Since the amount of DNA that can be packaged within the protein coat is limited, such viruses lack some or all of their own necessary genetic information. Although they may be able to infect a new host cell, they are usually not able to complete a lytic cycle. However, the genes they carry from their previous host may become incorporated into the chromosome of the new host. This process is called **general transduction** because virtually any gene can be transferred by this mechanism.

When prophages break loose from a host chromosome to initiate a lytic cycle, they may, similarly, take a fragment of the host chromosome with them. In this situation, however, the host DNA is *not* picked up at random. It is, instead, quite specifically restricted to the portions of the host chromosome next to the insertion site of the prophage. Hence, this process is known as **specialized transduction.**

The portions of the bacterial chromosome carried by a virus are replicated along with the viral chromosome through subsequent lytic cycles. If the DNA of one of these viruses should become part of the chromosome of a new host, the genes from the previous host may be inserted into the new host's chromosome and become part of its genetic equipment. For example, the temperate bacteriophage known as lambda (Figure 17–7) can carry the bacterial genes that code for the enzymes con-

17–6   *When certain types of bacteriophages infect their host cells, one of two processes may occur. (a) The viral DNA may enter the cell and set up a lytic cycle, as described in Chapter 14. The cell is lysed as the newly formed viral particles escape from it. (b) Alternatively, the viral DNA may become part of the bacterial chromosome, replicating with it and being passed on to the daughter cells through many cell divisions. A bacterial virus integrated into the chromosome is known as a prophage, and bacteria harboring such viruses are said to be lysogenic. From time to time, however, a prophage becomes activated (c) and is released from the bacterial chromosome, setting up a new lytic cycle (d). In these diagrams, the double-stranded viral and bacterial chromosomes are depicted by single lines.*

cerned with the breakdown of galactose—in other words, the galactose operon. When a lambda phage carrying the galactose operon becomes a prophage in a mutant bacterial cell that cannot synthesize one or more of these enzymes, the infected cell gains the capacity to utilize galactose as a nutrient.

Transduction resembles conjugation in that it involves the transfer of bacterial genes from one bacterial cell to another. It differs from conjugation in that in transduction the genes are carried by viruses rather than by plasmids.

17–7 *Lambda, a temperate bacteriophage that infects* E. coli *cells. Its chromosome is a double-stranded DNA molecule that contains about 45,000 base pairs. When the chromosome is within the capsid head, it is linear (with two free ends). However, when it is released into the cytoplasm of an* E. coli *cell, it immediately forms a circle (no free ends).*

Like a plasmid, the DNA of a lambda phage may become part of a host-cell chromosome. Insertion of the lambda DNA into the host-cell chromosome is brought about by an enzyme that recognizes specific nucleotide sequences on the two DNA molecules. This enzyme, which is coded by the phage DNA, brings the two molecules together and initiates the cutting and sealing reactions.

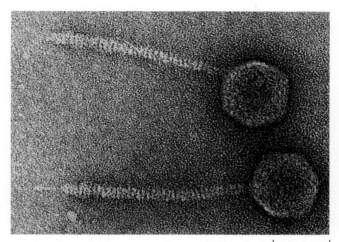

0.05 μm

Proviruses and Retroviruses

Some viruses of eukaryotes can, like the temperate bacteriophages, become integrated into the chromosomal DNA of the host cell. When integrated, these viruses are known as **proviruses.** Such viruses are of two general types: DNA viruses (analogous to the temperate bacteriophages) and RNA retroviruses.

A number of DNA viruses are known that can, depending on the type of cell they infect, either initiate a lytic cycle or insert themselves into the chromosomal DNA of the host cell. An example is simian virus 40, or SV40 (Figure 17–8). SV40 is a virus of monkeys that was first discovered in cells, growing in tissue culture, that were being used for the development of polio vaccines. It was subsequently found to cause cancers in newborn hamsters, though not in the monkeys that are its normal hosts. The cancers are caused by specific growth-promoting proteins produced in the cells by the viral genes. In short, SV40 can introduce new, functional genes into the DNA of the host cell, as can a number of other DNA viruses.

(a)

(b) 0.25 μm

17–8 *(a) A computer-generated model of the 20-sided capsid of simian virus 40 (SV40). The capsid encloses the SV40 chromosome (b), a circular molecule of double-stranded DNA that, like the DNA of the eukaryotic chromosome, is complexed with histone proteins to form nucleosomes. The histones are those of the host cell. Thus, SV40 not only uses host-cell machinery for its own synthetic activities but also directly incorporates products of the host cell's synthetic activities.*

The second group of eukaryotic viruses that can become integrated into host-cell chromosomes are the **RNA retroviruses.** Human immunodeficiency virus (HIV), the deadly virus that causes AIDS, is an example of an RNA retrovirus. In Chapter 33, we shall look more closely at this virus and its effects on the human immune system.

The integration of an RNA virus into a DNA chromosome poses some special problems that are solved by a remarkable enzyme known as **reverse transcriptase.** Molecules of reverse transcriptase are carried within the capsid of an RNA retrovirus, along with its RNA. In the host cell, the viral RNA is copied by reverse transcriptase to produce, after a complex series of events, a double-stranded DNA molecule (Figure 17–9).

Once integrated into a host chromosome, the DNA derived from the viral RNA utilizes the RNA polymerases and other resources of the host cell to synthesize new viral RNA and protein molecules, which are then packaged into new virus particles. Depending on the site of insertion into the host chromosome, DNA derived from a retrovirus may cause mutations by interfering with the expression of host-cell genes, either repressing them or releasing them from repression. Characteristically, however, most retroviral insertions do not damage or destroy their host cells but become permanent additions to the host-cell DNA. If cells destined to become eggs or sperm are infected with such a retrovirus, its genetic information will be transmitted to the next generation.

Genes, Viruses, and Cancer

Cancer is a disease in which cells escape the influence of the factors, still largely unknown, that regulate normal cell growth. As a consequence, the cells multiply out of control, crowding out, invading, and destroying other tissues. Cancer is often considered a group of diseases rather than a single disease because, with few exceptions, any one of the 200 or more cell types in the human body can become malignant.

Three lines of evidence have long linked the development of cancer with mutations. First, once a cell has become cancerous, all of its daughter cells are cancerous. In other words, cancer is an inherited property of cells. Second, gross chromosomal abnormalities, such as deletions and translocations, are often visible in cancer cells. Third, most carcinogens—agents known to cause cancer, such as x-rays, ultraviolet radiation, tobacco smoke, and a variety of chemicals—are also mutagens.

17–9   *Infection of an animal cell by a retrovirus (a). Contained within the capsid of a retrovirus are the viral RNA and one or more molecules of an enzyme, reverse transcriptase. The capsid is typically surrounded by an outer lipoprotein envelope formed from elements of the plasma membrane of its previous host. This envelope can fuse with the plasma membrane of a new host (b), allowing the virus to enter the cell by endocytosis (page 119). Once the retrovirus has gained entry to the cell, the viral RNA is (c) released from the capsid and (d) transcribed into a single strand of complementary DNA (cDNA). (e) Synthesis of the matching DNA strand follows immediately, producing a double-stranded molecule of cDNA. These two reactions, as well as the degradation of the original viral RNA molecule, are all catalyzed by reverse transcriptase.*

After the double-stranded cDNA has been inserted into a host-cell chromosome, new viral RNA molecules are transcribed from it, as are mRNA molecules that direct the synthesis of viral proteins.

Other lines of evidence, however, have linked the development of cancer with viruses. As long ago as 1911, a cancer-causing virus, the Rous sarcoma virus, was isolated from chicken tumors. Even though other cancer-causing viruses, particularly viruses affecting laboratory mice, were gradually discovered, a viral hypothesis of cancer was slow to emerge. For one thing, viruses could not be shown to be important as causes of human cancer. (Even today, after years of searching, only a few rare human cancers have been linked to viruses.) In addition, the fact that most of the known cancer-causing viruses, including the Rous sarcoma virus, are RNA viruses rather than DNA viruses seemed to make the viral hypothesis of cancer incompatible with the mutation hypothesis.

As we have seen, however, it is now known that viruses, like mutagens, can bring about changes in the cell's genetic makeup. All known cancer-causing viruses introduce information into host-cell chromosomes. These include both DNA viruses, like SV40, and RNA retroviruses. The discovery of the role of reverse transcriptase forged the crucial link between the RNA viruses and the chromosomes of eukaryotic cells.

New techniques have enabled molecular biologists to study the changes in the eukaryotic chromosome that lead to cancer. These studies are usually carried out in cells growing in tissue culture. When such cells are exposed to a carcinogenic agent, they undergo characteristic changes in their growth patterns and in their shape (Figure 17–10). Moreover, when these cells are transplanted into laboratory animals, they cause cancer.

Studies of such cells have uncovered a group of genes known as **oncogenes** (from the Greek word *onkos,* meaning "tumor"). Oncogenes closely resemble normal genes of the eukaryotic cells in which they are found.

According to the oncogene hypothesis, when something goes wrong in the expression of these normal genes, either because of mutations in the genes themselves or because of changes in gene regulation, they stimulate the uncontrolled cell division characteristic of cancer.

Other studies have revealed the presence of another group of genes that play a crucial role in determining whether cancer actually develops following oncogene activation. The function of these genes, known as **tumor suppressor genes,** is apparently to act as a brake on cell division, keeping it tightly regulated. As long as these genes are functioning normally, the development of cancer is blocked or, at the least, slowed. However, when they are disabled by mutation or actually lost from the chromosome, the way is clear for the rapid multiplication of cancer cells.

Collectively, these studies suggest that viruses can affect the development of cancer in three different ways. First, simply by their presence in the chromosome, viruses may disrupt the function of normal regulatory genes—converting normal genes into oncogenes, activating oncogenes, or disabling tumor suppressor genes. Second, viruses may encode proteins needed for viral replication that also affect the regulation of these cellular genes. Third, and most interesting of all, viruses may serve as vectors of oncogenes, enabling them to move from one cell to another or from one individual to another.

With these discoveries, the viral hypothesis and the mutation hypothesis of cancer are no longer regarded as incompatible but rather as mutually supportive. Step by step, this work is bringing us closer to the control of one of our oldest and ugliest enemies, and, at the same time, is yielding new information on the fundamental question of the regulation of cell growth.

(a)　　　　　　　　　(b)

17–10   (a) *Normal cells growing in tissue culture and* (**b**) *the same type of cells after transformation with a cancer-causing virus. The cancer cells not only show striking surface changes but also pile up on top of one another as they multiply out of control. The normal cells, by contrast, grow side by side in a single organized layer.*

17–11   (a) *A complex transposon "jumping" from one location to another in a bacterial chromosome. The transposon shown here includes the gene that codes for the enzyme transposase and another gene that codes for an enzyme that confers resistance to an antibiotic.* (b) *This transposon has inserted itself into a structural gene, interrupting the coding sequence and inactivating the gene.*

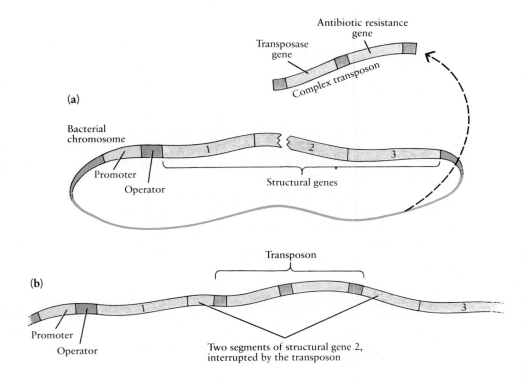

(a)

Antibiotic resistance gene

Transposase gene

Complex transposon

Bacterial chromosome

Promoter

Operator

1   2   3

Structural genes

Transposon

(b)

Promoter

Operator

1   3

Two segments of structural gene 2, interrupted by the transposon

Transposons

Both plasmids and viruses, as we have seen, can function as vectors, carrying genes from one chromosome to another. The chromosomes of both prokaryotes and eukaryotes contain yet another type of movable genetic element, one that incorporates within itself the mechanism for its movement. These movable genetic elements are known as **transposons,** and it was their activity that Barbara McClintock first detected in corn kernels more than 40 years ago.

Transposons are simply segments of DNA that move from one location to another in the chromosomes. They are characterized by a gene coding for an enzyme, **transposase,** that catalyzes their insertion into a new site. In many cases, a transposon does not disappear from its original site when it appears at a new location. Instead, the parent transposon gives rise to a copy that becomes inserted elsewhere in the chromosomes.

Two kinds of transposons—simple and complex—have been identified in prokaryotes. Simple transposons are relatively short and do not carry any genes beyond those essential for the process of transposition. They are detectable because they can cause mutations. If one of these transposons becomes inserted into a gene, it inactivates it. Simple transposons also contain promoter sequences, which may lead to the inappropriate transcription of previously inactive genes of the chromosome.

Complex transposons are much larger and carry genes that code for additional proteins (Figure 17–11). As is the case with simple transposons, complex transposons may cause mutations, but they are also detectable because of their gene products. Genes that are part of a complex transposon can move from place to place on a chromosome or from chromosome to chromosome, and are therefore known as "jumping genes." The drug-resistance genes of bacteria, for example, are often part of complex transposons, which explains why they can be transferred so readily from plasmid to plasmid and from plasmid to bacterial chromosome to plasmid again.

Complex transposons often have simple transposons flanking them—one at each end. This suggests that complex transposons are formed when two simple transposons jump at the same time, taking with them everything in between.

Eukaryotic transposons resemble their bacterial counterparts in structure, and like bacterial transposons, they can cause mutations when they become inserted into structural genes or promoter regions. However, they differ from bacterial transposons in one significant and interesting feature. In eukaryotes, many transposons are first copied into RNA—and then back into DNA—before their insertion into a new location in the chromosomal DNA. This discovery was surprising, for it was previously thought that reverse transcription was unique to retroviruses.

Although nucleotide sequences that resemble genes for reverse transcriptase have been identified in the DNA of some eukaryotes, there is, thus far, no evidence of reverse transcriptase activity in cells uninfected by retroviruses. It is, however, widely accepted that the action of reverse transcriptase in the course of retroviral infections has played an important role in the evolution of eukaryotic DNA and, in particular, in the evolution of transposons. There is also evidence that many of the repeated DNA sequences in eukaryotic chromosomes originated as transposons.

Expanding Triplet Repeats

In 1991, molecular biologists discovered yet another form of instability in eukaryotic genes. Although the affected genes were not moving from one location to another, they were expanding—often dramatically—from one generation to the next. In a few cases, the reverse process was occurring, and the genes were shrinking.

The initial discovery was made in studies of a human disorder known as **fragile X syndrome**. In this disorder, which is the most common form of inherited mental retardation, the tip of the long arm of the X chromosome hangs by a slender thread of DNA (Figure 17–12). Identification and analysis of the gene involved in fragile X syndrome revealed that it contains multiple copies of the nucleotide triplet CGG repeated in tandem, head to tail. The most astonishing discovery, however, was that the number of repeats varies from generation to generation and is correlated with the presence or absence of mental retardation and with its severity. The normal gene contains 60 or fewer CGG repeats. In healthy carriers of a fragile X chromosome, the gene may contain as many as 200 copies of the triplet. In individuals with fragile X syndrome, however, the number of copies balloons to hundreds or even thousands.

More recently, similar expanding triplet repeats have been identified in the genes responsible for Huntington's disease, for myotonic dystrophy (the most common type of muscular dystrophy in adults), and for several other rare disorders. An intensive search is now underway to identify other human genes that contain such repeats. At least 50 human genes are known to contain short triplet repeats, but it is not yet known if any of these repeats can expand.

The discovery of expanding (and occasionally contracting) triplet repeats has shed light on the incomplete

17–12 *A fragile X chromosome. The tip of the long arm is attached to the rest of the chromosome by a slender thread of DNA (arrow). Triplet repeats that increase in number from generation to generation occur in this stretch of the DNA. Like other X-linked disorders, fragile X syndrome occurs most often in males.*

penetrance and variable expressivity (page 227) observed in Huntington's disease, fragile X syndrome, and a number of other human genetic disorders. Below a certain threshold, which varies according to the disorder, the triplet repeats appear to have no harmful consequences. Above that threshold, however, symptoms of the disorder begin to appear, with their severity increasing as the number of repeats increases.

Like so many other discoveries, the discovery of expanding triplet repeats has raised many more questions than it has answered. At this time, no one knows what triggers the expansion or contraction of the repeats or how it occurs. Nor does anyone know why, in Huntington's disease, the expansion occurs most often when the affected gene is transmitted by the father, but in fragile X syndrome, it occurs most often when the gene is transmitted by the mother. The answers to these questions, when they come, may lead to new understandings of many previously inexplicable variations in human inheritance patterns. And, the answers may come very quickly. Stay tuned.

Summary

In addition to the bacterial chromosome, *E. coli* and other types of bacteria contain much smaller, also circular, double-stranded DNA molecules. These molecules are known as plasmids. Most plasmids can be transferred from cell to cell, often through the process of conjugation, in which a cytoplasmic bridge forms between two cells. Some plasmids can exist either independently in the cell or integrated into the bacterial chromosome. Plasmids often carry genes for drug resistance, with serious consequences for human health.

Viruses consist of either DNA or RNA wrapped in a protein coat. Some RNA viruses of animals, known as retroviruses, contain and code for the enzyme reverse transcriptase, which can transcribe DNA from an RNA template. Viruses can multiply only within a host cell, where the viral nucleic acid can use the cell's metabolic resources to synthesize more viral nucleic acid molecules and more viral proteins. The infection cycle of a virus is completed when newly synthesized virus particles break out of the host cell, ready to infect a new host cell.

Some DNA viruses of both prokaryotes and eukaryotes, as well as DNA transcribed from RNA retroviruses, can become integrated into a host-cell chromosome and replicate with the chromosome through many cell divisions. When integrated into a host chromosome, the DNA of a prokaryotic virus is known as a prophage. Integrated DNA from a eukaryotic virus is known as a provirus. From time to time, prophages and proviruses break loose from the chromosome and set up a new infection cycle.

Viruses can serve as vectors, transporting genes from cell to cell in a process known as transduction. General transduction occurs when host DNA, fragmented in the course of the viral infection, is incorporated into new virus particles that carry these fragments to a new host cell. Specialized transduction occurs when an integrated virus, on breaking away from the host chromosome, carries with it—as part of the viral chromosome—host genes, which are then transported to a new host cell.

According to present evidence, the development of cancer involves alterations in the function of certain normal cellular genes, converting them into oncogenes, as well as the inactivation or loss of tumor suppressor genes. The alterations in gene function may be caused by mutations or by changes in the regulation of gene expression. Viruses can cause cancer by disrupting the function of normal regulatory genes, by directing the synthesis of proteins that affect the regulation of host-cell genes, or by inserting oncogenes into a chromosome.

Transposons are another type of movable genetic element, found in both prokaryotic and eukaryotic cells. They carry a gene for the enzyme transposase, which catalyzes their transposition from one location to another in the chromosome. Transposons may cause mutations by interfering with the normal expression of host-cell genes. Simple transposons contain only genes involved in their transposition; complex transposons carry additional structural genes.

Certain human genes, including the genes responsible for fragile X syndrome and for Huntington's disease, contain triplet repeats that can increase (or sometimes decrease) in number from generation to generation. The cause or causes of these changes are not known, but it is clear that the severity of the disorders involved depends on the number of triplet repeats.

Questions

1. Distinguish among the following: transformation/conjugation/transduction; F^+ cell/F^- cell; plasmid/virus/transposon; prophage/provirus; general transduction/specialized transduction; oncogene/tumor suppressor gene.

2. As bacterial conjugation indicates, it is possible to separate the production of new genetic combinations from reproduction. Why do you think these two processes are combined in eukaryotic cells?

3. Before the structure of any virus was known, Crick and Watson predicted that the protein coats of viruses would prove to be made up of large numbers of identical subunits. Can you explain the basis of their prediction?

4. Describe two possible outcomes of infection of a bacterial cell by a temperate bacteriophage.

5. How do RNA retroviruses differ from other RNA viruses? What are the functions of the enzyme reverse transcriptase?

6. Most bacteriophages lack introns, but introns have been found in the DNA viruses of eukaryotes. The RNA viruses of eukaryotes, however, lack introns. What do these findings suggest about the origin of viruses?

7. Some biologists consider viruses to be living organisms. By what criteria might viruses be considered alive?

Suggestions for Further Reading

Alberts, Bruce, Dennis Bray, Julian Lewis, Martin Raff, Keith Roberts, and James D. Watson: *Molecular Biology of the Cell,* 3d ed., Garland Publishing, Inc., New York, 1994.

This outstanding cell biology text provides excellent discussions of plasmids, viruses, and transposons.

Amábile-Cuevas, Carlos F., and Marina E. Chicurel: "Horizontal Gene Transfer," *American Scientist,* vol. 81, pages 332–341, 1993.

Bishop, J. Michael: "Oncogenes," *Scientific American,* March 1982, pages 80–92.

Butler, P. Jonathan G., and Aaron Klug: "The Assembly of a Virus," *Scientific American,* November 1978, pages 62–69.

Campbell, A. M.: "How Viruses Insert Their DNA into the DNA of the Host Cell," *Scientific American,* December 1976, pages 102–113.

Cohen, S. N., and J. A. Shapiro: "Transposable Genetic Elements," *Scientific American,* February 1980, pages 40–49.

Doolittle, Russell F., and Peer Bork: "Evolutionarily Mobile Modules in Proteins," *Scientific American,* October 1993, pages 50–56.

Fedoroff, Nina V.: "Transposable Genetic Elements in Maize," *Scientific American,* June 1984, pages 85–98.

Gallo, Robert C.: "The AIDS Virus," *Scientific American,* January 1987, pages 46–56.

Gallo, Robert C.: "The First Human Retrovirus," *Scientific American,* December 1986, pages 88–98.

Hirsch, Martin S., and Joan C. Kaplan: "Antiviral Therapy," *Scientific American,* April 1987, pages 76–85.

Keller, Evelyn: *A Feeling for the Organism: The Life and Work of Barbara McClintock,* W. H. Freeman and Company, New York, 1993.*

An illuminating biography of McClintock, originally published in 1983, shortly before she was awarded the Nobel Prize.

Levine, Arnold J.: *Viruses,* Scientific American Library, New York, 1992.

This beautifully illustrated book begins with the bacterial viruses that have contributed so much to our understanding of genetics. Most of the book, however, is devoted to those human viruses—herpes viruses, tumor viruses, retroviruses, and the viruses responsible for influenza and hepatitis—that have, thus far, largely defied all efforts to bring them under control.

Morrell, Virginia: "The Puzzle of the Triple Repeats," *Science,* vol. 260, pages 1422–1423, 1993.

Rennie, John: "DNA's New Twists," *Scientific American,* March 1993, pages 122–132.

Simons, Kai, Henrik Garoff, and Ari Helenius: "How an Animal Virus Gets into and out of Its Host Cell," *Scientific American,* February 1982, pages 58–66.

Varmus, Harold: "Reverse Transcription," *Scientific American,* September 1987, pages 56–64.

Weinberg, Robert A.: "Finding the Anti-Oncogene," *Scientific American,* September 1988, pages 44–51.

*Available in paperback.

This is Cindy, playing catch in May of 1993. A few years earlier, such an activity would have been inconceivable, for Cindy was born with an inherited disorder known as SCID (severe combined immunodeficiency). In this disorder, the white blood cells—the cells that defend the body against infection—are defective. Until recently, Cindy and other victims of SCID were locked in a constant and losing battle against the multitude of infectious agents that surround us all.

The most common cause of SCID is a defect in the gene coding for an enzyme known as ADA. ADA is essential for the normal development and functioning of white blood cells. In 1988, doctors began to treat SCID children with synthetic ADA. Some of the children, including Cindy, improved. Others, however, did not.

In January of 1991, Cindy was the second child to receive a treatment that made a dramatic difference—and gave her the stamina to play baseball. White blood cells were taken from her body and treated with genetically engineered viruses that acted as vectors, inserting copies of the normal gene for ADA into the cells. The altered white blood cells were then reinfused into her body, where they synthesized the missing enzyme. As effective as this treatment was, however, it had to be repeated every few months as the altered white blood cells reached the end of their normal life span. It was an ordeal, and when asked about it, Cindy's response was, "I don't like to talk about it. It is too weird."

Shortly after this photo was taken, Cindy received a different treatment. This time, the gene for ADA was inserted into stem cells (page 180), the cells in the bone marrow that divide repeatedly, giving rise to red blood cells and to the white blood cells of the immune system. It is not yet known if the altered stem cells will reproduce in Cindy's bone marrow and make it possible for her to lead a long and normal life. If they do, it will mark the first successful cure of a human hereditary disorder—and an end to repeated trips to the hospital for this courageous pioneer.

18

New Frontiers in Genetics

In the last few years, advances in molecular genetics have come at an astonishing pace. One after another, new discoveries are revolutionizing our understanding of plant and animal genetics—and particularly our understanding of human genetic disorders. At the same time, new methods are being developed for the diagnosis of genetic disorders, and new hopes are being raised for actual cures. Human gene therapy, which a few years ago seemed only a distant dream, is becoming reality. And, that is only one of the many practical applications flowing from the new discoveries.

This burst of discovery and application has been made possible by the technology known as **recombinant DNA.** In recombinant DNA work, also known as **genetic engineering,** portions of DNA molecules from different sources are modified, recombined, and then inserted into other cells where the genes carried by the modified DNA are expressed.

Recombinant DNA: The Technology

In order to do recombinant DNA work, scientists must (1) have segments of DNA that are small enough to be analyzed and manipulated, (2) have large quantities of these small segments, (3) know the nucleotide sequence in the segments, and (4) be able to identify the specific segments in which they are interested. These are not, however, four sequential steps. The order in which these stages of the work are accomplished depends on the goals of a particular project. In some cases, for example, scientists know in advance the nucleotide sequences in which they are interested and where to find them. In other cases, learning the sequences and determining their location within the chromosome is the goal.

Recombinant DNA technology makes use of the natural mechanisms—plasmids, viruses, and transposons—by which genes can be moved from one location to another. It all began, however, with the discovery of a group of enzymes that protect bacterial cells against viral infection.

Obtaining Short DNA Segments

In the course of studies of the bacteriophages, it was found that viruses infecting one strain of *E. coli* were sometimes unable to infect another strain. This restriction of viral infection occurs, it was learned, because of the presence in the bacteria of enzymes that cut foreign DNA molecules into small pieces before they can be replicated or transcribed. These enzymes came to be known as **restriction enzymes.**

Restriction enzymes cut through foreign DNA molecules at specific nucleotide sequences, usually four to eight base pairs in length. These sequences are known as **recognition sequences,** since they are "recognized" by specific restriction enzymes. In the bacterial chromosome itself, the nucleotides in these sequences are chemically modified and thus protected from the enzymes.

Some restriction enzymes make straight cuts through both strands of the DNA molecule. Others, however, cleave the molecule by cutting through the two strands a few nucleotides apart. As a consequence, there is a short sequence of unpaired nucleotides at each end of the cut (Figure 18–1). These ends are said to be "sticky," because they can join again with each other when hydrogen bonds form spontaneously between the complementary bases. A type of enzyme known as a **DNA ligase,** which catalyzes the reaction that links the cut ends of each strand, is required to complete the resealing process.

From a practical standpoint, the most important thing about the "sticky" ends is that they can join with any other segments of DNA that have been cut by the same restriction enzyme and, as a result, have complementary "sticky" ends. This was the discovery that opened the way for molecular biologists to carry out their own genetic manipulations.

More than 200 different restriction enzymes have now been isolated from different types of bacteria, making it possible to cut a DNA molecule at any one of more than 90 recognition sequences. Because of the number of different restriction enzymes and recognition sequences, DNA molecules from virtually any organism can be cut into segments short enough to analyze and manipulate. These segments can be separated from one another on the basis of size by electrophoresis (Figure 18–2).

The particular segments produced by a restriction enzyme are determined by the locations at which the recognition sequence for the enzyme occurs—and these locations may be anywhere. Depending on where the recognition sequences occur, a given gene may be fragmented into a number of different segments, or, conversely, a particular segment may contain several different genes. However, short segments of DNA that represent specific genes (or known portions of specific genes) can be obtained by synthesizing them from an mRNA template, using the enzyme reverse transcriptase (page 294). If their nucleotide sequences are known, short DNA segments can also be synthesized by chemical means in the laboratory.

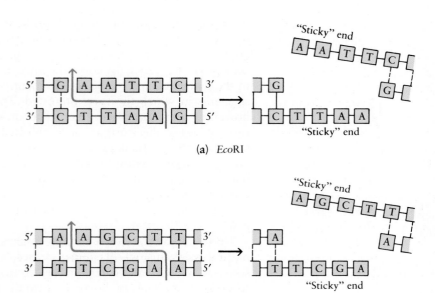

(a) *Eco*RI

(b) *Hind*III

18–1   *The DNA nucleotide sequences recognized by two widely used restriction enzymes,* **(a)** *EcoRI and* **(b)** *HindIII. These two enzymes cut the DNA so that "sticky" ends result. A "sticky" end can reattach to its complementary sequence at the end of any DNA molecule that has been cut by the same restriction enzyme. Restriction enzymes are named for the bacteria from which they are obtained. EcoRI is from* E. coli, *and* HindIII *is from* Hemophilus influenzae.

Mixture of DNA fragments

Gel

Large fragments move slowly in electrical field

Number of base pairs

— 350

— 150
— 140

— 54

— 39

(a) Gel at beginning of procedure

Small fragments move rapidly

(b) Gel at end of procedure

18–2   *Electrophoresis, the technique used by Pauling to separate normal and sickle-cell hemoglobins (page 257), can also be used to separate fragments of DNA. In electrophoresis, the electrical field separates molecules not only on the basis of their charge (the feature utilized by Pauling) but also on the basis of their size. Smaller molecules move faster than larger ones.* (a) *A mixture of DNA fragments containing different numbers of base pairs can be cleanly separated according to size* (b). *The gel can then be sliced into sections, and the separated, purified fragments washed out of the gel unharmed. This separation procedure is important in many aspects of recombinant DNA work.*

Obtaining Multiple Copies

DNA Cloning

Most recombinant DNA work requires multiple copies of the DNA segments being studied and manipulated. One way these copies can be obtained is by harnessing the replicative powers of plasmids and viruses.

Not long after the discovery of the restriction enzyme *Eco*RI, researchers isolated a small plasmid of *E. coli* that makes its host cells resistant to the antibiotic tetracycline. This plasmid has only one recognition sequence for *Eco*RI, and, as a consequence, it is cleaved at only one site by that enzyme. Thus it is possible to insert into the plasmid a small segment of foreign DNA that has also been cleaved by *Eco*RI (Figure 18–3).

Insertion of foreign DNA into the plasmid does not affect its ability to move into *E. coli* cells, its capacity to make the recipient cells tetracycline-resistant, or its ability to replicate. Typically, the plasmid replicates several times in its host cell, producing about 10 new plasmids per cell. Because the plasmids containing the foreign DNA are larger than those that do not contain it, after replication they can be readily isolated and collected. Treatment of the replicated plasmids by *Eco*RI

releases the foreign DNA, which can then be separated from the plasmid DNA by electrophoresis.

The discovery that plasmids could be used as vectors for foreign DNA opened the way to the production of large numbers of uniform, identical segments of DNA. Such multiple copies are known as **clones,** a term that has long been applied to genetically identical bacteria and other organisms produced asexually from a single parent cell or organism.

Plasmids are useful as vectors because they multiply rapidly and are easily taken up by bacteria through the plasma membrane (Figure 18–4). Modified strains of the bacteriophage lambda, which can gain ready entry to bacterial cells by way of its identifying protein coat, are also used in the laboratory to carry foreign DNA segments into bacterial cells for replication.

Polymerase Chain Reaction

In 1989, a new process was developed for producing multiple copies of DNA segments. This process, known as **polymerase chain reaction (PCR),** can take a tiny sample of DNA and, in a few hours, synthesize millions of copies of a specific segment of that DNA.

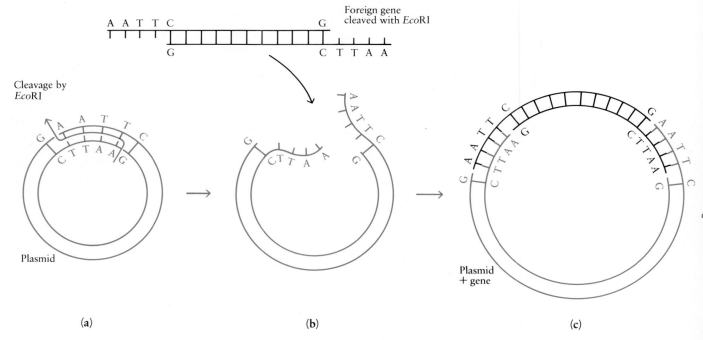

Foreign gene
cleaved with *EcoRI*

Cleavage by
EcoRI

Plasmid

Plasmid
+ gene

(a)

(b)

(c)

18–3   *The use of plasmids in DNA cloning. (a) The plasmid is cleaved by a restriction enzyme. In this example, the enzyme is EcoRI, which cleaves the plasmid at the sequence (5')-GAATTC-(3'), leaving "sticky" ends exposed. (b) These ends, consisting of TTAA and AATT sequences, can join with any other segment of DNA that has been cleaved by the same enzyme. Thus it is possible to splice a foreign gene into the plasmid (c). (In this illustration, the length of the GAATTC sequences is exaggerated and the lengths of the other portions of*
both the foreign gene and the plasmid are compressed.)

When plasmids incorporating a foreign gene are released into a medium in which bacteria are growing, (d) they are taken up by some of the bacterial cells. (e) As these cells multiply, the plasmids replicate. The result is an increasing number of cells, all making copies of the same plasmid. (f) The plasmids can then be separated from the other contents of the cells and treated with EcoRI to release the copies of the cloned gene.

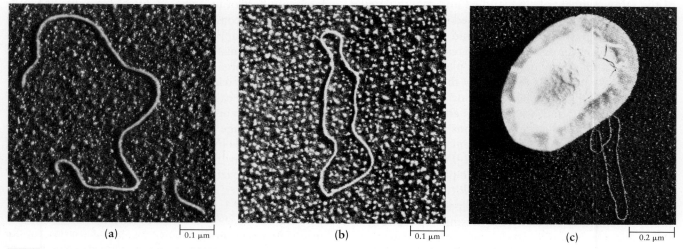

(a)   0.1 μm

(b)   0.1 μm

(c)   0.2 μm

18–4   *Insertion of a foreign gene into a bacterial cell. (a) A plasmid has been cut open with a restriction enzyme, leaving two "sticky" ends. A small segment of foreign DNA (lower right) also has "sticky" ends that can join with the "sticky" ends of the plasmid. (b) With the aid of DNA ligases, the for-*
eign DNA has been spliced into the plasmid. (c) The plasmid, now containing the foreign DNA, is about to enter a bacterial cell. (Huntington Potter and David Dressler, LIFE Magazine, 1980, Time, Inc.)

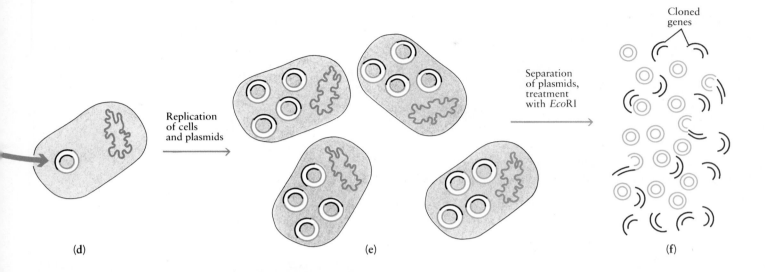

(d)   (e)   (f)

Replication of cells and plasmids

Separation of plasmids, treatment with *Eco*RI

Cloned genes

Unlike cloning, PCR requires knowledge of the nucleotide sequences at each end of the DNA segment that is to be copied. The DNA sample is added to a solution that contains DNA polymerase molecules, large quantities of the four nucleotides that occur in DNA, and "primer" molecules of DNA, about 20 nucleotides in length, that are complementary to the sequences at each end of the desired segment. When this solution is heated, the two strands of the DNA double helix separate. As the solution is cooled, the primers attach to their complementary sequences. DNA polymerase molecules "recognize" the primers and begin adding nucleotides at the locations where the primers are attached. Both strands of the DNA segment are copied simultaneously. Thus, at the end of the initial step, two copies of the segment are present, each consisting of one new strand and one old strand. The solution is then alternately heated and cooled, and with each cycle, the number of copies is doubled. If this is repeated for 20 cycles, about one million copies of the segment are produced.

Among its many applications, PCR is being used to speed up prenatal diagnosis of genetic disorders, to detect latent viral infections, and to produce large quantities of selected human DNA sequences at rates far faster than is possible with conventional DNA cloning.

Determining Nucleotide Sequences

With the development of techniques for cutting DNA molecules into smaller pieces and making multiple copies of those pieces, it is now possible, in principle, to determine the nucleotide sequence of any isolated gene. One of the most important features of restriction enzymes is that different enzymes cut DNA molecules at different sites. Cutting a DNA molecule with one restriction enzyme produces one particular set of short DNA fragments. Cutting an identical DNA molecule with a different restriction enzyme produces a different set of short DNA fragments. The fragments of each set can be separated from one another by electrophoresis and cloned into multiple copies.

These copies can then be analyzed to determine the exact nucleotide sequence of each fragment. The methods used in these analyses involve complex series of chemical or enzymatic reactions. However, the reactions have now been automated, greatly reducing the time required to determine the nucleotide sequences of short fragments.

Because the sets of fragments produced by different restriction enzymes overlap, the information obtained from sequencing the different sets can be pieced to-

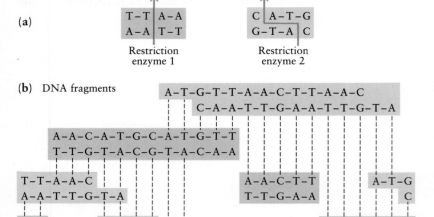

(a)

```
T - T | A - A        C | A - T - G
A - A | T - T        G - T - A | C
```
Restriction   Restriction
enzyme 1   enzyme 2

(b) DNA fragments

```
                    A-T-G-T-T-A-A-C-T-T-A-A-C
                      C-A-A-T-T-G-A-A-T-T-G-T-A

      A-A-C-A-T-G-C-A-T-G-T-T
      T-T-G-T-A-C-G-T-A-C-A-A

T-T-A-A-C                      A-A-C-T-T        A-T-G
A-A-T-T-G-T-A                  T-T-G-A-A          C

T-T        A-T-G-C                        A-A-C-A-T-G
A-A          C-G-T-A                      T-T-G-T-A-C
```

```
T-T-A-A-C-A-T-G-C-A-T-G-T-T-A-A-C-T-T-A-A-C-A-T-G
A-A-T-T-G-T-A-C-G-T-A-C-A-A-T-T-G-A-A-T-T-G-T-A-C
```

(c) Sequenced DNA molecule

18–5   A simplified example of DNA sequencing. Identical samples of the DNA molecule to be sequenced are treated with different restriction enzymes that cut the DNA at different sites (a). One sample is treated with one enzyme, producing one set of fragments, and another sample is treated with another enzyme, producing a different set of fragments. The fragments of each set are then separated from one another, cloned, and analyzed, revealing the nucleotide sequence of each individual fragment (b). As you can see, the fragments produced by the two restriction enzymes overlap, making it possible to determine the nucleotide sequence of the molecule as a whole (c).

gether like a puzzle to reveal the entire sequence of a DNA molecule (Figure 18–5). At the present time, the sequencing of genes from sources ranging from the smallest viruses to human cells is going forward in hundreds of laboratories around the world.

Locating Specific DNA Segments

Before a DNA segment of interest—such as a particular gene or portion of a gene—can be cloned, sequenced, or manipulated, it must first be located and isolated. The chromosomes of even the simplest eukaryotic cells contain an enormous quantity of DNA. Locating a specific segment of that DNA is like trying to find the proverbial needle in a haystack. The "magnet" that scientists use is nucleic acid hybridization, one of the earliest methods for studying DNA and RNA molecules (see essay, page 259).

The foundation of hybridization is the base-pairing properties of the nucleic acids. As you know, when dissolved DNA molecules are heated gently, the hydrogen bonds holding the two strands together are broken and the strands separate. When the solution is cooled slowly, the hydrogen bonds re-form, reconstituting the double helix.

When DNA molecules from different sources are mixed together and heated, the strands separate and undergo random collisions. If two strands with nearly complementary sequences find each other as the solution is cooled, they will form a hybrid double helix. The extent to which segments from two samples reassociate and the speed with which they do so provide an estimate of the similarity between their nucleotide sequences.

18–6   A set of human metaphase chromosomes in which the locations of specific nucleotide sequences are revealed by nucleic acid hybridization. The probes were labeled with fluorescent dyes. Each color indicates the location of a different DNA sequence: red = the location on the X chromosome of the gene for the most common form of childhood muscular dystrophy; green = the region of chromosome 21 where the gene for Down syndrome occurs; orange = the beta-globin gene on chromosome 11; white = an oncogene on chromosome 8; violet = a gene coding for a protein found on the surface of white blood cells; yellow = an unidentified sequence on chromosome 5. Because this is a diploid set of replicated chromosomes, each probe hybridized in four locations.

This procedure has been adapted to locate specific nucleotide sequences in either DNA or RNA. A probe can be prepared by incorporating a radioactive isotope into a short segment of single-stranded DNA or RNA that is complementary to the nucleotide sequence of interest. Alternatively, the probe can be labeled with a fluorescent dye (Figure 18–6). Probes may be mRNA molecules, DNA fragments produced by restriction enzymes, complementary DNAs synthesized from RNA by reverse transcriptase, or nucleotide sequences synthesized in the laboratory. After a probe has been prepared, it can be used to seek out segments of DNA or RNA that contain the complementary nucleotide sequence. Labeled RNA molecules are routinely used to find corresponding DNA segments, and vice versa.

Recombinant DNA: Some Applications

The "tools" of recombinant DNA technology are now being used in many different types of research studies. These studies are rapidly increasing our knowledge of chromosome structure and organization, of the regulation of gene expression, and of the ways in which things can go wrong, leading not only to inherited disorders but also to many health problems that develop later in life. An enormous number and variety of practical applications are also flowing from the new technology. In the pages that follow, we are going to look at a few representative examples.

Bacterial Synthesis of Useful Proteins

Early in the course of recombinant DNA research, biologists realized that if genes coding for proteins of medical or agricultural importance could be transferred into bacteria and expressed, the bacteria could function as "factories," producing a virtually limitless source of the proteins.

The first human protein synthesized in a bacterial cell was the hormone somatostatin (Figure 18–7). It is a small protein (only 14 amino acids long) and can be detected in very small amounts. The investigators knew the sequence of amino acids in somatostatin and thus could determine a sequence of nucleotides in DNA that would code for it. Using this information, they synthesized an artificial gene, including an initiation codon. Copies of the gene were spliced into plasmids carrying genes for drug resistance. Also spliced into the plasmids was a portion of the *lac* operon (page 274), including its regulatory sequences. The scientists then supplied the plasmids to *E. coli* cells. Only a few of the bacterial cells

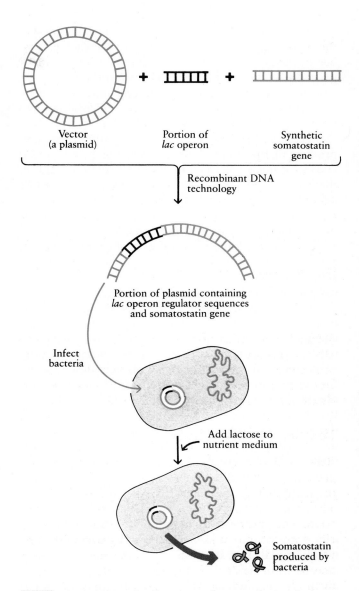

18–7   *How bacterial cells were made to synthesize the hormone somatostatin. The somatostatin gene and a portion of the* lac *operon were spliced into the plasmid. The addition of lactose to the nutrient medium turned the somatostatin gene on.*

became drug-resistant, indicating that they had taken up the plasmid. However, as these cells multiplied, the plasmids multiplied along with them. When the *lac* operon was turned on by the addition of lactose, proteins were synthesized. Chemical treatment of these proteins released pure somatostatin. When this somatostatin was tested in laboratory animals, it had the same biological activity as the natural hormone.

This experiment paved the way for the bacterial synthesis of medically useful proteins. An increasing number of medications are now produced using a similar procedure. They include human insulin (Figure 18–8), clotting factors for the treatment of hemophilia (page 231), and human growth hormone, which is used to treat some forms of dwarfism in children.

Vaccines against viral disease are another important product of this technology. As you know, all viruses consist of nucleic acid wrapped in a protein capsid. In the animal bloodstream, capsid proteins, recognized by cells of the immune system as foreign, evoke the formation of antibodies, molecules that play a crucial role in future immunity against the virus. Most vaccines are made from killed or altered forms of the virus particles. Vaccines produced from synthetic capsid proteins are safer, since without the viral nucleic acid, the vaccines cannot be contaminated by infectious particles.

The bacterial synthesis of other types of proteins is of increasing economic importance. For example, the enzyme rennin, extracted from calves' stomachs and used by the dairy industry for cheesemaking, can now be produced by recombinant DNA technology. Scientists have also succeeded in inducing bacteria to synthesize the enzyme cellulase, which is produced in nature by certain fungi. This enzyme converts cellulose, the plant polysaccharide that is indigestible by most organisms, into glucose, a food molecule of major importance in a hungry world.

Diagnosis of Genetic Disorders

Another application of recombinant DNA technology is in the diagnosis of human genetic disorders. Reliable tests have now been developed that, in conjunction with amniocentesis (page 202), make possible the prenatal diagnosis of a number of inherited disorders, including PKU, sickle-cell anemia, and the most common forms of hemophilia and muscular dystrophy. A test has also been developed that can identify those individuals at risk for Huntington's disease who are most likely to develop the disorder. All of these tests involve the use of restriction enzymes, nucleic acid probes, or both. Two examples illustrate the principles on which they are based.

Sickle-Cell Anemia

In order to develop a diagnostic test for sickle-cell anemia, radioactive copies of portions of the nucleotide sequence coding for the beta chain of hemoglobin were prepared. DNA from persons with normal hemoglobin and DNA from persons with sickle-cell anemia were

18–8   *Crystals of human insulin, produced by bacteria that have been modified by genetic engineering. Previously, human insulin was available only in very small quantities. Individuals with the type of diabetes that results from insulin deficiency were treated with insulin obtained from cows or pigs. These animal proteins were not as effective as human insulin, and they triggered allergic reactions in some individuals.*

18–9   *A test to detect the presence of the sickle-cell allele. Treating human DNA with the restriction enzyme HpaI produces three possible restriction fragments containing the gene for the beta chain of hemoglobin. Fragments containing the normal allele (light green) are either (a) 7,000 or (b) 7,600 nucleotides long. Fragments containing the sickle-cell allele (orange) are usually (c) 13,000 nucleotides in length. A recognition sequence for the restriction enzyme HpaI—present in the DNA of persons with the normal allele—is missing in the DNA of about 87 percent of the individuals carrying the sickle-cell allele.*

Some Ethical Dilemmas

Although the rapid advances in molecular genetics are contributing greatly to our understanding of human genetic disorders, the capacity to identify individuals at risk of developing certain disorders or of transmitting them to their children is creating perplexing ethical dilemmas both for individuals and for our society as a whole. The four questions below cover situations that have either arisen already or are likely to arise by the year 2000. They were posed by Dr. Eric Lander, a human geneticist at the Whitehead Institute for Biomedical Research and the Massachusetts Institute of Technology.

1. While the right to abortion is guaranteed by the U.S. Constitution, the choice is never easy. Imagine that you learned very early in a pregnancy that the child would certainly:
   (a) die within nine months from spinal muscular atrophy, a common fatal genetic disorder;
   (b) suffer throughout life from cystic fibrosis, a painful chronic disease, and die at about age 20;
   (c) suffer from Huntington's disease at age 40 and die at about 50;
   (d) suffer from Alzheimer's disease at about 60;
   (e) be congenitally deaf;
   (f) be a dwarf, but otherwise healthy; or,
   (g) be predisposed to severe manic depression, which could be partially controlled by medication.

   Would you choose to abort? (Assume that you are young enough that you may reasonably expect to have more children if you wish.) Regardless of your own choice, would you consider it *unethical* for another couple to do so? What principles underlie your choices?

2. Suppose you could learn with certainty whether you would suffer:
   (a) from Huntington's disease at about 40;
   (b) from Alzheimer's disease at about 60.

   Would you want to know?

3. Should insurance companies have the right to:
   (a) charge higher premiums to individuals with higher risk of inherited disease?
   (b) know the results of testing for genetic predispositions?
   (c) refuse to cover a child whom prenatal tests show will suffer a severe genetic disease?

4. Suppose that 10 percent of the work force is particularly prone to cancer induced by an industrial chemical. Should an employer:
   (a) have the obligation to make the work place safe for this minority?
   (b) have the right to require pre-employment genetic screening?
   (c) have the right to refuse employment to such workers?
   (d) have the right to require such workers to pay for supplementary insurance?

   Would your answer differ if the minority involved were 1 percent of the work force or 40 percent of the work force?

---

cleaved with the restriction enzyme *Hpa*I. The fragments produced were then exposed to the radioactive probe.

In persons with normal hemoglobin, the probe consistently hybridized with a DNA fragment that was either 7,000 or 7,600 nucleotides long. By contrast, in 87 percent of persons with sickle-cell anemia, the probe hybridized with a much longer fragment (Figure 18–9). The same result is seen with cells obtained by amniocentesis, thus providing a prenatal screening test for one of the most common serious genetic disorders.

Genetic markers of this type are known as **RFLPs,** pronounced "rif-lips" and translated as **restriction-fragment-length polymorphisms.** They occur because inherited variations—that is, mutations—in the nucleotide sequences in the DNA of different individuals lead to differences in the lengths of the fragments produced by restriction enzymes. Such differences occur when the recognition sequence for a restriction enzyme has been

eliminated or otherwise altered by a mutation. In the case of the sickle-cell allele, the mutation that causes production of the abnormal hemoglobin chain is usually accompanied by another mutation, some 5,000 nucleotides away, that eliminates a cleavage site for *Hpa*I.

The principle involved here is gene linkage, the same principle that made possible the genetic mapping studies described in Chapter 13. Those studies in *Drosophila* confirmed that two genes that are close together on the same chromosome tend to stay together. The updated version is that two nucleotide sequences close together on the same DNA molecule tend to stay together. In the *Drosophila* days, a marker was a gene that produced a detectable phenotypic change, such as a different wing shape or eye color. Today, however, a marker may be "anonymous"—detectable only in the pattern produced by restriction enzyme fragments. The closer the allele and its marker are, the more accurate the diagnosis.

Recombinant DNA techniques are now being used not only in the diagnosis of human genetic disorders but also in the identification of individuals and of their family relationships. Each individual's DNA is as distinctive as a fingerprint.

A relatively simple method for DNA fingerprinting has been devised. As we saw in Chapter 16, eukaryotic chromosomes contain many regions of simple-sequence DNA, identical short nucleotide sequences lined up in tandem and repeated thousands of times. The number of repeated units in such regions differs distinctively from individual to individual. These regions can be excised from the total DNA by the use of appropriate restriction enzymes, separated according to length by electrophoresis, and identified with radioactive probes. When the process is completed, the end result, visible on x-ray film, looks like the bar code on a supermarket package.

DNA fingerprinting is the most reliable method yet available for resolving cases of disputed parentage. It is also a precise method for identifying rapists and murderers, who do not always leave conventional fingerprints. However, semen is usually present on the person and clothing of a rape victim, and the blood of a murder victim is often on the clothing of the murderer. In some cases, samples of the murderer's blood are on the victim's body or clothing.

Often only very small samples of biological evidence are found at a crime scene—trace amounts of blood or semen, or even a single hair. Polymerase chain reaction (page 303) is used to synthesize multiple copies of the DNA present in these samples and thus make possible DNA fingerprinting.

The resulting DNA bar codes can be compared with the bar codes of blood samples from the victim and the suspects. When the tests are done correctly, the result is a reliable identification of the perpetrator, coupled with the exoneration of suspects who are innocent. To ensure the quality and admissibility in court of DNA fingerprinting evidence, uniform procedures and standards are now being developed for the laboratories that perform the tests.

The Russian royal family, photographed shortly before World War I. From left to right, Maria, Tsarina Alexandra, Tatiana, Olga, Tsar Nicholas II, Anastasia, with Alexis in front. Alexis, heir to the Romanov throne, was a victim of hemophilia (page 231).

DNA fingerprinting can also be used to identify the victims of violent death. In 1993, for example, it was used to identify the remains of Nicholas II and Alexandra, the last Tsar and Tsarina of Russia, and three of their five children. In 1918, the family, three servants, and their doctor were executed. According to the executioners' reports, the bodies of two of the children were burned, and the bodies of the others were placed in an abandoned salt mine in a forest in the Ural Mountains. Throughout the years of Communist rule, the location of the mine was a closely guarded state secret.

In 1991, however, the remains of nine bodies were recovered from the mine shaft. Forensic anthropologists in Russia tentatively identified five sets of bones as those of Nicholas, Alexandra, and their three oldest daughters, Olga, Tatiana, and Maria. The bones were then taken to the British government's Forensic Science Service Laboratory for testing. Samples of chromosomal and mitochondrial DNA from the bones were compared with samples from three living relatives—two cousins of the Tsar, who have not been publicly identified, and Prince Philip of Great Britain, whose maternal grandmother was the sister of the Tsarina. These tests confirmed that the remains are indeed those of the Tsar, the Tsarina, and three of their daughters.

In the years since the execution of the Russian royal family, rumors have circulated that the two youngest children, Anastasia and Alexis, survived. Over the years, a number of women claimed to be Anastasia. The most convincing case was made by one Anna Anderson, who was found in Berlin in a dazed state in 1920. Some believed her claim, others did not. At her death in 1984, however, she was buried in a Romanov family crypt in Europe. A lock of her hair is now being tested, and the mystery of Anna Anderson may soon be solved.

Huntington's Disease

In the early 1980s, James F. Gusella of the Massachusetts General Hospital, with a large team of co-workers, set out to develop a diagnostic test for Huntington's disease based on the RFLP technique. Huntington's presented a new problem because neither the defective gene nor its normal allele had been identified at that time. However, the workers had available a large library of cloned restriction fragments from human DNA, a family in the United States with a history of Huntington's disease, and, in Venezuela, another family of more than 3,000 individuals, also with a history of Huntington's (see page 206).

For three years, a team led by Nancy Wexler interviewed members of the Venezuelan family to obtain the information needed to construct a pedigree, performed neurological examinations of family members, and collected skin and blood samples from 570 people. As the samples were collected, they were flown to Gusella's laboratory, where the DNA was extracted, treated with restriction enzymes, and analyzed. Similar studies were conducted with the members of the family in the United States. Correlation of the restriction fragments with the family pedigrees and the results of the neurological examinations led ultimately to the identification of two RFLP patterns associated with the Huntington's gene. The presence or absence in an individual of the RFLP pattern present in a parent with Huntington's disease predicts with a high degree of certainty whether that person will or will not be similarly afflicted.

As it has turned out, however, many members of families with a history of Huntington's do not want to know their future in terms of the disease. With Huntington's, the most significant development may not be the possibility of earlier detection but rather the recent identification of the gene responsible for the disease.

Identification of Specific Human Genes

For some human genetic disorders, the underlying cause has been known for many years. In the case of sickle-cell anemia, for example, the mutation responsible for the abnormal hemoglobin molecules is known down to the last nucleotide. For the vast majority of disorders, however, the identities of the genes involved, as well as the nature and consequences of the mutations that lead to the disorders, have remained puzzles.

The search for any specific human gene is a complex and time-consuming endeavor, requiring the cooperative work of many researchers in different laboratories and the creative application of the methods of both classical and molecular genetics. One of the scientists in-

18–10   *Two members of the Venezuelan family afflicted by Huntington's disease: a man in his forties with Huntington's and his daughter, who has a 50 percent chance of developing the disorder.*

volved in the successful search for the cystic fibrosis gene, which was located in 1989, described the problem as the equivalent of looking for a broken faucet (the mutation that leads to cystic fibrosis) in a house (the gene) located somewhere in the United States (the 44 human autosomes). The problem is only slightly simpler when the gene being sought is known to be located on either the X or the Y chromosome. The enormous effort, however, is beginning to pay off.

In March of 1993, for example, some 10 years after the RFLP markers for Huntington's disease were first identified, the gene itself was located. Near the tip of chromosome 4, it consists of some 210,000 base pairs, coding for a protein that has not yet been identified. Like the gene responsible for fragile X syndrome (page 297), the Huntington's gene contains triplet repeats that vary in number. In individuals without Huntington's disease, the gene usually contains from 11 to 24 repeats of the sequence CAG. In individuals with Huntington's, the number ranges from 42 to 86 repeats. The fact that the repeats vary in number gave the gene a "now you see it, now you don't" quality that made its identification particularly difficult.

Crown Gall Disease: The Ti Plasmid

Agrobacterium tumefaciens is a common soil bacterium that infects plants, producing a lump, or tumor, of tissue known as a crown gall. Even if the bacteria are destroyed with antibiotics, the gall, once started, continues to grow. The cause of the gall, it turns out, is not the bacterium itself but rather a large plasmid it carries. A portion of this plasmid, which is known as Ti (for tumor-inducing), becomes integrated into the DNA of the host plant cell. Crown gall disease is the only naturally occurring genetic recombination yet recorded between prokaryotic and eukaryotic cells.

To try to understand how the Ti plasmid exerts its effects, recombinant DNA technology has been used to study its genes. Three of the genes direct the synthesis of plant hormones that act directly on the gall cells to promote their growth. Additional genes subvert the plant cell's metabolic machinery to produce unusual amino acids, which can be used by the gall cells but not by normal cells. Moreover, these substances appear to act as molecular aphrodisiacs, increasing bacterial conjugation and thus promoting the spread of the Ti plasmid to uninfected bacterial cells. In effect, Ti takes over and directs the activities of both of its hosts—the bacterial cells and the plant cells—to promote its own multiplication.

The Ti plasmid is of interest not only because of its remarkable powers but also because of its utility as a vector to ferry useful genes into crop plants. It is being used, for example, to transfer genes that confer resistance to major plant diseases. Other genes that are candidates for such transfer are those required for C_4 photosynthesis (page 171) and for nitrogen fixation (to be discussed in Chapter 41).

One of the technical problems in attempts at gene transfer, particularly in plants, is knowing whether a gene has actually been introduced into a new host cell and, if transferred, whether it is directing the synthesis of protein.

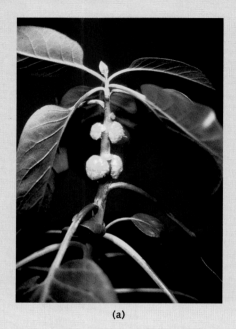

(a)

(a) *Crown galls growing on a tobacco plant. The galls, which are caused by a plasmid carried by* Agrobacterium tumefaciens, *are the large bulbous structures on the stem.*
(b) *Genes for the production of the enzyme luciferase were inserted into cells isolated*

(b)

from a tobacco plant, using Agrobacterium *as a vector. After the cells multiplied and developed into a new plant, the cells that had incorporated the luciferase gene into their DNA were luminescent in the presence of luciferin, ATP, and oxygen.*

An unusual "indicator" has now been devised, using the gene for an enzyme, known as luciferase, that is found in fireflies. The substrate of luciferase is a protein called luciferin. In the presence of oxygen, luciferin plus luciferase plus ATP produces bioluminescence, as seen in the flash of the firefly.

In the first demonstration of the utility of the luciferase gene, the gene was cloned in *E. coli* and then spliced into the chromosome of a plant virus, which provided a regulatory sequence for the gene. The altered viral chromosome was then inserted into Ti plasmids, the plasmids were transferred to *Agrobacte-*

rium, and the bacterial cells were incubated with tobacco leaf cells. The cells formed a mass of tissue from which, in a suitable growth medium, new plants were produced. The new plants were watered with a solution containing luciferin. You can see the result above: the plants shone!

Luciferase provides an extraordinary signal that gene transfer has taken place, and work is now in progress to combine its gene with other plant genes that are candidates for transfer. In the meantime, these experiments are a glowing example of the ingenuity of both molecular geneticists and the Ti plasmid.

With the gene located, work has begun to determine the exact nucleotide sequences of both the normal allele and the abnormal allele that is responsible for Huntington's disease. When these nucleotide sequences are determined, it should be possible to identify and characterize the protein coded by the gene. Ultimately, it is hoped, this work will not only reveal the underlying abnormality that causes brain cells to deteriorate some three to five decades after birth, but will also illuminate fundamental questions of the normal development and aging of the nervous system.

On the basis of previous experience, this hope seems well founded. For example, discovery of the gene responsible for the most common form of muscular dystrophy in children is shedding new light on muscle physiology. Similarly, discovery of the gene responsible for cystic fibrosis (page 212), which codes for a protein involved in the transport of chloride ions across the plasma membrane, is leading to new understandings of membrane transport processes.

Transfers of Genes between Eukaryotic Cells

As knowledge of the role of plasmids, viruses, and transposons in moving segments of DNA into and out of chromosomes increased, and as recombinant DNA techniques were refined, molecular biologists began to attempt to transfer functioning genes between eukaryotic cells. The hope was, of course, that it would someday be possible to cure human genetic disorders, substituting "good" genes for "bad" ones. As we saw at the beginning of the chapter, this hope appears on the verge of becoming reality. The shift from hope to reality has been made possible by a large body of work in other eukaryotic organisms.

The transfer of genes between eukaryotic cells is an extremely complex process. It requires, first, the preparation of a gene that will be taken up by the cells, become incorporated into a chromosome, and be expressed there—but this is only the beginning. The new gene must be established in a large number of cells of the appropriate type and be subject to the complicated regulatory controls of the normal gene.

The first stage proved easier than expected. Foreign genes will undergo recombination in eukaryotic cells growing in test tubes. In the first such experiment, SV40 virus was used as a vector to insert a rabbit gene for the beta-globin polypeptide into monkey cells. The recipient cells produced rabbit beta globin. The advantage of using viruses as vectors is that not only can viruses gain access to target cells but also they characteristically possess strong promoters (page 271), with the result that the gene is efficiently expressed. The gene coding for

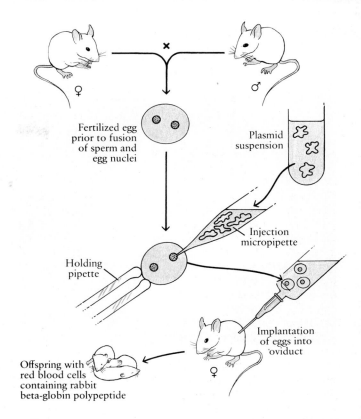

18–11 *The procedure by which the rabbit gene for the beta polypeptide chain of hemoglobin was inserted into mice. The gene was spliced into plasmids, which were then injected into the sperm nucleus within a fertilized egg before the egg and sperm nuclei fused to form a single diploid nucleus. A number of fertilized eggs were injected in this manner and then implanted in a female mouse, who subsequently gave birth. The rabbit beta-globin polypeptide was present in the red blood cells of the offspring. Nucleic acid hybridization revealed that the gene had been incorporated into their DNA.*

ADA (page 301), for example, was carried into white blood cells and, subsequently, into stem cells by a retrovirus.

Foreign genes have also been introduced into fertilized eggs and expressed in the organisms that developed from the eggs. For example, the gene for rabbit beta globin was transferred into the fertilized eggs of mice, using a plasmid as the vector (Figure 18–11). The red blood cells of the mice that developed from the eggs contained rabbit beta globin. The fact that the gene was expressed only in the red blood cells—and not in other tissues of the mouse—indicated that it had been incorporated in the "right" place and so had come under cellular control mechanisms. The gene was also passed, in a Mendelian distribution, to subsequent generations.

18–12   *The two female mice shown here are litter mates, approximately 24 weeks old. The fertilized egg from which the mouse on the left developed was injected with a gene consisting of the promoter and regulatory sequences of a mouse gene combined with the gene for human growth hormone. Following integration of the new gene into the DNA of this mouse, it was passed on to her offspring. Mice that express the new gene grow, on average, two to three times as fast as mice lacking the gene and, as adults, they are twice the normal size. (Those unable to see the utility of a giant mouse may be interested to learn that similar procedures have been successfully performed in fish.)*

Using the same technique, the gene for human growth hormone was combined with the regulatory portion of a mouse gene and injected into fertilized mouse eggs. The resultant "transgenic" mice—mice receiving the transferred human gene—grew to twice the normal size (Figure 18–12), indicating that the human gene was incorporated into the DNA of the mouse and was producing growth hormone. Because the DNA was injected into the egg cells, it was found in all cells, including the cells destined to become gametes. Thus, it could be passed on to the next generation. This type of procedure, using plasmids or transposons as vectors, has now been carried out with a number of cloned genes. From 10 to 30 percent of the eggs survive the manipulation, and the foreign gene functions in up to 40 percent of those eggs.

Such procedures are now making it possible to introduce desirable characteristics into crop plants (see essay) and domesticated animals, such as cows and sheep. With genetic engineering, the exact identity of the gene or genes being introduced is known, providing a significant degree of control over the characteristics that will be affected. By contrast, with the conventional breeding techniques that have been used for many years, enhancing particular desirable characteristics often leads to other, less desirable characteristics. Reduced fertility, for example, is frequently a consequence of the intensive inbreeding necessary to establish a particular characteristic in a population of plants or animals.

In economically important plants and animals, the reliable transmission of the introduced genes from generation to generation is an important goal. In the case of human gene therapy, however, there are serious moral and ethical questions about the possibility of changing the course of genetic history. Because of those questions—and also because of seemingly insurmountable technical difficulties—attempts at human gene therapy are, thus far, limited to therapies that affect only somatic cells and thus only the individual receiving the therapy.

Some observers, both inside and outside the scientific community, have expressed alarm at the newfound powers being made available by the triumphs of molecular biology. It is unlikely, however, that such doubts will slow the pace of discovery and experiment. The history of our species over the past 20,000 years suggests that curiosity, the search for knowledge, and the desire to improve and to heal are, for better or worse, a large part of what it means to be human. The best we can hope for is that we will use our knowledge and newly acquired skills with wisdom and compassion.

The Human Genome Project

Work is now underway to map and sequence the entire human genome—that is, all of the DNA in human cells. This endeavor has been compared to the Manhattan Project, which brought forth the atomic bomb, and the space program, which culminated with a human footprint on the moon. Although initially the subject of some controversy and dissent, which was also true of the projects to which it is compared, the human genome project is going forward.

There is action on two fronts. One multi-institutional enterprise is focused on mapping the human chromosomes. Data for chromosome maps can be generated in a number of ways. One method involves finding genes with different and detectable alleles and documenting their rate of recombination in breeding populations. For example, scientists at the University of Utah are tracing such marker genes through three generations of 60 Mormon families. (Mormon families are large and cohesive, and their genealogical records are meticulously maintained in the Church archives.) More than 500 genes have been mapped, and other investigators have located several hundred additional markers in other families.

The techniques of recombinant DNA are also being used in chromosome mapping. In addition to analyzing RFLP patterns to locate markers, scientists are identifying the sites at which specific restriction enzymes cut the DNA of particular chromosomes. In other studies, radioactive probes are being used to locate genes for which the protein products are known.

The types of data generated by these different techniques are, unfortunately, incompatible. Thus, it has been proposed that each time a marker is identified, by whatever technique, the researchers sequence a segment of 200 to 500 base pairs at one end of the marker, and then search both ends of the segment for short, unique sequences that identify the segment. In this way, the information generated by different techniques can be integrated into one cohesive map. The goal is to create a map of the human chromosomes in which markers are located at intervals of approximately 100,000 base pairs.

The second, and more controversial, effort involves the nucleotide sequencing of the entire genome. When it was first proposed years ago, many molecular biologists were against it, mainly because of its great expense. Much of the genome is "senseless," they pointed out, most funds available for research would have to be channeled into this one project, it would take decades to complete, and, perhaps most important of all, the creative talent of almost an entire generation of young scientists in the field would be wasted on this routine enterprise. However, the prevailing winds of opinion subsequently shifted. The crucial change came about as a result of the development of polymerase chain reaction and of methods for automatic sequencing of nucleotides, which have greatly decreased the estimates of time and money required for the project.

Also, as the sequencing of the genome comes closer to realization, there is mounting excitement about the new knowledge that may be generated. What is the mysterious role of the "senseless" DNA? How are genes regulated? (The answer to this question may hold the key to understanding and treating cancers and also be important in gene therapy.) What evolutionary relationships will be revealed among the human genes themselves? And among humans and other species?

A completion date for the project has been set: the year 2004. That is little more than a hundred years after the rediscovery of Mendel's work, which marks the beginning of genetics as a science, and only 51 years after the announcement by Watson and Crick that opened up the field of molecular biology. And you will see it happen.

Summary

Recombinant DNA technology includes methods for (1) obtaining DNA segments short enough to be analyzed and manipulated, (2) obtaining large quantities of identical DNA segments, (3) determining the exact order of nucleotides in a DNA segment, and (4) locating and identifying specific DNA segments of interest.

Short DNA segments can be obtained by transcribing mRNA into DNA with the enzyme reverse transcriptase, by chemical synthesis, or by cleaving DNA molecules with restriction enzymes. These are bacterial enzymes that cleave foreign DNA molecules. The DNA segments produced by restriction enzymes can be separated from one another on the basis of their size by electrophoresis.

Different restriction enzymes cut DNA at different specific nucleotide sequences. Instead of cutting the molecule straight across, some restriction enzymes leave "sticky" ends. Any DNA cleaved by such an enzyme can be joined readily to another DNA molecule cleaved by the same enzyme. The discovery of restriction enzymes made possible the development of recombinant DNA technology.

Two techniques, DNA cloning and polymerase chain reaction, are used to produce large quantities of identical DNA segments. In cloning, the segments to be copied are introduced into bacterial cells by means of plasmids or bacteriophages, which function as vectors. Once in the bacterial cell, the vector and the foreign DNA it carries are replicated, and the multiple copies can be harvested from the cells. Polymerase chain reaction is a much more rapid process but requires a greater knowledge of the segment that is to be copied.

The availability of multiple copies makes possible, in turn, the determination of the exact order of nucleotides in a DNA segment. By combining sequencing information for sets of short segments produced by different restriction enzymes, molecular biologists can determine the complete sequence of a long DNA segment (such as an entire gene).

Segments of DNA of interest for study and manipulation can be identified by nucleic acid hybridization, using single-stranded probes labeled with radioactive isotopes or fluorescent dyes. This technique is based on the capacity of a single strand of RNA or DNA to combine, or hybridize, with another strand with a complementary nucleotide sequence.

Recombinant DNA technology has many practical applications. For example, it is possible to incorporate specific genes into suitable vectors, introduce them into bacterial cells, and induce the bacterial cells to synthesize the proteins coded by the genes. Insulin, clotting factors, and growth hormone are among the medically important proteins now produced in this way.

Recombinant DNA technology is also providing new means for early diagnosis of genetic disorders. Among the most important tools for such diagnosis are RFLPs (restriction-fragment-length polymorphisms) and radioactive probes. RFLPs are the result of mutations that eliminate or alter the recognition sequence for a restriction enzyme. When such a mutation is associated with an allele causing a genetic disorder, it can provide a diagnostic marker for that allele. Labeled probes, which bind to either the normal or the mutant allele, can be used for detection and diagnosis when the nucleotide sequence of the allele is known or can be deduced.

A variety of techniques are being used to locate and identify the genes responsible for various human disorders. Recently, the genes responsible for Huntington's disease, cystic fibrosis, and the most common form of childhood muscular dystrophy have been identified.

Increasing success is being achieved in the transfer of genes between eukaryotic cells. Such transfers have been accomplished in both plants and animals. These experiments are leading to increased understanding of the regulatory factors governing the expression of eukaryotic genes and have made possible the first attempts at human gene therapy.

A massive effort is underway to map and sequence the entire human genome. This project, which is expected to be completed by the year 2004, may answer many puzzling questions. If past experience is any guide, it is also likely to raise many new questions.

Questions

1. What are restriction enzymes? What are their uses in recombinant DNA technology?

2. Describe the role of "sticky" ends in recombinant DNA technology. How are the "sticky" ends produced? What enzyme is required to complete the recombination?

3. What is electrophoresis? Why is it of such great value in recombinant DNA studies?

4. Suppose you treated a DNA molecule with a particular restriction enzyme and obtained five fragments, which you separated and cloned into multiple copies. Using the multiple copies, you then sequenced the five fragments. What would you do next to establish their order in the original molecule?

5. Suppose you wish to locate on the chromosome the gene coding for a small protein molecule. You know the amino acid sequence of the protein, and you have the technical skill to synthesize an artificial mRNA molecule with any nucleotide sequence you choose. How would you go about locating the gene? Suppose you wish to separate the gene from the rest of the chromosome. How would you proceed?

6. Why, in the somatostatin project, did the investigators link the synthetic somatostatin gene to regulatory elements of the *lac* operon?

7. *E. coli* cells used in recombinant DNA work are "disabled." That is, they lack the capacity to synthesize key components of their cell walls, for example, or to make a nitrogenous base, such as thymine. Consequently, they are able to survive only in an enriched laboratory medium. Why is such a precaution taken by molecular biologists?

8. In the first family in which the RFLP test for the sickle-cell allele was used, both parents were known to carry the allele, and they had previously had a baby with sickle-cell anemia. The woman was pregnant again, and the parents did not want to have another child with this extremely painful disease. DNA tests from each of the parents yielded both the short and the long fragments. Tests from their child with sickle-cell anemia yielded only long fragments. Tests of the fetal cells yielded both short and long fragments. What was the status of the unborn child with respect to the sickle-cell allele?

Suggestions for Further Reading

Alberts, Bruce, Dennis Bray, Julian Lewis, Martin Raff, Keith Roberts, and James D. Watson: *Molecular Biology of the Cell,* 3d ed., Garland Publishing, Inc., New York, 1994.

This outstanding cell biology text provides an excellent discussion of recombinant DNA technology.

Anderson, W. French: "Human Gene Therapy," *Science,* vol. 256, pages 808–813, 1992.

Anderson, W. R., and E. G. Diachmakos: "Genetic Engineering in Mammalian Cells," *Scientific American,* July 1981, pages 106–121.

Ayala, Francisco J., and Bert Black: "Science and the Courts," *American Scientist,* vol. 31, pages 230–239, 1993.

Chilton, Mary-Dell: "A Vector for Introducing New Genes into Plants," *Scientific American,* June 1983, pages 50–59.

DeLisi, Charles: "The Human Genome Project," *American Scientist,* vol. 76, pages 488–493, 1988.

Erickson, Deborah: "Hacking the Genome," *Scientific American,* April 1992, pages 128–137.

Franklin-Barbajosa, Cassandra: "DNA Profiling: The New Science of Identity," *National Geographic,* May 1992, pages 112–123.

Gasser, Charles S., and Robert T. Fraley: "Transgenic Crops," *Scientific American,* June 1992, pages 62–69.

Gilbert, Walter, and Lydia Villa-Kamaroff: "Useful Proteins from Recombinant Bacteria," *Scientific American,* April 1980, pages 74–94.

Lerner, Richard A.: "Synthetic Vaccines," *Scientific American,* February 1983, pages 66–74.

Mange, Arthur P., and Elaine Johansen Mange: *Genetics: Human Aspects,* 2d ed., Sinauer Associates, Inc., Sunderland, Mass., 1990.

A thorough account of all aspects of human genetics, including applications of recombinant DNA technology to the understanding and treatment of human genetic disorders.

Morell, Virginia: "Huntington's Gene Finally Found," *Science,* vol. 260, pages 28–30, 1993.

Moses, Phyllis B., and Nam-Hai Chua: "Light Switches for Plant Genes," *Scientific American,* April 1988, pages 88–93.

Mullis, Kary B.: "The Unusual Origin of the Polymerase Chain Reaction," *Scientific American,* April 1990, pages 56–65.

Nossal, G. J. V.: *Reshaping Life,* 2d ed., Cambridge University Press, New York, 1990.

A short, excellent, easily read introduction to recombinant DNA techniques and the biotechnology industry.

Scientific American: *Recombinant DNA,* W. H. Freeman and Company, New York, 1978.*

The articles in this collection from Scientific American *include discussions of the critical experiments that paved the way for modern molecular genetics.*

Thompson, Larry: "At Age 2, Gene Therapy Enters a Growth Phase," *Science,* vol. 258, pages 744–746, 1992.

Torrey, John G.: "The Development of Plant Biotechnology," *American Scientist,* vol. 73, pages 354–363, 1985.

Verma, Inder M.: "Gene Therapy," *Scientific American,* November 1990, pages 68–84.

Walters, LeRoy: "The Ethics of Human Gene Therapy," *Nature,* vol. 320, pages 225–227, 1986.

Watson, James D., John Tooze, and David T. Kurtz: *Recombinant DNA,* 2d ed., W. H. Freeman and Company, New York, 1992.

A short course in genetics, organized around the central theme of recombinant DNA. Clearly written and handsomely illustrated, it is accessible to anyone who enjoys reading Scientific American.

White, Ray, and Jean-Marc Lalouel: "Chromosome Mapping with DNA Markers," *Scientific American,* February 1988, pages 40–48.

* Available in paperback.

Evolution

19

The Genetic Basis of Evolution

In Charles Darwin's time, many naturalists were abandoning the concept of special creation and moving toward some theory of evolution. An intellectual revolution was in the making—one of the greatest in human history—but there was no strong framework to support it. None of the current ideas, of which there were many, were adequate until Darwin presented his theory in *The Origin of Species.*

What Darwin (and his younger colleague, Alfred Russel Wallace) grasped was that evolution is a two-part process. First, there are differences—variations—among individuals in any population, and some of these variations can be passed from generation to generation. Second, these variations, which occur by chance, may have consequences that affect survival and reproduction. It is this second step that Darwin called "natural selection."

Darwin received primary credit for the theory because of his carefully documented, closely argued presentation, drawing on his own wide-ranging experience and his voluminous reading and correspondence. One of his many examples was camouflage in insects. As is often the case, Darwin says it best:

"Insects often resemble for the sake of protection various objects, such as green or decayed leaves, dead twigs, bits of lichen. . . . The resemblance is often wonderfully close, and is not confined to colour, but extends to form, and even to the manner in which the insects hold themselves. . . . Assuming that an insect originally happened to resemble in some degree a dead twig or a decayed leaf, and that it varied slightly in many ways, then all the variations which rendered the insect at all more like any such object, and thus favoured its escape, would be preserved, whilst other variations . . . if they rendered the insect at all less like the imitated object, they would be eliminated."

These are walkingstick insects, photographed in the Ozark Mountains of Missouri. The male is smaller than the female. Note how carefully she is lined up with the twig on which she rests. He is slightly out of line, risking his life by mating.

A vast body of evidence indicates that Earth has had a long history and that all living organisms, including ourselves, arose in the course of that history from earlier, more primitive forms. This means that all species are descended from other species; in other words, all living things share common ancestors in the distant past. The thread that links together the enormous diversity of species that live on Earth today—and that have lived in the past—is **evolution,** a process of change through time.

In this section of our text (Chapters 19–22), we shall look at the mechanisms by which evolution occurs. In the next section (Chapters 23–27), we shall turn our attention to the diverse groups of organisms that have arisen in the course of evolutionary history.

Darwin's Theory

As we saw in the Introduction,* Charles Darwin was not the first to propose that the diversity of organisms is the result of historical processes. Credit for the theory of evolution is rightly his, however, for two reasons. First, his "long argument"—as he himself described *The Origin of Species*—left little doubt that evolution had actually occurred and so marked a turning point in the science of biology. The second reason, which is closely related to the first, is that Darwin correctly perceived the general mechanism by which evolution occurs.

Although it is now more than 130 years since the publication of *The Origin of Species*, Darwin's original concept of how evolution works still provides the basic framework for our understanding of the process. This concept rested on five premises:

1. Organisms beget like organisms—in other words, there is stability in the process of reproduction.

* If you have not read pages 1–12 of the Introduction, we suggest that you do so now.

2. In most species, the number of individuals in each generation that survive and reproduce is small compared with the number initially produced.

3. In any given population, there are variations among individual organisms, and some of these variations can be inherited.

4. Which individuals will survive and reproduce and which will not are determined to a significant degree by the interaction between these variations and the environment. Some variations enable individuals to survive longer and produce more offspring than other individuals. Darwin called these variations "favorable" and argued that inherited favorable variations tend to become more and more common from one generation to the next. This is the process that Darwin called **natural selection.**

5. Given sufficient time, natural selection can lead to the accumulation of changes that differentiate groups of organisms from one another.

As originally formulated by Darwin, however, the theory of evolution had a major weakness: the absence of any valid mechanism to explain heredity. Although Mendel was at work on his experiments with pea plants at the time Darwin was writing *The Origin of Species,* his paper was not delivered until 1865 and did not enter the mainstream of biological thinking until early in the twentieth century. The subsequent development of genetics made it possible to answer three questions that Darwin was never able to resolve: (1) how inherited characteristics are transmitted from one generation to the next; (2) why inherited characteristics are not "blended out" but can disappear and then reappear in later generations (like whiteness in pea flowers); and (3) how the variations arise on which natural selection acts.

The Concept of the Gene Pool

A new branch of biology, **population genetics,** emerged from the synthesis of Mendelian principles with Darwinian evolution. A **population** can be defined as an interbreeding group of organisms of the same species, living in the same place at the same time. For instance, all of the fish of one particular species in a pond constitute a population, as do all the fruit flies in one bottle.

A population is unified and defined by its **gene pool,** which is simply the sum total of all the alleles of all the genes of all the individuals in the population. From the viewpoint of the population geneticist, each individual organism is only a temporary vessel, holding a small

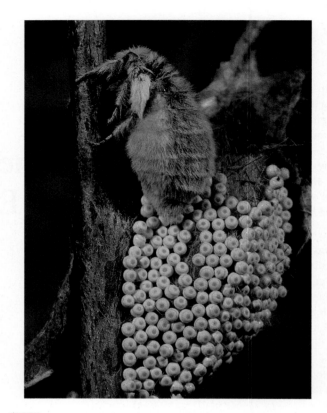

19–1 *A female rusty tussock moth, laying eggs on the empty cocoon from which she emerged after her transformation from caterpillar to adult moth. Despite the large number of eggs laid by each female who attracts a mate, there is, over a span of many generations, no increase in the number of rusty tussock moths.*

sampling of the gene pool for a moment in time. The subjects of interest to the population geneticist are gene pools, changes in their composition over time, and the forces causing these changes.

In natural populations, some alleles increase in frequency from generation to generation, and others decrease. If an individual has a favorable combination of alleles in its genotype, it is more likely to survive and reproduce. As a consequence, its alleles are likely to be present in an increased proportion in the next generation. Conversely, if the combination of alleles is not favorable, the individual is less likely to survive and reproduce. Representation of its alleles in the next generation will be reduced or perhaps eliminated. Evolution is the result of such accumulated changes in the gene pool over time.

In the context of population genetics, the **fitness** of an individual does not mean physical well-being or optimal adaptation to the environment. The only criterion—the sole measure—of an individual's fitness is the relative number of surviving offspring, that is, the extent to which the alleles in an individual's genotype are present in succeeding generations.

Survival of the Fittest

The phrase "survival of the fittest" is often used in describing the Darwinian theory. At various times in the twentieth century, the doctrine of survival of the fittest in natural populations has been used to defend gross social inequalities and ruthless competitive tactics in industry on the grounds that they are merely in accord with the "laws of nature." This philosophy is sometimes known as social Darwinism.

In fact, however, very little in the process of evolutionary change fits the concept of "nature red in tooth and claw." One plant with flowers a little brighter than those of its neighbors and so better able to catch the attention of a passing hummingbird is a more pertinent model of the struggle for survival. Fitness, as measured by population geneticists, is determined solely by the relative number of descendants of an individual in a future population.

Natural selection in operation. A female purple-throated hummingbird (Lampornis calolaema) *is making her choice. A resident of the* cloud forest of Costa Rica, this hummingbird is also known as the "mountain gem."

The Extent of Variation

Like begets like, we know now, because of the remarkable precision with which the DNA is copied and transmitted from each cell to its daughter cells. The DNA in the cells of any individual is, except for occasional mutations, a true replica of the DNA that individual received from its father and mother.

This fidelity of duplication is, of course, essential to the survival of the individual organisms of which a population is composed. However, if evolution is to occur, there must be variations among individuals. Such variations make it possible for populations to change as conditions change. They are the raw material on which evolutionary forces act.

As we have noted previously, Darwin was the first to recognize the importance of widespread, inherited variations in the process of evolution. One focus of research in population genetics has been to determine the extent of genetic variability and how it is preserved and fostered in gene pools. The extent of variation has been revealed in a number of different ways.

Artificial Selection

First-time readers of *The Origin of Species* are sometimes surprised to find themselves, in the very first chapter, caught up in a treatise on pigeon breeding. By selecting birds with particular characteristics—such as a larger beak or a tail with more feathers—for breeding, pigeon fanciers had been able, over the years, to produce a number of exotic breeds (see Figure I–9, page 10). All developed from the same wild species and still able to interbreed with one another, the new breeds differed widely in appearance, more widely in fact than many animals of different species.

Darwin coined the term **artificial selection** for this process of human choice in determining which individuals would be represented in the next generation—and which would not. He saw it as a direct analogy to the process of natural selection. As he was well aware, the selective breeding of pigeons is not an isolated example. Similar experiments in evolution have been carried out by breeders of plants and animals for centuries, producing some of our most familiar domestic varieties

Kale   Brussels sprouts   Broccoli   Kohlrabi   Cabbage   Cauliflower

19–2 *Six vegetables produced from a single species of plant* (Brassica oleracea), *a member of the mustard family. They are the result of selection for leaves (kale), lateral buds (brussels sprouts), flowers and stem (broccoli), stem (kohlrabi), enlarged terminal buds (cabbage), and flower clusters (cauliflower). Kale most resembles the ancestral wild plant. Artificial selection, as practiced by plant and animal breeders, gave Darwin the clue to the concept of natural selection.*

(Figure 19–2). These experiments have shown that there is a large amount of variability hidden in the gene pool and that this latent variability can be expressed when selection pressures are applied.

Breeding Experiments in the Laboratory

Breeding experiments that demonstrate the extent of latent variability have also been carried out in the laboratory. In studies with *Drosophila melanogaster,* for example, an easily observable hereditary trait, the number of bristles on the ventral surface of the fourth and fifth abdominal segments (Figure 19–3), was chosen for selection.

In the starting stock, the average number of bristles was 36. From this initial population, the investigators established two separate groups of fruit flies. In one group, selection was for a decrease in the number of bristles; in the other group, it was for an increase in the number of bristles. In every generation, those individ-

19–3 *Scanning electron micrographs of* (a) *a fruit fly* (Drosophila melanogaster) *and* (b) *the ventral surface of its posterior abdominal segments. The results of selection experiments to increase and decrease the number of bristles on the ventral surface of the fourth and fifth abdominal segments are shown in Figure 19–4.*

(a)   0.5 mm

(b)   0.1 mm

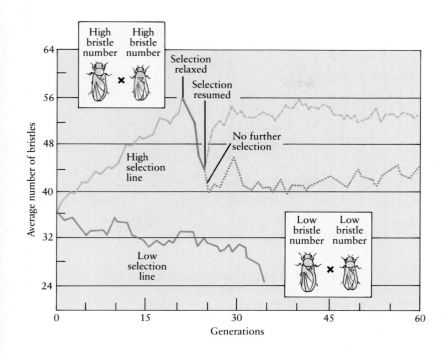

19–4 *The results of experiments with* Drosophila melanogaster, *demonstrating the extent of latent variability in a population. From a single parental stock, one group was selected for an increase in the number of bristles on the ventral surface (high selection line) and one for a decrease in the bristle number (low selection line).*

As you can see, the high selection line rapidly reached a peak at which the average number of bristles was 56, but then the stock began to become sterile. Selection was abandoned at generation 21 (dark blue line) and begun again at generation 24 (dashed line). This time, the previous high bristle number was regained, and there was no apparent loss in reproductive capacity. Note that after generation 24 the stock interbreeding without selection was also continued, as indicated by the dotted line. After 60 generations, members of the freely breeding group from the high selection line had an average of 45 bristles. The low selection line died out owing to sterility.

uals in the first group with the lowest number of bristles were selected and bred with each other. In the other group, the individuals with the highest number of bristles were selected and bred with each other.

Selection for low bristle number resulted in a drop over 30 generations from an average of 36 to an average of about 30 bristles. In the high-bristle-number line, progress was at first rapid and steady. In 21 generations, bristle number rose steadily from 36 to an average of about 56 (Figure 19–4). No new genetic material had been introduced. The potential for a wide range of bristle numbers had been already present—but hidden—within the single starting population. Subsequent experiments with *Drosophila* and other organisms have demonstrated a comparable range of natural variability in many different traits.

There is a second part to the bristle-number story. The low-bristle-number line soon died out because it became sterile. Although the flies mated, no offspring were produced. Presumably, changes in factors affecting fertility had also taken place during selection. When sterility started to become severe in the high-bristle line, the remaining flies of this line were permitted to interbreed freely without further selection. The average number of bristles fell sharply, and in five generations dropped from 56 to 40. Thereafter, as this line continued to breed without selection, the bristle number fluctuated up and down, usually between 40 and 45, which still was higher than the original 36. At generation 24, selection for high bristle number was begun again for a portion of this line. The previous high bristle number of 56 was regained, and this time there was no loss in

reproductive capacity. Apparently, the genotype had become reorganized in such a way that alleles producing a higher bristle number were present in more favorable combinations with alleles affecting fertility.

Mapping studies have shown that bristle number is controlled by a large number of genes, at least one on every chromosome and sometimes several at different sites on the same chromosome. Although we do not know how important bristle number is to the survival of the animal, we do know that selection for this trait in some way disrupted the entire genotype. Livestock breeders are well aware of this consequence of artificial selection. Loss of fertility is a major problem in virtually all circumstances in which animals have been purposely inbred for particular characteristics.

Quantifying Variability

Analysis at the molecular level provides another method for assessing variability. As you know, the amino acid sequences of proteins reflect the nucleotide sequences of the genes coding for them. In a classic study, J. L. Hubby and R. C. Lewontin ground up fruit flies from a natural population and extracted proteins from them. From these proteins, they were able to isolate 18 functionally different enzymes. Then they analyzed each of the 18 enzymes separately, to determine if its molecules existed in different structural forms in different flies or if the structure of the enzyme was uniform throughout the population. The method they used was electrophoresis—the same method used by Pauling to separate the variants of hemoglobin (see page 257).

Of the 18 enzymes studied in this way, nine were found to be composed of protein molecules that were indistinguishable by electrophoresis. In other words, the gene coding for each of these enzymes was the same throughout the entire population of fruit flies studied. However, each of the other nine enzymes was found to exist in two or more structurally different forms. Thus, without any direct analysis of the genes themselves, the investigators were able to conclude that among the fruit flies studied there were two or more alleles of the gene responsible for each of these nine enzymes. One enzyme had six slightly different structural forms (Figure 19–5). This meant that at least six alleles of the gene coding for that enzyme were present in the gene pool.

Each fruit-fly population examined was heterozygous for almost half of the genes tested. Each individual, it was estimated, was probably heterozygous for about 12 percent of its genes. Similar studies on humans, using accessible tissues such as blood or placenta, indicate that at least 25 percent of the genes in any given population are represented by two or more alleles, and individuals are heterozygous for at least 7 percent of their genes, on the average.

Now that methods for sequencing nucleotides in DNA have been developed, it is possible to make direct comparisons of the genetic material. Since all changes in nucleotides do not result in changes in amino acid sequences, and all changes in amino acid sequences are not detectable by electrophoresis, DNA sequencing is expected to reveal even more genetic variation.

Explaining the Extent of Variation

The findings of Hubby and Lewontin and others after them have, like most important scientific discoveries, raised major new questions. Many geneticists had previously thought that the individuals of a population should be close to genetic uniformity, as a result of a long history of selection for "optimal" genes. Yet, as these studies revealed, natural populations are far from uniform.

One school of geneticists, the selectionists, claim that even such small variations as those in enzyme structure are maintained by various forms of natural selection that favor some genotypes at some times and in some areas, and others at other times or in other localities. In the selectionist's view, all variation directly or indirectly affects fitness. An opposing school, the neutralists, claim that the observed variations in the protein molecules are so slight that they do not make any difference in the function of the organism and so are not acted upon by natural selection. Neutral alleles, according to this argument, accumulate as a result of random processes, including mutation. Although this difference of

19–5   *The method used by Hubby and Lewontin to analyze* Drosophila *enzymes was electrophoresis. In this process, as you know, the sample is dissolved, placed at the edge of a sheet of gel, and exposed to a weak electric field. The rate at which the molecules move in the electric field is determined by their size and electric charge. As a result, proteins with even very slight structural differences can be separated.*

This diagram shows the electrophoretic patterns of six different forms of one enzyme. The material in each column was obtained from flies homozygous for one of the six different alleles coding for the enzyme.

opinion is not yet resolved, we do have a growing understanding of the variety of processes that can act to maintain, increase, or decrease the variability in the gene pool of a population.

A Steady State: The Hardy-Weinberg Equilibrium

In the early 1900s, biologists raised an important question about the maintenance of variability in populations. How, they asked, can both dominant and recessive alleles remain in populations? Why don't dominants simply drive out recessives? For example, given that brachydactylism (Figure 19–6) is caused by a dominant allele, why don't most or even all people have short, fat fingers? This question was answered in 1908 by G. H. Hardy, an English mathematician, and G. Weinberg, a German physician.

Working independently, Hardy and Weinberg showed that the genetic recombination that occurs at each generation in diploid organisms does not *by itself*

19–6 *A dominant allele is responsible for the characteristic known as brachydactylism (short fingers). In the brachydactylous hands shown here, the first bones of the fingers are of normal length, but the second and third bones are abnormally short. If brachydactylism is caused by a dominant allele, why is it such a rare characteristic?*

change the overall composition of the gene pool. To demonstrate this, they examined the behavior of alleles in an idealized population in which five conditions hold:

1. No mutations occur.

2. There is no net movement of individuals—with their genes—into the population (immigration) or out of it (emigration).

3. The population is large enough that the laws of probability apply; that is, it is highly unlikely that chance alone can alter the **frequencies,** or relative proportions, of alleles.

4. Mating is random.

5. All alleles are equally viable. In other words, there is no difference in reproductive success. The offspring of all possible matings are equally likely to survive to reproduce in the next generation.

Consider a single gene, which has only two alleles, *A* and *a*. Hardy and Weinberg demonstrated mathematically* that if the five conditions listed above are met, the frequencies of alleles *A* and *a* in the population will not change from generation to generation. Moreover, the frequencies of the three possible combinations of these alleles—the genotypes *AA, Aa,* and *aa*—will not change from generation to generation. The gene pool will be in a steady state—an equilibrium—with respect to these alleles.

This equilibrium is expressed by the following equation:

$$p^2 + 2pq + q^2 = 1$$

* This mathematical demonstration and its application in specific examples are given in Appendix A.

In this equation, the letter *p* designates the frequency of one allele (for example, *A*), and the letter *q* designates the frequency of the other allele (*a*). The sum of *p* plus *q* must always equal 1 (that is, 100 percent of the alleles of that particular gene in the gene pool). The expression p^2 designates the frequency of individuals homozygous for one allele (*AA*), q^2 the frequency of individuals homozygous for the other allele (*aa*), and $2pq$ the frequency of heterozygotes (*Aa*).

Although normally any one diploid individual has no more than two alleles of the same gene, there may, of course, be more than two alleles for a given gene in the gene pool. For instance, in the case of the gene for coat color in rabbits, there are at least four alleles. Different diploid combinations of these alleles give rise to coat colors ranging from dark gray through medium and light gray to white, as well as a variety in which most of the coat is white but the feet, ears, and muzzle are black (see Figure 13–3, page 223). The Hardy-Weinberg equilibrium applies equally well to situations such as this, in which there are multiple alleles of the same gene, although the equation representing the equilibrium is more complex.

The Significance of the Hardy-Weinberg Equilibrium

The Hardy-Weinberg equilibrium and its mathematical formulation have proved as valuable a foundation for population genetics as Mendel's principles have been for classical genetics. At first glance, this seems hard to understand, since the five conditions specified for the equilibrium are seldom likely to be met in a natural population. An analogy from physics may be useful. Newton's first law says that a body remains at rest or maintains a constant velocity when not acted upon by external force. In the real world, bodies are always acted upon by external forces, but this first law is an essential premise for examining the nature of such forces. It provides a standard against which to measure.

Similarly, the Hardy-Weinberg equation provides a standard against which we can measure the changes in allele frequencies that are always occurring in natural populations. Without the Hardy-Weinberg equation, we would not be able to detect change, determine its magnitude and direction, or uncover the forces responsible for it. If, however, we can identify and count the individuals in a population that are homozygous for a particular allele of interest, we can then calculate the frequency of that allele—and thus all the other terms of the Hardy-Weinberg equation. If we do this over a period of generations, we can chart precisely the changes that are taking place in the gene pool—and then look for the causes.

19–7 *An example of the sometimes dramatic effects of mutation. The ewe in the middle is an Ancon, an unusually short-legged strain of sheep. The first Ancon on record was born in the late nineteenth century into the flock of a New England farmer. By inbreeding (the characteristic is transmitted as a recessive), it was possible to produce a strain of animals with legs too short to jump the low stone walls that traditionally enclosed New England sheep pastures. A similar strain was produced in northern Europe following an independent mutation there. At one time, it was proposed that evolution takes place in sudden, large jumps such as this—an idea sometimes referred to as the "hopeful monster" concept. One reason this idea has been abandoned is that nearly all mutations producing dramatic changes in the phenotype are harmful, as this one would be in a wild population.*

The Agents of Change

According to modern evolutionary theory, natural selection is the major force in changing allele frequencies. Because natural selection is of such great importance, we shall devote all of the next chapter to it. Here, let us look at some of the other agents that can change the frequencies of alleles in a population. There are four: mutations, gene flow, genetic drift, and nonrandom mating.

Mutations

Mutations, from the point of view of population genetics, are inheritable changes in the genotype. A mutation may involve the deletion, transposition, or duplication of a portion of a DNA molecule, or the substitution of one or more nucleotides in the molecule. Mutations can affect not only structural genes, such as the genes coding for the polypeptide chains of the hemoglobin molecule, but also regulatory genes, such as those responsible for turning on and off various processes in embryonic development.

As we noted in Chapter 15, mutations can be caused by a number of agents, including x-rays, ultraviolet rays, radioactive substances, and a variety of other chemicals. They can also be caused by the insertion of viral DNA into a chromosome or by the movement of transposons, as we saw in Chapter 17.

Most mutations occur "spontaneously"—meaning simply that we do not know the factors that trigger them. Mutations are generally said to occur at random, or by chance. This does not mean that mutations occur without cause but rather that the events triggering them are independent of their subsequent effects. Although the rate of mutation can be influenced by environmental factors, the specific mutations produced are independent of the environment—and independent of their potential for subsequent benefit or harm to the organism and its offspring.

The rate of spontaneous mutation is generally low. For mutations detectable in the phenotype, it varies from 1 in 1,000 to 1 in 1,000,000 gametes per generation, depending on the allele involved. It is estimated that each new human individual, with approximately 100,000 genes (pairs of alleles), carries two new mutations. Thus, although the incidence of mutation in any given gene or any given individual is low, the number of new mutations per population generation is very high. Mutations are regarded as the raw material for evolutionary change, since they provide the variation on which other evolutionary forces act. However, because mutation rates are so low, mutations seldom, if ever, determine the direction of evolutionary change.

Gene Flow

Gene flow is the movement of alleles into or out of a population. It can occur as a result of the immigration or emigration of individuals of reproductive age. In the case of plants and many aquatic invertebrates, it can also occur through the movement of gametes (for example, in the form of pollen) between populations.

Gene flow can introduce new alleles into a population, or it can change existing allele frequencies. Its overall effect is to decrease the difference between populations. Natural selection, by contrast, is more likely to increase differences, producing populations more suited for different local conditions. Thus, gene flow often counteracts natural selection. As we shall see in Chapter 21, geographic barriers that prevent gene flow are very important in the formation of new species.

Genetic Drift

As we stated previously, the Hardy-Weinberg equilibrium holds true only if the population is large. This qualification is necessary because the equilibrium depends on the laws of probability. Consider, for example, an allele, say *a*, that has a frequency of 1 percent. In a population of 1 million individuals, 20,000 *a* alleles would be present in the gene pool. (Remember that each diploid individual carries two alleles for any given gene. In the gene pool of this population there are 2 million alleles for this particular gene, of which 1 percent, or 20,000, are allele *a*). If a few individuals in this population were destroyed by chance before leaving offspring, the effect on the frequency of allele *a* would be negligible.

In a population of 50 individuals, however, the situation would be quite different. In this population, it is likely that only one copy of allele *a* would be present. If the lone individual carrying this allele failed to reproduce or were destroyed by chance before leaving offspring, allele *a* would be completely lost. Similarly, if 10 of the 49 individuals without allele *a* were lost, the frequency of *a* would jump from 1 in 100 to 1 in 80.

This phenomenon, a change in the gene pool that takes place as a result of chance, is **genetic drift.** Population geneticists and other evolutionary biologists generally agree that genetic drift plays a role in determining the evolutionary course of small populations. Its relative importance, however, as compared to that of natural selection, is a matter of debate. There are at least two situations in which it has been shown to be important.

The Founder Effect

A small population that becomes separated from a larger one may or may not be genetically representative of the larger population from which it was derived (Figure 19–8). Some rare alleles may be overrepresented or, conversely, may be completely absent in the small population. As a consequence, when and if the small population increases in size, it will continue to have a different genetic composition—a different gene pool—from that of the parent group. This phenomenon, a type of genetic drift, is known as the **founder effect.**

An example of the founder effect is provided by the Old Order Amish of Lancaster, Pennsylvania (Figure 19–9). Among these people, there is an unprecedented frequency of a recessive allele that, in the homozygous state, causes a combination of dwarfism and polydactylism (extra fingers). Since the group was founded in the early 1770s, some 61 cases of this rare congenital deformity have occurred, about as many as have been

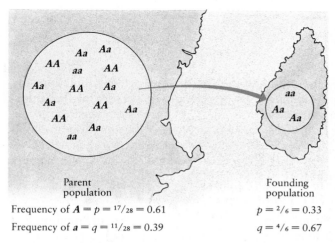

Parent population

Founding population

Frequency of $A = p = {}^{17}/_{28} = 0.61$ $p = {}^{2}/_{6} = 0.33$

Frequency of $a = q = {}^{11}/_{28} = 0.39$ $q = {}^{4}/_{6} = 0.67$

19–8 *When a small subset of a population founds a new colony (for example, on a previously uninhabited island), the allele frequencies within the founding group may be different from those within the parent population. Thus the gene pool of the new population will have a different composition than the gene pool of the parent population. This phenomenon is known as the founder effect.*

19–9 *Among the Old Order Amish, a group founded by only a few couples some 200 years ago, there is an unusually high frequency of a rare recessive allele. In its homozygous state, this allele results in extra fingers and dwarfism. This Amish child is a six-fingered dwarf.*

reported in all the rest of the world's population. Approximately 13 percent of the persons in the group, which numbers some 17,000, are estimated to carry this rare allele. The entire colony, which has kept itself virtually isolated from the rest of the world, is descended from only a few dozen individuals. By chance, one of those was a carrier of the allele.

Population Bottleneck

Population bottleneck is another type of situation that can lead to genetic drift. It occurs when a population is drastically reduced in numbers by an event that may have little or nothing to do with the usual forces of natural selection. For example, from the 1820s to the 1880s, the northern elephant seal was hunted so heavily along the coast of California and Baja California that it was rendered almost extinct, with perhaps as few as 20 individuals remaining. Since 1884, when the seal was placed under the protection of the United States and Mexican governments, the population has increased to more than 30,000, all presumably descendants of this small group. Studies of blood samples taken from 124 seal pups showed them to be homozygous for some 21 gene loci, indicating (by comparison with other mammalian groups) a drastic loss of genetic variability.

A population bottleneck is likely not only to eliminate some alleles entirely but also to cause others to become overrepresented in the gene pool. For example, the high rate of Tay-Sachs disease (page 212) among Jews of Eastern and Central European extraction has been attributed to a population bottleneck experienced by these people in the Middle Ages.

Nonrandom Mating

Disruption of the Hardy-Weinberg equilibrium can also be produced by nonrandom mating. A form of nonrandom mating particularly important in plants is self-pollination (as in the pea plants Mendel studied). In animals, nonrandom mating is often behavioral. For example, snow geese may be either white or blue (Figure 19–11). This is a type of variation known as **polymorphism,** in which two or more phenotypically distinct forms coexist in a population. White snow geese tend to mate preferentially with other white geese and blue with blue. Thus, assuming only two alleles are involved, there will be a decrease in the frequency of heterozygotes (represented as $2pq$ in the Hardy-Weinberg equation), with concomitant increases in the frequencies of the two homozygotes (p^2 and q^2).

Note that nonrandom mating can cause changes in

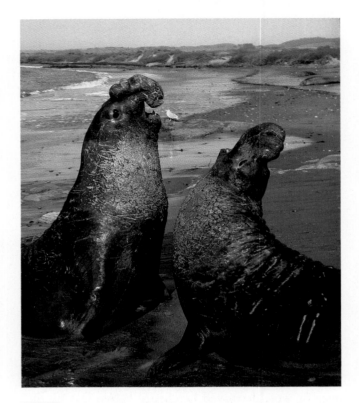

19–10   *Two northern elephant seal bulls fighting for supremacy of a harem, which may consist of as many as 50 females. Only a few males breed each year, and each breeding male fathers many offspring. This social system may be an additional factor contributing to the high degree of homozygosity found in the northern elephant seal population.*

19–11   *The lesser white snow goose ("the goose from beyond the north wind") and the blue goose were once believed to represent distinct species. In fact, they represent one species, with individuals of two colors. The allele for white is recessive; heterozygotes and homozygotes for blue are both the same dark blue color. However, birds mate preferentially with animals of their own color. As a result, there are more homozygotes than there would be if mating were random.*

the genotype frequencies without necessarily producing any changes in the frequencies of the alleles in question. Suppose, however, that all female snow geese, white or blue, preferred to mate only with blue snow geese. If, as a consequence, some of the blue males mated with more than one female, and some of the white males with none, changes in the frequencies of the two alleles and also of the phenotypes would occur. In fact, as we shall see in the next chapter, such nonrandom mating produced by female choice is an important agent of natural selection in some species.

Preservation and Promotion of Variability

Sexual Reproduction

By far the most important method by which eukaryotic organisms promote variation in their offspring is sexual reproduction. Sexual reproduction produces new genetic combinations in three ways: (1) by independent assortment at the time of meiosis—as diagrammed in Figure 11–16 on page 203, (2) by crossing over with genetic recombination, and (3) by the combination of two different parental genomes at fertilization. At every generation, alleles are assorted into new combinations.

In contrast, consider, for a moment, organisms that reproduce only asexually, by processes that involve mitosis and cytokinesis but not meiosis. Except when a mutation has occurred in the duplication process, the new organism will exactly resemble its only parent. In the course of time, various clones may form, each carrying one or more mutations, but potentially favorable combinations are unlikely to accumulate in one genotype.

Organisms that reproduce sexually can do so at only half the rate of asexually reproducing organisms. The only advantage to the organism of sexual reproduction, speaking strictly scientifically, appears to be the promotion of variation, the production of new combinations of alleles among the offspring. Why such variation is advantageous to the individual organism is a matter of controversy (see essay).

Mechanisms That Promote Outbreeding

Many means have evolved by which new genetic combinations are promoted in sexually reproducing populations. Among plants, a variety of mechanisms ensure that the sperm-bearing pollen from flowers on one plant is delivered to the stigmas of flowers on a different

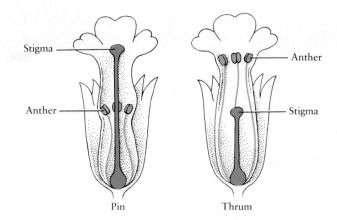

19–12   *Diagrams of two types of flowers ("pin" and "thrum") of the same species of primrose. Notice that the pollen-bearing anthers of the pin flower and the pollen-receiving stigma of the thrum flower are both situated about halfway up the length of the flower and that the pin stigma is level with the thrum anthers. An insect foraging for nectar in these flowers would collect pollen on different areas of its body, so that thrum pollen would be deposited on pin stigmas, and vice versa.*

plant. In some plants, such as the holly and the date palm, male flowers are on one tree and female flowers on another. In others, such as the avocado, the pollen of a particular plant matures at a time when its own stigma is not receptive. In some species, anatomical arrangements inhibit self-pollination (Figure 19–12).

Some plants have genes for self-sterility. Typically, such a gene has multiple alleles—s^1, s^2, s^3, and so on. A plant carrying the allele s^1 cannot fertilize a plant with an s^1 allele; one with an s^1/s^2 genotype cannot fertilize any plant with either of those alleles; and so forth. In one population of about 500 evening primrose plants, 37 different self-sterility alleles were found, and it has been estimated that there are more than 200 alleles for self-sterility in red clover. A plant with a rare self-sterility allele is more likely to be able to fertilize another plant than is a plant with a common self-sterility allele. As a consequence, the self-sterility system strongly encourages variability in a population. Selection for the rare allele makes it more common, whereas more common alleles become rarer.

Why Sex?

Sex is a very complicated and expensive way to reproduce. A sexually reproducing population wastes half of its reproductive potential on producing males and can therefore increase in numbers only half as fast as an asexually reproducing population. The same genetic penalty is paid by individual females: a sexually reproducing female will propagate her genes only half as fast as an asexually reproducing female. Moreover, sexual reproduction often involves considerable expenditure of time and effort and may expose an organism to predators while it is seeking a mate, courting, and copulating.

If sexual reproduction is so inefficient and so risky, why has it not been eliminated by natural selection? The answer appears to lie in the genetic variation promoted by sexual reproduction, but it is not easy to see how producing genetically variable offspring could compensate for producing, effectively, only half as many of them. Several hypotheses have been proposed to account for the success and persistence of sexual reproduction.

According to the "best man" hypothesis, the environment is very changeable. As a result, the offspring are likely to grow up in conditions quite different from those experienced by their parents. Offspring that are genetically identical to their parents will be poorly adapted to the changed conditions. Sexual reproduction is a way of creating a great variety of new genotypes, a few of which will happen to be well adapted to the unpredictable conditions experienced by the next generation.

The "tangled bank" hypothesis, by contrast, emphasizes that the environment is very diverse. It offers a great variety of different opportunities, but each opportunity is limited—that is, it can be exploited successfully by only a limited number of individuals of a particular genotype. Producing many uniform offspring is futile because, being very similar, they will all attempt to exploit the same opportunity. If you have 10 children, all of whom would like to stay in the same small town in which they grew up, you wouldn't want them all to be

The Red Queen from Through the Looking Glass:
"You have to run faster than that to stay in the same place."

doctors. Sexual reproduction is a way of avoiding competition between close relatives by suiting them to a variety of different ways of making a living.

Although often cited as correct, the "best man" hypothesis is, according to Graham Bell of McGill University, almost certainly wrong. If it were true, we would expect to find sexual reproduction in continually changing, disturbed, or novel conditions, with asexual reproduction prevailing in stable, long-established environments. The reverse appears to be true. For example, asexual reproduction is more common in freshwater environments, whereas sexual reproduction prevails in the more stable conditions provided by marine environments.

This ecological pattern is consistent with the "tangled bank" hypothesis but does not prove that it is correct. Other hypotheses may also be consistent with the observed pattern. For example, the "Red Queen" hypothesis—named after a character in Lewis Carroll's *Through the Looking Glass*—points out that many species (such as predators and prey, or parasites and hosts) are continually running an evolutionary race, each trying to increase its efficiency at the expense of others. Success requires a means of rapid genetic change in order to keep pace with the adaptations and

counter-adaptations of one's opponents. The continual reshuffling of alleles provided by sexual reproduction is such a means. If this hypothesis is correct, then sexual reproduction should be common wherever many species are found together, as in the sea and the tropics.

Correlation of the occurrence of sexual reproduction with ecological information appears to exclude the "best man" hypothesis. However, different or more detailed comparisons are needed to decide between the "tangled bank" and "Red Queen" hypotheses.

Other hypotheses have also been proposed to explain the success and persistence of sexual reproduction. For example, it has recently been suggested that the close pairing of homologous chromosomes at the beginning of meiosis makes possible not only the genetic recombinations that result from crossing over but also the repair of damaged DNA. Some biologists think that this "check" on the integrity of the DNA being transmitted to the next generation is of such importance that it could be a major factor in maintaining sexual reproduction.

The process of sexual reproduction, as it occurs in diverse organisms, is varied and complex. Understanding its evolution and function remains among the most challenging and perplexing problems in biology.

19–13   *Two giant African land snails* (Achatina) *in the process of mating. Like earthworms, many snails and slugs are hermaphrodites (each individual produces both sperm and egg cells). Hermaphroditism is advantageous for slow, solitary species because it doubles the chances of finding a mate. Every mature member of the same species that a hermaphrodite meets will always be of the opposite sex—and of the same sex too. As a result of the mating, each individual can produce new offspring.*

Among animals, even in those invertebrates that are **hermaphrodites,** producing both sperm and egg cells, an individual seldom fertilizes its own eggs (Figure 19–13). Among mammals, in particular, behavioral strategies promote outbreeding (mating with unrelated individuals). Often, for example, young males leave the family group as they reach reproductive age. This occurs among lions, gorillas, and baboons, to mention only a few. Among hunting dogs of the African plains, it is the young females who leave at reproductive age.

Diploidy

Another factor in the preservation of variability in eukaryotes is diploidy. In a haploid organism, genetic variations are immediately expressed in the phenotype and are therefore exposed to the selection process. In a diploid organism, however, such variations may be stored as recessives, as with the allele for white flowers in Mendel's pea plants. The extent to which a rare allele is protected is shown in Table 19–1.

As the table reveals, the lower the frequency of allele *a,* the smaller the proportion of it exposed in the *aa* homozygote becomes. The removal of the allele by natural selection slows down accordingly. This result should be of special interest to proponents of eugenics, who advocate improving the human gene pool through controlled breeding. For instance, consider a genetic disorder, such as PKU (page 212), that is expressed only in the homozygous recessive. If the frequency of allele *a* is about 0.01, the frequency of individuals with the *aa* genotype is 0.0001 (1 child for every 10,000 born). It would take 100 generations, roughly 2,500 years, of a program of sterilization of homozygous individuals with this condition to halve the allele frequency (to 0.005) and reduce the number born with this genetic disorder to 1 in 40,000, which should be enough to discourage even the most zealous eugenicist.

Heterozygote Advantage

Recessive alleles, even ones that may be harmful in the homozygous state, may not only be sheltered in the heterozygous state, but sometimes may actually be selected for. This phenomenon, in which the heterozygotes have greater reproductive success than either type of homozygote, is known as **heterozygote advantage.** It is another way that genetic variability is preserved.

One of the best-studied examples of heterozygote advantage involving a single gene locus is found in association with sickle-cell anemia. Until very recently, individuals homozygous for the sickle allele almost never lived long enough to become parents. Therefore, almost every time two sickle alleles came together in a homozygous individual, two sickle alleles were removed from the gene pool. At one time, it was thought that the sickle allele was maintained in the population by a steady in-

Frequency of Allele *a* in Gene Pool	Genotype Frequencies			Percentage of Allele *a* in Heterozygotes
	AA	*Aa*	*aa*	
0.9	0.01	0.18	0.81	10
0.1	0.81	0.18	0.01	90
0.01	0.9801	0.0198	0.0001	99

Table 19–1   Protection of Recessive Alleles by Diploidy

flux of new mutations. Yet in some African populations, as many as 45 percent of the individuals are heterozygous for the sickle allele, despite the loss of sickle alleles through homozygous individuals. To replace these alleles by mutations alone would require a rate of mutation about 1,000 times greater than any other known human mutation rate.

In the search for an alternative explanation, it was discovered that the sickle allele is maintained at high frequencies because the heterozygote has a selective advantage. In many regions of Africa, malaria is one of the leading causes of illness and death, especially among young children. Studies of malarial infections among young children showed that the severity of the illness is significantly lower in individuals heterozygous for the sickle allele than in normal homozygotes. Thus, although two alleles for sickle-cell hemoglobin are eliminated virtually every time they appear in the homozygous state, positive selection for the heterozygote maintains the allele in the gene pool. Moreover, for reasons that are not known, women who carry the sickle allele are more fertile than women who do not.

As this example makes clear, genetic variation not only is the raw material on which natural selection acts, but it can also be maintained by selection. In the next chapter, we shall consider other situations in which natural selection plays a key role in promoting and preserving genetic variation.

Summary

Population genetics is a synthesis of the Darwinian theory of evolution with the principles of Mendelian genetics. A population, for the population geneticist, is an interbreeding group of organisms, defined and united by its gene pool (the sum of all the alleles of all the genes of all the individuals in the population). Evolution is the result of accumulated changes in the composition of the gene pool.

The extent of genetic variability in a population is a major determinant of its capacity for evolutionary change. Natural populations can be shown by breeding experiments—artificial selection—to harbor a wide spectrum of genetic variations. The extent of genetic variation can be quantified by comparison of protein structures and, more recently, of DNA sequences. It has been estimated that at least 25 percent of the genes in any given human population are represented by two or more alleles.

The Hardy-Weinberg equilibrium describes the steady state in allele and genotype frequencies that would exist in an ideal population in which five conditions were met: (1) no mutations occurred, (2) no immigration or emigration occurred, (3) the population was large, (4) mating was random, and (5) there was no difference in the reproductive success of the offspring. The Hardy-Weinberg equilibrium demonstrates that the genetic recombination that results from meiosis and fertilization cannot, in itself, change the frequencies of alleles in the gene pool. The mathematical expression of the Hardy-Weinberg equilibrium provides a quantitative method for determining the extent and direction of change in allele and genotype frequencies.

The principal agent of change in the composition of the gene pool is natural selection. Other agents of change include mutation, gene flow, genetic drift, and nonrandom mating. Mutations provide the raw material for change, but mutation rates are usually so low that mutations, in themselves, do not determine the direction of evolutionary change. Gene flow, the movement of alleles into or out of the gene pool, may introduce new alleles or alter the proportions of alleles already present. It often has the effect of counteracting natural selection. Genetic drift is the phenomenon in which certain alleles increase or decrease in frequency, and sometimes even disappear, as a result of chance events. Circumstances that can lead to genetic drift, which is most likely to occur in small populations, include the founder effect and population bottleneck. Nonrandom mating causes changes in the proportions of genotypes but may or may not affect allele frequencies.

Sexual reproduction is the most important factor promoting genetic variability in populations. Mechanisms that promote outbreeding further promote variability. These mechanisms include self-sterility alleles in plants, anatomic adaptations that inhibit self-fertilization, and in animals, behavioral strategies. Variability is preserved by diploidy, which shelters rare, recessive alleles from selection. Natural selection can also act to promote and preserve variability. In cases of heterozygote advantage, for example, the heterozygote is selected over either homozygote, thus maintaining both the recessive and the dominant alleles in the population.

Questions

1. At the end of the experiment on bristle number in *Drosophila*, the flies in the high selection line had an average of 56 bristles. No fly at the beginning of the experiment had as many as 56, however. How do you explain this fact?

2. When domesticated animals, such as dogs, horses, pigs, and goats, are released into the wild, within a few generations their offspring closely resemble—in both behavior and appearance—the wild ancestors from which the animals were originally domesticated. What does this phenomenon reveal about the genetic variability latent in populations of domesticated animals? What does the speed with which the changes occur indicate about the strength of natural selection? About the long-term effectiveness of artificial selection?

3. As noted in the text, the Hardy-Weinberg equation was first formulated in response to a question: Why, if some alleles are dominant and some are recessive, don't the dominants drive out the recessives? What is the fallacy in the reasoning underlying that question? How does the Hardy-Weinberg equation answer that question?

4. What is the difference between gene flow and genetic drift? How does each affect the gene pool of a population?

5. What role might immigration play in promoting genetic variability? Compare the potential impact of immigration on the gene pool of a small island population versus that of a large mainland population.

6. How does self-fertilization, as in flowers, affect the Hardy-Weinberg equilibrium?

7. How do self-sterility alleles promote variability in a population?

8. In cases of heterozygote advantage involving a single gene with only two alleles, how does selection for the heterozygote affect the relative frequency of the two alleles? How would selection *against* the heterozygote affect this frequency?

9. Under what circumstances would two alleles be maintained in a population in exactly the same numbers?

Suggestions for Further Reading

Culotta, Elizabeth: "How Many Genes Had to Change to Produce Corn?" *Science*, vol. 252, pages 1792–1793, 1991.

Dawkins, Richard: *The Selfish Gene*, 2d ed., Oxford University Press, New York, 1989.*

> *Dawkins argues that we are "survival machines" programmed to preserve the "selfish molecules" known as genes. A wit-sharpening account and also an entertaining one.*

Koehn, Richard K., and Thomas J. Hilbish: "The Adaptive Importance of Genetic Variation," *American Scientist*, vol. 75, pages 134–141, 1987.

Lewin, Roger: "The Surprising Genetics of Bottlenecked Flies," *Science*, vol. 235, pages 1325–1327, 1987.

Li, Wen-Hsiung, and Dan Graur: *Fundamentals of Molecular Evolution*, Sinauer Associates, Inc., Sunderland, Mass., 1991.*

> *An authoritative introductory textbook on the new science that is emerging from the synthesis of evolutionary biology with the rapidly expanding discoveries of molecular genetics.*

Slatkin, Montgomery: "Gene Flow and the Geographic Structure of Natural Populations," *Science*, vol. 236, pages 787–792, 1987.

Strickberger, Monroe W.: *Evolution*, Jones and Bartlett Publishers, Boston, 1990.

> *An up-to-date textbook on evolutionary biology, written for undergraduate students who have completed a course in introductory biology.*

Travis, John: "Possible Evolutionary Role Explored for 'Jumping Genes,'" *Science*, vol. 257, pages 884–885, 1992.

Wills, Christopher: *The Wisdom of the Genes: New Pathways in Evolution*, Basic Books, Inc., New York, 1989.*

> *An introduction to modern evolutionary biology, written for a general audience and enriched with lively anecdotes and analogies. The author is particularly concerned with the evolutionary role of such features of DNA structure and organization as mobile genetic elements (transposons) and introns and exons.*

*Available in paperback.

20

Natural Selection

In species after species, the females are sensible in their size, attire, and accessories, while the males are often oversized, overdressed, and burdened with awkward appendages and seemingly foolish patterns of behavior. The male peacock is perhaps the most cited example; his resplendent tail not only attracts predators but also makes it more difficult for him to escape them. The enormous antlers of the male Irish elk, as preserved in the fossil record, would have been so cumbersome that some paleontologists believe they hurried the species toward extinction. Male bowerbirds spend much of their waking life collecting ornaments for elaborate nests that are never occupied, while the females quietly prepare homes for their young, which they will raise unassisted.

Darwin concluded that "this form of selection depends, not on the struggle for existence in relation to other organic beings or to external conditions, but on a struggle between the individuals of one sex, generally the males, for the possession of the other sex."

This phenomenon of sexual selection, as Darwin called it, was one of his most controversial proposals. One reason was a lack of understanding of how such characteristics could be passed through the female to the male, generation after generation (a question that has now been answered). Another reason was that it required "a sense of beauty," as Darwin called it, on the part of the selecting females, which some of his critics found difficult to imagine.

But beauty, as we are often reminded, lies not in our eyes alone. Among proboscis monkeys *(Nasalis larvatus)*, females have, as you can see, a dainty turned-up nose. By contrast, the nose of a mature male resembles a giant cucumber, drooping down over his mouth, almost reaching his chin. In order to eat, he has to push it out of the way. Like the tail of the peacock, however, this outsized appendage is believed to be his ticket to sexual success.

These wonderfully grotesque animals are found only in the swampy mangrove forests of Borneo. They are endangered because of destruction of their habitat.

According to Darwin's own account, the concept of natural selection came to him in 1838 upon reading Malthus's gloomy essay (see page 9). Darwin realized that all populations—not just the human population—are potentially doomed to exceed their resources. Only a small fraction of the individuals that might exist are born and survive to maturity. According to Darwin, those that do survive are those that are "favoured," to use his term, by reason of slight, advantageous variations. This process of survival of the "favoured" he termed natural selection, by analogy with the artificial selection practiced by breeders of domestic animals and plants.

In terms of population genetics, natural selection is now defined more rigorously as the differential rate of reproduction of different genotypes in a population. This differential reproductive success, which is a result of interactions between the individual organisms and their environment (including other organisms), can result in changes in the gene pool of a population—that is, in evolution. According to current theory, natural selection is the major force in evolution.

Natural Selection and the Maintenance of Variability

In the course of the controversies that led to the synthesis of evolutionary theory with Mendelian genetics, some biologists argued that natural selection would serve only to eliminate the "less fit." As a consequence, it would tend to reduce the genetic variation in a population and thus reduce the potential for further evolution. Modern population genetics has demonstrated this not to be true.

As we saw in the example of the sickle-cell allele (page 331), natural selection can be a critical factor in preserving and promoting variability in a population. There are many other examples of this effect. Here we shall consider just one, which illustrates the fact that a variety of selective forces can be at work simultaneously.

Balanced Polymorphism:
Color and Banding in Snail Shells

As you will recall (page 328), polymorphism is the co-existence within a population of two or more phenotypically distinct forms. In some cases, one phenotype gradually replaces the others, and the polymorphism is transient. In other cases, however, the phenotypes are maintained in fairly stable proportions by natural selection, and the polymorphism is said to be **balanced.**

One of the best-studied examples of a balanced polymorphism is found among land snails of the genus *Cepaea.* In one species, for instance, the shell of the snail may be yellow, brown, or any shade from pale fawn through pink and orange to red. The lip of the shell may be black or dark brown (normally) or pink or white (rarely). Up to five black or dark-brown longitudinal bands may decorate it (Figure 20–1a). Fossil evidence indicates that these different types of shells have coexisted for more than 10,000 years.

Studies among English colonies of *Cepaea* have revealed some of the selective forces at work on the snails, which occupy a variety of habitats. The snails are preyed upon by birds, among which are song thrushes. Song thrushes select snails from the colonies and take them to nearby rocks, where they break them open, eat the soft parts, and leave the shells (Figure 20–1b). By comparing the proportions of types of shells around the thrush "anvils" with the proportions in the nearby colonies, investigators have been able to correlate the shell patterns of snails seized by the thrushes with the habitats of the snails. In habitats, such as bogs, where the background is fairly uniform, unbanded snails blend in and are less likely to be preyed upon than banded ones. Conversely, in habitats, such as woodlands, where the backgrounds are mottled, unbanded snails are more likely to be the victims.

Studies of many different colonies have confirmed these correlations. In uniform environments, a higher proportion of snails are unbanded; in rough, tangled habitats, far more tend to be banded. Similarly, the greenest habitats have the highest proportion of yellow shells, but among snails living on dark backgrounds, the yellow shells are much more visible and are clearly disadvantageous, judging from the evidence conveniently assembled by the thrushes.

Many of the snail colonies studied were at distances so great from one another that the possibility of movement between populations could be ruled out. Why, then, are both shell types still present in these colonies? One would expect populations living on uniform backgrounds to be composed almost entirely of unbanded snails, and colonies on dark, mottled backgrounds to lose most of their yellow-shelled individuals. The answer to this problem is not fully worked out, but it seems that there are other factors that are correlated with the particular shell patterns and that form a part of the same group of genes that control color and banding. Experiments have shown, for instance, that unbanded snails (especially yellow ones) are more heat-resistant and cold-resistant than banded snails. In other words, several different selection pressures are at work, and they appear to maintain the genetic variations of color and banding.

(a)   (b)

20–1   (a) *Ten snails of the species* Cepaea hortensis. *The shells of snails of this species (and of other species in the genus* Cepaea) *exhibit a wide range of colors and banding patterns. Among the most common colors are yellow and pink. Shells of either color may be unbanded or may have dark bands of varying thickness and intensity. Banded snails living on dark, mottled backgrounds are less visible than unbanded snails and are therefore preyed upon less frequently.* (b) *A song thrush breaking open a yellow-shelled snail at an "anvil."*

The blood types A, B, AB, and O represent the most thoroughly studied polymorphism in human populations. Apparently, the three alleles associated with these blood types (page 224) are part of our ancestral legacy, since the same blood types are also found in other primates. A great deal is known about the chemistry of the different blood types and about the allele frequencies in different populations. Yet we know very little about how this polymorphism has been maintained.

Some population biologists regard the blood types as probably neutral in selective value. Others maintain that polymorphism in human blood types is a result of selection. For example, among Caucasian males, life expectancy is greatest for those with type O and least for those with type B; exactly the opposite is true for Caucasian females with these blood types. People with type A blood run a relatively higher risk of cancer of the stomach and of pernicious anemia. People with type O blood have a higher risk of duodenal ulcers and are more likely to contract Asian flu. With the exception of Asian flu, however, most of these conditions would not affect relative rates of reproduction—and thus act as selection forces—since they generally occur in individuals who are past reproductive age.

It has been suggested that there are correlations between the different blood types and susceptibility to such diseases as plague, leprosy, tuberculosis, syphilis, and smallpox—all diseases that, in the past, could have been powerful selection forces. For example, studies conducted in certain small towns and villages of India during a smallpox epidemic in 1965 and 1966 revealed that individuals carrying the A allele had a risk of contracting smallpox that was seven times higher than that of individuals who did not carry the A allele. Moreover, once ill with smallpox, such individuals were three times as likely to develop a severe case and twice as likely to die. These data would suggest that allele A should have been virtually eliminated in human populations long before the eradication of smallpox in the late 1970s. However, individuals with the A allele appear to be relatively resistant to cholera, a deadly disease that still takes a very high toll in India and many other parts of the world.

The geographic distributions of the A, B, AB, and O types are irregular. For example, there is a large predominance of type O in the Western Hemisphere and an increase in the B allele as one moves from Europe toward Central Asia. The B allele is totally absent in Native Americans and Australian aborigines who have not intermarried with individuals of European descent. Although most Native Americans are type O, the Blackfoot tribe has the highest frequency of type A blood found anywhere in the world (55 percent). Even within an area as small as the British Isles, there are significant variations in allele frequency, going both from north to south and from east to west. These differences may reflect some differences in the selective forces favoring particular blood types under particular conditions, they may be the result of population migrations and genetic drift, or some combination of both. At the present time, we do not know.

What Is Selected?

As the studies with *Cepaea* illustrate, natural selection acts on the complete phenotype. In the days of fruit-fly genetics, when populations were being scanned for white eyes and short wings, phenotype became synonymous with physical appearance. In terms of evolutionary theory, however, it must be regarded as including such important characteristics as the optimum temperature at which a particular enzyme works, or the speed of response to a stimulus. In short, the phenotype includes all the physical, physiological, and behavioral attributes of an organism.

A phenotype generally is the expression of many different genes. As a corollary, any particular phenotypic characteristic may be arrived at by a number of genotypic routes. For example, selection for high and low bristle number in *Drosophila* (page 322) led, in the first series of experiments, to sterility. The second attempt to select for high bristle number led to the same number of bristles (that is, the same phenotypic characteristic) but without a reduction in fertility. Obviously, different genotypes were responsible.

It is very rare that a single allele can determine a winning phenotype. In the peppered moth *Biston betularia* (pages 13–14), black wing color is determined by a single allele. However, the peppered coloring may be the result of several different genes, and, along with the mottled appearance, there are other characteristics important for escaping hungry predators. These include selecting the right background, lying very still while on the tree trunk, and positioning oneself correctly so as to enhance the camouflage effect, all of which involve genetic factors. Such underlying genetic complexity is typical not only of the variety of mechanisms by which animals avoid predators (Figure 20–2) but also of most characteristics of the whole organism.

As you will recall, however, the phenotype is not de-

(a)

(b)

20–2   *Coordinated anatomical and behavioral characteristics protect the peacock moth* (Automeris memusae) *from predators.* (a) *At rest, the moth positions itself in a way that enhances the camouflaging effect of its patterned wings.* (b) *Disturbed, the moth parts its wings, revealing a pair of "eyes," reported to be sufficiently menacing to scare away a bird.*

termined solely by the interactions of the multitude of alleles making up the genotype. It is also a product of the interaction of the genotype with the environment in the course of the individual's life (Figure 20–3). For example, among human identical twins (that is, twins produced from a single fertilized egg and therefore having the same genotype), noticeable differences—such as body weight—between the two are often apparent even at birth, owing to differences in the intrauterine environment.

Any particular phenotype lasts, in terms of evolutionary history, for as long as the blink of an eye. In the case of sexually reproducing organisms, the genotype is as unique and transient as the phenotype, shuffled and recombined at every generation. Only the individual genes survive. The great men and women of history have long since vanished, but each of us may carry a few of their alleles as part of our human legacy.

20–3   *Jeffrey pines usually grow tall and straight, as in* (a). *Environmental forces, however, can alter the normal growth patterns, as shown in* (b). *This tree is growing on a mountaintop in Yosemite National Park, California, where it is exposed to strong, constant winds.*

(a)

(b)

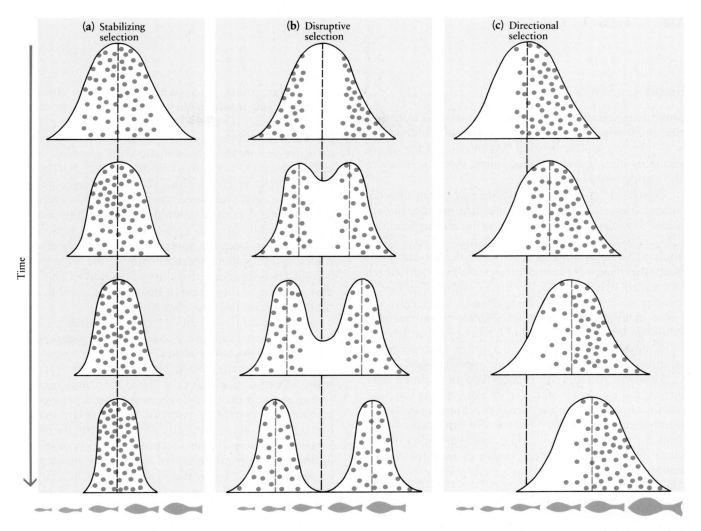

20–4 *A schematic representation of three types of natural selection acting on a trait, such as body size, that varies continuously throughout a population (see page 225 for a review of continuous variation). The horizontal axis of each small graph represents the range of values of the trait being considered, with one extreme at the left, the other extreme at the right, and intermediate states in between. The black curve in each graph summarizes the proportion of individuals in the population exhibiting a particular variant of the trait. The dots represent individuals in each generation that reproduced and left an average—or greater than average—number of offspring.*

Initially, each of the populations displays a normal bell-shaped curve. Most individuals exhibit intermediate variants of the trait, and only a few exhibit the extremes. As you can see, in (a) stabilizing selection and (b) disruptive selection, the shapes of the curves change in succeeding generations. Stabilizing selection involves the elimination of extremes, producing a more uniform population. In disruptive selection, intermediate forms are eliminated, producing two divergent populations. In (c) directional selection, one expression of the trait is gradually eliminated in favor of another. Note how the bell-shaped curve moves to the right in each succeeding generation.

Types of Selection

As we have seen, the results of natural selection in any given instance depend on the interaction of a variety of factors, both genetic and environmental. Selection processes can be classified into five broad categories on the basis of several different criteria. Depending on its effect on the distribution of characteristics within a population, natural selection can be described as stabilizing, disruptive, or directional (Figure 20–4). When selection is influenced by the relative proportions of different phenotypes within a population, it is said to be frequency-dependent. A fifth category, sexual selection, is defined by what is selected: characteristics of direct consequence in obtaining a mate and successfully reproducing.

Stabilizing Selection

Stabilizing selection, a process that is always in operation in all populations, is the elimination of individuals with extreme characteristics. Many mutant forms are probably immediately weeded out in this way, often in the zygote or embryo.

Clutch size in birds is an example of a trait affected by stabilizing selection. Clutch size (the number of eggs a bird lays) is determined genetically, although it also appears to be influenced by environmental factors. In a study of Swiss starlings (Table 20–1), the percentage of birds surviving was found to increase for each clutch size up to five. With a clutch size larger than five, a smaller percentage of birds survived—apparently because of inadequate nutrition. Female Swiss starlings whose genotypes lead to a clutch size of four or five will have more surviving young, on average, than members of the same species that lay more or fewer eggs.

Although the number of eggs laid in a clutch may appear to be a simple trait, rather like pushing the right number for copies on the Xerox machine, it involves a number of factors. They include the synthesis of proteins for the egg yolk and white, the availability of calcium for the shells, and the length of time the female will mate.

Disruptive Selection

A second type of selection, **disruptive selection,** increases the frequency of the extreme types in a population at the expense of intermediate forms. Disruptive selection in the laboratory gave rise to the high- and low-bristle-number lines of *Drosophila* described earlier.

A particularly clear instance of disruptive selection in nature has been demonstrated in studies of plants growing on soils that were previously contaminated by mining operations. The boundaries between contaminated and uncontaminated areas are often very sharp. Plants growing on uncontaminated soil are unable to survive on contaminated soil. Plants of the same species growing on contaminated soil are able to survive in the uncontaminated areas but cannot compete with those already growing there. Thus, the two extreme phenotypes have been "favoured" at the expense of intermediate forms. This has resulted in the development of very marked differences between the two groups in the 50 years or so since the mining operations were discontinued and the plants began to colonize the area. Under certain circumstances, disruptive selection of this sort may lead to the formation of new species, a subject we shall explore in the next chapter.

A more recently described example of disruptive selection is also an example of both frequency-dependent selection and sexual selection. As you may know, the coho salmon of the Pacific Northwest hatch in freshwater streams, in which they spend their first year before moving to the ocean, where they become sexually mature. Upon attaining maturity, the fish return to their home stream, breed, and die. Although females of this species attain sexual maturity and return to breed at age 3, the males may attain sexual maturity at either age 2 or age 3. The two-year-old males, known as "jacks," are about half the size of the three-year-old males, known as "hooknoses" (Figure 20–5). Whether a male will mature in two years and thus be a jack, or in three years and be a hooknose, is, in part, genetically determined.

Studies by Mart R. Gross of Simon Fraser University in British Columbia have shown that disruptive selection not only maintains these two types of males in the population but favors the smallest jacks and the largest hooknoses. When a female spawns, the males closest to the nest in which her eggs are laid are the first to deposit sperm over them. The hooknose males jockey for prox-

Table 20–1 Survival in Relation to Number of Young in Swiss Starling								
Number of young in clutch	1	2	3	4	5	6	7	8
Number of young marked	65	328	1,278	3,956	6,175	3,156	651	120
Number of marked birds recaptured after 3 months	0	6	26	82	128	53	10	1
Percentage of marked birds recaptured after 3 months	0	1.8	2.0	2.1	2.1	1.7	1.5	0.8

20–5   *Coho salmon, at the mouth of their home river, as they begin the journey upstream to the location where they will breed. The large fish at the bottom is a mature female, and the larger, more brightly colored fish above her is a hooknose male. The much smaller fish above him is a jack male. The small fish grouped together at the lower right appear to be holding back. They may be immature fish that will remain at the mouth of the river until they are mature, or they may be sexually mature jack males.*

20–6   *Scale insects on a branch of a citrus tree. With their specialized mouthparts, these insects suck fluid from the plant, debilitating it. If left unchecked, they eventually kill the plant.*

imity to the nests by fighting, and the largest males generally win. By contrast, the jacks sneak close to the nests by hiding among rocks or debris, or in shallow areas of the stream. The smaller the jack, the less likely it is to be discovered and chased away by a hooknose. Both the larger jacks and the smaller hooknoses seldom breed successfully, and the two extreme male types are thus maintained within the population.

Directional Selection

A third type of natural selection, **directional selection**, results in an increase in the proportion of individuals with an extreme phenotypic characteristic. It is therefore likely to result in the gradual replacement of one allele or group of alleles by another in the gene pool.

An example of directional selection is the development of insecticide resistance. Chemicals poisonous to insects, such as DDT, were originally hailed as major saviors of human health and property. They have fallen into disfavor not only because of their tendency to accumulate in the environment, but also because of the extraordinary increase in resistant strains of insects. At least 225 species of insects are now resistant to one or more insecticides. One species is even able to remove a chlorine atom from a DDT molecule and use the remainder as food.

A particularly striking example of insecticide resistance has been found in the scale insects (Figure 20–6) that attack citrus trees in California. In the early 1900s, a concentration of hydrocyanic gas sufficient to kill nearly 100 percent of the insects was applied to orange groves at regular intervals with great success. By 1914, orange growers near Corona, California, began to notice that the standard dose of the fumigant was no longer sufficient to destroy one type of scale insect, the red scale. A concentration of the gas that had left less than 1 survivor out of every 100 insects in the nonresistant strain left 22 survivors out of 100 in the resistant strain. By crossing resistant and nonresistant strains, it was possible to show that the difference between the two involved a single gene locus. The resistant insects were variants produced by chance in the original population. Subsequently, a new factor in the environment—a poisonous chemical—led to directional selection for these variants. Nonresistant insects died, but resistant insects survived and reproduced.

The mechanism for this insecticide resistance is not known. However, one group of experiments has shown that the resistant individual can keep its spiracles (the openings through which it draws air into its respiratory system) closed for 30 minutes under unfavorable conditions. The nonresistant insect can do so for only 60 seconds.

Another example of directional selection is provided by the peppered moth, *Biston betularia*. As we saw in the Introduction (pages 13–14), by the 1950s the black form of this moth had largely replaced the light-colored form in heavily polluted areas of the English countryside. In recent years, however, strong controls have been instituted in Great Britain on the particulate content of smoke, and the heavy soot accumulation has begun to decrease. The light-colored moths are already increasing in proportion to the black moths, but it is not yet known whether there will be a complete reversal either in pollution or in the direction of selection. There is a moral to this story. Note that the black moth is not absolutely superior to the light-colored one, or vice versa. Natural selection is, as Darwin realized, entirely a matter of time and place.

Frequency-Dependent Selection

In the examples of natural selection we have considered thus far, we have assumed that the fitness of a phenotype is independent of its relative proportion within a population. In some situations, however, a type of natural selection known as **frequency-dependent selection** acts to decrease the frequency of more common phenotypes and to increase the frequency of less common ones. Such selection is, for example, a factor in maintaining hooknose and jack males in coho salmon populations. As the frequency of either type increases, the competition between males of that type becomes more intense, allowing greater opportunities for successful reproduction by males of the other type.

Predator-prey interactions can also lead to frequency-dependent selection. If individuals of a prey species differ, for example, in color, some predators will prey disproportionately on individuals with the most common color (Figure 20–7). If, as a result, individuals with that color should become less common, selection pressure on them would be relaxed as predators switch their attention to individuals of alternative colors, which would now be more common. Thus frequency-dependent selection can be a factor in maintaining polymorphisms in prey populations.

Sexual Selection

As Darwin recognized, many of the conspicuous characteristics of animals have little to do with survival on a day-to-day basis but are instead the result of **sexual selection**, the "struggle between the members of one sex, generally the males, for the possession of the other sex."

(a)

(b)

20–7 *Frequency-dependent selection has been demonstrated in the aquatic insects known as water boatmen (a). These insects, which are preyed upon by fish, often exist in different color forms that provide different degrees of camouflage. (b) The results of experimental studies in which fish were provided with supplies of water boatmen of a species that exists in three distinct color forms. When the three forms were present in equal proportions (33 percent of each type), the least camouflaged form was the most likely to be eaten and the most camouflaged form the least likely to be eaten. However, when the proportions of the three forms were unequal, the form present in the highest proportion was the most likely to be eaten, regardless of its degree of camouflage.*

Male Ornamentation: The Role of Female Preference

Darwin hypothesized that in species in which females actively choose their mates, the preference of the females for particular characteristics plays a major role in the evolution of male ornamentation. This is, however, only one of several hypotheses that could explain the sometimes extreme characteristics observed in many male animals. For example, another hypothesis is that male ornamentation evolved as a device for signaling between competing males, with female preference having little or no influence. However, experiments performed by Malte Andersson with the long-tailed widowbirds of Kenya have provided dramatic support for Darwin's hypothesis.

Long-tailed widowbirds exhibit a striking sexual dimorphism. The females, which are smaller than the males, are mottled brown and have short tails. The larger males, by contrast, are a brilliant black, with red epaulets, and have tails that average 50 centimeters (about 20 inches) in length.

Males compete for territories, in which the females subsequently nest. Andersson observed as many as six females nesting in the territories of some males, while the territories of other males contained no females at all. As the males fly over their territories, which are in the grasslands of the African savanna, their long tails and black plumage are readily visible, presumably to both predators and other widowbirds, as well as to human observers.

Darwin suggested that selection of males by females on the basis of an ornament, such as the widowbird's tail, would produce directional selection for further elaboration of the ornament until selection in the opposite direction—as, for example, by predators—balanced the effects of the sexual selection. On the basis of this hypothesis, one would predict that, given the opportunity, females would show preference for males with still greater elaboration of the ornament, even in species in which the males are already highly ornamented.

A male long-tailed widowbird, flying over his territory.

Andersson tested this prediction by giving female widowbirds a choice between three groups of males: one group in which the tails were clipped to a length of 14 centimeters, one group in which the tails were the normal length (50 centimeters), and one group in which the tails were lengthened to 75 centimeters by gluing on feathers from the clipped tails of the first group. In the group in which the tails were the normal length, half of the birds were unaltered and half had their tails clipped and then reglued.

Before the experimental alteration of the tails, similar numbers of females were nesting in the territories of all the males. Following the alterations of tail length, the males with lengthened tails attracted four times as many females as the males with either shortened tails or tails of normal length. Moreover, the ability of the males to compete for and successfully acquire territories appeared to be unaffected by the alterations in tail length. As this study demonstrates, at least in the case of the long-tailed widowbird, female mate choice is a dominant selective agent in the evolution of ornamentation.

Andersson's experimental results were published in 1982, more than 100 years after Darwin's hypothesis about the role of female choice. The length of time between that hypothesis and these elegant experiments is indicative of the difficulty often facing evolutionary biologists as they try to devise appropriate experimental or observational tests of key hypotheses. The challenge is not only in the asking of questions but also in the framing of questions in ways that can provide clear answers.

Males produce many more gametes than do females. Thus the female has a larger investment, in terms of time, energy, and resources, in each fertilized egg. When parental care is involved, females are often the ones caring for the young. Hence, males, seeking to inseminate as many females as possible, are generally the competitors. Females, with a higher stake in each mating, seek the best possible genetic partner and so are the choosers. The competition among males may be either direct—for territories, harems, or privileges of consort—or indirect, as with nest-building and displays.

Sexual selection is thought to play a major role in establishing and maintaining **sexual dimorphism,** those differences between males and females that have to do not with the act of reproduction itself but rather with obtaining a mate (Figure 20–8). Examples of sexually selected characteristics are the nose of the male proboscis monkey (page 334), the extravagant plumage of many male birds, and the oversized antlers of male deer and elk. Among those few bird species in which the females do the courting and the males do the choosing, it is the female that is the gaudier, with the stay-at-home male the drab partner.

For those of us weary of comparisons between the resplendent peacock and the dowdy peahen, it is comforting to realize that it is the female whose size, color, and general behavior are considered optimal for the environment. The dimorphic characteristics of the male, by contrast, are useful principally for threat, display, and other bids for attention. In fact, as we noted earlier, these characteristics may well be maladaptive with respect to factors affecting survival, such as conspicuousness to predators.

Darwin, recognizing that such males were not "fitter to survive in the struggle for existence," categorized sexual selection as a force separate from that of natural selection. However, with fitness stringently redefined in terms of relative numbers of surviving offspring, many investigators now believe that such a distinction is invalid and that sexual selection should be considered simply as one of the forms natural selection may take.

The Result of Natural Selection: Adaptation

Natural selection results in **adaptation,** a term with several meanings in biology. First, it can mean a state of being adjusted to the environment. Every living organism is adapted in this sense, just as Abraham Lincoln's legs were, as he remarked, "just long enough to reach the ground." Second, adaptation can refer to a particular characteristic, that which is adapted—such as an eye or a hand—that aids in the adjustment of an organism to its environment. Third, adaptation can mean the evolutionary process, occurring over the course of many generations, that produces organisms better suited to their environment.

(a)

(b)

20–8   *Sexual selection is caused by competition for mates and often results in differences between the two sexes. (a) In many species of birds, such as the golden pheasants shown here, a few males father most of the offspring. The males direct their energies primarily to courtship and mating, and the females invest more heavily in parental care. In such species,* the females are colored in a way that blends with their surroundings, thus protecting them and their young, and the males have bright, conspicuous plumage. (b) In species in which mating couples form monogamous pairs, such as these albatrosses, males and females generally look alike. Both parents are involved in the care of the young.

20–9 *The woodpecker has a number of adaptations that enable it to obtain food. These include two toes pointing backward with which the woodpecker clings to the bark of the vertical tree trunk, strong tail feathers that prop it up, a strong beak that can chisel holes in the bark, strong neck muscles that make the beak work as a hammer, air spaces in the skull that cushion the brain during hammering, and a very long tongue that can reach insects under the bark.*

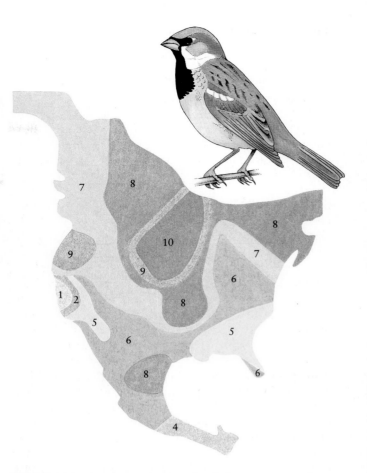

20–10 *The distribution by size of male house sparrows in North America. This artist's interpretation is adapted from a map generated by a computer program that processed data on 739 birds. The higher numbers indicate larger body sizes, based on a composite of 16 different measurements of the birds' skeletons.*

Adaptation has multiple manifestations. Consider, for example, a squirrel. Note how its tail serves as a counterbalance as the animal leaps and turns; in addition, the same marvelous structure serves as a parasol, a blanket, and an aerial rudder. Pluck a burr from your clothing and consider the artful contrivances by which it clings, ensuring that you—or some other animal—will carry the seeds within to a new location, far from the parent plant. Consider the love and devotion characteristic of the domesticated dog; these adaptations, related to the procurement of food and shelter, are as stringently selected as the beak of a woodpecker (Figure 20–9). Thread a needle; your capacity to do so represents the cumulative effect of millions of years of selection pressures for digital dexterity and eye-hand coordination. (The needle itself made its appearance a mere 20,000 years ago.)

Natural selection involves interactions between individual organisms, their physical environment, and their biological environment—that is, other organisms. In many cases, the adaptations that result from natural selection can be clearly correlated with environmental factors or with the selective forces exerted by other organisms.

Adaptation to the Physical Environment: Clines and Ecotypes

Sometimes phenotypic variations within the same species follow a geographic distribution and can be correlated with gradual changes in temperature, humidity, or some other environmental condition. Such a graded variation in a trait or a complex of traits is known as a **cline.**

Many species exhibit north-south clines of various traits. House sparrows, for example, tend to have a smaller body size in the warmer parts of the range of the species, and a larger body size in the cooler parts (Figure 20–10). As size increases, the surface-to-volume ratio decreases, thereby increasing the animal's ability to conserve heat in cold weather. House sparrows were first introduced into North America between 1852 and 1860 in the form of small founder populations drawn

20–11 *Ecotypes of* Potentilla glandulosa, *a relative of the strawberry. Notice the correlation between the size of the plant and the altitude at which it grows. There are other phenotypic differences among the plants as well. When plants from the four geographic areas are grown under identical conditions, many of these phenotypic differences persist and are passed on to the next generation, indicating that these plants are genotypically, as well as phenotypically, different.*

principally from central England and Germany. From an evolutionary standpoint, the time that has elapsed since their introduction is very brief indeed, yet a dramatic degree of size differentiation has occurred. To cite another example, plants growing in the south often have slightly different requirements for flowering or for ending dormancy than the same kind of plants growing in the north, although they may all belong to the same species.

A species that occupies many different habitats may appear to be slightly different in each one. Each group of distinct phenotypes is known as an **ecotype**. Are the differences among ecotypes determined entirely by the environment or do these differences represent adaptation resulting from the action of natural selection on genetic variation?

Figure 20–11 illustrates a series of experiments carried out with the perennial plant *Potentilla glandulosa*, a relative of the strawberry. Experimental gardens were established at various altitudes, and wild plants collected from near each of the experimental sites were grown in all of the gardens. Because *Potentilla glandulosa* reproduces asexually, the plants collected from each area were genetically uniform. The subsequent growth of the plants in the experimental gardens revealed that many of the phenotypic differences among the ecotypes of *P. glandulosa* were due to genetic differences. It is not surprising that in these very different environments, different characteristics had resulted from selection. Over time, the genetic differences between individual plants had been translated into genetic differences between subgroups of the *P. glandulosa* population. As we shall see in the next chapter, such a process is often the first step in the formation of new species.

Adaptation to the Biological Environment: Coevolution

When populations of two or more species interact so closely that each exerts a strong selective force on the other, simultaneous adjustments occur that result in **coevolution**. One of the most important, in terms of sheer numbers of species and individuals involved, is the coevolution of flowers and their pollinators, to be described in Chapter 25. Here we shall consider other examples of coevolution, also involving plants and insects, those ancient allies and enemies.

Milkweed, Monarchs, and Mimics

Various families of plants have evolved chemical defenses that, because they are toxic, bad-tasting, or both, deter insects and other animals that try to feed on the plants. The bitter white sap of plants of the milkweed family contains a toxic substance that acts as such a deterrent. In the course of the evolutionary race to stay in the same place (see essay, page 330), some species of insects, including monarch butterflies, have evolved enzymes that enable the caterpillars to feed on the milkweeds without being poisoned (Figure 20–12). Monarch caterpillars not only utilize the plant tissues for food, but they also ingest and store up the toxic substance, which is then present in the adult forms, the butterflies. The butterflies, in turn, are distasteful and poisonous to their predators (Figure 20–13).

Tasting bad, while useful, is not an ideal defense from the point of view of the individual, since making this fact known may demand a certain amount of personal sacrifice. (Although birds usually drop the monarch butterfly after the first bite, they often inflict fatal injury in the process.) For monarchs and other animals

(a)

(b)

(c)

20–12 (a) *A monarch butterfly egg (white speck in center of photograph) clings to the bud of a milkweed.* (b) *When the egg hatches, two to four days after it is laid, a* caterpillar emerges that feeds on the milk-weed, ingesting and storing the toxic com-pounds produced by the plant. The monarch caterpillar and also the monarch butterfly (c) thus become unpalatable and poisonous. The conspicuous coloration of caterpillar and butterfly warns would-be predators.

that have an effective protective device, it is advantageous to advertise. The more conspicuous or common such an animal is, the fewer the number of individuals that must be sacrificed before the bird or other predator learns to avoid it. Warning colors, such as those of the monarch, are common in the animal world. They are found not only among insects but also among poisonous reptiles and amphibians. The skunk is a familiar mammalian example of an animal whose distinctive markings remind us to keep our distance.

In many types of animals, especially insects, unrelated species with similar obnoxious defenses often come to resemble one another in their warning characteristics. This phenomenon is known as **Müllerian mimicry**, after F. Müller, who first described it. Bees, wasps, and hornets are probably the most familiar example of Müllerian mimicry (Figure 20–14). Even if we cannot tell which is which, we recognize them immediately as stinging insects and keep a respectful distance. Müllerian mimicry is adaptive for all the individuals involved because each benefits from a predator's experience with another. As you might expect, other insects that feed on milkweeds have the bright warning colors characteristic of the monarch.

(a)

(b)

20–13 *A blue jay that has never before tasted a monarch butterfly is fed one in a controlled experiment. Shortly after ingesting the butterfly, the blue jay begins to show signs of being uncomfortable and then vomits. Upon subsequently being offered a monarch, the bird refused it.*

Model (stings)

(a) Yellowjacket

20–14   (a) *A yellowjacket (the model) and some of its mimics. The* (b) *sand wasp,* (c) *masarid wasp, and* (d) *anthidine bee are Müllerian mimics; that is, they all sting. The* (e) *syrphid fly,* (f) Alcathae *moth, and* (g) *ornate checkered beetle, which do not sting, are all Batesian mimics. Müllerian mimics have been compared to reputable merchants who share a common advertisement and divide its costs. Batesian mimics, by contrast, are like unscrupulous retailers who copy the advertisements of successful firms.*

Müllerian mimics (sting)

(b) Sand wasp

(c) Masarid wasp

(d) Anthidine bee

Batesian mimics (do not sting)

(e) Syrphid fly

(f) *Alcathae* moth

(g) Ornate checkered beetle

Next to appear on the evolutionary scene, as events are reconstructed, were several species of butterflies that do not taste bad and are not poisonous but that have coloration similar to that of the monarch. These innocuous mimics escape predation by resembling the bad-tasting model that their would-be predators have previously learned to avoid. Such deception is known as **Batesian mimicry** because it was first described by the naturalist H. W. Bates, a friend and traveling companion of Alfred Russel Wallace (page 11).

Laboratory experiments have confirmed the selective value of Batesian mimicry. Jane Brower, working at Oxford University, made artificial models by dipping

mealworms in a solution of quinine, to give them a bitter taste, and then marking each one with a band of green cellulose paint. Other mealworms, which had first been dipped in distilled water, were painted green like the models, so as to produce mimics, and still others were painted orange to indicate another species. These colors were chosen deliberately: orange is a warning color, since it is clearly distinguishable, and green is usually found in species that are not repellent and that therefore benefit from being inconspicuous.

The painted mealworms were fed to caged starlings, which ordinarily eat mealworms voraciously. Each of the nine birds tested received models and mimics in

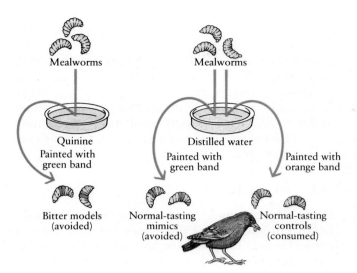

20–15 *An experimental test of Batesian mimicry. Models and mimics were produced by painting quinine-dipped worms (models) and water-dipped worms (mimics) with a green band. Other water-dipped worms were painted with an orange band, clearly distinguishing them from both the models and the mimics. After experiencing the bitter taste of a quinine-dipped model, starlings subsequently avoided not only the models but also their green-banded mimics. The starlings did, however, eagerly consume the orange-banded, water-dipped controls, demonstrating that the worms were not made unpalatable by the process of dipping or the taste of the paint.*

varying proportions. After initial tasting and violent rejection, the models were generally recognized by their appearance and avoided. In consequence, their mimics were also protected (Figure 20–15).

Batesian mimicry works only to the advantage of the mimic. The model suffers from attacks not only by inexperienced predators but also by predators who have had their first experience with the mimic rather than with the model. The mimetic pattern will be at its greatest advantage if the mimic is rare—that is, less likely to be encountered than the model. However, even when as many as 60 percent of the green-banded worms were mimics, 80 percent of the mimics escaped predation. A mimic also has a better chance if it times its emergence to appear after the model, thus reducing its chances of being encountered first. You will not be surprised to learn that in any given area, Batesian mimics do, indeed, generally emerge after the model.

The Imperfection of Adaptation

The concept of evolution has often carried with it some notion of progress. This view—regarded by most biologists as erroneous—has been reinforced by the evidence that evolution has, over time, produced organisms increasingly larger, more complex in structure, and more sophisticated in design, apparently culminating—as if it were preordained—in something as marvelous as ourselves.

Natural selection does tend to push populations toward better solutions to the particular "problems in living" with which they are confronted. However, evolutionary change does not necessarily mean improvement by human criteria, nor does it necessarily even result in organisms that are better adapted to their immediate environment. This is because natural selection acts on the here and now. As a result, populations—particularly those engaged in intense evolutionary races—are often a generation behind. Their capacity to "keep up," and thus maintain adaptation, depends primarily on the existence of sufficient genetic variability in the population.

Natural selection has to work with what is available, not only in the chance variations in the genetic material but also in the phenotypic expression of the genotype—a constraint that has been aptly described as "the law of used parts." To take a simple example, all terrestrial vertebrates and their descendants (including ourselves) are limited structurally to a body with a long vertebral column and four limbs, one on each corner, with which our ancestors waddled up on shore. A structural engineer setting out to build a flying machine or a submarine would scrap these blueprints and start from scratch, but evolution can only build on past history. Many of our human ailments, such as unusual difficulty in childbirth and a propensity to lower back pain, can be traced directly to the conversion of this basically quadrupedal form to an upright stance. In *The Panda's Thumb,* Stephen Jay Gould points out that such absence of perfection and the presence of jerry-built contraptions (the panda's thumb, for example) provide stronger evidence that evolution has occurred than the more usually cited examples of exquisite adaptation.

Patterns of Evolution

Natural selection, as we have seen, is a complex process, operating continuously in all populations. Its outcome, in any particular instance, depends on a great many factors. Viewed from a broader perspective, however, it has the effect of producing different patterns of evolution. One of these patterns, which we have already considered, is coevolution, in which organisms of different species act as selective forces on each other. In other patterns of evolution, natural selection may produce remarkably similar phenotypes in distantly related organisms and, conversely, widely different phenotypes in closely related organisms.

Convergent Evolution

Organisms that occupy similar environments often come to resemble one another even though they may be only very distantly related. This phenomenon is known as **convergent evolution.** When organisms are subjected to similar selection pressures, they show similar adaptations. The whales, a group that includes the dolphins and porpoises, are similar to sharks and other large fish in their streamlined shape and other external features, but the fins of whales conceal the remnants of a terrestrial hand. Whales are warm-blooded, like their land-dwelling ancestors, and they have lungs rather than gills. Similarly, two families of plants invaded deserts in different parts of the world, giving rise to the cacti and the euphorbs (Figure 20–16). Both evolved large fleshy stems with water-storage tissues and protective spines, and they appear superficially similar. However, their quite different flowers reveal their widely separate evolutionary origins.

Divergent Evolution

Divergent evolution occurs when a population becomes isolated from the rest of the species and, as a result of particular selection pressures, begins to follow a different evolutionary course. For example, *Ursus arctos,* the brown bear, is distributed throughout the Northern Hemisphere, ranging from the deciduous forests up through the coniferous forests and into the tundra. As is characteristic of such widespread species, there are many local ecotypes. *Ursus arctos* was similarly widespread some 1.5 million years ago, when the Northern Hemisphere was subjected to a series of glacial advances and retreats. During one of the massive glaciations, a population of *Ursus arctos* was split off from the main group, and, according to fossil evidence, this group, under selection pressure from the harsh environment, evolved into the polar bear, *Ursus maritimus* (Figure 20–17).

(a)

(b)

20–16   *An example of convergent evolution. Members of (a) the cactus family and (b) the euphorb family have been separated for millennia of evolutionary history, with the cacti evolving in the deserts of the New World, and the euphorbs in the desert regions of Asia and Africa. Members of both families have fleshy stems adapted for water storage, protective spines, and greatly reduced leaves.*

20–17 *The polar bear is one example of divergent evolution. The white coloring of the polar bear is probably related to its need for camouflage while hunting rather than to a need to hide from predators, since it is one of the world's largest carnivores. (Its greatest enemies are commercial fishermen who do not appreciate the polar bear's fishing activities and hunters who, prizing the skins for trophies, have taken to tracking the bears by helicopter.) Unlike their southern cousins, the brown bears, polar bears (except for females with young) do not become lethargic in the winter but continue to roam the ice and icy waters in search of fish, seals, young walruses, and other prey.*

Brown bears, although they are classified as carnivores and are closely related to dogs, are mostly vegetarians, supplementing their diet only occasionally with fish and game. The polar bear, however, is almost entirely carnivorous, with seals being its staple diet. The polar bear differs physically from the other bears in a number of ways, including its white color, its streamlined head and shoulders, and the stiff bristles that cover the soles of its feet, providing insulation and traction on the slippery ice.

As the example of the polar bear illustrates, divergent evolution can lead not only to the differentiation of locally adapted ecotypes but also to the formation of new species. In the next chapter, we shall examine this process in more detail.

Summary

Natural selection is the differential reproduction of genotypes resulting from interactions between individual organisms and their environment. According to current theory, it is the major force in evolution. Natural selection can act both to produce change and to maintain variability within a population, as exemplified by the color and banding patterns in snails.

Natural selection can act only on characteristics expressed in the phenotype. The unit of selection is the entire phenotype—the whole organism. In extreme cases, a single allele may be decisive in selection. More often, however, a successful phenotype is the result of the interaction of many genes.

Three major categories of natural selection are stabilizing selection, in which extreme phenotypes are eliminated from the population; disruptive selection, in which the extreme phenotypes are selected at the expense of intermediate forms; and directional selection, in which one of the extremes is favored, pushing the population along a particular evolutionary pathway. Another type of selection is frequency-dependent selection, in which the fitness of a phenotype decreases as it becomes more common in the population and increases as it becomes less common. A fifth category, sexual selection, results from the competition for mates; it can greatly increase reproductive success without improving adaptation to other environmental factors.

The result of natural selection is the adaptation—however imperfect—of populations to their environment. Evidence of adaptation to the physical environment can be seen in gradual variations that follow a geographic distribution (clines) and in distinct groups of phenotypes (ecotypes) of the same species occupying different habitats. Adaptation to the biological environment results from the selective forces exerted by interacting species of organisms on each other (coevolution). Examples include the relationship between milkweeds and monarchs and cases of mimicry, both Müllerian and Batesian.

Evolution by natural selection does not necessarily produce a population with the best possible relationship to its environment. Because natural selection operates in the here and now, many members of a population may be at least one generation behind optimal

adaptation. Moreover, the potential for evolutionary change is constrained by the extent of variability in the gene pool and by the limits on the phenotypic expression of the genotype imposed by past history.

Observed patterns of evolution resulting from natural selection include coevolution, convergent evolution, and divergent evolution. In convergent evolution, dissimilar populations, only distantly related, come to resemble one another as a result of similar selection pressures. In divergent evolution, similar, related populations become more dissimilar, a process that can lead to the formation of new species.

Questions

1. Distinguish between genotypic variability and phenotypic variability. Which is acted upon by natural selection? Which is necessary for evolution?

2. What is the difference between artificial selection and natural selection? Under what circumstances might they become indistinguishable?

3. How does stabilizing selection affect the gene pool? Disruptive selection? Directional selection?

4. What characteristics in the contemporary human population are probably being maintained by stabilizing selection? Which might be subject to directional selection?

5. Imagine a rapidly expanding population in an environment that allows every organism that is born to survive and reproduce to its maximum extent. Which organisms will make the greatest contribution to the gene pool of future generations? Is natural selection acting on such a population?

6. From an evolutionary perspective, longevity of animals past reproductive age is generally useless. Why is this so? What situations might be exceptions to this statement? Can

you think of any reason why longevity might be harmful from an evolutionary standpoint?

7. In his experiments with the long-tailed widowbird, Andersson clipped and reglued the tails of half of the males in the group with tails of normal length. Why did he perform this manipulation? Why was it important?

8. If you were a poisonous butterfly, would you rather have a Batesian mimic or a Müllerian one? Why?

9. In her experiments with quinine-dipped mealworms and their Batesian mimics, Brower dipped the mimics in distilled water before painting on the distinctive green band. Similarly, the mealworms on which orange bands were painted were first dipped in distilled water. Why?

10. Many of Darwin's arguments concerning evolution focused on imperfections rather than perfections. Can you explain why?

11. Distinguish between convergent evolution and divergent evolution. Which type of selection would you expect to be the major factor in convergent evolution? In divergent evolution?

Suggestions for Further Reading

Borgia, Gerald: "Sexual Selection in Bowerbirds," *Scientific American*, June 1986, pages 92–100.

Cook, L. M., G. S. Mani, and M. E. Varley: "Postindustrial Melanism in the Peppered Moth," *Science*, vol. 231, pages 611–613, 1986.

Dawkins, Richard: *The Blind Watchmaker*, W. W. Norton & Company, New York, 1986.*

A thorough and delightfully written explication of modern evolutionary theory, the evidence in its support, and the controversies that have surrounded it.

*Available in paperback.

Diamond, Jared: "A Pox upon Our Genes," *Natural History*, February 1990, pages 26–30.

Diamond, Jared M.: "Rapid Evolution of Urban Birds," *Nature*, vol. 324, pages 107–108, 1986.

Eldredge, Niles (ed.): *The* Natural History *Reader in Evolution*, Columbia University Press, New York, 1987.*

An anthology of articles from Natural History, *with emphasis on adaptation through natural selection, speciation, and the rise and fall of major groups of organisms.*

Gould, James L., and Carol Grant Gould: *Sexual Selection*, W. H. Freeman and Company, New York, 1989.

A clearly written, authoritative account of one of the most fascinating topics in evolutionary biology, enriched with a great variety of entertaining examples.

Gould, Stephen Jay: *Bully for Brontosaurus: Reflections in Natural History*, W. W. Norton & Company, New York, 1992.*

Gould, Stephen Jay: *Eight Little Piggies: Reflections in Natural History*, W. W. Norton & Company, New York, 1993.

Gould, Stephen Jay: *Ever Since Darwin: Reflections in Natural History*, W. W. Norton & Company, New York, 1992.*

Gould, Stephen Jay: *The Flamingo's Smile: Reflections in Natural History*, W. W. Norton & Company, New York, 1987.*

Gould, Stephen Jay: *Hen's Teeth and Horse's Toes: Further Reflections in Natural History*, W. W. Norton & Company, New York, 1984.*

Gould, Stephen Jay: *The Panda's Thumb: More Reflections in Natural History*, W. W. Norton & Company, New York, 1992.*

Collections of thoughtful, witty, and well-written essays from Natural History. *Evolution is the central theme.*

Gross, Mart R.: "Disruptive Selection for Alternative Life Histories in Salmon," *Nature*, vol. 313, pages 47–48, 1985.

Hutchinson, G. Evelyn: *The Ecological Theater and the Evolutionary Play*, Yale University Press, New Haven, Conn., 1965.

By one of the great modern experts on freshwater ecology, this is a charming and sophisticated collection of essays on the influence of environment in evolution— and also on an astonishing variety of other subjects.

May, Robert M.: "Evolution of Pesticide Resistance," *Nature*, vol. 315, pages 12–13, 1985.

Owen, Denis: *Camouflage and Mimicry*, University of Chicago Press, Chicago, 1982.*

A short, but thorough discussion of the diversity of ways in which organisms deceive other organisms. A multitude of wonderful photographs accompany the text.

Ryan, Michael J.: "Signals, Species, and Sexual Selection," *American Scientist*, vol. 78, pages 46–52, 1990.

Simpson, George Gaylord: *Penguins: Past and Present, Here and There*, Yale University Press, New Haven, Conn., 1983.*

As Simpson said, "Penguins are beautiful, interesting, inspiring, and funny." They are also excellent examples of evolution and adaptation, as this informal account demonstrates.

Small, Meredith F.: "Female Choice in Mating," *American Scientist*, vol. 80, pages 142–151, 1992.

Thornhill, Randy: "The Allure of Symmetry," *Natural History*, September 1993, pages 30–37.

Thornhill, Randy, and Darryl T. Gwynne: "The Evolution of Sexual Differences in Insects," *American Scientist*, vol. 74, pages 382–389, 1986.

Vane-Wright, R. I.: "A Case of Self-Deception," *Nature*, vol. 350, pages 460–461, 1991.

Walker, Tim: "Butterflies and Bad Taste: Rethinking a Classic Tale of Mimicry," *Science News*, June 1, 1991, pages 348–349.

*Available in paperback.

The Tasmanian devil, with its sharp teeth and powerful jaws, is both a hunter and a scavenger. Although it can kill living prey, such as lambs and chickens, it prefers carrion and can devour the entire carcass of a sheep, including the bones. Tasmanian devils are found today only on the island of Tasmania, just south of the continent of Australia. The introduction of the wild dog known as the dingo to the Australian mainland some 3,500 years ago apparently led to the extinction of the Tasmanian devil there.

Like kangaroos and koalas, Tasmanian devils are marsupials. Marsupials, unlike the more familiar placental mammals, are born at an extremely immature stage and then continue their development in a special protective pouch on the mother's body. They are native to Australia, South America, and a few neighboring islands, such as Tasmania.

The Tasmanian devil and other marsupials remind us that exotic lands and their inhabitants played an important role in raising the questions that led to the theory of evolution. Young, curious travelers, such as Darwin and Wallace, observed a profusion of different organisms and came to doubt the doctrine of special creation, which held that each species was created separately and remained fixed and immobilized in time. Madagascar, the Galapagos, Borneo—each of these faraway places has, as we have seen, its own inhabitants, many of them found nowhere else in the world.

Just as island biology raised the question of the origin of species for the early students of evolution, it has helped to answer many questions about that process for later observers. Islands are living laboratories in which organisms can become separated in time and space from their parents and their predecessors and thus set out on a new and different evolutionary adventure. Islands, as you will see, come in many different shapes, sizes, and disguises.

The Origin of Species

In the course of his voyage on H.M.S. *Beagle*, Charles Darwin was struck not only by the enormous number of species he encountered for the first time but also by their distribution. According to the doctrine of the time, accepted by Darwin when he set sail, each species had been created separately and was distinct from all other species. Moreover, each was specially created for a particular way of life and placed in the locality for which it was best suited.

It is no coincidence that Darwin first came to doubt the doctrine of special creation while traveling in the tropics, where the widest variety of species is found and where, even today, many species new to science are encountered. For example, the island continent of Australia is home to some 57 separate species of kangaroos. All of the kangaroos are marsupials. Had each kangaroo species been specially created and dropped off in Australia? And why only in Australia? Or, more plausibly, was there perhaps an ancestral marsupial that had given rise to all of these clearly related kangaroos?

Moreover, it became apparent to Darwin that species are not as distinct as had previously been thought. For example, as he traveled down the eastern coast and up the western coast of South America, he observed gradual changes in various characteristics of the plants and animals—evidence that organisms become modified with time, according to the different environments in which they live.

Darwin—like his friend and correspondent, Alfred Russel Wallace (page 11)—was also puzzled by the fact that places similar in climate and topography are often populated by very different organisms. For example, remote oceanic islands often have no terrestrial mammals at all but only peculiar species of bats. Moreover, bats are the only native placental mammals of Australia. All of the other diverse mammals indigenous to Australia are marsupials, all clearly related to one another and found only rarely elsewhere on the planet. Great Britain and continental Europe have rabbits galore, whereas similar areas in South America have only the Patagon-

(a)

(b)

(c)

21–1   *Although these three mammals are similar in appearance and have similar life styles, they are not closely related. (a) The European rabbit, a placental mammal, is classified as a lagomorph. (b) The Patagonian hare, or cavy, also a placental mammal,*   *is classified as a rodent. (c) The Australian "banded hare" is actually a wallaby, a member of the kangaroo family. Like almost all of the other native mammals of Australia, it is a marsupial.*   *The scientific study of the distribution of plants and animals in the various regions of the world, known as biogeography, was initially developed by Alfred Russel Wallace.*

ian hare—not a rabbit or hare at all but a rodent (Figure 21–1). A multitude of examples of this sort provided strong evidence for Darwin that living things are what and where they are because of events that occurred in the course of their previous history.

Although Darwin titled his great text *On the Origin of Species,* he was never really able to explain how species might originate. However, an enormous body of work, mostly in the twentieth century, has provided many insights into the process. A crucial starting point was the development of a clear definition of the term "species."

What Is a Species?

In Latin, *species* simply means "kind," and so species are, in the simplest sense, different kinds of organisms. A more precise definition is that a **species** is a group of natural populations whose members can interbreed with one another but cannot (or at least usually do not) interbreed with members of other such groups. The key concept in this definition is **genetic isolation:** if members of one species freely exchanged genes with members of another species, they could no longer retain those unique characteristics that identify them as different kinds of organisms.

From an evolutionary perspective, a species is a group or population of organisms, reproductively united but very probably changing as it moves through space and time. Splinter groups, reproductively isolated

from the population as a whole, can undergo sufficient change that they become new species. This process is known as **speciation.** Occurring repeatedly in the course of more than 3.5 billion years, it has given rise to the diversity of organisms that have lived in the past and that live today.

Modes of Speciation

The members of a species share a common gene pool effectively separated from the gene pools of other species. A central question, then, is how one pool of genes splits off from another to begin a separate evolutionary journey. A subsidiary question is how two species, often very similar to one another, inhabit the same place at the same time and yet remain reproductively isolated.

According to current perspectives, speciation is most commonly the result of the geographic separation of a population of organisms: this process is known as **allopatric** ("other country") speciation. Under certain circumstances, speciation may also occur without geographic isolation, in which case it is known as **sympatric** ("same country") speciation.

Allopatric Speciation

Every widespread species that has been carefully studied has been found to contain geographically representative populations that differ from each other to a greater or lesser extent. Examples are the ecotypes of

(a)

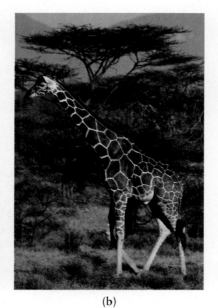
(b)

21–2 *Subspecies are variants of a species that are sufficiently different to have been formally recognized with a Latin name. Two subspecies that have formed on either side of a natural geographic barrier, the Tana River in Kenya, are* (**a**) *the common giraffe,* Giraffa camelopardalis rothschildi, *and* (**b**) *the reticulated giraffe,* Giraffa camelopardalis reticulata. *The two are considered subspecies because they are phenotypically distinct (note the patterns of their coats), but they are not considered different species because they can interbreed, producing fertile hybrids. Six other subspecies of giraffes are also recognized.*

(a)

(b)

(c)

Potentilla glandulosa (page 346) and the subspecies of giraffes (Figure 21–2). A species composed of such geographic variants is particularly susceptible to speciation if geographic barriers arise, preventing gene flow.

Geographic barriers are of many different types. Islands are frequently sites for the development of new species. The breakup of Pangaea (see essay on next page) profoundly altered the course of evolution, setting the island continent of Australia adrift as a veritable Noah's ark of marsupials. Populations of many organisms can become cut off from one another by barriers less obvious than oceans (Figure 21–3). For a plant, an island may be a mist-veiled mountaintop, and for a fish, a freshwater lake. A forest grove may be an island for a small mammal, and a few meters of dry ground can isolate two populations of snails. Islands may also form by the creation of barriers between formerly contiguous geographic zones. The Isthmus of Panama, for instance, has repeatedly submerged and reemerged in the course of geologic time. With each new emergence, the Atlantic and Pacific oceans became "islands," populations of marine organisms were separated from one another, and some new species formed. Then, when the oceans joined again (with the submergence of the Isthmus), North and South America were separated and became, in turn, the "islands."

(d)

21–3 *For different populations, the genetic isolation that can lead to speciation may take different geographic forms. Genetic "islands" may be* (**a**) *true islands,* (**b**) *mountaintops,* (**c**) *ponds, lakes, or even oceans, or* (**d**) *isolated clumps of vegetation.*

The Breakup of Pangaea

In the past 30 years, continental drift has become firmly established as part of the theory of plate tectonics. According to this theory, the outermost layer of our planet is divided into a number of segments, or plates. These plates, on which the continents and sea floors rest, move in relation to one another.

We see evidence of this relative motion mainly at the boundaries between adjacent plates. Where plates collide, volcanic islands such as the Aleutians may be formed or mountain belts such as the Andes or Himalayas may be uplifted. At the boundaries where plates are separating from one another, volcanic material wells up to fill the void. It is here that ocean basins are created. Plates may also move parallel to the boundary that joins them but in opposite directions, or in the same direction but at different speeds, as with the San Andreas fault.

About 200 million years ago, all the major continents were locked together in a supercontinent, Pangaea. Several reconstructions of Pangaea have been proposed, one of which is shown here. It is generally agreed that Pangaea began to break up about 190 million years ago, about the time the dinosaurs approached their zenith and the first mammals began to appear. First, the northern group of continents (Laurasia) split apart from the southern group (Gondwana). Subsequently, Gondwana broke into three parts: Africa-South America, Australia-Antarctica, and India. India drifted northward and collided with Asia about 50 million years ago. This collision initiated the uplift of the Himalayas, which continue to rise today as India still pushes northward into Asia.

By the end of the Cretaceous period, about 65 million years ago (for a geologic timetable, see the inside of the front cover), South America and Africa had separated sufficiently to have formed half the South Atlantic. Europe, North America, and Greenland had also begun to drift apart. However, final separation between Europe and North America–Greenland did not occur until about 43 million years ago, during the Eocene epoch of the Tertiary period. Also during the Eocene, Australia finally split from Antarctica and moved northward to its present position. Later, during the Pliocene epoch, the two Americas were joined by the Isthmus of Panama, which was created by volcanic action.

From the outset, the theory of continental drift has been closely interwoven with that of evolution. One of the earliest and most impressive pieces of evidence in favor of continental drift was the discovery of fossil remains of a small, snaggle-toothed reptile, *Mesosaurus*, found in the coastal regions of Brazil and South Africa but nowhere else. Early in 1982, a team of American scientists returned from Antarctica with the first fossil of a land mammal ever found there—a marsupial. This find supports the theory that marsupials migrated by land from South America across Antarctica to Australia before the two separated, some 55 million years ago. Fossils of many extinct species of marsupials have been found in South America, but only some 40 species have survived to the present.

(a)

(b)

(a) *Pangaea.* **(b)** *The San Andreas fault is a boundary between two giant moving plates.* *The fault, running through San Francisco and continuing southeast of Los Angeles, is* *responsible for California's notorious earthquakes.*

If the isolated population is a small, peripheral one, it is more likely to differ from the parent population (because of the founder effect), and gene flow is less likely to occur between the two populations. Once separated, the isolated population may begin to diverge genetically under the pressure of different selective forces. If enough time elapses and the differing selective forces are sufficiently great, the isolated population may diverge so much that, even if it were reunited with the parent population, interbreeding under natural conditions would no longer occur. At this point, speciation is said to have taken place. Every isolated population does not, of course, become a species. It may rejoin the parent group or it may perish, which is probably the fate of most small, isolated populations.

Sympatric Speciation

Polyploidy

A well-documented mechanism by which new species are produced through sympatric speciation is **polyploidy,** an increase in the number of chromosomes beyond the typical diploid (2n) complement. Polyploidy may arise as a result of nondisjunction (page 198) during mitosis or meiosis, or it may be generated when the chromosomes divide properly during mitosis or meiosis but cytokinesis does not subsequently occur. Polyploid individuals can be produced deliberately in the laboratory by the use of the drug colchicine, which prevents separation of chromosomes during mitosis.

Polyploidy leading to the formation of new species sometimes occurs as a result of a doubling of the chromosome number within individual organisms of a species (Figure 21–4). This process has, for example, produced several species of whiptail lizards in the American Southwest. More frequently, however, new species are generated by a doubling of the chromosome number in hybrid organisms.

A **hybrid** organism is the offspring of parents of different species. Hybrids can occur in animals (such as the mule, which is the offspring of a donkey and a horse), but they are far more common in plants. Kentucky bluegrass, for instance, is a promiscuous hybridizer and has crossbred with many related species, producing hundreds of hybrid strains, each well adapted to the conditions of the area in which it grows. Such hybrids, spreading by means of asexual reproduction, are sometimes able to outcompete both parents. Asexually reproducing strains of hybrid plants may be considered species in that they are genetically isolated from their parents and from each other.

21–4   *Polyploidy within individual organisms can lead to the formation of new species. If the chromosomes of a diploid organism do not separate during meiosis (nondisjunction), diploid (2n) gametes may result. Union of two such gametes, produced either by the same individual or by different individuals of the same species, will produce a tetraploid (4n) individual. Although this individual may be capable of sexual reproduction, it will be reproductively isolated from the parent species.*

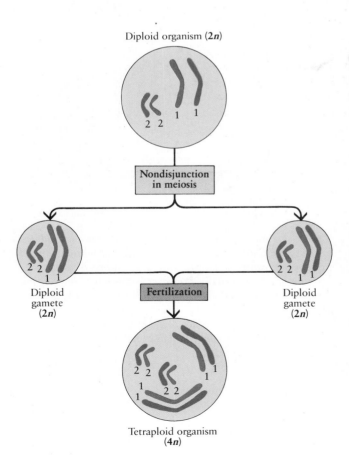

Diploid organism (**2n**)

Nondisjunction in meiosis

Diploid gamete (**2n**)

Diploid gamete (**2n**)

Fertilization

Tetraploid organism (**4n**)

(a)

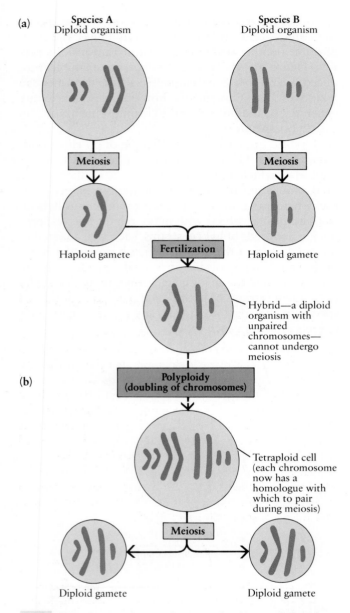

(b)

21–5 (a) *An organism, such as a mule, that is a hybrid between two different species and is produced from two haploid (n) gametes, can grow normally because mitosis is normal. It cannot reproduce sexually, however, because the chromosomes cannot pair at meiosis.* (b) *If polyploidy subsequently occurs and the chromosome number doubles, the hybrid can then produce viable gametes. Since each chromosome will have a partner, the chromosomes can pair at meiosis. The resultant gametes will be diploid (2n).*

Hybrids in both plants and animals are often sterile because the chromosomes cannot pair at meiosis (having no homologues), a necessary step for producing viable gametes (Figure 21–5a). If, however, polyploidy occurs in such a sterile hybrid and the resulting cells divide by further mitosis and cytokinesis so that they eventually produce a new individual asexually, that individual will have twice the number of chromosomes as its parent. As a consequence, it is reproductively isolated from the parental line. However, its chromosomes—now duplicated—can pair, meiosis can occur normally, and fertility is restored (Figure 21–5b). It is a new species, capable of sexual reproduction.

Sympatric speciation through polyploidy is an important, well-established phenomenon in plants. Approximately half of the 235,000 species of flowering plants have had a polyploid origin, and many important agricultural species, including wheat, are hybrid polyploids.

Disruptive Selection

Sympatric speciation as a result of disruptive selection (page 340) is a more controversial issue. Possible cases of this type of sympatric speciation are difficult to prove because of the time factor. Either (1) the two forms are in the process of diverging, as in the plants growing on contaminated soil, in which case it is not possible to prove that complete separation will ever occur, or (2) the two forms have separated completely, and it is not possible to prove that they were not formed as a consequence of geographic separation and have since been reunited in the same locale.

One often-cited example of sympatric speciation produced by disruptive selection involves a group of species of small fruit flies of the genus *Rhagoletis*. The larvae of each species of *Rhagoletis* feed on developing fruits of only one plant family. The host fruits of the family serve not only as food but also as the rendezvous for courtship and mating, followed by the deposition of eggs (Figure 21–6). *Rhagoletis pomonella,* for example, is a species that feeds on hawthorns. In 1865, farmers in the Hudson River Valley of New York reported that some of these flies had begun attacking their apples. The infestation then spread rapidly to apple orchards in adjacent areas of Massachusetts and Connecticut. There are now two distinct varieties of *R. pomonella,* with a number of genetic differences. One variety feeds and reproduces on hawthorns, and the other, known as the apple maggot fly, on apples. The two varieties are isolated by their reproductive behavior and so may be regarded as on the path to speciation.

Many biologists, however, question whether the two varieties of *R. pomonella* are truly sympatric. They

21–6 *A female apple maggot fly,* Rhagoletis pomonella. *Typically, the male lies in wait on the surface of an apple. When a female arrives, attracted by the ripening fruit, he will attempt to mate with her. The female subsequently lays her eggs, one at a time, on different apples. Before laying an egg, she probes the apple, as shown here, testing for the presence of nutrients and the absence of harmful chemicals. The eggs are deposited beneath the skin, where they hatch into small, white larvae that feed on the flesh of the apples. Apples infested by the larvae soon rot and fall to the ground.*

point out that this may instead be an example of allopatric processes operating on a greatly reduced scale. For these flies—and other small organisms that are parasitic or highly specialized for feeding and reproducing in a particular microhabitat—environmental differences that appear small to us may represent significant barriers, creating "mini-islands."

Maintaining Genetic Isolation

Once speciation has occurred, the now-separate species can live together without interbreeding, despite the fact that some are so similar phenotypically that only an expert with a microscope can tell them apart—*Dro-*

sophila offers several examples. What factors operate to maintain genetic isolation of closely related species?

Isolating mechanisms may be conveniently divided into two categories: **premating mechanisms,** which prevent mating between members of different species, and **postmating mechanisms,** which prevent the production of fertile offspring from such matings as are attempted or do occur. One of the most significant things about the postmating isolating mechanisms is that, in nature, they are rarely tested. The premating mechanisms alone usually prevent any interbreeding. This is an expected consequence of sexual selection. Choosing the best mate or mates from among the members of one's own species requires a finer level of discrimination than is required to distinguish members of one's own species from those of a different species (Figure 21–7).

21–7 *A male frigate bird, a resident of the Galapagos Islands, displays his crimson pouch. Throughout the courtship period, the pouch remains bright and inflated, even when the bird is flying or sleeping. The female will not mate with a male that does not have these species-specific characteristics.*

21–8   *Sticklebacks, small freshwater fish, have elaborate mating behavior. The male at breeding time, in response to increasing periods of sunlight, changes from dull brown to the radiant colors shown here (upper left). He builds a nest and (**a**) begins to court females, zigging toward them and zagging away from them. A female ready to lay eggs responds by displaying her swollen belly. (**b**) The male then leads her down to the tunnel-like nest. (**c**) After she enters the nest, he prods her tail and, in response, she lays her eggs and swims off. (**d**) He follows her through, fertilizes the eggs, and stays to tend the brood. If either partner fails in any step of this quite elaborate ritual, no young are produced.*

Premating Isolating Mechanisms

Premating isolating mechanisms in vertebrates often involve elaborate behavioral rituals, visual signals, or frequently, a combination of the two (Figure 21–8).

Visual and auditory signals are also important in many insects. Among the most familiar are the flashings of fireflies (which are actually beetles) and the chirping of crickets. In fireflies, each species has its own particular flashing pattern, which is different from that of other species both in duration of the flashes and in intervals between them. For example, one common flash pattern consists of two short pulses of light separated by about 2 seconds, with the phrase repeated every 4 to 7 seconds. The lights also vary among the species in intensity and color. The male flashes first, and the female answers, returning the species-specific signal. The predatory females of some firefly species not only produce the signal specific to their own species but also mimic the signals of other species, luring males of those species to the female's dinner table. This mimicry—and the subsequent feeding—generally occurs several days after the female has mated with a male of her own species.

Bird songs, frog calls, and the strident love notes of cicadas and crickets all serve to identify members of a species to one another. For example, studies of leopard frogs have revealed that some of the populations that can crossbreed under laboratory conditions do not do so in the wild. They are effectively isolated from one another by differences in their mating calls.

In many species of animals, chemical substances released by either the female or the male serve to bring the two sexes together. They may function as signals to attract the male, as in the case of the cecropia moth, or to trigger the release of gametes by the female, as with oysters. Because such substances are species-specific, they also act, in effect, as isolating mechanisms.

Temporal differences also play an important role in reproductive isolation. Species differences in flowering times are important isolating mechanisms in plants. Most vertebrates—we are a notable exception—have seasons for mating, often controlled by temperature or by day length. Figure 21–9 shows the mating calendar of different species of frogs and toads near Ithaca, New York.

Postmating Isolating Mechanisms

On the rare occasions when members of two different species attempt to mate or succeed in doing so, anatomical or physiological incompatibilities often maintain

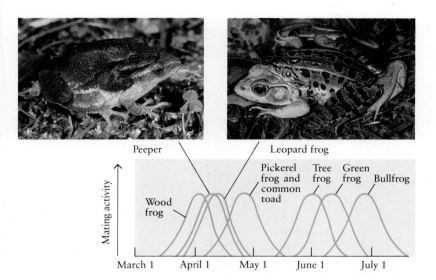

21–9 *Mating timetable for various frogs and toads that live near Ithaca, New York. In the two cases where two different species have mating seasons that coincide, the breeding sites differ. Peepers prefer woodland ponds and shallow water, whereas leopard frogs breed in swamps. Pickerel frogs mate in upland streams and ponds, and common toads use any ditch or puddle.*

genetic isolation. These postmating isolating mechanisms are of several different types. For instance:

1. Differences in the shape of the genitalia may prevent insemination, or differences in flower shape may prevent pollination.
2. The sperm may not be able to survive in the reproductive tract of the female, or the pollen tube may not be able to grow on the stigma.
3. The sperm cell may not be able to fuse with the egg cell.
4. The egg, once fertilized, may not develop.
5. The young may survive, but they may not become reproductively mature.
6. The mature offspring may be hardy but sterile—the mule, for example.

Postmating mechanisms reinforce premating ones. A female cricket that answers to the wrong song or a frog whose individual calendar is not synchronized with that of the rest of the species will contribute less (or nothing) to the gene pool. As a consequence, there is a steady selection for premating isolating mechanisms.

An Example of Speciation: Darwin's Finches

Admirers of Charles Darwin find it particularly appropriate that one of the best examples of speciation is provided by the finches observed by Darwin on his voyage to the Galapagos Islands. All the Galapagos finches are believed to have arisen from one common ancestral group—perhaps either a single pair or even a single female bearing a fertilized egg—transported from the South American mainland, some 950 kilometers away.

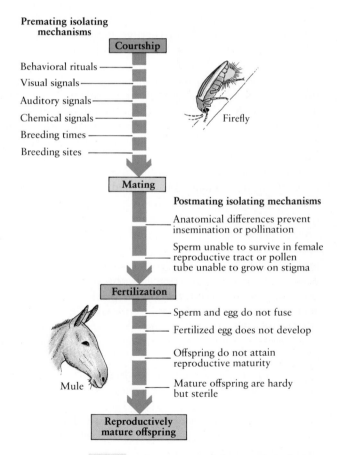

21–10 *A summary of the genetic isolating mechanisms that usually prevent successful interbreeding by organisms of different species. Depending on the species involved, the complex process from courtship (or flowering, in plants) to reproductively mature offspring can be interrupted at any one of a number of different stages.*

How she or they got there is, of course, not known, but it may have been the result of some particularly severe storm. (Periodically, for example, some North American birds and insects appear on the coasts of Ireland and England after having been blown across the North Atlantic.) It is very likely that finches were the first land birds to colonize the islands, which provide a highly diversified environment. Different groups of finches, occupying different habitats, would have been subjected to widely differing selective forces.

From the small ancestral group, 13 different species arose, plus one to the northeast on the Cocos Islands, 1,000 kilometers away. Apparently the various islands were near enough to one another that, over the years, small founding groups could emigrate. The islands were far enough apart, however, that once a group was established, there would be little or no gene flow between it and the parent group for a period of time long enough for significant differences to evolve as a result of natural selection and for genetic barriers to develop. (Finches are not very good long-distance fliers; if they were, this natural experiment in speciation would have been a failure.) Development of a new species in this way requires, it is estimated, at least 10,000 years of geographic isolation. However, since there are many islands, different species could have been evolving on different islands during the same time period.

The ancestral type was a finch, a smallish bird with a short, stout, conical bill adapted for seed-crushing. The ancestor is believed to have been a ground-feeding finch, and six of the Galapagos finches are ground finches. Four species of ground finches now live together on most of the islands. Three of them eat seeds and differ from one another mainly in the size of their beaks, which, in turn, influences the size of seeds they eat. The fourth lives largely on the prickly pear and has a much longer and more pointed beak. The other two species of ground finch are usually found only on outlying islands, where some supplement their diet with cactus.

In addition to the ground finches, there are six species of tree finches, also differing from one another mainly in beak size and shape. One has a parrotlike beak, suited to its diet of buds and fruit. Four of these tree finches have insect-eating beaks, each adapted to a different size range of insects. The sixth, and most remarkable of the insect eaters, is the woodpecker finch (Figure I–8, page 9).

By all ordinary standards of external appearance and behavior, the thirteenth species of Galapagos finch would be classified as a warbler. Its beak is thin and pointed like a warbler's, it even has the warblerlike habit of flicking its wings partly open, and, warblerlike,

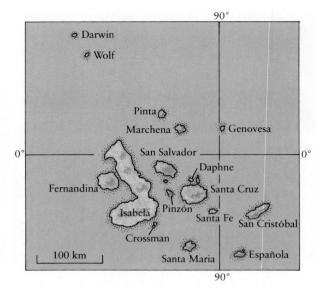

21–11   *The Galapagos archipelago, some 950 kilometers west of the coast of Ecuador, consists of 13 principal volcanic islands and many smaller islets and rocks. These islands have been called "a living laboratory of evolution." Species and subspecies of plants and animals that have been found nowhere else in the world inhabit the Galapagos. "One is astonished," wrote Charles Darwin in 1837, "at the amount of creative force . . . displayed on these small, barren, and rocky islands. . . ."*

it searches through leaves, twigs, and ground vegetation for small insects. However, its internal anatomy and other characteristics clearly place it among the finches. There is general agreement that it, too, is a descendant of the common ancestor or ancestors.

Studies directed by Peter R. Grant of Princeton University have demonstrated that reproductive isolation of the different finch species that now live together on a number of the islands depends on a combination of song and beak characteristics. It appears that young male and female finches learn to recognize their species-specific song from hearing the singing of their father and to recognize the beak of their species from observation of the beaks of both parents during feeding. Subsequently, songs, which are sung by the males only, serve as long-range cues to species identity, and beak size and shape serve as short-range cues. Females enter the territory of a courting male in response to his song but appear, at least in part, to base their decision on whether to mate on an inspection of his beak. The males are also known to actively pursue females. Sometimes a male finch will mistakenly pursue a female of another species—the finches often look very much alike from the rear—only to lose interest as soon as he sees her beak. He will rarely court a female of another species following a close inspection of her beak.

21-12   *Six of the 13 different species of Darwin's finches. Except for (a) the warbler finch (Certhidea olivacea), which resembles a warbler more than a finch, the species look very much alike. The birds are all small and dusky-brown or blackish, with stubby tails. The most obvious differences among them are in their bills, which vary from small, thin beaks to large, thick ones.*

(b) The small ground finch (Geospiza fuliginosa) and (c) the large ground finch (Geospiza magnirostris) are both seed eat- ers. G. magnirostris, *with a much larger beak than* G. fuliginosa, *is able to crack larger seeds.*

(d) The cactus finch (Geospiza scandens) lives on cactus blooms and fruit. Notice that its beak is more pointed than those of the two ground finch species.

(e) The small tree finch (Camarhynchus parvulus) and (f) the large tree finch (Camarhynchus psittacula), both insect eaters, take prey of different sizes.

The Record in the Rocks

The Earth's long history is recorded in the rocks that lie at or near its surface, layer piled upon layer, like the chapters in a book. These layers, or strata, are formed as rocks in upland areas are eroded by wind and water, and the resulting pebbles, sand, and clay are carried to the lowlands and the seas. Once deposited, they slowly become compacted and cemented into a solid form as new material is deposited above them. As continents and ocean basins change shape over the millennia, some strata sink below the surface of an ocean or a lake, others are forced upward into mountain ranges, and some are worn away, in turn, by water, wind, or ice or are deformed by heat or pressure.

Individual strata may be paper-thin or many meters thick. They can be distinguished from one another by the types of parent material from which they were laid down, the way the material was transported, and the environmental conditions under which the strata were formed, all of which leave their traces in the rock. They can be distinguished, moreover, by the types of fossils they contain. Small marine fossils, in particular, can be associated with specific periods in the Earth's history. The fossil record is seldom complete in any one place, but because of the identifying characteristics of the strata, it is possible to piece together the evidence from many different sources. It is somewhat like having many copies of the same book, all with chapters missing—but different chapters, so it is possible to reconstruct the whole.

The geologic eras—Precambrian, Paleozoic, Mesozoic, and Cenozoic—which are the major volumes of the geologic record, were identified and named in the early nineteenth century. As shown in the geologic timetable on the inside of the front cover, these eras are subdivided into periods, many of which are named, quite simply, for the areas in which the particular strata were first studied or studied most completely: the Devonian for Devonshire in southern England, the Permian for the province of Perm in Russia, the Jurassic for the Jura Mountains between France and Switzerland, and so on.

In the Badlands of South Dakota, the waters of the Missouri River and its tributaries have carved through the rock, revealing the horizontal bands of geologic strata. These formations are of particular interest because they encompass the boundary between the Cretaceous and Tertiary periods—a time of mass extinction, including that of the dinosaurs (see page 369).

Early attempts to date the various eras and periods were based on the ages of the strata in relation to each other. (A stratum occurring regularly above another is younger than the one below it.) The relative ages of the strata were then compared to the Earth's age. The first scientific estimate of the planet's age was made in the mid-1800s by the British physicist Lord Kelvin. On the basis of his calculations of the time necessary for the Earth to have cooled from its original molten state, Kelvin maintained that the planet was about 100 million years old, a calculation that posed considerable difficulties for Darwin. (Kelvin was not aware of the existence of radioactive materials under the surface that heat the planet from within.) In the last 50 years, however, new methods for determining the ages of strata have been developed using measurements of the decay of radioactive isotopes. As a result, the estimated age of the Earth has increased from 100 million years to about 4.6 billion years.

Geologic strata are now dated whenever possible by analysis of radioactive isotopes contained in crystals of igneous rock (rock formed from molten material) associated with particular strata. As we noted in Chapter 1, many naturally occurring isotopes are radioactive. All the heavier elements—atoms that have 84 or more protons in the nuclei—are unstable and, therefore, radioactive. All radioactive isotopes emit energy (as particles or rays) at a fixed rate; this process is known as radioactive "decay." The rate of decay is measured in terms of half-life: the half-life of a radioactive isotope is defined as the time in which half the atoms in a sample lose their radioactivity and become stable. Since the half-life of an isotope is constant, it is possible to calculate the fraction of decay that will take place for a given isotope in a given period of time. The radiometric clock starts to tick when the crystalline rock is formed.

Half-lives vary widely, depending on the isotope. The radioactive nitrogen isotope ^{13}N has a half-life of 10 minutes, and the most common isotope of uranium (^{238}U) has a half-life of 4.5 billion years. This uranium isotope undergoes a series of decays, eventually being transformed to an isotope of lead (^{206}Pb). Thus the proportion of ^{238}U to ^{206}Pb in a given rock sample is a good indication of how long ago that rock was formed. Five different isotopes are now commonly employed as radiometric clocks, and in thousands of instances, rocks have been dated by three or more independent clocks.

Any theory of evolution requires, as Darwin knew well, that the Earth have a long history. Thus, these radiometric clocks are doubly important to modern students of evolution. First, they demonstrate conclusively that the Earth's age is close to 5 billion years. In other words, the Earth is indeed old enough for evolution to have produced the observed diversity of organisms. Second, they provide the tools for estimating the ages of various rocks—and of the fossils within them—and so, for unraveling the details of the Earth's biological past.

The Evidence of the Fossil Record

Although we may make inferences about the course of evolution from the observation of living organisms, the proving ground of evolutionary theory is the fossil record itself. This record reveals a succession of living forms, with simpler forms generally preceding more complex forms (see essay).

Geologic studies, as well as the collecting of plant and animal specimens, were among Darwin's activities aboard the *Beagle*. The coasts of South America were of particular interest because they showed evidence of widespread upheaval, with many geologic strata exposed. In the course of his explorations, Darwin came across many fossils of extinct mammals, including those of giant armadillos. The fact that the remains of extinct armadillos were buried in the same South American plains where the only surviving species of these strange armored mammals lived provided tangible evidence of change and history.

Since Darwin's time, a steady stream of new discoveries has enormously increased our knowledge of the fossil record, which now extends back more than 3 billion years. In the case of many groups of organisms—plants and vertebrate animals, for example, as we shall see in Chapters 25 and 27—fossils have been found that exhibit a graded series of changes in anatomical characteristics, linking older forms with the modern forms and revealing pathways of divergence from common ancestors.

The fossil record provides overwhelming evidence that evolution has indeed occurred. It also reveals that the diversity of life on Earth is the result of three broad patterns of speciation, coupled with the effects of extinction.

Phyletic Change

One pattern observed in the fossil record is change within a single lineage of organisms. This is known as **phyletic change** (from the Greek *phylon,* meaning "race" or "tribe"). Under the pressures of directional selection, a species gradually accumulates changes until eventually it is so distinct from its predecessors that it may properly be considered a new species—a new kind—of organism. Darwin's concept of evolution emphasized the slow, gradual accumulation of change, and phyletic change is a large-scale version of the types of changes so ably documented by population geneticists.

Cladogenesis

A second pattern observed in the fossil record is the splitting of a lineage of organisms into two or more distinct lineages. This is known as **cladogenesis** (the forming of branches). Species formed by cladogenesis are the contemporaneous descendants of a common ancestor—like Darwin's finches.

Paleobiologists have tended to give more weight to the role of cladogenesis in evolution than to phyletic change. Ernst Mayr, an ornithologist and a leader in the formulation of modern evolutionary theory, agrees. He maintains that the formation of new species by the splitting off of small populations from the parent stock is responsible for almost all major evolutionary change. In such small populations, favorable genetic combinations, if present, can increase rapidly in number and frequency without being diluted out by gene flow to and from the larger parent population. Hence, evolution probably does not occur steadily and gradually but in spurts, accounting for the sudden increases of new species observed at many points in the fossil record. However, these two models—phyletic change and cladogenesis—have generally been viewed not as alternatives but rather as complementary to one another.

Adaptive Radiation

Another pattern observed in the fossil record is **adaptive radiation,** which is usually considered to result from the combined effects of both cladogenesis and phyletic change. Adaptive radiation is the sudden (in geologic time) diversification of a group of organisms that share a common ancestor, often itself newly evolved. It is associated with the opening up of a new biological frontier that may be as vast as the land or the air, or, as in the case of the Galapagos finches, as small as an archipelago. Adaptive radiation results in the almost simultaneous formation of many new species.

The late George Gaylord Simpson, a leading paleontologist, emphasized that adaptive radiation is the major pattern revealed by the fossil record. There are many examples. For instance, some 300 million years ago, the reptiles, liberated from an amphibian existence by the "invention" of an egg that retained its own water supply, diversified rapidly as they moved into a variety of different terrestrial environments. A similar, even more rapid burst of evolution later gave rise to the birds. As the dinosaurs became extinct, the mammals similarly burst forth upon the evolutionary scene, with many different kinds appearing simultaneously in the fossil record (Figure 21–13).

Extinction

A phenomenon particularly well documented by the fossil record is **extinction**—the complete demise not only of individual species but also of entire groups of

Placentals

Marsupials

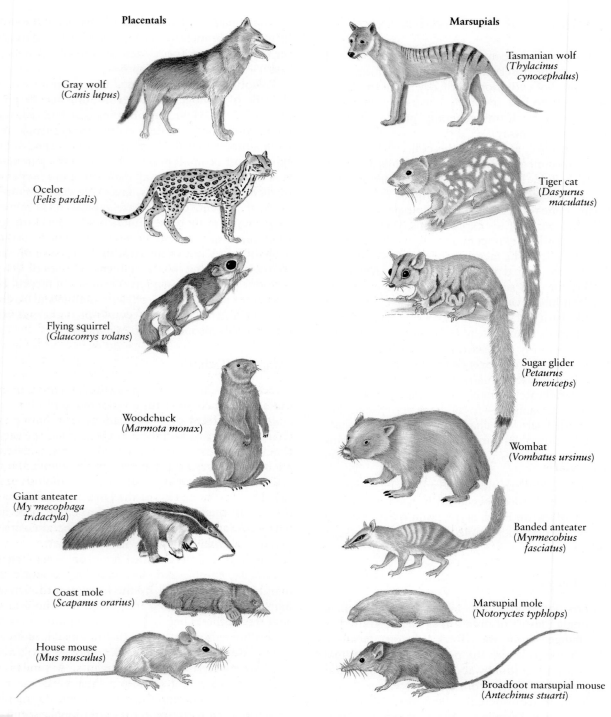

Gray wolf
(*Canis lupus*)

Tasmanian wolf
(*Thylacinus
cynocephalus*)

Ocelot
(*Felis pardalis*)

Tiger cat
(*Dasyurus
maculatus*)

Flying squirrel
(*Glaucomys volans*)

Sugar glider
(*Petaurus
breviceps*)

Woodchuck
(*Marmota monax*)

Wombat
(*Vombatus ursinus*)

Giant anteater
(*Myrmecophaga
tridactyla*)

Banded anteater
(*Myrmecobius
fasciatus*)

Coast mole
(*Scapanus orarius*)

Marsupial mole
(*Notoryctes typhlops*)

House mouse
(*Mus musculus*)

Broadfoot marsupial mouse
(*Antechinus stuarti*)

21–13 *Early in the evolutionary history of the mammals, the lineage gave rise to two principal branches, the marsupials and the placentals, which are distinguished primarily by the stage of development at which the young are born. Although the subsequent adaptive radiation of the marsupials in Australia and of the placentals on other continents were independent events, involving different immediate ancestors, striking simi-larities resulted in the descendant organisms. As shown in these examples, similar adaptations are exhibited by placentals and marsupials that occupy similar habitats and have similar life styles. All of the marsupials, however, are more closely related to each other than they are to any of the placentals, and vice versa.*

species. Only a small fraction of the species that have ever lived are presently in existence—certainly less than 1/10 of 1 percent, perhaps less than 1/1,000 of 1 percent.

In the 1980s, J. John Sepkoski, Jr., and David M. Raup of the University of Chicago gathered all available data on the extinction of marine organisms during the past 250 million years and analyzed the data statistically. Their analysis revealed a steady "background" extinction rate of about 180 to 300 species every one million years, interrupted every 26 million years by a period of mass extinctions. Some of these mass extinctions resulted in the elimination of enormous numbers of species. The greatest extinction of all occurred at the end of the Permian period, some 248 million years ago. It is estimated that 80 to 85 percent of all species then living became extinct, with the rate rising as high as 96 percent for marine species living in shallow waters.

Of all the organisms that once lived but live no more, the dinosaurs are perhaps the most intriguing. For some 150 million years these giant reptiles and their relatives dominated the land, air, and waters—a success story that indicates exquisite adaptation to their environments. And then, about 65 million years ago, at the end of the Cretaceous period, they vanished. Many other species of terrestrial animals died out at about the same time, including virtually all of those weighing more than 25 kilograms (about 55 pounds), as did a large proportion of the plants in temperate regions. Marine life was particularly hard hit. Paradoxically, tropical plants, small terrestrial animals, and freshwater organisms seem to have come through the general disaster relatively untouched.

In 1977, a group of scientists from the University of California at Berkeley, headed by geologist Walter Alvarez, made a most unexpected discovery. In the course of studies of sedimentary rocks in Italy, covering the transition from the Cretaceous period to the Tertiary period, they found that a layer of clay between the two sets of rocks contained unusually high levels of the metal iridium. This element, although relatively rare in the Earth's crust, is abundant in meteorites. Similar iridium deposits were subsequently found in clays at the Cretaceous-Tertiary boundary in Denmark, Spain, and New Zealand, and in deep-sea cores from both the Atlantic and Pacific Oceans.

The Berkeley scientists proposed that the cause of the iridium deposits—and the cause of the mass extinctions at the end of the Cretaceous period—was an asteroid, approximately 10 kilometers in diameter, that collided with Earth. They hypothesized that the impact and explosion of the asteroid kicked up a cloud of debris that circled Earth for a period of at least several months, producing continuous darkness, the cessation of photo-synthesis, and a subsequent collapse of food supplies for heterotrophic organisms, both terrestrial and marine. Such darkness would be expected to cause significant changes in climate, putting further stress on terrestrial organisms.

Enormous interest was generated by the discovery of the iridium deposits, by the asteroid hypothesis, and by the apparent periodicity of mass extinctions. As a consequence, scientists embarked on a variety of new studies. These have included a thorough reexamination of the fossil record, studies of the boundary layers associated with the different mass extinctions, explorations of possible terrestrial sources of high levels of iridium (for example, certain types of volcanic eruptions), new analyses of geologic phenomena known to have occurred in the periods marked by high rates of extinction, as well as a search for impact craters of the right size and age. In 1990, a buried impact crater of the right size—now estimated to be 300 kilometers in diameter—was located in the Caribbean, centered on the modern town of Chicxulub in the Yucatan Peninsula of Mexico. Different dating techniques have all placed its age at almost exactly 65 million years. A much smaller crater (32 kilometers in diameter) in Iowa has also been dated at 65 million years, suggesting that more than one impact may have been involved.

Many biologists, including those most familiar with the details of the fossil record on either side of the Cretaceous-Tertiary boundary, initially had serious doubts about the asteroid hypothesis. On the whole, the fossil evidence indicates that different groups did not become extinct simultaneously at the end of the Cretaceous period, but that the extinctions occurred over a period of tens of thousands—even hundreds of thousands—of years. This suggests more gradual processes at work. Nevertheless, there is a growing consensus that although asteroid impacts may not be the whole story, they played a major role in the extinctions that occurred 65 million years ago—and perhaps in other mass extinctions as well.

Whatever the cause or causes of mass extinctions, their effects on the subsequent course of evolutionary history are clear. When entire groups of organisms have died out, apparently without regard to the success of their particular solutions to the problems facing all living systems, new opportunities have opened up for the groups of organisms that survived. As the survivors have diversified in the course of exploiting newly available living space, new sets of solutions to the common problems have appeared, based on the survivors' genetic inheritance. At each mass extinction, the course of evolution has been dramatically altered, with some branches of the evolutionary tree permanently terminated and others undergoing vast new diversifications.

Equus: A Case Study

As a case study in evolution, let us consider horses, whose history is particularly well documented in the fossil record (Figure 21–14). At the beginning of the Tertiary period, shortly after the extinction of the dinosaurs, the horse lineage was represented by *Hyracotherium,* also known as Eohippus ("dawn horse"). It was a small herbivore (25 to 50 centimeters high at the shoulders) with three toes on its hind feet, four toes on its front feet, and doglike footpads on which its weight was carried. Its eyes were halfway between the top of its head and the tip of its nose. Its teeth had small grinding surfaces and low crowns, which probably could not have stood much wear. The teeth indicate that *Hyracotherium* did not eat grass; in fact, there was probably not much grass to eat then. It probably lived in forests and browsed on succulent leaves.

Slightly higher in the fossil-bearing strata, one finds larger horses, all three-toed, still browsing. Some of these clearly had molar teeth with large crowns that continued to grow as they were worn away (as do those of modern horses).

The strata of the Oligocene yield fossils of still larger horses. In these, the middle digit of each foot had expanded and was now the weight-bearing surface. In some, the two non-weight-bearing digits had become greatly reduced. These horses are thought to have been the direct ancestors of the genus *Equus,* which includes all modern horses.

These changes can be correlated with changes in the environment. In the time of *Hyracotherium,* much of the land was marshy and the chief vegetation was succulent leaves; the teeth of *Hyracotherium* were adapted for browsing. By the Miocene, the grasslands began to spread. Horses whose teeth became adapted to grinding grasses (which are very coarse and tough) survived, whereas those that remained browsers did not. The placement of the eye higher in the head may have facilitated watching for predators while grazing. The climate became drier, and the ground became harder; reduction of the number of toes, with the development of the spring-footed gait characteristic of modern horses, was an adaptation to harder ground and to larger size. An animal twice as high as another tends to weigh about eight times as much. Little *Hyracotherium* was probably as fast as the modern horse, but a larger, heavier horse with the foot and leg structure of *Hyracotherium* would have been too slow to escape from predators. (During this same period, predators were developing adaptations that rendered them better able to catch large herbivores, including horses.)

Reduction of toes   Increased grinding surface of molar teeth

Equus (Pleistocene)

Hoof

Pliohippus (Pliocene)

Merychippus (Miocene)

Mesohippus (Oligocene)

Hyracotherium (Eocene)

21–14   *The modern horse and some of its predecessors. Only one of the several branches represented in the fossil record is shown here. Over the past 60 million years, small several-toed browsers, such as* Hyracotherium, *were replaced in gradual stages by members of the genus* Equus, *characterized by, among other features, a larger size; broad molars adapted to grinding coarse grass blades; a single toe surrounded by a tough, protective keratin hoof; and a leg in which the bones of the lower leg had fused, with joints becoming more pulleylike and motion restricted to a single plane.*

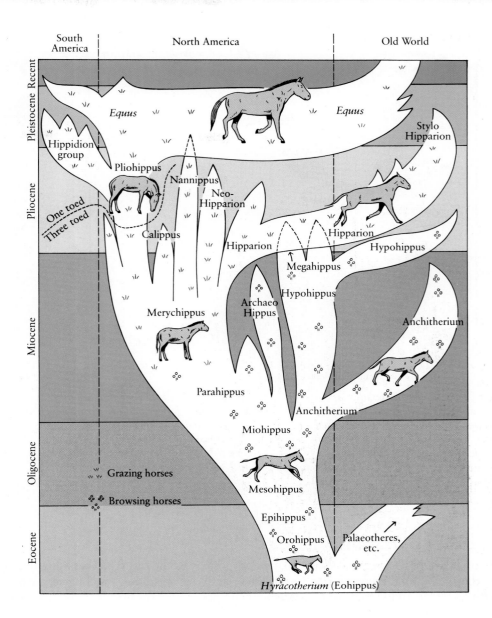

21–15 *The evolutionary history of the horse family, as summarized by George Gaylord Simpson in his classic book* Horses.

Thus the evolution of *Equus,* viewed from the long retrospective of the geologic record, represents an accumulation of adaptive changes related to pressure for increased size and for grazing. It might be considered— as it was for many years—as an example of phyletic change. As the number of fossil specimens has increased, however, it has become clear that at any given time many different species of horses coexisted, only some of which survived (Figure 21–15). Thus both phyletic change and cladogenesis are seen as contributing to the evolution of *Equus.*

Punctuated Equilibria

Although the fossil record documents many important stages in evolutionary history, there are numerous gaps. As Darwin noted in *The Origin of Species,* the geologic record is ". . . a history of the world imperfectly kept, and written in a changing dialect; of this history, we possess the last volume alone, relating only to two or three countries. Of this volume, only here and there a short chapter has been preserved; and of each page, only here and there a few lines."

Many more fossils have, of course, been discovered in the 100 years since Darwin's death. We are no longer limited to the most recent volume, and some of the volumes are now far more complete. Nevertheless, fewer examples of gradual change have been found than might have been expected. For many years, this discrepancy between the model of slow phyletic change and the poor documentation of such change in much of the fossil record was ascribed to the imperfection of the record itself.

In 1972, two young scientists, Niles Eldredge of the American Museum of Natural History and Stephen Jay

(a) Gradual phyletic change

(b) Punctuated equilibrium

21–16 *Two models of evolutionary change.* **(a)** *In the phyletic model, changes in anatomy and other characteristics within a lineage of organisms occur gradually and more or less continuously over relatively long periods of time.* **(b)** *In the punctuated equilibrium model, changes occur rapidly within short periods of time that are followed by much longer periods during which there are few detectable changes.*

The organisms shown here are African cichlid fishes. In Lake Victoria, which is less than 1 million years old, there are some 200 species of cichlid fishes, of many different sizes and shapes, adapted to the many different habitats and food sources available within the lake. Current evidence suggests that this diversity of species evolved from a common ancestor within the last 200,000 years.

Gould of Harvard University, ventured the proposal that perhaps the fossil record is not so imperfect after all. Both Eldredge and Gould have backgrounds in geology and invertebrate paleontology, and both were impressed with the fact that there was very little evidence of slow phyletic change in the fossil species they studied. Typically, a species would appear abruptly in fossil-bearing strata, last 5 million to 10 million years, and disappear, apparently not much different than when it first appeared. Another species, related but distinctly different, would take its place, persist with little change, and disappear equally abruptly. Suppose, Eldredge and Gould argued, that these long periods of no change, punctuated by gaps in the fossil record, are not flaws in the record but *are* the record, the evidence of what really happens.

How could new species make such sudden appearances? Eldredge and Gould found their answer in the model of allopatric speciation. If new species formed principally in small populations isolated from the parent population, if speciation occurred rapidly (by rapidly, paleobiologists mean in thousands rather than millions of years), and if the new species then outcompeted

the old species, taking over their geographic range, the resulting fossil pattern would be the one observed.

As first proposed, punctuated equilibria seemed to refer principally to the tempo of evolution. Population geneticists and Darwin before them had emphasized gradual change. There was clearly room, however, for the idea that populations would change more rapidly at some times than at others, particularly in periods of environmental stress, as in the case of the peppered moth (page 13).

As this model has become more fully developed, particularly by Steven M. Stanley of Johns Hopkins University, it has become more radical and more controversial. Its proponents now argue not only that cladogenesis is the principal mode of evolutionary change (as Mayr stated some 40 years ago) but that selection occurs among species as well as among individuals. Thus, at any one time in evolutionary history it may be possible to find a number of related species coexisting, each departing in a different way from the ancestral type. For example, as the horses evolved, some species became larger, others remained small. Some became grazers, others remained browsers. The overall trends in horse evolution resulted from the differential survival of species, rather than from the phyletic change that occurred within species. Thus, in this new formulation, species take the place of individuals, and speciation and extinction substitute for birth and death. The role of speciation is analogous to the role of mutation in population genetics; it provides the variation on which selection acts. According to this proposal, there are two levels of selection: in one, selection acts on the individual, and in the other, it acts on the species.

The punctuated equilibrium model has stimulated a vigorous and continuing debate among biologists, a reexamination of evolutionary mechanisms as currently understood, and a reappraisal of the evidence. These activities have at times been misinterpreted as a sign that Darwin's theory is "in trouble." In fact, they indicate that evolutionary biology is alive and well and that scientists are doing what they are supposed to be doing—asking questions. Darwin, we think, would have been delighted.

Summary

A species is defined as a group of natural populations whose members can interbreed with one another but cannot (or at least usually do not) interbreed with members of other such groups. In order for speciation—the formation of new species—to occur, populations that formerly shared a common gene pool must be reproductively isolated from one another and subsequently subjected to different selection pressures.

Two principal modes of speciation are recognized, allopatric ("other country") and sympatric ("same country"). Allopatric speciation occurs in geographically isolated populations. Sympatric speciation, which does not require geographic isolation, occurs principally in plants through polyploidy, often coupled with hybridization. It may also take place in some cases by disruptive selection.

The key event in speciation is genetic isolation. Once species have become genetically isolated, they can once again inhabit the same geographic area without interbreeding because of numerous behavioral, anatomical, and physiological mechanisms that prevent it.

The fossil record discloses four patterns of evolutionary change: phyletic change, cladogenesis, adaptive radiation, and extinction. Phyletic change is gradual change within a single lineage over time. Cladogenesis, by contrast, is evolutionary change produced by the branching off of populations from one another to form new species. Adaptive radiation is the rapid formation of many new species from a single ancestral group, characteristically to fill a new ecological zone. Extinction is the disappearance of a species from the Earth. The fossil record reveals a low, steady rate of extinction, interrupted periodically by mass extinctions involving enormous numbers of species. In modern evolutionary theory, large-scale evolution is regarded as the product of a combination of these patterns.

Paleontologists have presented evidence for an additional pattern of evolution known as punctuated equilibrium. They propose that new species are formed during bursts of rapid speciation among small isolated populations, that the new species outcompete many of the then-existing species (which become extinct), persist for long periods with little change, and then, in turn, abruptly become extinct. The punctuationalists propose that major changes in evolution take place as a result of selection acting on species as well as on individuals.

Questions

1. Distinguish among the following: allopatric speciation/sympatric speciation; hybridization/polyploidy; premating isolating mechanisms/postmating isolating mechanisms; phyletic change/cladogenesis/adaptive radiation.

2. Define genetic isolation. Why is it such an important factor in speciation?

3. Give three possible reasons why sympatric speciation by hybridization and polyploidy is more common in plants than in animals.

4. Two species of plants, each capable of reproducing both sexually and asexually, became distinct from their parental lineages as a result of polyploidy. Individual plants of these

two species then hybridized with each other. Would the resulting plants be capable of sexual reproduction? Would they constitute a new species? Explain your answers.

5. The genetic isolation of long-established, closely related species is generally maintained almost entirely by premating isolating mechanisms. Explain, in terms of natural selection, why this is to be expected.

6. Describe the separate steps involved in the formation of the distinct species of Galapagos finches. How does this example of speciation illustrate both the founder principle and adaptive radiation?

7. Darwin emphasized that when mammals (cats, goats, sheep, rodents, and so on) were introduced onto remote islands they often thrived there, yet there were no indigenous mammals on the islands except bats. Can you explain why he considered this important?

8. Could the fossil record disprove that evolution has occurred? If so, how?

9. Recent evidence supporting the periodic occurrence of mass extinctions suggests that Cuvier and Agassiz (page 6) were correct in their assertion that a series of catastrophes led to the demise of many previously existing forms of life. Neither of these eminent scientists, however, could provide a testable explanation for the proliferation of new forms that followed each of the proposed catastrophes. How would modern evolutionary theory explain the proliferation of new forms after mass extinctions?

Suggestions for Further Reading

Alvarez, Luis W., Walter Alvarez, Frank Asaro, and Helen V. Michel: "Extraterrestrial Cause for the Cretaceous-Tertiary Extinction," *Science,* vol. 208, pages 1095–1108, 1980.

Archibald, J. David, and William A. Clemens: "Late Cretaceous Extinctions," *American Scientist,* vol. 70, pages 377–385, 1982.

Carlquist, Sherwin: *Island Life: A Natural History of the Islands of the World,* Natural History Press, Garden City, N.Y., 1965.

An exploration of the nature of island life and the intricate and unexpected evolutionary patterns found in island plants and animals.

Carson, Hampton L., et al.: "Hawaii: Showcase of Evolution," *Natural History,* December 1982, pages 16–72.

A series of 10 articles on the evolution of the plants and animals of the Hawaiian Islands, which are, like the Galapagos Islands, a living laboratory of evolution.

Darwin, Charles: *The Origin of Species by Means of Natural Selection, or The Preservation of Favored Races in the Struggle for Life,* New American Library, Inc., New York, 1986.*

Darwin's "long argument." Every student of biology should, at the very least, browse through this book to catch its special flavor and to begin to understand its extraordinary force.

Darwin, Charles: *The Voyage of the Beagle,* New American Library, Inc., New York, 1988.*

Darwin's own chronicle of the expedition on which he made the discoveries and observations that eventually led him to his theory of evolution. The sensitive, eager young Darwin that emerges from these pages is very unlike the image many of us have formed of him from his later portraits.

Eldredge, Niles: *The Miner's Canary: Unravelling the Mysteries of Extinction,* Simon and Schuster, New York, 1991.

An account, for the general reader, of mass extinctions and their possible causes. Eldredge maintains that environmental degradation has played the major role in triggering past mass extinctions. An important aspect of this book is his analysis of what the currently accelerating rate of extinction reveals about the state of the global environment.

FitzGerald, Gerard J.: "The Reproductive Behavior of the Stickleback," *Scientific American,* April 1993, pages 80–85.

Gould, Stephen Jay: "Evolution and the Triumph of Homology, or Why History Matters," *American Scientist,* vol. 74, pages 60–69, 1986.

Gould, Stephen Jay: "Opus 200," *Natural History,* August 1991, pages 12–18.

Grant, Peter R.: *Ecology and Evolution of Darwin's Finches,* Princeton University Press, Princeton, N.J., 1986.*

A major work on the Galapagos finches, based on field studies carried out by Grant and his students over a period of some 15 years. In the course of these studies, an enormous quantity of data was gathered, which is analyzed from the perspective of modern ecological and evolutionary theory. Clearly written and beautifully illustrated, this book is already considered a classic.

Grant, Peter R.: "Speciation and the Adaptive Radiation of Darwin's Finches," *American Scientist,* vol. 69, pages 653–663, 1981.

Kerr, Richard A.: "Huge Impact Tied to Mass Extinction," *Science,* vol. 257, pages 878–880, 1992.

* Available in paperback.

Lack, David: *Darwin's Finches*, Cambridge University Press, New York, 1983.*

> *This short, readable book, first published in 1947, gives a marvelous account of the Galapagos, their finches and other inhabitants, and of the general process of evolution. Although in many respects superseded by Grant's 1986 book, it remains an excellent introduction to the Galapagos finches.*

Lewin, Roger: "A Lopsided Look at Evolution," *Science*, vol. 241, pages 291–293, 1988.

Marshall, Larry G.: "Land Mammals and the Great American Interchange," *American Scientist*, vol. 76, pages 380–388, 1988.

Mayr, Ernst: *The Growth of Biological Thought: Diversity, Evolution, and Inheritance*, Harvard University Press, Cambridge, Mass., 1982.*

> *This is the first of two volumes on the history of biology and its major ideas, written by one of the leading figures in the study of evolution. The introductory chapters provide an outstanding analysis of the philosophy and methodology of the biological sciences. This book, like Darwin's* The Origin of Species, *should at least be sampled by every serious student of biology.*

Mayr, Ernst: *Populations, Species, and Evolution: An Abridgement of Animal Species and Evolution*, Harvard University Press, Cambridge, Mass., 1970.*

> *A masterly, authoritative, and illuminating statement of contemporary thinking about species—how they arise and their role as units of evolution.*

Monastersky, Richard: "Closing in on the Killer," *Science News*, January 25, 1992, pages 56–58.

Monastersky, Richard: "Counting the Dead," *Science News*, February 1, 1992, pages 72–75.

Nance, R. Damian, Thomas R. Worsley, and Judith B. Moody: "The Supercontinent Cycle," *Scientific American*, July 1988, pages 72–79.

Prokopy, Ron, and Guy Bush: "Evolution in an Orchard," *Natural History*, September 1993, pages 4–10.

Raup, David M.: *Extinction: Bad Genes or Bad Luck?* W. W. Norton & Company, New York, 1992.*

Raup, David M.: *The Nemesis Affair: A Story of the Death of Dinosaurs and the Ways of Science*, W. W. Norton & Company, New York, 1986.*

> *Popular accounts of mass extinctions, the various hypotheses that have been proposed to explain them, and the debates within the scientific community over those hypotheses. Professor Raup, who has made major contributions to our knowledge of previous mass extinctions, is a lively and entertaining writer.*

Russell, Dale A.: "The Mass Extinctions of the Late Mesozoic," *Scientific American*, January 1982, pages 58–65.

Sagan, Carl, and Paul R. Ehrlich: *The Cold and the Dark: The World After Nuclear War*, W. W. Norton & Company, New York, 1985.*

> *A discussion, for the general reader, of the hypothesis that even a limited nuclear exchange would throw such quantities of smoke and debris into the atmosphere that a "nuclear winter"—analogous to the "winter" of the asteroid hypothesis—would result, with devastating biological consequences.*

Simpson, George Gaylord: *Horses*, Oxford University Press, New York, 1951.

> *The story of the horse family in the modern world and through 60 million years of history. Simpson was not only an outstanding vertebrate paleontologist but also a graceful and delightful writer.*

Simpson, George Gaylord: *Splendid Isolation: The Curious History of South American Mammals*, Yale University Press, New Haven, Conn., 1983.*

> *Charles Darwin found in the animals of South America some of the major clues that led to the formulation of* The Origin of Species. *Today, their history can be seen as an almost ideal natural experiment in evolution, told in Simpson's engaging style.*

Stanley, Steven M.: *Extinction*, W. H. Freeman and Company, New York, 1987.

> *A beautifully illustrated volume surveying the paleontological and geological evidence of the great extinctions of the past.*

Storch, Gerhard: "The Mammals of Island Europe," *Scientific American*, February 1992, pages 64–69.

Turco, Richard P., et al.: "The Climatic Effects of Nuclear War," *Scientific American*, August 1984, pages 33–43.

Vickers-Rich, Patricia, and Thomas Hewitt Rich: "Australia's Polar Dinosaurs," *Scientific American*, July 1993, pages 50–55.

Ward, Peter Douglas: *On Methuselah's Trail: Living Fossils and the Great Extinctions*, W. H. Freeman and Company, New York, 1993.*

> *A fascinating account of the ancient organisms, such as Nautilus,* the horseshoe crabs, *and the coelacanth, that survived the great extinctions of the past. As the author makes clear, these organisms have much to teach us.*

White, Mary E.: *The Flowering of Gondwana*, Princeton University Press, Princeton, N.J., 1990.

> *A marvelous study, enriched by some 400 photographs, of the plants of the supercontinent Gondwana and their descendant plants—both fossil and living—now found on the continent of Australia.*

* Available in paperback.

In many species of birds, the first behavior to appear in nestlings is a thrusting of the head upward toward the parent, with the mouth gaping open. A complementary behavior of the parent is the regurgitation of food into the open mouth.

The big bird in this nest is a baby European cuckoo. The little bird is an adult dunnock. Programmed by her evolutionary history, the dunnock finds the gaping beak of the nestling irresistible and spends all of her daylight hours feeding the greedy chick. The chick, however, is an imposter. His mother, following her own program, kept the dunnock's nest under surveillance until it had two or three eggs in it. She then waited until the nest was momentarily unguarded, swooped in, removed one of the eggs already there, and laid her own in its place—all in 10 seconds. As she was leaving, she thriftily swallowed the stolen egg.

Although cuckoos are much bigger than dunnocks, their eggs are not much larger than dunnock eggs and their incubation time is shorter. Consequently, the baby cuckoo hatched first. Once out of his shell, still blind and naked, he hoisted the dunnock eggs one by one into a special hollow on his back, shuffled to the edge of the nest, and heaved them over the rim. The cuckoo thus ensured himself a monopoly on all of the provisions that should have gone to the rightful progeny of his foster mother, the dunnock. In the meantime, his biological mother laid eight more eggs, each containing a murderous offspring, in eight other nests.

Brood parasitism, as this phenomenon is called, is found in some 80 species of birds. The most familiar to you may be the brown-headed cowbird, whose deceitful strategies are not quite as well developed as the cuckoo's, perhaps because they have a shorter evolutionary history. Note that although the behavior of the cuckoos seems wily and deliberate—and even subject to moral judgment—the cuckoos have no more choice about what they are doing than the seemingly industrious and virtuous dunnock has about feeding the chick. Indeed, they have no more choice about this behavior than they have about the colors of their feathers or the patterns of their flight.

The Evolution of Behavior

Virtually all organisms, from the simplest bacterial cells to the most complex animals, *act*. They seek out suitable environments in which homeostasis can be maintained. They obtain nutrients and other essential molecules, such as water and oxygen. Often they successfully avoid becoming nutrients for other organisms. And, they produce offspring, a process that can involve complex patterns of courtship, mating, and parenting. These activities, along with many others, constitute **behavior.**

The behavioral characteristics of an organism—its sensitivity to particular stimuli and its patterns of response to those stimuli—are as much the products of natural selection as the size and shape of a bird's beak or the process that regulates our blood cholesterol levels (see page 58). An organism's behavior is vitally important for its survival and the successful production of offspring.

The factors governing the evolution of behavioral characteristics are the same as those that apply to any other trait. First, there are variations among individual organisms in behavioral characteristics. Second, some of these variations are genetically determined; that is, they are influenced by the presence of particular alleles or combinations of alleles. Third, individuals with certain variations tend to have greater reproductive success than individuals with certain other variations. Fourth, as a result of this greater reproductive success, certain alleles tend to increase in the gene pool of a population, generation after generation.

The study of behavior and its evolution is currently one of the most active and exciting areas of biological research. The scope of the field is enormous, ranging from laboratory studies of bacteria to long-term studies of animals in the wild. We shall, of necessity, limit our discussion to a few of the most fascinating areas of this vast subject.

The Genetic Basis of Behavior

Behavior—whether that of a *Salmonella* cell navigating toward a food source (see page 113) or yours in reading and reacting to this text—has its roots in the genetic program carried in the DNA molecules of the individual. The steps between a sequence of nucleotides in a DNA molecule and the behavior of an organism, even a very simple one, are many and complex. The process involves, at the least, the synthesis of specific molecules, their organization into particular structures capable of receiving and responding to stimuli, the development of pathways for the transmission of information within the organism, and the modification of those structures and pathways as a result of interactions of the organism with its environment.

For some behaviors, the role of genetic components can be clearly demonstrated. A classic example is the behavior of honey bees when a pupa (Figure 22–1), undergoing metamorphosis within its cell in the hive, becomes diseased and dies. In some honey-bee strains, known as hygienic, worker bees uncap the cell and remove the corpse. In other strains, known as unhygienic, this behavior does not occur.

22–1   *Two honey-bee larvae (left) and two honey-bee pupae (right) in different stages of their development. Honey bees undergo their entire development from egg to adult in individual wax cells within the hive. During the early stages of development, the cells are open, enabling worker bees to feed the larvae continuously. When a larva has reached the size at which pupation begins, the workers add a wax cap to the cell, sealing in the pupa for its metamorphosis into an adult bee. When metamorphosis is complete, the adult bee emerges from the cell by gnawing through the wax cap.*

In a series of breeding experiments performed in the early 1960s, Walter Rothenbuhler first crossed unhygienic bees with hygienic bees. All of the offspring were unhygienic, indicating that this is a dominant characteristic. Rothenbuhler then made a testcross of these hybrid offspring with the original hygienic strain. Twenty-nine colonies of bees resulted. In eight colonies, the bees were unhygienic; they left the cells capped and did not remove dead pupae. Six colonies consisted of hygienic bees that uncapped the cells and removed dead pupae. In nine of the colonies, however, worker bees uncapped cells but left the dead pupae untouched, and in the remaining six colonies, they did not uncap cells but would remove the dead pupae if the cells were uncapped for them.

These results approximate a 1:1:1:1 ratio, which is the expected ratio in a testcross involving two independently assorted genes. They indicate that one gene, designated *U*, controls the uncapping behavior, whereas another gene, designated *R*, controls the removal behavior (Figure 22–2).

As this example illustrates, a single allele may determine a particular behavioral characteristic of an organism. For most behaviors, however, the underlying genetics are considerably more complicated. For some behaviors, genes with multiple effects (that is, pleiotropic genes) play an important role. Most often, however, behavioral characteristics are the result of polygenic inheritance—that is, they depend on the integrated action of the alleles of a large number of genes.

Proximate and Ultimate Causation

Imagine a pond on a spring day, with numerous tadpoles schooling near the water's edge. If you walk quietly along the bank of the pond and then make a sudden, loud noise, you will notice that the tadpoles immediately arch their bodies and shoot away into deeper water. Similarly, if you tap on the side of an aquarium in a pet shop, the fish also arch their backs, darting to the other side of the aquarium. Why do the tadpoles and the fish exhibit this behavior? This question, and similar questions about a vast diversity of behaviors exhibited by organisms, can be approached from two different points of view. One approach seeks an answer in terms of the **proximate cause**—that is, the immediate sequence of physiological events that lead to the observed behavior. Another approach seeks to explain the **ultimate cause** of the behavior—that is, its adaptive value and its evolutionary origins.

In the example of the startle behavior exhibited by many fish and larval amphibians, such as tadpoles, the

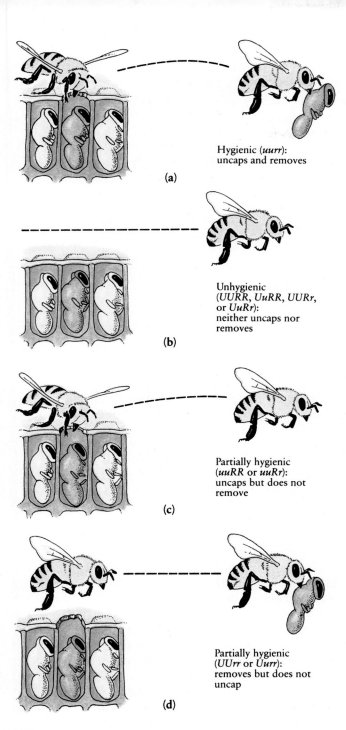

Hygienic (*uurr*):
uncaps and removes

(a)

Unhygienic
(*UURR, UuRR, UURr,*
or *UuRr*):
neither uncaps nor
removes

(b)

Partially hygienic
(*uuRR* or *uuRr*):
uncaps but does not
remove

(c)

Partially hygienic
(*UUrr* or *Uurr*):
removes but does not
uncap

(d)

22–2   *Hygienic behavior in honey bees is determined by two independently assorted genes, U and R. (a) Bees that are homozygous recessive for both genes* (uurr) *uncap the cells containing dead pupae and remove the corpses from the hive. (b) Bees that have at least one copy of the dominant allele for each gene (for example, UuRr) do neither. (c) Bees that are uuRR or uuRr uncap the cells but do not remove the corpses, while (d) bees that are UUrr or Uurr remove dead pupae from cells that have already been uncapped.*

proximate cause is stimulation of two giant neurons (nerve cells), known as Mauthner cells. Stimulation of the Mauthner cells by signals transmitted from the acoustic nerve leads to stimulation of other neurons on each side of the body. This, in turn, produces powerful contractions of the muscles of the body wall and tail that propel the animal forward.

The ultimate cause of the behavior, however, is to be found in its adaptive value in preserving the animal from predation. Individuals with a highly efficient response are more likely to avoid predation—and thus survive to reproduce—than those with a less efficient response.

Fixed Action Patterns

The comparative study of patterns of behavior and the construction of hypotheses concerning their evolutionary origins is known as **ethology,** from the Greek word *ethos,* meaning "character" or "custom." The pioneers in this field have been the European zoologists Konrad Lorenz, Niko Tinbergen, and Karl von Frisch.

Some behavior patterns—for example, the gaping of a baby bird's mouth (page 377) or the tongue flick of a toad attempting to capture a fly—develop with a minimum of sensory experience. Although the behavior may be subsequently refined, the pattern appears, essentially complete, the first time the organism encounters the relevant stimulus. Such an innate behavioral pattern, which tends to be highly stereotyped, rigid, and repetitive, is known as a **fixed action pattern.** The fixed action patterns of the members of a given species of the appropriate age, sex, and physiological condition are as specific and constant as their anatomical characteristics.

Fixed action patterns are initiated by external stimuli, known as **sign stimuli.** When these stimuli are communication signals exchanged between members of a species, they are known as **releasers.** Tinbergen hypothesized that certain circuits within the brain, which he termed **innate releasing mechanisms,** respond to the sign stimuli. A fixed action pattern remains blocked until the organism encounters the appropriate sign stimulus, which then stimulates the innate releasing mechanism, setting in action the sequence of movements that constitute the behavior.

An example of a fixed action pattern studied by Lorenz and Tinbergen is the egg-rolling behavior of greylag geese. The nests of these geese are shallow depressions in the ground. If an egg rolls out of her nest, a goose will retrieve it using a stereotyped sequence of movements (Figure 22–3). First, she stretches her neck toward the egg and then begins rolling it back under

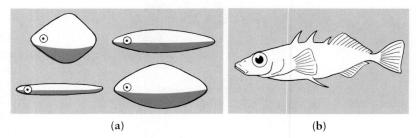

(a)   (b)

22–4 *Colors are the sign stimuli for many fixed action patterns in different animal species. In experiments by Niko Tinbergen with sticklebacks, (a) crude models, painted red on the lower surface,* *elicited much stronger reactions in both male and female sticklebacks (aggressiveness in the males and attraction in the females) than did (b) an exact replica of a male that was not colored.*

22–3 *The fixed action pattern by which a greylag goose retrieves an egg that has rolled from her nest. Because an egg is not a perfect sphere, a series of lateral movements of the goose's head are required to keep the egg on a straight course back to the nest.*

her chin. Through a series of lateral movements of her head, she keeps the egg rolling back on a straight line toward the nest. Occasionally, however, the egg slips away to the side—but the goose continues the retrieval movements all the way back to the nest, even though the egg is no longer there to retrieve. In this case, as in many others, once the fixed action pattern has been initiated, it cannot be altered but must be carried through to completion.

Through the use of physical models, ethologists have been able to identify the releasers for a number of fixed action patterns. For example, the red bellies of male sticklebacks in breeding condition are releasers not only for the stereotyped mating behavior of stickleback females (see page 362) but also for aggressive interactions between the males. Tinbergen constructed a series of models (Figure 22–4), by which he was able to show that a male stickleback reacts much more aggressively to a very crude model of a male with a red belly than to an exact replica of a stickleback without the red coloration.

Learning

All patterns of behavior, even those that seem relatively complete on their first appearance, depend not only on environmental cues but also on the normal physiological development of the animal. An enormous amount of animal behavior also requires learning—a process in which the responses of the organism are modified as a result of experience. The capacity for learning appears to be loosely correlated with the length of the life span and with the size and complexity of the brain. In small organisms, such as honey bees, with a short life span (and thus little available time for learning), most behavior appears to take the form of fixed action patterns. By contrast, in large organisms with a complex brain and a long life span, such as *Homo sapiens,* a large proportion of the behavioral repertoire is critically dependent on the prior experience of the individual.

One of the simplest forms of learning is **habituation,** in which an organism comes to ignore a persistent stimulus and go on about its other business, wasting neither time nor energy on a response. An example is provided by ground squirrels (Figure 22–5), which come to ignore the alarm calls of other ground squirrels *if* they are given frequently when there is, in fact, no danger.

Associative Learning

A type of learning familiar to us all is **association,** in which one stimulus comes to be linked, through experience, with another one. If you keep pets in your home, you will be able to cite many examples of associative learning, such as goldfish coming to the corner of the aquarium to be fed as you walk toward the tank or your dog becoming excited at the sight of a leash.

The first scientific studies of associative learning were performed in the 1920s by the Russian physiologist

22–5 *A female Belding's ground squirrel who has detected danger—most likely a potential predator—and is sounding the alarm. Other members of the species who hear the alarm usually retreat to the safety of their burrows. If, however, a particular squirrel repeatedly raises false alarms, the others stop responding to the calls of that squirrel.*

Ivan Pavlov. In his classic experiments, Pavlov restrained a hungry dog in a harness and offered it small portions of food at regular intervals. When he signaled the delivery of food by preceding it with an external stimulus, such as the sound of a bell or a signal light, the dog began to respond to the external stimulus by salivating. After every few trials, the external stimulus was presented without the food, and the amount of saliva produced by the dog was measured. Pavlov found that the number of drops of saliva triggered by the external stimulus alone was directly proportional to the number of previous trials in which it had been followed by food.

In natural settings, associative learning often involves trial and error. A newly hatched chick, for example, will peck at any spots that contrast with the background. As it grows older, it learns which of these spots represent edible objects, and its pecking accuracy improves. When young chicks first peck at grain, only about 15 percent of the pecks lead to obtaining a morsel worth swallowing. Accuracy improves with practice and also with maturity. Chicks that have been kept in the dark and fed by hand for a number of days peck more accurately than younger chicks, but they do not peck as accurately as chicks their own age that have had pecking experience.

Imprinting

Closely related to associative learning is the development of discrimination. Of vital importance for the ultimate reproductive success of many animals is the discrimination of members of one's own species from members of all other species, a discrimination that may be based on a variety of cues. In many species, particularly birds, this learning occurs very rapidly during a specific **critical period** in the early life of the individual and depends on exposure to particular characteristics of the parent or parents. This type of learning is known as **imprinting.**

The most familiar example of imprinting is the following response of birds that are physiologically mature enough to leave the nest soon after hatching (Figure 22–6). This keeps the young birds close behind and well within the protective range of the parent until the end of their juvenile period, when the following response is lost. Imprinting is also involved in song learning in birds, a process of considerable complexity.

22–6 *Many species of precocial birds (birds that are able to walk and feed as soon as they hatch) follow the first moving object they see after they have freed themselves from the egg. The object is usually mother goose (for goslings, that is), but it can be a matchbox on a string or a member of another species, such as zoologist Konrad Lorenz, shown here. In many species, this phenomenon, an example of the type of learning known as imprinting, also influences mate selection in the adult bird. Lorenz reported many examples of ducks and geese becoming sexually fixated on objects or on members of other species, including himself.*

CHAPTER 22 The Evolution of Behavior 381

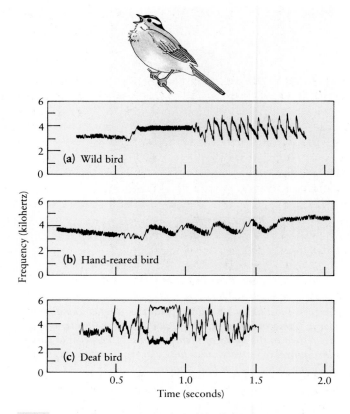

The Song of the White-Crowned Sparrow

About 150 days after hatching, a juvenile male white-crowned sparrow begins to utter a tentative, twittering call that has only a vague resemblance to the full song of the mature male. During the next 50 days or so, the juvenile's song gradually becomes more complex and sophisticated until, by the time the bird is 200 days old, he is singing a full song.

Laboratory experiments undertaken by ethologist Peter Marler and others have dissected the sequence of events in song learning in this species. Figure 22–7 shows sound spectrograms of (a) a wild bird, and (b) an adult, hand-reared bird that had never heard a song of its own species. Clearly, some form of learning is required for production of the full song. Exposing birds to tape-recorded songs at various stages of their development revealed that exposure to the full song of the white-crowned male in a critical period of 10 to 50 days after hatching is required for development of that song five to six months later. Even if a bird is kept in complete sound isolation after two months of age, he will produce a normal song if he was exposed to it during the critical period. If, however, he does not hear the song until about the time he begins to sing himself, he can never produce a full song. Exposure to the song of other sparrow species during the critical period has no effect.

If a bird is deafened after he has heard songs of his own species during the critical period—but before he himself has begun to sing—the full song is not normal (Figure 22–7c). It is, however, more complex than that of the bird that did not hear the song during the critical period. Apparently, then, for normal song, three requirements must be met: (1) the bird must have the genetic capacity to recognize and reproduce the song, (2) he must hear the song during the critical period for imprinting, and (3) he must be able to hear himself sing the song. As he sings, he apparently compares his own song to the song stored in his memory during the critical period, rehearsing until he gets it right. If a bird is deafened after he has mastered the full song, he continues to sing normally.

Imitative Learning

As the song of the white-crowned sparrow indicates, **imitation** is often a component of learning. A well-documented example of "monkey see, monkey do" is provided by the macaques (rhesus monkeys) of the island of Koshima, Japan. These animals lived and fed in the inland forest until about 40 years ago, when a group of Japanese researchers began throwing sweet potatoes on

22–7 *Sound spectrograms for adult male white-crowned sparrows: (a) wild; (b) reared by hand and isolated from songs of other white-crowned sparrows; (c) deafened prior to five months of age (before he himself began to sing), but after having heard normal white-crowned sparrow songs during the critical period. The unit in which the frequency (pitch) of the sound is expressed is the kilohertz; 1 kilohertz equals 1,000 cycles per second.*

the beach for them. The macaques quickly got used to venturing onto the beach, brushing sand off the potatoes, and eating them.

One year after the feeding started, a two-year-old female the scientists had named Imo was observed carrying a sweet potato to the water, dipping it in with one hand, and brushing off the sand with the other. Soon, other macaques began to wash their potatoes, too. Only macaques that had close associations with a potato washer took up the practice themselves. Thus it spread among close companions, siblings, and their mothers, but adult males, which were rarely part of these intimate groups, did not acquire the habit. However, when the young females that learned potato washing matured and had offspring of their own, those offspring learned potato washing from their mothers. Today all the macaques of Koshima dip their potatoes in the salt water to rinse them off, and many of them, having acquired a taste for salt, dip them between bites.

22–8   *A macaque of the island of Koshima, washing a sweet potato before eating it.*

This was only the beginning. Later, the scientists began scattering wheat kernels on the beach. Like the other macaques, Imo, then four years old, had been picking the grains one by one out of the sand. One day she began carrying handfuls of sand and wheat to the shore and throwing them into the water. The sand sank, the wheat kernels floated to the top, and Imo collected the wheat and ate it. The researchers were particularly intrigued by this new behavior since it involved throwing away food once collected, much less a part of the macaques' normal behavioral repertoire than holding on to food and cleaning it off. Washing wheat spread through the group in much the same way that washing potatoes had. Now the macaques, which had never even been seen on the beaches before the feeding program began, have taken up swimming. The youngsters splash in the water on hot days. Some of them dive and bring up seaweed, and at least one has left Koshima and swum to a neighboring island, perhaps as a cultural emissary.

Social Behavior: An Introduction

Of all behaviors, perhaps the most intriguing are the interactions that occur among animals, such as the macaques, living in structured societies. A **society** is a group of individuals of the same species, living together in an organized fashion, with divisions of resources, divisions of labor, and mutual dependence. Stimuli—that is, communications—exchanged among members of the group hold it together and maintain the social structure.

When biologists concerned with evolutionary theory began to analyze social behavior, some disturbing questions began to emerge. For instance, how can you explain, with a mechanism driven by differential reproductive success, the evolution of sterile castes—such as

worker bees—in insects? How is it that in many vertebrate societies only a few of the males breed, with their access to the females seldom successfully challenged by other males? Why, among animals in groups, do some utter warning cries or exhibit other forms of behavior that attract attention to the individual issuing the warning, thus threatening the warner's life? These behaviors all appear to be examples of **altruism**, which is, by definition, behavior that benefits others and is performed at some risk or cost to the doer. How can such acts of apparent altruism be explained in terms of natural selection acting on the individual organism?

Not all social behavior, of course, is altruistic. Table 22–1 shows a classification of social behaviors proposed by biologist W. D. Hamilton. As you can see, each of these behaviors has a different potential effect on the fitness—that is, the reproductive success—of both the individual performing the behavior (the donor) and the recipient of the behavior. Selfish, cooperative, and altruistic behaviors have all been well documented in numerous species of animals living in a natural setting. Spiteful behavior has thus far remained unobserved except in *Homo sapiens.*

Table 22–1	A Classification of Social Behaviors	
Type of Behavior	**Effect of the Behavior on**	
	The Donor	**The Recipient**
Selfish	Increases fitness	Decreases fitness
Cooperative	Increases fitness	Increases fitness
Altruistic	Decreases fitness	Increases fitness
Spiteful	Decreases fitness	Decreases fitness

Adapted from W. D. Hamilton, "The Evolution of Social Behavior," *Journal of Theoretical Biology,* vol. 7, pages 1–52, 1964.

The evolution of selfish behavior through the action of natural selection poses no problems for evolutionary theory and, indeed, is what would be expected. Although the maintenance of cooperation by natural selection is relatively easy to understand, the mechanism by which it could have evolved initially is almost as perplexing as the mechanism underlying the evolution of altruism.

In our consideration of social behavior, we shall look first at several different types of animal societies, paying particular attention to their organization and the behavior of their individual members. Then we shall return to the question of how the observed behaviors could have evolved.

Insect Societies

Insect societies are among the most ancient of all societies and, along with modern human societies, are among the most complex. Social insects include termites, ants, wasps, and bees.

Stages of Socialization

As with other animals, the social insects evolved from forms that were originally solitary. Among bees, for example, true sociality appears to have evolved on at least eight separate occasions, and among wasps four times.

Most living species of bees and wasps are solitary, and others show varying degrees of sociality. Thus it is possible to construct a scenario of the various stages of social evolution by the analysis of present-day species. Among the solitary species, the female builds a small nest, lays her eggs in it, stocks it with a food supply, and leaves it forever (Figure 22–9). She usually dies soon after, so there is no overlap between generations.

Among subsocial or presocial species, the mother returns to feed the larvae for some period of time, and the emerging young may subsequently lay their eggs in the same nest or comb. However, the colony is not permanent (usually being destroyed over the winter), there is no division of labor, and all females are fertile.

Eusocial, or "truly social," insects are characterized by cooperation in caring for the young and a division of labor, with sterile individuals working on behalf of reproductive ones. All ants and termites and some species of wasps and bees—for example, honey bees—are eusocial.

Honey Bees

A honey-bee society usually has a population of 30,000 to 40,000 workers and one adult queen. Each worker, always a diploid female, begins life as a fertilized egg deposited by the queen in a separate wax cell. (Drones, or male bees, develop from unfertilized eggs and are therefore haploid.) The fertilized egg hatches to produce a white, grublike larva (see Figure 22–1, page 378) that is fed almost continuously by the nurse workers; each larval bee eats about 1,300 "meals" a day. After the larva has grown until it fills the cell, a matter of about six days, the nurses cover the cell with a wax lid, sealing it in. Its metamorphosis requires about 12 days, after which the adult bee emerges.

(a)

(b)

22–9 *Wasps, most of which are not truly social insects, do not tend their young, but often provide for them.* (a) *Wasps of the genus* Apanteles, *all of which are solitary and parasitic, inject their eggs under the skins of caterpillars. The resultant larvae eat the internal tissues of the caterpillar, chew their way to* the surface, and spin the cocoons shown here. (b) *Female wasps of the genus* Ammophila *hunt caterpillars, which they then deposit in previously prepared burrows in the soil. The egg is laid on the caterpillar, which, paralyzed but not dead, provides up to 40 days' food for the larva.*

(a)

(b)

(c)

22–10   *Honey bees. (a) Workers tending honey and pollen storage cells. The honey is made from nectar processed by special enzymes in the workers' bodies. (b) "Undertaker" honey bees, removing the corpse of a dead pupa from its cell. (c) A foraging worker. The first segment of each of the three pairs of legs has a patch of bristles on its inner surface. Those of the first and second pairs are pollen brushes, which gather the pollen that sticks to the bee's hairy body. On the third pair of legs, the bristles form a pollen comb that collects pollen from the brushes and the abdomen. From the comb, the pollen is forced up into the pollen basket, a concave surface fringed with hairs on the upper segment of the third pair of legs. Transfer of pollen to the pollen basket occurs in midflight. The sting is at the tip of the abdomen.*

The newly emerged adult worker rests for a day or two and then begins successive phases of employment. She is first a nurse, bringing honey and pollen from storage cells to the queen, drones, and larvae. This occupation usually lasts about a week, but it may be extended or shortened, depending on the conditions of the colony. Then she begins to produce wax, which is exuded from the abdomen, passed forward by the hind legs to the front legs, chewed thoroughly, and then used to enlarge the comb. During this stage of employment as a houseworking bee, she may also remove sick or dead comrades from the hive, clean emptied cells for reuse, or serve as a guard at the hive entrance. During this period, she begins to make brief trips outside, seemingly to become familiar with the immediate neighborhood. It is only in the third and final phase of her existence that the worker bee forages for nectar and pollen. At about six weeks of age, she dies.

22–11   *Honey-bee workers in attendance on their queen (center). The workers feed the queen and also lick queen substance, which prevents their own sexual maturation, from the surface of her body.*

The Queen

Queens are raised in special cells that are larger than the ordinary cells and shaped somewhat like a peanut shell. Although all fertilized eggs have the genetic potential to become queens, new queens develop only at certain times and under specific circumstances. According to the best recent evidence, queens become queens because of a generally more nutritious diet in the larval stage, especially rich in protein, as compared to the mostly carbohydrate (honey) diet fed the worker larvae.

The queen influences her subjects by means of chemical substances that are synthesized in and secreted from specialized glands. One of these substances, known as "queen substance," inhibits sexual maturation in the worker bees and prevents them from becoming queens or from producing rival queens.

If a hive loses its queen, workers will notice her absence very quickly and will become quite agitated. Very shortly, they begin enlarging worker cells to form emergency queen cells, and the larvae in the enlarged cells are then fed the special diet. Any diploid larva so treated will become a queen.

The Annual Cycle

One of the important differences between subsocial and eusocial bees is that colonies of eusocial bees survive the winter. Within the wintering hive, bees maintain their temperature by clustering together in a dense ball; the lower the temperature, the denser the cluster.

In the spring, when the nectar supplies are at their peak, so many young may be raised that the group separates into two colonies. The new colony is always founded by the old queen, who leaves the hive, taking about half of the workers with her. The group stays together in a swarm for a few days, gathered around the queen, after which the swarm will settle in some suitable hollow tree or other shelter found by its scouts.

As the old queen is preparing to leave the hive, the new queens are getting ready to emerge. These two events are synchronized by sound signals transmitted through the comb. As these signals are exchanged, the workers remain motionless. During this period, sexual maturation begins in some of the workers, a few of which lay eggs. The unfertilized eggs develop into males, or drones. After the old queen leaves the hive, a new young queen emerges, and any other developing queens are destroyed. The young queen then goes on her nuptial flight, exuding a chemical (apparently queen substance) that entices the drones of neighboring colonies. She mates only on this one occasion (although she may mate with more than one male) and then returns to the hive to settle down to a life devoted to egg production.

During her nuptial flight, the queen receives enough sperm to last her entire life, which may be some five to seven years. The sperm are stored in a special organ in her reproductive tract and are released to fertilize each egg as it is being laid. The queen usually lays unfertilized eggs only in the spring, at the time males are required to inseminate the new queens.

The drones' only contribution to the life of the hive is their participation in the nuptial flight. Since they are unable to feed themselves, they become an increasing liability to the social group. As nectar supplies decrease in the fall, they are stung to death by their sisters or are driven out.

Vertebrate Societies

With rare exceptions (Figure 22–12), vertebrate societies do not have the rigid caste systems characteristic of the truly social insects. Many vertebrate societies are nonetheless highly structured, with social roles and access to resources determined by specific interactions that vary according to the species and the age and sex of the individuals.

Dominance Hierarchies

In many species of birds and mammals, dominance hierarchies, maintained by species-specific patterns of behavior, determine priority of access to resources and strongly influence relative reproductive success. One type of dominance hierarchy among vertebrates that has been studied in some detail is the pecking order in chickens.

A pecking order is established whenever a flock of hens is kept together over any period of time. In any one flock, one hen usually dominates all the others; she can peck any other hen without being pecked in return. A second hen can peck all hens but the first one; a third, all hens but the first two; and so on through the flock, down to the unfortunate pullet that is pecked by all and can peck none in return.

Hens that rank high in the pecking order have privileges such as first chance at the food trough, the roost, and the nest boxes. As a consequence, they can usually be recognized on sight by their sleek appearance and confident demeanor. Low-ranking hens tend to look dowdy and unpreened and to hover timidly on the fringes of the group.

During the period when a pecking order is being established, frequent and sometimes bloody battles may ensue, but once rank is fixed in the group, a mere raising or lowering of the head is sufficient to acknowledge the

22–12   *The only vertebrates known to have a social system similar to that of the eusocial insects are the naked mole rats, which live in underground tunnels in Kenya, Ethiopia, and Somalia. In each colony, only a single dominant female (the queen rat) and one or a few males breed. The other members of the colony, both male and female, are workers who forage for food (mostly root vegetables) and dig and maintain the tunnels in which the colony lives. When they are resting, naked mole rats—which typically weigh only 20 to 30 grams—huddle together to keep warm. As many workers as possible huddle under the queen rat, who, because of her large size, gives off large quantities of body heat.*

Naked mole rats can live more than 10 years. The queen rat usually breeds four times each year, producing an average of 10 pups per litter. It is thought that reproduction of the other females in the colony is suppressed by chemicals in the queen's urine.

dominance or submission of one hen in relation to another. The overt fighting behavior has become **ritualized,** and life subsequently proceeds in harmony. If, however, new members are added to a flock, the entire pecking order must be reestablished. The resulting disorganization leads to more fighting, less eating, and less tending to the essential business of growth and egg laying.

Pecking orders have the effect of reducing the breeding population. Cocks and hens low in the pecking order copulate much less frequently than socially dominant chickens. Thus the final outcome is probably the same as it would be if the social structure did not exist: the stronger and otherwise dominant animals eat better and leave the most offspring. However, because of the social hierarchy, this comes about with a minimum expenditure of lives and energy.

Territories and Territoriality

Many vertebrates stay close to their birthplaces, occupying a home range that is likely to be the same as that occupied by their parents. Even migratory birds that travel great distances are likely to return year after year to the same areas. Often these home ranges are defended, either by individuals or by groups, against other individuals or groups of the same species or closely related species that use the same resources. Areas so defended are known as **territories,** and the behavior of defending an area against rivals is known as **territoriality.**

22–13   *Wolf packs have dominance hierarchies of both males and females. Here a subordinate wolf greets a more dominant member of the pack by licking and smelling its mouth. This same muzzle-nuzzle gesture is used by pups begging for food. Usually only the dominant male and dominant female breed, and the rest of the pack cooperate in caring for the young. Caring includes guarding the den and providing food, which is swallowed at the kill and then regurgitated for the pups. (Some domestic dogs regularly vomit at the sight of a puppy, which should be recognized not as a sign of disapproval but as a social reflex.)*

(a)

(b)

(c)

(d)

22–14 *Territories come in many shapes and sizes.* (a) *The male Uganda kob displays on his stamping ground, which is about 15 meters in diameter and is surrounded by similar stamping grounds on which other males display. A female signifies her choice by entering one of the stamping grounds and grazing there. Only a small proportion of males possess stamping grounds, and those that do are the only ones that breed.*

(b) *A fiddler crab's territory is a burrow, from which he signals with his large claw, beckoning females and warning off other males.*

(c) *Howler monkeys shift their territories as they move through the jungle canopy but maintain spacing between groups by chorusing.*

(d) *Territoriality is common among reef fishes. For many species, a territory is a crevice in the coral, but for others, such as the clownfish, the territory is a sea anemone. The fish are covered by thick slime that partially protects them from the poison of the tentacles, but their acceptance by the anemone is chiefly a consequence of behavioral adaptations of the fish, which even mate and raise their brood among the tentacles. These two clownfish are guarding their fertilized eggs, which are on the rock just below the fish that is facing the camera. The eggs are the shiny, spherical objects that resemble small black marbles.*

Territoriality in Birds

Territoriality was first recognized by an English amateur naturalist and bird watcher, Eliot Howard. Howard observed that, in the spring, female birds were attracted to areas occupied by singing males of the same species, whereas other adult males of the species avoided those areas. In general, a breeding territory is established by a male. Courtship of the female, nest building, mating, raising of the young, and often feeding are carried out within this territory. Frequently the female also participates in territory defense.

By virtue of territoriality, a mating pair has a better chance of obtaining food and nesting material in the area and a safe place to carry on all the activities associated with reproduction and care of the young. Some pairs carry out all their domestic activities within the territory. Others perform mating and nesting activities in the territories, which are defended vigorously, but gather food on a nearby communal feeding ground, where the birds congregate amicably together. A third type of territory functions only for courtship and mating, as in the bower of the bowerbird or the courtship arena of the prairie chicken. In these territories, the males prance, strut, and posture—but very rarely fight—while the females look on and eventually indicate their choice of a mate by entering his territory. Males that have not been able to secure a territory for themselves are not able to reproduce. In fact, there is

22–15 *Conflict between two male blue-striped grunts at their territorial borderline. Each grasps the other by the lips, and the loser is the one who lets go first. Although defeated, the loser swims back into his own territory unharmed. This sort of "jaw wrestling" occurs in many species of fishes.*

evidence from studies of some territorial species, such as the Australian magpie, that adult males that do not secure territories do not mature sexually.

Territorial Defense

Even though territorial boundaries may be invisible, they are clearly defined and recognized by the territory owner. With birds, for example, it is not the mere proximity of another bird of the same species that elicits aggression, but its presence within a particular area. The territory owner patrols his territory by flying from tree to tree. He will ignore a nearby rival outside his territory, but he will fly off to attack a more distant one that has crossed the border. Animals of other species are generally ignored unless they are prey or predators or are in competition for some limited resource.

Once an animal has taken possession of a territory, he or she is usually undefeatable on it. Among territory owners, prancing, posturing, scent marking, and singing and other types of calls usually suffice to repel intruders, which are at a great disadvantage. For example, a male cichlid, a tropical freshwater fish (see page 372), will dart toward a rival male within his territory, but, as he chases the rival back into its own territory, he begins to swim more slowly, the tail fin seemingly working harder and harder, just as if he were making his way against a current that increases in strength the farther he pushes into the other male's home ground. The fish know just where the boundaries are and, after chasing each other back and forth across them, will usually end up with each one trembling and victorious on his own side of the line.

Similarly, the expulsion from communal territories is typically accomplished by ritual rather than by force. For example, among the red grouse of Scotland, the males crow and threaten only very early in the morning, and then only when the weather is good. This ceremony may become so threatening that weaker members of the group leave the moor. Those that leave often starve or are killed by predators. Once the early-morning contest is over, the remaining birds flock together and feed side by side for the rest of the day.

Kin Selection

In 1962, V. C. Wynne-Edwards proposed that individuals that failed to reproduce were doing so for the benefit of the society to which they belonged. In this way, a society could maintain its population at a level always slightly below its resources, he argued, and so the entire group would benefit. Such behavior was perpetuated by the increased survival of groups whose members behaved with such altruistic self-restraint. Although Wynne-Edwards's hypothesis of group selection has now been rejected by almost all biologists, this proposal served to galvanize a whole series of extremely productive studies that have revolutionized the way modern biologists view social behavior.

Group selection was rejected on fairly simple grounds: if there were genes with alleles for breeding and alleles for restraint-of-breeding, the alleles for breeding would soon overwhelm the alleles for restraint-of-breeding in the gene pool. Rejection of group selection led to another proposal, put forward by W. D. Hamilton and based largely on studies of social insects. As Darwin himself realized, the evolution of sterile castes of insects posed a special problem for evolutionary theory. If evolution is based on the number of surviving offspring, how can natural selection—acting on the individual organism—result in the development and persistence of sterility in a large proportion of a population? Darwin concluded that, in some cases, natural selection might act not only on individuals but also on families. Conceptually the idea is a simple one: members of families share inheritable characteristics. Thus there are variations among families as well as among individuals, and families that have favorable variations are likely to leave more survivors than other families.

Hamilton elaborated Darwin's suggestion, modifying it to the concept of the gene pool. In population genetics, as we noted earlier (page 320), the measure of fitness is the extent to which the alleles in an individual's genotype are present in the gene pool of succeeding generations. Viewed from this perspective, a mother takes risks and makes sacrifices for her offspring in order to increase the representation of her alleles in the gene pool. Or, to put it another way, she is programmed genetically to take certain risks or to make certain sacrifices for her offspring. To the extent that taking these risks or making these sacrifices increases her contribution to the gene pool, the alleles dictating this altruistic behavior will increase in subsequent generations.

However, it is not just mothers and offspring who share alleles. For example, consider siblings—that is, sons and daughters of the same mother. Siblings with the same father share, on an average, half of their alleles with one another. Therefore, any allele that favorably influences altruistic behavior among siblings could similarly increase its representation in the gene pool.

Hamilton's hypothesis, based on this principle, is called **kin selection.** According to this hypothesis, within a species different groups of related individuals (that is, kin) reproduce at different rates. The critical factor in kin selection is the effect of the individual on the reproductive success of its relatives.

The concept of kin selection served to establish a new evolutionary perspective, that of **inclusive fitness.** The single criterion of Darwinian fitness is the relative number of an individual's *offspring* that survive to reproduce. The criterion of inclusive fitness is the relative number of an individual's *alleles* that are passed from generation to generation, either as a result of his or her own reproductive success or as a result of the reproductive success of related individuals.

Tests of the Hypothesis

Kin selection is a useful hypothesis because it is testable. An example of such a test is the work of Patricia Moehlman, who studied 17 litters of silver-backed jackal pups near Ndutu Lodge in Tanzania. The family unit of the silver-backed jackal consists of a monogamous pair, young pups, and, often, one to three older siblings from a previous litter that serve as helpers (Figure 22–16).

In litters with no helpers, Moehlman found that an average of only one pup was raised. With a single helper, three pups survived, and in one family with three helpers, six pups survived. Thus, by being helpers, young jackals (with the same degree of relatedness to their siblings as they would have to their own offspring) produced more copies of their own alleles than if they had reproduced themselves with no helpers.

Among Florida scrub jays, the situation is somewhat different. Here again, breeding success is positively correlated with the number of helpers, which are usually close relatives of the breeding pair. However, the success is somewhat less than would be achieved by the young jay's setting up its own nest and rearing its own young. The reason that the young jays stay on as helpers is hypothesized to be the shortage of available territories for young males. This hypothesis is supported by the fact that as soon as a territory becomes available, the male helper leaves. One of the most common ways to get a new territory is to inherit part of the parents' territory. Thus, one is led to predict that females will help less than males and that older, more dominant males will help more than the younger males, both predictions that have been shown to be true.

22–16   (a) *Silver-backed jackal parents groom a helper. Being groomed is not only pleasurable but also removes insects and other parasites from the skin. Helpers are also provided food by parents.* (b) *This young silver-backed jackal, by helping raise his younger brother, probably increases his inclusive fitness more than if he tried to raise his own family.*

(a)

(b)

Kin Recognition in Tadpoles

For kin selection to be a factor in the evolution and maintenance of altruistic behavior, animals must be able to selectively direct their altruistic acts toward related individuals—thus increasing their own inclusive fitness. In many species, dispersal from the birth site is limited, and the probability is high that any individual of the same species in the vicinity—to whom assistance might be rendered—is a relative. In other species, however, dispersal is wider, and a group may consist of both related and unrelated individuals. How, then, can an animal distinguish relatives from nonrelatives, particularly if there has been no previous opportunity to get to know the relatives?

A series of studies conducted by Andrew Blaustein and Richard O'Hara of Oregon State University have revealed the mechanism involved in one organism, the Cascades frog (*Rana cascadae*). Members of this species are highly gregarious, particularly in the tadpole stage, during which they maintain cohesive schools. The schooling behavior is thought to aid in the defense against predators, principally the insect larvae that feed on these tadpoles. When a Cascades frog tadpole is captured and injured by an insect larva, it releases a chemical into the water that effectively warns the other tadpoles of the danger. Release of the chemical is followed by the flight of other members of the school.

In laboratory experiments, Blaustein and O'Hara allowed fertilized eggs to develop to the free-swimming tadpole stage under four different sets of conditions: (1) Eggs from the same clutch—all laid by the same female and fertilized by the same male—developed together in the same tank; these individuals were full siblings. (2) Eggs from one clutch developed at one end of a tank, with eggs from another, unrelated clutch developing at the other end of the tank; the two groups were separated by a mesh screen that allowed water to circulate freely throughout the tank. (3) A single egg developed within a mesh cylinder in a tank, while 12 eggs from an unrelated clutch developed in the surrounding water. (4) Individual eggs developed in small tanks, in total isolation from any other individuals, related or unrelated.

When the tadpoles hatched, they were given "preference tests," using tanks that

Rana cascadae *tadpoles schooling in a pond in the Cascade Range in Oregon. The preference of individuals of this species to associate with relatives rather than with unrelated individuals persists after their metamorphosis into frogs.*

were sectioned into three parts by mesh screens. Individual tadpoles were placed in the central section of the tank, with related individuals in the section at one end of the tank and unrelated individuals in the section at the other end. Regardless of the regimen under which the tadpole had developed, it spent most of its time closest to the end of the tank that contained its relatives. When tadpoles were mixed in tanks that were not sectioned and were allowed to school freely, time after time, they chose to school with relatives rather than with nonrelatives—even when they had been raised in a tank surrounded by nonrelatives.

In field experiments, when groups of related and unrelated individuals were released into ponds and lakes, they quickly sorted out into schools composed primarily of kin. Additional experiments, in which eggs were fertilized in the laboratory, with different males fertilizing the eggs of the same female and the same male fertilizing the eggs of different females, revealed that the tadpoles could distinguish between full siblings, half siblings, and

nonsiblings, almost always preferring the company of the most closely related individuals available.

A second set of preference tests, in which the tadpoles were separated from the related and unrelated groups by glass barriers that allowed the tadpoles to see one another but prevented any water circulation between compartments, revealed that the cue that identifies its relatives to a Cascades frog tadpole is a chemical substance. In the absence of water bearing the identifying substance, the tadpoles displayed no preference for either group.

Experiments by other investigators have demonstrated varying degrees of kin recognition in a vast diversity of animals, including social insects, birds, ground squirrels, mice, and monkeys, as well as tadpoles of other frog and toad species. In most species studied, the capacity to recognize kin appears early in development. It seems to function not only in determining the distribution of favors—such as warning signals—but also in enabling animals to avoid both fighting with kin and inbreeding.

The Selfish Gene

Many years ago the English satirist Samuel Butler remarked that a chicken was just an egg's way of making another egg. Updating this concept, some biologists see an organism as just a gene's way of making more genes.

The argument is simple: The individual organism is transient. The genotype is fragmented at every generation. All that can survive from generation to generation is the gene. The way it survives is in the form of replicas. The more replicas, the better its chance of survival. The organism is the gene's survival machine, and it so programs the machine that it will turn out gene copies at a maximum rate, regardless of personal cost. Thus, for example, there is a mite in which the female has a brood of one son and 20 daughters. The son mates with his sisters while still within the mother and dies before he is born—not much of a life for the male mite but a great low-risk, high-return strategy for his genes and a wise maternal investment.

This concept of the selfish gene is admitted, even by its articulate sponsor, Richard Dawkins, to be an oversimplification. It has served, however, to bring into sharper focus questions on the evolution of behavior and of survival strategies.

Conflicts of Interest

In directing their attention to the survival of genes rather than of individuals, biologists realized that there were opportunities for serious conflicts of interest. These conflicts could occur not only between those individuals that were in obvious direct competition for some limited resource—such as territory, a food supply, or a mate—but also among individuals previously envisaged as working together for some common good—such as members of the nuclear family.

For example, a mammalian female invests in her offspring by providing nourishment in the form of milk. Nursing in mammals, however, generally prevents new pregnancies. It is to the offspring's benefit to monopolize the milk supply for as long as possible. The mother, however, can maximize her fitness by weaning the offspring at an appropriate time and beginning a new investment.

Another example of intergenerational conflict occurs when a new male takes over a female with young. The new male's evolutionary goal, as programmed by his genes, is to produce his own young as rapidly as possible. The existing offspring (fathered by another male) may interfere with this objective, either by competing with their half-siblings or by preventing a new pregnancy in the female. Sarah Blaffer Hrdy was the first to

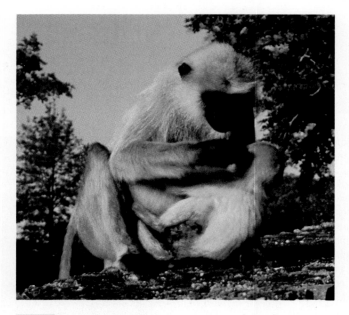

22–17   *Mother langur holding an infant that has been fatally wounded by an infanticidal male at the time of take-over of her troop.*

report the practice of infanticide by males at the time of take-over. Her observations, which were made on Hanuman langurs in India (Figure 22–17), were greeted with considerable skepticism when first reported in 1971, but similar acts have now been observed in other primates and also in birds, rodents, and lions. Infanticide can clearly represent a successful evolutionary strategy.

Male vs. Female

Among the most closely fought conflicts of interest are those that occur between the partners of a reproductive pair. A female, by definition, is the individual that produces the larger gametes. Female gametes are relatively expensive, metabolically speaking, and if the female must also carry the embryo in her body and care for the young after their birth, her investment in each reproductive effort may be, in terms of her total reproductive potential, very large indeed. Male gametes, by contrast, are usually cheap; a single insemination costs males of most species almost nothing.

In most animal species, all that the male contributes to the next generation is his genes. In such situations, it is to the interest of the female's selfish genes to find themselves in the best possible company, thus promoting their survival in the next generation. Hence, intense competition may arise among the males to prove to the females that they are the best endowed. Such proof may take the form of elaborate courtship displays, in which strength, vigor, stamina, and other desirable traits are displayed, or it may take the form of dominance over other males. Males also attract females by offers of food or other resources in the form of a territory or a nesting

22–18 *A great bowerbird, decorating his bower with hibiscus flowers and other red objects (note the bottle cap in his beak). The male spends much of his time building this elaborate nuptial structure, which is the focus of his territory, and decorating it with brightly colored objects. Male birds call from their bowers and display to females if they approach. The female selects one male, enters his bower, copulates briefly, and departs. Then she builds her own utilitarian nest some distance from the bower of her mate, lays her eggs, and cares for the young entirely on her own.*

site. Because the male that commands the most resources, or the best resources, is likely also to be the male that is superior genetically, the offering or the territory may be less important as an actual resource than as a symbol of the male's desirability (Figure 22–18).

In some few species, males make an equal contribution to the care of the young. Selection for such a reproductive strategy would depend on at least two factors: first, that the additional parental investment would result in a significant increase in surviving young and, second, that the male has some way of being sure that the offspring he cares for are his own. Females in such situations may give up the opportunity of mating with the highest-ranking male in return for a male that will provide care for the young. Selection for such a reproductive strategy would include a means for the female to be sure that the male she has selected will remain with her after her eggs have been fertilized.

A long courtship, such as is seen among some species of birds, serves the purposes of both male and female. It gives each of them a chance to assess the qualities of the other. For her, it symbolizes commitment and keeps him away from other females. It may also bring material benefits; for example, many female birds refuse to copulate until the nest is built. For him, the long courtship provides assurance that she has not already been inseminated. She loses nothing by cheating, whereas he loses his entire investment if she does.

The Advantage of Waiting

Dominance hierarchies and territories can also be viewed from the perspective of the selfish gene. An animal that is not breeding because it is low in the dominance hierarchy or lacks a territory is not sacrificing itself for the good of the group or of the species. It is more likely waiting for its chance. If the individual is very low in a hierarchy, such an opportunity may never come, but still its chances of perpetuating its genotype are better than if it challenges a superior and is killed or badly injured.

Studies of bird territories confirm the hypothesis that individuals that lack territories are waiting for their turn. An empty territory is almost always immediately filled, as in the example of Florida scrub jays. Another example is provided by lions. As young male lions approach maturity, they are driven from the pride by the established adult males. The young males tend to stay together as nomads (Figure 22–19) until they find a pride they can take over, often by defeating the resident males. Although their period of dominance may be limited, their reproductive success is greater than if they had been killed or maimed resisting expulsion from the pride in which they were born. As we noted in Chapter 19, this social system also promotes outbreeding.

22–19 *Young male lions, living as nomads. Such lions are typically found in small groups in which all members are about the same age and are probably brothers or, at least, from the same pride. In all likelihood, they will eventually take over a pride of their own. Like the males who preceded them, they will be transients, holding the females—and the territory occupied by the females—for only a few months, or, at most, a few years.*

The females of a pride (usually sisters or half-sisters) hunt together cooperatively, become sexually receptive at about the same time, give birth at about the same time, and nurse each other's cubs. New males taking over a pride often kill the young cubs. Following the death of the cubs, the females become sexually receptive again and so are available to perpetuate the genes of the newcomers, at least during the brief period of their reign.

Reciprocal Altruism

Seemingly altruistic acts may pay off, as we have seen, because they increase the probability of survival of genes shared by the performer of the act and the beneficiary. In 1971, Robert Trivers proposed another model for seemingly self-sacrificing behavior, which he termed **reciprocal altruism.** According to this model, an altruistic act is performed with the expectation that the favor will be returned. It might involve warning of danger, removing a tick from an inaccessible portion of the anatomy, or sharing food, to cite just a few possible examples. It also might involve acts of cooperation between organisms of different species, as well as between members of the same species. Reciprocal altruism may well have become established in early human societies, which is why this model is of special interest.

An example of reciprocal altruism is provided by vampire bats (Figure 22–20), which often have a difficult time locating suitable warm-blooded victims. Many individuals return each morning to the communal nesting sites with an empty stomach after an unsuc-

cessful night of searching. A study by Gerald Wilkinson of wild vampire bats in Costa Rica showed that the bats regularly regurgitate blood to feed each other and that individuals in dire need are fed preferentially. A starved bat that receives blood reciprocates on a later day when its hunt has been successful. Not only related animals but also unrelated animals reciprocate with one another.

For reciprocal altruism to be a successful strategy, it must be resistant to exploitation by individuals who accept favors but never return them (that is, cheaters). Two conditions are necessary. The first is that individuals meet more than once and are able to recognize each other. The second is that each individual cooperates on the first encounter and, on subsequent encounters, does whatever the other individual did on the preceding encounter. Under these particular conditions, cooperation pays off and cheating is punished.

A significant number of cooperators must be present in a population for the reciprocal behavior to become established. Thus, an important question is how reciprocal altruism gets started initially. One possible explanation is that cooperation originates in a group of related individuals—that is, as kin selection—and spreads from the original cluster to include unrelated individuals.

22–20 *Vampire bats, such as the females shown here, congregate during the day in communal nesting areas in caves or hollow trees. At night they emerge and begin their search for animals, preferably horses or cows, from which they can obtain a blood meal. Because of their high metabolic rate, each night these bats need to consume at least 50 percent of their body weight in blood (about 15 grams, which is a little more than half an ounce). If a bat's hunt is unsuccessful on two consecutive nights, it is in danger of death by starvation. In a striking example of reciprocal altruism, successful hunters share their food with the unsuccessful, a favor that is returned when their fortunes are reversed.*

The Biology of Human Behavior

It is tempting—indeed almost irresistible—to draw parallels between human behavior and the behavior observed in other species. The extent to which concepts concerning the evolution of behavior can be extrapolated to the human species is a matter of controversy. Some biologists maintain that the human species is basically no different from any other species, that our genes are as selfish as any, and that if we seek to modify human behavior for the common good, we need to understand its roots. Other biologists, however, maintain that while early human ancestors may have been governed by their genes, modern humans are so much a product of their culture and of their individual experience that such analyses are no longer valid. Moreover, they may be dangerous. The concept that biology determines human behavior lies at the root of all notions of racial superiority. Thus it has provided the rationale for slavery, exploitation, and genocide. More commonly, the notion that human behavior is, to some extent, biologically determined allows us to forgive ourselves for, and even to justify, violence, aggression, docility, and greed.

Summary

The behavioral characteristics of organisms are the products of natural selection. Like anatomical and physiological adaptations, they affect the fitness of organisms. Although all behavior has its roots in the underlying genetic program of the individual, in only a few instances has it been possible to correlate a particular behavior with the presence or absence of specific alleles. Behaviors are typically the result of the interactions of a large number of genes, the influence of which may be further modified by interactions of the individual with its environment.

A full explanation as to why an organism performs a particular behavior involves both proximate and ultimate causation. The proximate cause of a behavior is typically a series of physiological responses to a particular stimulus. The ultimate cause, however, is the adaptive value of the behavior for the survival and successful reproduction of the individual.

Ethology is the study of the behavior of animals in their natural environment, with particular emphasis on species-specific patterns of behavior and their evolutionary origins. Patterns of behavior that appear essentially complete the first time the organism encounters the relevant stimulus are known as fixed action patterns. These patterns are highly stereotyped and repetitive and, for members of a given species of a particular age, sex, and physiological condition, are as predictable and constant as the anatomical characteristics of the species. Communication signals among members of a species that act as stimuli for fixed action patterns are known as releasers, and the brain circuits that respond to them are known as innate releasing mechanisms.

Learning is the modification of behavior as a result of experience. Various categories of learning are recognized, of which habituation is one of the simplest. In associative learning, which often takes the form of trial-and-error learning, one stimulus comes to be associated with another. Imprinting is a form of learning of particular importance in the discrimination of members of one's own species from members of other species; it occurs quickly within a critical period in early life. Another form of learning, important in many species of birds and mammals, is imitation.

Of particular interest from an evolutionary standpoint is the behavior of animals in societies—groups of individuals of the same species, living together in an organized fashion, with divisions of resources and labor and with mutual dependence. Such behavior may be selfish, cooperative, altruistic, or spiteful—with differing consequences for the fitness of the donor and the recipient.

Insect societies are among the largest and most complex of animal societies. In the honey-bee colony (and among other eusocial insects as well), the queen is the only female reproductive form. She and her brood are tended by sterile workers. The behavior and physiology of the members of the hive are controlled by the exchange of chemical signals.

Vertebrate societies are often organized in terms of social dominance. Dominance hierarchies may take the form of pecking orders, in which higher-ranking animals have priority of access to food and other resources and reproduce more. Territoriality is a system of social dominance in which only animals with territories reproduce and animals with better territories reproduce more.

A central question in the study of social behavior is the mechanism by which natural selection can result in behavior that limits the reproductive potential of individuals in the society. An explanation is provided by the hypothesis of kin selection, which includes the concept of inclusive fitness. Previous concepts of evolutionary fitness focused on the relative number of surviving offspring in future generations. Inclusive fitness focuses on the relative number of an individual's alleles that are passed from generation to generation, as a result of the reproductive success either of the individual or of its relatives.

From the viewpoint of inclusive fitness, social behavior is regulated by the "selfish gene," which programs the individual not necessarily for the individual's own well-being, or even survival, but only for the perpetuation of its alleles. Competition for representation in the future gene pool can occur not only between individuals obviously competing for the same resources but also between parents and offspring and between members of a reproductive pair. The selfish-gene concept can also be used to explain the establishment of dominance hierarchies and territorial behavior.

Altruism is behavior that carries a cost to the individual that performs it and benefits some other individual or individuals. Some acts of altruism are thought to be based on inclusive fitness, and others may be based on reciprocal altruism, the performance of an unselfish act with the expectation that the favor will be returned. Reciprocal altruism is of particular interest in terms of the evolution of cooperative behavior in the ancestors of *Homo sapiens*.

Questions

1. Distinguish among the following: proximate cause/ultimate cause; fixed action pattern/innate releasing mechanism; sign stimulus/releaser; imprinting/imitation; subsocial/eusocial; queen/worker/drone; social dominance/territoriality; group selection/kin selection; Darwinian fitness/inclusive fitness; altruism/reciprocal altruism.

2. Define behavior. What distinguishes social behavior from other forms of behavior?

3. Territoriality and social dominance achieve the same results. What are they? How do the results differ in a population in which these forms of behavior are not present?

4. Among organisms in which both sexes are diploid, the degree of relatedness between parents and children is always 0.5 (that is, each child inherits one-half, or 50 percent, of its alleles from its mother and one-half, or 50 percent, from its father). The degree of relatedness between siblings has a theoretically possible range from 0 to 1 and averages 0.5. Explain.

5. The late J. B. S. Haldane, a mathematician who made major contributions to evolutionary biology, once remarked that he would lay down his life for two brothers or eight cousins. On what basis did he make such an offer? Actually, Haldane was being too generous. Explain why.

6. Animals that live within a dominance hierarchy may never breed, yet they may be as fit as those that do breed. How can this be explained?

7. In order to protect their young against infanticidal males, females might refuse to mate with murderers. Would this be a successful strategy?

8. A hypothesis known as the "superterritory hypothesis" suggests that an animal should defend a territory larger than it really needs. Could this be spiteful behavior? Explain.

9. In what ways are human societies different from insect societies? How are they similar?

Suggestions for Further Reading

Alcock, John: *Animal Behavior: An Evolutionary Approach,* 5th ed., Sinauer Associates, Inc., Sunderland, Mass., 1993.

> *An outstanding text on animal behavior. Highly recommended.*

Blaustein, Andrew R., and Richard K. O'Hara: "Kin Recognition in Tadpoles," *Scientific American,* January 1986, pages 108–116.

Dawkins, Richard: *The Selfish Gene,* 2d ed., Oxford University Press, New York, 1989.*

> *Dawkins argues that we are "survival machines" programmed to preserve the "selfish molecules" known as genes. A wit-sharpening account and also an entertaining one.*

Franks, Nigel R.: "Army Ants: A Collective Intelligence," *American Scientist,* vol. 77, pages 138–145, 1989.

Goodall, Jane: *In the Shadow of Man,* 2d ed., Houghton Mifflin Company, Boston, 1988.*

> *A personal account, first published in 1971, of 11 years spent observing the complex social organization of a single chimpanzee community in Tanzania.*

Goodall, Jane: *Through a Window: My Thirty Years with the Chimpanzees of Gombe,* Houghton Mifflin Company, Boston, 1991.*

> *The sequel to* In the Shadow of Man, *this fascinating and insightful book carries the story of the chimpanzee community through the subsequent 20 years. Highly recommended.*

Hölldobler, Bert, and Edward O. Wilson: *The Ants,* Harvard University Press, Cambridge, Mass., 1990.

> *An outstandingly detailed and richly illustrated compendium of information about ants, the dominant organisms in many terrestrial environments. This delightful resource covers all aspects of these fascinating insects, from their evolution to their anatomy and physiology, as well as their social organization and ecology.*

Honeycutt, Rodney L.: "Naked Mole-Rats," *American Scientist,* vol. 80, pages 43–53. 1992.

Hrdy, Sarah Blaffer: *The Langurs of Abu: Female and Male Strategies of Reproduction,* Harvard University Press, Cambridge, Mass., 1981.*

> *This book, which has been described as "spell-binding," is notable not only for its original observations but also for its contributions to the theoretical structure of sexual combat.*

* Available in paperback.

Hrdy, Sarah Blaffer: *The Woman that Never Evolved*, Harvard University Press, Cambridge, Mass., 1981.*

> *Both a sociobiologist and a feminist, Hrdy examines what it means to be female. Males are almost universally dominant over females in primate species, and Homo sapiens is no exception. Yet in studying our closest living relatives, the primates, she discovers that the female is competitive, independent, sexually assertive, and has just as much at stake in the evolutionary game as her male counterpart.*

Krebs, J. R., and N. B. Davies: *An Introduction to Behavioural Ecology*, 3d ed., Blackwell Scientific Publications, Boston, 1993.*

> *The best introduction to the general subject of social behavior. Well written with many good illustrations and an excellent balance between examples and theory.*

Mech, L. David: *The Wolf: The Ecology and Behavior of an Endangered Species*, University of Minnesota Press, Minneapolis, 1983.*

> *The definitive book on the social history of wolves, originally published in 1970. Mech, a wildlife biologist with the U.S. Fish and Wildlife Service, devoted more than 20 years to the study of wolves in their natural habitats.*

Mock, Douglas (ed.): *Behavior and Evolution of Birds*, W. H. Freeman and Company, New York, 1991.*

> *A collection of articles from* Scientific American *on various aspects of the behavior and evolution of these familiar and fascinating animals.*

Mock, Douglas W., Hugh Drummond, and Christopher H. Stinson: "Avian Siblicide," *American Scientist,* vol. 78, pages 438–449, 1990.

Moehlman, Patricia D.: "Social Organization in Jackals," *American Scientist,* vol. 75, pages 366–375, 1987.

Moss, Cynthia: *Portraits in the Wild: Animal Behavior in East Africa*, 2d ed., The University of Chicago Press, Chicago, 1982.*

> *A description of some of the studies carried out on various wild species. Full of fascinating bits of information and firsthand observations.*

Sapolsky, Robert M.: "Stress in the Wild," *Scientific American,* January 1990, pages 116–123.

Seeley, Thomas D.: "The Honey Bee Colony as a Superorganism," *American Scientist,* vol. 77, pages 546–553, 1989.

Seeley, Thomas D.: "How Honeybees Find a Home," *Scientific American,* October 1982, pages 158–168.

Sherman, Paul W., Jennifer U. M. Jarvis, and Stanton H. Braude: "Naked Mole Rats," *Scientific American,* September 1992, pages 78–86.

Tinbergen, Niko: *Curious Naturalists*, Doubleday & Company, Inc., Garden City, N.Y., 1984.*

> *Some charming descriptions of the activities and discoveries of scientists studying the behavior of animals in their natural environment, originally published in 1968.*

Topoff, Howard (ed.): *The* Natural History *Reader in Animal Behavior*, Columbia University Press, New York, 1987.*

> *A collection of* Natural History *articles covering four broad topics: sensory processes and orientation, evolution of behavior, social organization, and behavioral development.*

Topoff, Howard: "Slave-making Ants," *American Scientist,* vol. 78, pages 520–528, 1990.

Trivers, Robert: *Social Evolution*, The Benjamin/Cummings Publishing Company, Redwood City, Calif., 1985.

> *An excellent introductory text on the evolution of social behavior, this book assumes a minimal background on the part of the reader. Well illustrated, with many examples.*

Tumlinson, James H., W. Joe Lewis, and Louise E. M. Vet: "How Parasitic Wasps Find Their Hosts," *Scientific American,* March 1993, pages 100–106.

Wilkinson, Gerald S.: "Food Sharing in Vampire Bats," *Scientific American,* February 1990, pages 76–82.

Wilson, E. O.: *The Insect Societies*, Harvard University Press, Cambridge, Mass., 1971.*

> *A comprehensive and fascinating account of the social insects.*

Wilson, E. O.: *Sociobiology: The New Synthesis*, Harvard University Press, Cambridge, Mass., 1975.

> *In this extremely interesting, beautifully written, and beautifully illustrated book, Wilson undertakes to set forth the biological principles that govern social behavior in all kinds of animals. The last chapter, which concerns human sociobiology, became a focus of controversy about the inheritance of behavioral traits in Homo sapiens. Also available in an abridged, less technical version.*

Winston, Mark L., and Keith N. Slessor: "The Essence of Royalty: Honey Bee Queen Pheromone," *American Scientist,* vol. 80, pages 374–385, 1992.

Woolfenden, G. E., and J. W. Fitzpatrick: "The Inheritance of Territory in Group Breeding Birds," *BioScience,* vol. 28, pages 104–108, 1978.

---

* Available in paperback.

Ecology

Population Dynamics: The Numbers of Organisms

The California condor is North America's largest bird. This giant vulture weighs up to 25 pounds and has a wingspan of 8 to 9 feet. Although, by human standards, not very pretty up close, the condor is breathtaking to watch in its soaring flight, as it glides up to 10 miles on a single wing beat, traveling hundreds of miles in one day.

Condors do not breed until they are five or six years old. Each mating pair produces only one egg, incubates it for almost two months, and feeds and protects its solitary chick for the better part of another year. As a consequence, even under the best conditions, a mating pair produces only one offspring every two years.

The condor reminds us that just as feathers and fins, hooves and beaks, flower parts and fruit colors are shaped by an evolutionary past, so are life histories. The condor is a relic of 2 million years ago. In those days, condors ranged all over the North American continent. They had no enemies, and the giant cats, cave bears, and other carnivores of the time left ample carnage in their wake to support hundreds of pairs of condors and their chicks.

The California condor is now almost extinct. The reasons are many. They include shooting by hunters, poisoning by bait intended for coyotes, poisoning from feeding on carcasses containing lead shot, and contamination by DDT and other pesticides. The principal causes, however, are two. The first is the condor's own life history, especially its long, careful care of its young. Breeding so slowly, the condor cannot replace those young that are lost. The second is the loss of its habitat. Efforts to return condors to the wild are failing because there is no longer a "wild" for them to return to.

Ecologists are divided about the condor's future. Some support a captive breeding program, so future generations will at least be able to catch a glimpse of these homely giants. Others feel that only a free-flying condor should be preserved. In the words of writer Kenneth Brower, "For a bird that once fed on mastodons, it is too steep a fall from glory."

Ecology, the study of the interactions of organisms with their physical environment and with each other, is both the oldest and the youngest of the major subdivisions of biology. It is at least as old as the inquiring human mind—clearly not a subject early hominids could safely ignore. However, as a "hard" science, ecology is young. Only recently have biologists been able to devise ways to analyze the many variables that affect organisms in their natural environment, to study them quantitatively, and to construct models, pose hypotheses, and formulate and test predictions. The great ecologists of 40 years ago, who laid the foundations of modern ecology, were sensitive observers of nature. Their successors, equally sensitive observers as they gather the data needed to test their hypotheses, are often also wizards at calculus, statistics, and computer modeling.

As a science, ecology seeks to discover how organisms affect, and are affected by, their living and nonliving environment. Ecology also seeks to define how these interactions determine the kinds and numbers of organisms found in a particular place at a particular time. Because of the pressures of the growing human population on the environment and on the other organisms with which we share it, ecology has become, from a practical standpoint, one of the most important branches of biological science. We, even more than our ancestors, cannot safely ignore it.

We are going to begin these chapters on ecology by looking at some of the factors affecting the size of populations. Later chapters will be concerned with the interactions between and among the populations of different species in communities and with the interactions between communities and their physical environment in ecosystems. We shall end with a global overview of the biosphere.

Properties of Populations

As we noted in Chapter 1, different properties emerge at different levels of organization. A molecule has different properties from its constituent atoms, a cell different properties from its constituent molecules, and a multicellular organism different properties from its constituent cells. Similarly, a population has very different properties from the individuals it comprises.

As a simple illustration, picture a patch of woods, or a field, or a lake shore to which you return from time to time, which perhaps your parents and grandparents visited and where, with any luck, your children may walk someday. If you go there in the spring, you will see, for the sake of argument, squirrels, birds, wildflowers, and beetles—the same species you saw the preceding spring and the same species you expect to see in subsequent springs. Yet, year after year, it is unlikely that you see the same individual squirrels and birds, very unlikely that you encounter the same beetles, and seldom the same individual wildflowers. The individual is transitory, but—with a little luck—the population endures.

A **population** is a group of organisms of the same species living in the same place at the same time. Among the properties of populations that are not properties of individual organisms are growth patterns, mortality patterns, age structure, density, and dispersion.

Patterns of Population Growth

As Darwin noted some 150 years ago, the reproductive potential of all species is very high. It can be calculated, for instance, that a single female housefly producing 120 eggs (with half developing into females) can, within seven generations—the expected number of generations in a single year—give rise to almost 6×10^{12} houseflies (Table 43–1). With bacteria that have a generation time of 20 minutes, a single bacterium can give rise to eight bacteria within one hour, 512 in three hours, and 262,144 in six hours. Fortunately for us, few organisms ever attain their full reproductive potential.

The **per capita rate of increase** of a population is the increase in the number of individuals in a given unit of time per individual present. Thus, the annual per capita rate of increase of the human population in the United States is currently about 0.007 (which is equivalent to 7 per thousand), and in the world as a whole it is about 0.016 (or 16 per thousand). In the absence of net immigration (the movement of other individuals of the species into the population from elsewhere) or emigration (the departure of individuals from the population), the rate of increase is equal to the birth rate minus the death rate. Thus, the rate of increase can be equal to zero, it can be a positive figure, or it can be a negative figure—as it is now for the human population of some countries.

Table 43–1 Reproductive Potential of the Housefly (*Musca domestica*)*	
Generation	**Numbers If All Survive and Reproduce**
1	120
2	7,200
3	432,000
4	25,920,000
5	1,555,200,000
6	93,312,000,000
7	5,598,720,000,000

* In one year, about seven generations are produced. The numbers are based on the following assumptions: each female lays 120 eggs per generation, each fly survives just one generation, and half of the flies are females. Adapted from E. J. Kormondy, *Concepts of Ecology*, Prentice-Hall, Inc., Englewood Cliffs, N.J., 1969.

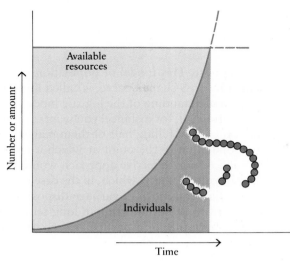

43–1   *An exponential growth curve. After an initial estab-lishment phase, the population increases in the same fashion as a savings account with compound interest. Exponential growth is characteristic of small populations, such as labora-tory cultures of bacteria, with access to abundant resources.*

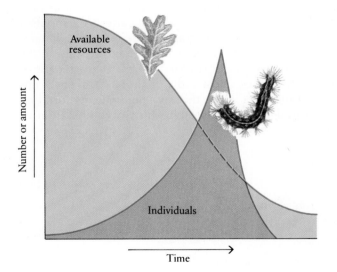

43–2   *A population "crash," as illustrated by the gypsy moth caterpillar. Infestations of these caterpillars can strip the leaves from trees so rapidly that the food supply is exhausted before the caterpillars have finished feeding and are ready to begin their metamorphosis. The result is a sudden drop in the number of individuals. Spraying the caterpillars with insecti-cides often reduces the size of the population enough to pre-vent a crash and thus maintains the population at a lower but fairly constant level.*

The **growth rate** of a population—that is, the change in the number of individuals over time—equals the per capita rate of increase times the number of individuals *already present*. In the type of population growth known as **exponential growth,** the number of individuals increases at a constant rate. As you can see in Figure 43–1, exponential growth starts out slowly, but then it shoots up very rapidly as the number of repro-ducing individuals increases in each succeeding gener-ation. The principle is the same as the compounding of interest on a savings account: the more you have, the more you get.

The exponential growth curve is most closely ap-proximated by microorganisms cultivated in the labo-ratory (where resources are constantly renewed), by the initial stages of seasonal "blooms" of algae, and by the recent growth of the human population. Under most circumstances, however, a population cannot long con-tinue to increase exponentially without reaching some environmental limits imposed by shortages of food, space, oxygen, nesting or hiding places, or accumula-tion of its own waste products. In nature, short-term exponential growth is characteristic of so-called fugitive or opportunistic species that invade an area, rapidly use up local resources, and then either enter a phase of dormancy or move on. Weeds and insects are typical opportunists.

Sometimes, a population that is growing very rapidly may hit an environmental limit suddenly and then "crash" to very low levels. An example can be found in the gypsy moth infestations that recurrently plague the northeastern United States (Figure 43–2). If a popula-tion grows rapidly enough, it can wipe out its food sup-ply before the caterpillars complete their metamorpho-sis and reach the reproductive stages. Or, alternatively, the population may reach such a high density that it provides a suitable environment for the exponential growth of an infectious microorganism (a virus, in the case of the gypsy moth) and is wiped out by disease.

The Effect of the Carrying Capacity

For many populations, the number of individuals is determined not by reproductive potential but by the environment. The number of individuals of a particular population that the environment can support under a particular set of conditions is known as the **carrying capacity** of the environment. For animal species, the carrying capacity may be determined by food supply or access to sheltered sites. For plants, the determining factor may be access to sunlight or the availability of water.

The factors limiting population size may vary seasonally. When the wildebeests migrate through the Serengeti plain of East Africa, all the lions are well fed, but when the big herd animals are gone, the lion cubs often starve to death, and some of the less capable adults as well. In North America, the carrying capacity for a herd of deer is not the number counted in the springtime when new fawns appear, but the number that can survive over several winters.

The patterns of population growth observed in nature are many and complex. One of the simplest patterns, which illustrates clearly the effect of the carrying capacity, is shown in Figure 43–3. When the number of organisms is very small and resources are abundant, the curve approximates the exponential growth curve of Figure 43–1. As the number of organisms increases and resources become scarce, population growth slows. If the number of organisms exceeds the carrying capacity, the growth rate becomes negative and the population declines. Eventually, the population stabilizes and oscillates around the maximum size the environment can support. This model of population growth, represented by an S-shaped curve, is called **logistic.**

An understanding of the logistic model has practical applications. If, for instance, you want to control a population of rats, killing half of them may merely reduce the population to the point at which it increases most rapidly. A more effective approach would be to reduce the carrying capacity, which, in the case of rats, usually means tighter control of garbage disposal. Similarly, if you want to achieve maximum long-term productivity in the harvesting of a particular type of economically valuable fish, you should not harvest the population below the level of rapid growth unless you are willing to wait a long time for the population to recover.

Mortality Patterns

Another important property, affecting both the size and composition of a population, is its **mortality pattern.** Figure 43–4 shows three different patterns of mortality, represented in the form of survivorship curves. In the first, exemplified by the oyster, a very high mortality rate in early life is followed by a lower rate of mortality among older individuals that have passed some critical stage in the life cycle. In the second pattern, exemplified by *Hydra*, the rate of mortality is the same for all members of the population, regardless of age or stage in the life cycle. In the third pattern, which is characteristic of our own species, most individuals reach a fairly advanced age, after which there is a high rate of mortality.

Natural populations often show a combination of

43–3   *One of the simplest growth patterns observed in natural populations is known as logistic growth and is represented by an S-shaped curve. As with exponential growth, there is an initial establishment phase when population growth is relatively slow (1), followed by a phase of rapid acceleration (2). Then, as the population approaches the carrying capacity of the environment, the growth rate slows (3 and 4), and finally stabilizes (5), though fluctuations around the carrying capacity may continue. This graph represents the growth pattern of a herd of deer introduced into a new area with limited resources. Other growth patterns observed in natural populations are considerably more complex.*

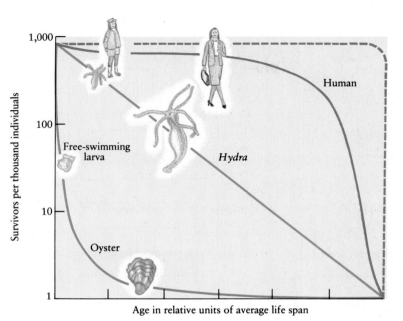

43–4   *Survivorship curves, showing three different patterns of mortality. In the oyster, mortality is extremely high during the free-swimming larval stage, but once the individual attaches itself to a favorable substrate, life expectancy levels off. Among* Hydra, *the mortality rate is the same at all ages.*

The curve at the top of the graph (dashed line) is for a hypothetical population in which all individuals live out the average life span of the species—a population, in other words, in which all individuals die at about the same age. The fact that the curve for humans approaches this hypothetical curve indicates that the human population as a whole is reaching some genetically programmed uniform age of mortality.

mortality patterns. For instance, in a study made of the saltmarsh song sparrows of San Francisco Bay, it was estimated that of every 100 eggs laid, 26 are lost before hatching. Of the 74 live nestlings, only 52 leave the nest, and of these, 42 (80 percent) die the first year. The remaining 10 breed the following season, but during the next year, 43 percent of these die, leaving only 6 of the original 100. Each subsequent year, mortality among the survivors amounts to 43 percent, apparently regardless of age. Once a bird survives its first, risk-laden year, the mortality rate for the subsequent years remains more or less constant. Thus, the early survivorship curve for the sparrows resembles that of the oyster, while the later curve is more *Hydra*-like.

Age Structure

The mortality pattern of a population, in turn, affects another important property of the population, its age structure. The **age structure** of a population is the proportion of individuals of different ages in the population.

In species in which the life span exceeds the reproductive span, knowledge of the age structure is useful in predicting future changes in population size. For example, if a large proportion of a population is of reproductive age or younger, as in the human population of India (Figure 43–5), population growth will continue at a high rate even though reproductive individuals only replace themselves. This is necessarily the case simply because the number of reproductive individuals continues to increase as younger individuals reach reproductive age.

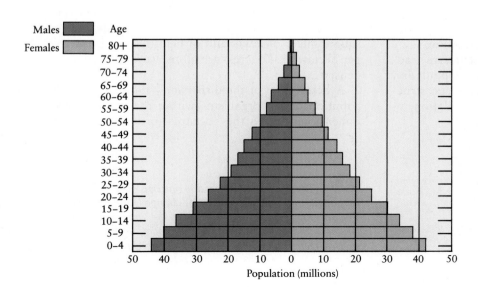

43–5   *The age structure of the human population of India. This pyramidal shape is characteristic of developing nations, with half the population under 20 years of age. In the absence of emigration, the population can stay the same size only if death rates are as high as birth rates. Even if members of the present generation in India limit their family size to only two children per couple (which means cutting the current birth rate in half), population growth will not level off until about the year 2040—and at a level of well over a billion.*

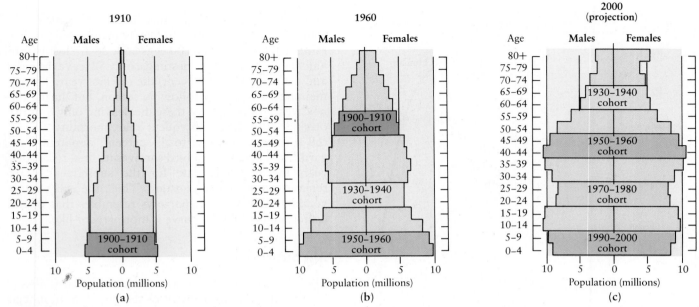

1910

Age	Males	Females

1900–1910
cohort

10   5   0   5   10
Population (millions)
(a)

1960

Age	Males	Females

1900–1910
cohort

1930–1940
cohort

1950–1960
cohort

10   5   0   5   10
Population (millions)
(b)

2000
(projection)

Age	Males	Females

1930–1940
cohort

1950–1960
cohort

1970–1980
cohort

1990–2000
cohort

10   5   0   5   10
Population (millions)
(c)

43–6   *Age structure of the United States population. (a) In 1910, the graph of age structure was shaped like a pyramid, although its base—that is, the number of persons in the youngest age groups—was not as large as that of India's. In subsequent years, as illustrated by (b) the graph of the age structure in 1960, and (c) the graph of the age structure projected for the year 2000, the proportion of the population that is more than 40 years of age has steadily increased, and the age structure of the population has become more uniform. The term "cohort" refers to the group of individuals born during the decade indicated.*

Note the decrease in the growth of the population during the depression years of 1930 to 1940 and the bulge produced by the "baby boom" of the 1950s. Although birth rates in the United States decreased in the 1970s and have remained low, the increase in the number of individuals during the 1950s was reflected in an increase in population growth in the 1980s as the boom babies reached reproductive age.

The age structure of the United States population (Figure 43–6) is one of the reasons why the population has continued to increase despite the fact that, on the average, young couples (between the ages of 20 and 30) are having slightly fewer than two children. Another reason for the increase is continued immigration.

As population growth slows, the age structure gradually becomes more uniform. Ultimately, a population that is not increasing attains a stationary age structure—that is, the age structure of the population remains the same over long periods of time.

Density and Dispersion

Two additional, interrelated properties of populations are density and dispersion. **Population density** is the number of individual organisms per unit area or volume. For example, the number of *Paramecium* cells per cubic centimeter of pond water, of dandelions per square meter of lawn, and of field mice and oak trees per hectare (2.47 acres) are all measures of population density.

A description of the **dispersion**—the pattern of distribution of the organisms within the two- or three-dimensional space they occupy—provides additional

(a)   **(b)**   **(c)**

43–7   *The three basic dispersion patterns observed in natural populations are (a) random, (b) clumped, and (c) regular. The dots may represent individuals of the same species, populations of the same species, or even populations of different species. Determination of dispersion patterns requires careful observation and precise mapping, repeated in a number of different areas and at different times.*

information about the population. The three basic dispersion patterns are shown in Figure 43–7. They are (1) **random,** in which the spacing between individuals is irregular; (2) **clumped,** in which the individuals are aggregated in patches; and (3) **regular,** in which individuals are evenly spaced within the area or volume.

A number of factors, both nonliving and living, may affect the spatial distribution of a particular population. Dispersion patterns are often dependent upon the distribution of essential resources. If, for example, water is available only in patches, members of a population of a particular plant species may be found clumped around these patches. In vertebrates, dispersion patterns often reflect social behavior. Individuals of a species that is highly territorial usually exhibit a regular spacing, whereas individuals of highly gregarious species tend to be clumped.

The dispersion pattern observed for a particular population or species depends upon the scale of the observation. On a local scale—a few square meters or kilometers, for example—a species may show a random or regular distribution. However, when viewed on a larger scale, such as the entire North American continent, the same species may show a clumped distribution.

(a)

(b)

(c)

43–8   *Some examples of dispersion patterns.* (a) *Many species of birds exhibit a regular dispersion when perched on telephone wires, power lines, or, as shown here, a dead tree branch. These birds are carmine bee-eaters* (Merops nubicoides). (b) *Fish, such as this school of French grunts, photographed in a shipwreck off the coast of Mexico, and* (c) *quaking aspen often exhibit clumped dispersion patterns.*

Climate Change and the Movement of Populations

The probability of a major global climate change within the next century is a fairly constant topic in the news. Although the speed of the predicted change and the agent of change—the activities of *Homo sapiens*—may be news, the occurrence of climate change itself is not. Periodic warming and cooling have characterized the global climate since shortly after the first living organisms made their appearance on Earth.

During most of our planet's history, its climate appears to have been warmer than it is at the present time. However, these long periods of milder temperatures have been interrupted periodically by Ice Ages, so called because they are characterized by **glaciations,** or persistent accumulations of ice and snow. Such glaciations occur whenever the summers are not hot enough and long enough to melt ice that has accumulated during the winter. In many parts of the world, an alteration of only a few degrees in the annual average temperature is enough to begin or end a glaciation.

An early Ice Age appears to have occurred at the beginning of the Paleozoic era, some 590 million years ago. Another, marked by extensive glaciations in the Southern Hemisphere, occurred at the end of the Paleozoic, some 248 million years ago. The conifers evolved during this period and possibly the angiosperms, as older forest types disappeared. A more recent, less severe cold, dry period occurred at the end of the Mesozoic, about 65 million years ago, and may be associated with the events causing the extinction of the dinosaurs (page 369).

The most recent Ice Age began during the Pleistocene epoch, about 1.5 million years ago. The Pleistocene has been marked by four extensive glaciations that have covered large areas of North America and Europe. Between the glaciations there have been intervals, called **interglacials,** during which the climate has become warmer.

In each of the four Pleistocene glaciations,

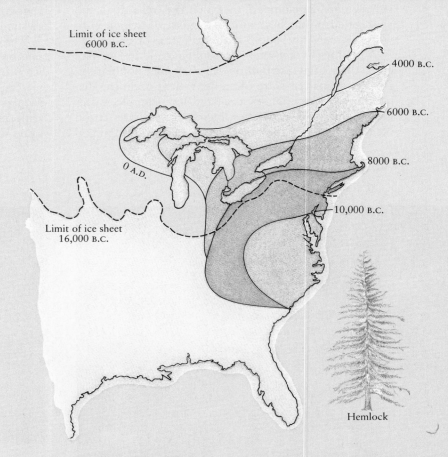

Hemlock

The northward migration of the eastern hemlock (Tsuga canadensis), *following the last glacial retreat. This retreat began about 18,000 years ago and ended some 8,000 years ago. The date at the right of each curve indicates the time of the first appearance of the leading edge of the hemlock population at the locations marked by the curve. This reconstruction of the northward movement of the hemlock population was compiled by Margaret B. Davis of the University of Minnesota, using records of pollen deposits that were preserved in lake sediments.*

sheets of ice, thicker than 3 kilometers in some regions, spread out from the poles, scraped their way over much of the continents—reaching as far south as southern Illinois in North America and covering Scandinavia, most of Great Britain, northern Germany, and northern Russia—and then receded again. We are living at the end of the fourth glaciation, which completed its retreat only some 8,000 years ago.

The fossil record shows that during periods

of dramatic climate change the populations of affected regions were under extraordinary ecological and evolutionary pressures. As conditions became too extreme, plant and animal populations moved, changed, or became extinct. In the interglacial periods, during which the average temperatures were at times warmer than those of today, the tropical forests and their inhabitants spread from the equatorial regions poleward through today's temperate zones. During the periods of glaciations, only animals of the polar regions could survive in these same locations.

In the colder periods, reindeer ranged as far south as southern France, while during the warmer periods, the hippopotamus reached England. Rhinoceroses, great herds of horses, large bears, and lions roamed Europe during the interglacials. In North America, as the fossil record shows, there were camels and horses, saber-toothed cats, and great ground sloths (one species as large as an elephant).

The reason for past climate changes is one of the most controversial issues in modern science. They have been variously ascribed to changes in the Earth's orbit, variations in the Earth's angle of inclination toward the sun, migration of the magnetic poles, fluctuations in solar energy, higher elevations of the continental masses, continental drift, and combinations of these and other causes. The predictions of a coming climate change are based upon a more obvious cause—the dramatic increase in the Earth's atmosphere of carbon dioxide and other by-products of human activity.

As we all wait to see the extent of this climate change, one of the greatest concerns of biologists is whether there will be time and opportunity for populations that are unable to adapt to changing conditions to move to new environments in which they can survive. If not, the resulting mass extinctions may well exceed those that marked the major climate upheavals of the past.

The Regulation of Population Size

Fifty years ago, the eminent ecologist Charles Elton noted that "no animal population remains the same for any great length of time, and . . . the numbers of most species are subject to violent fluctuations." The popular notion that "nature is in balance" and that populations generally reach an equilibrium state has also come under severe criticism from contemporary ecologists. As the two examples illustrated in Figure 43–9 show, the size of a population may vary greatly over a period of years.

(a)

(b)

43–9 *Two examples of the sometimes extreme fluctuations that occur in population size and density.* (a) *The population density of dormant larvae of the moth* Dendrolimus pini, *recorded over a period of 60 years in a coniferous forest in Germany.* (b) *Variations over a span of 30 years in the size of the breeding population of the great tit* (Parus major), *a common European bird of the same genus as the titmice and chickadees of North America, recorded at a single location in the Netherlands.*

Although it is often difficult to understand why fluctuations in population size occur, such knowledge can be important because population fluctuations in one species can have profound effects—for good or ill—on populations of other species, including ourselves. For example, earlier in this century an enormous increase occurred in the number of field mice in Kern County, California. The population density reached thousands of mice per hectare, leading not only to the devastation of crops but also to the infestation of playgrounds and living rooms. Abrupt declines in a population can also occur with staggering speed. A field plagued with mice in June may have practically none in July. A number of factors are thought to be involved in such fluctuations.

Limiting Factors

Among the influences affecting the size and density of a population are specific limiting factors that differ for different populations. Of critical importance is the organisms' range of tolerance for such factors as light, temperature, available water, salinity, nesting space, and shortages (or excesses) of required nutrients. If any essential requirement is in short supply or any environmental feature is too extreme, growth of the population is not possible, even though all other necessities are met.

Ecologists often divide the factors that influence the growth of a population into density-dependent and density-independent factors. Factors that cause changes in either the birth rate or the mortality rate as the density of the population changes are said to be **density-dependent.** Birth rates and mortality rates, however, may also vary without regard to changes in population size or density. Factors that cause such changes are said to be **density-independent,** and they often involve weather-related events.

Numerous factors operate on populations in a density-dependent fashion. As a population increases in size, it may deplete its food supply, leading to increased competition among members of the population that ultimately leads to a higher mortality rate or a lower birth rate. Predators may be attracted to areas in which the density of prey organisms becomes high, capturing an increased proportion of the population. Similarly, infectious diseases may spread more easily when population density is high.

Environmental disturbances often act as density-independent factors. For example, when Mount St. Helens erupted in 1980, populations of many species became locally extinct. Mortality was completely independent of the size and density of the populations. In many cases, however, the identification of density-independent factors is less clear-cut. The mortality resulting from a severe storm or cold spell may appear to be density-independent, yet animals that have secure places of shelter are more likely to survive than those that do not. And, the availability of good sheltering sites depends on the density of the population. As this example illustrates, there is often an interplay between density-dependent and density-independent factors, and populations are generally affected by both simultaneously.

Life-History Patterns

Among the most interesting and variable properties of populations are **life-history patterns,** which are groups of traits affecting reproduction and survival. Under what conditions, for example, will natural selection favor an organism that has two million almost microscopic young at one spawning, like an oyster, as opposed to an organism that has only one large infant, like an elephant?

In view of the fact that physiological constraints prevent an oyster from producing a 10-kilogram offspring, this question may sound silly. However, as a result of studies such as those on clutch size in birds (page 340), biologists have become aware that life-history patterns vary from individual to individual within a population and also from one population to another population of related organisms. In other words, these patterns involve genetically determined variations subject to natural selection.

Alternative life-history patterns have been given a variety of names: prodigal and prudent, opportunistic and equilibrium, and r-selected and K-selected. (In mathematical formulations of population growth, r represents the per capita rate of increase and K the carrying capacity.) Generally speaking, prodigal or opportunistic patterns would appear more adaptive for "weedy" species, colonizers of open fields, for instance, whereas prudent or equilibrium patterns would appear to be more adaptive for a population at the carrying capacity.

The characteristics associated with the alternative patterns are shown in Figure 43–10. Each pair of characteristics represents the extremes in a continuum of possibilities. The best choice in each continuum—the choice made by natural selection operating on a particular population—depends in large part on other properties of the population (such as its mortality pattern and its age structure) and on properties of the environment it occupies.

(a)

Prodigal or Opportunistic
Many young
Small young
Rapid maturation
Little or no parental care
Reproduction once ("Big Bang")

(b)

Prudent or Equilibrium
Few young
Large young
Slow maturation
Intensive parental care
Reproduction many times

43–10 *The characteristics associated with alternative life-history patterns. (a) Annual plants, such as the zinnia, have a prodigal, or opportunistic, life-history pattern. These plants reproduce only once, typically setting large numbers of seeds. The parent plants die soon after, and only the seeds survive. (b) Elephants, by contrast, have a prudent, or equilibrium, life-history pattern. The calves are suckled by their mothers for at least two years. The elephant social structure is a maternal hierarchy in which the young are zealously guarded by the mother and the sisters and the aunts.*

Some Examples of Life-History Patterns

Suppose you are a plant with a very short life expectancy—a flowering annual, for example. The choice that has been made for you, as a result of the action of natural selection in the past, is to put as much energy as possible, as rapidly as possible, into one reproductive effort. The result is a big burst of seeds, increasing the probability that at least a few will survive your short life span and, in turn, reproduce.

On the other hand, if you are an oak tree, once you have established a place for yourself, your life span is likely to be a very long one, more than 100 years. However, the likelihood of your offspring finding space on the forest floor is very small. You grow as tall as possible, for 20 years or more, since height aids in seed dispersal, and then put out moderate numbers of acorns per year, every year, over your entire life span. In this way, you are likely to produce more surviving offspring than if you had put the same amount of energy into one big burst of reproduction.

Bluefin tuna are like oak trees. Once a bluefin reaches a large size, its chances of continued survival are extremely good. However, the mortality among eggs and larvae is always very high, and, in years in which the plankton growth is low, there are no survivors at all. So it is better for a bluefin to concentrate its efforts first on reaching a safe size and thereafter to expend a relatively small amount of reproductive effort each season over a long period of time.

Whether breeding is early or late can greatly influence the rate of population growth. Reconsider the housefly of Table 43–1. If the generation time (that is, the interval between the time an egg is laid and the time that individual, in turn, mates and lays eggs) were shortened by only two days (from a minimum of 13 to a minimum of 11), there would be time for one more generation each season. The total number of possible offspring per season would increase by 330×10^{12}.

Early breeding does not always pay, however. Among many of the larger mammals, the rate of juvenile survival can be correlated with the experience of the mother, her size, or her social position, which are often determined by her age. In such situations, the action of natural selection favors delayed reproduction. A similar pattern is observed in the human population. Infant mortality is higher among teenage mothers in the United States than among women over 20, and higher still if the teenagers are from impoverished families, as they often are.

The Asexual Advantage

When we considered the evolution of sexual reproduction (page 329), we noted that an asexually reproducing population can increase in numbers much more rapidly than a sexually reproducing population. There are other advantages to asexual reproduction. Many plants, for example, reproduce by means of runners (page 734), and by so doing, may be able to grow to cover a very large area. All of the plants produced represent a single genotype. A young plant that develops in this way has a continued supply of resources from the parent plant and thus a much higher probability of survival.

Figure 43–11 shows the differences in survivorship in a species of buttercup between plants grown from seeds and those grown from runners. Note the very high mortality in the early stages of growth from seed—an oysterlike curve—compared to the uniform risk of death of the plantlets grown from runners.

Parthenogenesis

Another form of asexual reproduction is **parthenogenesis,** the development of an organism from an unfertilized egg. In species in which the male gamete determines the sex of the offspring, parthenogenesis always results in all female offspring. If the houseflies of Table 43–1 had reproduced parthenogenetically, each female would have had twice as many female young in every generation, and the population would have reached 358×10^{12} at the end of seven generations.

Parthenogenesis in plants lacks the advantage of the parental support system supplied by vegetative growth, which is traded off for the possibilities of larger numbers and, usually, wider dispersal of the young. Dandelions (Figure 43–12) reproduce parthenogenetically. They form conspicuous flowers and also some functionless pollen grains, which may be taken as evidence that the present asexual species of dandelions evolved from sexual ones.

Completely asexual species are also found among small invertebrates—some rotifers, for example—as well as among plants. Recently, several species of fishes and lizards have been found that apparently reproduce only parthenogenetically. Many other organisms alter-

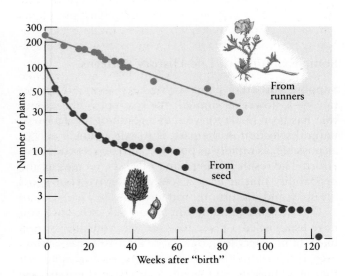

43–11 *Survivorship curves of populations of* Ranunculus repens, *a species of buttercup. Of 100 plants that started from seed (lower curve), only two (2 percent) were still alive 20 months later. Of 225 plants originating from runners, 30 (more than 15 percent) were still alive after 20 months (upper curve). Such plants receive support from the parent plant during early growth.*

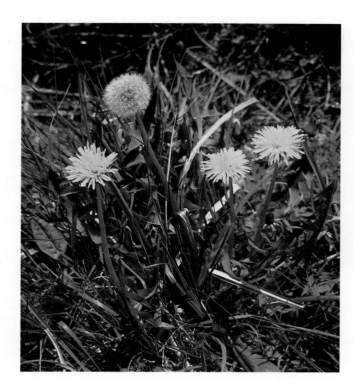

43–12 *One of the most widely distributed of all the flowering plants is the dandelion. Its flowers produce enormous numbers of seeds parthenogenetically. The seeds are enclosed in fruits bearing a plumelike structure that is caught by the wind, thereby aiding dispersal. Dandelions also reproduce vegetatively. If the plant is lopped off down to the root by either hominid or herbivore, three more grow in its place.*

nate sexual and asexual phases. Freshwater *Daphnia*, for instance, multiply by parthenogenesis when the plankton on which they feed is abundant. Then, in response to some environmental cue, they start producing both males and females. Typically, the asexual phase occurs when conditions are favorable for rapid local growth, and the sexual phase when the population is facing a less certain future and less homogeneous conditions.

Some Consequences of Life-History Patterns

Opportunistic organisms, which rapidly exploit an environment and then move on, would appear to lead risky lives, both as individuals and as species. There may not always be a new environment available for exploitation. However, populations of such organisms typically possess remarkable recuperative powers, because a population can be built up quickly from only a few individuals.

By contrast, populations composed of long-lived, slow-to-mature individuals, which would appear to have a higher probability of long-term survival, are very slow to recover when their size is reduced. Both the California condor (page 774) and the whooping crane are in this category. Each does not begin to reproduce until it is about five or six years old, and each raises only a single chick in each reproductive effort. As a result of a concerted conservation effort, the whooping crane population has made a remarkable recovery from the brink of extinction. For the condor, however, as we

saw at the beginning of this chapter, the prospects are not promising.

The Human Population Explosion

The California condor and the whooping crane are only two of the thousands of species of animals and plants whose continued existence has been made precarious by the extraordinary population growth of just one species—our own.

In the period from 25,000 to 10,000 years ago, when *Homo sapiens* lived as a hunter-gatherer, the rate of population increase was low. By the end of this period, the human population probably numbered a little more than 5 million, spread over the entire world. At this point, the establishment of agricultural communities began. During the next 5,000 years, as agriculture spread around the world, population growth increased dramatically. By 3000 B.C., the human population had reached about 100 million.

Population growth continued after an agricultural way of life had become established, but at a significantly lower rate. By 1650 A.D., the population had increased to approximately 500 million. At about this time, the rapid development of science, technology, and industry began, bringing profound changes in human life and its relationship to nature. In the 150 years between 1650 and 1800, the population doubled, to 1,000 million (1 billion), and then it doubled again by 1930, to 2 billion (Figure 43–13).

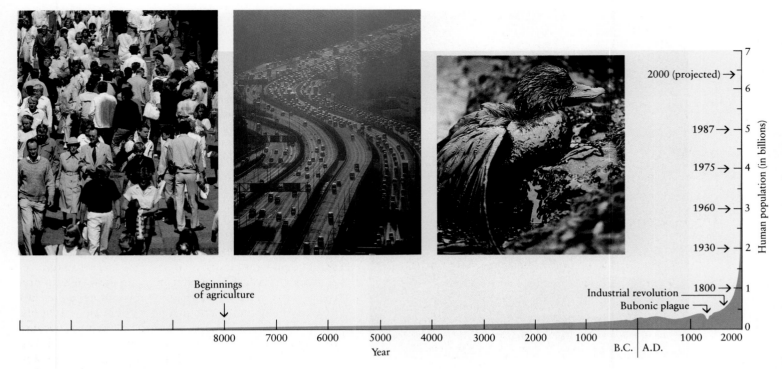

43–13 *Over the last 25,000 years, the human population has grown from an estimated 3 million to more than 5 billion. A significant increase in the rate of population growth occurred as a result of the agricultural revolution, and an even more dramatic increase began with the advent of the scientific and industrial revolution in the late 1600s.*

The consequences of the rapid growth of the human population are many and varied. In our country and other parts of the developed world, they include not only the sheer numbers of people but also heavy consumption of nonrenewable fossil fuels and the resulting pollution—both as the fuels are burned and as a result of accidents, such as oil spills, during their transport. In less de-

veloped parts of the world, the consequences include malnutrition and, all too often, starvation, coupled with a continuing vulnerability to infectious diseases. The consequences for other organisms include not only the direct effects of pollution but also—and most important—the loss of habitat.

By 1991, there were 5.42 billion people on our planet, and the population is growing at a rate of 1.6 percent per year. This means a net increase in the world population of more than 160 people every minute, about 230,000 each day, and almost 84 million every year. If this rate of increase is sustained, there will be about 6.2 billion people on Earth by the year 2000.

Birth Rates and Death Rates

For *Homo sapiens,* as for other species, the rate at which the population increases is the result of both the birth rate and the death rate. The relative influence of these two factors has varied at different times in human history.

In the period between 25,000 and 10,000 years ago, the low rate of increase is believed to have resulted from a low birth rate, in turn the result of physiological factors. For example, in some primitive hunter-gatherer societies, a woman is unable to conceive a child until she is between 19 and 20. Moreover, throughout the period when she is breast-feeding a child (which may be as long as three or four years), ovulation usually does

not occur. Compared to other contemporary human societies, the age at first reproduction is later (and thus the total reproductive span of the woman is shorter) and the interval between children is longer. These factors combine to produce a much lower birth rate in the population.

It is thought that the changes in culture and nutrition that accompanied the shift to an agricultural way of life may have led to a breakdown in the mechanisms that controlled the birth rate, resulting in the dramatic population increases that occurred between 10,000 and 5,000 years ago. The reasons for the subsequent slowing of population growth are not known with certainty. One contributing factor was probably an increased vulnerability to infectious diseases as people lived in closer proximity to each other within agricultural communities. Although the population continued to grow, increased mortality—particularly among the young—had a damping effect on the growth rate.

The enormous recent increase in the world's population is primarily a result of the decline in death rates, especially among the young, as shown in Figure 43–14. Some experts believe that further reductions in death

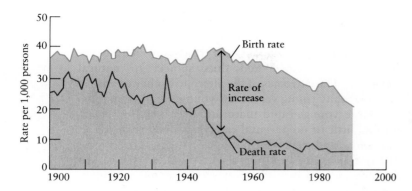

43–14 *In many tropical countries, the death rate has fallen rapidly since 1940, resulting in a rapid growth of the population. The drop in the death rate is primarily the result of increased medical services, better control of malaria-bearing mosquitoes, and the availability of new antibacterial drugs, especially the antibiotics. Note that the birth rate has also begun to fall. The area marked in color indicates the population growth in the absence of net immigration or emigration. The data shown here are for Sri Lanka.*

rates and a general increase in the standard of living will lead to a reduction in birth rates as well and thus to a reduction in the rate of population growth. Other experts, however, disagree. They point out that in many parts of the world a significant rise in the standard of living seems an almost unattainable goal and, in some cases, a rapidly receding one. Moreover, the age structure in many countries is such that even if birth rates were to drop dramatically, it would be many years before population size would begin to level off.

The current growth of the human population poses urgent and complex problems, made more difficult by the unequal distribution of growth and available resources. Births are occurring at the greatest rates in precisely those areas where the new arrivals have the least chance of an adequate diet, good housing, schools, medical care, or future employment. Moreover, because the affluent citizens of developed countries are not constantly reminded of the soaring population by experiencing firsthand the problems of hunger and crowding, they may feel it is not their problem. Yet a child born to a middle-class American will consume, in his or her lifetime, a far greater amount of the limited resources of the world—more than twice the amount of food, for instance—than a child born in a less developed country.

Bringing human population growth under control is an enormously complex undertaking, involving biological, economic, political, and social factors and requiring wisdom and commitment on the part of the diverse populations that constitute *Homo sapiens*.

Summary

Ecology is the study of the interactions of organisms with their physical environment and with each other. Ecologists seek to quantify the variables that affect organisms in nature, to construct explanatory hypotheses for the observed distribution and abundance of organisms, and to make and test predictions based on their hypotheses.

A population is a group of organisms of the same species living in the same place at the same time. Among the properties that characterize populations are patterns of growth, patterns of mortality, age structure, density, and dispersion.

In the absence of net immigration or emigration, the change in the number of individuals in a population over a given period of time—that is, the growth rate of the population—equals the per capita rate of increase (the birth rate minus the death rate) times the number of individuals present at the beginning of the period under study.

The reproductive potential of most populations is high. When the full reproductive potential of a population is achieved (a relatively infrequent occurrence in nature), exponential growth can occur. Exponential growth cannot continue long, however, without a "crash" in the population size.

The maximum number of individuals in a particular population that the local environment can support over a specified period of time is known as the carrying capacity of the environment. The logistic model, which takes the carrying capacity into account, describes one of the simpler patterns of population growth observed in nature. Growth is rapid when the population is small, gradually slows as it approaches the carrying capacity, and then oscillates as the population is maintained at or near the carrying capacity.

Populations also have characteristic mortality patterns, with varying risk of death at varying ages. A related property is the age structure of the population, that is, the proportion of individuals of different ages. Age structure is an important factor in predicting the future growth of a population.

Two additional, interrelated properties of a population are its density and its dispersion pattern. Density is the number of individual organisms per unit area or volume, whereas the dispersion pattern describes the two- or three-dimensional arrangement of the organisms in the environment.

A complex variety of environmental factors, both living and nonliving, play a role in the regulation of population size. Factors that influence birth rates or death rates independently of the population density are said to be density-independent; they often involve severe environmental disturbances. Factors that cause changes in birth rates or death rates as the density of the population changes are said to be density-dependent; these factors include numerous resources that are available in limited supply.

Populations are also characterized by their life-history patterns, which are groups of traits affecting reproduction and survival. These traits are generally genetically determined and thus subject to natural selection. The extremes of the alternative characteristics include many young versus few young, small young versus large young, rapid maturation versus slow maturation, little or no parental care versus intensive parental care, and reproducing once versus reproducing many times. In some organisms, the alternatives also include asexual reproduction versus sexual reproduction. Some combinations of these characteristics are favorable for organisms of opportunistic species undergoing exponential growth, whereas others will be selected for in more stable populations living at or near the carrying capacity.

About 25,000 years ago, the human population numbered perhaps 3 million worldwide. A surge in population growth began about 10,000 years ago with the development of agriculture. A second, more dramatic surge, still underway, began with the scientific and industrial revolution in the late 1600s. As the population now approaches 6 billion, this explosive growth is placing enormous pressure on the environment and the multitude of organisms with which we share it.

Questions

1. Distinguish between the following terms: exponential growth/logistic growth; limiting factors/carrying capacity; density-dependent factors/density-independent factors; opportunistic species/equilibrium species.

2. An old French riddle: "The pond lilies in a certain pond grow at a rate such that each day they cover twice as much of the pond as they did the day before. The pond is of a size that it will be completely covered at the end of 30 days. On what day is the pond half-covered? One-tenth covered? One-hundredth covered?"

What is the relevance of this riddle to human ecology?

3. Suppose that you have a "farm" on which you grow, harvest, and sell edible freshwater fish. The growth of the fish population is logistic. You, of course, wish to obtain maximum yields from your "farm" over a number of years. To ensure this, how large should you allow the population to become before you begin harvesting? Identify the point on the logistic growth curve (Figure 43–3) at which you should begin harvesting.

How large a population should you leave unharvested? Identify the point on the curve at which you should take no more fish from the population.

Factors in addition to the pattern of harvesting will affect the yields of fish obtained. What are some of these factors and how might you adjust them to further increase the yields?

4. What would be the shape of the survivorship curve of a population of annual plants? Of automobiles? Of salmon? Of butterflies? Of dishes in a dishwasher?

5. Note that in the three graphs in Figure 43–6 there is a marked difference in the proportion of individuals more than 80 years old, but the vertical axis has not become any longer. What do these data suggest? Does the age-structure graph of India (Figure 43–5) support your conclusion?

6. Distinguish among random, clumped, and regular dispersion patterns, and give an example of each. How does the scale from which organisms are viewed affect the dispersion pattern?

7. Explain how each of the following factors would affect the growth rate of a population: age at first reproduction; time between generations; pre-reproductive mortality; post-reproductive mortality; length of period of parental care.

8. Imagine a hypothetical species in which a particular individual lives only 48 hours and produces only two offspring. How is it possible that this individual may achieve greater fitness (see page 320) than a longer-lived individual that produces 100 offspring? Explain in evólutionary terms.

9. In an editorial entitled "Food, Overpopulation, and Irresponsibility" that appeared in a scientific journal, the authors concluded: "Because it creates a vicious cycle that compounds human suffering at a high rate, the provision of food to the malnourished nations of the world that cannot, or will not, take very substantial measures to control their own reproductive rates is inhuman, immoral and irresponsible." What is your opinion?

Suggestions for Further Reading

Begon, Michael, John L. Harper, and Colin R. Townsend: *Ecology: Individuals, Populations, and Communities,* 2d ed., Blackwell Scientific Publications, Boston, 1990.

An up-to-date, balanced, and thorough treatment of contemporary ecology.

Caldwell, John C., and Pat Caldwell: "High Fertility in Sub-Saharan Africa," *Scientific American,* May 1990, pages 118–125.

Diamond, Jared: "The Worst Mistake in the History of the Human Race," *Discover,* May 1987, pages 64–66.

Ehrlich, Paul R., and J. Roughgarden: *The Science of Ecology,* The Macmillan Company, New York, 1987.

This textbook provides a solid introduction to the many facets of ecology. It gives a broad overview of the science and also includes a great deal of information on behavioral ecology.

Grove, Richard H.: "Origins of Western Environmentalism," *Scientific American,* July 1992, pages 42–47.

Harper, John L.: *Population Biology of Plants,* Academic Press, Inc., New York, 1981.*

The world from a plant's perspective. A wonderful introduction to the ecology of populations and a must for any student of botany.

Holloway, Marguerite, and John Horgan: "Soiled Shores," *Scientific American,* October 1991, pages 88–95.

Hutchinson, G. Evelyn: *The Ecological Theater and the Evolutionary Play,* Yale University Press, New Haven, Conn., 1965.

By one of the great modern experts on freshwater ecology, this is a charming and sophisticated collection of essays on the influence of environment in evolution— and also on an astonishing variety of other subjects.

Hutchinson, G. Evelyn: *An Introduction to Population Ecology,* Yale University Press, New Haven, Conn., 1978.

A beautifully written and handsomely produced collection of lectures by one of the founders of the field, who was also a marvelous humanistic scholar and raconteur.

Keyfitz, Nathan: "The Growing Human Population," *Scientific American,* September 1989, pages 118–126.

Krebs, Charles J.: *Ecology: The Experimental Analysis of Distribution and Abundance,* 4th ed., HarperCollins Publishers, New York, 1992.

This excellent, modern text focuses on populations and on interactions among them. Requires some knowledge of mathematics.

Lents, James M., and William J. Kelly: "Clearing the Air in Los Angeles," *Scientific American,* October 1993, pages 32–39.

May, Robert M.: "Parasitic Infections as Regulators of Animal Populations," *American Scientist,* vol. 71, pages 36–45, 1983.

May, Robert M., and Jon Seger: "Ideas in Ecology," *American Scientist,* vol. 74, pages 256–267, 1986.

Myers, Judith H.: "Population Outbreaks in Forest Lepidoptera," *American Scientist,* vol. 81, pages 240–251, 1993.

Partridge, Linda, and Paul H. Harvey: "The Ecological Context of Life History Evolution," *Science,* vol. 241, pages 1449–1455, 1988.

Ricklefs, Robert E.: *Ecology,* 3d ed., W. H. Freeman and Company, New York, 1990.

This outstanding textbook provides a comprehensive introduction to the basic concepts of modern ecology. Beautifully written and rich with examples, it is highly recommended.

Robey, Bryant, Shea O. Rutstein, and Leo Morris, "The Fertility Decline in Developing Countries," *Scientific American,* December 1993, pages 60–67.

*Available in paperback.

44

Interactions in Communities

The Serengeti, the vast plain that spans northern Tanzania and southern Kenya, is home to the largest herds of grazing animals in the world. Zebras, wildebeest, and Thomson gazelles—affectionately known as "tommies"—move through the plains, following the annual cycle of rainfall. Ecologists have been able to demonstrate that species like these, which appear to be competitors, actually partition their resources—not only sharing but also, in some cases, facilitating each other's existence.

Zebras are the first animals to move into long-grass areas. Like horses, zebras have large incisors on both their top and bottom jaws and heavy grinding molars. They eat the long, tough grass stalks and the older leaves. By removing the coarse top stems, they make the more nutritious leaves accessible to the animals that follow.

The wildebeest—large, ungainly looking antelopes with long manes and a distinct beard—come next. A wildebeest feeds like a cow, wrapping its tongue around the long grass stems and drawing them across the saw-like ridge of its lower incisors. The great herds of wildebeest remove so much of the grass that new shoots of grass and tiny soft-stemmed dicots spring up.

The tommies come last. These exquisite little antelopes are only about 2 feet high at the shoulder and weigh about 40 pounds. With their tiny muzzles, they nibble at the new grass leaves and the small herbs that grow between them.

Grazing stimulates new growth, as do the trampling of the soil and the feces that the animals leave behind. Grazing also makes room for a variety of species, where otherwise only one or two species might dominate. Although each zebra, each wildebeest, each tommie, and every plant on the Serengeti is acting in its own self-interest, their combined activities have produced a vast, self-sustaining system that has operated to the benefit of all of its inhabitants for millennia—and perhaps will continue to do so for years to come.

Populations are made up of individual organisms. Communities are made up of populations. Ecologically speaking, a **community** consists of all the populations of organisms inhabiting a common environment and interacting with one another. These interactions are major forces of natural selection. They also influence the number of individuals in each population and the number and kinds of species in the community. The interactions among different populations are enormously varied and complex, but they can generally be categorized as competitive, predatory, or symbiotic.

Competition

Competition is the interaction between individual organisms of the same species (**intraspecific competition**) or of different species (**interspecific competition**) using the same resource, often present in limited supply. As a result of competition, the overall fitness—that is, the reproductive success—of the interacting individuals may be reduced. Among the many resources for which organisms may compete are food, water, light, or living space, including nesting sites or burrows.

Competition is generally greatest among organisms that have similar requirements and life styles. Plants often compete with other plants for sunlight and water. **Herbivores,** animals that eat plants and algae, may compete with other herbivores. **Carnivores,** animals that eat animals, may compete with other carnivores. Moreover, within these categories, the more similar two species are in their requirements and life style, the more intense the competition between them is likely to be.

For many years, competition has been invoked as a major force in determining the composition and structure of communities—that is, the number and kinds of species present and their arrangement in space and time within the community. Recently, however, a number of ecologists have come to question the importance of competition as an influence on community composition and structure. The debate—at times acrimonious—that has ensued concerns not only the role of competition but also the methods to be used in testing ecological hypotheses.

We shall begin our consideration of competition—and of the controversies surrounding it—with the principle that, until very recently, dominated the study of competition and the kinds of questions that ecologists asked about it.

The Principle of Competitive Exclusion

In 1934, the Russian biologist G. F. Gause formulated what became known as the principle of **competitive exclusion.** According to this principle, if two species are in competition for the same limited resource, one or the other will be more efficient at utilizing or controlling access to this resource and will eventually eliminate the other in situations in which the two species occur together.

Gause demonstrated the validity of his principle in his own, now classic, experiments involving laboratory cultures of two species of *Paramecium: Paramecium aurelia* and *Paramecium caudatum.* When the two species were grown under identical conditions in separate containers, both grew well. *Paramecium aurelia,* however, multiplied much more rapidly than *P. caudatum,* indicating that the former used the available food supply more efficiently than the latter. When the two were grown together, *P. aurelia* rapidly outmultiplied *P. caudatum,* which soon died out (Figure 44–1).

In laboratory experiments, the fastest growing species is not always the most successful competitor, how-ever, as observed with two species of duckweed, *Lemna gibba* and *Lemna polyrrhiza.* In pure culture, *L. gibba* grows more slowly than *L. polyrrhiza,* yet *L. gibba* always replaces *L. polyrrhiza* when they are grown together. The plant bodies of *L. gibba* have air-filled sacs that serve as little pontoons, so that these plants form a mass over the other species, cutting off the light. As a consequence, the shaded *L. polyrrhiza* dies out (Figure 44–2).

It is possible to devise different culture conditions so that the outcomes of both the *Paramecium* and *Lemna* experiments are reversed. However, as long as the conditions of a particular experiment are held constant, one species always wins and the other is always eventually eliminated.

44–1   *The results of Gause's experiments with two species of* Paramecium *demonstrate the principle that if two species are in direct competition for the same limited resource—in this case, food—one eliminates the other.* (a) Paramecium aurelia *and* (b) Paramecium caudatum *were first grown separately under controlled conditions and with a constant food supply. As you can see,* P. aurelia *grew much more rapidly than* P. caudatum, *indicating that* P. aurelia *uses available food supplies more efficiently.* (c) *When the two protists were grown together, the more rapidly growing species outmultiplied and eliminated the slower-growing species.*

(a)

(b)

(c)

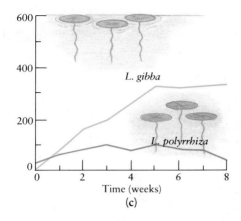

44–2 *An experiment with two species of floating duckweed, tiny angiosperms found in ponds and lakes. One species, Lemna polyrrhiza (a), grows more rapidly in pure culture than the other species, Lemna gibba (b). But L. gibba has tiny air-filled sacs that float it on the surface. When the two species are grown together, L. gibba shades L. polyrrhiza and is the victor in the competition for light (c).*

The Ecological Niche

Gause's competitive exclusion principle would lead you to expect that only dissimilar species would be found coexisting in natural communities. Yet, in fact, species with similar requirements and life styles are often found in the same community.

This observation raised the question of how similar two or more species can be and still continue to coexist in the same place at the same time, which led, in turn, to the concept of the **ecological niche.** This term is somewhat misleading because the word "niche" has the connotation of a physical space. An ecological niche, however, is not the space occupied by an organism but rather the role that it plays. The simplest analogy is that the niche is an organism's profession, as distinct from its habitat, which is its address.

A working definition of the niche is more complex, however, and includes many more factors than the way in which an organism makes its living. A niche is, in fact, the total environment and way of life of all the members of a particular species of organism in the community. Its description includes physical factors, such as the temperature limits within which the organisms can survive and their requirements for moisture. It includes biological factors, such as the nature and amount of required food sources. And, it also includes aspects of the behavior of the organisms, such as patterns of movement and daily and seasonal activity cycles. Although only a few of these factors can be studied at any one time, all are likely to influence the interactions of the members of a species with the members of other species in the community.

Resource Partitioning

The concept of the ecological niche suggested that when similar species are found coexisting together, a close examination will reveal that their niches are, in fact, different. An example is provided by the grazing animals of the Serengeti with which we began this chapter. Although the zebras, wildebeest, and gazelles appear to be sharing and competing for the same resources, they are not.

Numerous studies of different types of organisms have revealed that such dividing up, or partitioning, of resources occurs frequently among ecologically similar members of a community. The cause of the partitioning is a matter of considerable disagreement among ecologists, as we shall see shortly. First, however, let us consider two other examples.

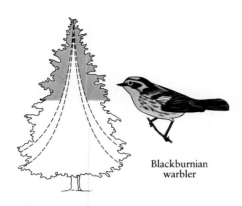

Cape May
warbler

Bay-breasted
warbler

Blackburnian
warbler

44–3 *The feeding zones in a spruce tree of five species of North American warblers. The colored areas in the tree indicate where each species spends at least half its feeding time. This partitioning of resources allows all five species to feed in the same trees.*

Black-throated
green warbler

Yellow-rumped
("myrtle")
warbler

44–4 *Mosses of the genus* Sphagnum, *which form extensive bogs in both temperate and cold regions of the world. Peat is formed from the accumulation and compression of the mosses themselves, as well as the sedges, reeds, and other grasses that grow among them. In Ireland, dried peat is widely used as an industrial fuel, as well as for home heating. It is estimated that peat bogs cover at least 1 percent of the Earth's land surface, an enormous area equivalent to half of the United States.*

Woodland Warblers

Some New England forests are inhabited by five closely related species of warblers, all about the same size and all insect eaters. The late Robert MacArthur, a brilliant and innovative young ecologist, meticulously observed and timed where the warblers fed within the trees. His data showed that the five species have different feeding zones (Figure 44–3). Because they exploit slightly different resources—that is, insects in different parts of the trees—the species can coexist.

Bog Mosses

In bogs, mosses of the genus *Sphagnum* (Figure 44–4) often appear to form a continuous cover, and several species are usually involved. How can these species, apparently very similar, continue to coexist?

When the situation is examined in detail, it is found that semiaquatic species grow along the bottoms of the wettest hollows. Other species grow in drier places on the sides of the hummocks, which they help to form. Still other species grow only in the driest conditions on the tops of the hummocks, where they are eventually succeeded by one or more species of flowering plants. Therefore, although all the species of *Sphagnum* coexist, in the sense that they are all present in the same bog, they actually occupy different microenvironments and continually replace one another as the characteristics of each microenvironment change.

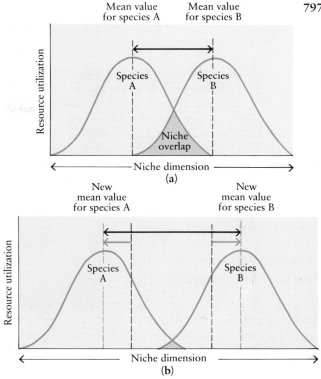

44–5 *Beak sizes in three species of ground finch found on the Galapagos Islands. Beak measurements are plotted horizontally, and the percentage of specimens of each species is shown vertically. Daphne and Crossman, which are very small islands, each have only one species of ground finch. These species have beak sizes halfway between those of the medium-sized and small finches on the larger islands.*

44–6 *Overlap in one dimension of an ecological niche. (a) The two bell-shaped curves represent utilization of a resource by two species in a community. The niche dimension might represent living space, as in the case of* Sphagnum *mosses; foraging space, as in the case of warblers; size of seeds eaten, as in the case of the Galapagos finches; and so on. Competition between the two species is potentially most intense in areas of niche overlap, leading to restriction of one or both species in living space, foraging space, or the size of seeds eaten, and so on. (b) It is hypothesized that such competition results in selection against individuals with overlapping characteristics, leading to divergence in the niches of the two species.*

The Role of Past Competition in Resource Partitioning

The resource partitioning observed among African herbivores, woodland warblers, *Sphagnum* mosses, and many other organisms was long considered to be the result of competition. In some cases, such as the warblers, the competition was thought to be occurring in the present. In other cases, such as the mosses and the herbivores, it was thought to have occurred in the evolutionary past, leading to the differing adaptations that enable the organisms to coexist. This phenomenon, in which species that live together in the same environment tend to diverge in those characteristics that overlap, is known as **character displacement.**

One of the most frequently cited examples of character displacement is provided by the beaks of Darwin's finches. As we saw in Chapter 21, the large, medium, and small ground finch species are very similar except for differences in overall body size and in the sizes and shapes of their beaks. The differences in the beaks are correlated with the sizes of the seeds the birds eat. On islands such as Pinta and Marchena (see page 364), where all three species of ground finch exist together, there are clear-cut differences in beak size (Figure 44–5). On Santa Maria and San Cristóbal Islands, the large species is not found, and the beak sizes of the medium ground finches on these islands are larger and overlap the beak sizes of the large finches found on

Pinta and Marchena. Daphne and Crossman, which are very small islands, each have only one species. Daphne has the medium-sized finch and Crossman, the small finch. These two populations have similar beak sizes, which are intermediate between those of the medium-sized and small finches on the larger islands.

These data have been interpreted in two different ways by ecologists. Some ecologists maintain that the observed differences in beak size are the result of the selection pressures exerted by interspecific competition. According to this interpretation, competition between organisms whose ecological niches overlap causes selection against individuals with overlapping characteristics, leading to the observed divergence between the species (Figure 44–6).

Other ecologists point out that it is impossible to determine if the differing beak sizes are the result of competition that occurred at times when the different species were coexisting on the same islands or if they are the result of adaptations to local conditions that occurred at times when the species were isolated from one another on different islands. Both groups of ecologists, however, agree that whatever their evolutionary cause, the differences in beak size and shape enable the different finch species to exploit different food sources and thus to coexist.

Experimental Approaches to the Study of Competition

Virtually all ecologists agree that competition does occur in nature, with the degree of intensity varying according to the particular species involved, the sizes of the interacting populations, and the abundance or scarcity of resources. Although the analysis of resource partitioning may provide clues as to the occurrence and importance of competition in a particular situation, experiments involving changes in community composition are required to demonstrate that competition is actually taking place. At their best, laboratory experiments, such as those performed by Gause, can only approximate natural conditions, which are invariably much more complex.

Barnacles in Scotland

One of the clearest experimental demonstrations of competition in a natural community was a study of barnacles performed by Joseph Connell. Barnacles are crustaceans. When they change from their free-swimming, larval form into their adult, sedentary form, they cement themselves to rocks and secrete protective calcium-containing plates. Once attached, barnacles remain fixed, so that by making careful observations over time, one can determine the dynamics of a particular population. One can identify exactly which barnacles have died and which new ones have arrived between visits to the study site.

Connell studied two barnacle species, *Chthamalus stellatus* and *Semibalanus balanoides*, that live on the coasts of Scotland. *Chthamalus* is found in the high part of the intertidal seashore. As the tides go in and out, these barnacles are exposed to wide fluctuations of temperature and salinity and to the hazards of desiccation. The other species, *Semibalanus,* occurs in the lower intertidal zone, where conditions are more constant. Although *Chthamalus* larvae, after their period of drifting in the plankton, often attach to rocks in the lower, *Semibalanus*-occupied zone, adults are rarely

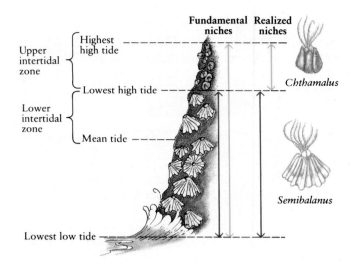

44–7   *Interspecific competition between* Semibalanus *and* Chthamalus *barnacles. The larvae of both species settle over a wide range, but the adults live in precisely restricted areas. The upper limits of the* Semibalanus *area are determined by physical factors such as exposure to air, which can lead to desiccation. Such exposure occurs at specific times during the month when the high tide covers only the lower portion of the intertidal zone, leaving the upper intertidal zone exposed.* Chthamalus *barnacles, however, are prevented from living in the* Semibalanus *area not by physical factors (they would probably thrive there since the area is less physically limiting) but by the* Semibalanus *barnacles.* Semibalanus *grows faster, and whenever it comes upon* Chthamalus *in the* Semibalanus *area, it either pries it off the rocks or grows right over it.*

found there. Connell showed that in the lower zone, *Semibalanus,* which grows faster, ousts *Chthamalus* by crowding it off the rocks and growing over it or undercutting it. When Connell experimentally removed *Semibalanus* from the lower portion of the intertidal zone, *Chthamalus* invaded the area and thrived there. In the control areas, where barnacles were not removed, each species remained in its own zone.

As this experiment demonstrates, *Chthamalus* is not restricted to the upper intertidal zone by a physiological inability to live elsewhere. It is restricted by competition with *Semibalanus.* There is no evidence, however, that competition with *Chthamalus* keeps *Semibalanus* in the lower zone. Because *Semibalanus* lacks the physiological adaptations required for life in the upper, drier zone, it cannot successfully invade it.

This study, and others like it, have generated the concepts of fundamental niche and realized niche. The **fundamental niche** describes the physiological limits of tolerance of the organism. It is the niche occupied by an organism in the absence of interactions with other organisms. The **realized niche** is that portion of the fundamental niche actually utilized. It is determined by physical factors and also by interactions with other organisms.

In its fundamental niche, *Chthamalus* can occupy both the high and low intertidal zones, but because of niche overlap with *Semibalanus*, *Chthamalus* actually occupies a smaller area, its realized niche (Figure 44–7). Because *Semibalanus* is restricted by physiological limits, it can occupy only the lower intertidal zone. Its fundamental niche is narrower than that of *Chthamalus* and is totally included within it. In such a situation, the species with the narrow, included fundamental niche must be a superior competitor or it will be driven to extinction. *Semibalanus* can survive in the same intertidal community as *Chthamalus* because it is the superior competitor. *Chthamalus* can survive, despite its competitive inferiority, because its fundamental niche is broader, providing it with a refuge that cannot be invaded by *Semibalanus*.

Winner Takes All

Most studies of competition have emphasized the adaptations and the partitionings of resources that make possible the coexistence of similar species within a community. This is, however, a biased view, for it is difficult to study the interactions between species after one of the protagonists has left. Just as competition within species leads to the elimination of the great majority of individual organisms, as Darwin observed, competition between species may lead to elimination of a species from the community.

One example is the disappearance from many localities of bluebirds (Figure 44–8). This is thought to have been caused, in part, by the usurpation of their nesting sites by starlings. Starlings were introduced into Central Park in New York City in 1891 and are now found throughout the United States, whereas some of us have never seen a bluebird.

Predation

Predation is the consumption of live organisms, including plants by animals, animals by animals, and even, as we have seen, animals by plants (page 741) or by fungi (page 437). Predators use a variety of techniques—known as **foraging strategies**—to obtain their food. Foraging strategies are under intense selection pressure. Those individuals that forage most efficiently are likely to leave the most offspring. From the standpoint of potential prey, those individuals that are most successful at avoiding predation are likely to leave the most offspring. Thus, predation affects the evolution of both predator and prey.

Predation also affects the number of organisms in a population and the diversity of species within a community.

Predation and Numbers

For many populations, predation is the major cause of death, yet, paradoxically, it is not at all clear that predation necessarily reduces the numbers of a prey population below the carrying capacity of its environment. However, when predation is heavier on certain age groups—juveniles versus adults, for instance—or certain life stages—such as caterpillars versus butterflies—it can alter the structure of a prey population and promote adjustments in life-history patterns.

44–8 *A male mountain bluebird at the entrance of a nesting box in Denver, Colorado. For more than 25 years, the U.S. Fish and Wildlife Service has been monitoring the breeding populations of the three bluebird species native to North America. A significant increase in the size of the populations began to occur in the mid-1980s. This increase is attributed, in large part, to a concerted effort by the North American Bluebird Society and others to encourage people to provide suitable nesting boxes for the birds in potential bluebird habitats. An important factor in the design of the nesting boxes is that the entrance holes be large enough to admit bluebirds but small enough to exclude starlings.*

44–9   (a) *Prickly-pear cactus growing on a pasture in Queensland, Australia, in November of 1926. Such rapid and environmentally destructive spread often occurs when alien organisms are introduced into a region where they have no competitors or predators.* (b) *The same pasture in October of 1929, slightly less than three years after the introduction of the cactus moth.*

(a)   (b)

Predation, especially on large herbivores, tends to cull animals in poor physical condition. Wolves, for instance, have great difficulty overtaking healthy adult moose or even healthy calves. A study of Isle Royale, an island in Lake Superior, showed that in some seasons more than 50 percent of the animals the wolves killed had lung disease, although the incidence of such individuals in the population was less than 2 percent. Thus, many of the animals killed by predators, according to this study, are animals that would have died soon anyway. (Modern human hunters, however, with their superior weapons and their desire for a "prize" specimen, are more likely to injure or destroy strong, healthy animals.)

In some situations, however, predators do limit their prey species. This has been most clearly demonstrated in cases involving the introduction of alien species into areas where they have no natural predators. When prickly-pear cactus, for instance, was brought to Australia from South America, it escaped from the garden of the gentleman who imported it and spread into fields and pastureland until more than 12 million hectares were so densely covered with prickly pears that they could support almost no other vegetation (Figure 44–9). The cactus then began to take over the rest of Australia at the rate of about 400,000 hectares a year. It was not brought under control until a natural predator was imported—a South American moth, whose caterpillars feed only on the cactus. Now only an occasional cactus and a few moths can be found. (Note, however, that the introduction of the alien moth was, in itself, risky.)

Few predator-prey relationships are this simple. Most predators have more than one prey species, although one prey species may be preferred. Characteristically, when one prey species becomes less abundant, predation on other species increases so the proportions of each in the predator's diet fluctuate.

Although predators may not always limit prey populations, the availability of prey constitutes a major component of the carrying capacity for predator populations, often stringently affecting their size. This is evident in relatively simple situations, such as when a bloom of phytoplankton (mostly microscopic algae) results from an upwelling of nutrients due to ocean currents and then is followed by a corresponding increase in zooplankton (animal plankton).

A more complex example is that of the lynx and the snowshoe hare. The data (Figure 44–10) are based on pelts received yearly by the Hudson's Bay Company over a period of almost 100 years. As you can see, there are oscillations in population density that occur about every 10 years. Generally speaking, a rise in the hare population is followed by a rise in the lynx population. The hare population then plummets, and the lynx population follows.

This example, which has been studied by ecologists over the last 50 years, can be interpreted in a variety of ways. The traditional explanation is that overpredation by the lynx reduces the snowshoe hare population. The lynx population, heavily dependent on the snowshoe hare as prey, is reduced in turn. The reduction in predation then permits the snowshoe hares to increase in number, followed by an increase in the number of lynx, and so on.

A second explanation is that the hare population undergoes a regular 10-year cycle caused, perhaps, by diseases associated with crowding or the effects of its own

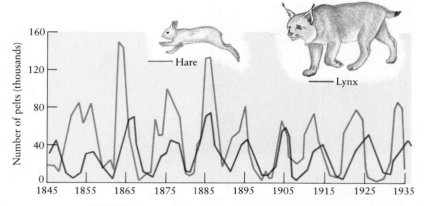

44–10 *The number of lynx and snowshoe hare pelts received yearly by the Hudson's Bay Company over a period of almost 100 years, indicating a pattern of 10-year oscillations in population density. The lynx reaches a population peak every 9 or 10 years, and these peaks are followed in each case by several years of sharp decline. The snowshoe hare follows a similar cycle, with a peak abundance generally preceding that of the lynx by a year or more.*

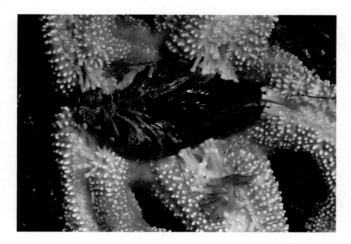

44–11 *Photographed off the coast of California, this starfish of the genus* Pisaster *is opening a mussel, one of its favorite foods. Studies have shown that predation by* Pisaster, *particularly on mussels and barnacles, is an important factor in maintaining species diversity in the rocky coastal communities in which it lives.*

predatory activities on the vegetation it consumes. This latter hypothesis is supported by two discoveries: (1) when overbrowsed, certain types of plants put out new shoots and leaves that contain chemicals toxic to hares, and (2) on an island where there are no lynx, the hare population undergoes similar cycles.

A third possibility is that the lynx undergo a regular cycle, independent of the hares, perhaps associated with some other factor, such as changes in the habits of their own predators, the human hunters. A decrease in the lynx population may permit growth of the hare population, or the two populations may oscillate independently. Thus, whereas the lynx and the hare used to serve as a simple model of predator-prey relationships, it is now perhaps more instructive as an example of the difficulties in dealing with ecological variables.

Predation and Species Diversity

The number and kinds of species in a community can be greatly influenced by predation. Although predation may occasionally eliminate a prey species, many experimental studies have shown that it is often an important factor in maintaining species diversity in a community.

For example, R. T. Paine studied a community on the rocky coast of Washington. In this community, the principal predator was the starfish *Pisaster* (Figure 44–11). At the beginning of the experiment, there were 15 prey species, including several species of barnacles and several kinds of mollusks, including mussels, limpets, and chitons. Paine systematically removed the starfish from one area of the community, 8 meters by 2 meters in size. By the end of the experiment, the number of prey species in the area from which the starfish were removed had declined to eight, and the community was dominated by one species of mussel. In the undisturbed community, starfish predation kept the densities of the prey populations low, reducing competition between the species and permitting all to survive. In the absence of the predator, the mussels were clear victors in the competition for living space.

Organisms, such as *Pisaster*, that are of exceptional importance in maintaining the diversity of a community are known as **keystone species.** When the keystone is removed from a stone arch, the arch falls apart. Similarly, when a keystone species is lost from a community—as a result, for example, of pollution, disease, or competition from an alien organism—the diversity of the community decreases and the structure of the community is significantly altered.

In another series of experiments, Jane Lubchenco showed that the herbivorous marine snail *Littorina lit-*

torea controls the abundance and type of algae in the higher intertidal pools on the New England coast. In such pools, the snails' preferred food (the green alga *Enteromorpha*) is competitively superior, and its removal by the snails permits the growth of other algal species (Figure 44–12). However, in areas exposed to the air at high tide as well as at low tide, *Enteromorpha* is competitively inferior. Its removal by the snails from those areas facilitates the growth of algal species that are competitively superior under exposed conditions. The result is domination by those species and a decrease in the total number of species in the exposed areas.

As this example illustrates, although predation often plays an important role in maintaining species diversity, it does not always do so. Even when the same predator and prey species are involved, the effect of predation on species diversity depends on the particular situation.

Symbiosis

As we saw in Chapter 25, **symbiosis** ("living together") is a close and long-term association between organisms of two different species. Long-continued symbiotic relationships can result in profound evolutionary changes in the organisms involved, as in the case of lichens (page 439), one of the oldest and most ecologically successful symbioses.

Although there is some disagreement among ecologists as to precisely what constitutes a symbiotic relationship, and the details of the relationship between two closely associated species are often difficult to determine, symbiotic relationships are generally considered to be of three kinds. If one species benefits and the other is harmed, the relationship is known as **parasitism.** If the relationship is beneficial to both species, it is called **mutualism.** Less common is **commensalism,** a relationship that is beneficial to one species and that neither benefits nor harms the other.

An example of commensalism is the relationship between the marine annelid *Chaetopterus* and tiny crabs of the genus *Pinnixa*, which live in the intertidal mud flats along the Atlantic coast of the United States. Each *Chaetopterus* constructs a U-shaped tube in which it lives (Figure 44–13), and the tube usually also contains two crabs, one male and one female. Both worm and crabs feed on particles of organic matter carried in water currents moved through the tube by fanlike appendages of the worm's body. The crabs obtain shelter and a steady food supply, and, as best anyone has been able to determine, their presence neither benefits nor harms the worm.

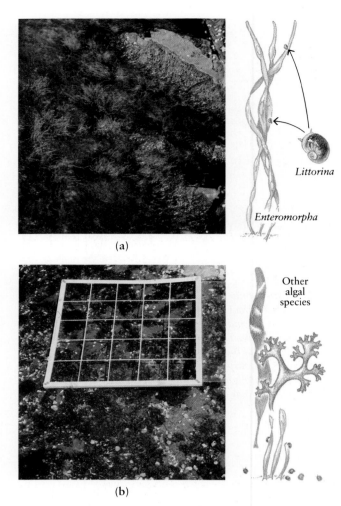

(a)

(b)

44–12   *An example of the effects of predation on species diversity.* (a) *In this high intertidal pool on the New England coast, the density of the herbivorous marine snail* Littorina littorea *is very low (between one and five individuals per square meter). The competitively superior green alga* Enteromorpha *dominates the pool, excluding other algae.* (b) *In a neighboring pool, less than a meter away, the density of* Littorina *is much higher—more than 250 individuals per square meter. The snails have grazed heavily on* Enteromorpha, *permitting the growth of other algal species that are, for* Littorina, *inedible. The aluminum grid at the top of the photograph was used to estimate the density of the snails.*

Parasitism

Parasitism may be considered a special form of predation in which the predator is considerably smaller than the prey. The plants and animals in a natural community support hundreds of parasitic species—in fact, certainly thousands and perhaps millions if one were to count nematodes and viruses.

As with more obvious forms of predation, parasitic diseases are most likely to wipe out the very young, the very old, and the disabled—either directly or, more often, indirectly, by making them more susceptible to other predators or to the effects of climate or food

44–13 Chaetopterus, *a marine annelid, lives in a parchment tube, constructed by the worm itself, with two commensal crabs of the genus* Pinnixa. *Modified parapodia ("side feet") of the worm function as fans, pulling water through the tube. Both the worm and the crabs are nourished by food particles carried into the tube with the moving water.*

44–14 *Rabbits crowd a water hole in Australia. Imported from Europe as potentially valuable herbivores, they soon overran the countryside. The myxoma virus was introduced to control them, and now host and parasite are coexisting.*

shortages. It is predictable that a parasite-caused disease should not be too virulent or too efficient. If a parasite were to kill all the hosts for which it is adapted, it too would perish.

This principle is particularly well illustrated by a series of further misadventures on the continent of Australia. There were no rabbits in Australia until 1859, when an English gentleman imported a dozen from Europe to grace his estate. Six years later he had killed a total of 20,000 on his own property and estimated he had 10,000 remaining. In 1887, in New South Wales alone, Australians killed 20 million rabbits. By 1950, Australia was being stripped of its vegetation by the rabbit hordes (Figure 44–14).

In that year, rabbits infected with myxoma virus were released on the continent. Myxoma virus, which is carried from host to host by mosquitoes, causes only a mild disease in South American rabbits, its normal hosts, but is usually fatal to the European rabbit. At first, the effects were spectacular, and the rabbit population steadily declined, yielding a share of pasture-

land once more to the sheep herds on which much of the economy of the country depends. But then occasional rabbits began to survive the disease, and their litters also showed resistance to the myxoma virus.

A double process of selection had taken place. The original virus was so rapidly fatal that often a rabbit died before there was time for it to be bitten by a mosquito and thereby infect another rabbit. The virus strain then died with the rabbit. Strains less drastic in their effects, on the other hand, had a better chance of survival since they had a greater opportunity to spread to a new host. So selection began to work in favor of a less virulent strain of the myxoma virus. Simultaneously, rabbits that were resistant to the original virus began to proliferate. Now, as a result of coevolution, the host-parasite relationship seems to be stabilizing.

Mutualism

If the current hypothesis as to the origin of eukaryotic cells (page 419) is correct, we owe our very origins to

(a)

(b)

(c)

(d)

44–15 *Mutualism.* (a) *Sea anemones on the back of a whelk shell occupied by a red hairy hermit crab. The anemones protect and camouflage the crab and, in turn, gain mobility—and so a wider feeding range—from their association with the crab. Hermit crabs, which periodically move into new, larger shells, will coax their anemones to move with them.*

(b) *Cleaner fish, removing parasites from a tiger grouper. Throughout the process of having its mouth cleaned, the grouper holds its mouth perfectly still. Cleaner fish are permitted to approach larger fish with impunity because of the service they render as they feed off the algae, fungi, and other microorganisms on the fish's body. Larger fish recognize the cleaners by their brilliant colors and distinctive markings. Other species of fish, by closely resembling the cleaners, are able to get close enough to the large fish to remove big bites of flesh. What would probably happen if the number of cleaner mimics began to approach the number of true cleaners?*

(c) *Aphids suck sap from the phloem, removing certain amino acids, sugars, and other nutrients from it and excreting most of it as "honeydew," or "sugar-lerp," as it is called in Australia, where it is harvested as food by the aborigines. Some species of aphids have been domesticated by some species of ants. These aphids do not excrete their honeydew at random, but only in response to caressing movements of the ants' antennae and forelimbs. The aphids involved in this symbiotic association have lost all their own natural defenses, including even their hard outer skeletons, relying upon their hosts for protection.*

(d) *Oxpeckers live on the ticks they remove from their hosts. Like cleaner fish, they perform an essential service. An oxpecker forms an association with one particular animal, such as the warthog shown here, conducting most of its activities, including courtship and mating, on the body of its host.*

(a)

(b)

44–16 *Ants and acacias.* (a) *A worker ant, drinking from a nectary of a bull's-horn acacia. At the left is an entrance hole into a thorn that was cut by the queen ant. After she had hollowed out the thorn, the queen raised her brood* within it. (b) *Worker ants collecting Beltian bodies from the tips of acacia leaflets. Rich in proteins and oils, the Beltian bodies are an important food source for both adult and larval ants.*

mutualism. Examples of present-day mutualistic associations are so abundant that we must limit ourselves to only a few. Perhaps among the most significant are those that take place underground, between roots and nitrogen-fixing bacteria (page 751) and in mycorrhizae (page 440).

Some particularly colorful examples of mutualism are illustrated in Figure 44–15. Another example is the interaction of ants and acacias.

Ants and Acacias

Trees and shrubs of the genus *Acacia* grow throughout the tropical and subtropical regions of the world. In Africa and tropical America, where plants are preyed upon by large herbivores, acacia species are protected by thorns. (Acacias that evolved where there are no large browsers—Australia, for instance—lack thorns.)

On one of the African species of *Acacia*, ants of the genus *Crematogaster* gnaw entrance holes in the walls of the thorns and live permanently inside them. Each colony of ants inhabits the thorns on one or more trees. The ants obtain food from nectar-secreting glands on the leaves of the acacias and eat caterpillars and other small herbivores they find on the trees. Both the ants and the acacias appear to benefit from this association.

In the lowlands of Mexico and Central America, the ant-acacia relationship has been extended to even greater lengths. The bull's-horn acacia is found frequently in cutover or disturbed areas, where it grows extremely rapidly. This species of acacia has a pair of greatly swollen thorns, several centimeters in length, at the base of most leaves. The petioles bear nectaries, and at the very tip of each leaflet is a small structure, rich in oils and proteins, known as a Beltian body. Thomas

Belt, the naturalist who first described these bodies, noted that their only apparent function was to nourish the ants. Ants live in the thorns, obtain sugars from the nectaries, eat the Beltian bodies, and feed them to their larvae.

Worker ants, which swarm over the surface of the plant, are very aggressive toward other insects and, indeed, toward animals of all sizes. They become alert at the mere rustle of a passing mammal, and when their tree is brushed by an animal, they swarm out and attack at once, inflicting painful, burning bites. Moreover, and even more surprising, other plants sprouting within as much as a meter of occupied acacias are chewed and mauled, and their bark is girdled. Twigs and branches of other trees that touch an occupied acacia are similarly destroyed. Not surprisingly, acacias inhabited by these ants grow very rapidly, soon overtopping other vegetation.

Daniel Janzen, who first analyzed the ant-acacia relationship in detail, removed ants from acacias either by insecticides or by removing thorns or entire occupied branches. Acacias without their ants grew slowly and usually suffered severe damage from insect herbivores. They were soon overshadowed by competing species of plants and vines. As for the ants, according to Janzen, these particular species live only on acacias.

There is an epilogue to the ant-acacia story. Three new species—a fly, a weevil, and a spider—have been discovered that mimic the ants that inhabit the acacias. So expert is their mimicry (probably involving chemical recognition signals as well as appearance) that the patrolling ants do not recognize them as interlopers, and so they enjoy the hospitality and protection of the ant-acacia complex.

Conservation Biology and the Island Biogeography Model

As the human demand for natural resources increases, ecological communities are being fragmented at an accelerating rate. Formerly large and continuous natural communities are being reduced to isolated "islands," often surrounded by areas that are unsuitable for most of the species involved. The most dramatic examples of this destruction of natural communities are occurring in tropical forests. However, the destruction of wildlife habitat is not unique to tropical forests. Whenever a marsh is cut by a new roadway, a forest is cleared for agriculture or a housing development, or a river is interrupted by a dam, the result is a subdivision of ecological communities into smaller, increasingly isolated habitat islands.

As discussed on the facing page, the equilibrium hypothesis of island biogeography predicts higher rates of extinction on and lower rates of immigration to small, isolated islands. According to this model, as natural communities are reduced to smaller and more isolated fragments, they can be expected to support fewer and fewer species.

This prediction of reduced species numbers in small, isolated habitat fragments has been supported by research in a number of different regions. One of the most thoroughly studied sites is Barro Colorado Island, an area of tropical rain forest located in Lake Gatun in Panama. The lake was created early in the twentieth century by the completion of the Panama Canal, isolating what is now Barro Colorado Island from the previously contiguous forest. Since that time, more than 50 species of birds have disappeared from the island, although they remain abundant on the mainland, only half a kilometer away.

A similar pattern of extinction has been observed in the 86-hectare woodland of the Bogor Botanical Garden in Java, Indonesia. Fifty years ago, this woodland was isolated by the destruction of the surrounding woodlands. Since that time, it has lost 20 of the 62 species of birds originally breeding there—more than

(a)

(b)

30 percent of the original community—and another four species are close to extinction.

It has been suggested that isolation of ecological communities may be reduced—and dispersal rates increased—by maintaining natural corridors between isolated communities. This possibility is being explored in the Netherlands, where studies are examining the role of hedgerows ("paths" of trees and shrubs, ranging from 1 to 10 meters in width) in increasing the rates of colonization of forest fragments. The hedgerows appear to increase the rate of dispersal of most forest species and may play a role in maintaining animal diversity in otherwise isolated patches of forest.

One of the most ambitious ecological projects currently underway was inspired by the equilibrium hypothesis of island biogeography. Known as the Minimum Critical Size of Ecosystems Project, it is a cooperative investigation by Brazilian and North American scientists in the Amazon rain forests. The project has revealed that forest fragmentation can have substantial effects on the physical environment. Hot, dry winds blowing across surrounding cleared areas reduce the relative humidity along the edges of the forest fragments

by as much as 20 percent. Moreover, increased light penetration at the edges of the forest fragments has elevated the temperatures there as much as 4.5°C (8.1°F) above the temperature in the forest interior. These physical changes may be responsible for some of the initial biological changes in the forest fragments: more trees die, more leaves drop from the remaining trees, and both the number of bird species and their population density decrease.

Although the equilibrium hypothesis of island biogeography began as a purely theoretical exercise to explain the maintenance of species diversity on islands, it has inspired research that is providing information vital for conservation biology, for the design and management of nature reserves, and for informed land-use planning. Some of the results of this research were not predicted by the original model. However, one of the most important predictions has held up: reductions in species diversity can be expected within nature reserves as they become increasingly isolated. For most groups of organisms, the numbers of species lost will depend upon both the size of the reserves and their degree of isolation.

(c)

(a) *The tropical rain forest of Barro Colorado Island, Panama, as seen from an observation tower above the forest canopy. The island is in the foreground, and the Panamanian mainland is in the background. Biologists affiliated with the Smithsonian Tropical Research Institute, assisted by Earthwatch volunteers, are conducting detailed studies of the forest and its occupants.* **(b)** *For North American visitors, one of the most familiar animals of the forest is the bay-breasted warbler, which summers in the forests of New England but winters on Barro Colorado Island— and nowhere else in the world.* **(c)** *The forest interior.*

Community Composition and the Question of Stability

Viewed from a global perspective, ecological communities often seem to be at equilibrium, with many species persisting for many generations over large areas. However, when communities are examined on a local scale, it becomes apparent that they, like the individual populations of which they are composed, undergo many fluctuations. Two questions concerning community composition have long perplexed ecologists. First, what determines the number of species in a community? And, second, what factors underlie the changes in a community that occur with the passage of time?

The Island Biogeography Model

Because of their size and relative isolation, small islands are often excellent natural laboratories for the study of evolutionary and ecological processes. In a classic study published in 1963, Robert MacArthur and E. O. Wilson used small islands as models to explore questions of community composition and stability.

MacArthur and Wilson hypothesized that the number of species on any given island remains relatively constant through time but that the identity of the species present is continually changing. According to their proposal, known as the **equilibrium hypothesis of island biogeography,** there is a balance between the rate at which new species immigrate to an island and the rate at which species already present become locally extinct. Although the number of species is in equilibrium, the species composition is *not* in equilibrium, because when a species becomes locally extinct it is usually replaced by a different species.

The island biogeography model was tested in an ingenious way by Wilson and Daniel Simberloff. They selected a number of small mangrove islands off the Florida Keys and counted the number of species of arthropods on each. They then destroyed all the animal life (mostly insects and other small arthropods) by covering the islands with plastic tents and fumigating them. Their plant life intact, the islands were soon colonized again from the mainland, and the recolonization process was monitored.

As predicted by the model, the total number of species present on an island after recolonization tended to be the same as the total number before the island was disturbed. However, and this is an important point, the species composition was often quite different from what it had been previously. Moreover, once the equilibrium number had been reached, the species composition con-

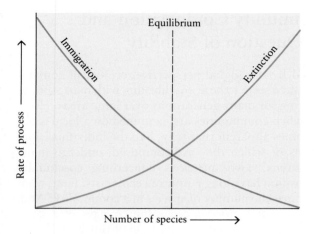

44–17 *Equilibrium model of the diversity of species on an island. The immigration rate declines as more species reach the island because species that arrived earlier have had time to become established. Thus, they are better able to compete against species that arrive later. Also, fewer immigrants will belong to new species. The extinction rate increases more rapidly at high species number because there is more competition between species. The equilibrium number of species (dashed line) is determined by the intersection of the immigration and extinction curves.*

44–18 **(a)** *Living sea fans, a coral that grows best in warm, clean water with a low level of nutrients.* **(b)** *The limestone remains of a sea fan of the same species that has been killed, most likely by parasites or polluted water. Reef-building corals are keystone species (page 801). When the living coral animals are destroyed, the structure of the reef community disintegrates. The diversity of other organisms that were once part of the community (page 460) must move elsewhere if they are to survive. When enormous areas of coral reef die, as is happening today in many parts of the world, there is no "elsewhere" for the other organisms.*

tinued to change, with extinction and immigration balancing one another out (Figure 44–17).

According to the island biogeography model, the two most important variables influencing species diversity are the size of the island and its distance from a source, usually the mainland, that can provide colonists. Distant islands tend to have fewer species than islands closer to the mainland. This is thought to be a result of lower rates of immigration, which appear to be a function of the distance that potential colonists must travel. Smaller islands are thought to reach equilibrium with a smaller number of species than larger islands, primarily because extinction rates are higher on smaller islands. This may be because populations tend to be smaller on such islands and thus are more susceptible to the effects of both predation and environmental disturbances.

This model applies not only to true islands but also to fragmented terrestrial areas. Thus it has important implications for conservation efforts (see essay).

The Intermediate Disturbance Hypothesis

Different types of communities vary widely in the number and diversity of species present. Among the most diverse are tropical rain forests and coral reefs. Until recently, it was thought that the species composition of these communities was relatively constant, and they were frequently cited as prime examples of an equilibrium state. Their high species diversity was thought to be a function of their stability. It now appears, however, that their diversity may be a function not of their stability but rather of the frequency and magnitude of the disturbances to which they are subjected.

Disturbances can take many forms. In the tropical forest, for example, trees are killed or severely damaged

(a)

(b)

by storms, landslides, lightning strikes, and insect outbreaks. The corals that form the basis of the coral reef community can be destroyed by predators or parasites, by the severe waves that accompany tropical storms, and by influxes of polluted water (Figure 44–18).

Soon after a disturbance, open areas—of forest or reef—are invaded by immature forms—seeds, spores, larvae, or even gametes. Initially, diversity in a newly colonized area is low. Only species that are close to the disturbed area and that are reproducing at the time are able to exploit the newly available area. If disturbances are frequent, the community will consist only of those species that can invade, mature, and reproduce before the next disturbance occurs.

According to the **intermediate disturbance hypothesis,** as the interval between disturbances increases, so does species diversity. Species that are excluded by frequent disturbances (because they are slow to mature or have limited dispersal abilities) now have an opportunity to colonize. However, if the interval between disturbances increases still further, species diversity may begin to decline (Figure 44–19). The primary factor in this decline is thought to be competition, but even if all species were competitively equal, the species most resistant to ill effects from physical extremes, predation, or disease would eventually come to dominate the community.

Among the smallest self-contained communities are those found on boulders located in the intertidal zones of rocky coasts. These communities, which are dominated by the algae growing on the rock surface, are often subjected to massive disturbances as a result of severe waves. The waves may either strip the algae away or actually overturn the boulders. As a result, bare rock becomes available for colonization. In a series of observations in intertidal zones of southern California, Wayne Sousa found that large boulders, which are infrequently overturned, and small boulders, which are frequently overturned, are typically dominated by one or a few algal species. By contrast, medium-sized boulders, which are subjected to an intermediate number of disturbances, tend to have a greater diversity of species.

Ecological Succession

Numerous observations have revealed that, if the interval between disturbances is relatively long, gradual changes occur in the composition of a community following the initial recolonization. The photosynthetic organisms that are usually the first colonists (Figure 44–20) are generally replaced in time by other types, which gradually crowd out the earliest species and

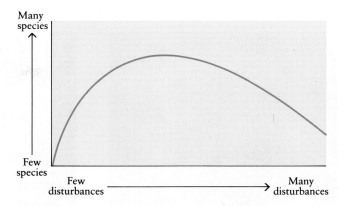

44–19   *According to the intermediate disturbance hypothesis, the diversity of species in a community is determined by the frequency of environmental disturbances. When disturbances occur either very seldom or very often, the number of species is low. By contrast, when the frequency of disturbances is somewhere in between, the number of species is high.*

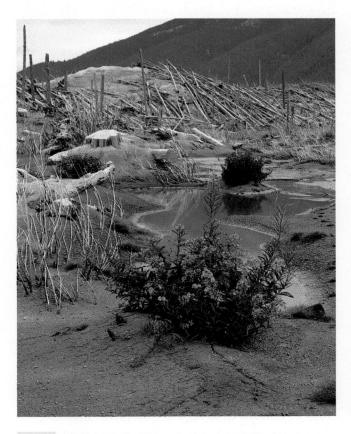

44–20   *When Mt. St. Helens in the state of Washington exploded on May 18, 1980, shock waves leveled all of the trees in an area of about 21,000 hectares, and a deep layer of ash was deposited. Fireweed* (Epilobium augustifolium) *and grasses, as seen here four months after the eruption, were among the first plants to recolonize the area.*

(a)

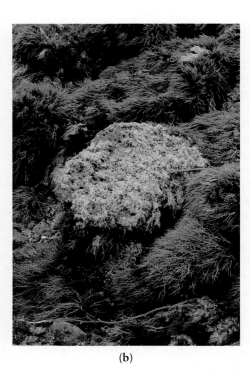

(b)

44–21   (a) *Boulders in a low intertidal zone on the southern California coast. Those with bare upper surfaces have recently been overturned by the waves, while those covered with algae have remained undisturbed.* (b) *A collection of intertidal boulders covered with different successional stages of algae. The boulder in the center was overturned by wave action during the winter before this photograph was taken and has been colonized by the green alga* Ulva. *The surrounding boulders, which are larger, have not been disturbed as recently and are covered with later successional stages that are dominated by red algae.*

which may themselves be subsequently replaced. As the photosynthetic components of the system change, the accompanying animal life changes as well. This process is known as **ecological succession.**

An example of succession is provided by an abandoned field surrounded by other vegetation. Such an area is bombarded by the seeds of numerous plants and is rapidly covered by seedlings of those species whose seeds germinate the most rapidly. In an open field, the survivors among the seedlings will be those plants that can survive the sunlight and drying winds—weeds and grasses and, in many regions of North America, trees such as cedars, white pines, poplars, and birch. For a while, these plants dominate the community, but eventually they are replaced by other trees—oaks, red maples, white ash, and tulip trees. These trees are, in turn, replaced by hemlock, beech, and sugar maple, which, in the absence of major disturbances, dominate the forest indefinitely.

In many communities, the sequence of photosynthetic organisms is so regular and predictable that ecological succession was at one time viewed as analogous to the developmental processes of a single organism, with each stage "preparing the way" for the next by altering the local conditions of temperature, light, humidity, soil content, and so on. Ultimately, the com-

munity would, according to this view, reach a "mature," stable state, known as the climax community. This concept of succession, formulated in 1916 by the plant ecologist F. E. Clements, is known as the **facilitation hypothesis.** For many years, it dominated the study of ecological succession.

Recent studies, however, have suggested that there are alternative mechanisms by which succession may occur. According to the **inhibition hypothesis,** the early species prevent—rather than assist—colonization by other species. Eventually, however, the earliest colonizers are replaced by later arrivals, and those species may, in turn, prevent colonization by others, until they too are replaced or a subsequent disturbance lowers their numbers.

Another model, the **tolerance hypothesis,** suggests that the earliest species neither facilitate nor inhibit colonization by later species. The species dominant at any given time are those that can best tolerate the existing physical conditions and availability of resources.

In a continuation of his earlier studies, Wayne Sousa charted the patterns of algal succession on intertidal boulders by scraping boulders clean of algae and by adding bare rocks to the intertidal zone. The first colonists on the bare rock were the green algae *Ulva* (Figure 24–18d, page 430) and *Enteromorpha* (Figure

44–12a). These pioneer species grow rapidly and, within a short period of time, fully occupy the available space on the boulders (Figure 44–21). Later, large perennial brown and red algae become established, replacing the green algae. Ultimately, the dominant organism is *Gigartina,* a red alga.

To determine the mechanism by which this successional sequence occurred, Sousa removed algae at various stages of the process. For example, when he removed *Ulva,* he found that *Gigartina* was then able to colonize the boulders. Similarly, when he removed the middle successional species of red algae, *Gigartina* was also able to colonize. These results support the inhibition hypothesis, but they raise the question as to how the later species ever gain a foothold. The answer, it turns out, is that *Ulva* and the other early colonizers are subjected to heavy predation and have high mortality rates. As a consequence, small areas of space open up on the boulders, which are then seized by the later species.

Other experiments, however, have supported the facilitation hypothesis, suggesting that there are no simple answers that apply in all situations. Each community is a unique assemblage of organisms, the product of a unique and ever-changing history involving both physical and biological factors.

The world of living organisms is so vast and diverse—and ecologists are, relatively speaking, so few—that it may be many years before enough information is available to provide a thorough understanding of community ecology. Paradoxically, the processes that occur at the next level of ecological complexity—the ecosystem—are in many respects more clearly understood than those occurring within either populations or communities. These processes, which involve the flow of energy and the recycling of minerals, will be the subject of the next chapter.

Summary

Populations live as part of a community—an assemblage of organisms of different species inhabiting a common environment and interacting with one another. Three major types of interaction in communities are competition, predation, and symbiosis.

The more similar organisms are in their requirements and life styles, the more intense the competition between them is likely to be. As a result of competition, the overall fitness of the interacting individuals may be reduced. The importance of competition as an influence on community composition and structure is currently a matter of debate among ecologists.

The competitive exclusion principle, formulated by Gause, states that if two species are in competition for the same limited resource, one or the other will eventually be eliminated in situations in which they occur together. Similar species are able to coexist in the same community because they occupy different ecological niches. The ecological niche of a species is defined by the resource requirements and total life style of the members of that species, including their interactions with members of other species. Analyses of situations in which similar species coexist have demonstrated that resources are often subdivided, or partitioned, by the coexisting species.

Niche overlap describes the situation in which members of more than one species utilize the same limited resource. In communities in which niche overlap occurs, natural selection may result in an increase in the differences between the competing species, a phenomenon known as character displacement. Although the adaptations that enable various organisms to partition resources have often been attributed to character displacement, it is difficult, if not impossible, to distinguish between adaptations that occurred in response to competition and those that were the result of differing local conditions.

Competition has been demonstrated in numerous experimental studies. Experiments in which potential competitors were removed from a study site have given rise to the concepts of the fundamental niche and the realized niche. A fundamental niche represents the resources that would be utilized by a species in the absence of interactions with other organisms. A realized niche describes the resources actually utilized.

Predation is the consumption of live organisms. Predator-prey interactions influence population dynamics and may increase species diversity by reducing competition among prey species. The size of a predator population is often limited by the availability of prey. However, predation is not necessarily the major factor in regulating the population size of prey organisms, which may be influenced more by their own food supply.

Symbiosis is a close and long-term association between organisms of different species. It may be beneficial to one species and harmless to the other (commensalism), beneficial to one and harmful to the other (parasitism), or beneficial to both species (mutualism).

Two important questions regarding community composition and structure remain unanswered. First, what determines the number of species in a community? And, second, what factors underlie the changes in community composition with time? According to the island biogeography hypothesis, the number of species on islands reaches an equilibrium determined by the balance between immigration and extinction. Species composition may vary widely over time but the number of species remains approximately the same. For new species to gain a foothold, established species must become extinct, leading to a continual turnover in species composition.

According to the intermediate disturbance hypothesis, the greatest species diversity is found in communities, such as tropical rain forests and coral reefs, that are subjected to environmental disturbances at an intermediate frequency. Communities in which disturbances occur either very seldom or very often generally have a lower species diversity.

Following environmental disturbances, communities are recolonized by dispersal of immature forms from neighboring communities. If enough time elapses before the next major disturbance, a community typically goes through a process of ecological succession in which the earliest colonizers are replaced by other species that may, in turn, be replaced by still others. The mechanism of succession appears to vary from community to community. Communities, like the populations of which they are composed, are dynamic, continually changing as conditions change.

Questions

1. Distinguish among the following: population/community; competitive exclusion/resource partitioning; habitat/niche; niche overlap/character displacement; fundamental niche/realized niche; symbiosis/commensalism/parasitism/mutualism.

2. Compare the effects of interspecific (between species) and intraspecific (within species) competition. What is the principal reason for the differences?

3. Compare the results of MacArthur's study of the warblers with Connell's experiment with the barnacles. What step did Connell perform that MacArthur did not? Why is the step important?

4. In the American southwest, grasses and mesquite compete with each other for dominance of the landscape. Mesquite, however, was rare before cattle were introduced into the western United States. How have cattle affected the competition between the two types of plant? Suppose all cattle were removed from a large area. What change would you predict in the competition between grasses and mesquite?

5. Introducing a new species into a community can have a number of possible effects. Name some of these possible consequences both to the community and to the introduced species. What types of studies should be done before the importation of an "alien" organism? Some states and many countries have laws restricting such importations. Has your own state adopted any such laws? Are they, in your opinion, ecologically sound?

6. In the long and ruthless war between coyotes and sheep herders, studies have shown that (a) coyotes kill sheep, and (b) the percentage of sheep lost from herds in areas where coyotes have been exterminated is about the same as the percentage lost in areas where coyotes are still present. How could you explain this finding?

7. In the opinion of some ecologists, animals that eat seeds, such as the Galapagos ground finches, should be regarded as predators, while animals that eat leaves, such as deer, should be regarded as parasites. Justify this classification of herbivores as either predators or parasites.

8. A national park or a game preserve can be considered an island, from the point of view of the species living there. What lesson in the management of such a park or preserve can be learned from the experiment carried out on the mangrove islands by Simberloff and Wilson?

9. Compare the three hypotheses (facilitation, inhibition, and tolerance) that have been proposed to explain the process of ecological succession.

Suggestions for Further Reading

Arehart-Treichel, Joan: "Science Helps the Bluebird," *Science News*, June 11, 1983, pages 376–377.

Bergerud, Arthur T.: "Prey Switching in a Simple Ecosystem," *Scientific American*, December 1983, pages 130–141.

Birkeland, Charles: "The Faustian Traits of the Crown-of-Thorns Starfish," *American Scientist*, vol. 77, pages 154–163, 1989.

Brown, Barbara E., and John C. Ogden: "Coral Bleaching," *Scientific American*, January 1993, pages 64–70.

Case, Ted J., and Martin L. Cody: "Testing Theories of Island Biogeography," *American Scientist*, vol. 75, pages 402–411, 1987.

Connell, Joseph H.: "Diversity and the Coevolution of Competitors, or the Ghost of Competition Past," *Oikos*, vol. 35, pages 131–138, 1980.

Connell, Joseph H., and Ralph O. Slatyer: "Mechanisms of Succession in Natural Communities and Their Role in Community Stability and Organization," *American Naturalist*, vol. 111, pages 1119–1144, 1977.

Connor, Edward F., and Daniel Simberloff: "Competition, Scientific Method, and Null Models in Ecology," *American Scientist*, vol. 74, pages 155–162, 1986.

DeVries, Philip J.: "Singing Caterpillars, Ants and Symbiosis," *Scientific American*, October 1992, pages 76–83.

Diamond, Jared M.: "Niche Shifts and the Rediscovery of Interspecific Competition," *American Scientist*, vol. 66, pages 322–331, 1978.

Diamond, Jared M., K. David Bishop, and S. Van Balen: "Bird Survival in an Isolated Javan Woodland: Island or Mirror?" *Conservation Biology*, vol. 1, pages 132–142, 1987.

Horn, Henry S.: "Forest Succession," *Scientific American*, May 1975, pages 90–98.

Karr, James R.: "Avian Extinction on Barro Colorado Island, Panama: A Reassessment," *American Naturalist*, vol. 119, pages 220–237, 1982.

Larson, Douglas: "The Recovery of Spirit Lake," *American Scientist*, vol. 81, pages 166–177, 1993.

Lubchenco, Jane: "Plant Species Diversity in a Marine Intertidal Community: Importance of Herbivore Food Preference and Algal Competitive Abilities," *American Naturalist*, vol. 112, pages 23–29, 1978.

O'Brien, W. John, Howard I. Browman, and Barbara I. Evans: "Search Strategies of Foraging Animals," *American Scientist*, vol. 78, pages 152–160, 1990.

Owen, Jennifer: *Feeding Strategy*, University of Chicago Press, Chicago, 1982.*

> *This short, beautifully illustrated book explores the variety of ways in which animals obtain food. It considers not only herbivorous and carnivorous animals but also filter-feeders, parasites, and scavengers.*

Peterson, Charles H.: "Intertidal Zonation of Marine Invertebrates in Sand and Mud," *American Scientist*, vol. 79, pages 236–249, 1991.

Rennie, John: "Living Together," *Scientific American*, January 1992, pages 122–133.

Rinderer, Thomas E., Benjamin P. Oldroyd, and Walter S. Sheppard: "Africanized Bees in the U.S.," *Scientific American*, December 1993, pages 84–90.

Romme, William H., and Don G. Despain: "The Yellowstone Fires," *Scientific American*, November 1989, pages 36–46.

Schaller, George: *The Serengeti Lion: A Study of Predator-Prey Relations*, University of Chicago Press, Chicago, 1976.*

> *One of the first, best, and most comprehensive of the modern field studies in animal ecology and behavior. Winner of the National Book Award.*

Schoener, Thomas W.: "The Controversy over Interspecific Competition," *American Scientist*, vol. 70, pages 586–595, 1982.

Sousa, Wayne P.: "Experimental Investigations of Disturbance and Ecological Succession in a Rocky Intertidal Algal Community," *Ecological Monographs*, vol. 49, pages 227–254, 1979.

Storer, John H.: *The Web of Life*, New American Library, Inc., New York, 1956.*

> *One of the first books ever written on ecology for the general public. In its simple presentation of the interdependence of living things, it remains a classic.*

Terborgh, John: "Why American Songbirds Are Vanishing," *Scientific American*, May 1992, pages 98–105.

Wilson, Edward O.: "Threats to Biodiversity," *Scientific American*, September 1989, pages 108–116.

Winkler, William G., and Konrad Bögel: "Control of Rabies in Wildlife," *Scientific American*, June 1992, pages 86–93.

Young, James A.: "Tumbleweed," *Scientific American*, March 1991, pages 82–87.

*Available in paperback.

Along the North Atlantic coast, from Labrador south to Maine, a cold current of water sweeps down from the North Pole, producing an upwelling of nutrients from the ocean depths to the sunlit surface. This upwelling of nitrates, phosphates, and other compounds profoundly affects the life and times of puffins, among others. These natural fertilizers spur the growth of phytoplankton, the "meadow" of the sea. The most abundant components of the phytoplankton are diatoms (page 427), which contain droplets of oil. Oil is less dense than water, and, as a consequence, the diatoms float near the surface, where they absorb the sun's rays.

Zooplankton graze on the phytoplankton. The most numerous organisms in the zooplankton are copepods (page 471), minute crustaceans less than 2 millimeters long. Shrimp and other larger invertebrates feed on the vast microscopic herds of copepods that move through the water. Fish larvae, herring, and other small fish are next in line, and they are eaten, in turn, by larger animals. Because there are, for reasons to be discussed in this chapter, fewer large animals than small animals, these interactions can be depicted in the form of a pyramid. This puffin, its mouth stuffed with herring, is at the top of one such pyramid.

Puffins look somewhat like penguins, because, like penguins, they are upright, with their feet placed well back under them. The reason they look like penguins, of course, is that they earn their living in a similar way. On land, puffins move with ease as they scramble over the rocks, using the claws at the ends of their webbed feet to grip the surface. In the water, they are superb swimmers, powered by their short, strong wings as they dive into and breast-stroke their way through schools of fish.

During the mating season, the puffin's bill is, as you see here, ornamented with vivid red, yellow, and black plates. Puffin pairs raise their chicks in huge, multispecies colonies of sea birds. At summer's end, they head to sea and spend the long, dark winter months harvesting the ocean's riches, never touching land until the following spring.

Ecosystems

As we saw in the last chapter, the populations within a community have numerous interactions with each other. Moreover, they interact with the **abiotic** (nonliving) environment. In all cases, these interactions have two consequences: (1) a one-way flow of energy through autotrophs (usually photosynthetic organisms) to heterotrophs, which consume either autotrophs or other heterotrophs; and (2) a cycling of materials, which move from the abiotic environment through the bodies of living organisms and back to the abiotic environment. This cycling of materials is dependent upon decomposers, organisms that break down organic materials into a form that can be used by autotrophs.

An assemblage of biotic (living) and abiotic (nonliving) components through which energy flows and materials cycle is known as an ecological system, or **ecosystem.** Taking a global view, the entire surface of the Earth can be seen as a single ecosystem. This view is useful when studying materials that are circulated on a worldwide basis, such as carbon dioxide, oxygen, and water.

A suitably stocked aquarium or terrarium is also an ecosystem, and such small-scale ecosystems can be useful models for studying certain ecological problems, such as the details of transfer of a particular mineral element. Most studies of ecosystems and their component communities have been made, however, on more or less self-contained natural units—for example, a pond, a swamp, or a meadow.

We shall begin our consideration of ecosystems with the principal physical factor affecting them—the energy received from the caldron of hydrogen and helium atoms that we know as the sun.

Solar Energy

Life on Earth is powered by the sun, which is also responsible for climate, wind, and weather. Every day, year in and year out, energy from the sun arrives at the

The living organisms of the Earth's ecosystems are shielded from the ultraviolet rays of the sun by a layer of ozone located in the stratosphere—the region of the atmosphere that begins about 10 kilometers above the Earth's surface and extends to an altitude of about 50 kilometers. Ozone (O_3) is formed when molecules of oxygen gas (O_2) are broken apart by radiant energy and then recombine.

Approximately 99 percent of the sun's ultraviolet radiation that reaches the stratosphere is converted to heat through a chemical reaction that continually recycles ozone molecules. When ultraviolet light strikes an ozone molecule, the high-energy radiation splits the molecule into highly reactive oxygen atoms. Almost immediately the atoms recombine, forming ozone once more and releasing energy in the form of heat. For hundreds of millions of years, the quantity of ozone in the atmosphere has apparently remained constant, as has the very small percentage of the ultraviolet radiation reaching the Earth's surface.

This is now changing, as a direct—although inadvertent—result of human activities. The most striking evidence of damage to the ozone layer has been the appearance of actual holes in the layer over both Antarctic and Arctic regions during their respective summers. Less dramatic, but of potentially far greater significance for living organisms, is growing evidence of a thinning of the ozone layer around the entire globe. Between 1978 and 1990, ozone levels over the temperate regions of the Northern Hemisphere dropped some 4.5 to 5 percent, and the loss since 1969 is estimated at about 10 percent.

The principal cause of the ozone loss is a group of synthetic chemicals known as chlorofluorocarbons. These chemicals are used throughout the world in refrigeration systems,

This computer-generated map of stratospheric ozone concentrations over the regions surrounding the South Pole is based on measurements made on October 6, 1993, by the Russian Meteor-3 satellite. The blue, pink, and white areas constitute the ozone hole. The ozone concentration in the center of the hole (white) is 60 percent lower than it was prior to 1975. This depletion of the ozone layer is more severe than in any year since measurements began. In this map, the outlines of the continents are shown in black and the concentric areas of color surrounding the hole indicate successively more substantial concentrations of ozone.

including home refrigerators, freezers, and air-conditioners. They are also used in fire extinguishers, particularly small household extinguishers, and they were previously used as propellants in aerosol cans.

For many years, chemists believed that chlorofluorocarbons released into the air diffused harmlessly into the upper atmosphere, where they were broken down by sunlight. In the 1970s, however, it was learned that, in the

presence of ultraviolet radiation, chlorine atoms break free of chlorofluorocarbon molecules and then react with ozone molecules. A single chlorine atom can react with—and destroy—as many as 100,000 ozone molecules. Current evidence indicates that if the addition of chlorofluorocarbons to the atmosphere were to continue unchecked, the ozone layer would eventually be entirely destroyed.

Depletion of the ozone layer leads directly and inevitably to an increase in the amount of ultraviolet radiation passing through the atmosphere. Each drop of 1 percent in the ozone layer leads to a 2 percent increase in the amount of ultraviolet radiation reaching the Earth's surface. For living systems, the consequences are many and serious. Ultraviolet radiation can trigger the dissociation of biologically important molecules, producing atoms and groups of atoms that are highly reactive.

Especially sensitive to ultraviolet radiation are the phytoplankton that form the base of aquatic food chains. Experimental studies have shown that even moderate levels of ultraviolet radiation are detrimental to these organisms and higher levels can cause mass mortality. Terrestrial photosynthetic organisms are also affected by increased levels of ultraviolet radiation. In a series of laboratory experiments in which some 100 different crop plants were exposed to slightly increased levels of ultraviolet radiation, the yields were reduced in 20 percent. Other experiments have demonstrated that the nitrogen-fixing bacteria that form symbiotic associations with the roots of legumes are killed by high levels of ultraviolet radiation.

For humans, one of the most immediate consequences of increased levels of ultraviolet radiation is an increased incidence of skin cancer. Calculations by the Environmental Protec-

upper surface of the Earth's atmosphere at an average rate of 1.94 calories per square centimeter per minute, for a total of about 1.3×10^{24} calories per year. This is known as the solar constant, and, although it is only a tiny fraction of the total energy radiated by the sun, it is a tremendous quantity of energy. However, because of the atmosphere, only a fraction of this energy reaches the surface of the Earth and becomes available to living organisms.

Of the incoming solar energy, about 30 percent is reflected back into space by clouds and atmospheric dust. Because of this reflected energy, Earth, as seen from outer space, is a shining planet, brighter than Venus. Another 20 percent of the energy is absorbed by the atmosphere.

The remaining 50 percent of the incoming radiation reaches the Earth's surface. A small amount of this is reflected from bright areas, but most is absorbed. En-tion Agency, based on 1990 satellite measurements of the ozone layer, project some 12 million additional cases of skin cancer in the United States in the next 50 years, of which more than 200,000 will be fatal. Also of concern is evidence that ultraviolet radiation can severely damage the immune system, the body's bulwark against infectious disease.

Intense efforts to find replacements for chlorofluorocarbons, particularly for refrigeration systems, are now underway. A landmark treaty, drafted in Montreal in 1987, bound its signatories to freeze chlorofluorocarbon production at 1986 levels by the end of 1989. Moreover, adherents to the treaty were required to cut production and use of these chemicals by at least half by the end of the century. It quickly became apparent, however, that the loss of stratospheric ozone is proceeding so rapidly that these measures were inadequate. Thus, in a subsequent treaty, drafted in London in 1990, the industrialized nations agreed to eliminate *all* production of chlorofluorocarbons by the turn of the century and the developing nations agreed to eliminate such production by the year 2010.

Current projections indicate that *if* the London agreement is faithfully observed, the concentration of ozone-destroying chlorine in the stratosphere should peak by the year 2005—at a level some 12 to 30 percent higher than at present. It is estimated that another 30 years would be required for the concentration of stratospheric chlorine to return to its current level, and perhaps another century for it to return to its preindustrial level. Although the London agreement cannot halt the loss of ozone that is already underway, it appears to be our best hope for slowing—and ultimately reversing—a process with potentially devastating effects for life on Earth.

Climate, Wind, and Weather

The amount of energy received by various parts of the Earth's surface is not uniform. This is the major factor determining the distribution of life on Earth. In the vicinity of the Equator, the sun's rays are almost perpendicular to the Earth's surface, and this sector receives more energy per unit area than the regions to the north and south, with the polar regions receiving the least (Figure 45–1). Moreover, because the Earth, which is tilted on its axis, rotates once every 24 hours and completes an orbit around the sun about once every 365 days, the amount of energy reaching different parts of the surface changes hour by hour and season by season (Figure 45–2).

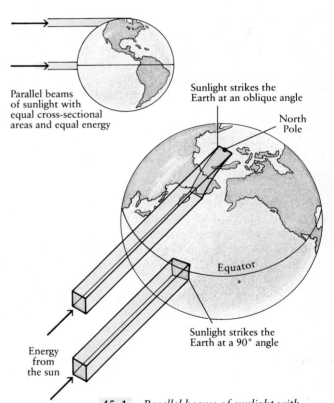

45–1 *Parallel beams of sunlight with equal cross-sectional areas contain equal amounts of energy. At the Equator, an incoming beam of sunlight strikes the Earth at a 90° angle (perpendicular) and its energy is concentrated into a small area. At higher latitudes, an identical incoming beam of sunlight strikes the curved surface of the Earth at an oblique angle. Thus, the same amount of energy is spread over a larger area. As a consequence, regions north and south of the Equator receive less solar energy per unit area than regions close to the Equator.*

ergy absorbed by the oceans warms the surface of the water, evaporating water molecules and powering the water cycle (page 45). Solar energy absorbed by the ground is reradiated from the surface in the form of infrared rays—that is, as heat. The gases of the atmosphere are transparent to visible light, but carbon dioxide and water, in particular, are not transparent to infrared rays. As a result, the heat is held in the atmosphere, warming the surface of the Earth.

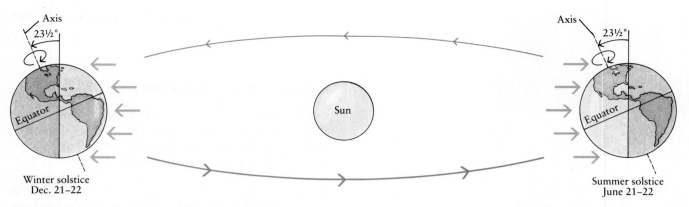

45–2 *In the Northern and Southern Hemispheres, temperatures change in an annual cycle because the Earth is slightly tilted on its axis of rotation in relation to its pathway around the sun. In winter in the Northern Hemisphere, the North Pole is tilted away from the sun, decreasing the angle at which the sun's rays strike the surface and also decreasing the duration of daylight, both of which result in lower temperatures. In summer in the Northern Hemisphere, the North pole is tilted toward the sun. Note that the polar region of the Northern Hemisphere is continuously dark during the winter and continuously light during the summer.*

Temperature variations over the surface of the Earth and the forces generated by the Earth's rotation establish major patterns of air circulation and rainfall. These patterns depend, to a large extent, on the fact that cold air is denser than warm air. As a consequence, warm air rises and cold air falls. As warm air rises, it encounters lower pressure and expands, and as a gas expands, it cools. Cooler air holds less moisture, so as warm air rises and then cools, it tends to lose its moisture in the form of rain or snow. As the air continues to cool, its density increases and it falls back toward the Earth's surface.

The air is warmest along the Equator, the region heated most intensely by the sun. This air rises, creating a low-pressure area (the doldrums) that draws in air from the surface north and south of the Equator. As the warm equatorial air rises, it cools, loses most of its moisture, moves poleward, and then falls again at latitudes of about 30° north and south, the regions where most of the great deserts of the world are found. This air then either moves toward the Equator, replacing the warm air rising there, or it moves poleward. The poleward-moving air warms, picks up moisture, and rises again at about 60° latitude (north and south). This is the polar front, another low-pressure area. The air rising at the polar front cools, loses moisture, and descends again at the poles, producing another region of descending air in which there is virtually no precipitation. The Earth's spinning movement twists the winds caused by this transfer of air from Equator to poles, creating the major wind patterns (Figure 45–3).

The worldwide patterns are modified locally by a variety of factors, such as the presence of mountains (Figure 45–4). For example, along the West Coast of the United States, where the prevailing winds are from the west, the western slopes of the Sierra Nevada have

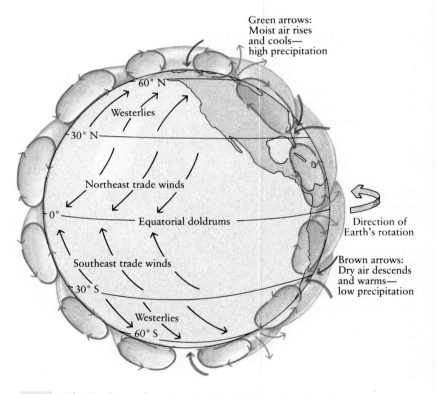

Green arrows: Moist air rises and cools— high precipitation

60° N
Westerlies
30° N
Northeast trade winds
0°
Equatorial doldrums
Southeast trade winds
30° S
Westerlies
60° S

Direction of Earth's rotation

Brown arrows: Dry air descends and warms— low precipitation

45–3 *The Earth's surface is covered by belts of air currents, which determine the major patterns of wind and rainfall. In this diagram, the dark blue arrows indicate the direction of movement of the air within the belts. The green arrows indicate regions of rising air, which are characterized by high precipitation, and the brown arrows indicate regions of descending air, which are characterized by low precipitation. The dry air descending at latitudes of 30° north and south is responsible for the great deserts of the world.*

The prevailing winds on the Earth's surface, indicated by the black arrows, are produced by the twisting effect of the Earth's rotation on the air currents within the individual belts.

45–4 *The effect of coastal mountains on patterns of precipitation. As winds come off the water and the air is forced upward by the contour of the land, it cools and releases its moisture in the form of rain or snow. As the air descends on the far side of the mountains and becomes warmer, its capacity to hold water increases, producing a rain shadow.*

abundant rainfall, while the eastern slopes are dry and desertlike. As air from the ocean hits the western slope, it rises, is cooled, and releases its water. Then, after passing the crest of the mountain range, the air descends again. As it becomes warmer, its water-holding capacity increases, producing a "rain shadow" on the eastern slope.

The Flow of Energy

The flow of energy through ecosystems is the most important factor in their organization. Of the solar energy that reaches the Earth's surface, only a very small fraction—an estimated 0.1 percent on a worldwide basis—is diverted into living systems. Even when light falls where vegetation is abundant, as in a forest, a cornfield, or a marsh, only 1 to 3 percent of that light (calculated on an annual basis) is used in photosynthesis. Yet this fraction, as small as it is, may result in the production—from carbon dioxide, water, and a few minerals—of several thousand grams (dry weight) of organic matter per year in a single square meter of field or forest. Worldwide, it results in the production of about 120 billion metric tons of organic matter per year.

Trophic Levels

The passage of energy from one organism to another takes place along a particular **food chain**—that is, a sequence of organisms related to one another as prey

45–5 *The energy on which life depends enters the living world in the form of light. It is converted to chemical energy by photosynthetic organisms, such as (a) wild barley, a terrestrial plant, and (b) giant kelp, a marine alga. Such photosynthetic organisms are, in turn, the energy source for all heterotrophs, including ourselves.*

(a)   (b)

and predator. The first is eaten by the second, the second by the third, and so on, in a series of feeding levels, or **trophic levels.** In most ecosystems, food chains are linked together in complex **food webs,** with many branches and interconnections (Figure 45–6).

A food web may involve more than 100 different species, with predators characteristically taking more than one type of prey, and each type of prey being exploited by several different species of predators. The relation of any species to others in its food web is an important dimension of its ecological niche.

Producers

The first trophic level of a food web is always occupied by a **primary producer.** On land, the primary producer is usually a plant; in aquatic ecosystems, it is usually an alga. These photosynthetic organisms use light energy to make carbohydrates and other compounds, which are sources of chemical energy. Producers far outweigh consumers. Some 99 percent of all the organic matter in the living world is made up of plants and algae. All heterotrophs combined account for only 1 percent of the organic matter.

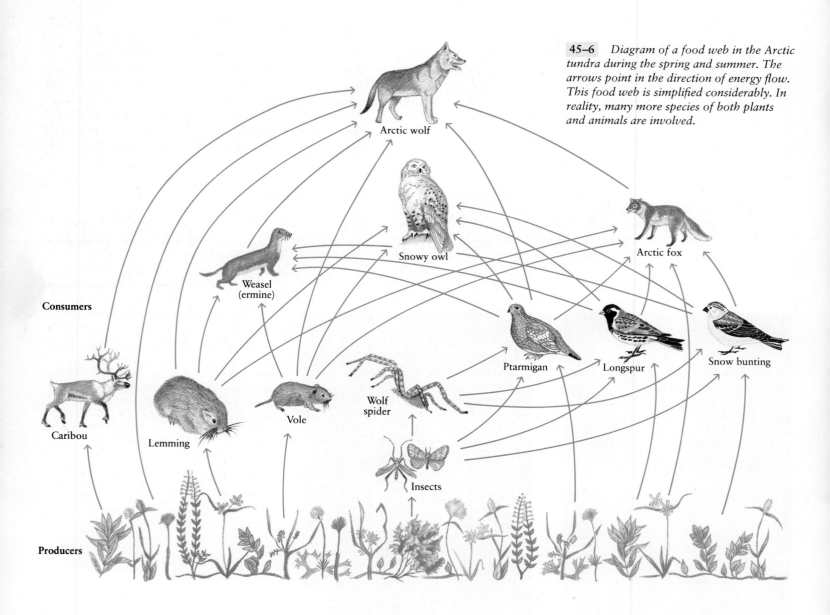

45–6   *Diagram of a food web in the Arctic tundra during the spring and summer. The arrows point in the direction of energy flow. This food web is simplified considerably. In reality, many more species of both plants and animals are involved.*

Consumers

Arctic wolf

Snowy owl

Arctic fox

Weasel (ermine)

Ptarmigan

Longspur

Snow bunting

Caribou

Lemming

Vole

Wolf spider

Insects

Producers

Life without the Sun

Throughout this text, we have emphasized that life on Earth depends on radiant energy from the sun. However, a series of discoveries beginning in 1977 have revealed the existence of entire ecosystems in which the primary producers are chemosynthetic bacteria (page 75) that are not dependent on the energy of sunlight to power their carbon-fixing reactions. Such ecosystems, which are typically located in the ocean depths, far beyond the reach of sunlight, are powered by geothermal energy—heat energy from within the Earth itself.

The first known chemosynthetic ecosystem was discovered by the *Alvin*, a research submarine from Woods Hole Oceanographic Institution, and its crew of scientific investigators. The place was the Galapagos Rift, a boundary between adjacent tectonic plates (see page 358). Sea water seeps into the fissures in the volcanic rocks erupting along this boundary, becomes heated, and rises again. As a consequence, oases of warmth are created in the near-freezing waters 2.5 kilometers below the surface. More important, the water reacts with the rocks deep within the Earth's crust. Here, under extreme heat and pressure, sulfate in the sea water is reduced to hydrogen sulfide, with heat supplying the necessary energy for the reaction. Chemosynthetic bacteria then oxidize the sulfide, obtaining the energy needed to extract carbon from carbon dioxide in the sea water and fix it in organic molecules.

More than a dozen of these deep-sea oases have now been located in the Pacific, each with chemosynthetic bacteria as the primary producers in a complex food web. The specific complement of animals varies from one chemosynthetic ecosystem to another and is thought to be the result of whatever floating larvae happened to colonize the particular area during its early stages of development.

Among the most spectacular inhabitants of the deep-sea oases are the clams, measuring some 20 centimeters in diameter, that cover the lava floor. Other permanent inhabitants include mussels, crabs, and octopuses. The bivalves feed on the bacteria, and the crustaceans and octopuses on the bivalves. One oasis, known as the Garden of Eden, is dominated by huge tube worms. These worms, which lack mouths and digestive tracts, apparently obtain energy from organic molecules synthesized by chemosynthetic bacteria living within their tissues. The bacteria have also been found living in the gills of some clams, suggesting that symbiosis plays an essential role in these ecosystems.

These discoveries triggered not only a search for other deep-sea chemosynthetic ecosystems but also a closer look at the biological interactions of chemosynthetic bacteria found in more accessible surroundings. Such ecosystems have been found deep in the Gulf of Mexico in cold waters and off the Palos Verde peninsula of California in very shallow waters.

The latter discovery was made by a young, skin-diving scientist who was puzzled by the black abalone he found grazing on mats of bacteria in sulfur-rich waters that would normally be toxic to animal life. Although the fissures through which the hot, sulfide-laden water escapes are only the size of teacups, they are surrounded by mats of chemosynthetic bacteria averaging 1 to 2 square meters in area. These mats are grazed not only by the black abalone but by other gastropods as well, including the California sea hare and the giant keyhole limpet.

Most recently, clams living in Los Angeles sewage outfalls have been found to contain symbiotic chemosynthetic bacteria, as have worms living in Bermuda. And, scientists have not yet examined the organisms of such sul-

Giant tube worms, 1.5 meters tall, growing in the depths of the Pacific Ocean at the Garden of Eden oasis. These tube worms are dependent on chemosynthetic bacteria as their energy source. Also visible are yellow-shelled clams and an eyeless, colorless crab, also participants in the food web for which the bacteria are the primary producers. Unlike the near-freezing temperatures normally found at these depths, it was 17°C (63°F) in this area.

fur-rich environments as pulpmill effluent zones, mangrove swamps, and eel grass beds. It is already clear, however, that although most life on Earth is based on radiant energy from the sun, there are highly viable alternatives—not only on this planet, but perhaps on others as well.

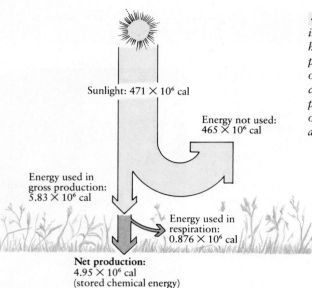

Sunlight: 471 × 10⁶ cal

Energy not used:
465 × 10⁶ cal

Energy used in
gross production:
5.83 × 10⁶ cal

Energy used in
respiration:
0.876 × 10⁶ cal

Net production:
4.95 × 10⁶ cal
(stored chemical energy)

45–7   *Calculation of the productivity of a field in Michigan in which the vegetation was mostly perennial grasses and herbs. Measurements are in terms of calories per square meter per year. In this field, the net primary production—the amount of chemical energy stored in plant material—was 4,950,000 calories per square meter per year. Thus, slightly more than 1 percent of the 471,000,000 calories per square meter per year of sunlight reaching the field was converted to chemical energy and stored in the plant tissues.*

Ecologists speak of the **productivity** of a trophic level, a community, or an ecosystem. **Gross productivity** is a measure of the rate at which energy is assimilated by the organisms in, for example, a particular trophic level. It might be considered analogous to the rate of gross income of a business.

A more useful—and often more easily measured—quantity is the **net productivity** (Figure 45–7). Net productivity is the gross productivity minus the cost of all the metabolic activities of the organisms in question. This cost might be considered analogous to the cost of doing business. Net productivity is thus comparable to the rate of net profit. It is usually expressed as the amount of energy (measured in calories) stored in chemical compounds or as the increase in biomass (measured in grams or metric tons) in a particular period of time. (**Biomass** is a convenient shorthand term meaning the total dry weight of all the organisms being measured at any one time.) Net productivity is a measure of the rate at which organisms store energy, which then becomes available to the organisms of the next trophic level.

At the first trophic level—the primary producers—the increase in the amount of plant or algal material between the beginning and the end of a specified time period represents the **net primary production** for that period. In agricultural ecosystems, for example, the standing crop at the end of the growing season represents the net primary production for that season.

As the data in Table 45–1 reveal, primary production varies enormously from one ecosystem type to another.

In terrestrial ecosystems, the key factors influencing productivity are the duration and intensity of sunlight, temperature, and precipitation. In aquatic ecosystems, the availability of essential mineral elements is more often the principal factor affecting productivity.

Consumers

Energy enters the animal world through the activities of herbivores, animals that eat plants or algae and are thus **primary consumers.** A herbivore may be a caterpillar, an elephant, a sea urchin, a snail, or a field mouse. Each type of ecosystem has its characteristic complement of herbivores.

Of the organic material consumed by herbivores, much is eliminated undigested. Most of the chemical energy from the digested food is used to maintain the metabolic processes of the animal and to power its daily activities: searching for food, eating and digesting it, mating and caring for the young, fleeing predators, and so on. Although this energy is generally described as "lost" through respiration, it is important to realize that, for the individual organism, this is the essential energy on which its life depends.

A fraction of the chemical energy consumed by the herbivore is converted to new animal biomass. The increase in animal biomass is the sum of the increase in the mass of individual animals plus the mass of new offspring. It represents energy available to the next trophic level.

Table 45-1	Average Primary Productivity of Selected Ecosystems	
Ecosystem	Net Primary Production (grams per meter² per year)	Biomass (kilograms per meter²)
Open ocean	125	0.003
Algal beds and reefs	2,500	2
Estuaries	1,500	1
Lakes and streams	250	0.02
Swamps and marshes	2,000	15
Tropical forest	2,000	12
Temperate forest	1,250	32
Taiga	800	20
Woodland and shrubland	700	6
Tropical grassland	900	4
Temperate grassland	600	1.6
Tundra and alpine	140	0.6
Desert and semidesert scrub	90	0.7
Cultivated land	650	1

Adapted from R. H. Whittaker and G. E. Likens, Table 5.2, in H. Lieth and R. H. Whittaker (eds.), *Primary Productivity of the Biosphere,* Springer-Verlag, Inc., New York, 1975. Tropical forest estimates based on Sandra Brown and Ariel E. Lugo, "Biomass of Tropical Forests: A New Estimate Based on Forest Volume," *Science,* vol. 223, pages 1290–1293, 1984.

This next level, the **secondary consumer** level, is made up of carnivores, animals that eat other animals. The carnivore that devours the herbivore may be a lion, a minnow, a starfish, a robin, or a spider. In every case, only a small part of the organic substance present in the body of the herbivore becomes incorporated into the body of the carnivore.

Some food chains have third and fourth consumer levels, but five links are usually the limit, regardless of the ecosystem. A study of 102 top predators (animals themselves free of predation) showed that there are usually only three links (four levels) in a food chain—plant to herbivore to carnivore to carnivore. For only one top predator were more than five links (six species) involved. At each higher trophic level, there is a decrease in the total amount of energy stored in animal biomass and therefore available to other consumers.

Detritivores

Detritivores are organisms that live on the refuse, or detritus, of a community—dead leaves, branches, and tree trunks, the roots of annual plants, feces, carcasses, even the discarded exoskeletons of insects. They include **scavengers,** such as vultures, jackals, crabs, and earthworms, as well as **decomposers,** such as fungi and bacteria.

Scavengers can be regarded as consumers that feed on dead prey rather than living prey. Decomposers are also consumers, but with a difference: they have evolved

(a)

(b)

45-8 *Representative consumers.* (a) *The hog deer is a herbivore of the Indian subcontinent. Its name derives from its relatively short stature. Adult hog deer generally range from 60 to* 72 *centimeters in height.* (b) *An Indian python, a carnivore, begins the four-hour process of swallowing an adult hog deer that it has killed by constriction.*

(a) (b) (c)

45–9 *Many birds, such as these vultures, are scavengers, living on dead animals. In East Africa, several species of vulture are often seen sharing the same carcass. (a) The lappet-faced vulture—the largest and most powerful, with a huge, heavy beak—is the one that can first break into a carcass. Other vultures are often seen standing by a carcass awaiting the arrival*

of a lappet-faced vulture. (b) Ruppell's griffon, also a vulture, specializes in reaching its long, snaky, featherless neck far inside the carcass to feed on the intestines and other internal organs. (c) The smaller Egyptian vulture, with its fine beak, can tear off scraps of flesh, such as from the skull, that the coarser beaks cannot grasp.

specializations that enable them to exploit sources of chemical energy, such as cellulose and nitrogenous waste products, that cannot be used by animals.

In a forest community, more than 90 percent of the net primary production is eventually consumed by detritivores rather than by herbivores. Some of this energy flows through the food web by way of consumers that feed on detritivores, while the rest of it is used in the metabolic processes of the detritivores themselves. As a result, essentially all of the energy stored by plants (and, in aquatic communities, by algae) is ultimately used to support life. Energy stored in organic matter goes unused only when it is trapped in an environment, such as a highly acidic peat bog, in which most detritivores cannot live.

Efficiency of Energy Transfer

The shortness of food chains is usually attributed to the inefficiency involved in the transfer of energy from one trophic level to another. In general, only about 10 percent of the energy stored in a plant is converted to animal biomass in the herbivore that eats the plant. A similar relationship is found at each succeeding level.

If, for instance, an average of 1,500 kilocalories of light energy per square meter fall per day on a land surface covered with plants, about 15 kilocalories are converted to plant material. (Remember that only 1 to 3 percent of the sunlight falling on vegetation is actually used in photosynthesis.) Of these 15 kilocalories, about

1.5 kilocalories are incorporated into the bodies of the herbivores that eat the plants, and about 0.15 kilocalorie is incorporated into the bodies of the carnivores that prey on the herbivores. Although meat is a more concentrated source of calories and nutrients than vegetation, carnivores must usually expend more energy in foraging than herbivores do, and so the net productivity of carnivores and herbivores may be roughly equivalent.

To give a concrete example, in studies of Cayuga Lake in western New York, Lamont Cole calculated that for every 1,000 calories of light energy utilized by algae in the lake, about 150 calories are reconstituted as the biomass of small aquatic animals. Of these 150 calories, 30 calories are converted to smelt biomass. If trout eat the smelt, about 6 calories are converted to trout biomass, and if we eat the trout, we gain about 1.2 calories from each 1,000 calories of light energy originally stored in organic compounds. (Note that if we ate smelt instead of trout, we would derive about five times as much food energy from the original 1,000 calories. This, in short, is the argument for living lower on the food chain.)

As these figures suggest, the 10 percent "rule of thumb" is only a crude estimate. Actual measurements show wide variations in transfer efficiencies, from less than 1 percent to over 20 percent, depending on the species involved. The flow of energy with large losses at each successive level can be depicted as a pyramid such as that shown in Figure 45–10.

	Respiration	Net production
	(kcal per m² per year)	
Secondary carnivore	13	6
Primary carnivore	316	67
Herbivore	1,890	1,478
Producer	11,977	8,833

45–10 *Pyramid of energy flow for a river ecosystem in Florida. A relatively small proportion of the energy in the system is transferred at each trophic level. Much of the energy is used metabolically and is measured as calories lost in respiration.*

Tertiary consumers

Secondary consumers

Primary consumers

Producers

Number of individuals
(a)

Tertiary consumers

Secondary consumers

Primary consumers

Producers

Number of individuals
(b)

45–11 *Pyramids of numbers for (a) a grassland ecosystem, in which the number of primary producers (grass plants) is large, and (b) a temperate forest, in which a single primary producer, a tree, can support a large number of herbivores.*

Energy Transfer and Ecosystem Structure

The energy relationships between the trophic levels determine the structure of an ecosystem in terms of both the numbers of organisms and the amount of biomass present. Figure 45–11a, for example, shows a pyramid of numbers for a grassland ecosystem. In this type of ecosystem, the primary producers (the individual grass plants) are small, and so a large number of them is required to support the primary consumers (the herbivores). In an ecosystem in which the primary producers are large (for instance, trees), one primary producer may support many herbivores, as indicated in Figure 45–11b.

Most pyramids of biomass take the form of an upright pyramid, as shown in Figure 45–12a, whether the producers are large or small. Pyramids of biomass are inverted only when the producers have very high reproduction rates. For example, in the ocean, the standing crop of phytoplankton may be smaller than the biomass of the zooplankton that feeds upon it (Figure 45–12b). Because the reproduction rate of the phytoplankton population equals or exceeds the rate of grazing by the zooplankton population, a small biomass of phytoplankton can supply food for a larger biomass of zooplankton.

Like pyramids of numbers, pyramids of biomass indicate only the quantity of organic material present at one time. They do not give the total amount of material produced or, as do pyramids of energy, the rate at which it is produced.

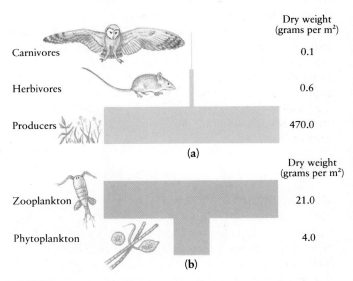

	Dry weight (grams per m²)
Carnivores	0.1
Herbivores	0.6
Producers	470.0

(a)

	Dry weight (grams per m²)
Zooplankton	21.0
Phytoplankton	4.0

(b)

45–12 *Pyramids of biomass for (a) the plants and animals in a field in Georgia and (b) plankton in the English Channel. Such pyramids reflect the mass present at any one time. The seemingly paradoxical relationship between the biomass of phytoplankton and zooplankton exists because the high reproduction rate of the smaller phytoplankton population is sufficient to support a larger zooplankton population.*

Biogeochemical Cycles

Energy takes a one-way course through an ecosystem, but many substances cycle through the system. Such substances include water, nitrogen, carbon, phosphorus, potassium, sulfur, magnesium, calcium, sodium, chlorine, and also a number of other minerals, such as iron and cobalt, that are required by living organisms in only very small amounts. The water cycle is shown on page 46, and the carbon cycle is on page 173.

Movements of inorganic substances are referred to as **biogeochemical cycles** because they involve geological as well as biological components of the ecosystem. The geological components are (1) the atmosphere, which is made up largely of gases, including water vapor; (2) the solid crust of the Earth; and (3) the oceans, lakes, and rivers, which cover three-fourths of the Earth's surface.

The biological components of biogeochemical cycles include the producers, consumers, and detritivores (both scavengers and decomposers). As a result of the metabolic work of the decomposers, inorganic substances are released from organic compounds and returned to the soil or water. From the soil or water, inorganic substances are taken back into the tissues of producers (plants or algae), passed along to consumers, and then released to the detritivores, from which they enter the producers again, repeating the cycle.

The cycling through the ecosystem of one important mineral, phosphorus, is shown in Figure 45–13. Other minerals in the ecosystem undergo similar cycles, differing only in details.

The Nitrogen Cycle

The chief reservoir of nitrogen is the atmosphere; in fact, nitrogen gas (N_2) makes up about 77 percent of the atmosphere. Since most living things, however, cannot use elemental atmospheric nitrogen to make amino acids and other nitrogen-containing compounds, they are dependent on nitrogen present in soil minerals. So, despite the abundance of nitrogen in the atmosphere, a shortage of nitrogen in the soil is often the major limiting factor in plant growth.

45–13   *The phosphorus cycle in a terrestrial ecosystem. Phosphorus is essential to all living organisms as a component of energy-carrier molecules, such as ATP, and also of the nucleotides of DNA and RNA. Like other minerals, it is released from dead tissues by the activities of decomposers, taken up from soil by plants, and cycled through the ecosystem.*

The phosphorus cycle in an aquatic ecosystem involves different organisms but is, in most respects, similar to the terrestrial cycle shown here. However, in aquatic ecosystems a considerable amount of phosphorus is incorporated into the shells and skeletons of aquatic organisms. This phosphorus, along with phosphates that precipitate out of the water, is subsequently incorporated into sedimentary rock. Such rock, returned to the land surface as a result of geological uplift, is the primary terrestrial reservoir of phosphorus.

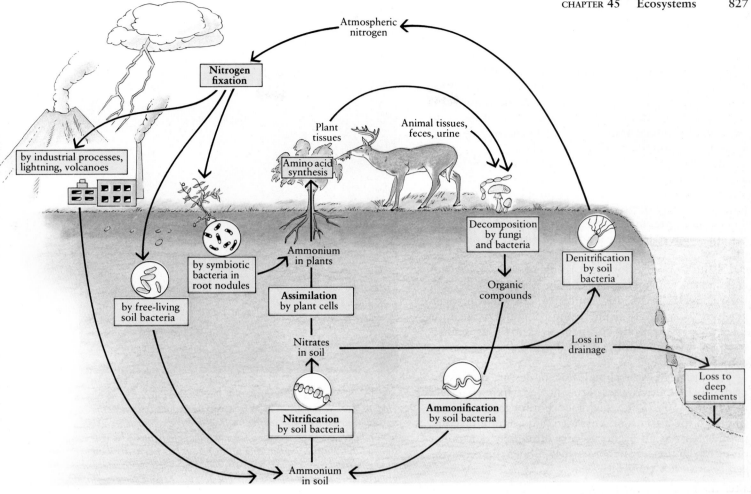

45–14 *The nitrogen cycle in a terrestrial ecosystem. The primary reservoir of nitrogen is the atmosphere, where it makes up about 77 percent of dry air. Only a few microorganisms, some symbiotic and others free-* *living, are capable of fixing nitrogen gas into inorganic compounds that can be used by plants in the synthesis of amino acids and other nitrogenous compounds.* *The nitrogen cycle in aquatic ecosystems is* *similar to the terrestrial cycle. Although the specific organisms differ, the processes are essentially the same, with specific groups of bacteria carrying out the various chemical reactions on which the cycle depends.*

The process by which this limited amount of nitrogen is circulated and recirculated throughout the world of living organisms is known as the **nitrogen cycle.** The three principal stages of this cycle are (1) ammonification, (2) nitrification, and (3) assimilation.

Much of the nitrogen found in the soil is the result of the decomposition of organic materials. Initially, it is in the form of complex organic compounds, such as proteins, amino acids, nucleic acids, and nucleotides. These nitrogenous compounds are usually rapidly decomposed into simple compounds by soil-dwelling organisms, chiefly bacteria and fungi. The microorganisms use the amino acids to build their own needed proteins and release the excess nitrogen in the form of ammonia (NH_3) or ammonium ion (NH_4^+). This process is known as **ammonification.**

Several species of bacteria common in soils are able to oxidize ammonia or ammonium ion. This oxidation,

known as **nitrification,** is an energy-yielding process, and the energy released is used by these bacteria as their primary energy source. One group of bacteria oxidizes ammonia (or ammonium) to nitrite (NO_2^-):

$$2NH_3 \ + \ 3O_2 \ \longrightarrow \ 2NO_2^- \ + \ 2H^+ \ + \ 2H_2O$$

Ammonia Oxygen Nitrite ion Hydrogen Water
           ion

Nitrite is toxic to many plants, but it rarely accumulates. Members of another group of bacteria oxidize the nitrite to nitrate (NO_3^-), again with a release of energy:

$$2NO_2^- \ + \ O_2 \ \longrightarrow \ 2NO_3^-$$

Nitrite ion Oxygen Nitrate ion

Although plants can utilize ammonium directly, nitrate is the form in which most nitrogen moves from the soil into the roots. Hence, these bacteria play a crucial role in the recycling of nitrogen.

Once the nitrate is within the plant cell, it is reduced back to ammonium. In contrast to nitrification, this **assimilation** process requires energy. The ammonium ions thus formed are transferred to carbon-containing compounds to produce amino acids and other nitrogenous organic compounds needed by the plant.

Although the nitrogen cycle appears complete and self-sustaining, nitrates are steadily lost from the soil and thus removed from the cycle. As we noted in Chapter 41, a variety of processes—the harvesting of plants, soil erosion, fire, and water percolating down through the soil—can reduce the amount of nitrate available to plants.

Nitrates are also lost as a result of the activities of certain types of soil bacteria. In the absence of oxygen, these bacteria break down nitrates, releasing nitrogen gas back into the atmosphere and using the oxygen for their own respiration. This process, known as **denitrification,** takes place in poorly drained (hence, poorly aerated) soils.

The nitrogen cycle is maintained, despite the losses, primarily by the activities of the nitrogen-fixing bacteria (see pages 750–751), which incorporate gaseous nitrogen from the air into organic nitrogen-containing compounds. Just as all organisms are ultimately dependent on photosynthesis for energy, they all depend on nitrogen fixation for their nitrogen.

45–15   *Weir at Hubbard Brook Experimental Forest in New Hampshire. Water from each of six experimental ecosystems was channeled through weirs, built where the water leaves the watershed, and was analyzed for chemical elements. The watershed behind the weir in this photograph has been stripped of vegetation. The experiments showed that such deforestation greatly increases the loss of nutrient elements from the system.*

Recycling in a Forest Ecosystem

Studies of a deciduous forest ecosystem in the White Mountains of New Hampshire have shown that the plant life of a community plays a major role in its retention of mineral elements. Investigators at the Hubbard Brook Experimental Forest first established a procedure for determining the mineral budget—input and output, "profit" and "loss"—of areas in the forest. By analyzing the content of rain and snow, they were able to estimate input, and by constructing concrete weirs (Figure 45–15) that channeled the water flowing out of selected areas, they were able to calculate output.

From these studies, the investigators discovered that the natural forest was extremely efficient in conserving its mineral elements. For example, the annual net loss of calcium represented only about 0.3 percent of the calcium in the ecosystem. In the case of nitrogen and potassium, the ecosystem was actually accumulating mineral nutrients.

In the winter of 1965–1966, all of the trees, saplings, and shrubs in one area of the forest were cut down completely. No organic materials were removed, however, and the soil was undisturbed. During the following summer, the area was sprayed with a herbicide to

inhibit regrowth. During the four months from June to September, 1966, the runoff of water from the area was four times higher than in previous years. Net losses of calcium were 10 times higher than in the undisturbed forest and of potassium, 21 times higher.

The most severe disturbance was seen in the nitrogen cycle. Plant and animal tissues continued to be decomposed to ammonia or ammonium, which then were acted upon by nitrifying bacteria to produce nitrates, the form in which nitrogen is usually taken up by plants. However, no plants were present, and the nitrate was leached from the soil. As a side effect, the stream draining the area became polluted with algal blooms, and its nitrate concentration exceeded the levels established by the U.S. Public Health Service for drinking water. Trees are now beginning to grow again on the devastated site, and the runoff of nutrients has dramatically decreased.

Studies in the tropical forests of Central and South America have further underlined the importance of plant life in mineral retention. In the tropical rain forest, in particular, virtually all mineral nutrients are held within the living components of the ecosystem—and none are retained by the soil for any length of time.

Agricultural Ecosystems and a Hungry World

The earliest traces of agriculture, dating back about 11,000 years, are found in areas of the Near East that are now parts of Iran, Iraq, and Turkey. Here were the raw materials required by an agricultural economy: cereals, which are grasses with seeds capable of being stored for long periods of time without serious deterioration, and herbivorous herd animals, which can be readily domesticated. The grasses in these areas were wild wheats and barley, which still grow wild in the foothills. The animals were wild sheep and goats.

By 8,100 years ago, agricultural communities were established in eastern Europe. By 7,000 years ago, agriculture had spread to the western Mediterranean and into central Europe. During this same period, agriculture originated independently in Central and South America, and perhaps slightly later in the Far East.

With the advent of widespread agriculture, *Homo sapiens* became a major actor in ecosystem dynamics. An area under cultivation is, in effect, an artificial ecosystem, maintained at a very early stage of ecological succession by human activities. Like other ecosystems subjected to frequent, large-scale disturbances, agricultural ecosystems are characterized by a small number of species, a relatively low total biomass, a high net productivity in relation to biomass, and a limited capacity to trap and retain nutrients.

In the earliest periods of agriculture, the most important of these features was undoubtedly the high net productivity, which provided the primitive farmers with a larger and more reliable food supply than they had previously known. As we saw in Chapter 43, an immediate consequence was a significant increase in the human population, a growth that has now become exponential. The consequences of this increase are many and serious—pollution, depletion of fuel supplies, destruction of natural resources, and extinction of other species as the human population expands into previously untouched ecosystems.

Ironically, by far the most difficult problem and also the most urgent is hunger and starvation. Of the world's more than 5.4 billion people, at least 1 billion are inadequately nourished. It is estimated that some 20 million people die each year as the direct result of malnutrition, and many more deaths are thought to be caused by its indirect effects. Perhaps even more important, in terms of the world's future, are the effects, both physical and psychological, of the prolonged chronic hunger of so large a proportion of our population.

A major effort to increase world food supplies by increasing agricultural productivity, known as the Green Revolution, has been under way for the past 40 years. The principal emphasis of this effort has been the development of new crop plants, especially grains.

Enormous progress has been made. From 1950 to 1970, for example, the production of wheat in Mexico increased from 270,000 metric tons per year to 2.35 million. In 1985, wheat production in Mexico reached an all-time high of 5.2 million metric tons, but by 1987, it had dropped back to 4.4 million. In the period from 1950 to 1970, the corn harvest in Mexico increased a more modest 250 percent, with yields per hectare almost doubling. In subsequent years, the corn harvest has continued to increase, reaching a record 14 million metric tons in 1985.

Similarly, between 1950 and the present, India has increased its production of food grains about 2.8 percent each year. (Its population during this period has increased an average of 2.1 percent a year, which is significantly less.) Since 1971, China, the most populous nation in the world, has become agriculturally self-sustaining. Most of this success—in Mexico, India, China, and elsewhere—has come about as a result of improved varieties of crop plants, combined with better techniques of irrigation and fertilization.

Despite its acknowledged success, the Green Revolution has come under criticism in recent years. Modern agricultural ecosystems consist of huge stands of single species, an open invitation to insects, weeds, and disease organisms. Such opportunistic species can be kept at bay only by constant attention and the application of pesticides.

This susceptibility to invasion was tragically illustrated by the great potato famine of 1845–1847 in Ireland, which was caused by a water mold (see page 426). This famine was responsible for more than a million deaths by starvation and initiated large-scale emigration from Ireland. Within a decade, the population of Ireland dropped from 8 million to 4 million. Virtually the entire Irish potato crop was wiped out in a single week in the summer of 1846. A number of plant geneticists have warned that the new strains of wheat and rice, which have made major contributions toward feeding the growing human population, are particularly susceptible to similar disasters because of their genetic uniformity and widespread distribution.

A related criticism concerns the enormous amounts of fertilizer required by the new plant varieties in order to achieve their high yields—fertilizer that must be repeatedly applied as nutrients are removed from the ecosystem by the harvesting of the crops. Fertilizer, pesticides, farm equipment, and fuel for the equipment are all increasingly expensive. Because large landowners are able to afford the necessary investment, whereas small-scale farmers are not, the Green Revolution is seen as accelerating the consolidation of farm lands into a few large holdings by the very wealthy.

Another problem is that although food production is still outstripping population growth, increased agricultural productivity will not be able to keep pace indefinitely with the rapid growth of the world's population.

Finally, there is a more fundamental though more elusive reason for the dissatisfaction with the Green Revolution. When it first began, it appeared to many to be an almost magical solution to problems so enormous and distressing that they had seemed insoluble. It is now clear, however, that poverty and famine—and the unrest and violence they may bring—will not be solved by a "technological fix." The Green Revolution must, of course, go forward. At the same time, we must recognize that the broader solutions are social, political, and ethical, involving not only the growth of crops but their distribution, not only the limiting of populations but the raising of living standards worldwide to tolerable levels.

When inorganic nutrients are released by the action of decomposers, they are immediately taken up again by the roots of plants. This has the effect of preventing leaching of nutrients from the soil by the frequent, and often torrential, rains. When forest is cleared to make way for farming, mineral nutrients are rapidly leached from the soil. The result is that it is often impossible to grow a second crop after the first has been harvested.

Concentration of Elements

The elements needed by living organisms often are present in their tissues in higher concentrations than in the surrounding air, soil, or water. This concentration of elements comes about as a result of the selective uptake of substances by living cells, amplified by the channeling effects of food chains.

Under natural circumstances, the concentration of elements is usually valuable. Animals generally have a greater requirement for minerals than do plants because so much of the biomass of plants is cellulose. In some special cases, the effect of the accumulation may be quite dramatic. For example, the thousands of kilometers of coral reefs in the world's warmer waters are composed of calcium compounds formed from ions extracted from sea water over the millennia by one tiny polyp after another (see page 460).

Foreign substances can also get caught up in biogeochemical cycles and, as they are passed from one living organism to another, reach high concentrations as they approach the top of the food chain. The insecticide DDT is probably the best known of the toxic substances whose effects were amplified in this way (Figure 45–16).

Another was strontium-90, a radioactive element released into the atmosphere by aboveground nuclear testing in the 1950s. Strontium is closely related to calcium and can take its place in many biochemical reactions. Strontium-90 from the fallout made its way through grasses into dairy cows and into milk; from there it became concentrated in the bones and teeth of children. By 1959 the bones of children in North America and Europe contained about six times as much strontium-90 as the bones of adults. Although the level of radioactivity present has not been proved dangerous, exposure to radioactive elements is known to cause leukemia, bone cancers, and genetic abnormalities and generally to shorten the life span. And, most important, the minimum exposure that can produce such effects is not established. Since the half-life of strontium-90 is 28 years (in other words, it takes 28 years for half of the atoms present to lose their radioactivity), the exposure continues.

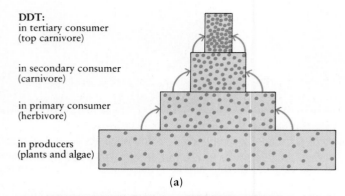

DDT:
in tertiary consumer
(top carnivore)

in secondary consumer
(carnivore)

in primary consumer
(herbivore)

in producers
(plants and algae)

(a)

(b)

45–16   (a) *Concentration of DDT residues being passed along a simple food chain. As organic matter is transferred from one level to the next in the chain, about 90 percent is usually respired or excreted. Only the remaining 10 percent forms new biomass. The losses of DDT residue, however, are small in proportion to the loss of other organic matter through respiration and excretion. Consequently, the concentration of DDT increases as the material passes up the chain, and high concentrations occur in the carnivores.*

(b) This dead peregrine falcon embryo, its development almost completed, was found with two broken, infertile eggs in a nest in southwestern Scotland in May of 1971. When its body was analyzed, it was found to contain one of the highest concentrations of DDT residue ever recorded in a bird in Great Britain. It is not known if the developing falcon died as a direct consequence of the levels of DDT residue in its body or as a result of the collapse of its shell. DDT causes a bird's liver to break down the hormones that, in females, mobilize calcium at the time of egg production. The result is thin-shelled, fragile eggs. Birds at the top of food chains, such as the peregrine falcon, the osprey, and the bald eagle, were the principal victims. With the banning of DDT in many countries, dramatic recoveries have been observed in populations of all three species.

The channeling of strontium-90 along the food chain could have been anticipated. Other consequences of the atomic tests were less predictable. For example, in the Arctic only slight radiation exposure was expected because the amount of fallout that reaches the ground at the poles is much less than it is in the temperate zones.

However, the Inuit peoples in the Arctic were discovered to have concentrations of radioactivity in their bodies that were much higher than those found in the inhabitants of temperate regions. The key link in the chain was the lichens. Lichens, which obtain their minerals largely from the rain, had absorbed a large amount of fallout material, little of which had time to decay and none of which was dissipated by absorption into the soil. In the winter, caribou live almost exclusively on lichens, and, at the top of the food chain, the Inuit live largely on caribou.

The Lessons of Chernobyl

In the early hours of April 26, 1986, the operators of a nuclear power plant at Chernobyl in what was then the Soviet Union began a series of tests of the operating capabilities of their newest reactor. According to official Soviet reports, without authorization the operators overrode the safety systems and deviated from standard procedures in order to conduct their tests. At some point, they lost control of the reactions occurring within the core of the reactor. Its output suddenly rose to 120 times its rated capacity, rupturing portions of the core. The cooling water in the system was instantly turned to steam, producing a powerful explosion that blew the 1,000-ton reactor cover aside and spewed hot nuclear fuel into the sky. The reactor was engulfed in fires that required more than 10 days to bring under control.

During this period, an enormous quantity of radioactive material was released into the environment—more than 1,000 times the amount released in the 1979 accident at Three Mile Island, near Harrisburg, Pennsylvania. A principal component of the material released in the Chernobyl accident was cesium-137, with a half-life of 30 years.

The consequences of the accident were most severe in the areas closest to Chernobyl, which included parts of what are now Belarus, Russia, and Ukraine (where Chernobyl is located). According to one unofficial estimate, as many as 10,000 people may have died in the five years following the accident, as a direct result of the radiation exposure. In 1986, some 135,000 people were evacuated from the most heavily contaminated areas, and in 1991 another 200,000 were still awaiting evacuation from other areas with significant contamination.

The radioactive cloud from the accident was moved in a northwesterly direction by the prevailing winds, and, when the rains subsequently came, the radioactive material fell back to the Earth. The fallout reached far beyond the Soviet Union, ultimately affecting some 100 million people in more than 20 European countries.

A substantial portion of the fallout was deposited

45–17   *In the winter of 1986–1987, as in other winters, the Sami peoples of Norway rounded up their reindeer herds in preparation for the annual slaughter—a process that previously provided them not only with meat for their own use but also with income for the next year. In central Norway, which was in the path of the radioactive cloud from the Chernobyl disaster, the levels of radioactivity in the reindeer averaged more than 10 times the level allowed in meat that is to be marketed. In some cases, the levels were more than 20 times the allowed limit. The contaminated carcasses were buried in large pits—in effect, nuclear waste disposal sites—in uninhabited regions in the far north. Although the Norwegian government has provided compensation for the financial losses, preventing economic collapse, the Sami people face an interval of some 20 to 30 years before they can hope to use and sell the milk and meat of their reindeer once more.*

more than 2,000 kilometers away in Norway, a country with no nuclear power plants. In the alpine tundra of Norway, the radioactive cesium-137 passed from rainwater to lichens and then to reindeer, in which its concentration increased to levels far exceeding those considered safe for human consumption. The highest concentrations occurred in the milk, muscles, and bones of the reindeer—the traditional subsistence for the Sami, or Lapp, peoples of central and southern Norway (Figure 45–17).

The accident at Chernobyl and its aftermath remind us of several important lessons. The first, and most obvious, is that the biological concentration of elements is a very real phenomenon with potentially severe consequences, especially for organisms at the top of the food chain, including ourselves. The second is that we dare not become complacent about the safeguards surrounding the use of potentially hazardous materials or technologies. The third, and perhaps most important, lesson is that the consequences of our misadventures do not respect international borders or local environmental regulations, no matter how well conceived or faithfully followed. We—and all other living organisms—are interconnected in one global ecosystem.

Summary

An ecosystem is a unit of biological organization made up of all the organisms in a given area and the environment in which they live. It is characterized by interactions between the living (biotic) and nonliving (abiotic) components that result in (1) a one-way flow of energy from the sun through autotrophs to heterotrophs, and (2) a cycling of mineral elements and other inorganic materials.

The ultimate source of energy for most ecosystems is the sun. Of the solar energy reaching the outer limits of the atmosphere, approximately half is reflected from or absorbed by the atmosphere. The distribution of the solar energy that reaches the Earth's surface is affected by the position and movements of the Earth in relation to the sun and by the movements of air and water over the surface. These factors can cause wide differences in temperature and rainfall from place to place and from season to season.

Within an ecosystem, there are at least three trophic (feeding) levels: primary producers, which are usually plants or algae; primary consumers, which are usually animals; and detritivores, which live on animal wastes and dead plant and animal tissues. The primary producers (the autotrophs) convert a small proportion (over 1 percent) of the sun's energy into chemical energy. The primary consumers (herbivores) eat the primary producers. A carnivore that eats a herbivore is a secondary consumer, and so on.

On average, about 10 percent of the energy transferred at each trophic level is stored in the body tissue of the consumers. Of the remaining 90 percent, part is used in the metabolism of the consumers and part is unassimilated. This unassimilated energy is ultimately utilized by detritivores.

The movements of water, carbon, nitrogen, and mineral elements through ecosystems are known as biogeochemical cycles. In such cycles, inorganic materials from the air, water, or soil are taken up by primary producers, passed on to consumers, and eventually transferred to the detritivores known as decomposers—bacteria and fungi. In the course of their metabolism, decomposers release the inorganic materials to the soil or water in a form in which they can again be taken up by primary producers.

The nitrogen cycle is of critical importance to all organisms. It involves three principal stages: ammonification, the breakdown of nitrogenous organic compounds to ammonia or ammonium ion; nitrification, the oxidation of ammonia or ammonium to nitrates, which are taken up by plants; and assimilation, the conversion of nitrates to ammonia and its incorporation into organic compounds. Nitrogen-containing organic compounds are eventually returned to the soil or water, completing the cycle. Nitrogen lost from the ecosystem is replaced by nitrogen fixation, the incorporation of elemental nitrogen into organic compounds.

In forest ecosystems, the principal reservoir of mineral elements is the vegetation. When a forest is stripped of its vegetation, the loss of mineral nutrients from the ecosystem increases dramatically.

Synthetic chemicals or radioactive elements released into the environment can be caught up in biogeochemical cycles, becoming concentrated in the tissues of organisms at higher trophic levels.

Questions

1. Distinguish among the following: community/ecosystem; biotic/abiotic; food chain/food web; gross productivity/ net productivity; producer/consumer/detritivore/ decomposer; ammonification/nitrification/assimilation/ denitrification.

2. Describe what happens to the light energy striking a temperate forest ecosystem. What happens when it strikes a cornfield? A pond? A field on which cattle are grazing?

3. Describe what happens to a mineral nutrient in each of the ecosystems in Question 2.

4. Arrange the following ecosystems in order of decreasing net primary production: coral reef, cultivated land, desert, estuaries, open ocean, temperate forest, temperate grassland, tropical forest, and tropical grassland. What environmental factors have the greatest influence on the net primary production in terrestrial ecosystems? In marine ecosystems?

5. Consider each of the organisms below and list the effects of each on its ecosystem. Consider how the organism receives its inputs of energy and nutrients, where its outputs

(metabolic wastes, offspring, dead carcasses) go, and its effects on other organisms.

(a) Earthworm
(b) Heterotrophic soil bacterium
(c) Oak tree or grass plant
(d) Deer or grasshopper
(e) Lion or wolf

6. Explain the different kinds of information provided by a pyramid of numbers, a pyramid of biomass, and a pyramid of energy flow. For what purposes might each type be more useful than the other two?

7. Among the highest energy-transfer efficiencies known occur when reptiles consume warm-blooded prey, such as birds or small mammals. Explain, in terms of characteristics of both prey and predator, why a high energy-transfer efficiency would be expected for this particular step in a food chain.

8. Although a carnivore at the top of a food chain is free of visible predation, it is—during its lifetime—a source of energy for many different species, representing as many as four kingdoms. Explain.

9. What are the implications for mineral cycling of the human practices of fertilization of land and harvesting of crops? How are these implications different for nutrients whose major inorganic reservoir is the atmosphere rather than the soil?

Suggestions for Further Reading

Barringer, Felicity: "Chernobyl: Five Years Later the Danger Persists," *The New York Times Magazine*, April 14, 1991, pages 28–39, 74.

Carson, Rachel: *Silent Spring*, Houghton Mifflin Company, Boston, 1987.*

This is the book, originally published in 1962, that awakened the nation to the dangers of pesticides. Once seen as highly controversial, it is now regarded as a classic.

Childress, James J., Horst Felbeck, and George N. Somero: "Symbiosis in the Deep Sea," *Scientific American*, May 1987, pages 114–120.

Clark, William C.: "Managing Planet Earth," *Scientific American*, September 1989, pages 46–54.

Colinvaux, Paul: *Why Big Fierce Animals Are Rare*, Princeton University Press, Princeton, N.J., 1978.*

A series of essays for the general reader.

Crosson, Pierre R., and Norman J. Rosenberg: "Strategies for Agriculture," *Scientific American*, September 1989, pages 128–135.

Firor, John: *The Changing Atmosphere: A Global Challenge*, Yale University Press, New Haven, Conn., 1992.*

This short book, by the director of the Advanced Study Program at the National Center for Atmospheric Research, provides a clear, accessible, and nonsensational introduction to the problems of acid rain, ozone depletion, and global warming. The author considers not only the origins of these problems but also practical approaches to their solution. Highly recommended.

Gosz, J. R., R. T. Holmes, G. E. Likens, and F. H. Bormann: "The Flow of Energy in a Forest Ecosystem," *Scientific American*, March 1978, pages 92–102.

Graedel, Thomas E., and Paul J. Crutzen: "The Changing Atmosphere," *Scientific American*, September 1989, pages 58–68.

Lappé, Frances Moore: *Diet for a Small Planet*, Ballantine Books, Inc., New York, 1991.*

In this classic book, first published in 1971, Lappé established the feasibility of eating "low on the food chain," thereby greatly increasing the availability of proteins to other human populations.

Nichol, Stephen, and William de la Mare: "Ecosystem Management and the Antarctic Krill," *American Scientist*, vol. 81, pages 36–47, 1993.

Power, J. F., and R. F. Follett: "Monoculture," *Scientific American*, March 1987, pages 78–86.

Reganold, John P., Robert I. Papendick, and James F. Parr: "Sustainable Agriculture," *Scientific American*, June 1990, pages 112–120.

Rowland, F. Sherwood: "Chlorofluorocarbons and the Depletion of Stratospheric Ozone," *American Scientist*, vol. 77, pages 36–45, 1989.

Simons, Ted, Steve K. Sherrod, Michael W. Collopy, and M. Alan Jenkins: "Restoring the Bald Eagle," *American Scientist*, vol. 76, pages 252–260, 1988.

Stephens, Sharon: "Lapp Life after Chernobyl," *Natural History*, December 1987, pages 32–41.

Stolarski, Richard S.: "The Antarctic Ozone Hole," *Scientific American*, January 1988, pages 30–36.

Toon, Owen B., and Richard P. Turco: "Polar Stratospheric Clouds and Ozone Depletion," *Scientific American*, June 1991, pages 68–75.

Tunnicliffe, Verena: "Hydrothermal-Vent Communities of the Deep Sea," *American Scientist*, vol. 80, pages 336–349, 1992.

* Available in paperback.

The Biosphere

"May I announce to you," French naturalist Philibert de Commerson wrote in 1771, "that Madagascar is the naturalists' promised land. Nature seems to have retreated there into a private sanctuary, where she could work on models different from any she has used elsewhere. There, you meet bizarre and marvelous forms at every step."

Most of Madagascar is a sanctuary no longer. The present inhabitants of the island arrived barely 1,500 years ago. Since that time, 14 species of lemurs, a pygmy hippopotamus, and tortoises as large as those in the Galapagos have all disappeared—not just from Madagascar but from this planet, forever. Vanished with them is the elephant bird, *Aepyornis maximus,* the largest bird that ever lived. It was a flightless, helpless giant that stood over 7 feet tall and weighed about 650 pounds. Its eggs, each of which weighed 20 pounds, still show up occasionally as fossils, exposed by heavy rains. Also vanished, without a doubt, are hundreds, perhaps thousands, of other species of plants and animals that were never seen, named, collected, described, or perhaps even imagined by any human eye or hand or mind.

Of the native vegetation that once covered the island, only 10 percent remains, and that is fast disappearing. On Madagascar, as in other tropical rain forests, when the land is cleared, the brick-red soil turns as hard as cement. Rainwater runs off the barren landscape, choking the rivers with silt and causing floods and mudslides. The sea around Madagascar is rust-red from the island's red soil. The rice paddies are being destroyed by the floods, and people are close to starving.

How do we feed a hungry planet without destroying its last sanctuaries? This question is being asked all over the world, in different ways and different languages. In our own Pacific Northwest, how do we balance the need for jobs against the life of a forest, and what do we do about jobs when the forest is gone? Biology can illuminate some of these problems, but it cannot do much to solve them. The solutions are ethical, political, economic, esthetic. And they are up to you.

The **biosphere,** that part of the Earth in which life exists, is only a thin film on the surface of this little planet. It extends about 8 to 10 kilometers above sea level and a few meters down into the soil, as far as roots penetrate and microorganisms are found. It includes the surface waters and the ocean depths. The biosphere is patchy, differing in both the depth and the density of life.

In this, the final chapter of our text, we shall survey the marvelously diverse communities of organisms that are found in different regions of the biosphere.

Life in the Waters

Life began in water, and although living organisms have long since conquered the land, by far the largest proportion of the biosphere consists of aquatic environments and their inhabitants. Freshwater environments can be conveniently classified as running water (rivers and streams) and standing water (lakes and ponds). Marine environments can be classified as oceans and seashores. Within these broad categories, there are a diversity of habitats, each with its own characteristic complement of organisms.

Rivers and Streams

Rivers and streams are characterized by continuously moving water. They may begin as outlets of ponds or lakes, as runoffs from melting ice or snow, or they may arise from springs (flows of groundwater emerging from below the surface of the land).

The character of life in a stream is determined to a large extent by the swiftness of the current, which generally changes as a stream moves downward from its source and, fed by tributaries, increases in volume. In swift streams, most organisms live in the riffles, or shallows, where small photosynthetic organisms—algae and mosses—cling to the rock surfaces. Many insects, both adult and immature forms, live on the underside of rocks and gravel in the riffles. For those small organ-

46–1 *A fast-moving stream in Michigan. The boulders within and beside the stream are constantly wetted by spray, making possible the luxuriant growth of mosses on their surfaces. What is a probable explanation for the total absence of mosses on some of the smaller boulders?*

46–2 *Some freshwater environments and their inhabitants. (a) A pond in the Pocono Mountains of Pennsylvania, covered with water lilies. (b) A great blue heron, with a bass that it has just caught. (c) A hungry muskrat.*

isms that can survive the swiftly moving current, there is an abundance of oxygen and nutrients swept along by the flowing waters.

As the stream travels along its course, the riffles are often interrupted by quieter pools, where organic materials may collect and be decomposed. Few plants can gain footholds on the shifting bottoms of stream pools, but some invertebrates, such as dragonflies and water striders, are typically found in or about the pools. Some organisms, notably trout, move back and forth between the riffles and the pools.

As streams broaden, they generally include peripheral habitats that have many of the characteristics of lakes and ponds.

Lakes and Ponds

Bodies of standing water vary in size from small ponds to very large lakes, covering thousands of square kilometers. They contain three distinct zones: littoral, limnetic, and profundal.

The **littoral zone,** at the edge of the lake, is the most richly inhabited. Here the most conspicuous plants are angiosperms, such as cattails and rushes, rooted to the bottom. Water lilies may grow farther out from the shore. There is often a green blanket of duckweed, a small, free-floating angiosperm. Other pond weeds grow entirely beneath the water. These plants lack a waxy outer cuticle and so can absorb minerals through their epidermis as well as through their roots. The tangle of submerged plant surfaces harbors large numbers of small organisms.

Snails, small arthropods, and mosquito larvae feed upon the plants and algae. Other insects that live among the submerged plants, such as the larvae of the dragonfly and damselfly, and the water scorpion, are carnivorous. Clams, worms, and the larvae of other insects

(a)

(b)

(c)

burrow in the mud. Frogs, salamanders, turtles, and water snakes are found almost exclusively in the littoral zone. The tadpoles of frogs and salamanders feed on algae and insect larvae, and the adults on insects. Most aquatic turtles feed on insects when young and on plants when fully grown. Water snakes are carnivores, feasting on amphibians and fishes, such as largemouth bass and bluegills, which are found in greater numbers along the lake margins. Ducks, geese, and herons feed on the plants, insects, mollusks, fishes, and amphibians abundant in this zone.

The shallow margins of some lakes and ponds are marshy. Among the inhabitants of the marshes are such animals as snails, frogs, ducks, herons, bitterns, muskrats, otters, and beavers.

In the **limnetic zone,** the zone of open water, small, floating algae—phytoplankton—are usually the dominant photosynthetic organisms. This zone, which extends down to the limits of light penetration, is the habitat, for example, of walleye, yellow perch, and, in colder waters, trout.

The deepwater **profundal zone,** extending down from the limnetic zone, has no plant life. Its principal occupants are detritivores, such as scavenging fishes, aquatic worms, insect larvae, crustaceans, fungi, and bacteria, that consume the organic debris filtering down from the overlying water. The nutrients released by bottom-dwelling decomposers are recirculated to the upper levels of the lake by the twice-yearly overturn of the water (see page 43).

The Oceans

The oceans cover almost three-fourths of the surface of the Earth. Although life extends to their deepest portions, photosynthesizing organisms are restricted to the upper, lighted zones. The sea has an average depth of nearly 4 kilometers and, except for a relatively shallow surface layer, is dark and cold. Most of its volume, therefore, is inhabited by bacteria, fungi, and animals, rather than by plants.

Sea water absorbs light readily. Even in clear water, less than 40 percent of the sunlight striking the surface reaches a depth of 1 meter, and less than 1 percent penetrates below 50 meters. Red, orange, and yellow wavelengths of light are absorbed first, so that only the shorter (higher energy) wavelengths, specifically blue and green, penetrate deeply. Thus, below depths of a few meters, only those photosynthetic organisms capable of utilizing the shorter wavelengths of light can grow.

There are two main divisions of life in the open ocean: **pelagic** (free-floating) and **benthic** (bottom-dwelling). A major component of the pelagic division is plankton. It is composed of rapidly dividing algae, intermingled with heterotrophic protists, small shrimp and other crustaceans, gelatinous invertebrates (such as jellyfish), and the eggs and larval forms of many fishes and invertebrates. These planktonic forms provide food for fishes and other relatively large pelagic animals, including some whales.

(a)

(b)

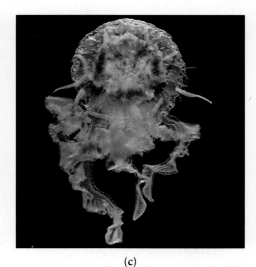

(c)

46–3   *Some ocean inhabitants.* (a) *This mixture of phytoplankton and zooplankton was photographed in the waters east of the Antarctic Peninsula. Planktonic organisms* form the base of the food webs in the pelagic division of the sea. (b) *A school of marine catfish, photographed off the* Philippine Islands. (c) *A purple stinger jellyfish* (Pelagia noctiluca), *swimming off Heron Island, Australia.*

The benthic division contains the sessile animals, such as sponges, sea anemones, and clams, and many motile animals, such as worms, starfish, snails, crustaceans, and fishes. A variety of fungi and bacteria also inhabit the benthic zone, subsisting on the accumulation of debris steadily drifting down from the more populated levels of the ocean.

Despite the fact that oceans cover three times more surface area of the planet than does the land, the total productivity of the open ocean—as measured by the amount of carbon converted to organic compounds by photosynthesis—is only about one-third as great. In fact, the open ocean is thought to be only slightly more productive per square meter than the desert (see Table 45–1, page 823), presumably because of the low concentration of mineral nutrients in those zones of the ocean where light penetration allows photosynthetic organisms to survive.

The major ocean currents, which are produced by a combination of winds and the Earth's rotation, profoundly affect life in the oceans and alter the climate along the ocean coasts (see essay). These patterns of water circulation—clockwise in the Northern Hemisphere and counterclockwise in the Southern Hemisphere—move currents of warm water north and south from the Equator (Figure 46–4). One such current, the Gulf Stream, warms a portion of the eastern coast of North America and the western shores of Europe, and another warms the eastern coast of South America. The same patterns of circulation bring cold waters to the western coasts of North and South America.

In regions where the winds move the surface water continuously away from the shores, as off the coasts of California, Portugal, and Peru, cold water rich in nutrients is brought to the surface from below, a process referred to as **upwelling.** Such areas contain high densities of pelagic life and traditionally have supported highly profitable fishing industries.

46–4   *The major currents of the ocean have profound effects on climate. Because of the warming effects of the Gulf Stream, Europe is milder in temperature than is North America at similar latitudes. The eastern coast of South America is warmed by water from the Equator, and the Humboldt Current brings cooler weather to the western coast of South America.*

El Niño

The importance of ocean currents to weather patterns and to both marine and terrestrial organisms is dramatically illustrated by the disastrous consequences that can follow disturbance of the normal patterns of water circulation. The most frequent disturbance occurs in the Pacific Ocean and is known as El Niño (Spanish for "the Child") because its effects are usually first felt in South America around Christmastime. Major El Niño events occur approximately every 8 to 10 years, and smaller ones every 2 to 3 years. The most severe El Niño in the past century occurred between the fall of 1982 and the summer of 1983.

Normally, the patterns of weather and ocean currents in the Pacific are determined by a large high-pressure system in the eastern Pacific and a large low-pressure system over Indonesia. Air moves from the region of high pressure to the region of low pressure, creating winds that blow from east to west. These winds move warm surface water toward the western Pacific, and cold water moves in along the western coasts of North and South America as an undercurrent, as shown in Figure 46–4.

An El Niño is apparently triggered by a rise in the surface temperature of equatorial regions of the eastern Pacific, causing changes in the location and strength of the atmospheric pressure systems. These changes, in turn, cause a reversal in the wind direction and an accompanying reversal in the surface current. Over a period of two to three months, warm water surges back toward North and South America, trapping the cold undercurrent well below the surface. With higher water temperatures and a cutoff of nutrient upwelling, planktonic organisms die off—followed by the fish that feed on the plankton, followed by the seabirds that feed on the fish.

The 1982–1983 El Niño led to a collapse of the plankton-fish-seabird food chain and thus a collapse of the fishing industries in Peru, Ecuador, and the Pacific Northwest. It also apparently disrupted weather patterns in many parts of the world. Massive rainstorms battered the usually dry coasts of California, Ecuador, and Peru that winter, and record snows fell in the Sierra Nevada and Rocky Mountains, followed by record flooding in the spring. Indonesia, the Philippines, India, Australia, Mexico, and southern Africa were stricken with droughts that destroyed vegetation and millions upon millions of animals, both domestic and wild, that had neither food nor water.

The complex, interacting events occurring in an El Niño are still poorly understood, but they appear to be part of a natural, self-sustaining cycle of warming and cooling. In a 10-year international study begun in 1985, satellites, ships, and ground stations across the Pacific are continuously monitoring air pressure, water and air temperatures, and the direction and velocity of winds and water currents. This study made possible the accurate prediction, a year in advance, of a moderate El Niño in 1986–1987 and of another, somewhat stronger El Niño that occurred in 1991–1992. With each succeeding El Niño, our knowledge of the forces involved increases, as does our understanding of the pattern that occurs as an El Niño becomes established and then declines.

The power of the oceans and the atmosphere is far beyond our meager capacity to control. However, better understanding of El Niños may make possible more accurate predictions and thus action to minimize their impact on both human activities and natural ecosystems.

The Seashore

Depending on the specific area, the submerged edges of the continents extend anywhere from 10 to several hundred kilometers out into the sea. Along these shallow borders, known as the **continental shelves,** nutrients are washed out from the land, and life is much denser than in the open seas. In temperate latitudes at the edge of the sea, where the large primary producers are brown algae (such as the rockweed *Fucus* and the kelp *Laminaria),* net primary productivity is as high as anywhere else in the biosphere. Primary productivity is also extremely high in the coral reefs (page 460) of tropical regions.

Sessile animals, such as sponges and anemones, are found all over the ocean bottom, but they are most abundant in relatively shallow areas that are close to the shores. Predators, such as mollusks, echinoderms, crustaceans, and many kinds of fishes, roam over the bottoms of the continental shelves. Eel grass, turtle grass, and seaweeds provide shelter for many animals, increase the supply of oxygen, and produce large amounts of detritus that forms the base of many food webs. Snails, sea slugs, and worms crawl over the surfaces of the plants and algae, feeding as they go.

Seashores—where the sea and the land join—are of three general types along most of the shores of the temperate zones: rocky, sandy, and muddy. Organisms that live on rocky coasts, like those that live in the riffles of fast-moving streams, often have adaptations for clinging to rocks. The algae have strong holdfasts. The starfishes of the rocky coast lie spread-eagled on the rocks,

46–5   *Some examples of life at the seashore. (a) Railroad vine, photographed on a Georgia beach, plays a role in stabilizing the shifting sand. (b) Sanderlings, seen here basking in the winter sun in Santa Cruz, California, are familiar residents of both the Atlantic and Pacific coasts of North America. When not sleeping, they race along the tide line, snatching up mollusks and crustaceans exposed as the waves retreat. (c) Sally lightfoot crabs are common inhabitants of sandy tropical beaches. Also known as red lava crabs, they run sideways at high speeds across the sand and are even able to skip across short stretches of water. (d) Among the inhabitants of this rocky tidal pool on Rialto Beach, Washington, are kelp, sea anemones, and starfish. (e) California sea lions, sunning themselves after a swim.*

clinging with their tube feet. The abalone holds tight with its well-developed muscular foot. Mussels secrete coarse, ropelike strands that anchor them to rocky surfaces.

The organisms of the rocky coast face the additional problem of the rising and falling tides. The supratidal zone, which is wetted only by waves and spray, is a zone of dark algal and lichen growth. The intertidal zone, alternately submerged and exposed by the tides, is commonly characterized by *Fucus,* often intermixed with many other species of brown and red algae. Animal life includes barnacles, oysters, mussels, limpets, and periwinkles. The subtidal zone, always submerged, often contains forests of kelp *(Laminaria* and *Macrocystis),* sea squirts, starfish, and other invertebrates. This zonation of the organisms of rocky shores is due, in part, to gradients of light, temperature, and wave action, partly to competitive interactions, and also to predation by both herbivores, such as sea urchins, and carnivores, such as starfish.

Sandy beaches have fewer inhabitants because of the constantly shifting sands. Clams, ghost crabs, sand fleas, lugworms, and other small invertebrates live below the surface of the sand, feeding on the debris washed in and out by the tide. Beach grasses, which spread by means of underground stems, are important for stabilizing the shifting dunes behind the sandy beaches.

The mud flat, while not so rich in species as the rocky coast, supports a large number of organisms, with many animals living not only on but also beneath its surface. A mud flat can support tens of thousands of individuals per cubic meter.

Mud flats, salt marshes, and **estuaries** (areas where the fresh water of streams and rivers drains into the sea) are the receiving grounds for a constant flow of nutrients drained off from the land. As a consequence, they are extremely rich in animal life. They function as the spawning places and nurseries for many forms of marine life, including many commercially important species of fish and crustaceans. In the tropics and subtropics (including parts of Florida, Puerto Rico, and Hawaii), mangrove forests are important tideland communities, serving as spawning grounds for marine organisms and exporting minerals and nutrients to the sea.

Mud flats, salt marshes, and mangrove forests are often located in prime recreational and commercial areas. Because they cannot be directly exploited for agriculture or lumbering, each year thousands of square kilometers of these wetlands are destroyed as they are filled and paved for human habitation—and thereby rendered sterile for many other species. Their protection is of special importance because of their role in nurturing life in the oceans.

(a)   (b)

46–6 (a) *A salt marsh on the North Carolina coast. The bright green areas, which are flooded at every high tide, are primarily the marsh grass* Spartina. *The dark areas, which are* flooded only during the highest high tides, consist principally of black needlerush. (b) *A snowy egret in a salt marsh in New Jersey in the autumn.*

Life on the Land

The characteristic patterns of life on the land are determined principally by physical factors. Of the most immediate consequence to terrestrial organisms are temperature and precipitation (Figure 46–7). As we saw in the last chapter, these critical factors are influenced by the angle of the Earth's axis in relation to its orbit around the sun and by the Earth's rotation, which affect the duration and intensity of sunlight, the prevailing wind patterns, and the major ocean currents. Temperature and precipitation are also affected by the structure of the continents themselves.

The surfaces of the continents change slowly but constantly. They are crumpled by contractions and collisions as the continents rise, sink, and collide because of the motion of the plates on which they are carried (page 358). As a consequence, the Earth's surface is not at all uniform but varies widely from place to place in its composition and in its height above sea level. The mineral content of the Earth's surface is a major factor affecting the growth of plants and other living organisms, and, as we have seen (page 819), the mountain ranges of the continents do much to determine the patterns of rainfall.

Average temperatures decrease about 0.5°C for each degree of increase in latitude. A similar effect occurs with each rise of about 100 meters in elevation. One consequence of this relationship is that plants and animals characteristic of Arctic regions may also be found in mountain ranges near the Equator (Figure 46–8).

The Concept of the Biome

The land surface of the Earth can be seen as divided into a number of geographic areas distinguished by particular types of dominant vegetation. Each major continent has, for example, deserts, grasslands, and deciduous forests. These categories of characteristic plant life are called **biomes**.

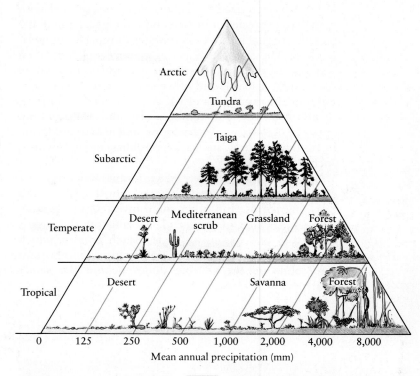

46–7   *The characteristic patterns of plant life in different regions of the biosphere are determined principally by temperature, which decreases with increasing latitude, and precipitation.*

46–8   *In the Northern Hemisphere, we can experience a similar sequence of dominant plant life and its associated animal life by either traveling north for hundreds of kilometers or ascending a mountain. To experience a similar sequence in the Southern Hemisphere, we could ascend a mountain. However, by simply traveling south, we would never encounter vegetation corresponding to the taiga and the tundra of the Northern Hemisphere. Can you explain why?*

Acid Rain and Forest Decline

The average pH of normal rainfall is about 5.6 (mildly acidic), a result of the combination of carbon dioxide with water vapor to produce carbonic acid. In the 1920s, however, the pH of rain and snow in Scandinavia began to drop, and by the 1950s, the pH had begun to drop elsewhere in Europe and in the northeastern United States. It was subsequently learned that, in certain geographic areas, the average annual pH of precipitation is between 4.0 and 4.5. Occasional storms release rain with a pH as low as 2.1, which is extremely acidic.

The low pH results primarily from two acids in the rainfall: sulfuric acid and nitric acid, both of which ionize almost completely in water, releasing hydrogen ions. These acids are formed when the gaseous oxides of sulfur and nitrogen react with water vapor and other gases in the air. Sulfur and nitrogen oxides are released into the atmosphere by some natural processes, but far greater quantities are released as a result of human activities. Sulfur oxides are produced by the combustion of high-sulfur coal and oil and by the smelting of sulfur-containing ores. Nitrogen oxides are by-products of gasoline combustion and of some generating processes for electricity.

That sulfur oxides could be damaging to vegetation was evident in the early 1900s, when a large copper smelter was opened in a mountainous area of Tennessee. Within a few years, all vegetation had been killed in the formerly luxuriant forest surrounding the smelter. The solution devised for the problem—still used today—was to build very tall smokestacks, so the wind would carry the pollutants away from the immediate area. It was assumed, incorrectly as it turned out, that the pollutants would be so widely dispersed that they would be rendered harmless.

By the 1960s, it was clear that sulfur oxides released from tall smokestacks are transported hundreds or thousands of miles by the prevailing winds and then returned to the Earth in rain and snow. Nitrogen oxides released from automobiles are also carried off by the wind. What was once a local problem has now become an international problem, in which the pollutants respect no boundaries.

The biological consequences of acid rain

(a)

(a) In the mountains of New England, red spruce trees are dying at a high rate. Only their skeletons remain, standing silent sentinel over the forest. Such scenes are also an

(b)

increasingly familiar sight in the forests of central and northwestern Europe. **(b)** In a forest in Vermont, a student sets up an apparatus to collect rainwater that will be tested for acidity.

are numerous. In plants, it causes reduced germination of seeds, a decrease in the number of seedlings that mature, reduced growth, and lowered resistance to disease. Moreover, acid rain lowers the pH of soil, increasing the leaching of essential nutrients from the soil. Recently, it has become apparent that the forests of the eastern United States, from Maine through Georgia, are in serious decline. A dramatic slowing of growth has occurred during the past 20 years, and in some locations at high altitude, red spruce trees are dying in large numbers. Current evidence suggests that, among its other effects, acid rain reduces the cold tolerance of this species.

Acid rain also affects freshwater ecosystems, particularly lakes in mountainous regions. A 1977 study of the lakes at high elevations in the western Adirondack Mountains of New York found that 51 percent had a pH below 5.0, of which 90 percent were devoid of fish life. By contrast, a similar study carried out from 1929 to 1937 found that only 4 percent of the lakes were acidic and without fish. More recent studies indicate increasingly acidic conditions in lakes at lower levels in the Adirondacks (which were previously unaffected) and in many of the lakes in the Cas-

cade Mountains in the Pacific Northwest. The effects of low pH on fish include the depletion of calcium in their bodies, leading to weakened and deformed bones; the failure of many eggs to hatch and deformed fish from those that do hatch; and the clogging of the gills by aluminum, which is released from the soil by acid.

The accumulating evidence indicates that acid rain is one of the more serious worldwide pollution problems confronting us today. The potential consequences of its effects on biological systems include lowered crop yields, decreased timber production, the need for greater amounts of fertilizer to compensate for nutrient losses, the loss of important freshwater fishing areas, and, possibly, of the eastern forests as well.

Scientists from many fields are engaged in research to gain a greater understanding of the causes and effects of acid rain and the likely consequences and costs of proposed solutions. As is the case with the other environmental problems we face, scientists can provide information on which decisions can be based, but the choices that lie ahead are essentially social and economic, to be made through political processes.

844

46–9 *Biomes of the world. The information in these maps and the accompanying key was originally supplied by A. W. Küchler of the University of Kansas, one of the leading authorities on the distribution of biomes. Because of the global coverage of the maps, the scale is relatively small and the content is generalized. Any given biome is not always uniform, and all of the biomes include considerable variations in vegetation. The boundaries between the biomes may be sharp, but more often they are blurred, consisting of broad zones of transition from one type of vegetation to another.*

1. Temperate deciduous forests
2. Temperate mixed forests
3. Subtropical mixed forests
4. Taiga
4. Northwestern coniferous forest
5. Mountain forests and alpine tundra
6. Mixed west-coast forests
7. Arctic tundra
8. Ice desert
9. Grasslands
10. Savannas
11. Mediterranean scrub
12. Deserts and semideserts
13. Juniper savanna
14. Southern woodland and scrub
15. Tropical mixed forests
16. Monsoon forests
17. Rain forests

(a)

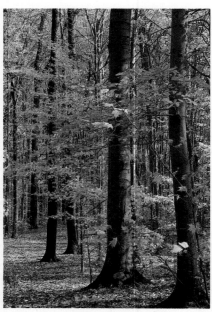

(b)

46–10 *The temperate deciduous forest of North America grades into temperate mixed forest in northern and upland regions and into subtropical mixed forest in the southeastern United States. (a) A beech and maple forest in Michigan, photographed in the spring. The forest floor is carpeted with large-flowered trillium (Trillium grandiflorum). (b) A beech and maple forest in the autumn. (c) Scarlet flycap mushrooms. The spots on the caps and the "skirt" near the top of each stalk identify these mushrooms as members of the deadly genus Amanita. Primary consumers in the forest include (d) chipmunks, and (e) white-tailed deer. (f) This red fox, which has just killed a cottontail rabbit, is a secondary consumer. All of these organisms are common inhabitants of the North American deciduous forest.*

As Figure 46–9 on pages 844–845 shows, the communities of plants and associated animal life that make up a biome are discontinuous, but a community may closely resemble another community on the other side of the planet. For instance, many of the deserts of the world look remarkably similar. However, when you look more closely, you see that although the physical features of the environment—temperature and rainfall—are the same, the particular organisms are not. But they look and act alike. The same is true of the Mediterranean scrub of California and Spain, the grasslands of North and South America, and so on.

A biome is a class, or category, of similar communities—not a place. When we speak of the tropical forest biome, we are not speaking of one particular geographic region, but rather of all the tropical forests on the planet. As with most abstractions, important details are omitted. For example, the boundaries are not as sharp as shown on biome maps nor are all areas of the world easy to categorize. However, the biome concept emphasizes one important truth: Where the climate is the same, the organisms are also very similar, even though they may not be related genetically and may be far apart in their evolutionary history. Organisms of geographically separate patches of the same biome provide many examples of convergent evolution (page 350).

Temperate Forests

Temperate forests once covered most of eastern North America, as well as most of Europe, part of Japan and Australia, and the tip of South America. In the United States, only scattered patches of the original forest remain.

Temperate deciduous forests occupy areas where there is a warm, mild growing season with moderate precipitation, followed by a colder period less suited to plant growth. Leaf-shedding in the deciduous forest evolved as a protection against water loss. (As you will recall, most of the water lost in transpiration escapes through the stomata of the leaves.) However, discarding leaves is expensive, in terms of both energy and nu-

(c)

(d)

(e)

Temperate
diciduous
forest

Temperate
mixed
forest

Subtropical
mixed forest

(f)

trients. For the water-saving benefits of defoliation to exceed the costs in energy and lost nutrients, deciduous trees must have a growing season of at least four months.

In deciduous woodlands there are up to four layers of plant growth: (1) a tree layer, in which the crowns form a continuous canopy that is usually between 10 and 35 meters high; (2) a shrub layer, which grows to a height of about 5 meters; (3) a field layer, made up of ferns, grasses, and herbaceous plants, including annual flowering plants that typically bloom in the spring before the trees regain their leaves; and (4) a ground layer, which consists of mosses and liverworts and is often covered with leaf litter.

The dominant trees of temperate forests vary from region to region, depending largely on the local rainfall. In the northern and upland regions of the United States, oak, birch, beech, and maple are the most prominent trees. Maple and basswood predominate in Wisconsin and Minnesota, and maple and beech in southern Michigan, becoming mixed with hemlock and white pine as the deciduous forest gives way to mixed forest. The southern and lowland regions have forests of oak and hickory. Along the southeastern coast of the United States, the warm, seasonally wet climate supports a mixed subtropical forest of pine, oak, and magnolia.

Temperate forests support an abundance of animal life. In North America, smaller mammals, such as chipmunks, voles, squirrels, raccoons, opossums, and white-footed mice, live mainly on nuts and other fruits, mushrooms, and insects. Deer typically live on the forest borders, where they browse on shrubs and seedlings. Wolves, bobcats, gray foxes, and mountain lions, in the areas where they have not been driven out by the encroachments of civilization, feed on the other mammals.

Beneath the ground layer is the soil of the forest, often a rich, gray-brown topsoil. Such soil is composed largely of organic material—decomposing leaves and other plant parts and decaying insects and other animals—and the bacteria, protists, fungi, worms, and arthropods that live on this organic matter. The roots of plants penetrate the soil to depths measurable in meters and add organic matter to the soil when they die. Carnivorous arthropods carry fragments of their prey to considerable depths in the soil. The myriad passageways left by dead roots and fungi and by the earthworms and other small animals that inhabit the forest make the soil a sponge that holds water and nutrients. Land where deciduous forests have stood often makes good farmland.

Coniferous Forests

Most conifers are evergreens, with small, compact leaves protected against water loss by a thick cuticle. They usually cannot compete successfully with deciduous trees in temperate zones with adequate summer rainfall and rich soil. Coniferous forest biomes include the **taiga** (also known as the northern boreal forest), the northwestern coniferous forest (located in the Pacific Northwest of the United States and Canada), mountain forests, and the mixed west-coast forests found along portions of the coastlines of Chile and California.

The Taiga

The boundary between the temperate mixed forest and the taiga occurs where summers become too short and winters too long for deciduous trees to grow well. The taiga is characterized by long, severe winters and a constant cover of winter snow. It is composed chiefly of evergreen needle-leaved trees such as pine, fir, spruce, and hemlock. A thick layer of needles and dead twigs, matted together by fungal mycelia, covers the ground. Along the stream banks grow deciduous trees, such as tamarack, willow, birch, alder, and poplar. There are very few annual plants.

The principal large animals of the North American taiga are elk, moose, mule deer, black bears, and grizzlies. Among the smaller animals are porcupines, red-backed voles, snowshoe hares, wolverines, lynxes, warblers, and grouse. The small animals use the dense

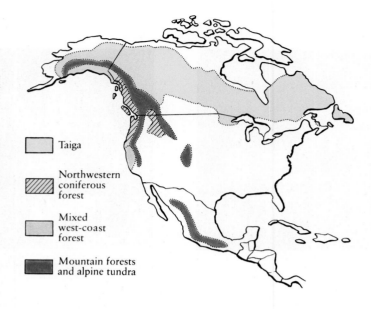

Taiga

Northwestern coniferous forest

Mixed west-coast forest

Mountain forests and alpine tundra

growths of the evergreens for breeding and for shelter. Wolves feed upon the mammals, particularly the larger ones. The black bear and grizzly bear eat everything—leaves, buds, fruits, nuts, berries, fish, the supplies of campers, and occasionally the flesh of other mammals. Porcupines are bark eaters and may seriously damage trees by girdling them. Moose and mule deer are largely browsers.

The ground layer of the coniferous forest is less richly populated by invertebrates than that of the deciduous forest, and the accumulated litter is slower to decompose.

The Forests of the Pacific Northwest

The massive evergreen coniferous forests of the Pacific Northwest are adaptations to the winter-wet/summer-dry environment of that region. Because photosynthesis is limited by lack of moisture during the warm season, deciduous trees are at a disadvantage. They are usually found only along stream banks, where water is plentiful. The evergreen conifers, however, can synthesize carbohydrates all year round. Also, these trees, because of their massive size, can store water and nutrients for use during the dry season, and their thick barks and high crowns protect them from the fires characteristic of the region.

These forests, which were untouched by the Pleistocene glaciation but have been seriously disturbed by recent human activities—primarily logging—are among the most ancient in North America, in terms of both individuals and populations. Their fate, and that of the animals, such as the northern spotted owl, that live only in the old growth forest, is currently the subject of intense controversy.

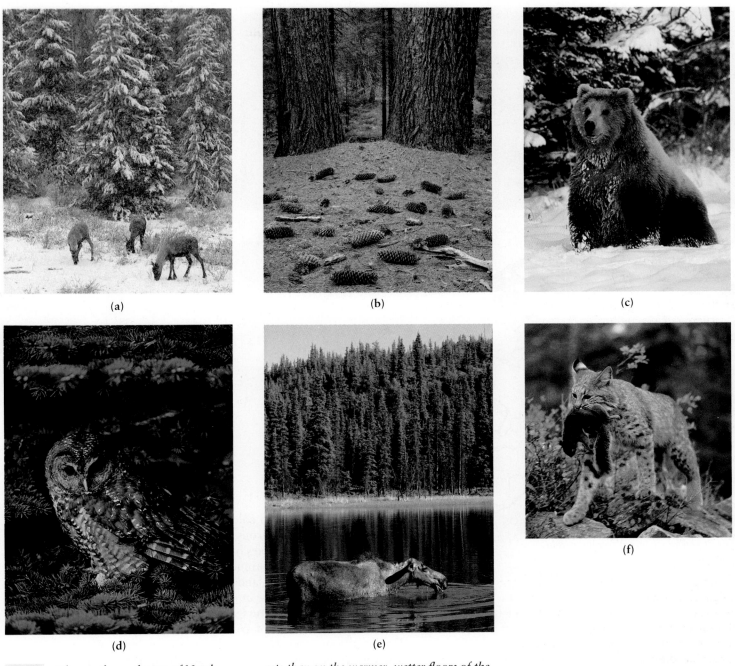

(a)

(b)

(c)

(d)

(e)

(f)

46–11   *The coniferous forests of North America include the taiga, the northwestern coniferous forest, mixed west-coast forests, and mountain forests.* (a) *Woodland caribou, searching for food in a clearing in the taiga of Jasper National Park in Alberta, Canada.* (b) *The floor of Buena Vista Grove in Kings Canyon National Park in California is carpeted with the needles and cones of sweet pines. Decay is slower in the coniferous for-*ests *than on the warmer, wetter floors of the deciduous forest, and, because of the shade cast by the mature trees, there is no undergrowth.*

Coniferous forests support a diversity of animals, including (c) *grizzly bears,* (d) *the northern spotted owls of the old growth forest of the Pacific Northwest,* (e) *moose, and* (f) *lynx and red squirrels.*

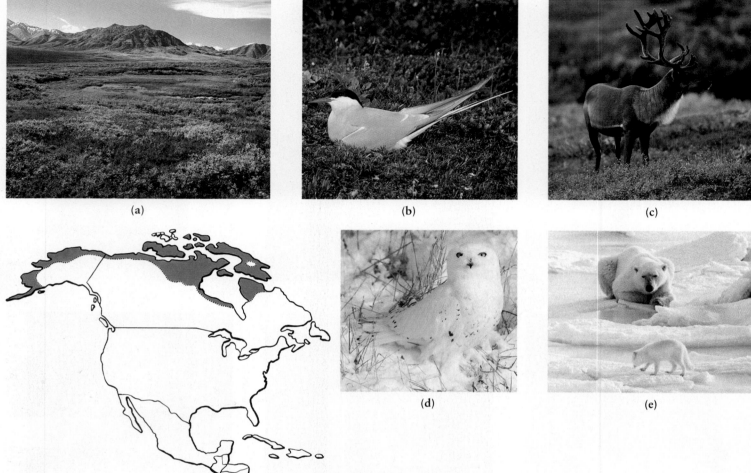

46–12 (a) *Tundra of North America on a long Arctic day. Kettle holes, formed by chunks of glacial ice, are a common sight.* (b) *The Arctic tern is one of a number of bird species that breed in the tundra, taking advantage of the long summer days to gather food for their nestlings. The terns winter in the Antarctic, following migration routes of 13,000 to 18,000 kilometers. Within three months after hatching, the young must be ready to migrate.* (c) *A barren-ground caribou, photographed in Alaska in the autumn.* (d) *A snowy owl in its winter plumage.* (e) *An Arctic fox (foreground) that has ventured from the tundra onto the ice floes, where polar bears spend the winter.*

The Tundra

Where the climate is too cold and the winters too long even for conifers, coniferous forest grades into **tundra.** The Arctic tundra occupies one-fifth of the Earth's land surface, forming a continuous belt across northern North America, Europe, and Asia. Similar vegetation is found above the tree line in mountainous regions.

The most characteristic feature of the Arctic tundra is **permafrost,** a layer of permanently frozen subsoil. During the summer the ground thaws to a depth of a few centimeters and becomes wet and soggy; in winter it freezes again. This freeze-thaw process, which tears and crushes the roots, keeps the plants small and stunted. Drying winter winds and abrasive, driven snow further reduce the growth of tundra plants.

The virtually treeless vegetation of the tundra is dominated by herbaceous plants, such as grasses, sedges, and rushes, and woody shrubs like heather. Beneath these is a well-developed ground layer of mosses and lichens, particularly the lichen known as reindeer moss. The growing season in many areas of the tundra is less

Grasslands

North American savannas

than two months, and all of the flowering plants are perennials, except for one native annual species.

The largest animals of the Arctic tundra are the herbivorous musk oxen and caribou of North America and the reindeer of the Old World. Other herbivores of the tundra include lemmings (small rodents with short tails) and ptarmigans (pigeon-sized grouse). The white fox and the snowy owl are among the principal predators, feeding largely on lemmings.

During the brief Arctic summer, insects emerge in great numbers. Migratory birds visit, taking advantage of the insect hordes and the long periods of daylight to feed their young.

Temperate Grasslands

Temperate grasslands, which are transitional areas between temperate forests and deserts, are usually found in the interior areas of continents. They are characterized by rolling to flat terrain, hot-cold seasons (rather than the wet-dry seasons of the savannas), periodic droughts, and fires. The temperate grasslands of the

46–13 *The grasslands of North America include large regions of both temperate grassland and savanna. (a) A June day on a tall-grass prairie in North Dakota. The cottonwood grove by the prairie creek is characteristic of this biome. A thunderstorm is gathering on the horizon. (b) A female bison nursing her calf in the short-grass prairie of Custer State Park, South Dakota. Among the other native animals of the grasslands are (c) the black-tailed prairie dog and (d) the redshouldered hawk and its prey, the garter snake.*

46–14   *A savanna in East Africa. The giraffes are surrounded by a herd of impala. The trees are acacias.*

world include the plains and prairies of North America, the steppes of Asia, the veld of South Africa, and the pampas of Argentina.

The vegetation is largely bunch or sod-forming grasses, often mixed with legumes (clovers and wild indigo) and a variety of annuals. In North America, there is a transition from the eastern temperate deciduous woodland, through the rich, moist, tall-grass prairie (the Corn Belt), to the more desertlike, western short-grass prairie (the Great Plains). Grasslands are drier to the west, where they are in the rain shadow of the Rockies. Periodic fires serve to maintain the nature of the grasslands, destroying tree seedlings and preventing their encroachment.

The temperate grasslands of the world support small, seed-eating rodents and also large herbivores, such as the bison of early America, the gazelles and zebras of the African veld, the wild horses, wild sheep, and saiga antelope of the Asiatic steppes, and now the domestic herbivores. These large, grass-eating mammals, in turn, support carnivores, such as lions and wolves, as well as omnivorous humans.

Savannas

Savannas are grasslands with scattered clumps of trees (Figure 46–14). They occur in regions characterized by wet-dry seasons. The transition from open forest with grassy undergrowth to savanna is gradual. It is determined by the duration and severity of the dry season and, often, by fire and by grazing and browsing animals.

In the savanna, the critical competition is for water, and grasses are the victors. Grasses are well suited to a fine, sandy soil with seasonal rain because their roots form a dense network capable of extracting the maximum amount of water during the rainy period. During dry seasons, the aboveground portions of the plants die, but the deep roots are able to survive even many months of drought.

The balance between woody plants and grasses is a delicate one. If rainfall decreases, the trees die. If rainfall increases, the trees increase in number until they shade the grasses, which, in turn, die. If the grasses are overgrazed (which often happens when people begin to use the savanna for agricultural purposes and introduce livestock), enough water is left in the soil so that the woody plants can increase in number, and the grassland is eventually destroyed.

The best-known savannas are those of Africa, which are inhabited by the most abundant and diverse group of large herbivores in the world, including the gazelles, the impala, the eland, the buffalo, the giraffe, the zebra, and the wildebeest.

852

(a)

(b)

(c)

46–15 (a) *The Mediterranean scrub, or chaparral, of North America. The bushy vegetation that characterizes the chaparral grows as dense as a mat on the foothills of southern California. It is the result of long, dry summers, during which much of the plant life is semidormant, followed by a cool, rainy season. The name comes from* chaparro, *the Indian word for the scrub oak that is one of the prominent components of the chaparral.* Chaps, *the leather leggings worn by cowboys making their way through this dense, dry growth, has the same derivation.* (b) *A cacomistle, or ring-tailed cat, an inhabitant of the chaparral.*

(c) *Mediterranean scrub, or* maquis, *on the island of Crete. Note the limestone on which the plants are growing. It is characteristic of much of the Mediterranean coastline and of the islands within the Mediterranean.*

Mediterranean Scrub

Regions with mild, moist winters and long, dry summers, such as the southern coast of California and southern Spain, are dominated by small trees or, often, by spiny shrubs with broad, thick evergreen leaves.

This vegetation, known formally as **Mediterranean scrub,** has been given a variety of local names. In the United States, it is known as the **chaparral,** whereas in Mediterranean regions, it is the *maquis,* and in Chile it is the *matorral.* Mediterranean scrub is also found in southern Africa and along portions of the coast of Australia. Although the plants of these various areas are unrelated, they closely resemble one another in their growth patterns and characteristic appearance.

Mule deer live in the North American chaparral during the spring growing season, moving out to cooler regions during the summer. Many of the resident vertebrates—lizards, wren-tits, brown towhees, brush rabbits, and a diversity of rodents—are small and dull-colored, matching the dull-colored vegetation.

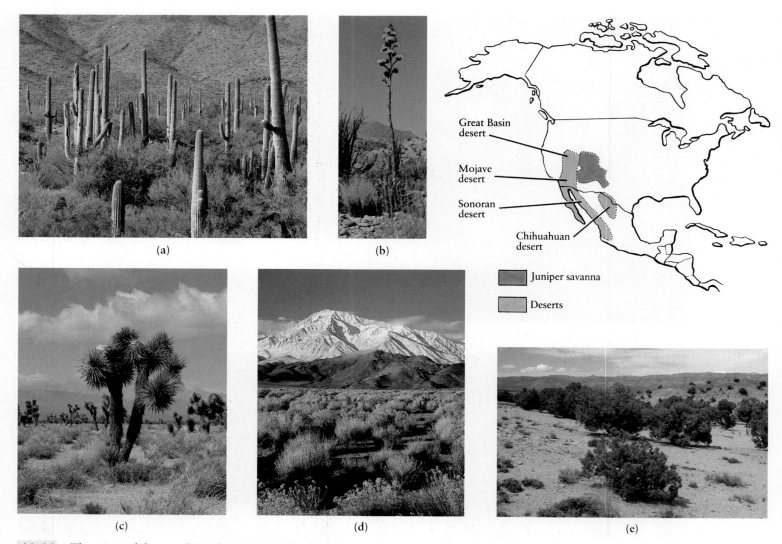

46–16 *The principal deserts of North America, adjacent to which are large areas of juniper savanna and semidesert.*

The Sonoran stretches from southern California to western Arizona and down into Mexico. A dominant plant, the giant saguaro cactus (a) is often as much as 15 meters high, with a widespreading network of shallow roots. Water is stored in a thickened stem, which expands, accordionlike, after a rainfall.

To the east is the Chihuahuan desert, one of whose principal plants is the agave (b), or century plant, a monocot.

North of the Sonoran is the Mojave, whose characteristic plant is the Joshua tree (c). This plant was named by early Mormon colonists who thought that its form resembled that of a bearded patriarch gesticulating in prayer. The Mojave contains Death Valley, the lowest point on the continent (90 meters below sea level), only 130 kilometers from Mt. Whitney, whose elevation is more than 4,000 meters.

The Mojave blends into the Great Basin, a cold desert bounded by the Sierra Nevada to the west and the Rockies to the east. It is the largest and bleakest of the American deserts. The dominant plant is sagebrush (d), shown here with the snow-covered Sierra Nevada mountains in the background.

Juniper savanna (e) is a high-altitude, cold biome named for its dominant plant species, Juniperus osteosperma.

The Desert

The great deserts of the world are located at latitudes of about 30°, both north and south, and extend poleward in the interiors of the continents. These are areas of falling, warming air and, consequently, little rainfall (see Figure 45–3, page 818). Only about 5 percent of North America is desert.

The Sahara, which stretches all the way from the Atlantic coast of Africa to the Arabian Peninsula, is the largest desert in the world (almost equal to the size of the United States) and is increasing in size, spreading along its southern boundaries. This spread is due in large part to the growth of the human population, resulting, in turn, in intensified grazing by domestic animals along the margins of the desert.

Desert regions are characterized by less than 25 centimeters of rain a year. Because there is little water vapor in the air to moderate the temperature, the nights are often extremely cold. The temperature may drop as much as 30°C at night, in contrast to the humid tropics,

(a)

(b)

(c)

(d)

46–17 *Inhabitants of the North American deserts.* (a) *With a few precisely targeted blows, this roadrunner will dispatch the bull snake held in its sharp beak. The roadrunner will then eat the snake headfirst.* (b) *A collared lizard, raising its body so cooling air currents can pass beneath it.* (c) *A cougar, or mountain lion.* (d) *A jackrabbit, which is actually a hare, resting in the shade during the hottest part of the day. Other familiar animals of the North American deserts include the horned lizard (Figure 32–8, page 576) and the kangaroo rat (Figure 32–12, page 579).*

where day and night temperatures vary by only a few degrees.

Many desert plants are annuals that race from seed to flower to seed during periods when water is available. During the brief growing seasons, the desert may be carpeted with flowers. Many of the perennials are succulents (adapted for water storage). Some are drought-deciduous (dropping their leaves in dry seasons), some have small, leathery water-conserving leaves, and some, like cacti, have leaves that are modified as hard, dry spines. C₄ and CAM photosynthesis (pages 171 and 745), both of which conserve water, are common among desert plants, as are extensive root systems that are able to trap large amounts of water during the brief periods it is available.

The animals that live in the desert are similarly adapted to this extreme climate (see page 579). Reptiles and insects, for example, have waterproof outer coverings and dry, and therefore water-conserving, excretions. Many of the mammals of the desert are small and nocturnal and obtain what little water they require from the plants they eat.

46–18   *Tropical rain forest.* (a) *The interior of a rain forest in Costa Rica. The broad-leaved plants with red flowers are Heliconia irrasa.* (b) *The diversity of the trees in the forest, which may reach several hundred species per hectare, is revealed when individual trees burst into bloom, like these trees in the coastal rain forest of Brazil.* (c) *A buttressed tree in the rain forest of Ecuador. Note the lianas on the trunk.*

(a)

(b)

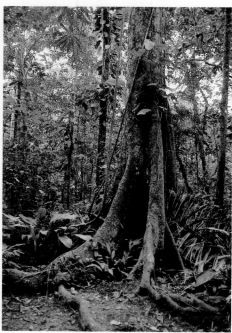

(c)

Tropical Forests

In the equatorial zone, where most of the tropical forests of the world are found, the average daily temperature is the same throughout the year, and the length of daylight varies by less than one hour. Rainfall, however, is often seasonal, and variations in total rainfall from one area to another are also caused by mountains and their rain shadows.

Areas in which rainfall is limited all year are characterized by southern woodland and scrub, in which the trees have small, water-conserving leaves. In areas with distinct wet and dry seasons, tropical mixed forests and monsoon forests occur. These forests are dominated by trees that lose their leaves during the dry seasons. The tropical rain forest, the most complex of all ecosystems, is found in areas where rainfall is abundant all year round.

Tropical Rain Forest

In the tropical rain forest, total rainfall is between 200 and 400 centimeters per year, and a month with less than 10 centimeters of rain is considered relatively dry. There are more species of plants and animals in the tropical rain forest than in all the rest of the biomes of the world combined. As many as 100 species of trees can be counted on 1 hectare, but each species may be represented by only one tree. By contrast, a comparable area in a deciduous forest in the northeastern United States typically contains only a few tree species, but each species is represented by many individual trees.

The critical competition among plants of the tropical forest is for light. About 70 percent of all species of plants are trees. The upper tree story consists of solitary giants 50 to 60 meters tall. A lower story of trees characteristically forms a continuous canopy. The trees forming the canopy are remarkably similar in appearance. Their trunks are usually slender and branch only near the crowns, which are very high and relatively small as a result of crowding. Because the soil is perpetually wet, their roots do not reach deep into it, and trunks often end in thick buttresses that provide broad anchorage. Their leaves are large, leathery, and dark green; their bark is thin and smooth; and their flowers are often inconspicuous and greenish or whitish in color.

Woody vines, or **lianas**, are abundant, especially where an opening has appeared in the forest, as, for example, where a tree has fallen. Vines as long as 240 meters have been measured. There are also many **epiphytes**, which are plants that grow on other plants, often high above the forest floor. The epiphytes of the tropical rain forest germinate in the branches of trees and obtain water from the humid air of the canopy. Unlike the plants that have contact with the moist floor, epiphytes need to conserve water between rainfalls. Some epiphytes resemble desert succulents, having fleshy water-storing leaves and stems. Others have spongy roots or cup-shaped leaves that capture moisture and organic debris. Many epiphytes can take up nutrients from decaying organisms in these storage tanks. A variety of plants, including ferns, orchids, mosses, and bromeliads (page 719), have exploited the epiphyte life style.

An extraordinary variety of insects, birds, and other animals, including mammals, have moved into the forest canopy along with the vines and epiphytes to make it the most abundantly and diversely populated area of the tropical rain forest (Figure 46–19).

Little light reaches the forest floor (from 0.1 to 1 percent of the total), and the relatively few plants that are found there are adapted to growing at low light intensities. Many of these, such as the African violet, are familiar to us as house plants.

Plants also compete for nutrients. The nutrient cycles are tight, and turnover is rapid. There is almost no accumulation of leaf litter on the forest floor, such as we find in our northern forests. Everything that touches the ground disappears almost immediately—carried off, consumed, or rapidly decomposed. As a consequence, the soils of tropical rain forests are relatively infertile. Many are chiefly composed of a red clay. When they are cleared, they generally either erode rapidly or form thick, impenetrable crusts that cannot be cultivated after a season or two.

Tropical soils are generally deficient in nutrients, and what nutrients are found there are likely to be leached out by heavy rainfall. Most of the nitrogen, phosphorus, calcium, and other nutrients are found in the plants rather than the soil. Trees that most effectively store these nutrients may be the ones that win the competition for light.

Tropical Forests, Mass Extinction, and Human Responsibility

Although tropical forests currently form about half of the forested area of the Earth, they are being rapidly destroyed. If the current rate of destruction continues, almost all the tropical forests will have disappeared by the turn of the century, and with them the thousands of plants and animals found nowhere else in the biosphere.

The destruction of the forests results directly from the rapid growth of the human population in the tropics and, as we saw at the beginning of this chapter, indirectly from the nature of tropical forest soils. Because

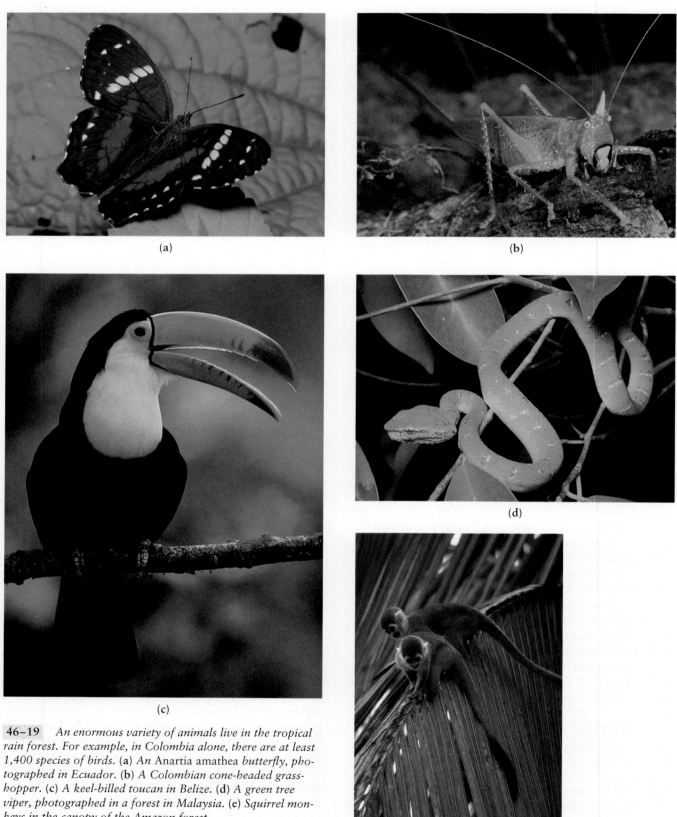

46–19 *An enormous variety of animals live in the tropical rain forest. For example, in Colombia alone, there are at least 1,400 species of birds.* (a) *An* Anartia amathea *butterfly, photographed in Ecuador.* (b) *A Colombian cone-headed grasshopper.* (c) *A keel-billed toucan in Belize.* (d) *A green tree viper, photographed in a forest in Malaysia.* (e) *Squirrel monkeys in the canopy of the Amazon forest.*

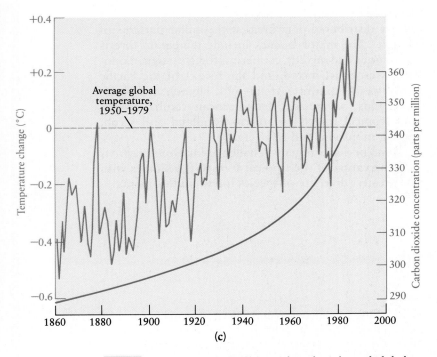

46–20 *The greenhouse effect, carbon dioxide, and global warming. (a) Light rays penetrate the glass of a greenhouse, are absorbed by the plants and soil, and are then reradiated as longer-waved infrared radiation (heat). The glass does not permit these rays to escape, and so the heat remains within the greenhouse. (b) Carbon dioxide, like glass, is transparent to light but absorbs infrared rays, preventing their escape.*

(c) Over the past 140 years, there has been a clear correlation between changes in average global temperature (red line) and the concentration of carbon dioxide in the atmosphere (blue line). As atmospheric carbon dioxide continues to increase, due to the burning of fossil fuels and forest destruction, it is expected to lead to further temperature increases. A study by the Environmental Protection Agency predicts that global temperatures could increase an average of 2°C by the year 2040 and 5°C by the year 2100.

such soils retain few nutrients, they are very poor for agricultural purposes and will support crops for only a few years after the trees have been cleared. Thus, the traditional agricultural practices in the tropics have been those of clearing and short-term cultivation. With a rapidly expanding, hungry human population, such practices have become immensely destructive as they are carried out on an ever-increasing scale. Commercial logging also contributes to the demise of the forests, as does the need of the local populations for firewood.

The loss of the organisms of the tropical forests would, in itself, represent a mass extinction on the scale of previous extinctions (see page 367). Unlike the earlier extinctions, however, it would be the direct result of the activities of a single species—our own. Moreover, many biologists are concerned that the destruction of the tropical forests will lead indirectly to extinctions in other biomes.

During the past 100 years, the concentration of carbon dioxide in the atmosphere has been rising steadily, largely as a result of increased burning of fossil fuels. Recent analyses have shown that forest destruction is also a major contributor to increased atmospheric carbon dioxide levels. As forest is cleared, the oxidation of its biomass—either by burning or by the metabolic processes of decomposers—releases large amounts of carbon dioxide. Moreover, carbon dioxide that would previously have been used by the trees and other plants of the forest in photosynthesis remains in the atmosphere.

As a result of the **greenhouse effect** (Figure 46–20), the increased carbon dioxide levels in the atmosphere are expected to result in a significant increase in global temperatures, beginning within the next 20 years. Among the expected consequences of this temperature increase are rising sea levels (as a result of the melting of polar ice), lengthened growing seasons and more precipitation in some areas, reduced precipitation and primary productivity in other areas, and the extinction of a variety of species as local environmental conditions change beyond the organisms' limits of adaptability (see page 782).

Halting, or even slowing, the destruction of the tropical forests is an enormously complex undertaking that cannot be separated from the needs of the human population of the tropics. Setting aside areas of forest as undisturbed preserves and slowing the rate of human population growth are only part of the answer. There is a pressing need for developing more productive and appropriate agricultural practices for the tropics and devising techniques for harvesting the richness of the forest—providing a long-overdue improvement in the standard of living for local populations—without simultaneously destroying the forest.

In the age of the dinosaurs, the earliest primates survived, it would appear, largely by their wits. Now, if we and other species are to survive the consequences of our own evolution, we will have to use well the contents of our own skulls. For within the human brain—that complex collection of neurons and synapses—resides the uniquely human capacity to accumulate knowledge, to plan with foresight, and so to act with enlightened self-interest and even, on occasion, with compassion for others and reverence for the diversity of life on our planet.

Summary

The biosphere is that part of the Earth that contains living organisms. It is a thin film on the surface of the planet, irregular in its thickness and its density.

The biosphere is affected by the position and movements of the Earth in relation to the sun and by the movements of air and water over the Earth's surface. These factors cause wide differences in temperature and rainfall from place to place and from season to season. There are also differences in the surfaces of the continents. These differences are reflected in the kinds of plant and animal life found in different parts of the biosphere.

Distinctive communities of organisms are found in running fresh waters (rivers and streams), standing fresh waters (lakes and ponds), at different depths of the open oceans, and along rocky, sandy, and muddy seashores.

On land, biomes are communities of organisms characterized by distinctive patterns of vegetation. They are widely distributed over areas with similar patterns of climate. The major biomes include temperate forests (deciduous and mixed), coniferous forests (taiga, mountain, west-coast mixed, and the forests of the Pacific Northwest), tundra (Arctic and alpine), temperate grasslands, savannas, Mediterranean scrub, deserts, and tropical forests (southern woodland and scrub, mixed, monsoon, and rain forest).

A major—and urgent—challenge for *Homo sapiens* is preservation of the tropical forests, the biomes with the greatest diversity of species found on this planet.

Questions

1. Mud flats are extremely rich in animal life, yet few plants are found there. What reasonable explanation can you give for the scarcity of plant life? How can the mud flats support such a profusion of animal life in the absence of plants?

2. What are the eight major biomes? Describe the principal abiotic features of each.

3. Name a plant and an animal associated with each of the eight major biomes, and describe their adaptations.

4. Although each of the biomes we have considered is sufficiently different from all the others to warrant its identification as a distinct biome, there are important similarities among some of the biomes. Consider the following groups of biomes: tropical rain forest/monsoon forest; monsoon forest/temperate deciduous forest/taiga; savanna/temperate grasslands/tundra. Describe the essential similarities and the most significant differences in the environmental factors affecting the members of each group. How do these factors affect the types of plants characterizing each biome?

5. The rate of decomposition of plant litter, animal wastes, and dead plants and animals varies from biome to biome. Describe the differences in decomposition rates in the following biomes: tropical rain forest, temperate deciduous forest, taiga. What factors in each biome are important in causing these differences? What are the consequences of these differing decomposition rates for nutrient recycling, soil quality, and the size and diversity of detritivore populations?

6. A leading ecologist has stated: "The plough is the most deadly weapon of extinction ever devised; not even thermonuclear weapons pose such a threat to the beauty and diversity of life on Earth." Explain.

Suggestions for Further Reading

Bazzaz, Fakhri A., and Eric D. Fajer: "Plant Life in a CO_2-Rich World," *Scientific American*, January 1992, pages 68–74.

Berner, Robert A., and Antonio C. Lasaga: "Modeling the Geochemical Carbon Cycle," *Scientific American*, March 1989, pages 74–81.

Bertness, Mark D.: "The Ecology of a New England Salt Marsh," *American Scientist*, vol. 80, pages 260–268, 1992.

Burney, David A.: "Recent Animal Extinctions: Recipes for Disaster," *American Scientist*, vol. 81, pages 530–541, 1993.

Colinvaux, Paul A.: "The Past and Future Amazon," *Scientific American*, May 1989, pages 102–108.

Crutzen, Paul J., and Meinrat O. Andreae: "Biomass Burning in the Tropics: Impact on Atmospheric Chemistry and Biogeochemical Cycles," *Science*, vol. 250, pages 1669–1678, 1990.

Detwiler, R. P., and C. A. S. Hall: "Tropical Forests and the Global Carbon Cycle," *Science*, vol. 239, pages 42–47, 1988.

Ellis, Gerry, and Karen Kane: *America's Rain Forest*, North Word Press, Inc., Minocqua, Wis., 1991.

> This beautiful book, filled with almost 200 outstanding color photographs, provides a valuable overview of the diversity and complexity of the forests of the Pacific Northwest. As the authors make clear, the very fabric of the ecosystem—not just individual species, such as the northern spotted owl—is endangered by logging practices that destroy the forests' capacity for renewal.

Forsyth, Adrian, and Ken Miyata: *Tropical Nature: Life and Death in the Rain Forests of Central and South America*, Charles Scribner's Sons, New York, 1984.*

> This book, for the general reader, consists of 17 essays on different aspects of the complex—and still poorly understood—ecology of tropical rain forests. It is filled with a multitude of marvelous examples of the diverse organisms of the forests and their many interactions with each other. The authors, both specialists in tropical vertebrates, spent seven years working in the forests, principally in Costa Rica and Peru, and bring vast personal experience to their essays.

Goulding, Michael: "Flooded Forests of the Amazon," *Scientific American*, March 1993, pages 114–120.

Holloway, Marguerite: "Sustaining the Amazon," *Scientific American*, July 1993, pages 90–99.

Homer-Dixon, Thomas F., Jeffrey H. Boutwell, and George W. Rathjeus: "Environmental Change and Violent Conflict," *Scientific American*, February 1993, pages 38–45.

Horn, Michael S., and Robin H. Gibson: "Intertidal Fishes," *Scientific American*, January 1988, pages 64–70.

Houghton, Richard A., and George M. Woodwell: "Global Climatic Change," *Scientific American*, April 1989, pages 36–44.

Janzen, Daniel H.: "Tropical Ecological and Biocultural Restoration," *Science*, vol. 239, pages 243–244, 1988.

Jones, Philip D., and Tom M. L. Wigley: "Global Warming Trends," *Scientific American*, August 1990, pages 84–91.

Krutch, J. W.: *The Desert Year*, Viking Press, Inc., New York, 1977.*

> A captivating, beautifully written description, first published in 1960, of the animal and plant life of the North American desert.

La Rivière, J. W. Maurits: "Threats to the World's Water," *Scientific American*, September 1989, pages 80–94.

Laws, Richard M.: "The Ecology of the Southern Ocean," *American Scientist*, vol. 73, pages 26–40, 1985.

Lopez, Barry: *Arctic Dreams: Imagination and Desire in a Northern Landscape*, Bantam Books, New York, 1987.*

> A brilliant evocation of life in the Arctic by a master nature writer. "This is a land," says Lopez, "where airplanes track icebergs the size of Cleveland and polar bears fly down out of the stars."

Mares, Michael A.: "Conservation in South America: Problems, Consequences, and Solutions," *Science*, vol. 233, pages 734–739, 1986.

Mohnen, Volker A.: "The Challenge of Acid Rain," *Scientific American*, August 1988, pages 30–38.

Moran, Joseph M., Michael D. Morgan, and James H. Wiersma: *An Introduction to Environmental Sciences*, 2d ed., W. H. Freeman and Company, New York, 1985.

> A clear, matter-of-fact presentation of the basic features of our planet and, in particular, of the biosphere. It provides a good background for the study of ecology in general and problems of environmental pollution and technological change in particular.

Newell, Reginald E., Henry G. Reichle, Jr., and Wolfgang Seiler: "Carbon Monoxide and the Burning Earth," *Scientific American*, October 1989, pages 82–88.

Perry, Donald R.: "The Canopy of the Tropical Rain Forest," *Scientific American*, November 1984, pages 138–147.

---

* Available in paperback.

Philander, George: "El Niño and La Niña," *American Scientist*, vol. 72, pages 451–459, 1989.

Pollack, Henry N., and David S. Chapman: "Underground Records of Changing Climate," *Scientific American,* June 1993, pages 44–50.

Post, Wilfred M., et al.: "The Global Carbon Cycle," *American Scientist,* vol. 78, pages 310–326, 1990.

Ramage, Colin S.: "El Niño," *Scientific American,* June 1986, pages 76–83.

Repetto, Robert: "Accounting for Environmental Assets," *Scientific American,* June 1992, pages 94–101.

Repetto, Robert: "Deforestation in the Tropics," *Scientific American,* April 1990, pages 36–42.

Richardson, Philip L.: "Tracking Ocean Eddies," *American Scientist,* vol. 81, pages 261–271, 1993.

Schneider, Stephen H.: "The Changing Climate," *Scientific American,* September 1989, pages 70–79.

Sebens, Kenneth P.: "The Ecology of the Rocky Subtidal Zone," *American Scientist,* vol. 73, pages 548–557, 1985.

Smith, Robert L.: *Ecology and Field Biology,* 4th ed., Harper & Row, Publishers, Inc., New York, 1990.

The outstanding sections of this text are those that deal with the biomes and the plants and animals within them. Although meant to accompany a course in field biology, this book is an asset and a pleasure to the amateur naturalist.

Terborgh, John: *Diversity and the Tropical Rain Forest,* W. H. Freeman and Company, New York, 1992.

This beautifully illustrated volume explores the incredible diversity of organisms found in the tropical forests, as well as the complex interactions of ecology and evolution that have given rise to that diversity. The author, who is the director of Duke University's Center for Tropical Conservation, also discusses current scientific efforts to counteract the devastating effects of the destruction of the tropical forests.

Waring, Richard H.: "Land of the Giant Conifers," *Natural History,* October 1982, pages 54–63.

White, Robert M.: "The Great Climate Debate," *Scientific American,* July 1990, pages 36–43.

Wilson, Edward O.: *The Diversity of Life,* Harvard University Press, Cambridge, Mass., 1992.

A presentation of the diversity of life and of the evolutionary processes that have produced it, by one of the world's leading authorities on both diversity and evolution. Filled with marvelous stories of many fascinating organisms, Wilson's text is also infused with examples of how rapidly species are being driven to extinction throughout the world and the obstacles that must be overcome if the loss of diversity is to be stemmed.

Wilson, Edward O.: "Threats to Biodiversity," *Scientific American,* September 1989, pages 108–116.

* Available in paperback.

The Hardy-Weinberg Equation

As we noted in the text (page 325), Hardy and Weinberg demonstrated the constancy of allele and genotype frequencies in an idealized population in which five conditions hold:

1. No mutations occur.

2. There is no net movement of individuals—with their genes—into the population (immigration) or out of it (emigration).

3. The population is large enough that the laws of probability apply; that is, it is highly unlikely that chance alone can alter the frequencies, or relative proportions, of alleles.

4. Mating is random.

5. All alleles are equally viable. In other words, there is no difference in reproductive success. The offspring of all possible matings are equally likely to survive to reproduce in the next generation.

To understand how Hardy and Weinberg arrived at their equation expressing the equilibrium of allele and genotype frequencies in a population meeting these five conditions, let us look at just one gene. For simplicity, we select one that has only two alleles, which we will call A and a. We are interested in the relative proportions—that is, the frequencies—of A and a from one generation to the next. By convention, the letter p is used to designate the frequency of one allele and the letter q to designate the frequency of the other. When there are only two alleles, p and q together must equal one: $p + q = 1$.

These proportions—p and q—could be expressed in terms of fractions, as done by Mendel, but since the proportions of the two alleles will probably not be equal, as they were in Mendel's carefully controlled experiments, it is more convenient to express the numbers as decimals. For example, suppose that in a particular population 80 percent of the alleles of the gene under study are allele A. The frequency of A is 0.8, or $p = 0.8$. Given that there are only two alleles, we then know that the frequency of allele a is 0.2. In other words, $q = 1 - p$.

Let us assume that the relative frequencies of A and a are the same in both male and female (as they are for most alleles in natural populations). Now suppose that males and females mate at random with respect to alleles A and a. We can calculate the frequencies of the resulting genotypes by drawing a Punnett square. As you can see from the illustration, the genotypes in the population produced by this random mating would consist of 64 percent AA, 32 percent Aa, and 4 percent aa.

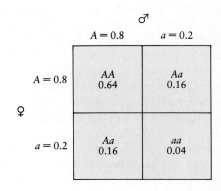

Results of random mating in a population in which the frequency (p) of allele A equals 0.8 and the frequency (q) of allele a equals 0.2. In setting up this Punnett square, we are assuming that the gene is not sex-linked and that the alleles appear in the same frequencies in males as in females.

Instead of drawing a Punnett square, we can do the same thing algebraically. Because $p + q = 1$, it follows that:

$$(p + q)(p + q) = 1 \times 1 = 1$$

Or, as you probably remember from algebra:

$$p^2 + 2pq + q^2 = 1$$

This algebraic expression of the genotype frequencies is the Hardy-Weinberg equation.

Let us apply this equation to the random mating that just occurred in our population. Taking the initial values for the frequencies of the two alleles, we obtain the following results:

$$p^2 = 0.8 \times 0.8$$
$$= 0.64 \text{ (the frequency of } AA \text{ genotypes)}$$
$$2pq = 2 \times 0.8 \times 0.2$$
$$= 0.32 \text{ (the frequency of } Aa \text{ genotypes)}$$
$$q^2 = 0.2 \times 0.2$$
$$= 0.04 \text{ (the frequency of } aa \text{ genotypes)}$$

What has happened to the frequencies of the two alleles in the gene pool as a result of this round of mating? We know from our calculations that the frequency of AA is 0.64. In addition, half of the alleles in the heterozygotes (Aa) are A, so the total frequency of allele A is 0.64 plus one-half of 0.32—that is, 0.64 plus 0.16, to give a total of 0.8. The frequency of allele A (p) has not changed. Similarly, the total frequency of allele a is 0.04 (in the homozygotes) plus 0.16 (half the alleles in the heterozygotes), or 0.2. The frequency of allele a (q) has also remained the same.

If another round of mating occurs, the proportion of AA, of Aa, and of aa genotypes in our population will again be 64 percent, 32 percent, and 4 percent, respectively. Again the frequency of allele A will be 0.8 and of allele a 0.2. And so on, and so on, generation after generation. In an ideal population in which the five conditions are met, neither the allele frequencies nor the genotype frequencies change from generation to generation.

The Hardy-Weinberg equilibrium also applies to situations in which there are more than two alleles of the same gene. However, the equation representing the equilibrium is more complex. For example, the genotype frequencies of three alleles are expressed by the algebraic expansion of $(p + q + r)^2 = 1$, with r representing the frequency in the gene pool of the third allele.

An Application of the Hardy-Weinberg Equation

As we have noted, the Hardy-Weinberg equilibrium holds only in populations in which the five stated conditions are met. What happens if one of these conditions is not met? Let us imagine another pair of alleles in which allele a has a damaging effect in the homozygous state, reducing the likelihood that a homozygous aa individual will survive to reproduce. We can estimate the number of aa genotypes in the population, perhaps by screening tests on newborn infants. Suppose, for instance, that the condition shows up in 1 in 10,000 infants. In other words, $q^2 = 1/10,000$, or 0.0001. Thus, $q = \sqrt{0.0001}$, or 0.01. If $q = 0.01$, then $p = 0.99$ and $2pq = 0.0198$, or almost 0.02. Thus about 2 percent of the population—one person in every 50—can be estimated to be a heterozygous carrier of this allele.

Now suppose that we do the same screening tests five years later and find that q is not 0.01 but 0.009, and then we repeat it a few years later and find that it has once again decreased very slightly, perhaps to 0.008. In other words, evolution is occurring. One allele is decreasing while the other is increasing. Using the Hardy-Weinberg equilibrium as our yardstick, we know not only that change has taken place and its direction but also that there must be some reason for it. We can then look for the factors causing the change.

Some Problems

1. Suppose that in a breeding experiment, 7,000 AA individuals and 3,000 aa individuals mate at random. In the first generation of offspring, what would be the frequencies of the three genotypes (AA, Aa, and aa)? What would be the frequencies of the two alleles? What would be the values in the second generation?

Solution: Initially, there are a total of 20,000 alleles in the gene pool: 14,000 A alleles and 6,000 a alleles. The frequency of allele A (that is, the value for p) is 14,000/20,000, or 0.7. The frequency of allele a (that is, the value for q) is $1 - p$, or 0.3.

In the first generation of offspring, the values of the three terms of the Hardy-Weinberg equation would be as follows:

$$p^2 = (0.7)^2$$
$$= 0.49 \text{ (frequency of } AA \text{ genotypes)}$$
$$2pq = 2(0.7)(0.3)$$
$$= 0.42 \text{ (frequency of } Aa \text{ genotypes)}$$
$$q^2 = (0.3)^2$$
$$= 0.09 \text{ (frequency of } aa \text{ genotypes)}$$

The allele frequencies would be as follows:

$$p = \sqrt{0.49} = 0.7 \text{ (frequency of allele } A\text{)}$$

$$q = \sqrt{0.09} = 0.3 \text{ (frequency of allele } a\text{)}$$

Assuming that the five conditions for the Hardy-Weinberg equilibrium continue to be met, the values will be the same in the second generation. Work through the calculations to satisfy yourself that this is the case.

2. Among African-Americans, the frequency of sickle-cell anemia (which, as you will recall, is a homozygous recessive condition) is about 0.0025. What is the frequency of heterozygotes? When one African-American marries another, what is the probability that both will be heterozygotes? If both are heterozygotes, what is the probability that their first child will have sickle-cell anemia?

B

Classification of Organisms

There are several ways to classify organisms. The one presented here follows the overall scheme described at the end of Chapter 23. Organisms are divided into five major groups, or kingdoms: Monera, Protista, Fungi, Plantae, and Animalia.

The chief taxonomic categories are kingdom, division or phylum, class, order, family, genus, and species. (The taxonomic categories of division and phylum are equivalent. The term "division" is generally used in the classification of prokaryotes, algae, fungi, and plants, whereas "phylum" is used in the classification of protozoa and animals.) The following classification includes all of the generally accepted divisions and phyla. Certain classes and orders, particularly those mentioned in this book, are also included, but the listing is far from complete. The number of species given for each group is an estimated number of living (that is, contemporary) species described and named to date.

KINGDOM MONERA

Monerans (prokaryotes) are cells that lack a nuclear envelope, chloroplasts and other plastids, mitochondria, and 9 + 2 flagella. Prokaryotes are unicellular but sometimes occur as filaments or other superficially multicellular bodies. Their predominant mode of nutrition is heterotrophic, by absorption, but some groups are autotrophic, either photosynthetic or chemosynthetic. Reproduction is primarily asexual, by cell division or budding, but genetic exchanges occur in some as a result of conjugation, transformation, transduction, and exchanges of plasmids. Motile forms move by means of bacterial flagella or by gliding.

The classification of prokaryotes is not hierarchical, and the schemes that are most widely used do not reflect evolutionary relationships. Kingdom Monera contains representatives of two distinct lineages, the Archaebacteria and the Eubacteria. Archaebacteria include the methanogens, thermoacidophiles, and extreme halophiles. Among the principal lineages of Eubacteria are the green bacteria (sulfur and nonsulfur), purple bacteria (sulfur and nonsulfur) and related forms, spirochetes, cyanobacteria, and gram-positive bacteria. About 2,700 distinct species of prokaryotes are recognized.

KINGDOM PROTISTA

Eukaryotic organisms, including unicellular or simple colonial heterotrophs (protozoa), multinucleate or multicellular heterotrophs (slime molds and water molds), and unicellular and multicellular photosynthetic autotrophs (algae). Their modes of nutrition include ingestion, absorption, and photosynthesis. Reproduction is both sexual (in some forms) and asexual. They move by 9 + 2 flagella (or cilia) or pseudopodia or are nonmotile. About 60,000 living species and another 60,000 extinct species known only from their fossils.

Phylum

Phylum Mastigophora: mastigophorans (flagellates). Unicellular heterotrophs with flagella, most of which are symbiotic forms such as *Trichonympha* and *Trypanosoma,* the cause of sleeping sickness. Some free-living species. Reproduction usually asexual by cell division. About 1,500 species.

Phylum Sarcodina: sarcodines. Unicellular heterotrophs with pseudopodia, such as amoebas. No stiffening pellicle; some produce shells. Reproduction may be asexual or sexual. About 11,500 living species and some 33,000 fossil species.

Phylum Ciliophora: ciliates. Unicellular heterotrophs with cilia, including *Paramecium* and *Stentor.* Reproduction is asexual, but genetic exchange through the phenomenon of conjugation is also common. About 8,000 species.

Phylum Opalinida: opalinids. Parasitic protists found mainly in the digestive tracts of frogs and toads. Covered uniformly with cilia or flagella. Reproduction asexual by cell division or sexual, with flagellated gametes. About 400 species.

Phylum Sporozoa: sporozoans. Parasitic protists; usually without locomotor organelles during a major part of their complex life cycle. Includes *Plasmodium,* several species of which cause malaria. About 5,000 species.

Division

Division Myxomycota: plasmodial slime molds. Heterotrophic amoeboid organisms that form a multinucleate plasmodium that creeps along as a mass and eventually differentiates into sporangia, each of which is multinucleate and eventually gives rise to many spores. Predominant mode of nutrition is by ingestion. More than 550 species.

Division Acrasiomycota: cellular slime molds. Heterotrophic organisms in which there are separate amoebas that eventually swarm together to form a mass but retain their identity within this mass, which eventually differentiates into a compound sporangium. Principal mode of nutrition is by ingestion. Seven genera and about 65 species.

Division Chytridiomycota: chytrids (water molds). Multinucleate aquatic heterotrophs with a vegetative body, or thallus, usually differentiated into rhizoids and a sporangium. Nutrition by absorption. Cell walls contain chitin. Asexual reproduction, with sexual reproduction in some forms. Both spores and gametes are flagellated. About 900 species.

Division Oomycota: oomycetes (water molds). Multinucleate filamentous heterotrophs, primarily aquatic. Nutrition by absorption. Their cell walls are composed of glucose polymers, including cellulose. Both asexual and sexual reproduction, with flagellated asexual spores and nonmotile gametes. The multinucleate filaments (hyphae) are diploid, and the haploid gametes are produced by meiosis. About 800 species.

Division

Division Euglenophyta: euglenoids. Unicellular photosynthetic (or sometimes secondarily heterotrophic) organisms with chlorophylls *a* and *b*. They store food as paramylon, an unusual carbohydrate. Euglenoids usually have a single apical flagellum and a contractile vacuole. Sexual reproduction is unknown. Euglenoids occur mostly in fresh water. There are some 1,000 species.

Division Chrysophyta: diatoms, golden-brown algae, and yellow-green algae. Unicellular photosynthetic organisms with chlorophylls *a* and *c* and the accessory pigment fucoxanthin. Food stored as the carbohydrate laminarin or as large oil droplets. Cell walls consisting mainly of cellulose, sometimes heavily impregnated with siliceous materials. There may be as many as 13,000 living species.

Class

Class Bacillariophyceae: diatoms. Chrysophyta with double siliceous shells, the two halves of which fit together like a pillbox. They are sometimes motile by the secretion of mucilage fibrils along a specialized groove, the raphe. There are nearly 10,000 living species and at least 15,000 extinct species.

Class Chrysophyceae: golden-brown algae. A diverse group of organisms, including flagellated, amoeboid, and nonmotile forms, some naked and others with a cell wall that may be ornamented with siliceous scales. About 3,000 species.

Division

Division Dinoflagellata: "spinning" flagellates. Unicellular photosynthetic organisms with chlorophylls *a* and *c*. Food is stored as starch. Cell walls contain cellulose. Most of the organisms in this division have two lateral flagella, one of which beats in a groove that encircles the organism. They probably have no form of sexual reproduction, and their mitosis is unique. Nearly 2,000 living species.

Division Chlorophyta: green algae. Unicellular, colonial, multinucleate, or multicellular, characterized by chlorophylls *a* and *b* and various carotenoids. The carbohydrate food reserve is starch. The cell walls consist of polysaccharides, including cellulose in some forms. Motile cells have two lateral or apical flagella. True multicellular genera do not exhibit complex patterns of differentiation. Multicellularity has arisen at least three times, and quite possibly more often. There are about 9,000 known species and possibly many more.

Class

Class Chlorophyceae: Unicellular, colonial, or multicellular green algae, found predominantly in fresh water. Cell division involves a system of microtubules parallel to the plane of cell division, and the nuclear envelope persists during mitosis. Sexual reproduction involves the formation of a dormant zygote that subsequently undergoes meiosis, producing the haploid cells in which the organisms spend most of the life cycle.

Class Charophyceae: Unicellular or multicellular green algae, found predominantly in fresh water. Cell division involves a system of microtubules perpendicular to the plane of cell division; the nuclear envelope breaks down during the course of mitosis. Sexual reproduction involves the formation of a dormant zygote that subsequently undergoes meiosis, producing the haploid cells in which the organisms spend most of the life cycle. Certain members of this class resemble plants more closely than do any other organisms.

Class Ulvophyceae: Multinucleate or multicellular green algae, found predominantly in salt water. Cell division involves a system of microtubules perpendicular to the plane of cell division, but the nuclear envelope persists during mitosis. Sexual reproduction often involves alternation of generations, with meiosis in the diploid sporophyte producing haploid spores that germinate to produce the haploid gametophyte.

Division

Division Phaeophyta: brown algae. Multicellular marine organisms characterized by the presence of chlorophylls *a* and *c* and the pigment fucoxanthin. Their food reserve is the carbohydrate laminarin. Motile cells are biflagellate, with one forward flagellum and one trailing flagellum. A considerable amount of tissue differentiation is found in some of the kelps, with specialized conducting cells in some genera that transport the products of photosynthesis to dimly lighted regions of the alga. There is, however, no differentiation into leaves, roots, and stem, as in plants. About 1,500 species.

Division Rhodophyta: red algae. Primarily marine organisms characterized by the presence of chlorophyll *a* and red pigments known as phycobilins. Their carbohydrate reserve is a special type of starch (floridean). No motile cells are present at any stage in the complex life cycle. The algal body is built up of closely packed filaments in a gelatinous matrix and is not differentiated into leaves, roots, and stem. It lacks specialized conducting cells. There are some 4,000 species.

KINGDOM FUNGI

Eukaryotic filamentous or, rarely, unicellular organisms. The filamentous forms consist basically of a continuous mycelium; this mycelium becomes septate (partitioned off) in certain groups and at certain stages of the life cycle. Chitin is present in the cell walls of all fungi. Fungi are saprobic or parasitic heterotrophs, with nutrition by absorption. Reproductive cycles often include both sexual and asexual phases. Most fungi are haploid, with the zygote the only diploid stage in the life cycle. No flagella or cilia at any stage of the life cycle. Some 100,000 species of fungi have been named.

Division

Division Zygomycota: terrestrial fungi, such as black bread mold, with the hyphae septate only during the formation of reproductive bodies. The division includes about 600 described species, of which about 30 occur as components of the mycorrhizae that are found in about 80 percent of all vascular plants.

Division Ascomycota: terrestrial and aquatic fungi, including *Neurospora,* powdery mildews, morels, and truffles. The hyphae are septate but the septa perforated; complete septa cut off the reproductive bodies, such as spores or gametangia. Sexual reproduction involves the formation of a characteristic cell, the ascus, in which meiosis takes place and within which spores are formed. The hyphae in many ascomycetes are packed together into complex "fruiting bodies." Yeasts are unicellular ascomycetes that reproduce asexually by budding. About 30,000 species, in addition to 25,000 species that occur in lichens.

Division Basidiomycota: terrestrial fungi, including the mushrooms and toadstools, with the hyphae septate but the septa perforated; complete septa cut off reproductive bodies. Sexual reproduction involves formation of basidia, in which meiosis takes place and on which the spores are borne. There are some 25,000 species.

Division Deuteromycota: Fungi Imperfecti. Mainly fungi in which the sexual cycle has not been observed. The deuteromycetes are classified by their asexual spore-bearing structures. There are some 25,000 species, including *Penicillium* (the original source of penicillin), fungi that cause athlete's foot and other skin diseases, and many of the molds that give cheeses, such as Roquefort and Camembert, their special flavor.

KINGDOM PLANTAE

Multicellular photosynthetic eukaryotes, primarily adapted to life on land. The photosynthetic pigment is chlorophyll *a,* with chlorophyll *b* and a number of carotenoids serving as accessory pigments. The cell walls contain cellulose. There is considerable differentiation of tissues and organs. Reproduction is primarily sexual with alternating gametophytic and sporophytic phases; the gametophytic phase has been progressively reduced in the course of evolution. The egg- and sperm-producing structures are multicellular and are surrounded by a sterile (nonreproductive) jacket layer; the zygote develops into an embryo, or young sporophyte, while encased in the archegonium (seedless plants) or embryo sac (seed plants). The living members of the plant kingdom include the bryophytes and nine divisions of vascular plants—plants with complex differentiation of the sporophyte into leaves, roots, and stems and with well-developed strands of conducting tissue for the transport of water and organic materials.

Division

Division Bryophyta: liverworts, hornworts, and mosses. Multicellular plants with photosynthetic pigments and food reserves similar to those of the green algae. They have multicellular gametangia with a sterile jacket one-cell-layer thick. The sperm are biflagellate and motile. Gametophytes and sporophytes both exhibit complex multicellular patterns of development, but conducting tissues are usually completely absent and are not well differentiated when present; true roots, leaves, and stems are absent. Most of the photosynthesis is carried out by the gametophyte, upon which the sporophyte is nutritionally dependent, at least initially. There are some 16,000 species.

Class

Class Hepaticae: liverworts. The gametophytes are either thallose (not differentiated into roots, leaves, and stem) or "leafy," and the sporophytes are relatively simple in construction. About 6,000 species.

Class Anthocerotae: hornworts. The gametophytes are thallose. The sporophyte grows from a basal meristem for as long as conditions are favorable. Stomata are present on the sporophyte. About 100 species.

Class Musci: mosses. The gametophytes are "leafy." The sporophytes have complex patterns of spore discharge. Stomata are present on the sporophyte. About 9,500 species.

Division

Division Psilotophyta: whisk ferns. Homosporous vascular plants with or without microphylls. (Homosporous plants produce only one kind of spore, giving rise to gametophytes with both male and female structures; microphylls are primitive leaves.) The sporophytes are extremely simple, and there is no differentiation between root and shoot. The sperm are motile. Two genera, with several species.

Division Lycophyta: club mosses. Homosporous and heterosporous vascular plants with microphylls; extremely diverse in appearance. (Heterosporous plants produce two kinds of spores, one kind giving rise to male gametophytes, the other to female gametophytes.) All lycophytes have motile sperm. There are five genera, with about 1,000 species.

Division Sphenophyta: horsetails. Homosporous vascular plants with jointed stems marked by conspicuous nodes and elevated siliceous ribs. Sporangia are borne in a cone at the apex of the stem. Leaves are scalelike. Sperm are motile. Although now thought to have evolved from a megaphyll (a complex leaf), the leaves of the horsetails are structurally indistinguishable from microphylls. There is one genus, *Equisetum,* with 15 living species.

Division Pterophyta: ferns. They are mostly homosporous, although some are heterosporous. All possess megaphylls. The gametophyte is more or less free-living and usually photosynthetic. Multicellular gametangia and free-swimming sperm are present. About 12,000 species.

Division Coniferophyta: conifers. Seed plants with active cambial growth and simple, needle-like leaves; the ovules are not enclosed and the sperm are not flagellated. There are some 50 genera, with about 550 species; the most familiar group of gymnosperms.

Division Cycadophyta: cycads. Seed plants with sluggish cambial growth and pinnately compound, palmlike or fernlike leaves. The ovules are not enclosed. The sperm are flagellated and motile but are carried to the ovule in a pollen tube. Cycads are gymnosperms. There are 10 genera, with about 100 species.

Division Ginkgophyta: ginkgo. Seed plants with active cambial growth and fan-shaped leaves with open dichotomous venation. The ovules are not enclosed and are fleshy at maturity. Sperm are carried to the ovule in a pollen tube but are flagellated and motile. They are gymnosperms. There is one species only.

Division Gnetophyta: gnetophytes. Seed plants with many angiospermlike features. They are the only gymnosperms in which vessels are present in the xylem. Motile sperm are absent. There are three very distinctive genera, with about 70 species.

Division Anthophyta: flowering plants (angiosperms). Seed plants in which the ovules are enclosed in a carpel (in all but a very few genera), and the seeds at maturity are borne within fruits. They are extremely diverse vegetatively but are characterized by the flower, which is basically insect-pollinated. Other modes of pollination, such as wind pollination, have been derived in a number of different lines. The gametophytes are much reduced, with the female gametophyte often consisting of only seven cells at maturity. Double fertilization involving the two nonmotile sperm nuclei of the mature male gametophyte gives rise to the zygote and to the primary endosperm nucleus; the former becomes the embryo and the latter a special nutritive tissue, the endosperm. About 235,000 species.

Class

Class Monocotyledones: monocots. Flower parts are usually in threes, leaf venation is usually parallel, vascular bundles in the young stem are scattered, true secondary growth is not present, and there is one cotyledon. About 65,000 species.

Class Dicotyledones: dicots. Flower parts are usually in fours or fives, leaf venation is usually netlike, the vascular bundles in the young stem are in a ring, there is true secondary growth with vascular cambium commonly present, and there are two cotyledons. About 170,000 species.

KINGDOM ANIMALIA

Eukaryotic multicellular organisms. Their principal mode of nutrition is by ingestion. Many animals are motile, and they generally lack the rigid cell walls characteristic of plants. Considerable cellular migration and reorganization of tissues often occur during the course of embryonic development. Their reproduction is primarily sexual, with male and female diploid organisms producing haploid gametes that fuse to form the zygote. More than 1.5 million living species have been described, and the actual number may be more than 50 million.

Phylum

Phylum Porifera: sponges. Simple multicellular animals, largely marine, with stiff skeletons, and bodies perforated by many pores that admit water containing food particles. All have choanocytes, "collar cells." About 5,000 species.

Phylum Mesozoa: extremely simple wormlike animals, all parasites of marine invertebrates. The body consists of 20 to 30 cells, organized in two layers. About 50 species.

Phylum Cnidaria: polyps and jellyfishes. Radially symmetrical animals with a gastrovascular cavity and two-layered bodies of a jellylike consistency. Reproduction is asexual or sexual. They are the only organisms with cnidocytes, special stinging cells. All are aquatic and most are marine. About 9,000 species.

Class

Class Hydrozoa: Hydra, Obelia, and other *Hydra*-like animals. They are often colonial and frequently have a regular alternation of asexual and sexual forms. The polyp form is dominant.

Class Scyphozoa ("cup animals"): marine jellyfishes, including *Aurelia.* The medusa form is dominant. They have true muscle cells.

Class Anthozoa ("flower animals"): sea anemones, colonial corals, and related forms. They have no medusa stage.

Phylum

Phylum Ctenophora: comb jellies and sea walnuts. They are free-swimming, often almost spherical animals, with a gastrovascular cavity. They are translucent, gelatinous, delicately colored, and often bioluminescent. They possess eight bands of cilia, for locomotion. About 90 species.

Phylum Platyhelminthes: flatworms. Bilaterally symmetrical with three embryonic tissue layers. The digestive cavity has only one opening. They have no coelom or pseudocoelom and no circulatory system. They have complex hermaphroditic reproductive systems and excrete by means of special flame cells. About 13,000 species.

Class

Class Turbellaria: planarians and other nonparasitic flatworms. They are ciliated, carnivorous, and have ocelli ("eyespots").

Class Trematoda: flukes. They are parasitic flatworms with digestive cavities.

Class Cestoda: tapeworms. They are parasitic flatworms with no digestive cavities; they absorb nourishment through body surfaces.

Phylum

Phylum Gnathostomulida: tiny acoelomate marine worms, characterized by a unique pair of hard jaws. The digestive cavity has only one opening. About 80 species.

Phylum Rhynchocoela: proboscis, nemertine, or ribbon worms. These acoelomate worms are nonparasitic, usually marine, and have a tubelike gut with mouth and anus, a retractile proboscis armed with a hook for capturing prey, and simple circulatory and reproductive systems. About 650 species.

Phylum Nematoda: roundworms. The phylum includes minute free-living forms, such as vinegar eels, and plant and animal parasites, such as hookworms. They have a pseudocoelom and are characterized by elongated, cylindrical, bilaterally symmetrical bodies. About 12,000 species have been described and named, and there may be as many as 400,000 to 500,000 species.

Phylum Nematomorpha: horsehair worms. They are extremely slender, brown or black pseudocoelomate worms up to 1 meter long. Adults are free-living, but the larvae are parasitic in arthropods. About 230 species.

Phylum Acanthocephala: spiny-headed worms. They are parasitic pseudocoelomate worms with no digestive tract and a head armed with many recurved spines. About 500 species.

Phylum Kinorhyncha: tiny pseudocoelomate worms that burrow in muddy ocean shores. They are short-bodied, covered with spines, and have a spiny retractile proboscis. About 100 species.

Phylum Gastrotricha: microscopic, wormlike pseudocoelomate animals that move by longitudinal bands of cilia. About 400 species.

Phylum Loricifera: microscopic pseudocoelomate animals of the ocean bottom, characterized by a plate-covered body, numerous spines projecting from the head, and a retractile tube for a mouth. The first of the three known species was discovered in 1982; there may be many more.

Phylum Rotifera: microscopic, wormlike or spherical pseudocoelomate animals, "wheel animalcules." They have a complete digestive tract, flame cells, and a circle of cilia on the head, the beating of which suggests a wheel; males are minute and either degenerate or unknown in many species. About 1,500 to 2,000 species.

Phylum Entoprocta: microscopic, stalked, sessile animals that superficially resemble hydrozoans but are much more complex, having three embryonic tissue layers, a pseudocoelom, and a complete digestive tract. They were long misclassified with the coelomate bryozoans (ectoprocts), which they also resemble. About 75 species.

Phylum Mollusca: unsegmented coelomate animals, with a head, a mantle, and a muscular foot, variously modified. They are mostly aquatic; soft-bodied, often with one or more hard shells, and a heart with three chambers. All mollusks, except bivalves, have a radula (rasplike organ used for scraping or marine drilling). At least 47,000 living species, and perhaps many more; an additional 35,000 species are known only from their fossils.

Class

Class Aplacophora: solenogasters. Wormlike marine animals, with no clearly defined shell, mantle, or foot. The presence of a radula identifies them as mollusks. About 250 species.

Class Polyplacophora: chitons. The mollusks with the closest resemblance to the hypothetical primitive form, they have an elongated body covered with a mantle in which are embedded eight dorsal shell plates. About 600 species.

Class Monoplacophora: Neopilina. Mostly deep-sea mollusks with a large, single dorsal shell and multiple pairs of gills, nephridia, and retractor muscles. Two genera with eight species.

Class Scaphopoda: tooth or tusk shells. They are marine mollusks with a conical tubular shell. About 350 species.

Class Bivalvia: two-shelled mollusks, including clams, oysters, mussels, scallops. They usually have a hatchet-shaped foot and no distinct head. Generally sessile. At least 7,500 species.

Class Gastropoda: asymmetrical mollusks, including snails, whelks, slugs. They usually have a spiral shell and a head with one or two pairs of tentacles. At least 37,500 living species and about 15,000 fossil species.

Class Cephalopoda: octopuses, squids, *Nautilus.* They are characterized by a "head-foot" with eight or ten arms or many tentacles, mouth with two horny jaws, and well-developed eyes and nervous system. The shell is external *(Nautilus),* internal (squid), or absent (octopus). All except *Nautilus* have ink glands. About 600 species.

Phylum

Phylum Annelida: segmented worms. They usually have a well-developed coelom, a one-way digestive tract, head, circulatory system, nephridia, and well-defined nervous system. About 9,000 species.

Class

Class Oligochaeta: soil, freshwater, and marine annelids, including *Lumbricus* and other earthworms. They have scanty bristles and usually a poorly differentiated head. About 3,000 species.

Class Polychaeta: mainly marine worms, such as *Nereis.* They have a distinct head with tentacles, antennae, and specialized mouthparts. Parapodia are often brightly colored. About 6,000 species.

Class Hirudinea: leeches. They have a posterior sucker and usually an anterior sucker surrounding the mouth. They are freshwater, marine, and terrestrial; either free-living or parasitic. About 300 species.

Phylum

Phylum Sipuncula: peanut worms. Unsegmented marine worms with a stout body, a long, retractile proboscis, and a trochophore larva that resembles those of the polychaete annelids. About 300 species.

Phylum Echiura: spoon worms. Marine worms with a nonretractile proboscis that contracts to form a structure resembling a spoon. Their embryonic development and trochophore larvae resemble those of the polychaete annelids. About 100 species.

Phylum Priapulida: predatory burrowing marine worms, characterized by a retractile, spine-bearing proboscis. They resemble the pseudocoelomate kinorhynchs, but may have a true coelom. Only 15 species.

Phylum Pogonophora: beard worms. Unsegmented except at the posterior end, these slender marine worms live in long tubes in deep-sea sediments. Although they have no mouth or digestive tract, the anterior region of the body bears a crown of tentacles. About 100 species.

Phylum Pentastomida: wormlike parasites of vertebrate respiratory systems, sometimes with two pairs of short appendages at the anterior end of the body. They lack circulatory, respiratory, and excretory systems, but the nervous system resembles those of annelids and arthropods. About 70 species.

Phylum Tardigrada: water bears. Tiny segmented animals, with a thin cuticle and four pairs of stubby legs. They are common in fresh water and in the moisture film on mosses; when water is unavailable, they enter a state of suspended animation. About 350 species.

Phylum Onychophora: *Peripatus.* Caterpillarlike animals with many short, unjointed pairs of legs. Their relatively soft bodies, segmentally arranged nephridia, muscular body walls, and ciliated reproductive tracts resemble those of the annelids; their antennae and eyes resemble those of both the polychaete annelids and the arthropods; their jaws, protective cuticle, brain, and circulatory and respiratory systems resemble those of arthropods. Most give birth to live young, and in some species the embryo is nourished through a placenta. About 70 species.

Phylum Arthropoda: The largest phylum in the animal kingdom, arthropods are segmented animals with paired, jointed appendages, a hard jointed exoskeleton, a complete digestive tract, reduced coelom, no nephridia, a dorsal brain, and a ventral nerve cord with paired ganglia in each segment. More than 1 million species have been classified to date.

Class

Class Merostomata: horseshoe crabs. Aquatic arthropods with chelicerae (pincers or fangs), pedipalps, compound eyes, four pairs of walking legs, and book gills. Only four species.

Class Pycnogonida: sea spiders. Aquatic chelicerates with slender bodies and four or, rarely, five pairs of legs, which are often very long. About 500 species.

Class Arachnida: spiders, mites, ticks, scorpions. Most members are terrestrial, air-breathing; four pairs of walking legs; chelicerae may be pincers or fangs; pedipalps are usually sensory. About 57,000 species.

Class Crustacea: lobsters, crayfish, crabs, shrimps. Crustaceans are mostly aquatic, with compound eyes, two pairs of antennae, one pair of mandibles, and typically two pairs of maxillae. The thoracic segments have appendages, and the abdominal segments are with or without appendages. About 25,000 species.

Class Chilopoda: centipedes. They have a head and 15 to 177 trunk segments, each with one pair of jointed appendages. About 3,000 species.

Class Diplopoda: millipedes. They have a head and a trunk with 20 to 200 body rings, each with two pairs of appendages. About 7,500 species.

Class Pauropoda: tiny, soft-bodied arthropods that resemble millipedes but have only 11 or 12 segments and 9 or 10 pairs of legs. They have branched antennae. About 300 species.

Class Symphyla: garden centipedes and their relatives. Soft-bodied arthropods with one pair of antennae, 12 pairs of jointed legs with claws, and a pair of unjointed posterior appendages. About 130 species.

Class Insecta: insects, including bees, ants, beetles, butterflies, fleas, lice, flies, etc. Most insects are terrestrial and breathe by means of tracheae. The body has three distinct parts (head, thorax, and abdomen); the head bears compound eyes and one pair of antennae; the thorax bears three pairs of legs and usually two pairs of wings. About 1 million species.

Phylum

Phylum Bryozoa (Ectoprocta): "moss" animals. These microscopic aquatic organisms are characterized by the lophophore, a crown of hollow, ciliated tentacles, with which they feed and by a U-shaped digestive tract. They usually form fixed and branching colonies. Coelomates, they superficially resemble the pseudocoelomate entoprocts. Some species retain larvae in a special brood pouch. About 4,000 species.

Phylum Phoronida: sedentary, elongated worms that secrete and live in a leathery tube. They have a U-shaped digestive tract and a ring of ciliated tentacles (the lophophore) with which they feed. Marine. Only 18 species.

Phylum Brachiopoda: lamp shells. Marine animals with two hard shells (one dorsal and one ventral), they superficially resemble clams. Fixed by a stalk or one shell in adult life, they feed by means of a lophophore. About 250 living species; 30,000 extinct species.

Phylum Echinodermata: starfishes and sea urchins. Echinoderms are radially symmetrical in adult stage, with well-developed coelomic cavities, an endoskeleton of calcareous ossicles and spines, and a unique water vascular system. They have tube feet. All are marine. About 6,000 living species and 20,000 species known only from their fossils.

Class

Class Crinoidea: sea lilies and feather stars. Sessile animals, they often have a jointed stalk for attachment, and they have 10 arms bearing many slender lateral branches. Most species are fossils; only 20 living species.

Class Stelleroidea: starfishes and brittle stars. They have 5 to 50 arms, an oral surface directed downward, and rows of tube feet on each arm. Brittle stars have greatly elongated, highly flexible slender arms and rapid horizontal locomotion.

Class Echinoidea: sea urchins and sand dollars. Skeletal plates form rigid external covering that bears many movable spines.

Class Concentricycloidea: sea daisies. The microscopic members of this newly established class have five plates on the dorsal surface, two rings of tube feet on the ventral surface, and no digestive tract.

Class Holothuroidea: sea cucumbers. They have a sausage-shaped or wormlike elongated body.

Phylum

Phylum Chaetognatha: arrow worms. Free-swimming, planktonic marine worms, they have a coelom, a complete digestive tract, and a mouth with strong sickle-shaped hooks on each side. About 60 species.

Phylum Hemichordata: acorn worms. The body is divided into three regions—proboscis, collar, and trunk. The coelomic cavities provide a hydrostatic skeleton similar to the water vascular system of echinoderms, and their larvae resemble starfish larvae. They have both ventral and dorsal nerve cords, and the anterior portion of the dorsal cord is hollow in some species. They also have a pharynx with gill slits. About 80 species.

Phylum Chordata: animals having at some stage a notochord, pharyngeal gill slits (or pouches), a hollow nerve cord on the dorsal side, and a tail. More than 45,000 species.

Subphylum

Subphylum Cephalochordata: lancelets. This small subphylum contains only *Branchiostoma* and related forms. They are somewhat fishlike marine animals with a permanent notochord the whole length of the body, a nerve cord, a pharynx with gill slits, and no cartilage or bone. About 28 species.

Subphylum Urochordata: tunicates. Adults are saclike, usually sessile, often forming branching colonies. They feed by ciliary currents, have gill slits, a reduced nervous system, and no notochord. Larvae are active, with well-developed nervous system and notochord. They are marine. About 1,300 species.

Subphylum Vertebrata: vertebrates. The major subphylum of Chordata. In the vertebrates the notochord is an embryonic structure; it is typically replaced in the course of development by cartilage or bone, forming the segmented vertebral column, or backbone. A cranium, or skull, surrounds a well-developed brain. Vertebrates usually have a tail. About 44,000 species.

Class

Class Agnatha: lampreys and hagfish. These are eel-like aquatic vertebrates without limbs, with a jawless sucking mouth, and no bones, scales, or fins. About 60 species.

Class Chondrichthyes: sharks, rays, skates, and other cartilaginous fishes. They have complicated copulatory organs, scales, and no swim bladders. They are almost exclusively marine. About 625 species.

Class Osteichthyes: bony fishes, including nearly all modern freshwater fishes, such as sturgeon, trout, perch, anglerfish, lungfish, and some almost extinct groups. They usually have a swim bladder or (rarely) a lung. More than 21,000 species.

Class Amphibia: salamanders, frogs, and toads. They usually respire by gills in the larval stage and by lungs in the adult stage. They have incomplete double circulation and a usually naked skin. The limbs are legs. They were the first vertebrates to inhabit the land and were ancestors of the reptiles. Their eggs are unprotected by a shell and embryonic membranes. About 2,500 species.

Class Reptilia: turtles, lizards, snakes, crocodiles; includes many extinct species such as the dinosaurs. Reptiles breathe by lungs and have incomplete double circulation. Their skin is usually covered with scales. The four limbs are legs (absent in snakes and some lizards). Almost all are ectotherms. Most live and reproduce on land, although some are aquatic. The embryo is enclosed in an egg shell and has protective membranes. About 6,000 species.

Class Aves: birds. Birds are homeothermic animals with complete double circulation and a skin covered with feathers. The forelimbs are wings. The embryo is enclosed in an egg shell with protective membranes. About 9,000 living species; an estimated 14,000 species are extinct.

Class Mammalia: mammals. Mammals are homeothermic animals with complete double circulation. Their skin is usually covered with hair. The young are nourished with milk secreted by the mother. They have four limbs, usually legs (forelimbs sometimes arms, wings, or fins), a diaphragm used in respiration, a lower jaw made up of a single pair of bones, three bones in each middle ear connecting eardrum and inner ear, and almost always seven vertebrae in the neck. About 4,500 species.

Subclass

Subclass Prototheria: monotremes. These are the oviparous (egg-laying) mammals with imperfect temperature regulation. There are only three living species: the duckbilled platypus and spiny anteaters of Australia and New Guinea.

Subclass Metatheria: marsupials, including kangaroos, opossums, and others. Marsupials are viviparous mammals, usually with a yolk-sac placenta; the young are born in a very immature state and are carried in an external pouch of the mother for some time after birth. They are found chiefly in Australia and South America. About 260 species.

Subclass Eutheria: mammals with a well-developed chorioallantoic placenta. This subclass comprises the great majority of living mammals. There are 16 orders of Eutheria:

Order

Insectivora: shrews, moles, hedgehogs, etc.

Chiroptera: bats. Aerial mammals with the forelimbs wings.

Dermoptera: "flying" lemurs.

Edentata: toothless mammals—anteaters, sloths, armadillos, etc.

Lagomorpha: rabbits and hares.

Carnivora: carnivorous animals—cats, dogs, bears, weasels, seals, etc.

Tubulidentata: aardvarks.

Pholidota: pangolins.

Rodentia: rodents—rats, mice, squirrels, etc.

Artiodactyla: even-toed ungulates (hoofed mammals)—cattle, deer, camels, hippopotamuses, etc.

Perissodactyla: odd-toed ungulates—horses, zebras, rhinoceroses, etc.

Proboscidea: elephants.

Hyracoidea: hyraxes.

Sirenia: manatees, dugongs, and sea cows. Large aquatic mammals with the forelimbs finlike, the hind limbs absent.

Cetacea: whales, dolphins, and porpoises. Aquatic mammals with the forelimbs fins, the hind limbs absent.

Primates: prosimians, monkeys, apes, and humans.

Glossary

This list does not include units of measure, which can be found inside the back cover, or names of taxonomic groups, which can be found in Appendix B, or terms that are used only once in the text and defined there.

abdomen: In vertebrates, the portion of the trunk containing visceral organs other than heart and lungs; in arthropods, the posterior portion of the body, made up of similar segments and containing the reproductive organs and part of the digestive tract.

abiotic [Gk. *a,* not, without + *bios,* life]: Nonliving; specifically, the nonliving components of an ecosystem, such as temperature, humidity, the mineral content of the soil, etc.

abscisic acid [L. *ab,* away, off + *scissio,* dividing]: A plant hormone with a variety of inhibitory effects; brings about dormancy in buds, maintains dormancy in many types of seeds, and effects stomatal closing; also known as the "stress hormone."

abscission [L. *ab,* away, off + *scissio,* dividing]: In plants, the dropping of leaves, flowers, fruits, or stems at the end of a growing season, as the result of formation of a two-layered zone of specialized cells (the abscission zone) and the action of a hormone (ethylene).

absorption [L. *absorbere,* to swallow down]: The movement of water and dissolved substances into a cell, tissue, or organism.

acetylcholine (asset-ill-**coal**-een): One of the principal chemicals (neurotransmitters) responsible for the transmission of nerve impulses across synapses.

acid [L. *acidus,* sour]: A substance that causes an increase in the relative number of hydrogen ions (H^+) in a solution and a decrease in the relative number of hydroxide ions (OH^-); having a pH of less than 7; the opposite of a base.

ACTH: Abbreviation of adrenocorticotropic hormone.

actin [Gk. *aktis,* a ray]: A protein, composed of globular subunits, that forms filaments that are among the principal components of the cytoskeleton. Also one of the two major proteins of muscle (the other is myosin); the principal constituent of the thin filaments.

action potential: A transient change in electric potential across a membrane; in nerve cells, results in conduction of a nerve impulse; in muscle cells, results in contraction.

activation energy: The energy that must be possessed by atoms or molecules in order to react.

active site: The region of an enzyme surface that binds the substrate during the reaction catalyzed by the enzyme.

active transport: The energy-requiring transport of a solute across a plasma membrane (or a membrane of an organelle) from a region of lower concentration to a region of higher concentration (that is, against a concentration gradient).

adaptation [L. *adaptare,* to fit]: (1) The evolution of features that make a group of organisms better suited to live and reproduce in their environment. (2) A peculiarity of structure, physiology, or behavior that aids the organism in its environment.

adaptive radiation: The evolution from a primitive and unspecialized ancestor of a number of divergent forms, each specialized to fill a distinct ecological niche; associated with the opening up of a new biological frontier.

adenosine diphosphate (ADP): A nucleotide consisting of adenine, ribose, and two phosphate groups; formed by the removal of one phosphate from an ATP molecule.

adenosine monophosphate (AMP): A nucleotide consisting of adenine, ribose, and one phosphate group; can be formed by the removal of two phosphates from an ATP molecule; in its cyclic form, functions as a "second messenger" for a number of vertebrate hormones and neurotransmitters.

adenosine triphosphate (ATP): The nucleotide that provides the energy currency for cell metabolism; composed of adenine, ribose, and three phosphate groups. On hydrolysis, ATP loses one phosphate group to become adenosine diphosphate (ADP), releasing energy in the process. ATP is formed from ADP and phosphate in an enzymatic reaction that traps the energy released by the breakdown of glucose and other molecules or the energy captured in photosynthesis.

ADH: Abbreviation of antidiuretic hormone.

adhesion [L. *adhaerere,* to stick to]: The holding together of molecules of different substances.

ADP: Abbreviation of adenosine diphosphate.

adrenal gland [L. *ad,* near + *renes,* kidney]: A vertebrate endocrine gland. The cortex (outer surface) is the source of cortisol, aldosterone, and other steroid hormones; the medulla (inner core) secretes adrenaline and noradrenaline.

adrenaline: A hormone, produced by the medulla of the adrenal gland, that increases the concentration of glucose in the blood, raises blood pressure and heartbeat rate, and increases muscular power and resistance to fatigue; also a neurotransmitter across synaptic junctions. Also called epinephrine.

adrenocorticotropic hormone (ACTH): A hormone, produced by the anterior lobe of the pituitary gland, that stimulates the production of cortisol by the adrenal cortex.

adventitious [L. *adventicius,* not properly belonging to]: Referring to a structure arising from an unusual place, such as roots growing from stems or leaves.

aerobic [Gk. *aēr,* air + *bios,* life]: Any biological process that can occur in the presence of molecular oxygen (O_2).

afferent [L. *ad,* near + *ferre,* to carry]: Bringing inward to a central part, applied to nerves and blood vessels.

agar: A gelatinous material prepared from certain red algae that is used to solidify nutrient media for growing microorganisms.

aldosterone [Gk. *aldainō,* to nourish + *stereō,* solid]: A hormone produced by the adrenal cortex that affects the concentration of ions in the blood; it stimulates the reabsorption of sodium and the excretion of potassium by the kidney.

aleurone layer [Gk. *aleuron*, flour]: The outermost cell layer of the endosperm of the grains (seeds) of wheat and other grasses; when acted upon by gibberellin, the aleurone layer releases enzymes that digest the stored food of the endosperm into small nutrient molecules that can be taken up by the embryo.

alga, *pl.* **algae** (al-gah, al-jee): A unicellular or simple multicellular eukaryotic photosynthetic organism lacking multicellular sex organs.

alkaline: Pertaining to substances that increase the relative number of hydroxide ions (OH$^-$) in a solution; having a pH greater than 7; basic; opposite of acidic.

allantois [Gk. *allant*, sausage]: One of the four extraembryonic membranes that form during the development of reptiles, birds, and mammals.

allele frequency: The proportion of a particular allele in a population.

alleles (al-eels) [Gk. *allelon*, of one another]: Two or more different forms of a gene. Alleles occupy the same position (locus) on homologous chromosomes and are separated from each other at meiosis.

allergic reaction: An inflammatory response triggered by a weak antigen (an allergen) to which most individuals do not react; involves the release of large amounts of histamine from mast cells.

allopatric speciation [Gk. *allos*, other + *patra*, fatherland, country]: Speciation that occurs as the result of the geographic separation of a population of organisms.

alternation of generations: A sexual life cycle in which a haploid *(n)* phase alternates with a diploid *(2n)* phase. The gametophyte *(n)* produces gametes *(n)* by mitosis. The fusion of gametes yields zygotes *(2n)*. Each zygote develops into a sporophyte *(2n)* that forms haploid spores *(n)* by meiosis. Each haploid spore forms a new gametophyte, completing the cycle.

altruism: Self-sacrifice for the benefit of others; any form of behavior that increases the fitness of the recipient while reducing the fitness of the altruistic individual.

alveolus, *pl.* **alveoli** [L. dim. of *alveus*, cavity, hollow]: One of the many small air sacs within the lungs in which the bronchioles terminate. The thin walls of the alveoli contain numerous capillaries and are the site of gas exchange between the air in the alveoli and the blood in the capillaries.

amino acids (am-ee-no) [Gk. *Ammon*, referring to the Egyptian sun god, near whose temple ammonium salts were first prepared from camel dung]: Organic molecules containing nitrogen in the form of —NH$_2$ and a carboxyl group, —COOH, bonded to the same carbon atom; the "building blocks" of protein molecules.

aminoacyl–tRNA synthetase: In protein synthesis, one of a class of enzymes that catalyze the attachment of an amino acid to its specific tRNA molecule.

ammonification: The process by which decomposers break down proteins and amino acids, releasing the excess nitrogen in the form of ammonia (NH$_3$) or ammonium ion (NH$_4{}^+$).

amnion (am-neon) [Gk. dim. of *amnos*, lamb]: One of the four extraembryonic membranes that form during the development of reptiles, birds, and mammals; it encloses a fluid-filled space, the amniotic cavity, that surrounds the developing embryo.

amniote egg: An egg that is isolated and protected from the environment during the period of its development by a series of extraembryonic membranes and, often, a more or less impervious shell; the amniote eggs of birds and many reptiles are completely self-sufficient, requiring only oxygen from the outside.

amoeboid [Gk. *amoibē*, change]: Moving or feeding by means of pseudopodia (temporary cytoplasmic protrusions from the cell body).

AMP: Abbreviation of adenosine monophosphate.

anabolic steroids: Synthetic chemical variants of the male sex hormone testosterone; they produce increased muscle mass but also suppress testosterone production, leading to shrinkage of the testes, growth of the breasts, and premature baldness; long-term use increases the risk of kidney and liver damage and of liver cancer.

anabolism [Gk. *ana*, up + -*bolism* (as in metabolism)]: Within a cell or organism, the sum of all biosynthetic reactions (that is, chemical reactions in which larger molecules are formed from smaller ones).

anaerobic [Gk. *an*, without + *aēr*, air + *bios*, life]: Applied to a process that can occur without oxygen, such as fermentation; also applied to organisms that can live without free oxygen.

analogous [Gk. *analogos*, proportionate]: Applied to structures similar in function but different in evolutionary origin, such as the wing of a bird and the wing of an insect.

anaphase (anna-phase) [Gk. *ana*, up + *phasis*, form]: In mitosis and meiosis II, the stage in which the chromatids of each chromosome separate and move to opposite poles; in meiosis I, the stage in which homologous chromosomes separate and move to opposite poles.

androgens [Gk. *andros*, man + *genos*, origin, descent]: Male sex hormones; any chemical with actions similar to those of testosterone.

angiosperms (an-jee-o-sperms) [Gk. *angeion*, vessel + *sperma*, seed]: The flowering plants. Literally, a seed borne in a vessel; thus, any plant whose seeds are borne within a matured ovary (fruit).

annual plant [L. *annus*, year]: A plant that completes its life cycle (from seed germination to seed production) and dies within a single growing season.

antennae: Long, paired sensory appendages on the head of many arthropods.

anterior [L. *ante*, before, toward, in front of]: The front end of an organism.

anther [Gk. *anthos*, flower]: In flowering plants, the pollen-bearing portion of a stamen.

antheridium, *pl.* **antheridia:** In bryophytes and some vascular plants, the multicellular sperm-producing organ.

anthropoid [Gk. *anthropos*, man, human]: A higher primate; includes monkeys, apes, and humans.

antibiotic [Gk. *anti*, against + *bios*, life]: An organic compound, inhibitory or toxic to other species, that is formed and secreted by an organism.

antibody [Gk. *anti*, against]: A globular protein, synthesized by a B lymphocyte, that is complementary to a foreign substance (antigen) with which it combines specifically.

anticodon: In a tRNA molecule, the three-nucleotide sequence that base pairs with the mRNA codon for the amino acid carried by that particular tRNA; the anticodon is complementary to the mRNA codon.

antidiuretic hormone (ADH) [Gk. *anti*, against + *diurgos*, thoroughly wet + *hormaein*, to excite]: A peptide hormone synthesized in the hypothalamus that inhibits urine excretion by inducing the reabsorption of water from the nephrons of the kidneys; also called vasopressin.

antigen [Gk. *anti*, against + *genos*, origin, descent]: A foreign substance, usually a protein or polysaccharide, that, when bound to a complementary antibody displayed on the surface of a B lymphocyte or to a complementary T-cell receptor, stimulates an immune response.

aorta (a-ore-ta) [Gk. *aeirein*, to lift, heave]: The major artery in blood-circulating systems; the aorta sends blood to the other body tissues.

apical dominance [L. *apex*, top]: In plants, the hormone-mediated influence of a terminal bud in suppressing the growth of axillary buds.

apical meristem [L. *apex*, top + Gk. *meristos*, divided]: In vascular plants, the growing point at the tip of the root or stem.

arboreal [L. *arbor*, tree]: Tree-dwelling.

archegonium, *pl.* **archegonia** [Gk. *archegonos*, first of a race]: In bryophytes and some vascular plants, the multicellular egg-producing organ.

archenteron [Gk. *arch*, first, or main + *enteron*, gut]: The main cavity within the early embryo (gastrula) of many animals; lined with endoderm, it opens to the outside by means of the blastopore and ultimately becomes the digestive tract.

artery: A vessel carrying blood from the heart to the tissues; arteries are usually thick-walled, elastic, and muscular. A small artery is known as an arteriole.

artificial selection: The breeding of selected organisms for the purpose of producing descendants with desired characteristics.

asexual reproduction: Any reproductive process, such as budding or the division of a cell or body into two or more approximately equal parts, that does not involve the union of gametes.

assimilation: The energy-requiring process by which plant cells convert nitrate ions (NO_3^-) taken up by the roots of plants into ammonium ions (NH_4^+), which can then be used in the synthesis of amino acids and other nitrogenous compounds.

atmospheric pressure [Gk. *atmos*, vapor + *sphaira*, globe]: The weight of the Earth's atmosphere over a unit area of the Earth's surface.

atom [Gk. *atomos*, indivisible]: The smallest particle into which a chemical element can be divided and still retain the properties characteristic of the element; consists of a central core, the nucleus, containing protons and neutrons, and electrons that move around the nucleus.

atomic number: The number of protons in the nucleus of an atom; equal to the number of electrons in the neutral atom.

atomic weight: The number of protons plus neutrons in the nucleus of an atom.

ATP: Abbreviation of adenosine triphosphate, the principal energy-carrying compound of the cell.

ATP synthase: The enzyme complex in the inner membrane of the mitochondrion and the thylakoid membrane of the chloroplast through which protons flow down the gradient established in the first stage of chemiosmotic coupling; the site of formation of ATP from ADP and phosphate during oxidative phosphorylation and photophosphorylation.

atrioventricular node [L. *atrium*, yard, court, hall + *ventriculus*, the stomach + *nodus*, knot]: A group of slow-conducting fibers in the atrium of the vertebrate heart that are stimulated by impulses originating in the sinoatrial node (the pacemaker) and that conduct impulses to the bundle of His, a group of fibers that stimulate contraction of the ventricles.

atrium, *pl.* **atria** (a-tree-um) [L., yard, court, hall]: A thin-walled chamber of the heart that receives blood and passes it on to a thick, muscular ventricle.

autoimmune disease: A disorder, such as multiple sclerosis, in which an individual's immune system makes antibodies against his or her own cells.

autonomic nervous system [Gk. *autos*, self + *nomos*, usage, law]: In the peripheral nervous system of vertebrates, the neurons and ganglia that are not ordinarily under voluntary control; innervates the heart, glands, visceral organs, and smooth muscle; subdivided into the sympathetic and parasympathetic divisions.

autosome [Gk. *autos*, self + *soma*, body]: Any chromosome other than the sex chromosomes. Humans have 22 pairs of autosomes and one pair of sex chromosomes.

autotroph [Gk. *autos*, self + *trophos*, feeder]: An organism that is able to synthesize all needed organic molecules from simple inorganic substances (e.g., H_2O, CO_2, NH_3) and some energy source (e.g., sunlight); in contrast to heterotroph. Plants, algae, and some groups of prokaryotes are autotrophs.

auxin [Gk. *auxein*, to increase + *in*, of, or belonging to]: One of a group of plant hormones with a variety of growth-regulating effects, including promotion of cell elongation.

axillary bud [Gk. *axilla*, armpit]: In plants, a bud located just above the point at which a leaf is (or was) attached to the stem; axillary buds give rise to lateral branches or specialized shoots, such as flowers.

axis: An imaginary line passing through a body or organ around which parts are symmetrically aligned.

axon [Gk. *axon*, axle]: A long process of a neuron, or nerve cell, that is capable of rapidly conducting nerve impulses over great distances; a nerve fiber.

B cell: A type of white blood cell capable of becoming an antibody-secreting plasma cell; a B lymphocyte.

bacteriophage [L. *bacterium* + Gk. *phagein*, to eat]: A virus that parasitizes a bacterial cell.

bark: In plants, all tissues outside the vascular cambium in a woody stem.

basal body [Gk. *basis*, foundation]: A cytoplasmic organelle of animals and some protists, from which cilia or flagella arise; identical in structure to the centriole, which is involved in mitosis and meiosis in animals and some protists.

basal metabolism [Gk. *basis*, foundation + *metabole*, change]: The energy expenditure of an animal at rest.

base: A substance that causes an increase in the relative number of hydroxide ions (OH^-) in a solution and a decrease in the relative number of hydrogen ions (H^+); having a pH of more than 7; the opposite of an acid. *See* Alkaline.

base-pairing principle: In the formation of nucleic acids, the requirement that adenine must always pair with thymine (or uracil) and guanine with cytosine.

behavior: All of the acts an organism performs, as in, for example, seeking a suitable habitat, obtaining food, avoiding predators, and seeking a mate and reproducing.

biennial [L. *biennium*, a space of two years; *bi*, twice + *annus*, year]: Occurring once in two years; a plant that requires two years to complete its reproductive cycle; vegetative growth occurs in the first year, sexual reproduction and death in the second.

bilateral symmetry [L. *bi*, twice, two + *lateris*, side; Gk. *summetros*, symmetry]: A body form in which the right and left halves of an organism are approximate mirror images of each other.

bile: A yellow secretion of the vertebrate liver, temporarily stored in the gallbladder and composed of organic salts that emulsify fats in the small intestine.

binomial system [L. *bi*, twice, two + Gk. *nomos*, usage, law]: A system of naming organisms in which the name consists of two parts, with the first designating genus and the second, species; originated by Linnaeus.

biochemical pathway [Gk. *bios*, life + *chēmeia*, alchemy]: An ordered series of chemical reactions in a living cell, in which each step is catalyzed by a specific enzyme; different biochemical pathways serve different functions in the life of the cell.

biogeochemical cycle [Gk. *bios*, life + *geō*, earth + *chēmeia*, alchemy; *kyklos*, circle, wheel]: The cyclic path of an inorganic substance, such as carbon or nitrogen, through an ecosystem. Its geological components are the atmosphere, the crust of the Earth, and the oceans, lakes, and rivers; its biological components are producers, consumers, and detritivores, including decomposers.

biological clock [Gk. *bios*, life + *logos*, discourse]: Proposed internal factor(s) in organisms that governs functions that occur rhythmically in the absence of external stimuli.

biomass [Gk. *bios*, life]: Total dry weight of all organisms (or some group of organisms) living in a particular habitat or place.

biome: One of the major types of distinctive plant formations; for example, the temperate grassland biome, the tropical rain forest biome, etc.

biosphere [Gk. *bios*, life + *sphaira*, globe]: The zones of air, land, and water at the surface of the Earth occupied by living things.

biosynthesis [Gk. *bios*, life + *synthesis*, a putting together]: Formation by living organisms of organic compounds from elements or simple compounds.

biotic [Gk. *bios*, life]: Of or pertaining to life; specifically, the living components of an ecosystem.

bipedal [L. *bi*, twice, two + *pes*, foot]: Walking upright on two feet.

blade: (1) The broad, expanded part of a leaf. (2) The broad, expanded photosynthetic part of the thallus of a multicellular alga or a simple plant.

blastocoel [Gk. *blastos*, sprout + *koilos*, a hollow]: The fluid-filled cavity in the interior of a blastula.

blastocyst [Gk. *blastos*, sprout + *kystis*, sac]: The blastula stage of a developing mammal; consists of an inner cell mass that will give rise to the embryo proper and a double layer of cells, the trophoblast, that is the precursor of the chorion.

blastodisc [Gk. *blastos*, sprout + *discos*, a round plate]: Disklike area on the surface of a large, yolky egg that undergoes cleavage and gives rise to the embryo.

blastopore [Gk. *blastos*, sprout + *poros*, a way, means, path]: In the gastrula stage of an embryo, the opening that connects the archenteron with the outside; represents the future mouth in some animals (protostomes), the future anus in others (deuterostomes).

blastula [Gk. *blastos*, sprout]: An animal embryo after cleavage and before gastrulation; usually consists of a fluid-filled sphere, the walls of which are composed of a single layer of cells.

bond strength: The strength with which a chemical bond holds two atoms together; conventionally measured in terms of the amount of energy, in kilocalories per mole, required to break the bond.

botany [Gk. *botanikos*, of herbs]: The study of plants.

Bowman's capsule: In the vertebrate kidney, the bulbous unit of the nephron, which surrounds the glomerulus. In filtration, the initial process in urine formation, blood plasma is forced from the glomerular capillaries into Bowman's capsule.

brainstem: The most posterior portion of the vertebrate brain; includes medulla, pons, and midbrain.

bronchus, *pl.* **bronchi** (bronk-us, bronk-eye) [Gk. *bronchos*, windpipe]: One of a pair of respiratory tubes branching into either lung at the lower end of the trachea; it subdivides into progressively finer passageways, the bronchioles, culminating in the alveoli.

bud: (1) In plants, an embryonic shoot, including rudimentary leaves, often protected by special bud scales. (2) In animals, an asexually produced outgrowth that develops into a new individual.

bulb: A modified bud with thickened leaves adapted for underground food storage.

bulk flow: The overall movement of a fluid induced by gravity, pressure, or an interplay of both.

bundle of His: In the vertebrate heart, a group of muscle fibers that carry impulses from the atrioventricular node to the walls of the ventricles; the only electrical bridge between the atria and the ventricles.

C₃ pathway: See Calvin cycle.

C₄ pathway: The set of reactions by which some plants initially fix carbon in the four-carbon compound oxaloacetic acid; the carbon dioxide is later released in the interior of the leaf and enters the Calvin cycle.

callus [L. *callos*, hard skin]: In plants, undifferentiated tissue; a term used in tissue culture, grafting, and wound healing.

calorie [L. *calor*, heat]: The amount of energy in the form of heat required to raise the temperature of 1 gram of water 1°C; in making metabolic measurements the kilocalorie (Calorie) is generally used. A Calorie is the amount of heat required to raise the temperature of 1 kilogram of water 1°C.

Calvin cycle: The set of reactions in which carbon dioxide is reduced to carbohydrate during the second stage of photosynthesis; the light-independent reactions.

calyx [Gk. *kalyx*, a husk, cup]: Collectively, the sepals of a flower.

CAM photosynthesis: *See* Crassulacean acid metabolism.

capillaries [L. *capillaris*, relating to hair]: Smallest thin-walled blood vessels through which exchanges between blood and the tissues occur; connect arteries with veins.

capillary action: The movement of water or any liquid along a surface; results from the combined effect of cohesion and adhesion.

capsid: The protein coat surrounding the nucleic acid core of a virus.

capsule (kap-sul) [L. *capsula*, a little chest]: (1) A slimy layer around the cells of certain bacteria. (2) The sporangium of a bryophyte.

carbohydrate [L. *carbo*, charcoal + *hydro*, water]: An organic compound consisting of a chain or ring of carbon atoms to which hydrogen and oxygen are attached in a ratio of approximately 2:1; carbohydrates include sugars, starch, glycogen, cellulose, etc.

carbon cycle: Worldwide circulation and reutilization of carbon atoms, chiefly due to metabolic processes of living organisms. Inorganic carbon, in the form of carbon dioxide, is incorporated into organic compounds by photosynthetic organisms; when the organic compounds are broken down in respiration, carbon dioxide is released. Large quantities of carbon are "stored" in the seas and the atmosphere, as well as in fossil fuel deposits.

carbon fixation: The light-independent reactions of the second stage of photosynthesis; energy stored in ATP and NADPH by the light-dependent reactions of the first stage is used to reduce carbon from carbon dioxide to simple sugars.

cardiovascular system [Gk. *kardio*, heart + L. *vasculum*, a small vessel]: In animals, the heart and blood vessels.

carnivore [L. *caro, carnis*, flesh + *voro*, to devour]: Predator that obtains its nutrients and energy by eating flesh.

carotenoids [L. *carota*, carrot]: A class of pigments that includes the carotenes (yellows, oranges, and reds) and the xanthophylls (yellow); accessory pigments in photosynthesis.

carpel [Gk. *karpos*, fruit]: A leaflike floral structure enclosing the ovule or ovules of angiosperms, typically divided into ovary, style, and stigma; a flower may have one or more carpels, either single or fused.

carrying capacity: In ecology, the average number of individuals of a particular population that the environment can support under a particular set of conditions.

cartilage [L. *cartilago*, gristle]: A connective tissue in skeletons of vertebrates; forms much of the skeleton of adult lower vertebrates and immature higher vertebrates.

Casparian strip (after Robert Caspary, German botanist): In the roots of plants, a thickened, waxy strip that extends around and seals the walls of endodermal cells, restricting the diffusion of solutes across the endodermis into the vascular tissues of the root.

catabolism [Gk. *katabole*, throwing down]: Within a cell or organism, the sum of all chemical reactions in which large molecules are broken down into smaller parts.

catalyst [Gk. *katalysis*, dissolution]: A substance that lowers the activation energy of a chemical reaction by forming a temporary association with the reacting molecules; as a result, the rate of the reaction is accelerated. Enzymes are catalysts.

category [Gk. *katēgoria*, category]: In a hierarchical classification system, the level at which a particular group is ranked.

cell [L. *cella*, a chamber]: The structural unit of organisms, surrounded by a plasma membrane and composed of cytoplasm and, in eukaryotes, one or more nuclei. In most plants, fungi, and bacteria, there is a cell wall outside the plasma membrane.

cell cycle: A regular, timed sequence of the events of cell growth and division through which dividing cells pass.

cell membrane: The outer membrane of the cell; the plasma membrane.

cell plate: In the dividing cells of most plants (and in some algae), a flattened structure that forms at the equator of the mitotic spindle in early telophase; gives rise to the middle lamella.

cell theory: All living things are composed of cells; cells arise only from other cells. No exception has been found to these two principles since they were first proposed well over a century ago.

cellulose [L. *cellula*, a little cell]: The chief constituent of the cell wall in all plants and some protists; an insoluble complex carbohydrate formed of microfibrils of glucose molecules.

cell wall: A plastic or rigid structure, produced by the cell and located outside the plasma membrane in most plants, algae, fungi, and prokaryotes; in plant cells, it consists mostly of cellulose.

central nervous system: In vertebrates, the brain and spinal cord; in invertebrates it usually consists of one or more cords of nervous tissue plus their associated ganglia.

centriole (sen-tree-ole) [Gk. *kentron*, center]: A cytoplasmic organelle identical in structure to a basal body; flagellated cells and all animal cells, including those without flagella, have centrioles at the spindle poles during division.

centromere (sen-tro-mere) [Gk. *kentron*, center + *meros*, a part]: Region of constriction of chromosome that holds sister chromatids together.

cerebellum [L. dim. of *cerebrum*, brain]: A subdivision of the vertebrate brain that lies above the brainstem and behind and below the cerebrum; functions in coordinating muscular activities and maintaining equilibrium.

cerebral cortex [L. *cerebrum*, brain]: A thin layer of neurons and glial cells forming the upper surface of the cerebrum, well developed only in mammals; the seat of conscious sensations and voluntary muscular activity.

cerebrum [L., brain]: The portion of the vertebrate brain occupying the upper part of the skull, consisting of two cerebral hemispheres united by the corpus callosum; coordinates most activities.

character displacement: A phenomenon in which species that live together in the same environment tend to diverge in those characteristics that overlap; exemplified by Darwin's finches.

chemical reaction: An interaction among atoms, ions, or molecules that results in the formation of new combinations of atoms, ions, or molecules; the making or breaking of chemical bonds.

chemiosmotic coupling: The mechanism by which ADP is phosphorylated to ATP in mitochondria and chloroplasts. The energy released as electrons pass down an electron transport chain is used to establish a proton gradient across an inner membrane of the organelle; when protons subsequently flow down this electrochemical gradient, the potential energy released is captured in the terminal phosphate bonds of ATP.

chemoreceptor: A sensory cell or organ that responds to the presence of a specific chemical stimulus; includes smell and taste receptors.

chemosynthetic: Applied to autotrophic bacteria that use the energy released by specific inorganic reactions to power their life processes, including the synthesis of organic molecules.

chitin (kye-tin) [Gk. *chitōn*, a tunic, undergarment]: A tough, resistant, nitrogen-containing polysaccharide present in the exoskeleton of arthropods, the epidermal cuticle or other surface structures of many other invertebrates, and in the cell walls of fungi.

chlorophyll [Gk. *chloros*, green + *phyllon*, leaf]: A class of green pigments that are the receptors of light energy in photosynthesis.

chloroplast [Gk. *chloros*, green + *plastos*, formed]: A membrane-bound, chlorophyll-containing organelle in eukaryotes (algae and plants) that is the site of photosynthesis.

chorion (core-ee-on) [Gk., skin, leather]: The outermost extraembryonic membrane of developing reptiles, birds, and mammals; in placental mammals it contributes to the structure of the placenta.

chromatid (crow-ma-tid) [Gk. *chrōma*, color]: Either of the two strands of a replicated chromosome, which are joined at the centromere.

chromatin [Gk. *chrōma*, color]: The deeply staining complex of DNA and histone proteins of which eukaryotic chromosomes are composed.

chromosome [Gk. *chrōma*, color + *soma*, body]: The structure that carries the genes. Eukaryotic chromosomes are visualized as threads or rods of chromatin, which appear in a contracted form during mitosis and meiosis, and are otherwise enclosed in a nucleus. Prokaryotic chromosomes consist of a closed circle of DNA with which a variety of proteins are associated. Viral chromosomes are linear or circular molecules of DNA or RNA.

chromosome map: A diagram of the linear order of the genes on a chromosome.

cilium, *pl.* **cilia** (silly-um) [L., eyelash]: A short, thin structure embedded in the surface of some eukaryotic cells, usually in large numbers and arranged in rows; has a highly characteristic internal structure of two inner microtubules surrounded by nine pairs of outer microtubules; involved in locomotion and the movement of substances across the cell surface.

circadian rhythms [L. *circa*, about + *dies*, day]: Regular rhythms of growth or activity that occur on an approximately 24-hour cycle.

cladogenesis [Gk. *clados*, branch + *genesis*, origin]: The splitting of an evolutionary lineage into two or more separate lineages; one of the principal patterns of evolutionary change.

class: A taxonomic grouping of related, similar orders; category above order and below phylum.

cleavage: The successive cell divisions of a fertilized egg of an animal to form a multicellular blastula.

cline [Gk. *klinein*, to lean]: A graded series of changes in some characteristic within a species, correlated with some gradual change in temperature, humidity, or other environmental factor over the geographic range of the species.

clone [Gk. *klon*, twig]: A line of cells, all of which have arisen from the same single cell by repeated cell divisions; a population of individuals derived by asexual reproduction from a single ancestor.

cnidocyte (ni-do-site) [Gk. *knide*, nettle + *kytos*, vessel]: A stinging cell containing a nematocyst; characteristic of cnidarians.

cochlea [Gk. *kochlias*, snail]: Part of the inner ear of mammals; concerned with hearing.

codominance: In genetics, the phenomenon in which the effects of both alleles at a particular locus are apparent in the phenotype of the heterozygote.

codon (code-on): Basic unit ("letter") of the genetic code; three adjacent nucleotides in a molecule of DNA or mRNA that form the code for a specific amino acid or for polypeptide chain termination.

coelom (see-loam) [Gk. *koilos*, a hollow]: A body cavity formed between layers of mesoderm and in which the digestive tract and other internal organs are suspended.

coenzyme [L. *co*, together + Gk. *en*, in + *zyme*, leaven]: A nonprotein organic molecule that plays an accessory role in enzyme-catalyzed processes, often by acting as a donor or acceptor of a substance involved in the reaction. NAD$^+$, FAD, and coenzyme A are common coenzymes.

coevolution [L. *co*, together + *e-*, out + *volvere*, to roll]: The simultaneous evolution of adaptations in two or more populations that interact so closely that each is a strong selective force on the other.

cofactor: A nonprotein component that plays an accessory role in enzyme-catalyzed processes; some cofactors are ions, and others are coenzymes.

cohesion [L. *cohaerere*, to stick together]: The attraction or holding together of molecules of the same substance.

cohesion-tension theory: A theory accounting for the upward movement of water in plants. According to this theory, transpiration of a water molecule results in a negative (below 1 atmosphere) pressure in the leaf cells, inducing the entrance from the vascular tissue of another water molecule, which, because of the cohesive property of water, pulls with it a chain of water molecules extending up from the cells of the root tip.

coleoptile (coal-ee-**op**-tile) [Gk. *koleon*, sheath + *ptilon*, feather]: The sheath enclosing the apical meristem and leaf primordia of a germinating monocot.

collagen [Gk. *kolla*, glue]: A fibrous protein in bones, tendons, and other connective tissues.

collenchyma [Gk. *kolla*, glue]: In plants, a type of supporting cell with an irregularly thickened primary cell wall; alive at maturity.

colony: A group of organisms of the same species living together in close association.

commensalism [L. *com*, together + *mensa*, table]: *See* Symbiosis.

community: All of the populations of organisms inhabiting a common environment and interacting with one another.

companion cell: In angiosperms, a specialized parenchyma cell associated with a sieve-tube member and arising from the same mother cell as the sieve-tube member.

competition: Interaction between members of the same population or of two or more populations using the same resource, often present in limited supply.

competitive exclusion: The hypothesis that two species with identical ecological requirements cannot coexist stably in the same locality and the species that is more efficient in utilizing the available resources will exclude the other; also known as Gause's principle, after the Russian biologist G. F. Gause.

complementary DNA (cDNA): DNA molecules synthesized by reverse transcriptase from an RNA template.

compound [L. *componere*, to put together]: A chemical substance composed of two or more kinds of atoms in definite ratios.

compound eye: In arthropods, a complex eye composed of many separate elements, each with light-sensitive cells and a lens that can form an image.

condensation: *See* Dehydration synthesis.

cone: (1) In plants, the reproductive structure of a conifer. (2) In vertebrates, a type of photoreceptor cell in the retina, concerned with the perception of color and with the most acute discrimination of detail.

conjugation [L. *conjugatio*, a joining, connection]: The sexual process in some unicellular organisms by which genetic material is transferred from one cell to another by cell-to-cell contact.

connective tissues: Supporting or packing tissues that lie between groups of nerves, glands, and muscle cells, and beneath epithelial cells, in which the cells are irregularly distributed through a relatively large amount of extracellular material; include bone, cartilage, blood, and lymph.

consumer, in ecological systems: A heterotroph that derives its energy from living or freshly killed organisms or parts thereof. Primary consumers are herbivores; higher-level consumers are carnivores.

continental drift: The gradual movement of the Earth's continents that has occurred over hundreds of millions of years.

continuous variation: A gradation of small differences in a particular trait, such as height, within a population; occurs in traits that are controlled by a number of genes.

convergent evolution [L. *convergere*, to turn together; *evolutio*, to unfold]: The independent development of similarities between unrelated groups, such as porpoises and sharks, resulting from adaptation to similar environments.

cork [L. *cortex*, bark]: A secondary tissue that is a major constituent of bark in woody and some herbaceous plants; made up of flattened cells, dead at maturity; restricts gas and water exchange and protects the vascular tissues from injury.

cork cambium [L. *cortex*, bark + *cambium*, exchange]: The lateral meristem that produces cork.

corolla (ko-**role**-a) [L. dim. of *corona*, wreath, crown]: Petals, collectively; usually the conspicuously colored flower parts.

corpus callosum [L., callous body]: In the vertebrate brain, a tightly packed mass of myelinated nerve fibers connecting the two cerebral hemispheres.

corpus luteum [L., yellowish body]: An ovarian structure that secretes estrogens and progesterone, which maintain the uterus during pregnancy. It develops from the remaining cells of the ruptured follicle following ovulation.

cortex [L., bark]: (1) The outer, as opposed to the inner, part of an organ, as in the adrenal gland. (2) In a stem or root, the primary tissue bounded externally by the epidermis and internally by the central cylinder of vascular tissue.

cortisol: A steroid hormone, produced by the adrenal cortex, that promotes the formation of glucose from protein and fat; also suppresses the inflammatory and immune responses.

cotyledon (cottle-ee-don) [Gk. *kotyledon*, a cup-shaped hollow]: A leaf-like structure of the embryo of a seed plant; contains stored food used during germination.

countercurrent exchange: An anatomical device for manipulating gradients so as to maximize uptake (or minimize loss) of O_2, heat, etc.

coupled reactions: In cells, the linking of endergonic (energy-requiring) reactions to exergonic (energy-releasing) reactions that provide enough energy to drive the endergonic reactions forward.

covalent bond [L. *con*, together + *valere*, to be strong]: A chemical bond formed as a result of the sharing of one or more pairs of electrons.

Crassulacean acid metabolism: A process by which some species of plants in hot, dry climates take in carbon dioxide during the night, fixing it in organic acids; the carbon dioxide is released during the day and used immediately in the Calvin cycle.

cristae: The "shelves" formed by the intricate folding of the inner membrane of the mitochondrion.

cross-fertilization: Fusion of gametes formed by different individuals; as opposed to self-fertilization.

crossing over: During meiosis, the exchange of genetic material between paired chromatids of homologous chromosomes.

cuticle (ku-tik-l) [L. *cuticula*, dim. of *cutis*, the skin]: (1) In plants, a layer of waxy substance (cutin) on the outer surface of epidermal cell walls. (2) In animals, the noncellular, outermost layer of many invertebrates.

cyclic AMP: A form of adenosine monophosphate (AMP) in which the atoms of the phosphate group form a ring; functions in chemical communication in slime molds and as a "second messenger" for a number of vertebrate hormones and neurotransmitters.

cytochromes [Gk. *kytos*, vessel + *chrōma*, color]: Heme-containing proteins that participate in electron transport chains; involved in cellular respiration and photosynthesis.

cytokinesis [Gk. *kytos*, vessel + *kinesis*, motion]: Division of the cytoplasm of a cell following nuclear division.

cytokinin [Gk. *kytos*, vessel + *kinesis*, motion]: One of a group of chemically related plant hormones that promote cell division, among other effects.

cytoplasm (**sight**-o-plazm) [Gk. *kytos*, vessel + *plasma*, anything molded]: The living matter within a cell, excluding the genetic material.

cytoskeleton: A network of filamentous protein structures within the cytoplasm that maintains the shape of the cell, anchors its organelles, and is involved in cell motility; includes microtubules, actin filaments, and intermediate filaments.

cytosol: A concentrated solution of ions, small molecules (such as amino acids, sugars, and ATP), and proteins that constitutes the fluid portion of the cytoplasm.

day-neutral plant: A plant that flowers without regard to the length of the light period to which it is exposed.

deciduous [L. *decidere*, to fall off]: Refers to plants that shed their leaves at a certain season.

decomposers: Specialized detritivores, usually bacteria or fungi, that consume such substances as cellulose and nitrogenous waste products. Their metabolic processes release inorganic nutrients, which are then available for reuse by plants and other organisms.

dehydration synthesis [L. *de,* from + *hydro,* water; Gk. *synthesis,* a putting together]: A type of chemical reaction in which two molecules join to form one larger molecule, simultaneously splitting out a molecule of water. The biosynthetic reactions in which monomers (e.g., monosaccharides, amino acids) are joined to form polymers (e.g., polysaccharides, polypeptides) are dehydration synthesis reactions.

dendrite [Gk. *dendron,* tree]: A process of a neuron, typically branched, that receives stimuli from other cells.

denitrification: The process by which certain bacteria living in poorly aerated soils break down nitrates, using the oxygen for their own respiration and releasing nitrogen back into the atmosphere.

density-dependent factors: Factors affecting the birth rate or death rate of a population, the effects of which vary with the density (number of individuals per unit area or volume) of the population; include resources for which members of the same or different populations compete, predation, and disease.

density-independent factors: Factors affecting the birth rate or death rate of a population, the effects of which are independent of the density of the population; often involve weather-related events.

deoxyribonucleic acid (DNA) (dee-ox-y-rye-bo-new-**clay**-ick): The carrier of genetic information in cells, composed of two complementary chains of nucleotides wound in a double helix; capable of self-replication as well as coding for RNA synthesis.

dermis [Gk. *derma,* skin]: The inner layer of the skin, beneath the epidermis.

desmosome [Gk. *desmos,* bond + *soma,* body]: A type of cell-cell junction that provides mechanical strength in animal tissues; consists of a plaque of dense fibrous material between adjacent cells, with clusters of filaments looping in and out from the cytoplasm of the two cells.

detritivores [L. *detritus,* worn down, worn away + *voro,* to devour]: Organisms that live on dead and discarded organic matter; include large scavengers, smaller animals such as earthworms and some insects, as well as decomposers (fungi and bacteria).

deuterostome [Gk. *deuteros,* second + *stoma,* mouth]: An animal in which the anus forms at or near the blastopore in the developing embryo and the mouth forms secondarily elsewhere; echinoderms and chordates are deuterostomes.

development: The progressive production of the phenotypic characteristics of a multicellular organism, beginning with the fertilization of an egg.

diaphragm [Gk. *diaphrassein,* to barricade]: In mammals, a sheetlike tissue (tendon and muscle) forming the partition between the abdominal and thoracic cavities; functions in breathing.

dicotyledon (dye-cottle-ee-don) [Gk. *di,* double, two + *kotyledon,* a cup-shaped hollow]: A member of the class of flowering plants having two seed leaves, or cotyledons, among other distinguishing features; often abbreviated as dicot.

differentiation: The developmental process by which a relatively unspecialized cell or tissue undergoes a progressive (usually irreversible) change to a more specialized cell or tissue.

diffusion [L. *diffundere,* to pour out]: The net movement of suspended or dissolved particles down a concentration gradient as a result of the random spontaneous movements of individual particles; the process tends to distribute the particles uniformly throughout a medium.

digestion [L. *digestio,* separating out, dividing]: The breakdown of complex, usually insoluble foods into molecules that can be absorbed into the body and used by the cells.

dioecious (dye-ee-shus) [Gk. *di,* two + *oikos,* house]: In angiosperms, having the male (staminate) and female (carpellate) flowers on different individuals of the same species.

diploid [Gk. *di,* double, two + *ploion,* vessel]: The condition in which each autosome is represented twice (2*n*); in contrast to haploid *(n).*

disaccharide [Gk. *di,* two + *sakcharon,* sugar]: A carbohydrate molecule composed of two monosaccharide subunits; examples are sucrose, maltose, and lactose.

diurnal [L. *diurnus,* of the day]: Applied to organisms that are active during the daylight hours.

division: A taxonomic grouping of related, similar classes; a high-level category below kingdom and above class. Division is generally used in the classification of prokaryotes, algae, fungi, and plants, whereas an equivalent category, phylum, is used in the classification of protozoa and animals.

DNA: Abbreviation of deoxyribonucleic acid.

dominant allele: An allele whose phenotypic effect is the same in both the heterozygous and homozygous conditions.

dormancy [L. *dormire,* to sleep]: A period during which growth ceases and metabolic activity is greatly reduced; dormancy is broken when certain requirements, for example, of temperature, moisture, or day length, are met.

dorsal [L. *dorsum,* the back]: Pertaining to or situated near the back; opposite of ventral.

double fertilization: A phenomenon unique to the angiosperms, in which the egg and one sperm nucleus fuse (resulting in a 2*n* fertilized egg, the zygote) and simultaneously the second sperm nucleus fuses with the two polar nuclei (resulting in a 3*n* endosperm nucleus).

duodenum (duo-dee-num) [L. *duodeni,* twelve each—from its length, about 12 fingers' breadth]: The upper portion of the small intestine in vertebrates, where food is digested into molecules that can be absorbed by intestinal cells.

ecological niche: A description of the roles and associations of a particular species in the community of which it is a part; the way in which an organism interacts with all of the biotic (living) and abiotic (nonliving) factors in its environment.

ecological pyramid: A graphic representation of the quantitative relationships of numbers of organisms, biomass, or energy flow between the trophic levels of an ecosystem. Because large amounts of energy and biomass are dissipated at every trophic level, these diagrams nearly always take the form of pyramids.

ecological succession: The gradual process by which the species composition of a community changes.

ecology [Gk. *oikos,* home + *logos,* a discourse]: The study of the interactions of organisms with their physical environment and with each other and of the results of such interactions.

ecosystem [Gk. *oikos,* home + *systema,* that which is put together]: The organisms in a community plus the associated abiotic (nonliving) factors with which they interact.

ecotype [Gk. *oikos,* home + L. *typus,* image]: A locally adapted variant of a species, differing genetically from other ecotypes of the same species.

ectoderm [Gk. *ecto,* outside + *derma,* skin]: One of the three embryonic tissue layers of animals; it gives rise to the outer covering of the body, the sensory receptors, and the nervous system.

ectotherm [Gk. *ecto,* outside + *therme,* heat]: An organism, such as a reptile, that maintains its body temperature by taking in heat from the environment or giving it off to the environment.

effector [L. *ex,* out of + *facere,* to make]: Cell, tissue, or organ (such as muscle or gland) capable of producing a response to a stimulus.

efferent [L. *ex,* out of + *ferre,* to bear]: Carrying away from a center, applied to nerves and blood vessels.

egg: A female gamete, which usually contains abundant cytoplasm and yolk; nonmotile and often larger than a male gamete.

electric potential: The difference in the amount of electric charge between a region of positive charge and a region of negative charge. The establishment of electric potentials across the plasma membrane and across organelle membranes makes possible a number of phenomena, including the chemiosmotic synthesis of ATP, the conduction of nerve impulses, and muscle contraction.

electron: A subatomic particle with a negative electric charge equal in magnitude to the positive charge of the proton but with a much smaller mass; the electrons of an atom move around its positively charged nucleus at almost the speed of light.

electron acceptor: Substance that accepts or receives electrons in an oxidation-reduction reaction, becoming reduced in the process.

electron carrier: A specialized molecule, such as a cytochrome, that can lose and gain electrons reversibly, alternately becoming oxidized and reduced.

electron donor: Substance that donates or gives up electrons in an oxidation-reduction reaction, becoming oxidized in the process.

electron transport: The movement of electrons down a series of electron-carrier molecules that hold electrons at slightly different energy levels; as electrons move down the chain, the energy released is used to form ATP from ADP and phosphate. Electron transport plays an essential role in the final stage of cellular respiration and in the light-dependent reactions of photosynthesis.

element: A substance composed only of atoms of the same atomic number and that cannot be decomposed by ordinary chemical means.

embryo [Gk. *en*, in + *bryein*, to swell]: The early developmental stage of an organism produced from a fertilized egg; a young organism before it emerges from the seed, egg, or body of its mother. In humans, refers to the first two months of intrauterine life. *See* Fetus.

embryo sac: The female gametophyte of a flowering plant, contained within an ovule; typically consists of seven cells with a total of eight haploid nuclei.

endergonic [Gk. *endon*, within + *ergon*, work]: Energy-requiring, as in a chemical reaction; applied to an "uphill" process.

endocrine gland [Gk. *endon*, within + *krinein*, to separate]: Ductless gland whose secretions (hormones) are released into the extracellular spaces, from which they diffuse into the circulatory system; in vertebrates, includes pituitary, sex glands, adrenal, thyroid, and others.

endocytosis [Gk. *endon*, within + *kytos*, vessel]: A cellular process in which material to be taken into the cell induces the plasma membrane to form a vacuole enclosing the material; the vacuole is released into the cytoplasm. Includes phagocytosis (endocytosis of solid particles), pinocytosis (endocytosis of liquids), and receptor-mediated endocytosis.

endoderm [Gk. *endon*, within + *derma*, skin]: One of the three embryonic tissue layers of animals; it gives rise to the epithelium that lines certain internal structures, such as most of the digestive tract and its outgrowths, most of the respiratory tract, and the urinary bladder, liver, pancreas, and some endocrine glands.

endodermis [Gk. *endon*, within + *derma*, skin]: In plants, a layer of specialized cells, one cell thick, that lies between the cortex and the vascular tissues in young roots. The Casparian strip of the endodermis prevents diffusion of solutes across the root.

endometrium [Gk. *endon*, within + *metrios*, of the womb]: The glandular lining of the uterus in mammals; thickens in response to secretion of estrogens and progesterone; one of its two principal layers is sloughed off in menstruation.

endoplasmic reticulum [Gk. *endon*, within + *plasma*, from cytoplasm; L. *reticulum*, network]: An extensive system of membranes present in most eukaryotic cells, dividing the cytoplasm into compartments and channels; often coated with ribosomes.

endorphin: One of a group of small peptides with morphine-like properties; produced by the vertebrate brain.

endosperm [Gk. *endon*, within + *sperma*, seed]: In plants, a triploid (3*n*) tissue containing stored food; develops from the union of a sperm nucleus and the two nuclei (the polar nuclei) of the central cell of the female gametophyte; found only in angiosperms.

endothelium [Gk. *endon*, + *thele*, nipple]: A type of epithelial tissue that forms the walls of the capillaries and the inner lining of arteries and veins.

endotherm [Gk. *endon*, within + *therme*, heat]: An organism, such as a bird or a mammal, that maintains its body temperature internally through metabolic processes. *See also* Homeotherm.

entropy [Gk. *en*, in + *trope*, turning]: A measure of the randomness or disorder of a system.

enzyme [Gk. *en*, in + *zyme*, leaven]: A globular protein molecule that accelerates a specific chemical reaction.

epidermis [Gk. *epi*, on or over + *derma*, skin]: In plants and animals, the outermost layers of cells.

epinephrine: *See* Adrenaline.

epistasis [Gk., a stopping]: Interaction between two nonallelic genes in which one of them interferes with or modifies the phenotypic expression of the other.

epithelial tissue [Gk. *epi*, on or over + *thele*, nipple]: In animals, a type of tissue that covers a body or structure or lines a cavity; epithelial cells form one or more regular layers with little intercellular material.

equilibrium [L. *aequus*, equal + *libra*, balance]: The state of a system in which no further net change is occurring; result of counterbalancing forward and backward processes.

equilibrium species: Species characterized by low reproduction rates, long development times, large body size, and long adult life with repeated reproductions.

erythrocyte (eh-**rith**-ro-site) [Gk. *erythros*, red + *kytos*, vessel]: Red blood cell, the carrier of hemoglobin.

estrogens [Gk. *oistros*, frenzy + *genos*, origin, descent]: Female sex hormones, which are the predominant secretions of the ovarian follicle during the preovulatory phase of the menstrual cycle; also produced by the corpus luteum and the placenta.

ethology [Gk. *ethos*, habit, custom + *logos*, discourse]: The comparative study of patterns of animal behavior, with emphasis on their adaptive significance and evolutionary origin.

ethylene: A simple hydrocarbon ($H_2C{=}CH_2$) that functions as a plant hormone; plays a role in fruit ripening and leaf abscission.

etiolation [Fr. *etioler*, to blanch]: In plants, a condition characterized by stem elongation, poor leaf development, and lack of chlorophyll; occurs in plants growing in the dark or with greatly reduced light.

eukaryote (you-**car**-ry-oat) [Gk. *eu*, good + *karyon*, nut, kernel]: A cell having a membrane-bound nucleus, membrane-bound organelles, and chromosomes in which DNA is combined with histone proteins; an organism composed of such cells.

eusocial [Gk. *eu*, good + L. *socius*, companion]: Applied to animal societies, such as those of certain insects, in which sterile individuals work on behalf of reproductive individuals.

evolution [L. *e-*, out + *volvere*, to roll]: Changes in the gene pool from one generation to the next as a consequence of processes such as mutation, natural selection, nonrandom mating, and genetic drift.

exergonic [Gk. *ex*, out of + *ergon*, work]: Energy-yielding, as in a chemical reaction; applied to a "downhill" process.

exocrine glands [Gk. *ex*, out of + *krinein*, to separate]: Glands, such as sweat glands and digestive glands, that secrete their products into ducts that empty onto surfaces, such as the skin, or into cavities, such as the interior of the stomach.

exocytosis [Gk. *ex*, out of + *kytos*, vessel]: A cellular process in which particulate matter or dissolved substances are enclosed in a vacuole and transported to the cell surface; there, the membrane of the vacuole fuses with the plasma membrane, expelling the vacuole's contents to the outside.

exon: A segment of DNA that is transcribed into RNA and expressed; dictates the amino acid sequence of part of a polypeptide.

exoskeleton: The outer supporting covering of the body; common in arthropods.

exponential growth: In populations, the increasingly accelerated rate of growth due to the increasing number of individuals being added to the reproductive base. Exponential growth is very seldom approached or sustained in natural populations.

expressivity: In genetics, the degree to which a particular genotype is expressed in the phenotype of individuals with that genotype.

extinct [L. *exstinctus*, to be extinguished]: No longer existing.

extraembryonic membranes: In reptiles, birds, and mammals, membranes formed from embryonic tissues that lie outside the embryo proper, protecting it and aiding metabolism; include amnion, chorion, allantois, and yolk sac.

F_1 (first filial generation): The offspring resulting from the crossing of plants or animals of a parental generation.

F_2 (second filial generation): The offspring resulting from crossing members of the F_1 generation among themselves.

facilitated diffusion: The transport of substances across the plasma membrane or across an organelle membrane from a region of higher concentration to a region of lower concentration by protein molecules embedded in the membrane; driven by the concentration gradient.

FAD: Abbreviation of flavin adenine dinucleotide, a coenzyme that functions as an electron acceptor in the Krebs cycle.

Fallopian tube: *See* Oviduct.

family: A taxonomic grouping of related, similar genera; the category below order and above genus.

fatty acid: A molecule consisting of a —COOH group and a long hydrocarbon chain; fatty acids are components of fats, oils, phospholipids, glycolipids, and waxes.

feedback systems: Control mechanisms whereby an increase or decrease in the level of a particular factor inhibits or stimulates the production, utilization, or release of that factor; important in the regulation of enzyme and hormone levels, ion concentrations, temperature, and many other factors.

fermentation: The breakdown of organic compounds in the absence of oxygen; yields less energy than aerobic processes.

fertilization: The fusion of two haploid gamete nuclei to form a diploid zygote nucleus.

fetus [L., pregnant]: An unborn or unhatched vertebrate that has passed through the earliest developmental stages; a developing human from about the second month of gestation until birth.

fibril [L. *fibra*, fiber]: Any minute, threadlike structure within a cell.

fibrous protein: Insoluble structural protein in which the polypeptide chain is coiled along one dimension. Fibrous proteins constitute the main structural elements of many animal tissues.

filament [L. *filare*, to spin]: (1) A chain of cells. (2) In flowers, the stalk of a stamen.

filtration: The first stage of kidney function; blood plasma is forced, under pressure, out of the glomerular capillaries into Bowman's capsule, through which it enters the renal tubule.

fitness: The genetic contribution of an individual to succeeding generations relative to the contributions of other individuals in the population.

fixed action pattern: A behavior that appears substantially complete the first time the organism encounters the relevant stimulus; tends to be highly stereotyped, rigid, and repetitive.

flagellum, *pl.* **flagella** (fla-jell-um) [L. *flagellum*, whip]: A long, threadlike organelle found in eukaryotes and used in locomotion and feeding; has an internal structure of nine pairs of microtubules encircling two central microtubules.

flower: The reproductive structure of angiosperms; a complete flower includes sepals, petals, stamens (male structures), and carpels (female structures).

food chain: A sequence of organisms related to one another as prey and predator.

food web: A set of interactions among organisms, including producers, consumers (herbivores and carnivores), and detritivores, through which energy and materials move within a community or ecosystem.

fossil [L. *fossilis*, dug up]: The remains of an organism, or direct evidence of its presence (such as tracks). May be an unaltered hard part (tooth or bone), a mold in a rock, petrification (wood or bone), unaltered or partially altered soft parts (a frozen mammoth).

founder effect: Type of genetic drift that occurs as the result of the founding of a population by a small number of individuals.

fovea [L., pit]: A small area in the center of the retina in which cones are concentrated; the area of sharpest vision.

frequency-dependent selection: A type of natural selection that decreases the frequency of more common phenotypes in a population and increases the frequency of less common phenotypes.

fruit [L. *fructus*, fruit]: In angiosperms, a matured, ripened ovary or group of ovaries and associated structures; contains the seeds.

function [L. *fungor*, to busy oneself]: Characteristic role or action of a structure or process in the normal metabolism or behavior of an organism.

gametangium, *pl.* **gametangia** [Gk. *gamein*, to marry + L. *tangere*, to touch]: A unicellular or multicellular structure in which gametes are produced.

gamete (gam-meet) [Gk., wife]: A haploid reproductive cell whose nucleus fuses with that of another gamete of an opposite mating type or sex (fertilization); the resulting cell (zygote) may develop into a new diploid individual or, in some protists and fungi, may undergo meiosis to form haploid somatic cells.

gametophyte: In organisms that have alternation of haploid and diploid generations (all plants and some green algae), the haploid *(n)* gamete-producing generation.

ganglion, *pl.* **ganglia** (gang-lee-on) [Gk. *ganglion*, a swelling]: Aggregation of nerve cell bodies; in vertebrates, refers to an aggregation of nerve cell bodies located outside the central nervous system.

gap junction: A junction between adjacent animal cells that allows the passage of materials between the cells.

gap phases: In the cell cycle, the phases that precede (G_1) and follow (G_2) the synthesis (S) phase in which DNA is replicated; in the G_1 phase, the cell doubles in size, and its enzymes, ribosomes, and other cytoplasmic molecules and structures increase in number; in the G_2 phase, the replicated chromosomes begin to condense and the structures required for mitosis or meiosis are assembled.

gastric [Gk. *gaster*, stomach]: Pertaining to the stomach.

gastrovascular cavity [Gk. *gaster*, stomach + L. *vasculum*, a small vessel]: A digestive cavity with only one opening, characteristic of the phyla Cnidaria (jellyfishes, *Hydra*, corals, etc.) and Ctenophora (comb jellies, sea walnuts); water circulating through the cavity supplies dissolved oxygen and carries away carbon dioxide and other waste products.

gastrula [Gk. *gaster*, stomach]: An animal embryo in the process of gastrulation; the stage of development during which the blastula, with its single layer of cells, turns into a three-layered embryo, made up of ectoderm, mesoderm, and endoderm, often enclosing an archenteron.

gated channel: In the plasma membrane of the axon, a protein that functions as a channel for the movement of ions into or out of the axon and that opens or closes as a result of changes in its shape.

Gause's principle: *See* Competitive exclusion.

gene [Gk. *genos*, birth, race; L. *genus*, birth, race, origin]: A unit of heredity in the chromosome; a sequence of nucleotides in a DNA molecule that performs a specific function, such as coding for an RNA molecule or a polypeptide.

gene flow: The movement of alleles into or out of a population.

gene pool: All the alleles of all the genes of all the individuals in a population.

genetic code: The system of nucleotide triplets in DNA and RNA that carries genetic information; referred to as a code because it determines the amino acid sequence in the enzymes and other protein molecules synthesized by the organism.

genetic drift: Evolution (change in allele frequencies) owing to chance processes.

genetic isolation: The absence of genetic exchange between populations or species as a result of geographic separation or of premating or postmating mechanisms (behavioral, anatomical, or physiological) that prevent reproduction.

genome: The complete set of chromosomes, with their associated genes.

genotype (jean-o-type): The genetic constitution of an individual cell or organism with reference to a single trait or a set of traits; the sum total of all the genes present in an individual.

genus, *pl.* **genera (jean-us)** [L. *genus,* race, origin]: A taxonomic grouping of closely related species.

geologic eras: See the inside of the front cover.

germ cells [L. *germinare,* to bud]: Gametes or the cells that give rise to gametes

germination [L. *germinare,* to bud]: In plants, the resumption of growth or the development from seed or spore.

gibberellins (jibb-e-rell-ins) [Fr. *gibberella,* genus of fungi]: A group of chemically related plant growth hormones, whose most characteristic effect is stem elongation in dwarf plants and bolting.

gill: The respiratory organ of aquatic animals, usually a thin-walled projection from some part of the external body surface or, in vertebrates, from some part of the digestive tract.

gland [L. *glans, glandis,* acorn]: A structure composed of modified epithelial cells specialized to produce one or more secretions that are discharged to the outside of the gland.

glial cell [Gk. *glia,* glue + L. *cella,* a chamber]: In nerve tissue, a type of cell that supports and insulates the neurons; Schwann cells, which form the myelin sheaths around axons in the peripheral nervous system of vertebrates, are glial cells.

globular protein [L. dim. of *globus,* a ball]: A polypeptide chain folded into a roughly spherical shape.

glomerulus (glom-mare-u-lus) [L. *glomus,* ball]: In the vertebrate kidney, a cluster of capillaries enclosed by Bowman's capsule; blood plasma minus large molecules filters through the walls of the glomerular capillaries into the renal tubule.

glucagon [Gk. *glykys,* sweet + *agō,* to lead toward]: A hormone, produced by the vertebrate pancreas, that acts to raise the concentration of glucose in the blood.

glucose [Gk. *glykys,* sweet]: A six-carbon sugar ($C_6H_{12}O_6$); the most common monosaccharide in animals.

glycerol: A three-carbon molecule with three hydroxyl (—OH) groups attached; a glycerol molecule can combine with three fatty acid molecules to form a fat or an oil.

glycogen [Gk. *glykys,* sweet + *genos,* race or descent]: A complex carbohydrate (polysaccharide); one of the main stored food substances of most animals and fungi; it is converted into glucose by hydrolysis.

glycolipids [Gk. *glykys,* sweet + *lipos,* fat]: Organic molecules similar in structure to fats, but in which a short carbohydrate chain rather than a fatty acid is attached to the third carbon of the glycerol molecule; as a result, the molecule has a hydrophilic "head" and a hydrophobic "tail." Glycolipids are important constituents of the plasma membrane and of organelle membranes.

glycolysis (gly-coll-y-sis) [Gk. *glykys,* sweet + *lysis,* loosening]: The process by which a glucose molecule is split anaerobically into two molecules of pyruvic acid with the liberation of a small amount of useful energy; catalyzed by enzymes in the cytosol.

glycoprotein [Gk. *glykys,* sweet + *proteios,* primary]: A large organic molecule consisting of a short carbohydrate chain covalently bonded to a protein; glycoproteins are important components of the plasma membrane and of organelle membranes; the carbohydrate chains protrude to the outside of the cell or organelle.

Golgi complex (goal-jee): An organelle present in many eukaryotic cells; consists of flat, membrane-bound sacs, tubules, and vesicles. It functions as a processing, packaging, and distribution center for substances that the cell manufactures.

gonad [Gk. *gone,* seed]: Gamete-producing organ of multicellular animals; ovary or testis.

gonadotropin [Gk. *gone,* seed + *trope,* a turning]: A vertebrate hormone, produced by the anterior lobe of the pituitary gland, that stimulates the gonads; follicle-stimulating hormone and luteinizing hormone.

granum, *pl.* **grana** [L., grain or seed]: In chloroplasts, stacked membrane-bound disks (thylakoids) that contain chlorophylls and carotenoids and are the sites of the light-dependent reactions of photosynthesis.

gravitropism [L. *gravis,* heavy + Gk. *trope,* turning]: The direction of growth or movement in which the force of gravity is the determining factor; also called geotropism.

gross productivity: A measure of the rate at which energy is assimilated by the organisms in a trophic level, a community, or an ecosystem.

ground meristem [Gk. *merizein,* to divide]: In plants, a primary meristematic tissue; gives rise to the ground tissues of the plant body.

ground tissues: In leaves and young roots and stems, all tissues other than the epidermis and the vascular tissues.

guard cells: Specialized epidermal cells surrounding a pore, or stoma, in a leaf or green stem; changes in turgor of a pair of guard cells cause opening and closing of the pore.

gymnosperm [Gk. *gymnos,* naked + *sperma,* seed]: A seed plant in which the seeds are not enclosed in an ovary; the conifers are the most familiar group.

habitat [L. *habitare,* to live in]: The place in which individuals of a particular species can usually be found.

habituation [L. *habitus,* condition]: A response to a repeated stimulus in which the stimulus comes to be ignored and a previous behavior pattern is restored; one of the simplest forms of learning.

half-life: The average time required for the disappearance or decay of one-half of any amount of a given substance.

haploid [Gk. *haploos,* single + *ploion,* vessel]: Having only one set of chromosomes (*n*), in contrast to diploid (*2n*); characteristic of eukaryotic gametes, of gametophytes in plants, and of some protists and fungi.

Hardy-Weinberg equilibrium: The steady-state relationship between relative frequencies of two or more alleles in an idealized population; both the allele frequencies and the genotype frequencies will remain constant from generation to generation in a population breeding at random in the absence of evolutionary forces.

heat of vaporization: The amount of heat required to change a given amount of a liquid into a gas; 540 calories are required to change 1 gram of liquid water into vapor.

heme [Gk. *haima,* blood]: The iron-containing group of heme proteins such as hemoglobin and the cytochromes.

hemoglobin [Gk. *haima,* blood + L. *globus,* a ball]: The iron-containing protein in vertebrate blood that carries oxygen.

hemophilia [Gk. *haima,* blood + *philios,* friendly]: A group of hereditary disorders characterized by failure of the blood to clot and consequent excessive bleeding from even minor wounds.

hepatic [Gk. *hēpatikos,* liver]: Pertaining to the liver.

herbaceous (her-bay-shus) [L. *herba,* grass]: In plants, nonwoody.

herbivore [L. *herba,* grass + *vorare,* to devour]: A consumer that eats plants or other photosynthetic organisms to obtain its food and energy.

heredity [L. *herres, heredis,* heir]: The transmission of characteristics from parent to offspring.

hermaphrodite [Gk. *Hermes* and *Aphrodite*]: An organism possessing both male and female reproductive organs; hermaphrodites may or may not be self-fertilizing.

heterotroph [Gk. *heteros,* other, different + *trophos,* feeder]: An organism that must feed on organic materials formed by other organisms in order to obtain energy and small building-block molecules; in contrast to autotroph. Animals, fungi, and many unicellular organisms are heterotrophs.

heterozygote [Gk. *heteros,* other + *zygōtos,* a pair]: A diploid organism that carries two different alleles at one or more genetic loci.

heterozygote advantage: The greater fitness of an organism heterozygous at a given genetic locus as compared with either homozygote.

hibernation [L. *hiberna,* winter]: A period of dormancy and inactivity, varying in length, depending on the species, and occurring in dry or cold seasons. During hibernation, metabolic processes are greatly slowed and, even in mammals, body temperature may drop to just above freezing.

histamine: A chemical, released by injured body cells and by certain types of white blood cells, that plays a role in the inflammatory response, increasing blood flow into the area and increasing the permeability of nearby capillaries; also released in large amounts in allergic reactions.

histones: A group of five relatively small, basic polypeptide molecules found bound to the DNA of eukaryotic cells.

HIV: Abbreviation of human immunodeficiency virus, the causative agent of AIDS.

homeobox: A sequence of some 180 nucleotides that has been identified in the DNA of a wide variety of animals that exhibit a segmental pattern at some stage of their development; codes for a polypeptide hypothesized to function as a regulatory molecule, altering the course of gene expression during development.

homeostasis (home-e-o-**stay**-sis) [Gk. *homos,* same or similar + *stasis,* standing]: Maintenance of a relatively stable internal physiological environment or internal equilibrium in an organism.

homeotherm [Gk. *homos,* same or similar + *therme,* heat]: An organism, such as a bird or mammal, capable of maintaining a stable body temperature independent of the environment.

hominid [L. *homo,* man]: Humans and closely related primates; includes modern and fossil forms, such as the australopithecines, but not the apes.

hominoid [L. *homo,* man]: Hominids and the apes.

homologues [Gk. *homologia,* agreement]: Chromosomes that carry corresponding genes and associate in pairs in the first stage of meiosis; each member of the pair is derived from a different parent.

homology [Gk. *homologia,* agreement]: Similarity in structure and/or position, assumed to result from a common ancestry, regardless of function, such as the wing of a bird and the foreleg of a mammal.

homozygote [Gk. *homos,* same or similar + *zygōtos,* a pair]: A diploid organism that carries identical alleles at one or more genetic loci.

hormone [Gk. *hormaein,* to excite]: An organic molecule secreted, usually in minute amounts, in one part of an organism that regulates the function of another tissue or organ.

host: (1) An organism on or in which a parasite lives. (2) A recipient of grafted tissue.

human immunodeficiency virus (HIV): The RNA retrovirus that is the causative agent of acquired immune deficiency syndrome (AIDS).

hybrid [L. *hybrida,* the offspring of a tame sow and a wild boar]: (1) Offspring of two parents that differ in one or more inheritable characteristics. (2) Offspring of two different varieties or of two different species.

hydrocarbon [L. *hydro,* water + *carbo,* charcoal]: An organic compound consisting of only carbon and hydrogen.

hydrogen bond: A weak molecular bond linking a hydrogen atom that is covalently bonded to another atom (usually oxygen, nitrogen, or fluorine) to another oxygen, nitrogen, or fluorine atom of the same or another molecule.

hydrolysis [L. *hydro,* water + Gk. *lysis,* loosening]: Splitting of one molecule into two by addition of H^+ and OH^- ions from water.

hydrophilic [L. *hydro,* water + Gk. *philios,* friendly]: Having an affinity for water; applied to polar molecules or polar regions of large molecules.

hydrophobic [L. *hydro,* water + Gk. *phobos,* fearing]: Having no affinity for water; applied to nonpolar molecules or nonpolar regions of molecules.

hypertonic [Gk. *hyper,* above + *tonos,* tension]: Of two solutions of different concentration, the solution that contains the higher concentration of solute particles; water moves across a selectively permeable membrane into a hypertonic solution.

hypha [Gk. *hyphe,* web]: A single tubular filament of a fungus or an oomycete; the hyphae together make up the mycelium, the matlike "body" of a fungus.

hypothalamus [Gk. *hypo,* under + *thalamos,* inner room]: The region of the vertebrate brain just below the thalamus; responsible for the integration of many basic behavioral patterns that involve correlation of neural and endocrine functions.

hypothesis [Gk. *hypo,* under + *tithenai,* to put]: A temporary working explanation or supposition based on accumulated facts and suggesting some general principle or relation of cause and effect; a postulated solution to a scientific problem that must be tested and if not validated, discarded.

hypotonic [Gk. *hypo,* under + *tonos,* tension]: Of two solutions of different concentration, the solution that contains the lower concentration of solute particles; water moves across a selectively permeable membrane from a hypotonic solution.

imbibition [L. *imbibere,* to drink in]: The capillary movement of water into germinating seeds and into substances such as wood and gelatin, which swell as a result.

immune response: A highly specific defensive reaction of the body to invasion by a foreign substance or organism; consists of a primary response in which the invader is recognized as foreign, or "not-self," and eliminated and a secondary response to subsequent attacks by the same invader. Mediated by two types of lymphocytes: B cells, which mature in the bone marrow and are responsible for antibody production, and T cells, which mature in the thymus and are responsible for cell-mediated immunity.

immunoglobulins: Complex, highly specific globular proteins synthesized by B cells; include both circulating antibodies and antibodies displayed on the surface of B cells prior to activation.

imprinting: A rapid and extremely narrow form of learning, common in birds and important in species recognition, that occurs during a very short critical period in the early life of an animal; depends on exposure to particular characteristics of the parent or parents.

inbreeding: The mating of individuals that are closely related genetically.

inclusive fitness: The relative number of an individual's alleles that are passed on from generation to generation, either as a result of his or her own reproductive success, or that of related individuals.

incomplete dominance: In genetics, the phenomenon in which the effects of both alleles at a particular locus are apparent in the phenotype of the heterozygote.

independent assortment: *See* Mendel's second law.

induction [L. *inducere,* to induce]: (1) In genetics, the phenomenon in which the presence of a substrate (the inducer) initiates transcription and translation of the genes coding for the enzymes required for its metabolism. (2) In embryonic development, the process in which one tissue or body part causes the differentiation of another tissue or body part.

inflammatory response: A nonspecific defensive reaction of the body to invasion by a foreign substance or organism; involves phagocytosis by white blood cells and is often accompanied by accumulation of pus and an increase in the local temperature.

innate releasing mechanism: In ethology, a circuit within an animal's brain that is hypothesized to respond to a specific stimulus, setting in motion, or "releasing," the sequence of movements that constitute a fixed action pattern.

insulin: A peptide hormone, produced by the vertebrate pancreas, that acts to lower the concentration of glucose in the blood.

interferon: A protein made by virus-infected cells that inhibits viral multiplication.

interleukin: A protein, produced by an activated helper T cell, that acts as a hormone, stimulating the differentiation and proliferation of both B cells and cytotoxic T cells following their activation.

intermediate filaments: Fibrous protein filaments that form part of the cytoskeleton; found in greatest density in cells subject to mechanical stress.

interneuron: Neuron that transmits signals from one neuron to another within a local region of the central nervous system; may receive signals from and transmit signals to many different neurons.

interphase: The portion of the cell cycle that occurs before mitosis or meiosis can take place; includes the G_1, S, and G_2 phases.

intron: A segment of DNA that is transcribed into RNA but is removed enzymatically from the RNA molecule before the mRNA enters the cytoplasm and is translated; also known as an intervening sequence.

invagination [L. *in,* in + *vagina,* sheath]: The local infolding of a layer of tissue, especially in animal embryos, so as to form a depression or pocket opening to the outside.

inversion: A chromosomal aberration in which a double break occurs and a segment is turned 180° before it is reincorporated into the chromosome.

ion (**eye**-on): Any atom or small molecule containing an unequal number of electrons and protons and therefore carrying a net positive or net negative charge.

ionic bond: A chemical bond formed as a result of the mutual attraction of ions of opposite charge.

isolating mechanisms: Mechanisms that prevent genetic exchange between individuals of different populations or species; they prevent mating or successful reproduction even when mating occurs; may be behavioral, anatomical, or physiological.

isotonic [Gk. *isos,* equal + *tonos,* tension]: Having the same concentration of solutes as another solution. If two isotonic solutions are separated by a selectively permeable membrane, there will be no net flow of water across the membrane.

isotope [Gk. *isos,* equal + *topos,* place]: Atom of an element that differs from other atoms of the same element in the number of neutrons in the atomic nucleus; isotopes thus differ in atomic weight. Some isotopes are unstable and emit radiation.

karyotype [Gk. *kara,* the head + *typos,* stamp or print]: The general appearance of the chromosomes of an organism with regard to number, size, and shape.

keratin [Gk. *karas,* horn]: One of a group of tough, fibrous proteins formed by certain epidermal tissues and especially abundant in skin, claws, hair, feathers, and hooves.

keystone species: A species that is of exceptional importance in maintaining the species diversity of a community; when a keystone species is lost, the diversity of the community decreases and its structure is significantly altered.

kidney: In vertebrates, the organ that regulates the balance of water and solutes in the blood and the excretion of nitrogenous wastes in the form of urine.

kinetic energy [Gk. *kinetikos,* putting in motion]: Energy of motion.

kinetochore [Gk. *kinetikos,* putting in motion + *choros,* chorus]: Disk-shaped protein structure within the centromere to which spindle fibers are attached during mitosis or meiosis.

kingdom: A taxonomic grouping of related, similar phyla or divisions; the highest-level category in biological classification.

kin selection: The differential reproduction of lineages of related individuals—that is, different groups of related individuals of a species reproduce at different rates; leads to an increase in the frequency of alleles shared by members of the groups with the greatest reproductive success.

Krebs cycle: Stage of cellular respiration in which acetyl groups are broken down into carbon dioxide; molecules reduced in the process can be used in ATP formation; also known as the citric acid cycle.

lamella (lah-**mell**-ah) [L. dim. of *lamina,* plate or leaf]: Layer, thin sheet.

larva [L., ghost]: An immature animal that is anatomically very different from the adult; examples are caterpillars and tadpoles.

lateral meristem [L. *latus, lateris,* side + Gk. *meristos,* divided]: In vascular plants, one of the two rings of tissue (vascular cambium and cork cambium) that produce new cells for secondary growth.

leaching: The dissolving of minerals and other elements in soil or rocks by the downward movement of water.

learning: The process that leads to modification in individual behavior as the result of experience.

leukocyte [Gk. *leukos,* white + *kytos,* vessel]: White blood cell; two principal types are phagocytic cells, such as macrophages, and lymphocytes (B cells and T cells).

leukotriene: A type of prostaglandin produced by various white blood cells involved in the inflammatory and immune responses and in allergic reactions.

lichen: Organism composed of a fungus and a green alga or a cyanobacterium that are symbiotically associated.

life cycle: The entire span of existence of any organism from time of zygote formation (or asexual reproduction) until it itself reproduces.

life-history pattern: A group of traits, such as size and number of offspring, length of maturation, age at first reproduction, and the number of times reproduction occurs, that affect reproduction, survival, and the rate of population growth.

light-dependent reactions: The reactions of the first stage of photosynthesis, in which light energy is captured by chlorophyll molecules and converted to chemical energy stored in ATP and NADPH molecules.

light-independent reactions: The carbon-fixing reactions of the second stage of photosynthesis; energy stored in ATP and NADPH by the light-dependent reactions is used to reduce carbon from carbon dioxide to simple sugars; light is not required for these reactions.

lignin: An organic molecule, found in the secondary cell walls of many vascular plants, that adds strength and rigidity to the walls.

limbic system [L. *limbus,* border]: Neuron network forming a loop around the inside of the brain and connecting the hypothalamus to the cerebral cortex; thought to be circuit by which drives and emotions are translated into complex actions and to play a role in the consolidation of memory.

linkage: The tendency for certain alleles to be inherited together because they are located on the same chromosome.

linkage group: A pair of homologous chromosomes.

lipid [Gk. *lipos,* fat]: One of a large variety of organic substances that are insoluble in polar solvents, such as water, but that dissolve readily in nonpolar organic solvents; includes fats, oils, waxes, steroids, phospholipids, glycolipids, and carotenes.

lipoprotein [Gk. *lipos,* fat + *proteios,* primary]: A large organic molecule consisting of a lipid covalently bonded to a protein.

locus, *pl.* **loci** [L., place]: In genetics, the position of a gene in a chromosome. For any given locus, there may be a number of possible alleles.

logistic growth: A pattern of population growth in which growth is rapid when the population is small, gradually slows as the population approaches the carrying capacity of its environment, and then oscillates as the population stabilizes at or near its maximum size; it is one of the simplest growth patterns observed for populations in nature.

long-day plant: A plant that must be exposed to a light period longer than some critical length for flowering to occur; long-day plants flower in spring or summer.

loop of Henle (after F. G. J. Henle, German pathologist): A hairpin-shaped portion of the renal tubule of mammals in which a hypertonic urine is formed by processes of diffusion and active transport.

lumen [L., light]: The cavity of a tubular structure, such as endoplasmic reticulum or a blood vessel.

lymph [L. *lympha*, water]: Colorless fluid derived from blood by filtration through capillary walls in the tissues; carried in special lymph ducts.

lymphatic system: The system through which lymph circulates; consists of lymph capillaries, which begin blindly in the tissues, and a network of progressively larger vessels that empty into the vena cava; also includes the lymph nodes, spleen, thymus, and tonsils.

lymph node [L. *lympha*, water + *nodus*, knot]: A mass of spongy tissues, separated into compartments; located throughout the lymphatic system, lymph nodes remove dead cells, debris, and foreign particles from the circulation; also are sites at which foreign antigens are displayed to immunologically active cells.

lymphocyte [L. *lympha*, water + Gk. *kytos*, vessel]: A type of white blood cell involved in the immune response; B cells differentiate into antibody-producing plasma cells, whereas cytotoxic T cells lyse diseased eukaryotic cells; other T cells interact with both cytotoxic T cells and with B cells.

lymphokine: A chemical, released by an activated cytotoxic T cell, that attracts macrophages and stimulates phagocytosis.

lysis [Gk. *lysis*, a loosening]: Disintegration of a cell by rupture of its plasma membrane.

lysogenic bacteria (lye-so-**jenn**-ick) [Gk. *lysis*, a loosening + *genos*, race or descent]: Bacteria carrying a bacteriophage integrated into the bacterial chromosome. The virus may subsequently set up an active cycle of infection, causing lysis of the bacterial cells.

lysosome [Gk. *lysis*, loosening + *soma*, body]: A membrane-bound organelle in which hydrolytic enzymes are sequestered.

macromolecule [Gk. *makros*, large + L. dim. of *moles*, mass]: An extremely large molecule; refers specifically to proteins, nucleic acids, polysaccharides, and complexes of these.

macronutrient [Gk. *makros*, large + L. *nutrire*, to nourish]: An inorganic nutrient required in large amounts for plant growth, such as nitrogen, potassium, calcium, phosphorus, magnesium, and sulfur.

macrophage [Gk. *makros*, large + *phagein*, to eat]: A type of phagocytic white blood cell important in both the inflammatory and immune responses.

major histocompatibility complex (MHC): In mammals, a group of at least 20 different genes, each with multiple alleles, coding for the protein components of the antigens that are displayed on nucleated cells and that serve to identify "self."

mantle: In mollusks, the outermost layer of the body wall or a soft extension of it; usually secretes a shell.

marine [L. *marini(us)*, from *mare*, the sea]: Living in salt water.

marsupial [Gk. *marsypos*, pouch, little bag]: A mammal in which the female has a ventral pouch or folds surrounding the nipples; the premature young leave the uterus and crawl into the pouch, where each one attaches itself by the mouth to a nipple until development is completed.

mast cell: A type of noncirculating white blood cell, found in connective tissue, that is the major protagonist in allergic reactions; when an allergen binds to complementary antibodies on the surface of a mast cell, large amounts of histamine are released from the cell.

matrix: The dense solution in the interior of the mitochondrion, surrounding the cristae; contains enzymes, phosphates, coenzymes, and other molecules involved in cellular respiration.

mechanoreceptor: A sensory cell or organ that receives mechanical stimuli such as those involved in touch, pressure, hearing, and balance.

medulla (med-**dull**-a) [L., the innermost part]: (1) The inner, as opposed to the outer, part of an organ, as in the adrenal gland. (2) The most posterior region of the vertebrate brain; connects with the spinal cord.

medusa: The free-swimming, bell- or umbrella-shaped stage in the life cycle of many cnidarians; a jellyfish.

megaspore [Gk. *megas*, great, large + *spora*, a sowing]: In plants, a haploid *(n)* spore that develops into a female gametophyte.

meiosis (my-o-sis) [Gk. *meioun*, to make smaller]: The two successive nuclear divisions in which a single diploid *(2n)* cell forms four haploid *(n)* nuclei, and segregation, crossing over, and reassortment of the alleles occur; gametes or spores may be produced as a result of meiosis.

memory cell: A type of white blood cell, produced by the differentiation of a B cell or a T cell following stimulation by a complementary antigen, that remains in the circulation indefinitely; enables the immune system to respond rapidly to any subsequent attack by the same invader.

Mendel's first law: The factors for a pair of alternative characters are separate, and only one may be carried in a particular gamete (genetic segregation). In modern form: Alleles segregate in meiosis.

Mendel's second law: The inheritance of a pair of factors for one trait is independent of the simultaneous inheritance of factors for other traits, such factors "assorting independently" as though there were no other factors present (later modified by the discovery of linkage). Modern form: The alleles of unlinked genes assort independently.

menstrual cycle [L. *mensis*, month]: In humans and certain other primates, the cyclic, hormone-regulated changes in the condition of the uterine lining; marked by the periodic discharge of blood and disintegrated uterine lining through the vagina.

meristem [Gk. *merizein*, to divide]: Undifferentiated plant tissue, including a mass of rapidly dividing cells, from which new tissues arise.

mesenteries [Gk. *mesos*, middle + *enteron*, gut]: Double layers of mesoderm that suspend the digestive tract and other internal organs within the coelom.

mesoderm [Gk. *mesos*, middle + *derma*, skin]: In animals, the middle layer of the three embryonic tissue layers. In vertebrates, includes the chordamesoderm, which gives rise to the notochord and skeletal muscle, and the lateral plate mesoderm, which gives rise to the circulatory system, most of the excretory and reproductive systems, the lining of the coelom, and the outer covering of the internal organs.

mesophyll [Gk. *mesos*, middle + *phyllon*, leaf]: The internal tissue of a leaf, sandwiched between two layers of epidermal cells; consists of palisade parenchyma and spongy parenchyma cells.

messenger RNA (mRNA): A class of RNA molecules, each of which is complementary to one strand of DNA and which serves to carry the genetic information from the chromosome to the ribosomes, where it is translated into protein.

metabolism [Gk. *metabole*, change]: The sum of all chemical reactions occurring within a cell or organism.

metamorphosis [Gk. *metamorphoun*, to transform]: Abrupt transition from larval to adult form, such as the transition from tadpole to adult frog.

metaphase [Gk. *meta*, middle + *phasis*, form]: The stage of mitosis or meiosis during which the chromosomes lie in the equatorial plane of the spindle.

MHC: Abbreviation of major histocompatibility complex.

microbe [Gk. *mikros*, small + *bios*, life]: A microscopic organism.

micronutrient [Gk. *mikros*, small + L. *nutrire*, to nourish]: An inorganic nutrient required in only minute amounts for plant growth, such as iron, chlorine, copper, manganese, zinc, molybdenum, and boron.

microspore [Gk. *mikros*, small + *spora*, a sowing]: In plants, a haploid (*n*) spore that develops into a male gametophyte; in seed plants, it becomes a pollen grain.

microtubule [Gk. *mikros*, small + L. dim. of *tubus*, tube]: An extremely small hollow tube composed of two types of globular protein subunits. Among their many functions, microtubules make up the internal structure of cilia and flagella.

middle lamella: In plants, distinct layer between adjacent cell walls, rich in pectins and other polysaccharides; derived from the cell plate.

mimicry [Gk. *mimos*, mime]: The superficial resemblance in form, color, or behavior of certain organisms (mimics) to other more powerful or more protected ones (models), resulting in protection, concealment, or some other advantage for the mimic.

mineral: A naturally occurring element or inorganic compound.

mitochondrion, *pl.* **mitochondria** [Gk. *mitos*, thread + *chondros*, cartilage or grain]: An organelle, bound by a double membrane, in which the reactions of the Krebs cycle, terminal electron transport, and oxidative phosphorylation take place, resulting in the formation of CO_2, H_2O, and ATP from acetyl CoA and ADP. Mitochondria are the organelles in which most of the ATP of the eukaryotic cell is produced.

mitosis [Gk. *mitos*, thread]: Nuclear division characterized by chromosome replication and formation of two identical daughter nuclei.

mole [L. *moles*, mass]: The amount of an element equivalent to its atomic weight expressed in grams, or the amount of a substance equivalent to its molecular weight expressed in grams.

molecular weight: The sum of the atomic weights of the constituent atoms in a molecule.

molecule [L. dim. of *moles*, mass]: A particle consisting of two or more atoms held together by chemical bonds; the smallest unit of a compound that displays the properties of the compound.

molting: Shedding of all or part of an organism's outer covering; in arthropods, periodic shedding of the exoskeleton to permit an increase in size.

monocotyledon [Gk. *monos*, single + *kotyledon*, a cup-shaped hollow]: A member of the class of flowering plants having one seed leaf, or cotyledon, among other distinguishing features; often abbreviated as monocot.

monoecious (mo-nee-shus) [Gk. *monos*, single + *oikos*, house]: In angiosperms, having the male and female structures (the stamens and the carpels, respectively) on the same individual but on different flowers.

monomer [Gk. *monos*, single + *meros*, part]: A simple, relatively small molecule that can be linked to others to form a polymer.

monosaccharide [Gk. *monos*, single + *sakcharon*, sugar]: A simple sugar, such as glucose, fructose, ribose.

monotreme [Gk. *monos*, single + *trēma*, hole]: A nonplacental mammal, such as the duckbilled platypus, in which the female lays shelled eggs and nurses the young.

morphogenesis [Gk. *morphe*, form + *genesis*, origin]: The development of size, form, and other structural features of organisms.

morphological [Gk. *morphe*, form + *logos*, discourse]: Pertaining to form and structure, at any level of organization.

mortality rate [L. *mors*, death]: Death rate.

motor neuron: Neuron that conducts nerve impulses from the central nervous system to an effector, which is typically a muscle or a gland.

muscle fiber: Muscle cell; a long, cylindrical, multinucleated cell containing numerous myofibrils, which is capable of contraction when stimulated.

mutagen [L. *mutare*, to change + *genus*, source or origin]: A chemical or physical agent that increases the mutation rate.

mutant [L. *mutare*, to change]: An organism carrying a gene that has undergone a mutation.

mutation [L. *mutare*, to change]: The change of a gene from one allelic form to another; an inheritable change in the DNA sequence of a chromosome.

mutualism [L. *mutuus*, lent, borrowed]: *See* Symbiosis.

mycelium [Gk. *mykes*, fungus]: The mass of hyphae forming the body of a fungus.

mycorrhizae [Gk. *mykes*, fungus + *rhiza*, root]: Symbiotic associations between particular species of fungi and the roots of vascular plants.

myelin sheath [Gk. *myelinos*, full of marrow]: A lipid-rich layer surrounding the long axons of neurons in the vertebrate nervous system; in the peripheral nervous system, made up of the plasma membranes of Schwann cells.

myofibril [Gk. *mys*, muscle + L. *fibra*, fiber]: Contractile element of a muscle fiber, made up of thick and thin filaments arranged in sarcomeres.

myoglobin [Gk. *mys*, muscle + L. *globus*, a ball]: An oxygen-binding, heme-containing globular protein found in muscles.

myosin [Gk. *mys*, muscle]: One of the principal proteins in muscle; makes up the thick filaments.

NAD: Abbreviation of nicotinamide adenine dinucleotide, a coenzyme that functions as an electron acceptor in glycolysis and the Krebs cycle.

NADP: Abbreviation of nicotinamide adenine dinucleotide phosphate, a coenzyme that functions as an electron acceptor in the light-dependent reactions of photosynthesis.

natural selection: A process of interaction between organisms and their environment that results in a differential rate of reproduction of different phenotypes in the population; can result in changes in the relative frequencies of alleles and genotypes in the population—that is, in evolution.

nectar [Gk. *nektar*, the drink of the gods]: A sugary fluid that attracts insects to plants.

negative feedback: A control mechanism whereby an increase in some substance inhibits the process leading to the increase; also known as feedback inhibition.

nematocyst [Gk. *nema, nematos*, thread + *kyst*, bladder]: A threadlike stinger, containing a poisonous or paralyzing substance, found in the cnidocyte of cnidarians.

nephridium, *pl.* **nephridia** [Gk. *nephros*, kidney]: A tubular excretory structure found in many invertebrates.

nephron [Gk. *nephros*, kidney]: The functional unit of the kidney in reptiles, birds, and mammals; a human kidney contains about 1 million nephrons.

nerve: A group or bundle of nerve fibers with accompanying connective tissue, located in the peripheral nervous system. A bundle of nerve fibers within the central nervous system is known as a tract.

nerve fiber: A filamentous process extending from the cell body of a neuron and conducting the nerve impulse; an axon.

nerve impulse: A rapid, transient, self-propagating change in electric potential across the membrane of an axon.

nervous system: All the nerve cells of an animal; the receptor-conductor-effector system; in humans, the nervous system consists of the central nervous system (brain and spinal cord) and the peripheral nervous system.

net primary production: In a community or an ecosystem, the increase in the amount of plant or algal material between the beginning and end of a specified time period, such as a growing season.

net productivity: In a trophic level, a community, or an ecosystem, the amount of energy (in calories) stored in chemical compounds or the increase in biomass (in grams or metric tons) in a particular period of time; it is the difference between gross productivity and the energy used by the organisms in respiration.

neural groove: Dorsal, longitudinal groove that forms in a vertebrate embryo; bordered by two neural folds; preceded by the neural-plate stage and followed by the neural-tube stage.

neural plate: Thickened strip of ectoderm in early vertebrate embryos that forms along the dorsal side of the body and gives rise to the central nervous system.

neural tube: Primitive, hollow, dorsal nervous system of the early vertebrate embryo; formed by fusion of neural folds around the neural groove.

neuromodulator: A chemical agent that is released by a neuron and diffuses through a local region of the central nervous system, acting on neurons within that region; generally has the effect of modulating the response to neurotransmitters.

neuromuscular junction: The junction between an axon terminal of a motor neuron and a muscle fiber innervated by that motor neuron; the axon terminal of a motor neuron is typically branched, forming neuromuscular junctions with a number of different muscle fibers.

neuron [Gk., nerve]: Nerve cell, including cell body, dendrites, and axon.

neurosecretory cell: A neuron that releases one or more hormones into the circulatory system.

neurotransmitter: A chemical agent that is released by a neuron at a synapse, diffuses across the synaptic cleft, and acts upon a postsynaptic neuron or muscle or gland cell and alters its electrical state or activity.

neutron (new-tron): An uncharged particle with a mass slightly greater than that of a proton. Found in the atomic nucleus of all elements except hydrogen, in which the nucleus consists of a single proton.

niche: *See* Ecological niche.

nitrification: The oxidation of ammonia or ammonium to nitrites and nitrates, as by nitrifying bacteria.

nitrogen cycle: Worldwide circulation and reutilization of nitrogen atoms, chiefly due to metabolic processes of living organisms; plants take up inorganic nitrogen and convert it into organic compounds (chiefly proteins), which are assimilated into the bodies of one or more animals; bacterial and fungal action on nitrogenous waste products and dead organisms return nitrogen atoms to the inorganic state.

nitrogen fixation: Incorporation of atmospheric nitrogen into inorganic nitrogen compounds available to plants, a process that can be carried out only by some soil bacteria, many free-living and symbiotic cyanobacteria, and certain symbiotic bacteria in association with legumes.

nitrogenous base: A nitrogen-containing molecule having basic properties (tendency to acquire an H^+ ion); a purine or pyrimidine.

nocturnal [L. *nocturnus*, of night]: Applied to organisms that are active during the hours of darkness.

node [L. *nodus*, knot]: (1) In plants, a joint of a stem; the place where branches and leaves are joined to the stem. (2) In the vertebrate axon, the uninsulated openings between segments of the myelin sheath; the nerve impulse jumps from node to node, greatly speeding its conduction.

nondisjunction [L. *non*, not + *disjungere*, to separate]: The failure of homologous chromosomes or of chromatids to separate during meiosis, resulting in one or more extra chromosomes in some gametes and correspondingly fewer in others.

noradrenaline: A hormone, produced by the medulla of the adrenal gland, that increases the concentration of glucose in the blood, raises blood pressure and heartbeat rate, and increases muscular power and resistance to fatigue; also one of the principal neurotransmitters; also called norepinephrine.

norepinephrine: *See* noradrenaline.

notochord [Gk. *noto*, back + L. *chorda*, cord]: A dorsal, rodlike structure that runs the length of the body and serves as the internal skeleton in the embryos of all chordates; in most adult chordates the notochord is replaced by a vertebral column that forms around (but not from) the notochord.

nuclear envelope [L. *nucleus*, a kernel]: The double membrane surrounding the nucleus within a eukaryotic cell.

nucleic acid (new-clay-ick): A macromolecule consisting of nucleotides; the principal types are deoxyribonucleic acid (DNA) and ribonucleic acid (RNA).

nucleoid: In prokaryotic cells, the region of the cell in which the chromosome is localized.

nucleolus (new-klee-o-lus) [L., a small kernel]: A small, dense region visible in the nucleus of nondividing eukaryotic cells; consists of rRNA molecules, ribosomal proteins, and loops of chromatin from which the rRNA molecules are transcribed.

nucleosome [L. *nucleus*, a kernel + Gk. *soma*, body]: A complex of DNA and histone proteins that forms the fundamental packaging unit of eukaryotic DNA; its structure resembles a bead on a string.

nucleotide [L. *nucleus*, a kernel]: A molecule composed of a phosphate group, a five-carbon sugar (either ribose or deoxyribose), and a purine or pyrimidine base; nucleotides are the building blocks of nucleic acids.

nucleus [L., a kernel]: (1) The central core of an atom, containing protons and neutrons, around which electrons move. (2) The membrane-bound structure characteristic of eukaryotic cells that contains the genetic information in the form of DNA organized into chromosomes. (3) A group of nerve cell bodies within the central nervous system.

omnivore [L. *omnis*, all + *vorare*, to devour]: An organism that "eats everything"; for example, an animal that eats both plants and meat.

oncogene [Gk. *onkos*, tumor + *genos*, birth, race]: One of a group of eukaryotic genes that closely resemble normal genes of the cells in which they are found and that are thought to play a role in the development of cancer.

oocyte (o-uh-sight) [Gk. *oion*, egg + *kytos*, vessel]: A cell that gives rise by meiosis to an ovum.

operator: A segment of DNA that interacts with a repressor protein to regulate the transcription of the structural genes of an operon.

operon [L. *opus*, *operis*, work]: In the bacterial chromosome, a segment of DNA consisting of a promoter, an operator, and a group of adjacent structural genes; the structural genes, which code for products related to a particular biochemical pathway, are transcribed onto a single mRNA molecule, and their transcription is regulated by a single repressor protein.

opportunistic species: Species characterized by high reproduction rates, rapid development, early reproduction, small body size, and uncertain adult survival.

orbital [L. *orbis*, circle, disk]: In the current model of atomic structure, the volume of space surrounding the atomic nucleus in which an electron will be found 90 percent of the time.

order: A taxonomic grouping of related, similar families; the category below class and above family.

organ [Gk. *organon*, tool]: A body part composed of several tissues grouped together in a structural and functional unit.

organelle [Gk. *organon*, instrument, tool]: A membrane-bound compartment in the cytoplasm of a cell.

organic [Gk. *organon*, instrument, tool]: Pertaining to (1) organisms or living things generally, or (2) compounds formed by living organisms, or (3) the chemistry of compounds containing carbon.

organism [Gk. *organon*, instrument, tool]: Any living creature, either unicellular or multicellular.

organogenesis [Gk. *organon*, instrument, tool + *genos*, origin, descent]: In vertebrates, the later stages of development, following cleavage and gastrulation, during which the organ systems form.

osmosis [Gk. *osmos*, impulse, thrust]: The diffusion of water across a selectively permeable membrane (a membrane that permits the free passage of water but prevents or retards the passage of a solute). In the absence of other factors that affect the water potential, the net movement of water is from the side containing a lower concentration of solute to the side containing a higher concentration.

osmotic potential [Gk. *osmos*, impulse, thrust]: The tendency of water to move across a selectively permeable membrane into a solution; it is determined by measuring the pressure required to stop the osmotic movement of water into the solution.

ovarian follicle [L. *ovum*, egg + *folliculus*, small ball]: A developing oocyte and the specialized cells surrounding it; located near the surface of the ovary; following ovulation, forms the corpus luteum.

ovary [L. *ovum*, egg]: (1) In flowering plants, the enlarged basal portion of a carpel or a fused carpel, containing the ovule or ovules; the ovary matures to become the fruit. (2) In animals, the egg-producing organ.

oviduct [L. *ovum*, egg + *ductus*, duct]: The tube serving to transport the eggs to the outside or to the uterus; also called uterine tube or Fallopian tube (in humans).

ovulation: In animals, release of an egg or eggs from the ovary.

ovule [L. dim. of *ovum*, egg]: In seed plants, a structure composed of a protective outer coat, a tissue specialized for food storage, and a female gametophyte with an egg cell; becomes a seed after fertilization.

ovum, *pl.* **ova** [L., egg]: The egg cell; female gamete.

oxidation: Gain of oxygen, loss of hydrogen, or loss of an electron by an atom, ion, or molecule. Oxidation and reduction take place simultaneously, with the electron lost by one reactant being transferred to another reactant.

oxidative phosphorylation: The process by which the energy released as electrons pass down the mitochondrial electron transport chain in the final stage of cellular respiration is used to phosphorylate (add a phosphate group to) ADP molecules, thereby yielding ATP molecules.

oxygen debt: In muscle, the cumulative deficit of oxygen that develops during strenuous exercise when the supply of oxygen is inadequate for the demand; ATP is produced anaerobically by glycolysis, and the resulting pyruvic acid is converted to lactic acid, which is subsequently metabolized when adequate oxygen is available.

pacemaker: *See* Sinoatrial node.

paleontology [Gk. *palaios*, old + *onta*, things that exist + *logos*, discourse]: The study of the life of past geologic times, principally by means of fossils.

palisade cells [L. *palus*, stake + *cella*, a chamber]: In plant leaves, the columnar, chloroplast-containing parenchyma cells of the mesophyll.

pancreas (pang-kree-us) [Gk. *pan*, all + *kreas*, meat, flesh]: In vertebrates, a small, complex gland located between the stomach and the duodenum, which produces digestive enzymes and the hormones insulin and glucagon.

parasite [Gk. *para*, beside, akin to + *sitos*, food]: An organism that lives on or in an organism of a different species and derives nutrients from it.

parasitism: *See* Symbiosis.

parasympathetic division [Gk. *para*, beside, akin to]: A subdivision of the autonomic nervous system of vertebrates, with centers located in the brain and in the most anterior and most posterior parts of the spinal cord; stimulates digestion; generally inhibits other functions and restores the body to normal following emergencies.

parathyroid glands [Gk. *para*, beside, akin to + *thyra*, a door]: Four pea-sized endocrine glands of vertebrates, located behind or within the thyroid gland; source of parathyroid hormone, which acts to increase the concentration of calcium in the blood.

parenchyma (pah-renk-ee-ma) [Gk. *para*, beside, akin to + *en*, in + *chein*, to pour]: A plant tissue composed of living, thin-walled, randomly arranged cells with large vacuoles; usually photosynthetic or storage tissue.

parthenogenesis [Gk. *parthenon*, virgin + *genesis*, birth]: The development of an organism from an unfertilized egg.

pathogen [Gk. *pathos*, suffering + *genos*, origin, descent]: An organism or a virus that causes disease.

pattern formation: The developmental process that gives rise to the shape and structure of specific body parts, such as the limbs.

penetrance: In genetics, the proportion of individuals with a particular genotype that show the phenotype ascribed to that genotype.

peptide bond [Gk. *pepto*, to soften, digest]: The type of bond formed when two amino acids are joined end to end; the acidic group (—COOH) of one amino acid is linked covalently to the basic group (—NH$_2$) of the next, and a molecule of water (H$_2$O) is removed.

perennial [L. *per*, through + *annus*, year]: A plant that persists in whole or in part from year to year and usually produces reproductive structures in more than one year.

pericycle [Gk. *peri*, around + *kyklos*, circle]: One or more layers of cells completely surrounding the vascular tissues of the root; branch roots arise from the pericycle.

peripheral nervous system [Gk. *peripherein*, to carry around]: All of the neurons and axons outside the central nervous system, including both motor neurons and sensory neurons; consists of the somatic nervous system and the autonomic nervous system.

peristalsis [Gk. *peristellein*, to wrap around]: Successive waves of muscular contraction in the walls of a tubular structure, such as the digestive tract or an oviduct; moves the contents, such as food or an egg cell, through the tube.

peritoneum [Gk. *peritonos*, stretched over]: A membrane that lines the body cavity and forms the external covering of the visceral organs.

peritubular capillaries [Gk. *peri*, around + L. *tubus*, tube]: In the vertebrate kidney, the capillaries that surround the renal tubule; water and solutes are reabsorbed into the bloodstream through the peritubular capillaries and some substances are secreted from them into the renal tubule.

permeable [L. *permeare*, to pass through]: Penetrable by molecules, ions, or atoms; usually applied to membranes that let given solutes pass through.

peroxisome: A membrane-bound organelle in which enzymes catalyzing peroxide-forming and peroxide-destroying reactions are segregated; in plant cells, the site of photorespiration.

petiole (pet-ee-ole) [Fr., from L. *petiolus*, dim. of *pes*, *pedis*, a foot]: The stalk of a leaf, connecting the blade of the leaf with the branch or stem.

pH: A symbol denoting the concentration of hydrogen ions in a solution; pH values range from 0 to 14; the lower the value, the more acidic a solution, that is, the more hydrogen ions it contains; pH 7 is neutral, less than 7 is acidic, more than 7 is alkaline.

phage: *See* Bacteriophage.

phagocytosis [Gk. *phagein*, to eat + *kytos*, vessel]: Cell "eating." *See* Endocytosis.

phenotype [Gk. *phainein*, to show + *typos*, stamp, print]: Observable characteristics of an organism, resulting from interactions between the genotype and the environment.

pheromone (fair-o-moan) [Gk. *phero*, to bear, carry]: Substance secreted by an animal that influences the behavior or development of other animals of the same species, such as the sex attractants of moths, the queen substance of honey bees.

phloem (flow-em) [Gk. *phloos*, bark]: Vascular tissue of higher plants; conducts sugars and other organic molecules from the leaves to other parts of the plant; in angiosperms, composed of sieve-tube members, companion cells, other parenchyma cells, and fibers.

phospholipids: Organic molecules similar in structure to fats, but in which a phosphate group rather than a fatty acid is attached to the third carbon of the glycerol molecule; as a result, the molecule has a hydrophilic "head" and a hydrophobic "tail." A double layer (bilayer) of phospholipids forms the fundamental structure of the plasma membrane and of organelle membranes.

phosphorylation: Addition of a phosphate group or groups to a molecule.

photoperiodism [Gk. *photos*, light]: The response to relative day and night length, a mechanism by which organisms measure seasonal change.

photophosphorylation [Gk. *photos*, light + *phosphoros*, bringing light]: The process by which the energy released as electrons pass down the electron transport chain between photosystems II and I during photosynthesis is used to phosphorylate ADP to ATP.

photoreceptor [Gk. *photos*, light]: A cell or organ capable of detecting light.

photorespiration [Gk. *photos*, light + L. *respirare*, to breathe]: The oxidation of carbohydrates in the presence of light and oxygen; occurs when the carbon dioxide concentration in the leaf is low in relation to the oxygen concentration.

photosynthesis [Gk. *photos*, light + *syn*, together + *tithenai*, to place]: The conversion of light energy to chemical energy; the synthesis of organic compounds from carbon dioxide and water in the presence of chlorophyll, using light energy.

photosystem [Gk. *photos*, light + *systema*, that which is put together]: A discrete unit of organization of chlorophyll and accessory pigments, embedded in the thylakoids of chloroplasts; the site of the light-dependent reactions of photosynthesis.

phototropism [Gk. *photos*, light + *trope*, turning]: Movement in which the direction of the light is the determining factor, such as the growth of a plant toward a light source; a curving response to light.

phyletic change [Gk. *phylon*, race, tribe]: The changes taking place in a single lineage of organisms over a long period of time; one of the principal patterns of evolutionary change.

phylogeny [Gk. *phylon*, race, tribe]: Evolutionary history of a taxonomic group. Phylogenies are often depicted as "evolutionary trees."

phylum, *pl.* **phyla** [Gk. *phylon*, race, tribe]: A taxonomic grouping of related, similar classes; a high-level category below kingdom and above class. Phylum is generally used in the classification of protozoa and animals, whereas an equivalent category, division, is used in the classification of prokaryotes, algae, fungi, and plants.

physiology [Gk. *physis*, nature + *logos*, a discourse]: The study of function in cells, organs, or entire organisms; the processes of life.

phytochrome [Gk. *phyton*, plant + *chrōma*, color]: A plant pigment that is a photoreceptor for red or far-red light and is involved with a number of developmental processes, such as flowering, dormancy, leaf formation, and seed germination.

phytoplankton [Gk. *phyton*, plant + *planktos*, wandering]: Aquatic, free-floating, microscopic, photosynthetic organisms.

pigment [L. *pigmentum*, paint]: A colored substance that absorbs light over a narrow band of wavelengths.

pinocytosis [Gk. *pinein*, to drink + *kytos*, vessel]: Cell "drinking." *See* Endocytosis.

pith: In the roots and stems of plants, the ground tissue occupying the region to the interior of the vascular cylinder; usually consists of parenchyma cells.

pituitary [L. *pituita*, phlegm]: Endocrine gland in vertebrates; the anterior lobe is the source of tropic hormones, growth hormone, and prolactin and is regulated by secretions of the hypothalamus; the posterior lobe stores and releases oxytocin and ADH produced by the hypothalamus.

placenta [Gk. *plax*, a flat object]: A tissue formed as the result of interactions between the inner lining of the mammalian uterus and the extraembryonic chorion; serves as the connection through which exchanges of nutrients and wastes occur between the blood of the mother and that of the embryo.

plankton [Gk. *planktos*, wandering]: Small (mostly microscopic) aquatic and marine organisms found in the upper levels of the water, where light is abundant; includes both photosynthetic (phytoplankton) and heterotrophic (zooplankton) forms.

planula [L. dim. of *planus*, a wanderer]: The ciliated, free-swimming type of larva formed by many cnidarians.

plasma [Gk., form or mold]: The clear, colorless fluid component of vertebrate blood, containing dissolved ions, molecules, and plasma proteins; blood minus the blood cells.

plasma cell: An antibody-producing cell resulting from the differentiation and proliferation of a B lymphocyte that has interacted with an antigen complementary to the antibodies displayed on its surface; a mature plasma cell can produce from 3,000 to 30,000 antibody molecules per second.

plasma membrane [Gk. *plasma*, form or mold + L. *membrana*, skin, parchment]: The membrane surrounding the cytoplasm of a cell; the cell membrane. Its fundamental structure consists of a bilayer of phospholipids (and a smaller number of glycolipids) in which protein molecules are embedded.

plasmid: In prokaryotes, an extrachromosomal, independently replicating, small, circular DNA molecule.

plasmodesma, *pl.* **plasmodesmata** [Gk. *plassein*, to mold + *desmos*, band, bond]: In plants, a minute, cytoplasmic thread that extends through pores in cell walls and connects the cytoplasm of adjacent cells.

plastid [Gk. *plastos*, formed or molded]: A membrane-bound, often pigmented, organelle in the cytoplasm of plant cells; includes chloroplasts and the starch-filled plastids in the core cells of the root cap.

platelet (**plate**-let) [Gk. *platus*, flat]: In mammals, a round or biconcave disk suspended in the blood and involved in the formation of blood clots.

pleiotropy (ply-o-**trop**-ee) [Gk. *pleios*, more + *trope*, a turning]: The capacity of a gene to affect a number of different phenotypic characteristics.

polar [L. *polus*, end of axis]: Having parts or areas with opposed or contrasting properties, such as positive and negative charges, head and tail.

polar body: Minute, nonfunctioning cell produced during those meiotic divisions that lead to egg cells; contains a nucleus but very little cytoplasm.

polar covalent bond: A covalent bond in which the electrons are shared unequally between the two atoms; the resulting polar molecule has regions of slightly negative and slightly positive charge.

polar nuclei: In angiosperms, the two nuclei of the central cell of the female gametophyte; they fuse with a sperm nucleus to form the triploid ($3n$) endosperm nucleus.

pollen [L., fine dust]: In seed plants, spores consisting of an immature male gametophyte and a protective outer covering.

pollination [L. *pollen*, fine dust]: The transfer of pollen from the anther to a receptive surface of a flower.

polygenic inheritance [Gk. *polus*, many + *genos*, race, descent]: The determination of a given characteristic, such as weight or height, by the interaction of many genes.

polymer [Gk. *polus*, many + *meris*, part or portion]: A large molecule composed of many similar or identical molecular subunits.

polymerase: An enzyme, such as DNA polymerase or RNA polymerase, that catalyzes the synthesis of a polymer from its subunits.

polymerase chain reaction (**PCR**): A recently developed chemical process in which millions of copies of a specific segment of DNA can be rapidly synthesized from a tiny sample of DNA containing that segment.

polymorphism [Gk. *polus*, many + *morphe*, form]: The presence in a single population of two or more phenotypically distinct forms of a trait.

polyp [Gk. *polus*, many + *pous*, foot]: The sessile stage in the life cycle of cnidarians.

polypeptide [Gk. *polus*, many + *pepto*, to soften, digest]: A molecule consisting of a long chain of amino acids linked together by peptide bonds.

polyploid [Gk. *polus*, many + *ploion*, vessel]: Cell with more than two complete sets of chromosomes per nucleus.

polyribosome: Two or more ribosomes together with a molecule of mRNA that they are simultaneously translating; a polysome.

polysaccharide [Gk. *polus*, many + *sakcharon*, sugar]: A carbohydrate polymer composed of monosaccharide subunits in long chains; includes starch, cellulose.

polysome: *See* Polyribosome.

population: Any group of individuals of one species that occupy a given area at the same time; in genetic terms, an interbreeding group of organisms.

population bottleneck: Type of genetic drift that occurs as the result of a population being drastically reduced in numbers by an event having little to do with the usual forces of natural selection.

population density: The number of individuals of a population per unit area or volume of living space.

posterior: Of or pertaining to the rear, or tail, end.

potential energy: Energy in a potentially usable form that is not, for the moment, being used; often called "energy of position."

predator [L. *praedari*, to prey upon; from *prehendere*, to grasp, seize]: An organism that eats other living organisms.

pressure-flow hypothesis: A hypothesis accounting for sap flow through the phloem system. According to this hypothesis, the solution containing nutrient sugars moves through the sieve tubes by bulk flow, moving into and out of the sieve tubes by active transport and diffusion.

prey [L. *prehendere*, to grasp, seize]: An organism eaten by another organism.

primary growth: In plants, growth originating in the apical meristem of the shoots and roots, as contrasted with secondary growth; results in an increase in length.

primary structure of a protein: The amino acid sequence of a protein.

primate: A member of the order of mammals that includes anthropoids and prosimians.

primitive [L. *primus*, first]: Not specialized; at an early stage of evolution or development.

primordium, *pl.* **primordia** [L. *primus*, first + *ordiri*, to begin to weave]: A cell or organ in its earliest stage of differentiation.

procambium [L. *pro*, before + *cambium*, exchange]: In plants, a primary meristematic tissue; gives rise to vascular tissues of the primary plant body and to the vascular cambium.

producer, in ecological systems: An autotrophic organism, usually a photosynthesizer, that contributes to the net primary productivity of a community.

progesterone [L. *progerere*, to carry forth or out + *steiras*, barren]: In mammals, a steroid hormone produced by the corpus luteum that prepares the uterus for implantation of the embryo; also produced by the placenta during pregnancy.

prokaryote [L. *pro*, before + Gk. *karyon*, nut, kernel]: A cell lacking a membrane-bound nucleus and membrane-bound organelles; a bacterium or a cyanobacterium.

promoter: Specific segment of DNA to which RNA polymerase attaches to initiate transcription of mRNA from an operon.

prophage: A bacterial virus (bacteriophage) integrated into a host chromosome.

prophase [Gk. *pro*, before + *phasis*, form]: An early stage in nuclear division, characterized by the condensing of the chromosomes and their movement toward the equator of the spindle. Homologous chromosomes pair up during meiotic prophase.

prosimian [L. *pro*, before + *simia*, ape]: A lower primate; includes lemurs, lorises, tarsiers, and bush babies, as well as many fossil forms.

prostaglandins [Gk. *prostas*, a porch or vestibule + L. *glans*, acorn]: A group of fatty acids that function as chemical messengers; synthesized in most, possibly all, cells of the body; thought to play key roles in fertilization and in triggering the onset of both menstruation and labor.

prostate gland [Gk. *prostas*, a porch or vestibule + L. *glans*, acorn]: A mass of muscle and glandular tissue surrounding the base of the urethra in male mammals; the vasa deferentia merge with ducts from the seminal vesicles, enter the prostate gland, and there merge with the urethra. The prostate gland secretes an alkaline fluid that has a stimulating effect on the sperm as they are released.

protein [Gk. *proteios*, primary]: A complex organic compound composed of one or more polypeptide chains, each made up of many (about 100 or more) amino acids linked together by peptide bonds.

protoderm [Gk. *protos*, first + *derma*, skin]: In plants, a primary meristematic tissue; gives rise to the epidermis.

proton: A subatomic particle with a single positive charge equal in magnitude to the charge of an electron and with a mass slightly less than that of a neutron; a component of every atomic nucleus.

protoplasm [Gk. *protos*, first + *plasma*, anything molded]: Living matter.

protostome [Gk. *protos*, first + *stoma*, mouth]: An animal in which the mouth forms at or near the blastopore in the developing embryo; mollusks, annelids, and arthropods are protostomes.

provirus: A virus of a eukaryote that has become integrated into a host chromosome.

pseudocoelom [Gk. *pseudes*, false + *koilos*, a hollow]: A body cavity consisting of a fluid-filled space between the endoderm and the mesoderm; characteristic of the nematodes.

pseudopodium [Gk. *pseudes*, false + *pous, pod-*, foot]: A temporary cytoplasmic protrusion from an amoeboid cell, which functions in locomotion or in feeding by phagocytosis.

pulmonary [L. *pulmonis*, lung]: Pertaining to the lungs.

pulmonary artery [L. *pulmonis*, lung]: In birds and mammals, an artery that carries deoxygenated blood from the right ventricle of the heart to the lungs, where it is oxygenated.

pulmonary vein [L. *pulmonis*, lung]: In birds and mammals, a vein that carries oxygenated blood from the lungs to the left atrium of the heart, from which blood is pumped into the left ventricle and from there to the body tissues.

punctuated equilibrium: A model of the mechanism of evolutionary change that proposes that long periods of no change ("stasis") are punctuated by periods of rapid speciation, with natural selection acting on species as well as on individuals.

Punnett square: The checkerboard diagram used for analysis of allele segregation.

pupa [L., girl, doll]: A developmental stage of some insects, in which the organism is nonfeeding, immotile, and sometimes encapsulated or in a cocoon; the pupal stage occurs between the larval and adult phases.

purine [Gk. *purinos*, fiery, sparkling]: A nitrogenous base, such as adenine or guanine, with a characteristic two-ring structure; one of the components of nucleic acids.

pyramid, ecological: *See* Ecological pyramid.

pyramid of energy: A diagram of the energy flow between the trophic levels of an ecosystem; plants or other autotrophs (at the base of the pyramid) represent the greatest amount of energy, herbivores next, then primary carnivores, secondary carnivores, etc.

pyrimidine: A nitrogenous base, such as cytosine, thymine, or uracil, with a characteristic single-ring structure; one of the components of nucleic acids.

quaternary structure of a protein: The overall structure of a globular protein molecule that consists of two or more polypeptide chains.

queen: In social insects (ants, termites, and some species of bees and wasps), the fertile, or fully developed, female whose function is to lay eggs.

radial symmetry [L. *radius*, a spoke of a wheel + Gk. *summetros*, symmetry]: The regular arrangement of parts around a central axis such that any plane passing through the central axis divides the organism into halves that are approximate mirror images; seen in cnidarians and adult echinoderms.

radiation [L. *radius*, a spoke of a wheel, hence, a ray]: Energy emitted in the form of waves or particles.

radioactive isotope: An isotope with an unstable nucleus that stabilizes itself by emitting radiation.

receptor: A protein or glycoprotein molecule with a specific three-dimensional structure, to which a substance (for example, a hormone, a neurotransmitter, or an antigen) with a complementary structure can bind; typically displayed on the surface of a membrane. Binding of a complementary molecule to a receptor may trigger a transport process or a change in processes occurring within the cell.

recessive allele [L. *recedere*, to recede]: An allele whose phenotypic effect is masked in the heterozygote by that of another, dominant allele.

reciprocal altruism: Performance of an altruistic act with the expectation that the favor will be returned.

recognition sequence: A specific sequence of nucleotides at which a restriction enzyme cleaves a DNA molecule.

recombinant DNA: DNA formed either naturally or in the laboratory by the joining of segments of DNA from different sources.

recombination: The formation of new gene combinations; in eukaryotes, may be accomplished by new associations of chromosomes produced during sexual reproduction or crossing over; in prokaryotes, may be accomplished through transformation, conjugation, or transduction.

reduction [L. *reducere*, to lead back]: Loss of oxygen, gain of hydrogen, or gain of an electron by an atom, ion, or molecule; oxidation and reduction take place simultaneously, with the electron lost by one reactant being transferred to another.

reflex [L. *reflectere*, to bend back]: Unit of action of the nervous system involving a sensory neuron, often one or more interneurons, and one or more motor neurons.

relay neuron: Neuron that transmits signals between different regions of the central nervous system.

releaser: In ethology, a stimulus that functions as a communication signal between members of the same species and that sets in motion, or "releases," the sequence of movements that constitute a fixed action pattern.

renal [L. *renes*, kidneys]: Pertaining to the kidney.

replication fork: In DNA synthesis, the Y-shaped structure formed at the point where the two strands of the original molecule are being separated and the complementary strands are being synthesized.

repressor [L. *reprimere*, to press back, keep back]: In genetics, a protein that binds to the operator, preventing RNA polymerase from attaching to the promoter and transcribing the structural genes of the operon; coded by a segment of DNA known as the regulator.

resolving power [L. *resolvere*, to loosen, unbind]: The ability of a lens to distinguish two lines as separate.

respiration [L. *respirare*, to breathe]: (1) In aerobic organisms, the intake of oxygen and the liberation of carbon dioxide. (2) In cells, the oxygen-requiring stage in the breakdown and release of energy from fuel molecules.

resting potential: The difference in electric potential (about 70 millivolts) across the membrane of an axon at rest.

restriction enzymes: Enzymes that cleave the DNA double helix at specific nucleotide sequences.

reticular formation [L. *reticulum*, a network]: A brain circuit involved with alertness and direction of attention to selected events; consists of a loose network of interneurons running through the brainstem, plus certain neurons in the thalamus that function as an extension of this network.

reticulum [L., network]: A fine network (e.g., endoplasmic reticulum).

retina [L. dim. of *rete*, net]: The light-sensitive layer of the vertebrate eye; contains several layers of neurons and photoreceptor cells (rods and cones); receives the image formed by the lens and transmits it to the brain via the optic nerve.

retrovirus [L., turning back]: An RNA virus that codes for an enzyme, reverse transcriptase, that transcribes the RNA into DNA and catalyzes the insertion of that DNA into a host-cell chromosome.

reverse transcriptase: An enzyme that transcribes RNA into DNA; found only in association with retroviruses.

rhizoid [Gk. *rhiza*, root]: Rootlike anchoring structure in fungi and nonvascular plants.

rhizome [Gk. *rhizoma*, mass of roots]: In vascular plants, a horizontal stem growing along or below the surface of the soil; may be enlarged for storage or may function in vegetative reproduction.

ribonucleic acid (RNA) (rye-bo-new-clay-ick): A class of nucleic acids characterized by the presence of the sugar ribose and the pyrimidine uracil; includes mRNA, tRNA, and rRNA. RNA is the genetic material of many viruses.

ribosomal RNA (rRNA): A class of RNA molecules found, along with characteristic proteins, in ribosomes; transcribed from DNA of the chromatin loops that form the nucleolus.

ribosome: A complex of specific protein and ribonucleic acid molecules; the site of translation in protein synthesis; in eukaryotic cells, often bound to the endoplasmic reticulum. Many ribosomes attached to a single strand of mRNA are called a polyribosome, or polysome.

RNA: Abbreviation of ribonucleic acid.

rod: Photoreceptor cell found in the vertebrate retina; sensitive to very dim light, responsible for "night vision."

root: The descending axis of a plant, normally below ground and serving both to anchor the plant and to take up and conduct water and dissolved minerals.

root cap: A thimblelike mass of cells that covers and protects the growing tip of a root.

root hair: An extremely fine cytoplasmic extension of an epidermal cell of a young root; root hairs greatly increase the surface area for the uptake of water and dissolved minerals.

saprobe [Gk. *sapros*, rotten, putrid + *bios*, life]: An organism that feeds on nonliving organic matter.

sarcolemma [Gk. *sarx*, the flesh + *lemma*, husk]: The specialized plasma membrane surrounding a muscle cell (muscle fiber); capable of propagating action potentials.

sarcomere [Gk. *sarx*, the flesh + *meris*, part of, portion]: Functional and structural unit of contraction in striated muscle.

sarcoplasmic reticulum [Gk. *sarx*, the flesh + *plasma*, from cytoplasm + L. *reticulum*, network]: The specialized endoplasmic reticulum that encases each myofibril of a muscle cell.

saturated fatty acid: A fatty acid in which there are no double bonds between carbon atoms; fats containing saturated fatty acids tend to be solid at room temperature.

sclerenchyma [Gk. *skleros*, hard]: In plants, a type of supporting cell with thick, often lignified, secondary walls; may be alive or dead at maturity; includes fibers and sclereids.

secondary growth: In the stems and roots of woody plants, growth originating in the lateral meristems (cork cambium and vascular cambium); as contrasted with primary growth; results in an increase in girth.

secondary sex characteristics: Characteristics of animals that distinguish between the two sexes but that do not produce or convey gametes; includes facial hair of the human male and enlarged hips and breasts of the female.

secondary structure of a protein: The simple structure (often a helix, a sheet, or a cable) resulting from the spontaneous folding of a polypeptide chain as it is formed; maintained by hydrogen bonds and other weak forces.

secretion [L. *secermere*, to sever, separate]: (1) Product of any cell, gland, or tissue that is released through the plasma membrane and that performs its function outside the cell that produced it. (2) The stage of kidney function in which, through active transport processes, molecules remaining in the blood plasma are selectively removed from the peritubular capillaries and pumped into the filtrate in the renal tubule.

seed: A complex structure formed by the maturation of the ovule of seed plants following fertilization; upon germination, a seed develops into a new sporophyte; generally consists of seed coat, embryo, and a food reserve.

segregation: *See* Mendel's first law.

selectively permeable [L. *seligere*, to gather apart + *permeare*, to go through]: Applied to membranes that permit passage of water and some solutes but block passage of most solutes; semipermeable.

self-fertilization: The union of egg and sperm produced by a single hermaphroditic organism.

self-pollination: The transfer of pollen from anther to stigma in the same flower or to another flower of the same plant, leading to self-fertilization.

semen [L., seed]: Product of the male reproductive system; includes sperm and the sperm-carrying fluids.

seminal vesicles [L. *semen*, seed + *vesicula*, a little bladder]: In male mammals, small vesicles, the ducts of which merge with the vasa deferentia as they enter the prostate gland; they produce an alkaline, fructose-containing fluid that suspends and nourishes the sperm cells.

seminiferous tubules [L. *semen*, seed + *ferre*, to bear or carry + dim. of *tubus*, tube]: The structures within the testes in which sperm cells are produced; contain spermatogonia, which give rise to the sperm cells, and Sertoli cells, which provide nourishment to the developing sperm cells.

sensory neuron: A neuron that conducts impulses from a sensory receptor to the central nervous system or a central ganglion.

sensory receptor: A cell, tissue, or organ that detects internal or external stimuli.

septum [L., fence]: A partition, or cross wall, that divides a structure, such as a fungal hypha, into compartments.

sessile [L. *sedere*, to sit]: Attached; not free to move about.

sex chromosomes: Chromosomes that are different in the two sexes and that are involved in sex determination.

sex-linked trait: An inherited trait, such as color discrimination, determined by a gene located on a sex chromosome and that therefore shows a different pattern of inheritance in males and females.

sexual reproduction: Reproduction involving meiosis and fertilization.

sexual selection: A type of natural selection that acts on characteristics of direct consequence in obtaining a mate and successfully reproducing; thought to be the chief cause of sexual dimorphism, the striking phenotypic differences between the males and females of many species.

shoot: The aboveground portions, such as the stem and leaves, of a vascular plant.

short-day plant: A plant that must be exposed to a light period shorter than some critical length for flowering to occur; short-day plants usually flower in autumn.

sieve tube: A series of sugar-conducting cells (sieve-tube members) found in the phloem of angiosperms.

sinoatrial node [L. *sinus*, fold, hollow + *atrium*, yard, court, hall + *nodus*, knot]: Area of the vertebrate heart that initiates the heartbeat; located where the superior vena cava enters the right atrium; the pacemaker.

smooth muscle: Nonstriated muscle; lines the walls of internal organs and arteries and is under involuntary control.

social dominance: A hierarchical pattern of social organization involving domination of some members of a group by other members in a relatively orderly and long-lasting pattern.

society [L. *socius*, companion]: An organization of individuals of the same species in which there are divisions of resources, divisions of labor, and mutual dependence; a society is held together by stimuli exchanged among members of the group.

sodium-potassium pump: In the plasma membrane of animal cells, a transport protein that pumps sodium ions (Na^+) out of the cell and potassium ions (K^+) into the cell in a process powered by ATP; establishes and maintains concentration gradients that are essential for osmotic balance and, in neurons, for conduction of the nerve impulse.

solution: A homogeneous mixture of the molecules of two or more substances; the substance present in the greatest amount (usually a liquid) is called the solvent, and the substances present in lesser amounts are called solutes.

somatic cells [Gk. *soma*, body]: The differentiated cells composing body tissues of multicellular plants and animals; all body cells except those giving rise to gametes.

somatic nervous system [Gk. *soma*, body]: In vertebrates, the motor and sensory neurons of the peripheral nervous system that control skeletal muscle; the "voluntary" system, as contrasted with the "involuntary," or autonomic, nervous system.

somatotropin [Gk. *soma*, body + *trope*, a turning]: A hormone, produced by the pituitary gland, that stimulates protein synthesis and promotes the growth of bone; also known as growth hormone.

specialized: (1) Of cells, having particular functions in a multicellular organism. (2) Of organisms, having special adaptations to a particular habitat or mode of life.

speciation: The process by which new species are formed.

species, *pl.* **species** [L., kind, sort]: A group of organisms that actually (or potentially) interbreed in nature and are reproductively isolated from all other such groups; a taxonomic grouping of anatomically similar individuals (the category below genus).

species-specific: Characteristic of (and limited to) a particular species.

specific: Unique; for example, the proteins in a given organism, the enzyme catalyzing a given reaction, or the antibody to a given antigen.

specific heat: The amount of heat (in calories) required to raise the temperature of 1 gram of a substance 1)C. The specific heat of water is 1 calorie per gram.

sperm [Gk. *sperma*, seed]: A mature male sex cell, or gamete, usually motile and smaller than the female gamete.

spermatid [Gk. *sperma*, seed]: Each of four haploid (*n*) cells resulting from the meiotic divisions of a spermatocyte; each spermatid becomes differentiated into a sperm cell.

spermatocytes [Gk. *sperma*, seed + *kytos*, vessel]: The diploid (2*n*) cells formed by the enlargement and differentiation of the spermatogonia; they give rise by meiotic division to the spermatids.

spermatogonia [Gk. *sperma*, seed + *gonos*, a child, the young]: The unspecialized diploid (2*n*) cells on the walls of the seminiferous tubules that, by enlargement, differentiation, and meiotic division, become spermatocytes, then spermatids, then sperm cells.

sphincter [Gk. *sphinktēr*, a band]: A circular muscle surrounding the opening of a tubular structure or the juncture of different regions of a tubular structure (e.g., the pyloric sphincter, at the juncture of the stomach and the small intestine); contraction of the sphincter closes the passageway, and relaxation opens it.

spinal cord: Part of the vertebrate central nervous system; consists of a thick, dorsal, longitudinal bundle of nerve fibers extending posteriorly from the brain.

spindle: In dividing cells, the structure formed of microtubules that extends from pole to pole; the spindle fibers appear to maneuver the chromosomes into position during metaphase and to pull the newly separated chromosomes toward the poles during anaphase.

spiracle [L. *spirare*, to breathe]: One of the external openings of the respiratory system in terrestrial arthropods.

splitting evolution: *See* Cladogenesis.

spongy parenchyma: In plant leaves, a tissue composed of loosely arranged chloroplast-containing parenchyma cells.

sporangiophore (spo-ran-ji-o-for) [Gk. *spora*, seed + *phore*, from *phorein*, to bear]: A specialized hypha or a branch bearing one or more sporangia.

sporangium, *pl.* **sporangia** [Gk. *spora*, seed]: A unicellular or multicellular structure in which spores are produced.

spore [Gk. *spora*, seed]: An asexual reproductive or resting cell capable of developing into a new organism without fusion with another cell; in contrast to a gamete.

sporophyte [Gk. *spora*, seed + *phytos*, growing]: In organisms that have alternation of haploid and diploid generations (all plants and some green algae), the diploid (*2n*) spore-producing generation.

stamen [L., a thread]: The male structure of a flower, which produces microspores or pollen; usually consists of a stalk, the filament, bearing a pollen-producing anther at its tip.

starch [M.E. *sterchen*, to stiffen]: A class of complex, insoluble carbohydrates, the chief food-storage substances of plants; composed of 1,000 or more glucose units and readily broken down enzymatically into these units.

statocyst [Gk. *statos*, standing + *kystis*, sac]: An organ of balance, consisting of a vesicle containing granules of sand (statoliths) or some other material that stimulates sensory cells when the organism moves.

stem: The aboveground part of the axis of vascular plants, as well as anatomically similar portions below ground (such as rhizomes).

stem cells: The common, self-regenerating cells in the marrow of long bones that give rise, by differentiation and division, to red blood cells and all of the different types of white blood cells.

stereoscopic vision [Gk. *stereōs*, solid + *optikos*, pertaining to the eye]: Ability to perceive a single, three-dimensional image from the simultaneous but separate images delivered to the brain by each eye.

steroid: One of a group of lipids having four linked carbon rings and, often, a hydrocarbon tail; cholesterol, sex hormones, and the hormones of the adrenal cortex are steroids.

stigma [Gk. *stigme*, a prick mark, puncture]: In plants, the region of a carpel serving as a receptive surface for pollen grains, which germinate on it.

stimulus [L., goad, incentive]: Any internal or external change or signal that influences the activity of an organism or of part of an organism.

stoma, *pl.* **stomata** [Gk., mouth]: A minute opening in the epidermis of leaves and stems, bordered by guard cells, through which gases pass.

strategy [Gk. *strategein*, to maneuver]: A group of related traits, evolved under the influence of natural selection, that solve particular problems encountered by living organisms; often includes anatomical, physiological, and behavioral characteristics.

striated muscle [L., from *striare*, to groove]: Skeletal voluntary muscle and cardiac muscle. The name derives from the striped appearance, which reflects the arrangement of contractile elements.

stroma [Gk., a bed, from *stronnymi*, to spread out]: A dense solution that fills the interior of the chloroplast and surrounds the thylakoids; the site of the light-independent reactions of photosynthesis.

structural gene: Any gene that codes for a protein; in distinction to regulatory genes.

style [L. *stilus*, stake, stalk]: In angiosperms, the stalk of a carpel, down which the pollen tube grows.

substrate [L. *substratus*, strewn under]: (1) The foundation to which an organism is attached. (2) A substance on which an enzyme acts.

succession: *See* Ecological succession.

sucrose: Cane sugar; a common disaccharide found in many plants; a molecule of glucose linked to a molecule of fructose.

sugar: Any monosaccharide or disaccharide.

surface tension: A tautness of the surface of a liquid, caused by the cohesion of the molecules of liquid. Water has an extremely high surface tension.

symbiosis [Gk. *syn*, together with + *bioonai*, to live]: An intimate and protracted association between two or more organisms of different species. Includes mutualism, in which the association is beneficial to both; commensalism, in which one benefits and the other is neither harmed nor benefited; and parasitism, in which one benefits and the other is harmed.

sympathetic division: A subdivision of the autonomic nervous system, with centers in the midportion of the spinal cord; slows digestion; generally excites other functions.

sympatric speciation [Gk. *syn*, together with + *patra*, fatherland, country]: Speciation that occurs without geographic isolation of a population of organisms; usually occurs as the result of hybridization accompanied by polyploidy; may occur in some cases as a result of disruptive selection.

synapse [Gk. *synapsis*, a union]: A specialized junction between two neurons where the activity in one influences the activity in the other; in mammals, transmission across the synaptic cleft is by means of chemical substances (neurotransmitters) and may be excitatory or inhibitory.

synthesis [Gk. *syntheke*, a putting together]: The formation of a more complex substance from simpler ones.

synthesis phase: In the cell cycle, the phase in which the DNA of the chromosomes is replicated and DNA-associated proteins, such as histones, are synthesized.

T cell: A type of white blood cell involved in cell-mediated immunity and in interactions with B cells; a T lymphocyte.

taxon, *pl.* **taxa** [Gk. *taxis*, arrange, put in order]: A particular group, ranked at a particular categorical level, in a hierarchical classification scheme; for example, *Drosophila* is a taxon at the categorical level of genus.

taxonomy [Gk. *taxis*, arrange, put in order + *nomos*, law]: The study of the classification of organisms; the ordering of organisms into a hierarchy that reflects their essential similarities and differences.

telophase [Gk. *telos*, end + *phasis*, form]: The last stage in mitosis and meiosis, during which the chromosomes become reorganized into two new nuclei.

temperate bacteriophage: A bacterial virus that may become incorporated into the host-cell chromosome.

template: A pattern or mold guiding the formation of a negative or complementary copy.

tentacles [L. *tentare*, to touch]: Long, flexible protrusions located about the mouth of many invertebrates; usually prehensile or tactile.

territory: An area or space occupied and defended by an individual or a group; trespassers are attacked (and usually defeated); may be the site of breeding, nesting, food gathering, or any combination thereof.

tertiary structure of a protein: A complex structure, usually globular, resulting from further folding of the secondary structure of a protein; forms spontaneously due to attractions and repulsions among amino acids with different charges on their R groups.

testcross: A mating between a phenotypically dominant individual and a homozygous recessive "tester" to determine the genetic constitution of the dominant phenotype, that is, whether it is homozygous or heterozygous for the relevant gene.

testis, *pl.* **testes** [L., witness]: The sperm-producing organ; also the source of the male sex hormone testosterone.

testosterone [Gk. *testis*, testicle + *steiras*, barren]: A steroid hormone secreted by the testes in higher vertebrates and stimulating the development and maintenance of male sex characteristics and the production of sperm; the principal androgen.

tetrad [Gk. *tetras,* four]: In genetics, a pair of homologous chromosomes that have replicated and come together in prophase I of meiosis; consists of four chromatids.

thalamus [Gk. *thalamos,* chamber]: A part of the vertebrate brain tucked below the cerebrum; the main relay center between the brainstem and the higher brain centers.

thallus [Gk. *thallos,* a young twig]: A simple plant or algal body without true roots, leaves, or stems.

theory [Gk. *theorein,* to look at]: A generalization based on many observations and experiments; a verified hypothesis.

thermodynamics [Gk. *therme,* heat + *dynamis,* power]: The study of transformations of energy. The first law of thermodynamics states that, in all processes, the total energy of a system plus its surroundings remains constant. The second law states that all natural processes tend to proceed in such a direction that the disorder or randomness of the system increases.

thorax [Gk., breastplate]: (1) In vertebrates, that portion of the trunk containing the heart and lungs. (2) In crustaceans and insects, the fused, leg-bearing segments between head and abdomen.

thylakoid [Gk. *thylakos,* a small bag]: A flattened sac, or vesicle, that forms part of the internal membrane structure of the chloroplast; the site of the light-dependent reactions of photosynthesis and of photophosphorylation; stacks of thylakoids collectively form the grana.

thyroid [Gk. *thyra,* a door]: An endocrine gland of vertebrates, located in the neck; source of an iodine-containing hormone (thyroxine) that increases the metabolic rate and affects growth.

tight junction: A junction between adjacent animal cells that prevents materials from leaking through the tissue; for example, intestinal epithelial cells are surrounded by tight junctions.

tissue [L. *texere,* to weave]: A group of similar cells organized into a structural and functional unit.

trachea, *pl.* **tracheae** (**trake-ee-a**) [Gk. *tracheia,* rough]: An air-conducting tube. (1) In insects and some other terrestrial arthropods, a system of chitin-lined air ducts. (2) In terrestrial vertebrates, the windpipe.

tracheid (**tray-key-idd**) [Gk. *tracheia,* rough]: In vascular plants, an elongated, thick-walled conducting and supporting cell of xylem, characterized by tapering ends and pitted walls without true perforations.

tract: A group or bundle of nerve fibers with accompanying connective tissue, located within the central nervous system.

transcription [L. *trans,* across + *scribere,* to write]: The enzymatic process by which the genetic information contained in one strand of DNA is used to specify a complementary sequence of bases in an RNA molecule.

transduction [L. *trans,* across + *ducere,* to lead]: The transfer of genetic material (DNA) from one cell to another by a virus.

transfer RNA (tRNA) [L. *trans,* across + *ferre,* to bear or carry]: A class of small RNAs (about 80 nucleotides each) with three functional sites: one is the attachment site for a specific amino acid, another is the nucleotide triplet (anticodon) for that amino acid, and yet another is the recognition site for the enzyme (an aminoacyl-tRNA synthetase) that catalyzes the attachment of the amino acid to the tRNA molecule. Each type of tRNA accepts a specific amino acid and transfers it to a growing polypeptide chain as specified by the nucleotide sequence of the messenger RNA being translated.

transformation [L. *trans,* across + *formare,* to shape]: A genetic change produced by the incorporation into a cell of DNA from the external medium.

translation [L. *trans,* across + *latus,* that which is carried]: The process by which the genetic information present in a strand of messenger RNA directs the sequence of amino acids during protein synthesis.

translocation [L. *trans,* across + *locare,* to put or place]: (1) In plants, the transport of the products of photosynthesis from a leaf to another part of the plant. (2) In genetics, the breaking off of a piece of chromosome with its reattachment to a nonhomologous chromosome.

transpiration [L. *trans,* across + *spirare,* to breathe]: In plants, the loss of water vapor from the stomata.

transposon [L. *transponere,* to change the position of]: A DNA sequence carrying one or more genes that is capable of moving from one location in the chromosomes to another. Simple transposons carry only the genes essential for transposition; complex transposons carry genes that code for additional proteins.

trophic level [Gk. *trophos,* feeder]: The position of a species in the food web or chain, that is, its feeding level; a step in the movement of biomass or energy through an ecosystem.

trophoblast [Gk. *trophos,* feeder + *blastos,* sprout]: In the early mammalian embryo (the blastocyst), a double layer of cells that surrounds the inner cell mass and subsequently gives rise to the chorion.

tropic [Gk. *trope,* a turning]: Pertaining to behavior or action brought about by specific stimuli, for example, phototropic ("light-oriented") motion, gonadotropic ("stimulating the gonads") hormone.

tuber [L. *tuber,* bump, swelling]: A much-enlarged, short, fleshy underground stem, such as that of the potato.

tumor suppressor gene: One of a group of eukaryotic genes that apparently act as a brake on cell division, keeping it tightly regulated; when such genes are lost or disabled by mutation, the rapid multiplication of cancer cells can occur.

turgor [L. *turgere,* to swell]: The pressure exerted on the inside of a plant cell wall by the fluid contents of the cell; the interior of the cell is hypertonic in relation to the fluids surrounding it and so gains water by osmosis.

unsaturated fatty acid: A fatty acid in which there are double bonds between some of the carbon atoms; fats containing unsaturated fatty acids tend to be liquid at room temperature.

urea [Gk. *ouron,* urine]: An organic compound formed in the vertebrate liver; principal form of disposal of nitrogenous wastes by mammals.

ureter [Gk. from *ourein,* to urinate]: The tube carrying urine from the kidney; in mammals, it empties into the bladder.

urethra [Gk. from *ourein,* to urinate]: The tube carrying urine from the bladder to the exterior of mammals.

uric acid [Gk. *ouron,* urine]: An insoluble nitrogenous waste product that is the principal excretory product in birds, reptiles, and insects.

urine [Gk. *ouron,* urine]: The liquid waste filtered from the blood by the kidney and stored in the bladder pending elimination through the urethra.

uterus [L., womb]: The muscular, expanded portion of the female reproductive tract modified for the storage of eggs or for housing and nourishing the developing embryo.

vacuole [L. *vacuus,* empty]: A membrane-bound, fluid-filled sac within the cytoplasm of a cell.

vaporization [L. *vapor,* steam]: The change from a liquid to a gas; evaporation.

vascular [L. *vasculum,* a small vessel]: Containing or concerning vessels that conduct fluid.

vascular bundle: In plants, a group of longitudinal supporting and conducting tissues (xylem and phloem).

vascular cambium [L. *vasculum,* a small vessel + *cambium,* exchange]: In plants, a cylindrical sheath of meristematic cells that divide mitotically, producing secondary phloem to one side and secondary xylem to the other, but always with a cambial cell remaining.

vas deferens, *pl.* **vasa deferentia** (**vass deff-er-ens**) [L. *vas,* a vessel + *deferre,* to carry down]: In mammals, the tube carrying sperm from a testis to the urethra.

vector [L., carrier]: In recombinant DNA, a small, self-replicating DNA molecule, or a portion thereof, into which a DNA segment can be spliced and introduced into a cell; generally a plasmid or a virus.

vein [L. *vena*, a blood vessel]: (1) In plants, a vascular bundle forming part of the framework of the conducting and supporting tissue of a leaf. (2) In animals, a blood vessel carrying blood from the tissues to the heart. A small vein is known as a venule.

vena cava (vee-na **cah**-va) [L., blood vessel + hollow]: A large vein that brings blood from the tissues to the right atrium of the four-chambered mammalian heart. The superior vena cava collects blood from the forelimbs, head, and anterior or upper trunk; the inferior vena cava collects blood from the posterior body region.

ventral [L. *venter*, belly]: Pertaining to the undersurface of an animal that holds its body in a horizontal position; to the front surface of an animal that holds its body erect.

ventricle [L. *ventriculus*, the stomach]: A muscular chamber of the heart that receives blood from an atrium and pumps blood out of the heart, either to the lungs or to the body tissues.

vertebral column [L. *vertebra*, joint]: The backbone; in nearly all vertebrates, it forms the supporting axis of the body and protects the spinal cord.

vesicle [L. *vesicula*, a little bladder]: A small, intracellular membrane-bound sac.

vessel [L. *vas*, a vessel]: A tubelike element of the xylem of angiosperms; composed of dead cells (vessel members) arranged end to end. Its function is to conduct water and dissolved minerals from the soil.

viable [L. *vita*, life]: Able to live.

villus, *pl.* **villi** [L., a tuft of hair]: In vertebrates, one of the minute, fingerlike projections lining the small intestine that serve to increase the absorptive surface area of the intestine.

virus [L., slimy, liquid, poison]: A submicroscopic, noncellular particle composed of a nucleic acid core and a protein coat (capsid); parasitic; reproduces only within a host cell.

viscera [L., internal organs]: The collective term for the internal organs of an animal.

vitamin [L. *vita*, life]: Any of a number of unrelated organic substances that cannot be synthesized by a particular organism and are essential in minute quantities for normal growth and function.

water cycle: Worldwide circulation of water molecules, powered by the sun. Water evaporates from oceans, lakes, rivers, and, in smaller amounts, soil surfaces and bodies of organisms; water returns to the Earth in the form of rain and snow. Of the water falling on land, some flows into rivers that pour water back into the oceans and some percolates down through the soil until it reaches a zone where all pores and cracks in the rock are filled with water (groundwater); the deep groundwater eventually reaches the oceans, completing the cycle.

water potential: The potential energy of water molecules; regardless of the reason (e.g., gravity, pressure, concentration of solute particles) for the water potential, water moves from a region where water potential is greater to a region where water potential is lower.

wild type: In genetics, the phenotype that is characteristic of the vast majority of individuals of a species in a natural environment.

worker: A member of the nonreproductive laboring caste of social insects.

xylem [Gk. *xylon*, wood]: A complex vascular tissue through which most of the water and dissolved minerals are conducted from the roots to other parts of the plant; consists of tracheids or vessel members, parenchyma cells, and fibers; constitutes the wood of trees and shrubs.

yolk: The stored food in egg cells that nourishes the embryo.

yolk sac: In developing reptiles and birds, the extraembryonic membrane that surrounds and encloses the yolk; performs a nutritive function. In mammals, the extraembryonic membrane in which the germ cells are set aside very early in development.

zoology [Gk. *zoe*, life + *logos*, a discourse]: The study of animals.

zooplankton [Gk. *zoe*, life + *plankton*, wanderer]: A collective term for the nonphotosynthetic organisms present in plankton.

zygote (**zi**-goat) [Gk. *zygon*, yolk, pair]: The diploid (2n) cell resulting from the fusion of male and female gametes (fertilization); a zygote may either develop into a diploid individual by mitotic divisions or may undergo meiosis to form haploid (n) individuals that divide mitotically to form a population of cells.

Illustration Credits

Page xii Frans Lanting/Minden Pictures; **Page xiv** Eric Gravé/Science Source/Photo Researchers, Inc.; **Page xvi** (left) Neil Leifer/Camera 5; (right) Dr. Gopal/MURTI–CNRI/Phototake; **Page xviii** Steven Dalton/NHPA; **Page xx** Jeff Lepore; **Page xxi** Helen Rhode; **Page xxiii** Chuck Place; **Page xxiv** Art Wolfe

Page xxvi Frans Lanting/Minden Pictures; **I–2** (a),(c)–(f) Frans Lanting/Minden Pictures; (b) Konrad Woethe/Bruce Coleman, Inc.; **I–3** George Richmond, *Portrait of Charles Darwin*, 1840, Down House, Downe, Kent, Bridgeman/Art Resource; **I–4** (a) E. R. Degginger/Bruce Coleman, Inc.; (b) George O. Poinar, Jr., University of California, Berkeley; (c) T. E. Whitely; **I–5** Frans Lanting/Minden Pictures; **I–7** (a) Mervin W. Larson/Bruce Coleman, Inc.; (b) Frans Lanting/Minden Pictures; **I–8** (a),(b) Tui De Roy **I–9** Picture Collection/New York Public Library; **Page 11** The Royal College of Surgeons; **I–10** Frans Lanting/Minden Pictures; **I–11** M. W. F. Tweedie/Bruce Coleman, Inc.; **I–14** (a) Stephen Dalton/NHPA; (b) M. K. and I. M. Morcombe/NHPA

Page 18 ©1961 California Institute of Technology and Carnegie Institution of Washington; **1–3** W. H. Hodge/Peter Arnold, Inc.; **1–6** (b) Runk and Schoenberger/Grant Heilman Photography; **1–7** Stephen Dalton/NHPA; **Pages 30–31** (a) John D. Cunningham/Visuals Unlimited; (b) M. Walker/NHPA; (c) Mary M. Thacher/Photo Researchers, Inc.; (d) John Cancalosi; (e) Silvestris/R. Gross/Peter Arnold, Inc.; (f) Fred Bruemmer; (g) Stephen J. Krasemann/DRK Photo; **1–12** (a) M. Abbey/Visuals Unlimited; (b) M. I. Walker/Photo Researchers, Inc.; (c) David Scharf/Peter Arnold, Inc.; (d) Cabisco/Visuals Unlimited

Page 36 Michael and Patricia Fogden; **2–3** (a) G. I. Bernard/Animals Animals; (b) Runk and Schoenberger/Grant Heilman Photography; **2–5** Richard Smith/Tom Stack and Associates; **2–6** (a) Richard Kolar/Animals Animals; (b) Luciano Viti/Retna, Ltd.; (c) Anthony Bannister/Animals Animals; **2–7** (a) adapted from Roger H. Pain: "Helices of Antifreeze," *Nature*, vol. 333, page 207, 1988; (b) Steve McCutcheon

Page 48 Anthony Bannister/ABPL; **3–2** Avril Ramage/Oxford Scientific Films; **3–4** (c) Dr. Lloyd M. Beidler/Science Photo Library/Photo Researchers, Inc.; **3–5** (b) Don W. Fawcett/Visuals Unlimited; **3–6** (b) Biophoto Associates/Photo Researchers, Inc.; **3–7** David P. Maitland/Seaphot Ltd./Planet Earth Pictures; **Page 58** (a)–(c) Lennart Nilsson/©Boehringer Ingelheim International GmbH; **3–11** Dr. Jeremy Burgess/Science Photo Library/Photo Researchers, Inc.; **3–16** sequence information from A. L. Lehninger, *Principles of Biochemistry*, Worth Publishers, Inc., New York, 1982; **3–19** (a) Stephen J. Krasemann/DRK Photo; (b) Gregory G. Dimijian/Photo Researchers, Inc.; **3–20** David Goodsell, *The Machinery of Life*, Springer-Verlag, New York, 1993; **3–21** adapted from R. E. Dickerson and I. Geis, *The Structure and Action of Proteins*, W. A. Benjamin, Inc., Menlo Park, Calif., 1969, copyright 1969 by Dickerson and Geis; **3–22** (a),(b) Stanley Flegler/Visuals Unlimited

Page 70 Courtesy of Kosuke Kamo, Sakurajima Volcanological Observatory; **4–1** Tokai University Research Center; **4–3** (a) Sidney W. Fox; (b) David Deamer; **4–4** S. M. Awramik/Biological Photo Service; **4–5** (b) Ralph Robinson/Visuals Unlimited; **4–6** (a) Dr. Jeremy Burgess/Science Photo Library/Photo Researchers, Inc.; (b) Courtesy of the National Library of Medicine; (c) John Mais; **4–7** (b) A. Ryter; **4–8** (b) Norma J. Lang/Biological Photo Service; **4–9** (b) George Palade; **4–11** (b) Michael A. Walsh; **4–12** (b) Keith Porter; **4–14** (d) Biological Photo Service; **4–15** (a)–(c) David M. Phillips/Visuals Unlimited

Page 86 Lennart Nilsson et al., *The Body Victorious*, Delacorte Press, New York, 1987; **5–2** J. D. Robertson; **5–4** (b) Biophoto Associates/Photo Researchers, Inc.; **5–5** after P. Albersheim, *Scientific American*, April 1975; **5–6** (a) Daniel Friend; (b) Nigel Unwin; **5–7** Biophoto Associates/Science Source/Photo Researchers, Inc.; **5–8** (b) Mia Tegner and David Epel; **5–12** (b) Peter Webster; **5–13** (b) Don W. Fawcett/Visuals Unlimited; **5–14** (b) C. J. Flickinger, *Journal of Cell Biology*, vol. 49, page 221, 1975; **5–16** (b) George B. Chapman, Priscilla Devadoss/Visuals Unlimited; **5–17** (b) Bill Longcore/Photo Researchers, Inc.; **5–18** (a)–(c) Mary Osborn, Max Planck Institute for Biophysical Chemistry; **5–20** Manfred Kage/Peter Arnold, Inc.; **5–21** (a) H. Cante-Lund/NHPA; (b) Karl Aufderheide/Visuals Unlimited; **5–22** (b) Lewis Tilney

Page 108 Eric Gravé/Science Source/Photo Researchers, Inc.; **Page 113** USDA/Science Source/Photo Researchers, Inc.; **6–5** (a) Robert Brons/Biological Photo Service; (b),(c) Thomas Eisner; **6–6** after A. L. Lehninger, *Biochemistry*, 2d ed., Worth Publishers, Inc., New York, 1975; **6–11** Birgit Satir; **6–12** adapted from *Scientific American*, May 1984, page 54; **6–13** (a)–(d) M. M. Perry and A. B. Gilbert, *Journal of Cell Science*, vol. 39, pages 257–272, 1979; **Page 122** (a),(b) London Scientific Films/Oxford Scientific Films; (c)–(e) Robert Kay; **6–14** (a) Ray F. Evert; **6–15** adapted in part from James Darnell, Harvey Lodish, and David Baltimore, *Molecular Cell Biology*, W. H. Freeman and Company, New York, 1986; in part from *Scientific American*, May 1978, page 150; and in part from *Scientific American*, October 1985, page 106

Page 126 Jeff Foott/Bruce Coleman, Inc.; **7–1** Alain Eurard/Photo Researchers, Inc.; **7–5** (a)–(c) Fred Bavendam; **7–8** after A. L. Lehninger, *Biochemistry*, 2d ed., Worth Publishers, Inc., New York, 1975; **7–10** Clive Freeman, The Royal Institution/Science Photo Library/Photo Researchers, Inc.; **7–11** after A. L. Lehninger, *op. cit.*; **7–16** (a) Larry West/Bruce Coleman, Inc.; (b) Ken Lucas/Planet Earth Pictures; **Page 141** K. Amman/Bruce Coleman, Inc.

Page 142 Neil Leifer/Camera 5; **8–2** ©1989 Egyptian Expedition of the Metropolitan Museum of Art, Rogers Fund, 1915 (15.5.19e); **8–6** adapted from James Darnell et al., *Molecular Cell Biology*, W. H. Freeman and Company, New York, 1986; **8–10** adapted from A. L. Lehninger, David Nelson, and Michael Cox, *Principles of Biochemistry*, 2d ed., Worth Publishers, Inc., New York, 1993, page 554; **8–11** (a) adapted from Lehninger, Nelson, and Cox, *op. cit.*, page 557; (b) John N. Telford; **8–12** adapted from Lehninger, Nelson, and Cox, *op. cit.*, page 559; **8–14** (b) Grant Heilman Photography; **8–15** (b) Ken Regan/Camera 5; **Page 156** (a),(b) Bruce Iverson; **8–16** after A. L. Lehninger, *Biochemistry*, 2d ed., Worth Publishers, Inc., New York, 1975

Page 160 Victor Englebert; **9–1** Colin Milkins/Oxford Scientific Films; **9–4** absorption spectra determined by Govindjee; **9–6** (a) Dr. Kenneth Miller/Science Photo Library/Photo Researchers, Inc.; (b) R. Howard Berg/Visuals Unlimited; **9–9** adapted from Bruce Alberts et al., *Molecular Biology of the Cell*, 2d ed., Garland Publishing, Inc., New York, 1989, page 378; **9–10** Dr. Jeremy Burgess/Science Photo Library/Photo Researchers, Inc.

Page 176 Edward Hicks, *Noah's Ark* (detail), 1846, Philadelphia Museum of Art, bequest of Lisa Norris Elkins (50.92.7); **10–3** (a),(b) David Phillips/Visuals Unlimited; **10–4** G. Shih and R. Kessel/Visuals Unlimited; **10–5** adapted from Bruce Alberts et al., *Molecular Biology of the Cell*, Garland Publishing, Inc., New York, 1983, page 619; **10–6** (a) Andrew S. Bajer; (b),(c) adapted from Alberts et al., *op. cit.*, page 652; **10–7** (a)–(f) James Solliday/Biological Photo Service; **10–8** (a)–(d) Andrew S. Bajer; **10–9** David Phillips/Visuals Unlimited; **10–10** (d) B. A. Palevitz/E. H. Newcomb/Biological Photo Service; **10–11** M. I. Walker/Photo Researchers, Inc.

Page 188 Michael and Patricia Fogden; **11–8** (a)–(j) C. A. Hasenkampf/Biological Photo Service; **11–12** (a) Dr. Ram Verma/Phototake; **11–13** (a) Terry Mendoza/Picture Cube; **11–15** (b) Custom Medical Stock; **11–16** after J. L. Moore, *Heredity and Development,* Oxford University Press, New York, 1963; **Page 204** Ripon Microslides

Page 206 Steve Uzzell; **12–2** adapted from K. von Frisch, *Biology,* translated by Jane Oppenheimer, Harper and Row, Publishers, Inc., New York, 1964; **Page 212** John Chiasson/Gamma Liaison; **12–7** Moravian Museum, Brno

Page 220 Biophoto Associates/Photo Researchers, Inc.; **13–1** Joe McDonald/Bruce Coleman, Inc.; **13–3** (a),(c) Hans Reinhard/Bruce Coleman, Inc.; (b) Grant Heilman Photography; (d) Jane Burton/Bruce Coleman, Inc.; **13–4** (a)–(d) Ralph Somes/University of Connecticut; **13–6** Heather Angel; **13–9** Jeremy Burgess/Science Photo Library/Photo Researchers, Inc.; **Page 231** (a) Richmond Products, Boca Raton; **13–15** B. John/Cabisco/Visuals Unlimited; **13–18** P. J. Bryant/Biological Photo Service

Page 240 Manfred Kage/Peter Arnold, Inc.; **14–1** (a),(b) Bruce Iverson; **14–2** after D. D. Koob and W. E. Boggs, *The Nature of Life,* Addison-Wesley Publishing Co., Inc., Reading, Mass., 1972; **14–4** Michel Wurtz; **14–6** Lee D. Simon/Science Photo Library/Photo Researchers, Inc.; **Page 247** (a) from J. D. Watson, *The Double Helix*, Atheneum Publishers, New York, 1968; (b) Will and Deni McIntyre/Photo Researchers, Inc.; **14–11** A. B. Blumenthal, H. J. Kreigstein, and D. S. Hognes, *Cold Spring Harbor Symposium on Quantitative Biology*, vol. 38, page 205, 1973

Page 255 Dr. Gopal/MURTI-CNRI/Phototake; **15–1** Ray Nelson/Phototake; **15–2** after L. Pauling et al., "Sickle Cell Anemia, A Molecular Disease," *Science*, vol. 110, pages 543–548, 1949; **15–4** (a) after K. Namba, D. L. D. Casper, and G. J. Stubbs, *Science*, vol. 227, pages 773–776, 1985; (b) K. G. Murti/Visuals Unlimited; **15–7** adapted from J. A. Lake, *Annual Review of Biochemistry*, vol. 54, pages 507–530, 1985; **15–10** Science Photo Library/Custom Medical Stock

Page 270 M. Bruno/Jacana/Photo Researchers, Inc.; **16–5** David Goodsell, *The Machinery of Life*, Springer-Verlag, New York, 1993; **16–6** (a) Victoria Foe; **16–7** adapted from Bruce Alberts et al., *Molecular Biology of the Cell*, 2d ed., Garland Publishing, Inc., New York, 1989; **16–9** (a) George T. Rudkin; **16–10** (a),(b) Ullrich Scheer, W. W. Franke, M. F. Trendelenberg; **16–11** adapted from U. Goodenough, *Genetics*, 3d ed., Saunders College/Holt, Rinehart and Winston, Inc., New York, 1983; **Page 280** Don Fawcett/Visuals Unlimited; **16–12** (a) James German; (b) adapted from P. Chambon, "Split Genes," *Scientific American*, May 1981, pages 60–71; **Page 282** P. J. Grabowski and T. R. Cech, *Cell*, vol. 23, pages 467–476, 1981; **16–14** adapted from R. Lewin, *Science*, vol. 212, pages 28–32, 1981

Page 286 Nik Kleinberg; **17–1** K. G. Murti/Visuals Unlimited; **17–2** Dr. Dennis Kunkel/Phototake; **17–3** adapted from F. J. Ayala, and J. A. Kiger, *Modern Genetics*, Benjamin/Cummings Publishing Co. Inc., Menlo Park, Calif., 1980; **17–4** S. N. Cohen, *Nature*, 1969; **17–5** (a) M. Wurtz/Biozentrum, University of Basel/Science Photo Library/Photo Researchers, Inc.; (b) Philippe Plailly/Science Photo Library/Photo Researchers, Inc.; (c) CNRI/Phototake; (d) Biozentrum/Science Photo Library/Photo Researchers, Inc.; **17–7** Michel Wurtz/Biozentrum/University of Basel; **17–8** (a) Dr. Robert Liddington, Dana Farber Cancer Institute; (b) Jack Griffith; **17–9** adapted from A. L. Lehninger, *Principles of Biochemistry*, Worth Publishers, Inc., New York, 1982, page 863; **17–10** (a),(b) Dr. David L. Spector, Cold Spring Harbor Laboratory; **17–12** Christine J. Harrison, from Harrison et al., "The Fragile X: A Scanning Electron Microscope Study," *Journal of Medical Genetics*, vol. 20, pages 280–285, 1983

Page 300 AP/Wide World Photos (Elizabeth Fulford, 1993); **18–1** adapted from Bruce Alberts et al., *Molecular Biology of the Cell*, Garland Publishing, Inc., New York, 1983; **18–2** adapted from a photo by John T. Fiddles and Howard M. Goodman; **18–4** (a)–(c) Huntington Potter and David Dressler, LIFE Magazine, © Time Warner; **18–6** David Ward; **18–7** adapted from *City of Hope Quarterly*, vol. 7, page 2, Winter 1978; **18–8** Eli Lilly and Company; **Page 310** ITAR/TASS/Sovfoto; **18–10** Steve Uzzell III; **Page 312** (a) Eugene W. Nester; (b) Keith Wood, University of California at San Diego; **18–11** after Jon W. Gordon and Frank Ruddle; **18–12** R. L. Brinster

Page 318 Lynn Stone; **19–1** Michael Ederegger/DRK Photo; **Page 321** Michael and Patricia Fogden; **19–3** (a),(b) John Mais; **19–4** after Mather and Harrison, *Heredity*, vol. 3, page 977, 1949; **19–5** R. C. Lewontin; **19–6** Dr. Agnes Bankier, Murdoch Institute, Royal Children's Hospital, Melbourne; **19–7** Edmund Gerard; **19–9** Victor McKusick; **19–10** Kennan Ward/DRK Photo; **19–11** Jim Brandenburg/Minden Pictures; **Page 330** from Lewis Carroll, *Through the Looking Glass*, illustrated by John Tenniel; **19–13** E. S. Ross

Page 334 Mark D. DeMello/New York Zoological Society; **20–1** (a) Jane Burton/Bruce Coleman, Ltd.; (b) Kim Taylor/Bruce Coleman, Ltd.; **20–2** (a),(b) James L. Castner; **20–3** (a) M. and C. Ederegger/Peter Arnold, Inc.; (b) Kent and Donna Dannen/Photo Researchers, Inc.; **20–5** Chris Huss/The Wildlife Collection; **20–6** Steve Hopkin/Planet Earth Pictures; **20–7** (a) E. R. Degginger/Animals Animals; (b) adapted from D. Futuyma, *Evolutionary Biology,* 1979 (after Clarke 1962, based on data of Popham 1942); **Page 343** M. Andersson/VIREO; **20–8** (a) Russ Kinne/Comstock; (b) Frans Lanting/Minden Pictures; **20–9** David Kjaer/Planet Earth Pictures; **20–10** (a),(b) adapted from S. J. Gould, *Ever Since Darwin*, W. W. Norton and Company, New York, 1977, page 235; **20–11** after J. Clausen and W. M. Hiesen, Publication 615, Carnegie Institute of Washington, 1950; **20–12** (a) Lincoln Brower; (b) Frans Lanting/Minden Pictures; (c) Dwight Kuhn; **20–13** (a),(b) Lincoln Brower; **20–14** (a)–(d) E. S. Ross; (e) James L. Castner; (f) D. Wilder/Tom Stack and Associates; (g) Peter J. Bryant/Biological Photo Service; **20–16** (a) S. J. Krasemann/Peter Arnold, Inc.; (b) E. S. Ross; **20–17** E. R. Degginger/Bruce Coleman, Inc.

Page 354 Erwin and Peggy Bauer/Bruce Coleman, Inc.; **21–1** (a) C. A. Henley/Biofotos; (b) Kenneth W. Fink/Photo Researchers, Inc.; (c) Tom McHugh/Photo Researchers, Inc.; **21–2** (a) Belinda Wright/DRK Photo; (b) Stephen J. Krasemann/DRK Photo; **21–3** adapted from E. O. Wilson and W. H. Burrert, *A Primer of Population Biology*, Sinauer Associates, Sunderland, Mass., 1971; **Page 358** (b) USGS; **21–6** Ron Prokopy, University of Massachussetts–Amherst; **21–7** Sid Bahrt/Photo Researchers, Inc.; **21–9** (a) Scott Camazine/Photo Researchers, Inc.; (b) Jeff Lepore/Photo Researchers, Inc.; (c) after V. Wallace and A. Srb, *Adaptation*, Prentice-Hall, Inc., Englewood Cliffs, N. J., 1964; **21–11** after D. Lack, *Darwin's Finches*, Harper and Row, Publishers, Inc., New York, 1961; **21–12** (a)–(f) Tui De Roy; **Page 366** Michael Franis/The Wildlife Collection; **21–14, 21–15** redrawn from G. G. Simpson, *Horses*, Oxford University Press, New York, 1951

Page 376 Steven Dalton/NHPA; **22–1** E. S. Ross; **22–5** Richard Hansen/Photo Researchers, Inc.; **22–6** Nina Leen/Time Warner, Inc.; **22–7** after Mazudazo Konish; **22–8** Heather Angel; **22–9** (a) E. S.

University Press, New York, 1983; **32–2** From Lewis Carroll, *Alice's Adventure in Wonderland*, illustrated by John Tenniel; **32–3** Gray Mortimore/Tony Stone Images; **32–7** Doug Allen/Oxford Scientific Films/Animals Animals; **32–8** John Cancalosi/Auscape International; **32–9** after A. J. Vander, J. H. Sherman and D. S. Luciano, *Human Physiology*, McGraw-Hill Publishing Company, New York, 1969; **32–10** (a) Ken Lucas/Planet Earth Pictures; (b) Robert Franz/Planet Earth Pictures; (c) Leonard Lee Rue III/Animals Animals; **32–12** John Cancalosi/DRK Photo; **32–13** Richard T. Nowitz

Page 582 Lennart Nilsson/©Boehringer Ingelheim International GmbH, *The Incredible Machine*, National Geographic Society, Washington, D.C., 1986 **Page 588** Bernard Pierre Wolff/Photo Researchers, Inc.; **33–6** (b) Leonard Lessin/Peter Arnold, Inc.; **33–10** Don Fawcett/Science Source/Photo Researchers, Inc.; **33–11** (a),(b) Paul Travers, University of London; **33–15** (a)–(c) Lennart Nilsson/©Boehringer Ingelheim International GmbH; **33–18** Lennart Nilsson/©Boehringer Ingelheim International GmbH, *The Body Victorious*, Delacorte Press, New York, 1987; **33–20** Lennart Nilsson/©Boehringer Ingelheim International GmbH, *The Incredible Machine*, National Geographic Society, Washington, D.C., 1986; **33–21** Stephen Turner/Oxford Scientific Films

Page 604 E. S. Ross; **34–1** adapted from Guttman and Hopkins; **34–10** Astrid and Hanns-Frieder Michler/Science Photo Library/Photo Researchers, Inc.; **34–13** adapted from A. J. Vander, J. H. Sherman, and D. S. Luciano, *Human Physiology*, 4th ed., McGraw-Hill Publishing Company, New York, 1985

Page 620 Janis Carter; **35–4** (b) Cedric S. Raine/Visuals Unlimited; **35–6** courtesy Dr. Peter S. Amenta, Hahneman University; **35–9** after R. Eckert and D. Randall, *Animal Physiology*, W. H. Freeman and Company, New York, 1978; **35–11** after David Hubel; **35–16** adapted from J. Darnell, H. Lodish, and D. Baltimore, *Molecular Cell Biology*, W. H. Freeman and Company, 1986; **35–17** adapted from Williams and Warwick, *Functional Neuroanatomy of Man*, W. B. Saunders, Philadelphia, 1975; **Page 637** Comstock

Page 640 Michael S. Quinton; **36–1** after W. S. Beck, *Human Design*, Harcourt Brace Jovanovich, New York, 1971; **36–5** Lennart Nilsson/©Boehringer Ingelheim International GmbH, *Behold Man*, Little, Brown and Company, Boston, 1974; **Page 646** Stephen Dalton/NHPA; **36–7** CNRI/Science Photo Library/Photo Researchers, Inc.; **36–9** (a) Martin M. Rotker/Science Source/Photo Researchers, Inc.; (b) CNRI/Science Photo Library/Photo Researchers, Inc.; **36–10** Bassett/Photo Researchers, Inc.; **36–15** adapted from *Scientific American*, June 1987, page 83; **36–17** adapted from *Scientific American*, June 1987, page 82; **Page 654** (a) Christine Gall/University of California, Irvine; (b) Daniel P. Perl; **36–18** Jim Greenfield/Planet Earth Pictures; **36–20** (a) Hugh E. Huxley; **36–24** (a) Lennart Nilsson et al., *Behold Man*, Little, Brown and Company, Boston, 1974; (b),(c) after K. R. Porter and M. Bonneville, *Fine Structure of Cells and Tissues*, Lea and Febiger, Philadelphia, 1968; **Page 662** Hugh E. Huxley

Page 664 Helen Rhode; **37–2** (b) G. Shih and R. Kessel/Visuals Unlimited; **37–3** after M. B. V. Roberts, *Biology: A Functional Approach*, The Ronald Press Company, New York, 1971; **37–6** (a) Doug Allen/Oxford Scientific Films; (b) Jonathan Scott/Seaphot Ltd./Planet Earth Pictures; (c) Lady Philippa Scott/NHPA; (d) Oxford Scientific Films/Animals Animals; **Pages 670–671** (a) NIAID/NIH/Peter Arnold, Inc.; (b) CNRI/Science Photo Library/Photo Researchers, Inc.; (c) Institut Pasteur/CNRI/Phototake; (d) NIAID/NIH/Peter Arnold, Inc.; (e) Science VU/Visuals Unlimited; **37–10** (a)–(d) C. Edelmann/La Villette/Photo Researchers, Inc.; **37–12** after A. J. Vander, J. H. Sherman, and D. S. Luciano, *Human Physiology*, 3d ed., McGraw-Hill Publishing Company, New York, 1980

Page 680 Keith A. Szafaranski; **38–1** (a) William Byrd; (b),(c) Mia Tegner/Scripps Institution of Oceanography; **38–2** adapted from Bruce Alberts et al., *Molecular Biology of the Cell*, Garland Publishing, Inc., New York, 1983; **38–3** (a)–(h) Tryggve Gustafson; **38–4** after Huettner, 1949; **38–5** M. P. Olsen; **38–6** after E. O. Wilson et al., *Life on Earth*, 2d ed., Sinauer Associates, Sunderland, Mass., 1978; **38–7** (a)–(d) Tryggve Gustafson; **38–8** after Wilson et

al., *op. cit.*; **38–9** after Huettner, 1949; **38–10** Hans Pfletschinger/Peter Arnold, Inc.; **38–11** after Huettner, 1949; **38–12** (a) after J. J. W. Baker and G. E. Allen, *The Study of Biology*, 3d ed., Addison-Wesley Publishing Co., Inc., Reading, Mass., 1977; (b) after Maurice Sussman, *Animal Growth and Development*, 2d ed., Prentice-Hall, Inc., Englewood Cliffs, N. J., 1964; **38–13** *Ibid.*; **38–15** S. R. Hilfer; **38–16** (a)–(c) adapted from G. Karp and N. J. Berrill, *Development*, 2d ed., McGraw-Hill Publishing Company, New York, 1981; (d) Kathryn W. Tosney; **38–17** (a) Martin Raff; (b) adapted from Alberts et al., *op. cit.*; **38–18** Lewis Wolpert; **38–20** (a)–(d) Runk and Schoenberger/Grant Heilman Photography; **Page 693** (a),(b) Laboratory of Edward B. Lewis, California Institute of Technology; **38–21** Lennart Nilsson, *The Incredible Machine*, National Geographic Society, Washington, D.C., 1986; **38–22** (a)–(c) Petit Format, Nestle/Science Source/Photo Researchers, Inc.; **38–27** (a)–(c) Lennart Nilsson et al., *Behold Man*, Little, Brown and Company, Boston, 1973; **38–28** (a),(b) Lennart Nilsson/©Boehringer Ingelheim International GmbH, *The Incredible Machine*, National Geographic Society, Washington, D.C., 1986; **38–29, 38–30** Lennart Nilsson et al., *Behold Man*, Little, Brown and Company, Boston, 1974; **38–32** (a)–(c) Jeffrey Reed/Medichrome; **38–33** David Austen/Tony Stone Images

Page 706 Merlin D. Tuttle/Bat Conservation International, Inc.; **39–2** (a) Kenneth W. Fink/Bruce Coleman, Inc.; (b) E. R. Degginger/Earth Scenes; (c) Thomas C. Boyden; (d) John C. Coulter/Visuals Unlimited; (e) Doug Wechsler; (f) E. S. Ross; (g) John Tiszler/Peter Arnold, Inc.; **39–3** (a),(b) Dr. Dennis Kunkel/Phototake; (c) Dr. Jeremy Burgess/Science Photo Library/Photo Researchers, Inc.; **39–4** (b) David L. Muleahy; **39–9** (a) W. H. Hodge/Peter Arnold, Inc.; (b) Larry West; (c) Rence Purse/Photo Researchers, Inc.; **Page 717** Heather Angel

Page 718 Tom Bean/Allstock; **40–4** Dr. Jeremy Burgess/Science Photo Library/Photo Researchers, Inc.; **40–5** (a) G. J. James/Biological Photo Service; (b) Dwight R. Kuhn; (c) James L. Castner; **40–7** (b) Ray F. Evert; **40–8** G. R. Roberts; **40–9** after P. Ray, *The Living Plant*, 2d ed., Holt, Rinehart and Winston, Inc., New York, 1971; **40–10** (a),(b) Ray F. Evert; **40–11** after Ray, *op. cit.*; **40–13** (a) Jack Dermid; (b) Robert Mitchell; **40–14** (a) Biophoto Associates/Photo Researchers, Inc.; (b) Ray F. Evert; (c) Randy Moore/Visuals Unlimited; **40–15** (c) Ray F. Evert; **40–16** (d) I. N. A. Lott, McMaster University/Biological Photo Service; **40–17** (a)–(c), **40–18** (a)–(c) Ray F. Evert; **40–22** from R. Ketchum, *The Secret Life of the Forest*, American Heritage Press, 1970.

Page 740 Kim Taylor/Bruce Coleman, Inc.; **41–2** Walter H. Hodge/Peter Arnold, Inc.; **41–3** after M. Richardson, *Translocation in Plants*, St. Martin's Press, Inc., New York, 1968 (after Stoat and Hoagland, 1939); **41–4** adapted from D. K. Northington and J. R. Goodin, *The Botanical World*, Mosby-Year Book, St. Louis, 1984; **41–5** Jerry D. Davis; **41–6** (a),(b) Biological Photo Service; **41–7** (a) Science Photo; (b) M. H. Zimmerman; **41–10** (a) Runk and Schoenberger/Grant Heilman Photography; (b) C. P. Vanae/Visuals Unlimited

Page 754 Chuck Place; **42–1** Dwight Kuhn; **42–4** Runk and Schoenberger/Grant Heilman Photography; **42–7** (a)–(c) Bruce Iverson; **42–8** E. Webber/Visuals Unlimited; **Page 761** Philip Harrington; **42–10** (a) Ray F. Evert; (b) Heather Angel; **42–11** Larry West; **42–12** Carolina Biological Supply Company; **42–13** after E. O. Wilson et al., *Life: Cells, Organisms, Populations*, Sinauer Associates, Inc., Sunderland, Mass., 1977; **42–14** (a),(b) Randy Moore; **42–15, 42–16** after A. W. Naylor, "The Control of Flowering," *Scientific American*, May 1952; **42–17** John R. MacGregor/Peter Arnold, Inc.; **42–20** Breck P. Kent/Earth Scenes; **42–21** (a),(b) Jane Burton/Bruce Coleman, Inc.; **42–22** Stephen P. Parker/Photo Researchers, Inc.; **42–23** (a),(b) James L. Castner; **42–24** (a),(b) Runk and Schoenberger/Grant Heilman Photography

Page 774 Kenneth W. Fink/Photo Researchers, Inc.; **43–6** after *Nature*, vol. 298, 1982; **43–7** adapted from McNaughton and Wolfe, *General Ecology*, 2d ed., Holt, Rinehart, and Winston, Inc., New York, 1979; **43–8** (a) Peter Johnson/NHPA; (b) Doug Perrine/

Index

Page numbers in italic type indicate references to illustrations, captions, tables, and boxed essays; boldface type is used to indicate major text discussions.